Chapter 12

1-9, 11-18, 21-25, 30, 31

Chapter
6-7-11-12

BASIC ELECTRIC CIRCUITS

BASIC ELECTRIC CIRCUITS

SECOND EDITION

DONALD P. LEACH, Chairman

Engineering/Technology Division
Foothill College
Los Altos Hills, California

John Wiley & Sons New York London Sydney Toronto

***Library of Congress Cataloging in Publication Data*:**

Leach, Donald P
Basic electric circuits.

1. Electric circuits. I. Title

TK454.L4 1976 621.319'2 75-35751
ISBN 0-471-52003-9

Printed in the United States of America

10 9 8 7 6 5 4

To Fafou and Skeeter

PREFACE

This second edition of *Basic Electric Circuits* incorporates many ideas and suggestions taken from a rather extensive survey of instructors and students familiar with the first edition. Two completely new chapters have been included. Chapter 19 has been added to expand and improve the presentation on transformers and transformer circuits. The new Chapter 21 illustrates the applications of circuit laws and theorems to simple electronic circuits, and it provides a smooth transition between a basic circuits course and a first course using active electronic devices. The widespread use of electronic hand calculators has made it possible to reduce the arithmetic detail somewhat.

This book is intended to provide a firm foundation in the fundamentals of electricity and electric circuits. For the electrical/electronics technician it provides the background necessary to compete successfully in industry and to study advanced courses in fields such as power, electronics, industrial electronics, or communications. For other students it provides a basic understanding to participate effectively in these fields. No previous background in chemistry, physics, or electronics is assumed. A course in college algebra should be taken concurrently or previously.

The material has been successfully used in a one-semester (7-hour) course, and it is now being taught in two quarters (5 hours each). It could be easily used in a one-year course sequence consisting of two semesters or three quarters.

The subject matter can be divided into two main categories: the first ten chapters deal with the fundamentals of electricity and steady dc circuits; the second eleven chapters deal with circuits in which the currents and voltages vary with time. The book is designed so that numerous sections can be deemphasized or omitted entirely without any loss of continuity, and the remaining material can then be expanded to accommodate individual needs or preferences.

There are over 325 example problems worked out in detail. More than 440 Review Questions with answers given in the appendix provide an immediate self-check of individual progress. The Important Relationships listed at the end of each chapter summarize key points. Finally, there are over 575 problems chosen from a wide variety of applications. Answers to approximately one half of the problems are given in the appendix. A complete solutions manual is available.

Los Altos Hills, California *Donald P. Leach*

September 1975

CONTENTS

BASIC ELECTRIC CIRCUITS

INTRODUCTION

The study of electricity and electronics can be very rewarding, for the various electrical devices we daily encounter have a profound effect on our lives. Common electrical devices, such as lights, phones, radios, TV, auto ignition systems, we take for granted, but there are more complex electrical systems which equally affect our lives: electrical power for industry and homes, telephone communications and the array of computer installations used by business, industry, and government. To realize their importance we need only to consider living one day without electrical energy!

Your reading this book indicates you have encountered one, or both, of the following situations: (*a*) Observing an electrical device in operation, you ask "I wonder how it works?"; (*b*) observing an electrical device that has failed, you ask "I wonder how it can be fixed?" This book is intended to help you answer these questions. The first few chapters tell about the nature of electricity and its effects. The remainder of the book is devoted to the basic laws of electricity and their application.

1.1 ELECTRICAL PROPERTIES OF MATTER

Significant advancements in science were made when man began investigating the substances from which objects were made rather than studying the objects themselves. It may be informative and useful to study all the properties of some particular object, such as an iron knifeblade. However, there are many kinds of knifeblades; they may be long or short, thick or thin, sharp or dull. Surely it is of greater importance to study the properties of iron itself, since this leads to a more general body of knowledge applicable to all iron objects—nails, knifeblades, or bridge girders. Over the years man has gained a large volume of information by studying the properties of different basic substances. This knowledge is summarized in tables and charts contained in handbooks and includes mechanical characteristics, melting points, boiling points, and electrical characteristics.

This general knowledge has in turn been extended to obtain an even more general body of knowledge which applies to all substances, which we call matter. We can describe *matter* as any substance which: (*a*) has *inertia*, (*b*) exerts a *gravitational force* on other matter, (*c*) exerts *electric forces* on other matter, and (*d*) exerts a *nuclear force* on other matter.

Inertia is the attribute of any piece of matter, which we call a *body*, to resist any change in its motion. Thus an automobile at rest tends to remain at rest and requires an applied force to cause it to move. Similarly, an automobile traveling at a certain velocity will tend to continue traveling in the same direction, at the same velocity, and an applied force is necessary to slow it or to change its direction.

It is commonly accepted that the *weight* of any body is actually due to the *gravitational force* between the earth and the body. The weight of a baseball, for example, is due to an interaction between the ball and the earth which produces a gravitational force of attraction. It is a basic property of all matter to exert a gravitational force on all other matter anywhere in the universe. There is in fact a gravitational force of attraction between two baseballs held some distance apart. At the surface of the earth, this force is too small to be noticed since it is masked by the much greater force of attraction between each baseball and the earth. However, if two baseballs were taken out into space, far removed from the gravitational forces of any other bodies, the force of attraction between them would be readily apparent. The gravitational force of attraction F between two bodies is given by Newton's law of gravitation

$$F = k\frac{m_1 m_2}{d^2} \tag{1.1}$$

where m_1 and m_2 are the masses of the two bodies, k is a constant of proportionality, and d is the distance separating the two bodies. Using Newton's law of gravitation, the concept of a *gravitational field* surrounding the earth can be developed. The force of attraction between the earth and other bodies can then be accounted for by considering their behavior relative to the earth's gravitational field. Thus the trajectories of projectiles or missiles can be accurately predicted, as well as the behavior of other bodies such as the orbits of satellites. Although the effects of gravitational forces can be accurately calculated, it is not known exactly what a gravitational force is.

A third attribute of matter is that it exerts electric forces on most other matter (with the possible exceptions of neutral subatomic particles such as neutrons or neutral mesons; these particles are unstable and shortlived when far removed from other particles and we will not consider them here). The electric forces present in matter are somewhat varied and complex, but we will consider some of the effects. First of all we will consider the forces between stationary charged bodies. These are called *electrostatic forces,* or *coulomb forces,* and are quite similar to the gravitational forces discussed above. We can draw an analogy between the gravitational field surrounding the earth and the *electric field* surrounding a charged body. The force F between the two charged bodies is given by Coulomb's law:

$$F = k\frac{Q_1Q_2}{d^2} \tag{1.2}$$

where Q_1 and Q_2 are the charges on the two bodies, k is a proportionality constant, and d is the distance separating the two bodies. You will notice the similarity between the equations for electrostatic force of attraction and for gravitational force of attraction, Eq. 1.1. Our analogy breaks down at this point, however, since gravitational forces are always forces of attraction (as far as we know). On the other hand, electrostatic forces may be forces of attraction or repulsion, since there are both positive and negative charges. You may recall that *like* charges *repel* and *unlike* charges *attract.* Using the concept of an electric field surrounding a charged body and Coulomb's law, the effects of the field can be accurately predicted; however, as in the case of a gravitational field, it is not known exactly what an electric field is. We will not be overly concerned about this, however, since our goal is to recognize electrical phenomena and the laws that govern them and to gain competence in the practical applications. It is interesting to note that electrostatic forces are far stronger than gravitational forces. It is only because most matter is neutral (i.e., contains an equal number of positive and negative charges) that the electrostatic forces cancel and we are thus able to detect such weak gravitational forces as the weight of a 20-ton truck.

A second type of electric force existing in matter is the *magnetic force* exerted between charges in motion. These forces are present in electromagnets, electric generators, electric motors, and other electrical machines.

Two other important types of electric forces present in matter are *elastic forces* and *chemical forces.* Both are *short range* forces since they are effective only when atoms are in close contact with one another; at greater distances their effects are negligible compared to electrostatic or magnetic forces. Elastic forces are a result of an overlapping of charged portions of atoms and are present in a stretched rubber band or in a long bow. Chemical forces are present in the various types of chemical bonding (covalent, ionic, and metallic bonding) which we will investigate in a future section. It is important to remember that some electric force exists in virtually all matter, because application of these various forces has led to all of the modern electric and electronic devices. It is interesting to note that even though gravitational and electric forces are quite similar, their effects are independent and there seems to be no precise connection between them. Albert Einstein (1879–1955) devoted a good portion of his later years searching for such a connection between the two—but was unsuccessful.

The fourth attribute of matter is the presence of nuclear forces. These are extremely short-range forces and we will not be concerned with them. Suffice it to say that these forces account for the *binding energy* holding the subatomic particles of matter together to form atoms.

1.2 THE STRUCTURE OF MATTER

In the process of examining a piece of iron, let us suppose that it is possible to continually divide the iron into smaller and smaller pieces. The smallest piece of iron which we could obtain would be a single *atom*; if this single atom could somehow be divided, we would no longer have iron. Similarly, the smallest piece of aluminum or copper is one atom of the substance. The atom can thus be considered as the basic building block of matter, and the different characteristics of different substances can be attributed to the structure of individual atoms.

The *planetary model* of the atom, proposed by Niels Bohr in 1913, is very useful for investigating the structure of different atoms*. At the center of each atom is a *nucleus* which accounts for about 99.9% of its total mass. The remainder of the atom is composed of *electrons* which travel around the nucleus in orbits similar to those of the planets around the sun in our solar system.

The nucleus contains two basic particles called *nucleons.* The two particles have almost identical masses, but one is electrically neutral and is called a *neutron,* while the other has a *positive* charge and is called a *proton.*

Each electron has a *negative* charge equal in magnitude to the positive charge of a proton. The mass of the electron is, however, only about 1/1840 that of either a neutron or a proton as shown in Table 1.1.

Table 1.1. Basic Atomic Particles.

Particle	Charge (coulomb)	Mass (gram)
Electron	-1.6×10^{-19}	9.108×10^{-28}
Proton	$+1.6 \times 10^{-19}$	1.672×10^{-24}
Neutron	none	1.675×10^{-24}

Since each atom is electrically neutral, there must be the same number of protons in the nucleus as there are planetary electrons orbiting the nucleus. The *atomic number* of each element is equal to the number of electrons (or protons) in each atom of that element. Referring to the Periodic Table of the Elements (Appendix *C*) the element with atomic number 1 is hydrogen. The Bohr model of this atom can be drawn as shown in Fig. 1.1*a*. Notice that there is one proton in the nucleus and one planetary electron. Since the charges of these two particles are equal in magnitude, but opposite in sign, the atom is electrically neutral. As a matter of notation the symbol for hydrogen is written ${}^{1}_{1}\mathrm{H}$. The lower left subscript is the atomic number (the total number of electrons), and the upper left superscript is the *mass number* and is equal to the total number of nucleons in the atom.

The second element in the periodic table has the atomic number 2 and is helium. The Bohr model for the helium atom is shown in Fig. 1.1*b*. There are two planetary electrons, and two protons in the nucleus and the atom is seen to be electrically neutral. The two neutrons are placed in the nucleus to account for the total mass of the helium atom. The symbol for helium is ${}^{4}_{2}\mathrm{He}$.

*Deficiencies in the Bohr model of the atom were noted as early as 1920; however, it is sufficiently accurate for our purposes.

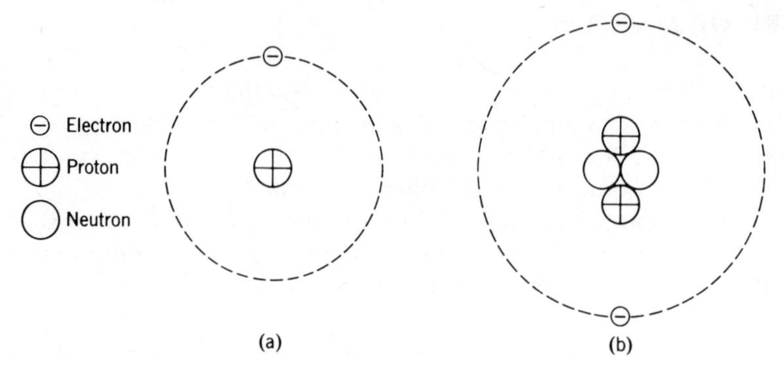

Figure 1.1 Planetary models of atoms (*a*) Hydrogen, ${}^{1}_{1}H$ (*b*) Helium, ${}^{4}_{2}He$

The subscript shows that there are two electrons and the superscript tells us there are four nucleons in the nucleus.

We could continue to construct a planetary model of each element in the periodic table. We could not do it randomly however, since the atoms are constructed in a very definite pattern. This pattern is governed by the total number of electrons permissible in various orbits around each nucleus. It is convenient to talk about *orbital rings* or *shells* surrounding each nucleus. Each shell contains a certain number of *different* orbits in which electrons can exist. The first four shells, called the K, L, M, and N shells, are shown in Fig. 1.2. The K shell can contain *at most* 2 electrons; the L shell can contain at most 8; and the M shell can contain at most 18 electrons. The various shells and the corresponding maximum number of electrons which can exist in each shell are summarized in Table 1.2. The outer shell of any atom can never contain more than 8 electrons.

The differences between elements can be explained on the basis of the total

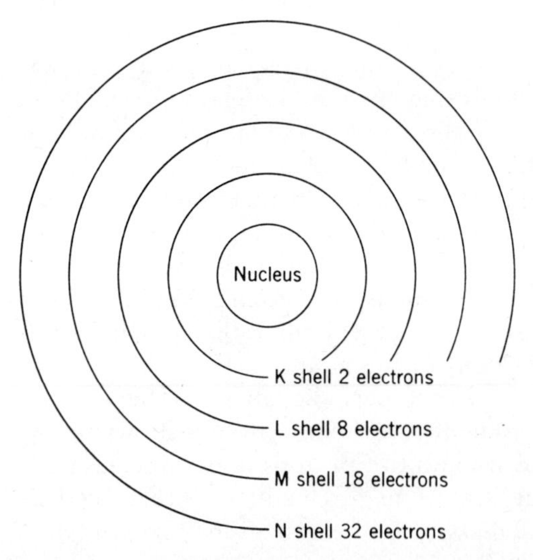

Figure 1.2 The first four shells of an atom and the maximum number of electrons in each shell

Table 1.2. Shells of Orbital Electrons.

Shell	Maximum number of electrons
K	2
L	8
M	18
N	32
O	18
P	18
Q	8

number of electrons surrounding the nucleus *and* on the number of electrons in the outermost shell. The outermost shell is filled when it contains either 2 or 8 electrons. All atoms tend to stabilize themselves under these conditions. The Bohr model for the *inert* gas neon, Fig. 1.3, shows that the outermost shell does contain 8 electrons. Helium is another inert gas (Fig. 1.1*b*) and its outermost shell contains 2 electrons. The reason that these gases are inert (do not combine readily with other elements) is that their outermost shells are filled.

The *valence* of an atom is determined by the number of electrons required to fill the outermost shell. For example, the hydrogen atom would require 2 electrons to fill its shell; however, it has only one and, therefore, hydrogen has a valence of +1. The helium atom has its shell filled with 2 electrons and, therefore, has a valence of zero. The Bohr model of a copper atom is shown in Fig. 1.4; since it has only one electron in its outermost shell its valence is +1. On the other hand, the carbon atom which has 4 electrons in its outermost shell has a valence of +4. Since the goal of all atoms, other than hydrogen or helium, is to fill the outer shell with 8 electrons, the carbon atom can be considered to have a valence of −4. The negative valence shows a deficit of electrons and if the carbon atom could "borrow" 4 electrons, its outer shell would be filled. The carbon atom does in fact *share* 4 electrons with neighboring atoms to form stable structures such as diamond. With the concept of valence we can show how atoms combine with one another to form molecules and crystalline structures.

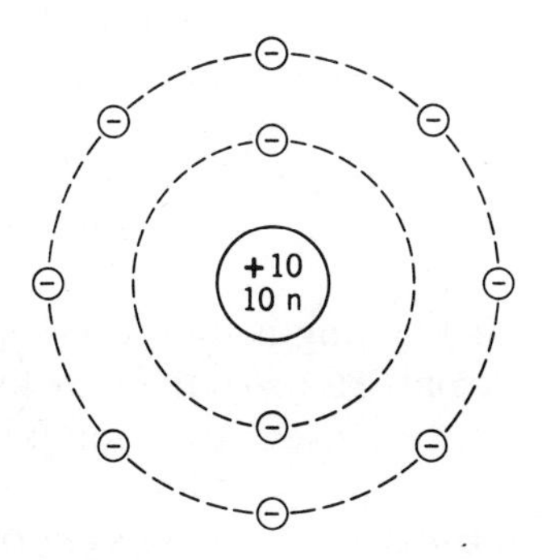

Figure 1.3 The Bohr model of the inert gas neon

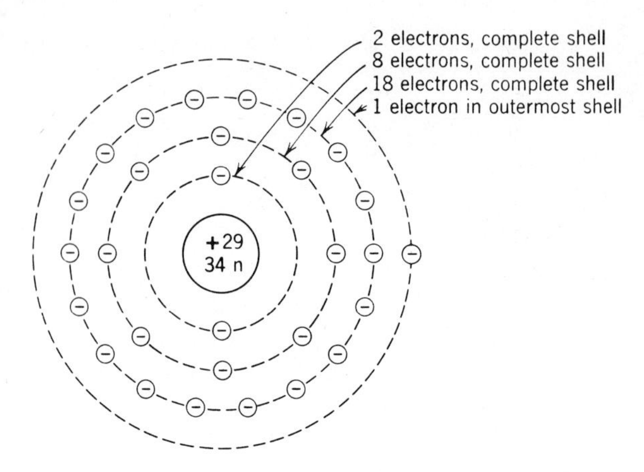

Figure 1.4 The Bohr model of a copper atom

1.3 COMBINATIONS OF ATOMS

Atoms combine with other atoms in an effort to fill their outermost shells and in the process *molecules* are formed. Thus, molecules are combinations of two or more atoms and are the smallest particles of a substance which retain the same chemical characteristics. For example, hydrogen combines with oxygen in the ratio of two hydrogen atoms to one oxygen atom to form water. A single molecule of water thus consists of two hydrogen atoms and one oxygen atom (H_2O). There are in general three types of *atomic bonding* in the formation of molecules.

One form of atomic bonding is found in the formation of sodium chloride (ordinary table salt) NaCl. The sodium atom has a total of 11 electrons as shown in Fig. 1.5*a*. It has only one electron in its outermost shell, or *valence shell*, and therefore has a valence of +1. The chlorine atom shown in the same figure has 7 electrons in its valence shell. It has a valence of −1 since it would like to add an electron to complete the shell. The valence electron in the outermost shell of the sodium atom is rather weakly bound to its nucleus and the chlorine atom will "borrow" this electron in order to fill its outer shell. Both atoms now have completely filled outer shells, but they are no longer electrically neutral. The sodium atom has lost an electron and therefore has a net charge of +1; similarly, the chlorine atom has gained an electron and thus has a net charge of −1. Such *charged atoms* are called *ions* and appear as shown in Fig. 1.5*b*. Since these two ions have unlike charges of equal magnitude, there is an electric force of attraction which tends to bind them together and they form a stable molecule of sodium chloride as shown in Fig. 1.5*c*. This form of atomic bonding is called *ionic bonding*. It is the recognition of ionic bonding which leads to such practical applications as *chemical cells, batteries*, and *electroplating*.

Another form of atomic bonding is found in molecules which are formed when atoms "share" electrons. Hydrogen atoms for example each have one electron. In order to fill its shell, each atom would like to have one additional electron. Hence two hydrogen atoms will combine to form one molecule of hydrogen gas. Each atom will then "share" its electron with the other atom in such a way that each nucleus appears to have two electrons in its outer shell. A

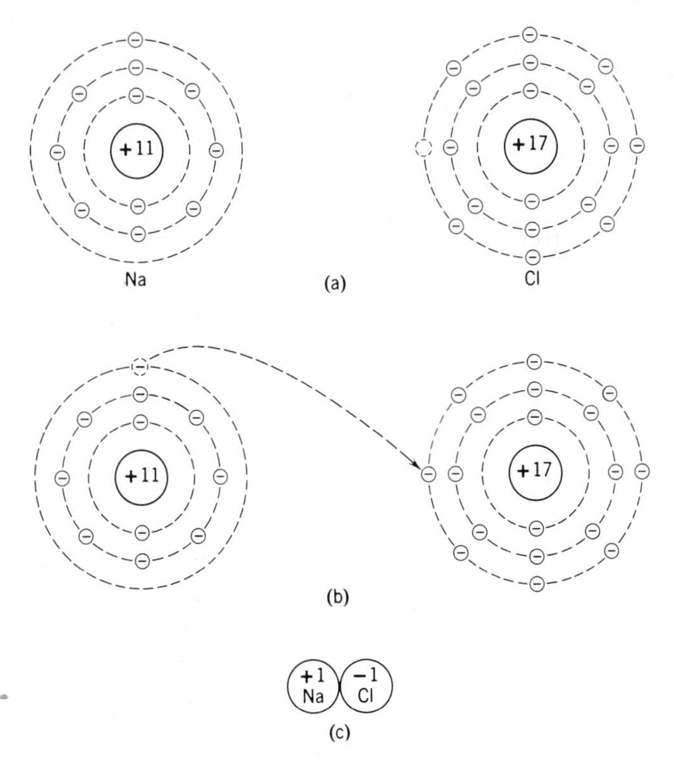

Figure 1.5 (*a*) Atoms of sodium and chlorine (*b*) Sodium and chlorine ions (*c*) A sodium chloride molecule

pictorial representation of a hydrogen molecule might appear as shown in Fig. 1.6. There is an important group of elements which form very stable crystalline structures by means of covalent bonding. The group includes carbon (diamond), germanium, and silicon. Recognition of the covalent bonding in these crystalline structures has led to all of the important semiconductor devices (transistors, diodes, etc.).

A third form of atomic bonding occurs in the group of elements called metals (iron, copper, gold, silver, aluminum, etc.). The copper atom in Fig. 1.4 has only one electron in its valence shell. The pictorial representation of the copper atom in Fig. 1.7 is intended to show that the valence electron in the copper atom is very weakly bound to the nucleus of the atom, that is, there are 28 electrons in the inner orbits tightly bound to the nucleus that are not easily removed. However, the valence electron is at a further distance from the nucleus, and the nucleus itself is somewhat "masked" from the valence electron by the other 28 electrons. As a result, the valence electron of copper is easily removed from its parent nucleus. In a solid piece of copper, virtually all

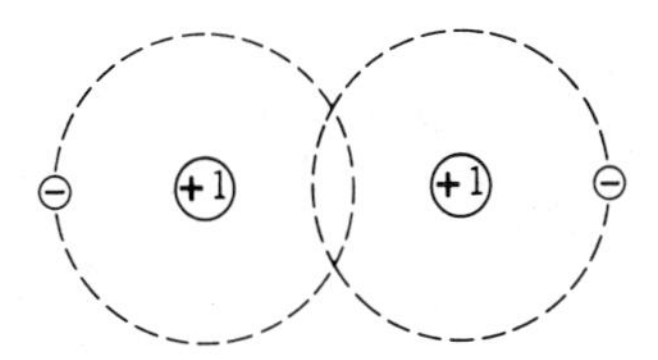

Figure 1.6 Hydrogen molecule

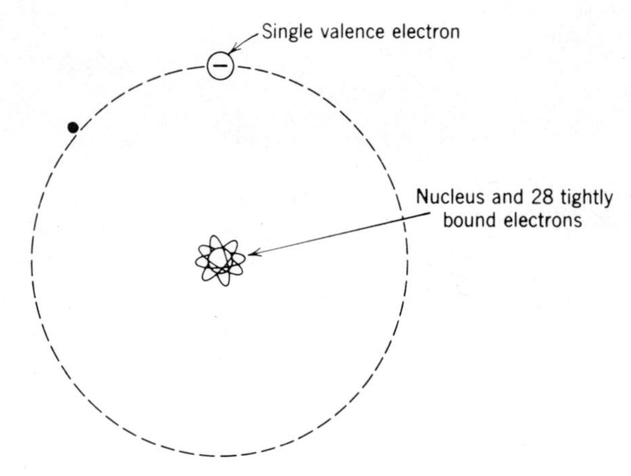

Figure 1.7 A copper atom

of the valence electrons are free from their parent atoms and they form a cloud of *free electrons* which drift randomly throughout the piece of copper. Each copper atom then becomes a positively charged ion. As a result of the force of repulsion between these positive metal ions, they will arrange themselves in a fixed geometrical pattern which is the shape of the piece of copper itself. The atoms are bound together by the force of attraction between the negative cloud of electrons and the positive metal ions. This is called *metallic bonding.* The cloud of free electrons in a metal accounts for the fact that it is a good conductor of electric current.

1.4 CONDUCTORS, INSULATORS, AND SEMICONDUCTORS

For a material to be a good *conductor* of electric current, it must contain a large number of electrons free to move about in the material, such as the copper discussed in the previous section. Most all metals have this property, but copper remains the most widely used metal in the construction of electrical conductors. Copper has many desirable physical properties, is easily drawn into wire, is not prohibitively expensive, and is second only to silver as a conductor of electrical current.

Silver is a somewhat better conductor than copper since it has about 5% more free electrons per unit volume than copper. However, silver is prohibitively expensive.

Aluminum is increasing in popularity as an electrical conductor since it has many desirable physical characteristics. It is much lighter and somewhat less expensive than copper, but it contains only about 60% as many free electrons per unit volume as copper and is, therefore, not quite as good a conductor. The approximate number of free electrons per cubic inch for these three metals is given in Table 1.3. The number 1.4×10^{24} means 1.4 million, million, million, million. A very large number indeed!

An *insulator,* as might be expected, is just the opposite of a conductor. Insulators are characterized as materials whose electrons are tightly bound to their nuclei. Within the insulator there are no electrons that are free to move about and an insulator is, therefore, a very poor conductor of electric current. In a perfect insulator, there would be absolutely no free electrons and the

Table 1.3.

Material	Approximate number of free electrons per cubic inch
Copper	1.40×10^{24}
Silver	1.47×10^{24}
Aluminum	8.36×10^{23}

material would conduct absolutely no electric current. In practice however, it would be virtually impossible to build a perfect insulator since there are nearly always impurities in any material. Therefore, even the best insulators will conduct some electric current. Examples of good insulating materials are paper, rubber, glass, plastic, and mica. These materials are useful in preventing electric current flow. They are also frequently referred to as *dielectrics* since they can be used to store electric charge.

The elements carbon, germanium, and silicon which form covalent bonds (discussed in the previous section) are neither good insulators nor good conductors. They are in a special class and are called *semiconductors.* The electrical characteristics of pure crystalline germanium and silicon can be altered with great precision by the controlled addition of certain elements such as arsenic or aluminum. In this fashion all the modern semiconductor devices such as transistors and diodes are constructed.

1.5 SCIENTIFIC NOTATION

In the previous section, the need for a concise way of writing very large numbers became apparent. It is certainly much easier to write 1.4×10^{24} than 1.4 million, million, million, million or 1,400,000,000,000,000,000,000,000. Numbers written in the form of 1.4×10^{24} or 7×10^{-5} are said to be written in *scientific notation,* which permits expressing any quantity as a number between 1 and 10 multiplied by a power of 10. Scientific notation is widely used in many fields and when mastered will prove an invaluable labor saving device.

Some of the powers of ten we will be using are shown in Table 1.4. You will

Table 1.4.

Quantity	Scientific notation
10,000	1.0×10^{4}
1,000	1.0×10^{3}
100	1.0×10^{2}
10	1.0×10^{1}
1	1.0×10^{0}
0.1	1.0×10^{-1}
0.01	1.0×10^{-2}
0.001	1.0×10^{-3}
0.0001	1.0×10^{-4}

notice that the power simply shows the number of places the decimal must be moved in order to write the original number. A *positive* power shows that the decimal must be shifted to the *right*, while a *negative* power shows that the decimal must be shifted to the *left*. The power of 10 is then simply a means for keeping track of the decimal point. You will also notice that $10^0 = 1$, so that $1.0 \times 10^0 = 1.0 \times 1 = 1$.

• • •

Example 1.1 Write the following numbers in scientific notation.

a. 121
b. 5116
c. 0.092
d. 4.76

Solution

a. $121 = 1.21 \times 10^2$
b. $5116 = 5.116 \times 10^3$
c. $0.092 = 9.2 \times 10^{-2}$
d. $4.76 = 4.76 \times 10^0 = 4.76$

• • •

In *d* the number 4.76 is already in scientific notation since it is a number between 1 and 10. Because the power 10^0 indicates no movement of the decimal it is customarily omitted.

You will notice in the solution to Example 1.1 that there are four figures preceding the times sign in *b* (5.116) and there are only two figures in *c* (9.2). This brings up the question, how many figures should there be? Would it, for example, have been correct to write 9.20×10^{-2} instead of 9.2×10^{-2}? In scientific notation all numbers preceding the times sign are *significant figures.* The problem is, *how many* significant figures should there be?

If there are exactly 120 students in a certain classroom, all three figures have significance; that is, there are not 119 students, nor 121, nor $119\frac{1}{2}$ students (hopefully), but there are *exactly* 120 students. This number could therefore be correctly written in scientific notation using three significant figures and would appear as 1.20×10^2.

On the other hand, if you are told that the population of a certain city is 42,248, you might question the last three digits in this number. Even if the number were the result of a very recent census, there may have been a number of births and deaths since then, not to mention families who have moved into or away from the city. If the census were not too recent, it might be safer and more correct to give the population as 42,000 by rounding to the nearest thousand. This number might then be more correctly written in scientific notation as 4.2×10^4.

A similar situation would occur is someone told you his height is 68.187493 inches, or his weight is 186.48395 pounds. You would question the numbers and the methods by which they were obtained. Depending on the situation, these two numbers would be more correctly stated by rounding them to two or perhaps three significant figures. It should be apparent then that the number of significant figures depends entirely upon the situation. In this text we will generally use *three* significant figures. Situations where more or less significant figures are warranted will be noted.

SUMMARY

One of the most important attributes of matter is the existence of various electric forces: the electrostatic, or coulomb, forces present between stationary electric charges; the magnetic forces associated with moving charges; and the elastic and chemical forces associated with individual atoms. Electrostatic and magnetic forces are long-range forces and their effects can be explained by the use of electric and magnetic fields. Elastic and chemical forces are short-range forces and are used to explain atomic bonding. The three types of atomic bonding which lead to the formation of molecules are ionic, covalent, and metallic. Recognition and understanding of these types of atomic bonding leads to the practical applications of electric forces. The classification of a material as a conductor, semiconductor, or insulator depends upon the structure of the atom, the type of bonding, and the presence (or absence) of free electrons. Scientific notation is a useful method of mathematical shorthand.

GLOSSARY

Atomic number. *A number that identifies each element in the periodic table; it is equal to the number of orbital electrons surrounding the nucleus.*

Bohr model. *A planetary model of the atom in which the central nucleus, composed of neutrons and protons, is surrounded by a number of orbiting electrons.*

Conductor. *Any material which contains an abundance of electrons that are free to move about in the material.*

Covalent bonding. *The formation of molecules or crystalline structures through sharing of electrons by atoms.*

Electrostatic force. *The force of attraction or repulsion between stationary charged bodies; also known as coulomb forces.*

Inertia. *The tendency for a body to resist any change in its motion.*

Insulator. *A body that does not contain free electrons.*

Ionic bonding. *The formation of molecules by the attraction of oppositely charged ions.*

Magnetic force. *Electric forces exerted between charges in motion.*

Mass number. *The total number of nucleons in the nucleus of an atom.*

Matter. *Any substance which (a) has inertia, (b) exerts gravitational forces on other matter, (c) exerts electric forces on other matter, and (d) exerts nuclear forces on other matter.*

Metallic bonding. *The type of atomic bonding found in most metals whereby the positive metal ions are held together by attraction to the cloud of free electrons.*

Scientific notation. *A method of mathematical shorthand.*

Semiconductor. *A material which has too many free electrons to be considered a good insulator, but not enough to be considered a good conductor.*

Significant figures. *All numbers preceding the times sign in scientific notation.*

Valence. *The number of electrons in the valence shell of an atom, or the number of electrons needed to fill the valence shell.*

IMPORTANT RELATIONSHIPS

- Newton's law for the gravitational force of attraction between two bodies:

$$F = k\frac{m_1 m_2}{d^2}$$

- Coulomb's law for the electrostatic force of attraction between two bodies:

$$F = k\frac{Q_1Q_2}{d^2}$$

REVIEW QUESTIONS

These questions give you the opportunity to determine whether you have learned the material in this chapter.

1. One of the most important attributes of matter is that it exerts electric forces on other matter. (TF)
2. Although the effects of gravitational forces can be accurately calculated, it is not known exactly what a gravitational field is. (TF)
3. Four types of electric forces present in matter are _____, _____, _____, and _____.
4. The force of attraction between two oppositely charged bodies is inversely proportional to the distance separating the bodies. (TF)
5. The electric force between charges in motion is called a _____ force.
6. The electric force between stationary charged bodies is called an electrostatic force. (TF)
7. The atomic number of an element is equal to the (*a*) valence, (*b*) number of planetary electrons, (*c*) number of neutrons in the nucleus, (*d*) total number of neutrons and protons in the nucleus.
8. The mass number of an element is equal to the total number of protons and neutrons in the nucleus of one atom of that element. (TF)
9. Electrons and protons have the same amount of charge but they are of opposite polarities. The charge of the electron has a _____ polarity while the charge of the proton has a _____ polarity.
10. The three types of atomic bonding are _____, _____, and _____.
11. Practical insulators contain an abundance of free electrons. (TF)
12. In scientific notation a positive power of ten means that the decimal must be shifted to the right in order to write the original number. (TF)
13. The number of significant figures which can be reasonably read from a ten inch slide rule is _____.

PROBLEMS

1. Make a sketch of the Bohr model of lithium.
2. Make sketches of the Bohr models of carbon and magnesium.
3. Using the results of Problem 1, determine the valence of lithium.
4. Determine the valence of oxygen.
5. Write the following numbers in scientific notation: (*a*) 857,400,000 (*b*) 0.00000000428 (*c*) 0.0120 (*d*) 7.35.
6. Change the following numbers from scientific notation to conventional decimal form: (*a*) 7.94×10^{13} (*b*) 1.20×10^{-11} (*c*) 4.81×10^{22}.
7. If you measured the width of a door with an ordinary ruler, explain how many significant figures you would use to express this measurement.
8. Use scientific notation to express the following: (*a*) $147{,}000 \times 386$ (*b*) 0.00074×682 (*c*) 0.0038×0.00084 (*d*) $2{,}693/73.9$ (*e*) $0.00843/144{,}000$ (*f*) $94{,}000/0.0025$ (*g*) $14{,}000 \times 843/0.007 \times 4{,}380$ (*h*) $221 \times 0.0073 \times 84/483 \times 0.00073 \times 0.43$
(*i*) $(73.4 + 19.7)0.0076 + 683 \times 0.0091/4{,}940{,}000 \times 0.00073(21.3 - 85.2)$.

FUNDAMENTALS OF ELECTRICITY

In the previous chapter, it was shown that the nuclei of atoms contain positively charged particles called protons while the planetary electrons surrounding the nuclei are negatively charged. Since the charge on the proton and that on the electron are equal in magnitude but opposite in sign, atoms which have an equal number of electrons and protons are electrically neutral. You will recall that atoms that have lost electrons, or have gained electrons, have a net positive, or a net negative charge and are called ions. Thus it is possible to produce a charged body by creating an excess or a deficit of electrons. In the process of producing a charged body, electrons must be transported or moved from one place to another. Furthermore, there must be some force applied to move these electrons. In this chapter we will consider the effects of static charge, moving charge (which is electric current), and the forces necessary to move the electrons.

2.1 ELECTRIC CHARGE

It has been known for many years that charged bodies can be produced by friction. The Greeks observed the phenomenon by rubbing pieces of amber* with fur or cloth and noting the forces exerted between the amber and bits of

*Amber is a yellowish to brownish translucent fossil resin.

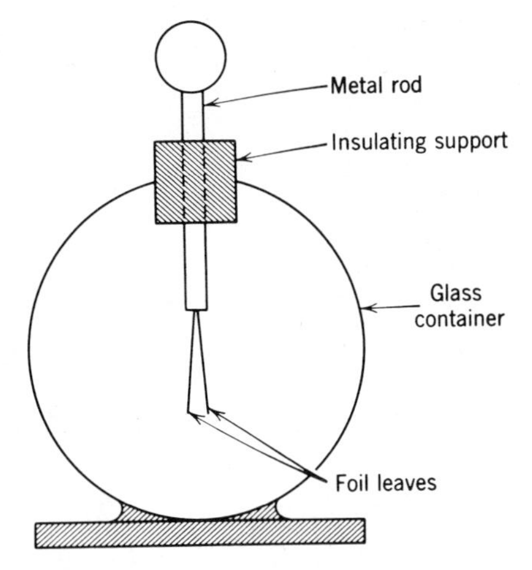

Figure 2.1 An electroscope

string or paper. The Greek word for amber was "elektron" and such charged bodies were said to be *electrified.* A glass rod can be similarly electrified by rubbing it with a piece of silk. There are in fact two ways by which a body can be charged: by gaining electrons and by losing electrons. The amber gains electrons from the fur and is said to be negatively charged. The glass rod actually gives up electrons to the silk and thus has a net positive charge. The two types of charge were recognized by early experimenters and the charge on the amber was called *resinous* while the charge on the glass rod was called *vitreous.** A very simple instrument called an *electroscope* can be used to detect charged bodies and to demonstrate a number of fundamental concepts.

The electroscope shown in Fig. 2.1 consists of a central rod with two gold or aluminum foil leaves attached to one end, a structure made of good insulating material to support the rod, and a glass container to protect the leaves from air currents. The foil leaves are attached to the rod in such a way that they move very easily. In Fig. 2.2*a* a piece of amber which has been previously

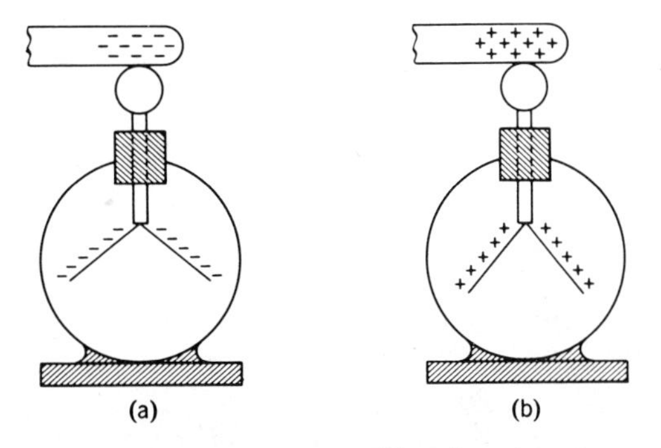

Figure 2.2 Force of repulsion between like charges

*It is interesting to note that the electron is said to have a negative charge and the proton is said to have a positive charge simply because early experimenters recognized the two types of charge (resinous and vitreous) and arbitrarily attached a negative sign to the charge on the amber and a positive sign to the charge on the glass. This sign convention is now universally accepted.

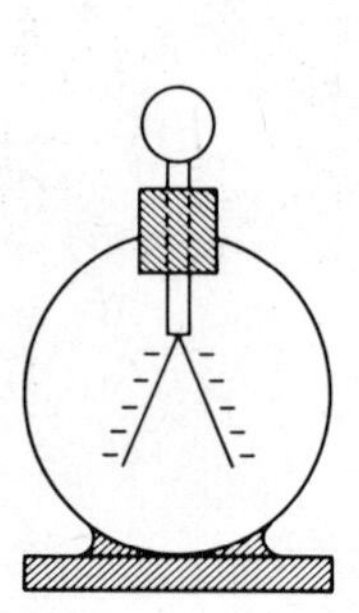

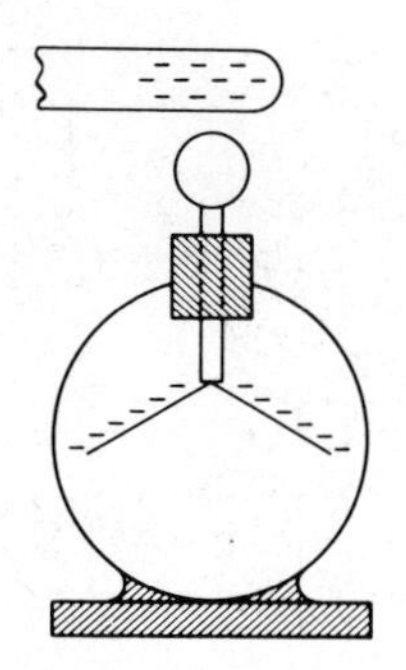

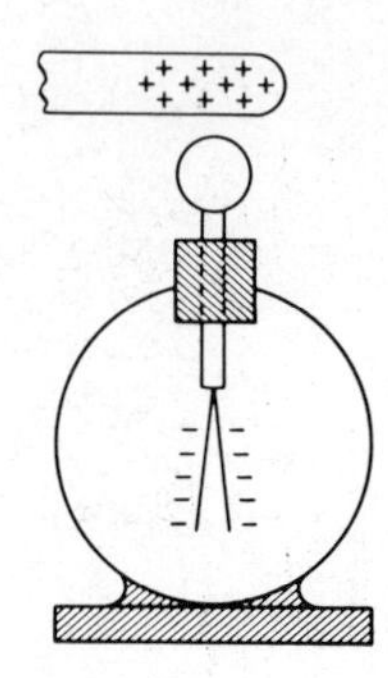

Figure 2.3 Demonstration of two different kinds of charge

charged is touched to the rod of the electroscope. Some of the excess electrons from the amber have traveled down the rod and spread themselves evenly over the foil leaves. The fact that the leaves are spread apart shows the repulsive force between charges of like sign.

In Fig. 2.2*b* a glass rod which has been charged is touched to the electroscope. Since the glass rod has a deficit of electrons (a positive charge), some of the electrons from the foil leaves travel up to the glass rod leaving the leaves with a net positive charge. Again, the fact that the leaves spread themselves shows the repulsive force present between charges of like sign.

If the electroscope is charged as in Fig. 2.2 and then left standing, the leaves will eventually fall back together again. This is because the support for the rod is not a perfect insulator and the charge on the leaves is eventually neutralized. If the support were a perfect insulator, the leaves would remain extended indefinitely.

The existence of two different types of charge is demonstrated in Fig. 2.3. The electroscope is initially charged negatively as shown. If a negatively charged rod is now brought close to the electroscope without touching it, the foil leaves will move even further apart. This is a result of the negatively charged rod forcing even more electrons down onto the leaves. On the other hand, a positively charged rod brought close to the electroscope will cause the leaves to come together slightly. This is a result of the positive charge on the rod attracting some of the electrons upward away from the foil leaves. There is then a force of attraction between charges of *unlike* sign; in this case the force of attraction is between the positive charge on the glass rod and the negatively charged electrons on the foil leaves.

The basic unit for measuring the *quantity* of electric charge is the *coulomb,* named after the famous French engineer and scientist, Charles-Augustin Coulomb (1736–1806). The most natural unit for the amount of charge is the electron (or proton). Since 1900 many ingenious experiments have been conducted to determine the amount of charge present on one electron. It is now known to three significant figures that the charge on one electron is 1.60×10^{-19} coulomb. We could on the other hand say that one coulomb of charge is equivalent to an excess of $1/(1.60 \times 10^{-19}) = 6.24 \times 10^{18}$ electrons. Since charge can be thought of as a *quantity* of electrons, the symbol Q is used to designate a charge measured in coulombs.

• • •

Example 2.1 A certain dielectric has an excess of 3.12×10^{18} electrons. What is the charge stored on the dielectric?

Solution Since the charge on one electron is -1.60×10^{-19} coulomb, the charge on 3.12×10^{18} electrons is

$$Q = -1.60 \times 10^{-19} \times 3.12 \times 10^{18} = -0.500 \text{ coulomb} = -0.500 \text{ C}$$

The minus sign in the answer demonstrates that there is an excess of electrons.

• • •

Example 2.2 What must be done to a certain dielectric in order to give it a net charge of +2.00 C?

Solution In order to achieve a positive charge, electrons must be removed from the dielectric. Since one C is equivalent to 6.24×10^{18} electrons, we must remove

$$2.00 \times 6.24 \times 10^{18} = 1.25 \times 10^{19} \text{ electrons}$$

• • •

.2 MOVING CHARGE—ELECTRIC CURRENT

The movement of electric charge can be demonstrated by connecting a charged body to a previously discharged electroscope using a rod as shown in Fig. 2.4. The fact that the leaves of the electroscope diverge shows that some of the charge from the charged body has moved through the rod down to the foil leaves. If the connecting rod is a good insulating material, the leaves will move apart rather slowly, showing that the charge does not readily move through the insulating material. If the rod were made of some semiconductor material the leaves would move apart much more rapidly showing a greater ease of movement of electrons through the material. Finally, if the rod were made of copper (a good conductor) the leaves would fly apart almost instantaneously.

In the experiment shown in Fig. 2.4, it is obvious that some of the electrons from the charged body have moved through the connecting rod down to the leaves of the electroscope. Let us suppose that the rod is made of copper and investigate the movement of the electrons through the rod. We know from our previous work that there is a cloud of free electrons contained within the copper. If we could trace the movements of one single electron within a piece of copper it might appear as shown in Fig. 2.5*a*. We notice that the electron

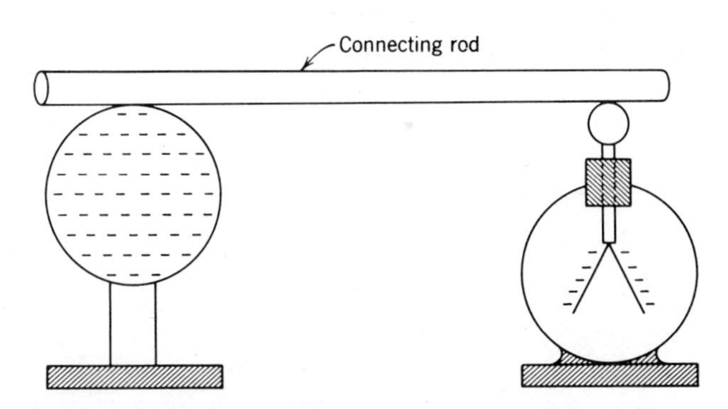

Figure 2.4 The movement of charge

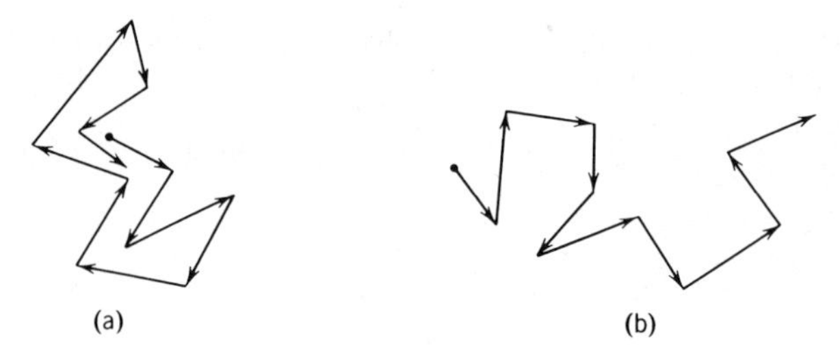

Figure 2.5 Movement of free electrons (*a*) Random drift (*b*) Net drift to right

moves randomly within the copper with no net movement in any particular direction. However, when the copper is used as the connecting rod between the charged body and the electroscope, the same electron might have a movement as shown in Fig. 2.5*b*. In this case, the electron is seen to have a net movement to the right. Since the movements of this one electron will affect the movements of its neighboring electrons, and they in turn will affect the movements of their neighbors, it is easily seen that the entire cloud of free electrons in the copper rod will drift toward the right. This net movement or flow of electrons through the copper is called *electric current.*

In order to obtain a specific definition for electric current, we note two things: (1) there is a certain number of electrons and therefore a certain quantity of electric charge involved, and (2) this charge moves in different periods of time as shown by the experiment in Fig. 2.4. In measuring electric current, we are interested in the quantity of charge which flows during a certain period of time, or the *rate of flow of charge.* Therefore,

electric current is defined as the rate of flow of electric charge.

The symbol for electric current is *I*, from the French phrase, "*Intensité* of electron flow."

It is important to note that in this case the sign of the moving charge is negative since the electrons are in motion. There are instances, however, where the sign of the moving charge is positive, such as the positive ions in the electrolyte (Chapter 4) of a battery or in a gas-filled tube. This does not alter our definition, however, since we are concerned simply with the total quantity of charge and the rate at which it moves.

Consider now the length of copper wire shown in Fig. 2.6. It is assumed that the electrons in the wire are flowing toward the right at a constant rate. Let us suppose that somehow it is possible to count the total number of electrons passing through the plane labeled A. If we were to count exactly 6.24×10^{18}

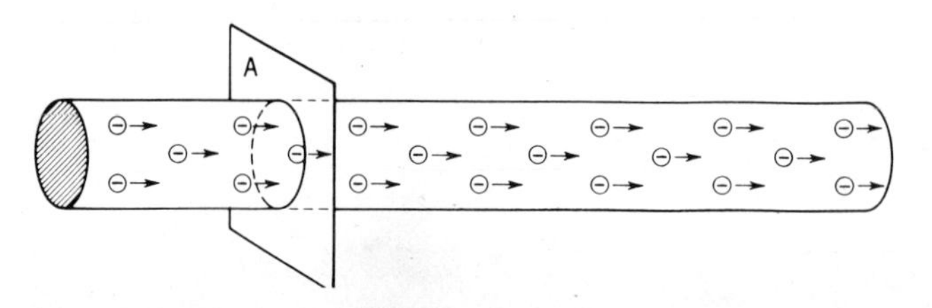

Figure 2.6

electrons passing through the plane in one second, this would represent a rate of flow of electric charge of 1 C per second. The electric current would therefore be 1 C/second. If the number of electrons passing through the plane in one second were doubled, the electric current would then be 2 C/second. It is, therefore, quite proper to measure electric current in units of coulombs/second. However, a special unit of electric current (the ampere) has been defined in honor of the French physicist, André Marie Ampère (1775–1836). Thus,

an electric current of one ampere is defined as the flow of one coulomb of electric charge through a plane in one second.

Based on this definition, electric current can be conveniently expressed by

$$I(\text{amperes}) = \frac{Q(\text{coulombs})}{t(\text{seconds})} \tag{2.1}$$

A convenient way to remember this definition and the proper units is

one ampere equals one coulomb per second.

Equation 2.1 can be rearranged to show that the total charge in coulombs is equal to the product of the current in amperes multiplied by the time in seconds:

$$Q(\text{coulombs}) = I(\text{amperes}) \times t(\text{seconds}) \tag{2.2}$$

Alternatively, Eq. 2.1 can be rearranged to solve for the time in terms of the current and the charge:

$$t(\text{sec}) = \frac{Q(\text{coulombs})}{I(\text{amperes})} \tag{2.3}$$

Finally we note that if the plane in Fig. 2.6 were moved to any other point on the wire, the total number of electrons passing through it would remain the same. Thus the electric current is seen to be the same across any section of the wire. This is reasonable since we assume that we can neither gain nor lose electrons along the length of the wire.

• • •

Example 2.3 If 2.00 C of charge pass through the plane in the copper wire in Fig. 2.6 in 0.100 sec, what is the current?

Solution The current is found using Eq. 2.1 and thus

$$I = \frac{Q}{t} = \frac{2.00\text{ C}}{0.100\text{ sec}} = 20.0\text{ amperes} = 20.0\text{ A}$$

• • •

Example 2.4 If 1.20 C of charge pass a certain point in a copper wire every minute, what is the electric current?

Solution Using Eq. 2.1

$$I = \frac{Q}{t} = \frac{1.20\text{ C}}{60.0\text{ sec}} = 0.0200\text{ A}$$

• • •

Example 2.5 How long will it take for 0.00500 C of charge to pass through a lamp if the current is 0.100 A?

Solution From Eq. 2.3

$$t = \frac{Q}{I} = \frac{0.00500\ \text{C}}{0.100\ \text{A}} = 5.00 \times 10^{-2}\ \text{sec}$$

• • •

2.3 ELECTRON VELOCITY

Since current is the flow of electric charge during a period of time, it is interesting to consider the *average velocity* of the moving charge. There is a current of 10 A in the copper conductor shown in Fig. 2.7. Since a current of 10 A represents a flow of charge of 10 C/sec, the conductor has been divided into equal sections, each of which contains 10 C of electrons. All of the electrons in any one section must move into the next section in exactly one second in order to maintain a current of 10 A. We can therefore solve for the average velocity with which the electrons move by determining the length L of each section of wire.

The number of electrons required for one C is 6.24×10^{18}, and thus 6.24×10^{19} electrons are equivalent to 10 C of charge. Since one cubic inch of copper contains about 1.40×10^{24} electrons, the volume of copper necessary to contain 6.24×10^{19} electrons (10 C) is

$$\text{Volume} = V = \frac{6.24 \times 10^{19}\ \text{electrons}}{1.40 \times 10^{24}\ \text{electrons/in.}^3} = 4.46 \times 10^{-5}\ \text{in.}^3$$

The volume of each section of conductor in the figure is the product of the cross-sectional area A and the length L, or $V = LA$. Thus the L of each section is found to be

$$L = \frac{V}{A} = \frac{4.46 \times 10^{-5}\ \text{in.}^3}{4.46 \times 10^{-2}\ \text{in.}^2} = 1.00 \times 10^{-3}\ \text{in.} = 0.001\ \text{in.}$$

Thus the electrons must move at an average velocity of 0.001 in./sec.

This low velocity of electron flow is surprising considering that the effect of turning on a light switch is almost instantaneous. This brings us to a very important point. Even though the effect of the 10 A current is instantaneous* at all points along the conductor, it would require about 1000 sec for any one electron to travel only one inch along the wire. An analogy to this situation

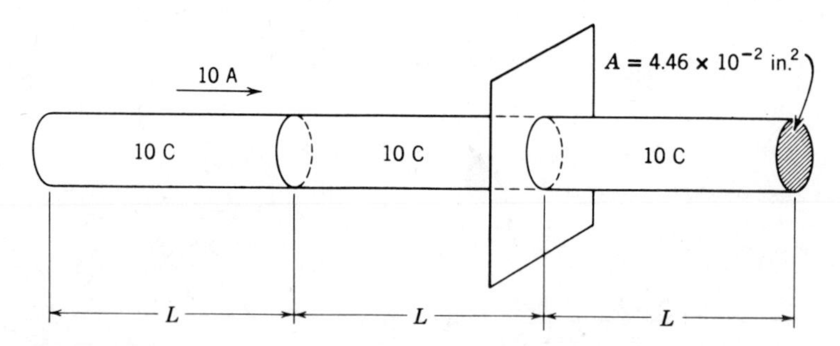

Figure 2.7

*The effect occurs at almost the speed of light, 186,000 mi/sec.

might be a long tube which is completely filled with marbles. If one marble is forced into one end of the tube, one marble must leave the other end of the tube. Thus, even though it might take a long period of time for any one marble to travel completely through the tube, the effect of inserting a marble at one end is felt almost immediately at the other end.

2.4 POTENTIAL DIFFERENCE—EMF

We have seen it is possible to produce a charged body by friction. To produce a charged body in this fashion, it is necessary to perform a certain amount of *work* (e.g., the act of rubbing the fur and the amber, or the silk and the glass). As we shall see in the next chapter, the exertion of a force is one of the basic requirements in performing work. Thus the separated charges have the ability to do work since any charged body exerts forces on other charged bodies. The ability to do work is often referred to as *potential.*

If we examine the electrons in a piece of copper wire as shown in Fig. 2.8, we see that each electron exerts a repulsive force on every other electron. Therefore each electron has the ability to do work, or has potential. However, the electrons will arrange themselves so that the mutual forces of repulsion between them are balanced and the cloud of free electrons is said to be in *equilibrium.* An analogy is the tendency of the molecules of a gas to distribute themselves evenly throughout the interior of a closed container. As a result, the potential of every electron is the same. Even though the electrons move about randomly within the wire, equilibrium is maintained and no net movement of electrons occurs; in other words, there is no electric current.

The two bodies in Fig. 2.9*a* both contain an excess of electrons. However, body A contains a much greater charge than body B. It is easily seen that all electrons on A have the same potential and thus there is no electric current. The same is true for all electrons on B. However, since the charge on A is much greater than that on B, the electrons on A have a much greater potential. This is true since the mutual forces of repulsion between the electrons on A are larger. As a result, there is a *potential difference between* A and B. It is important to note that a *difference* in potential is determined between two different points (in this case it exists between the two bodies).

If a copper rod is now connected between the two bodies as shown in Fig. 2.9*b*, it is obvious that electrons will travel from A to B. Thus there is an electric current. It is precisely this difference in potential which is necessary to produce electric current. After a period of time, the electrons will have established equilibrium as shown in Fig. 2.9*c* and they will all be at the same potential. There is, therefore, no potential difference, and thus no electric current exists. In summary,

there must be a potential difference between two points in order to sustain an electric current between them.

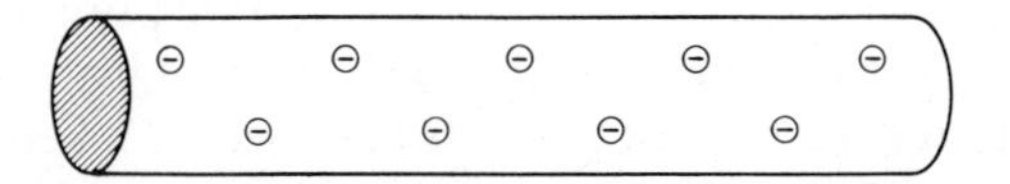

Figure 2.8 Electrons at the same potential

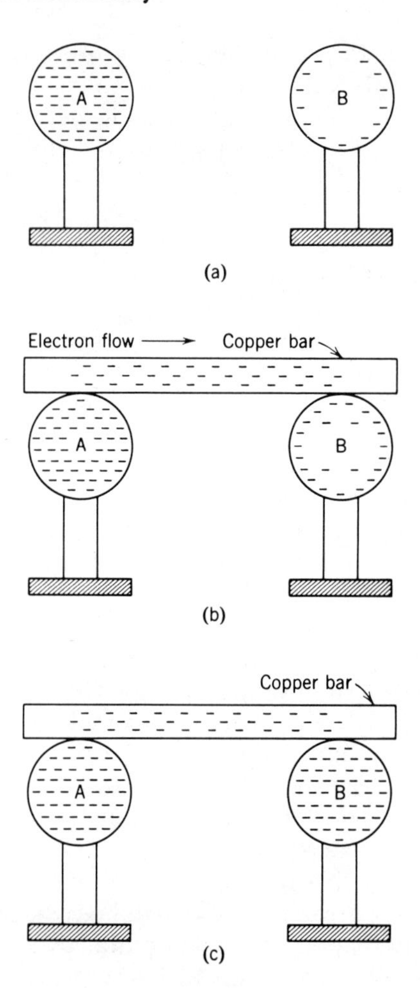

Figure 2.9 (*a*) A difference of potential between two bodies (*b*) Electric current caused by potential difference (*c*) Zero potential difference. No current flow

Referring again to Fig. 2.9, there is a certain potential difference between A and B *before* the copper wire is connected. A short time *after* the wire is connected, some electrons have traveled away from A and additional electrons have accumulated on B. The potential difference between A and B is, therefore, reduced, resulting in a smaller electric current. After a sufficient period of time, equilibrium is reached (the electrons are evenly distributed throughout the system), the potential difference is zero, and there is no electric current. Thus the electric current in the wire changes with time (is not constant) and, in fact, exists for only a very short period of time. This was one of the main problems faced by early experimentors in electricity. The problem was not overcome until the discovery of the *voltaic pile* by Alessandro Volta (1745–1827).

Volta, a professor at the University of Pavia, Italy, noted that the muscles and nerves of a dead frog would twitch when touched by two dissimilar metals which were separated by a paper which had been soaked in salt water. In order to increase the effect, Volta piled up alternate disks of silver and zinc separated by paper which had been soaked in acid; thus the name voltaic pile

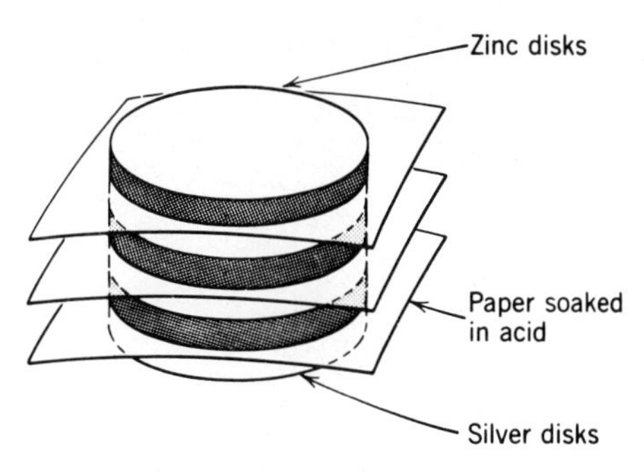

Figure 2.10 Voltaic pile

(Fig. 2.10). Finally he devised the scheme of submersing two dissimilar metals in a glass cup containing acid. Such an arrangement is shown in Fig. 2.11 and is known as a *voltaic cell*; we will discuss such cells in detail in Chapter 4. Any solution which will sustain an electric current is defined as an *electrolyte,* and thus the acid in the voltaic cell is called the electrolyte. The metallic bars are called *electrodes.* Other experimentors soon constructed voltaic cells and noted some of the effects of electric current such as the decomposition of water into hydrogen and oxygen gases, the heating effect in a wire, the brilliant light given off by an electric spark, and electroplating.

Even though these early researchers did not fully understand the process, it is now known that the *chemical action* between the electrolyte and the two metals results in a separation of charge. The zinc bar then accumulates an excess of electrons and becomes negatively charged while the copper bar loses electrons thus becoming positively charged. As a result of this chemical action, a potential difference is developed between the two electrodes. If a conductor is now connected between the electrodes as shown in Fig. 2.12, an electric current will be produced in the conductor (i.e., electrons will flow from the zinc to the copper). Unlike the experiment shown in Fig. 2.9, however, the electric current produced by the cell will be *continuous*; that is, the electrons will continue to flow through the wire at a constant rate since the chemical action in the cell will continually replenish the charge on the electrodes.

Since a constant potential difference is developed between the electrodes of the cell, and since this potential difference is the force which causes the

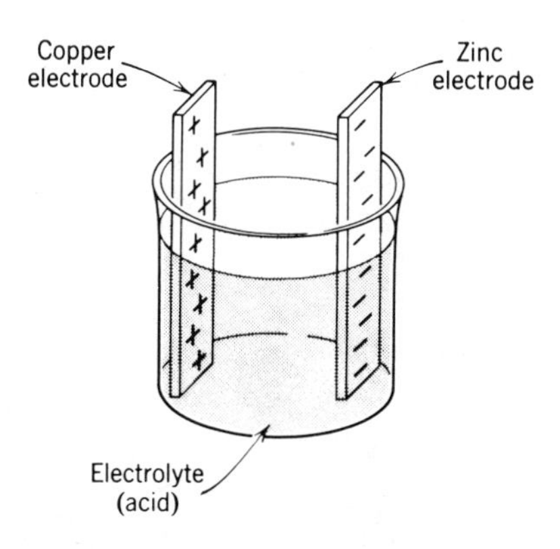

Figure 2.11 A voltaic cell

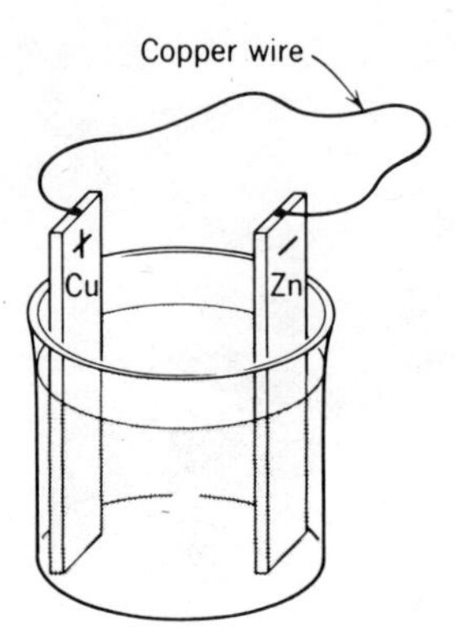

Figure 2.12 Electric current produced by a voltaic cell

electrons to move through the conductor, the potential of a cell is given the name *electromotive force,* or EMF. The unit of measurement of potential difference or electromotive force is the *volt,* named in honor of Alessandro Volta, and the symbol used is V. The EMF of a cell is essentially a measure of the amount of potential available to move a quantity of charge. Thus in the familiar English system of units,

the potential difference between two points is defined as one volt when 0.738 ft-lb (foot-pound) of work is necessary to move 1 C of charge from one point to the other.

The unit of work (or energy) in the MKS (meter-kilogram-second) system of units, which we will discuss in the next chapter, is the *joule,* and since one joule is equal to 0.738 ft-lb,

one volt is the potential difference between two points when one joule of work is required to move 1 C of charge from one point to the other.

The definition for the volt can be conveniently expressed in equation form:

$$V\,(\text{volts}) = \frac{W\,(\text{joules})}{Q\,(\text{coulombs})} \tag{2.4}$$

where V is the potential difference in volts, W is the work (or energy) in joules, and Q is the charge in coulombs. A convenient way to remember this definition and the proper units is

one volt equals one joule per coulomb.

The copper-zinc cell in Fig. 2.11 has an EMF of 1.1 V. Another word which is synonymous with potential difference is *voltage.* This term is widely used and thus it is meaningful to say "the voltage across the cell is 1.1 V," or to say "the cell has a voltage of 1.1 V."

• • •

Example 2.6 How many joules of work must be performed by the cell in Fig. 2.11 in order to transfer 10 C of charge from the copper to the zinc? How many ft-lb of work does this represent?

Solution Equation 2.4 can be rearranged as

$$W\,(\text{joules}) = V\,(\text{volts}) \times Q\,(\text{coulombs})$$

Therefore

$$W = 1.1\ \text{V} \times 10\ \text{C} = 11\ \text{joules} = 11\ \text{J}$$

Since there are 0.738 ft-lb/J, 11 joules is equivalent to

$$W = 11\ \text{J} \times 0.738\frac{\text{ft-lb}}{\text{J}} = 8.12\ \text{ft-lb}$$

• • •

Example 2.7 How many joules of work are performed by the cell in Fig. 2.12 in a period of 10 seconds if the current in the conductor is 2 A?

Solution The total charge moved is found using Eq. 2.2. Thus

$$Q = I \times t = 2\ \text{A} \times 10\ \text{sec} = 20\ \text{C}$$

The work performed is then found using Eq. 2.4,

$$W = V \times Q = 1.1\ \text{V} \times 20\ \text{C} = 22\ \text{J}$$

• • •

2.5 OHM'S LAW

In the previous section we have seen that a potential difference is necessary to produce an electric current. Thus, in Fig. 2.12 the EMF of the cell causes the cloud of free electrons in the wire to drift from the zinc electrode towards the copper electrode. If the wire were disconnected from the cell, the electric current in the wire would cease, that is, the cloud of free electrons in the wire would cease to drift. From these observations we can deduce that the free electrons in the wire exhibit an opposition to movement, and thus the wire itself is said to exhibit an opposition to electric current.

The nature of this opposition can be explained by examining the section of copper conductor shown in Fig. 2.13. There is an electric current in the wire, and the cloud of free electrons drifts from left to right. We have somewhat idealistically traced the path of one single electron. The applied EMF accelerates the electron towards the right. It does not travel far however before it suffers a collision with one of the fixed copper ions. At the point of collision, the electron decelerates rapidly (it is acted upon by the copper ion) and undergoes a change in direction. It begins to accelerate again because of the applied EMF, but soon suffers another collision. This process is repeated and the electron thus travels along, alternately undergoing accelerations due to the applied EMF and decelerations due to collisions with the fixed copper ions. The many collisions suffered by all of the free electrons as they drift

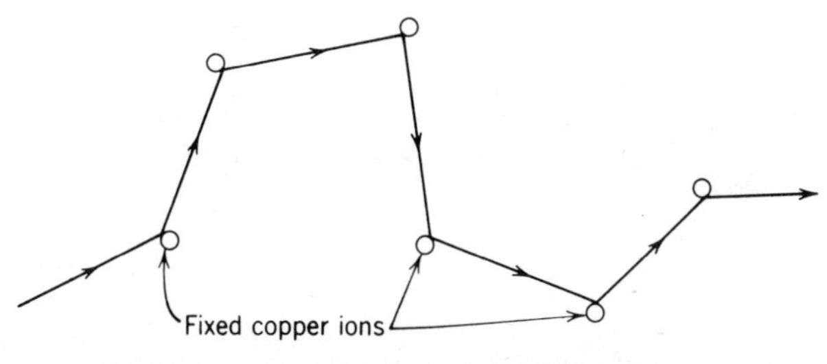

Figure 2.13 Idealized electron drift showing collisions

under the influence of the applied EMF constitute the opposition to electric current. Each time an electron suffers a collision, energy in the form of heat is imparted to a copper ion. This increase in heat energy will cause the temperature of a piece of copper wire to increase when an electric current is forced through it. This *opposition* to electric current is the property known as electric *resistance*, and is a property common to many different materials.

The electric resistance of a conductor was noted by Georg Simon Ohm (1787–1854). Using an apparatus similar to that shown in Fig. 2.14, Ohm discovered a very fundamental relationship. First, he noted that when the free end of the wire was connected to point A, a certain electric current was produced. Furthermore, every time he tried it, the current was the same. Secondly, he noted that the current doubled when the wire was connected to point B, and that it tripled when connected to point C.* From these observations, and others using different sizes of conductors, he determined that the value of the EMF divided by the current produced was equal to a constant. In equation form,

$$\frac{\text{EMF}}{\text{current}} = \text{constant}$$

Thus if an EMF of 1 V causes a current of 1 A, the constant is 1.0. Therefore, an EMF of 2 V must cause a current of 2 A in order to maintain the constant at 1.0. Similarly, an EMF of 3 V would cause a current of 3 A and so on. These experiments were quite difficult to perform at the time and Ohm was ridiculed for many years for his work. It was finally realized however that the constant proposed by Ohm was, in fact, a measure of the electric resistance of a circuit. The relationship then became known as *Ohm's Law*. The unit of resistance is the *Ohm*, the symbol for resistance is R, and the abbreviation for Ohm is the Greek letter *omega* Ω. In equation form Ohm's Law is

$$R(\text{ohms}) = \frac{V(\text{volts})}{I(\text{amperes})} \tag{2.5}$$

Since the volt and the ampere have previously been defined, the ohm can be defined as follows:

One ohm is the electric resistance when a potential difference of one volt produces a current of one ampere.

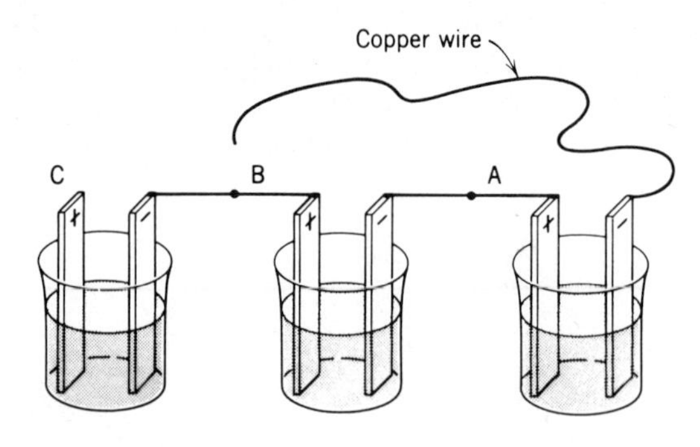

Figure 2.14

*It was necessary to maintain the temperature of the wire at a constant value.

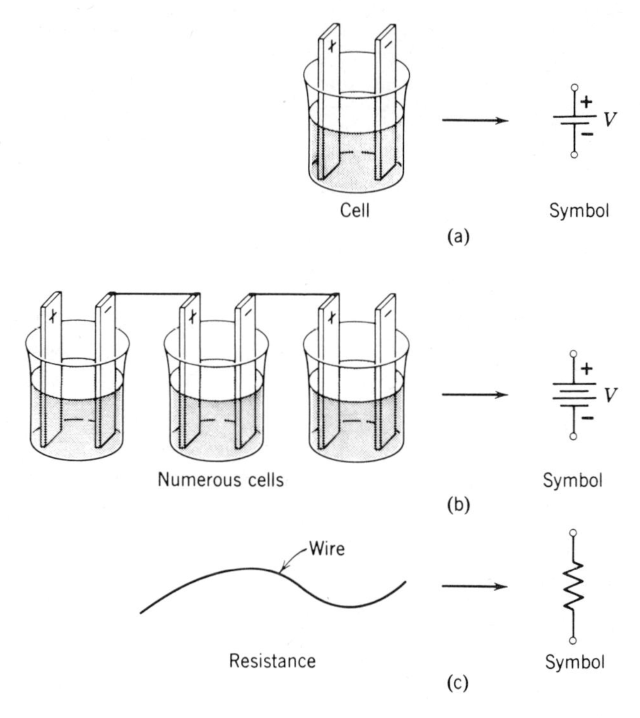

Figure 2.15 Common circuit symbols (*a*) Chemical cell (*b*) Battery (*c*) Resistor

Another definition of the ohm that has been established by international agreement is:

One ohm is the resistance of a column of mercury 106.3 cm long and 1 mm² in cross-sectional area held at a constant temperature of 0°C.

As it is not convenient to draw cells and wires as shown in Fig. 2.14, a set of standard symbols has been established to ease this problem. The symbols we are interested in at the moment are shown in Fig. 2.15 (a more complete group of symbols is given in the appendix). Figure 2.15*a* shows the symbol for a single cell. A number of cells connected as shown in Fig. 2.15*b* is defined as a *battery*. The resistance of a wire or other conductor is represented by the symbol shown in Fig. 2.15*c*. Note that this element is called a resistor.

We can then use the proper symbols to redraw the apparatus shown in Fig. 2.14. The result is shown in Fig. 2.16. You will notice that there is a complete

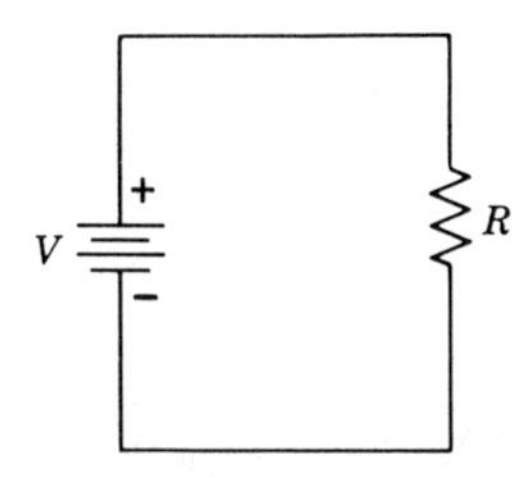

Figure 2.16

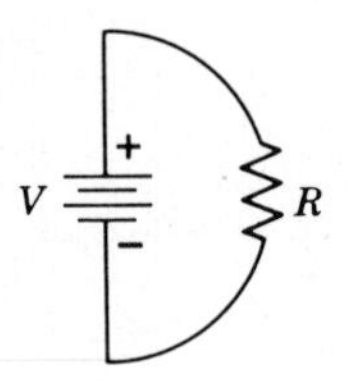

Figure 2.17

closed path around which electric current can exist. Such a closed path is called a *loop* or an *electric circuit*, or more simply, just a circuit. It is very important to notice that there must be a circuit in order for electric current to exist, since the current would cease if one of the components in the circuit were removed. We also assume that the wires used to connect the battery and the resistor are perfect, that is, they have no resistance. This is legitimate since the resistor and the battery could be connected as shown in Fig. 2.17; this is seen to be exactly the same circuit, but Fig. 2.16 is easier to draw, easier to work with, and more pleasant to look at.

• • •

Example 2.8 In the circuit shown in Fig. 2.16, $V = 10$ V, and there is a current of 5 A. What is the resistance of the circuit?

Solution Using Ohm's Law, Eq. 2.5,

$$R = \frac{V}{I} = \frac{10\text{ V}}{5\text{ A}} = 2\text{ ohms} = 2\ \Omega$$

• • •

Example 2.9 If the resistance in Fig. 2.16 is 5 Ω, make a table showing the values of current for a number of different values of voltage. Plot the results.

Solution Ohm's Law can be rearranged as

$$I\text{(amperes)} = \frac{V\text{(volts)}}{R\text{(ohms)}}$$

for $V = 5$ V, $I = 5\text{ V}/5\ \Omega = 1$ A; for $V = 10$ V, $I = 10\text{ V}/5\ \Omega = 2$ A, and so forth. This plot emphasizes the linear relationship between voltage and current (i.e., $I = 1/R \times V$ is the equation of a straight line).

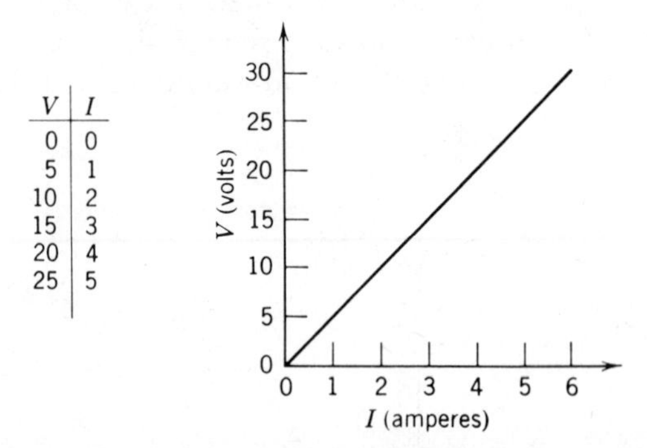

V	I
0	0
5	1
10	2
15	3
20	4
25	5

Figure (Example 2.9)

• • •

Example 2.10 If the current in Fig. 2.16 is 0.5 A and $R = 100\ \Omega$, what must be the applied EMF?

Solution Ohm's Law can also be arranged as

$$V\,(\text{volts}) = I\,(\text{amperes}) \times R\,(\text{ohms})$$

thus

$$V = 0.5\ \text{A} \times 100\ \Omega = 50\ \text{V}$$

• • •

2.6 PRACTICAL UNITS—PREFIXES

You will notice in the examples of the preceding section that three significant figures were not used. There are two very practical reasons for doing this. First, it is much easier to label the components in a circuit using only one or two significant figures as shown in Fig. 2.18*b*. Second, and more important, is the fact that most of the resistors used in practical circuits have tolerances of ±20%, ±10%, ±5%, and ±1%. A tolerance of ±20% on the 25-Ω resistor shown in Fig. 2.18 means that the actual resistance may be as low as 20 Ω or as high as 30 Ω. This certainly does not justify using three significant figures. In the remainder of this book we will label circuits as shown in Fig. 2.18*b*; we will however do all calculations to three significant figures. We are then making the assumption that the resistors and batteries in the circuits are ideal and have values *exactly* as shown. If resistor tolerances must be taken into account, we can then return to the circuit and make *worst case* calculations.

• • •

Example 2.11 Calculate the current in the circuit of Fig. 2.18*b*.

Solution

$$I = \frac{V}{R} = \frac{12\ \text{V}}{25\ \Omega} = 0.480\ \text{A}$$

• • •

Example 2.12 If the resistor in Fig. 2.18 has a ±10% tolerance, what will be the maximum and minimum values of current?

Solution 10% of 25 Ω = 2.5 Ω. Thus the maximum resistance is

$$R_{\text{MAX}} = 25\ \Omega + 2.5\ \Omega = 27.5\ \Omega$$

The minimum resistance is

$$R_{\text{MIN}} = 25\ \Omega - 2.5\ \Omega = 22.5\ \Omega$$

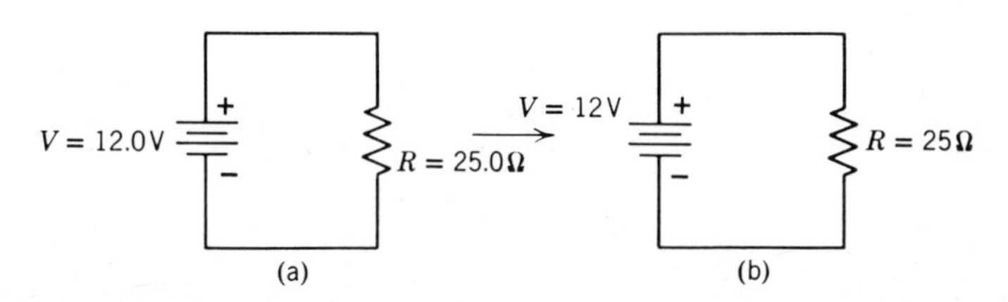

Figure 2.18 (*a*) Actual values (*b*) Ideal values

Thus, the minimum current is

$$I_{MIN} = \frac{V}{R_{MAX}} = \frac{12 \text{ V}}{27.5 \, \Omega} = 0.437 \text{ A}$$

and the maximum current is

$$I_{MAX} = \frac{V}{R_{MIN}} = \frac{12 \text{ V}}{22.5 \, \Omega} = 0.533 \text{ A}$$

• • •

Example 2.13 Calculate the current in the circuit shown in Fig. 2.19.

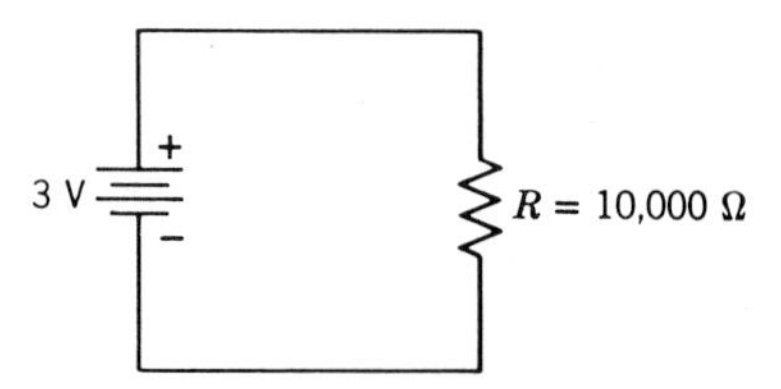

Figure 2.19

Solution Using Ohm's Law,

$$I = \frac{V}{R} = \frac{3 \text{ V}}{10{,}000 \, \Omega} = 0.000300 \text{ A}$$

• • •

Example 2.13 illustrates the value of scientific notation in solving electric circuit problems. Scientific notation is so widely used that a number of special symbols called *prefixes* have been developed. These prefixes are simply symbols for the various powers of ten, and some of the more common ones are shown in Table 2.1 along with the proper abbreviations. The prefixes most commonly encountered in electronics are shown in bold face type.

Table 2.1.

Prefix	Abbreviation	Power of Ten
Tera	T	10^{12}
Giga	G	10^{9}
Mega	M	10^{6}
kilo	k	10^{3}
hecto	h	10^{2}
deka	da	10^{1}
—	—	—
deci	d	10^{-1}
centi	c	10^{-2}
Milli	m	10^{-3}
Micro	μ	10^{-6}
Nano	n	10^{-9}
Pico	p	10^{-12}
femto	f	10^{-15}
atto	a	10^{-18}

A few examples will serve to illustrate the use of these prefixes.

$$9{,}000\ \Omega = 9 \times 10^3\ \Omega = 9 \text{ kilohms} = 9\ \text{k}\Omega$$
$$2{,}000{,}000\ \Omega = 2 \times 10^6\ \Omega = 2 \text{ Megohms} = 2\ \text{M}\Omega$$
$$4{,}100\ \text{V} = 4.1 \times 10^3\ \text{V} = 4.1 \text{ kilovolts} = 4.1\ \text{kV}$$
$$0.003\ \text{A} = 3 \times 10^{-3}\ \text{A} = 3 \text{ milliamperes} = 3\ \text{mA}$$
$$0.000{,}004\ \text{A} = 4 \times 10^{-6}\ \text{A} = 4 \text{ microamperes} = 4\ \mu\text{A}$$

Let us now make use of these prefixes by applying them to an electric circuit.

• • •

Example 2.14 There is a current of 20 μA in the circuit shown in Fig. 2.20. What must be the battery EMF?

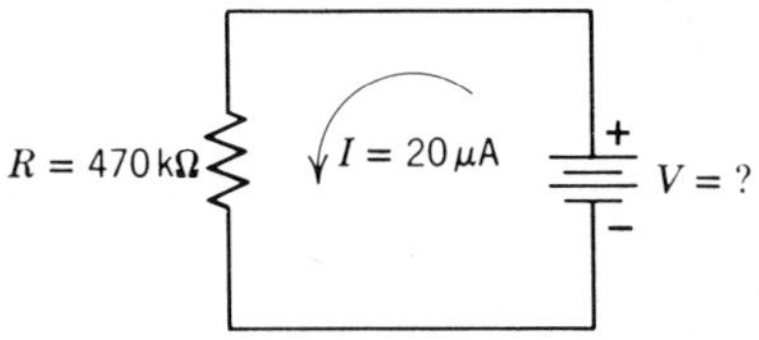

$R = 470\ \text{k}\Omega$ $I = 20\ \mu\text{A}$ + − $V = ?$

Figure 2.20

Solution Using Ohm's Law, and observing the proper units,

$$V = I \times R = 20\ \mu\text{A} \times 470\ \text{k}\Omega = 20 \times 10^{-6}\ \text{A} \times 470 \times 10^3\ \Omega = 9.40\ \text{V}$$

• • •

Example 2.15 How long will it take for 300 μC of charge to pass through the lamp in Fig. 2.21 if the lamp has a resistance of 2.4 kΩ?

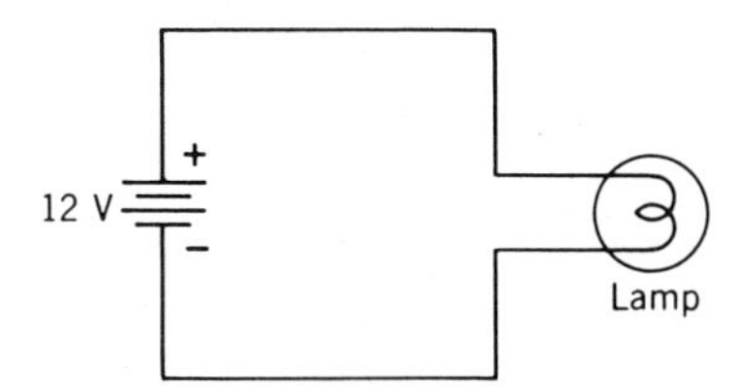

Figure 2.21

Solution We must first find the current in the lamp. Using Ohm's law,

$$I = \frac{V}{R} = \frac{12\ \text{V}}{2.4\ \text{k}\Omega} = 5.00\ \text{mA}$$

The time can then be found using Eq. 2.3

$$t = \frac{Q}{I} = \frac{300\ \mu\text{C}}{5.00\ \text{mA}} = 60.0\ \text{msec}$$

• • •

2.7 TYPES OF ELECTRIC CURRENT

At this point we should discuss a number of practical matters regarding electric circuits. First, notice that we have not discussed the direction of current in an electric circuit. The actual charge carriers which constitute electric current are shown in Fig. 2.22*a*. From our previous discussions we know that the charged particles which move in the copper wire are electrons. The electrons drift through the wire from the negative electrode of the cell (zinc rod) towards the positive electrode (copper rod). The charge carriers in the electrolyte of the cell are ions, some of which are positive, some negative. These carriers move in opposite directions through the electrolyte as shown (we will discuss cells in Chapter 4). Thus the current in this circuit is composed of charge carriers of unlike signs moving in opposite directions. There are numerous other examples in which current is due to the movement of both positive and negative charges. For example, semiconductors (electrons and fictitious positive particles called holes), gas-filled tubes (electrons and positive ions in a neon light or tube), and high energy nuclear accelerators (electrons, protons, ions, and other charged bodies).

It would be convenient if we could simply consider electric current in one direction only around an electric circuit; either clockwise or counterclockwise. This would be satisfactory since the definition of the ampere is simply

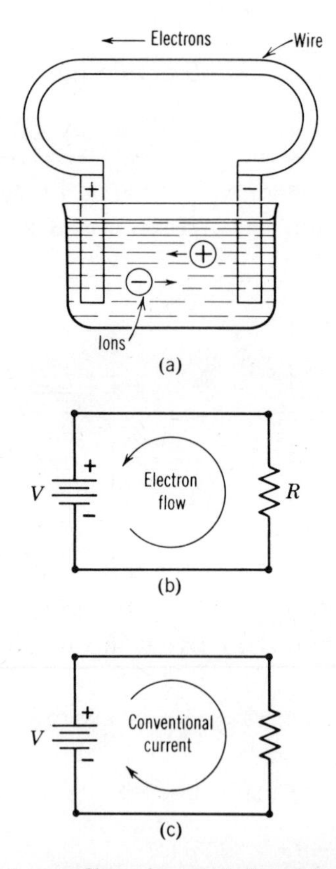

Figure 2.22 Electric current direction (*a*) Actual current carriers (*b*) Electron flow direction (*c*) Conventional current direction

concerned with the *total amount* of charge moved in a period of time. Since there are carriers moving in both directions through the circuit, two different conventions have been adopted—unfortunately. One convention considers current as carriers moving around the circuit in a counterclockwise direction as shown in Fig. 2.22*b*. This convention is based on a *negative* test charge and is referred to as *electron flow*.

The second convention considers current as carriers moving around the circuit in a clockwise direction as shown in Fig. 2.22*c*. This convention is based on a *positive* test charge and is called *conventional current direction*. This convention is in agreement with the idea of the positive test charge used in discussing electric fields, is widely used in other electrical and electronic textbooks, and is adopted almost exclusively in the field of electrical engineering. It really makes little difference which method we use so long as we are *consistent. In this text* we will adhere strictly to *conventional current direction*, and we will consider the actual current carriers only in cases where we are concerned with the physical operation of a device (e.g., sources in Chapter 4).

Electric current then is in a direction such that the carriers *leave* the *positive* terminal of a source, and *enter* the *negative* terminal. The carriers thus flow from a point of *high* potential (more positive) to a point of *lower* potential (more negative). This fits well with the analogy of water flowing from a point of higher elevation to a point of lower elevation.

In Fig. 2.22*c*, the current is constant and always has the same direction. A graph of the current in this circuit versus time is shown in Fig. 2.23 and it is seen to have a constant value. That is, it changes neither in *magnitude* nor *direction* as time passes. Such an electric current is called *direct current*, and the abbreviation dc is used.

The electric current available in a home or business, however, is quite different. A graph of the current available from a commercial power system is shown in Fig. 2.24. In this case the current changes with time *both* in magnitude and in direction. From time t_0 to t_1, the current is in one direction (called the positive direction for reference), and from time t_1 to t_2 the current has reversed and is in the opposite direction (called the negative direction). Since the current *alternately* changes direction from positive to negative, it is called *alternating current*, and is abbreviated ac. A number of other types of electric currents are shown in Fig. 2.25. We will encounter and discuss some of these types of currents in this book; others we will not encounter until we study electronic circuits.

A few points on *terminology*. First, an electric current is due to the *flow* of charge carriers; thus it is redundant to speak of *current flow*—only of *current*. It is, however, common practice to speak of a current flowing through a resistor, or a current flowing through a circuit, and since no loss of meaning or

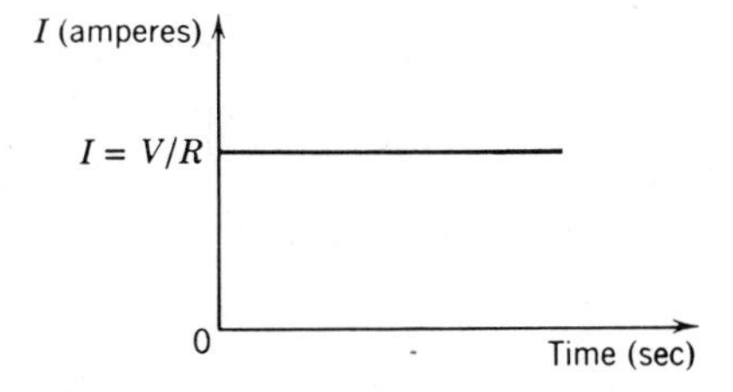

Figure 2.23 Direct current—dc

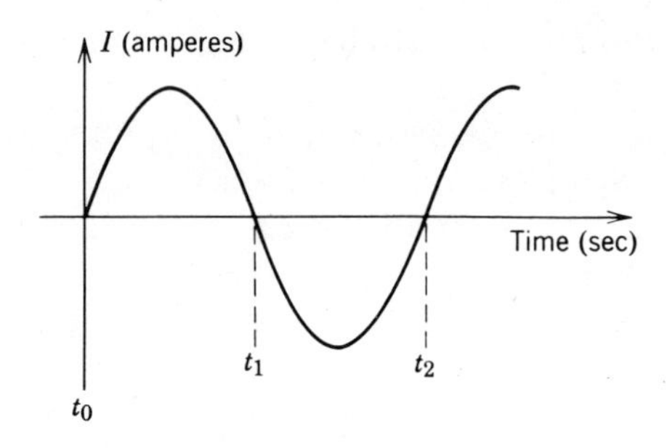

Figure 2.24 Alternating current—ac

ambiguity results, we will use such terms freely. It is important to note, however, that electric current is a flow of charge carriers *through* an element. It is, therefore, completely erroneous to speak of the current *across* an element and one must always speak of the *current through* an element.

Similarly, a voltage is a potential difference *between* two points. It is, therefore, incorrect to speak of voltage *through* an element; one must *always* refer to the *voltage across* an element or between two points.

A similar situation exists in the common applications of the terms direct current dc and alternating current ac. Strictly speaking it is redundant to say a *dc current* or an *ac current*; we need only say ac or dc.

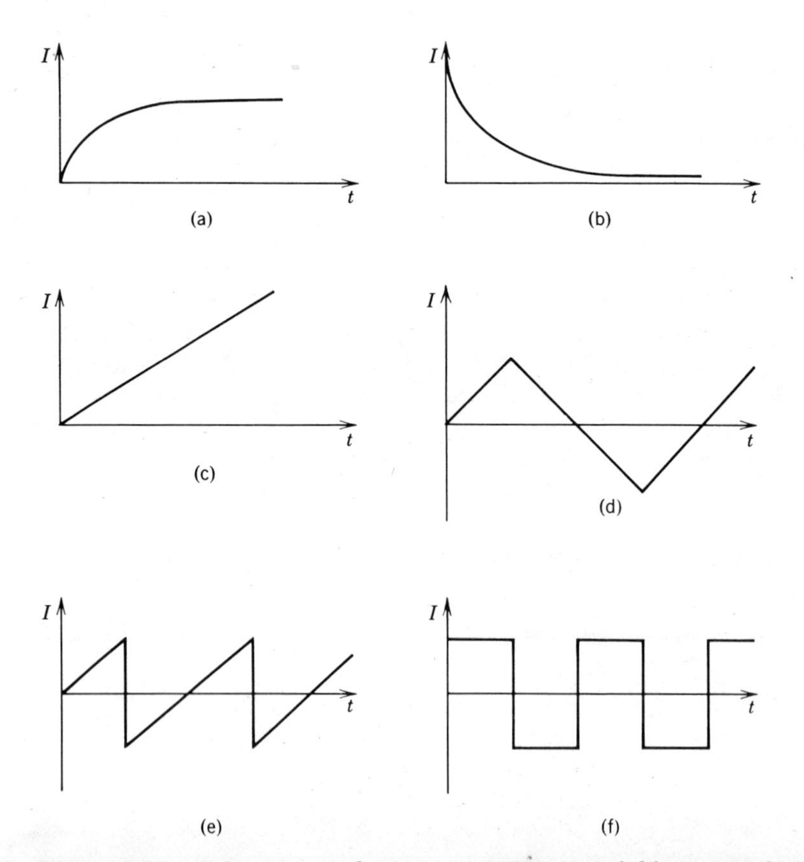

Figure 2.25 Other types of current (*a*) Exponential increase (*b*) Exponential decrease (*c*) Ramp (*d*) Triangular (*e*) Sawtooth (*f*) Square wave

SUMMARY

A charged body can be produced by accumulating an excess of electrons or by creating a deficit of electrons. The charging process requires a certain amount of work, but the charged bodies are capable of doing work in return as a result of electric forces. Making use of the potential difference between charged bodies, an electric current can be produced. This current is composed of charged carriers drifting under the influence of a potential difference, or EMF. A constant EMF can be produced by the chemical action in a voltaic cell. Ohm's Law gives the relationship between current, voltage, and resistance and provides a means of calculating any of these quantities when the other two are known.

GLOSSARY

ac. *Alternating current.*
Ampere. *The basic unit of electric current.*
Circuit. *A closed path around which charge carriers can drift.*
Coulomb. *The basic unit of electric charge.*
dc. *Direct current.*
Electric current. *The rate of flow of electric charge.*
Electrode. *The conductors immersed in the electrolyte of a voltaic cell.*
Electrolyte. *The chemically active solution in a voltaic cell.*
EMF. *Electromotive force.*
Joule. *The unit of work in the metric system. Equal to 0.738 ft-lb.*
Ohm. *The basic unit of resistance.*
Potential. *The ability to do work.*
Potential difference. *The difference in potential between two points.*
Resistance. *Opposition to the flow of electric current.*
Volt. *The basic unit of potential difference.*
Voltaic cell. *The different conductors immersed in an electrolyte in order to produce an EMF by chemical action.*

IMPORTANT RELATIONSHIPS

- One coulomb of charge is equivalent to 6.24×10^{18} electrons.
- One ampere equals one coulomb per second. $I = Q/t$.
- One volt equals one joule per coulomb. $V = W/Q$.
- Ohm's Law: $I = V/R$, or $V = I \times R$, or $R = V/I$.
- The prefixes in Table 2.1.

REVIEW QUESTIONS

1. Two ways in which a body can be charged are by creating an _____ of electrons, or by creating a _____ of electrons.
2. Why is the sign of the electron negative?
3. When an electroscope is charged, the leaves will (*a*) move apart, (*b*) move together, (*c*) stay together.
4. The charge on one electron is -1.00 C. (TF)

5. The drift of free electrons through a copper conductor is called ____ ____.
6. The electric current in the electrolyte of a voltaic cell is due to the drift of both positive and negative ____.
7. The flow of 5 C of electric charge past a point in one second is equivalent to a current of ____.
8. Electrons move through a copper conductor with the speed of light. (TF)
9. The ability to do work is often referred to as (*a*) resistance, (*b*) potential, (*c*) rate of flow.
10. Another word for potential difference is ____.
11. Three different types of metals are required to construct a voltaic pile. (TF)
12. The three necessary parts of a voltaic cell are ____, ____ ____, and ____ ____.
13. EMF is the abbreviation for ____________.
14. One volt equals one ____ per ____.
15. One ampere equals one ____ per ____.
16. Electrical resistance of a conductor is accounted for by the large number of collisions between free electrons and fixed atoms. (TF)
17. Explain the meaning of an *electric circuit.*
18. The prefix pico represents (*a*) 10^{12}, (*b*) 10^{6}, (*c*) 10^{-6}, (*d*) 10^{-12}.
19. The abbreviation ac stands for ____ ____.
20. It is proper to speak of the voltage (across, through) a resistor.

PROBLEMS

1. Briefly define each of the following terms. Give the proper symbol and units of measurement. (*a*) charge, (*b*) current, (*c*) potential difference, (*d*) resistance.
2. A charged body has an excess of 2.49×10^{19} electrons. What is the stored charge in C?
3. A charged body has a deficit of 1.87×10^{16} electrons. What is the stored charge in C?
4. How many electrons must be removed from a body in order to give it a charge of 1.00 μC?
5. If 0.900 C of charge pass a point in a circuit in $1\frac{1}{2}$ minutes, what is the value of the electric current?
6. If the current through a certain circuit is 5.00 A, how long will it take to transport 0.255 C of charge?
7. What is the average drift velocity of the electrons in a copper wire having a cross-sectional area of 4.46×10^{-2} in.^{2} if the current in the wire is 1.00 A?
8. If 2.158 J of work are required to move 2.00 C in a certain voltaic cell, what must be the EMF of the cell?
9. How many seconds will be required for a cell to expend 0.738 J of work, if the voltage of the cell is 10.0 V and the current is 100 mA?
10. What must be the voltage of a cell which is capable of expending work at the rate of 1.00 J/sec when the current is 1.00 A?
11. The current flowing through the resistor in Fig. 2.16 is 0.125 A. What must be the value of the resistance if the EMF of the battery is (*a*) 1.00 V, (*b*) 10.0 V, (*c*) 50.0 V, (*d*) 800 V?

12. The battery in Fig. 2.16 has an EMF of 12 V. What current will flow in the circuit if the resistor has a value of (*a*) 5 Ω, (*b*) 240 Ω, (*c*) 12 kΩ, (*d*) 2 MΩ?
13. The resistor in Fig. 2.16 has a value of 1 kΩ. What must be the EMF of the battery to provide a current of (*a*) 1.00 mA, (*b*) 20.0 mA, (*c*) 500 mA, (*d*) 2.00 A?
14. The battery in Fig. 2.20 has an EMF of 470 V. What are the minimum and maximum values of current if the 470 kΩ resistor has a tolerance of ±5%?
15. If the lamp in Fig. 2.21 draws a current of 75 mA, what must be the resistance of the lamp?
16. What must be the tolerance of the resistor in Fig. 2.18*b* if the current must never exceed 0.485 A?
17. Rewrite the following resistor values using the proper prefixes: (*a*) 21 Ω, (*b*) 2000 Ω, (*c*) 680,000 Ω, (*d*) 1,100,000 Ω, (*e*) 3300 Ω.
18. Rewrite the following voltages using the proper prefixes: (*a*) 16 V, (*b*) 1480 V, (*c*) 0.0843 V, (*d*) 0.000,072 V.
19. Change the following values as shown: (*a*) 0.100 MΩ to kΩ, (*b*) 7.3 mV to μV, (*c*) 3400 mA to A, (*d*) 4.5 kΩ to Ω.
20. In the circuit shown in Fig. 2.16, the battery has an EMF of 3 V and the resistance of the resistor is 2.2 MΩ. What is the value of the current and what is its direction?

3

WORK, ENERGY, AND POWER

Over the years man has been very clever in adapting natural forces to suit his needs. An example is the water wheel. The falling water caused the wheel to turn and the rotating axle of the wheel was then used to perform some useful task such as grinding wheat into flour. Using today's scientific terms, we would say that the *mechanical energy* of the falling water was used to perform *mechanical work.* Another example is the voltaic cell where the *chemical energy* of the cell is converted into *electric energy* which again can be used to perform useful work. The primary advantage of electric energy is that it can be easily transported over great distances. Consider for example the difficulty of replacing a 150-mile electric power transmission line with a water wheel having a rotating shaft 150 miles long! In this chapter we will be interested in the meaning of work, and the utilization of various forms of energy to perform useful work. We will begin with familiar mechanical systems and progress to electrical systems.

3.1 WORK AND ENERGY

Mechanical work is accomplished when a *force* is applied to a body causing the body to be moved a certain *distance.* The amount of work W is defined as the product of the applied force F and the displacement d. Thus

work equals force times distance.

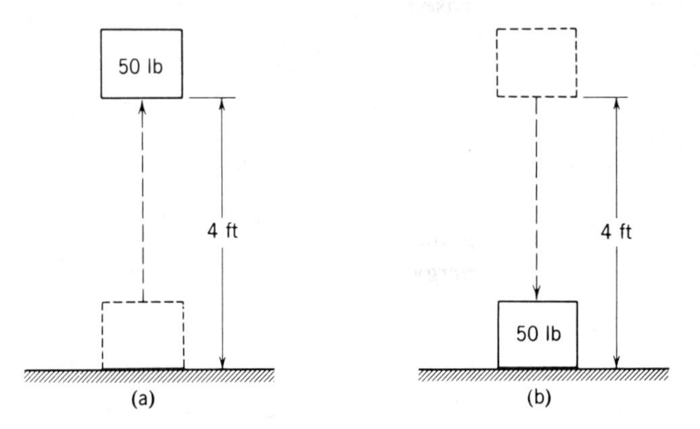

Figure 3.1 Reversible system (*a*) Doing work (*b*) Recovering work

In equation form

$$W = F \times d \tag{3.1}$$

In the English system of units, if the force is measured in pounds and the distance is measured in feet, the unit of work is the foot-pound (ft-lb). If a 10-lb weight is raised a vertical distance of 5 ft the work done is

$$\begin{aligned} W \text{ (ft-lb)} &= F \text{ (lb)} \times d \text{ (ft)} \\ &= 10 \text{ lb} \times 5 \text{ ft} = 50 \text{ ft-lb} \end{aligned}$$

Similarly, if the 10-lb weight were raised 20 ft the work done would be 200 ft-lb. Mechanical work can be precisely defined and measured.

In Fig. 3.1*a*, the amount of work required to raise the 50-lb block a distance of 4 ft is clearly 50 lb × 4 ft = 200 ft-lb. In this case, work must be done on the block in order to raise it. If the block is now lowered a distance of 4 ft as shown in Fig. 3.1*b*, work is still accomplished since the 50-lb block has moved a distance of 4 ft; however, the block has done the work. If the raising and lowering were done by hand, the block would do the same amount of work on the hand when lowered as the hand did on the block when raised. In this system, all work in raising the block is returned when the block is lowered. Such a system is said to be *reversible.*

In the system shown in Fig. 3.2 a force of 10 lb is needed to move the block a distance of 4 ft. The work done on the block is

$$W \text{ (ft-lb)} = 10 \text{ lb} \times 4 \text{ ft} = 40 \text{ ft-lb}.$$

In this case the 10-lb force is necessary to overcome the sliding friction between the block and the surface on which it rests. Another 40 ft-lb of work would be required to move the block back to its original position since the

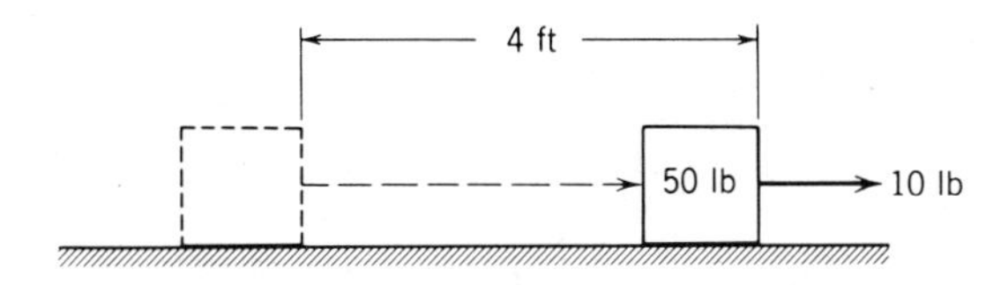

Figure 3.2 Irreversible system

block is not capable of moving back by itself, that is, the block is not capable of doing work. Thus the 40 ft-lb of work input is spent in overcoming the sliding friction and is not recoverable. Such a system is said to be *irreversible.* The differences between these two systems can be explained in terms of energy.

Energy is defined as the *capacity to do work.* Thus work and energy are measured in the same units and we will use the symbol W for both. The elevated block in Fig. 3.1 has the capacity to do work and it is therefore said to have *potential energy.* In this case its potential energy is 200 ft-lb since this is the amount of work it could perform in moving back down a distance of 4 ft. It could for example be connected to a rope and pulley system and do 200 ft-lb of work in raising another block. We would then say that 200 ft-lb of energy have been expended by the system. Note that the amount of work returned can *never* exceed the amount of work put into the system. This is the *principle of conservation of energy* which states that energy can neither be created nor destroyed.*

In the system shown in Fig. 3.2 the 40 ft-lb of work used in moving the block is spent in the form of *heat energy* due to the friction between the block and the support. Consider for example the heat generated when a piece of sandpaper is firmly and rapidly used to sand a piece of wood.

These principles of work and energy can be applied to electric charge since we know there are electric forces of attraction and repulsion between charged bodies. In Fig. 3.3*a*, a certain force must be applied to the electron in order to move it towards the negatively charged body. The work done in moving the electron can be found in terms of the applied force and the distance moved. It is however more convenient to use the units of charge and voltage which we have previously discussed. The basic unit of work or energy used with individual electrons is the electron-volt eV. Thus

one electron volt is the amount of work required to move one electron through a potential difference of one volt.

Clearly in Fig. 3.3*a* it requires one eV of work to move the electron from A to B. On the other hand, the electron in Fig. 3.3*b* is capable of expending one eV of energy in moving from B to A.

We have seen that the charge on one electron is quite small and we have defined one C as the charge equivalent to 6.24×10^{18} electrons. It is then convenient to refer to the amount of work required to move one C of

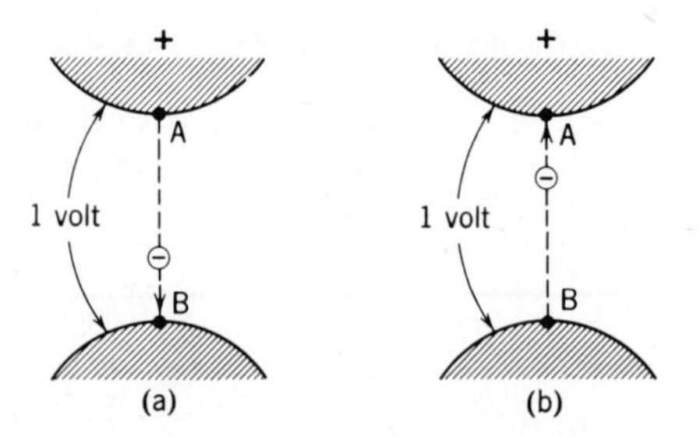

Figure 3.3 One electron volt (*a*) Work (*b*) Energy

*We will not be concerned with $E = mc^2$, Einstein's equation, which provides the relationship for changing mass to energy and vice-versa (the atomic bomb).

electrons through a potential difference of one volt. This is nothing new, however, since we have already made such a definition in Eq. 2.4 which is repeated here for convenience.

$$V \text{ (volts)} = \frac{W \text{ (joules)}}{Q \text{ (coulombs)}} \tag{2.4}$$

This equation can be rearranged as

$$W \text{ (joules)} = V \text{ (volts)} \times Q \text{ (coulombs)} \tag{3.2}$$

Thus from Eq. 3.2,

one joule is the amount of work required to move one coulomb of charge through a potential difference of one volt.

• • •

Example 3.1 What amount of work must be done in moving an electron through a potential difference of 100 V? How many J of energy does this represent?

Solution Since it requires 1 eV to move an electron through a potential difference of 1 V, it will require 100 eV to move it through 100 V. There are 6.24×10^{18} eV/J, and thus there must be 1.60×10^{-19} J/eV. Therefore,

$$\begin{aligned} 100 \text{ eV} &= 100 \text{ eV} \times 1.60 \times 10^{-19} \frac{\text{J}}{\text{eV}} \\ &= 1.60 \times 10^{-17} \text{ J} \end{aligned}$$

• • •

Example 3.2 At what rate must the battery shown in Fig. 3.4 convert energy?

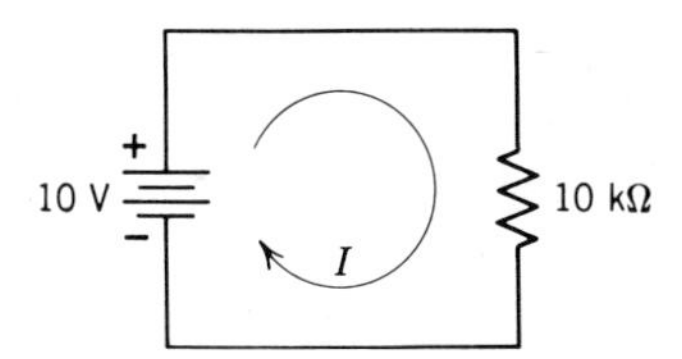

Figure 3.4

Solution Since the problem asks *at what rate* must the battery do work, we must determine the work done in a period of time. We will then determine the work done in 1 sec, or the rate of work in J/sec. Using Ohm's Law,

$$I = \frac{V}{R} = \frac{10 \text{ V}}{10 \text{ k}\Omega} = 10^{-3} \text{ A}$$

Then using Eq. 2.2

$$Q = I \times t = 10^{-3} \text{ A} \times 1 \text{ sec} = 10^{-3} \text{ C}$$

Finally, using Eq. 3.2

$$W = Q \times V = 10^{-3} \text{ C} \times 10 \text{ V} = 10^{-2} \text{ J}$$

Alternatively we could have combined Eq. 3.2 and 2.2 with Ohm's Law to obtain

$$W = Q \times V \qquad Q = I \times t \qquad I = V/R$$

$$W\ (\mathrm{J}) = Q \times V = I \times t \times V = \frac{V}{R} \times t \times V = V^2 t/R$$

$$= \frac{(10\ \mathrm{V})^2 \times 1\ (\mathrm{sec})}{10^4\ \Omega} = 10^{-2}\ \mathrm{J}$$

• • •

In discussing the voltaic cell in the previous chapter, we saw that the cell converts chemical energy into electric energy by separating electric charge. When the charges flow through an external resistance, the electric energy is changed into heat energy in the resistor. The English units for measuring the amount of heat energy in the resistor are *British Thermal Units* (*Btu*). By definition,

one Btu is the amount of heat energy required to raise the temperature of one pound of water one degree Fahrenheit (°F).

It can be shown that 1 Btu is equivalent to 1054.8 J. We now have the proper relationships and units to discuss such problems as the transformation of electric energy into mechanical work (motors) or the transformation of electric energy into heat energy (heaters, stoves, etc.). The necessary conversion factors are given in Table 3.1.

Table 3.1.

There are:	There are:
1.356 J/ft-lb	0.738 ft-lb/J
1.60×10^{-19} J/eV	6.24×10^{18} eV/J
1054.8 J/Btu	9.49×10^{-4} Btu/J

• • •

Example 3.3 How many joules of electric energy are required to increase the temperature of 1 ft^3 of water by 30°F?

Solution One cubic foot of water weighs 62.4 lb; thus there are 62.4 $\mathrm{lb/ft}^3$. Since it requires 1 Btu to raise the temperature of 1 lb of water by 1°F, it will take 30 Btu to raise the temperature of 1 lb by 30°F. There is a total of 62.4 lb, and it will then require

$$30\,\frac{\mathrm{Btu}}{\mathrm{lb}} \times 62.4\ \mathrm{lb} = 1870\ \mathrm{Btu}$$

From Table 3.1, there are 1054.8 J/Btu, and thus

$$W\ (\mathrm{J}) = 1870\ \mathrm{Btu} \times 1054.8\,\frac{\mathrm{J}}{\mathrm{Btu}} = 1.97 \times 10^6\ \mathrm{J}$$

• • •

Example 3.4 How many joules of electric energy are required to raise a 5-ton elevator a vertical distance of 10 ft?

Solution Using Eq. 3.1

$$W \text{ (ft-lb)} = F \text{ (lb)} \times d \text{ (ft)} = 5 \text{ ton} \times 2000 \frac{\text{lb}}{\text{ton}} \times 10 \text{ ft}$$
$$= 10^5 \text{ ft-lb}$$

Using Table 3.1

$$W(\text{J}) = 10^5 \text{ ft-lb} \times 1.356 \frac{\text{J}}{\text{ft-lb}} = 1.356 \times 10^5 \text{ J}$$

• • •

.2 THE MKS SYSTEM

In the previous sections we have calculated ft-lb of work in the familiar English system (foot-pound-second, fps). We have also rather freely calculated joules of work by simply stating that there are 0.738 ft-lb/J. The joule is in fact the basic unit of work in the *Meter-Kilogram-Second* (*MKS*) system of units. This system is widely used throughout the world and offers definite advantages over the English system as we shall see. If it seems a burden to learn another system of measurement, one only needs to attempt to recall how many grains in a troy ounce, or in an ounce avoirdupois, how many ounces in a pound (troy or avoir.), how many quarts in a peck or a bushel, how many rods in a chain or a mile! The list seems endless, and the *metric system* will indeed prove a welcome relief. The metric system is a method of measurement based on powers of 10 and the MKS system is included in the metric system.

The basic unit of *length* in the MKS system is the *meter*, the unit of *mass* is the *kilogram*, and the unit of *time* is the *second*. All other units in the MKS system, either larger or smaller, are found by multiplying or dividing by powers of ten. The prefixes given in Chapter 2 can then be applied to provide the different units of measurement, and some of the more common ones are given in Table 3.2. The basic unit of work in the metric system is the joule, and one joule is defined as one newton-meter.*

Table 3.2. Metric System.

Prefix	Power	Length		Mass		Time	
kilo	10^3	kilometer	km	kilogram	kg	—	—
hecto	10^2	hectometer	hm	hectogram	hg	—	—
deka	10^1	dekameter	dam	dekagram	dag	—	—
—	$10^0 = 1$	meter	m	gram	g	second	sec
deci	10^{-1}	decimeter	dm	decigram	dg	—	—
centi	10^{-2}	centimeter	cm	centigram	cg	—	—
milli	10^{-3}	millimeter	mm	milligram	mg	millisecond	msec
micro	10^{-6}	micrometer	μm	microgram	μg	microsecond	μsec
nano	10^{-9}	—	—	—	—	nanosecond	nsec

*One joule of work is performed when a force of one newton displaces a body a distance of one meter. One newton is equal to 0.102 kilogram.

• • •

Example 3.5 Express the following quantities in a more convenient form: (a) 0.000,37 m; (b) 1280 g; (c) 0.000,000,087 sec; (d) 0.000,21 sec

Solution Using Table 3.2:
(a) 0.37 mm or 370 μm; (b) 1.28 kg; (c) 87 nsec; (d) 0.21 msec or 210 μsec.

• • •

Historically, the French established the metric system in 1791. It was intended that the meter be exactly 1/10,000,000 the distance from the equator of the earth to either pole, and that the gram be exactly the mass of one cubic centimeter of water at zero degrees *centigrade* (Celsius) °C. These goals were not quite achieved however, and today the meter is defined as 1,650,763.73 wavelengths of orange-red light emitted by ionized krypton-86 operating at the triple point of nitrogen—a somewhat more precise definition based on an atomic standard. The standard for a kilogram of mass is based on a certain block of platinum-iridium alloy preserved at Sèvres, France. The second was originally defined as 1/86,400 of a mean solar day. There are however variations in the solar day and in 1955 the second was redefined as exactly 1/31,556,925.9747 of a mean tropical year. This will more than likely be permanently replaced in the near future with another atomic standard based on the radiation from cesium-133 (it is now the temporary standard and is accurate to $1/10^{11}$).

Since it may be necessary for us to work in either the English system or the MKS system, we must be able to easily convert from one to the other. In spite of the very precise standards mentioned above, we will be content with three (sometimes four) significant figures. The important conversion factors needed for this purpose are summarized in Table 3.3.

Table 3.3. Conversion Factors.

English to Metric	Metric to English
There are:	There are:
2.54×10^{-2} m/in.	39.37 in./m
0.454 kg/lb	2.20 lb/kg

• • •

Example 3.6 Make the following conversions:
(a) 46.3 in. to m; (b) 1800 kg to lb; (c) 21 mils to m.

Solution Using Table 3.3:

(a) $\text{m} = 46.3 \text{ in.} \times 2.54 \times 10^{-2} \text{ m/in.} = 1.18 \text{ m}$
(b) $\text{lb} = 1800 \text{ kg} \times 2.20 \text{ lb/kg} = 3{,}960 \text{ lb}$
(c) 1 mil = 0.001 in. Thus, $\text{m} = 21 \times 10^{-3} \text{ in.} \times 2.54 \times 10^{-2} \text{ m/in.}$
$= 5.33 \times 10^{-4} \text{ m} = (0.533 \text{ mm})$.

• • •

One metric system unit which we may encounter is the measurement of temperature in degrees centigrade (Celsius) °C. In this system, water freezes at 0°C and boils at 100°C. Thus 0°C is the same as 32°F and 100°C is equivalent to 212°F. The conversion from °C to °F can be accomplished using Eq. 3.3.

$$°F = \frac{9}{5}°C + 32 \tag{3.3}$$

Alternatively, one can convert from °F to °C using Eq. 3.4.

$$°C = \frac{5}{9}(°F - 32) \tag{3.4}$$

• • •

Example 3.7 Make the following temperature conversions: (*a*) 50°C to °F; (*b*) 68°F to °C.

Solution (*a*) Using Eq. 3.3,

$$°F = \frac{9}{5} \times 50 + 32 = 90 + 32 = 122°F$$

(*b*) Using Eq. 3.4

$$°C = \frac{5}{9}(68 - 32) = \frac{5}{9} \times 36 = 20°C$$

3.3 POWER

Suppose you were given the task of raising a 30-lb box from a point on the ground to a height of ten feet using the pulley system shown in Fig. 3.5. The box must be moved a vertical distance of 10 ft and you would clearly be required to expend 30 lb × 10 ft = 300 ft-lb of work in performing this task. If this work were performed in a time of 1 min, or in 30 sec, or even 10 sec, it would not be too difficult. On the other hand, you might find it difficult, or perhaps impossible, to do this 300 ft-lb of work in only 1 sec. It is thus apparent that the *rate of doing work* is just as important as the total amount of work to be done. The amount of work accomplished in a period of time is called *power*. Thus,

power is defined as the rate of doing work.

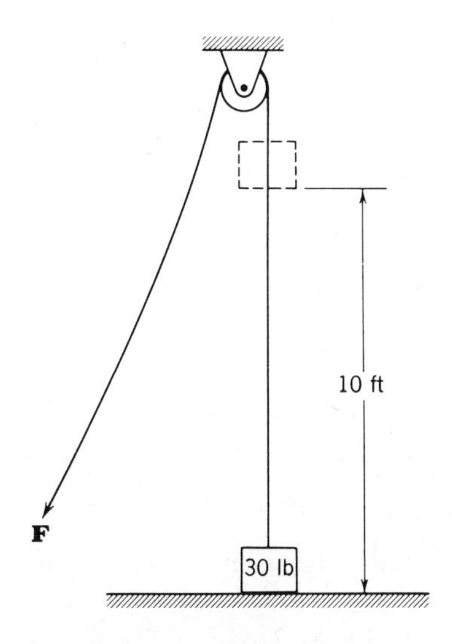

Figure 3.5

The symbol used for power is P, and power can be expressed in equation form as

$$P = W/t \tag{3.5}$$

In the MKS system the unit of work is the J and the unit of time is the sec. Power can therefore be expressed in units of J/sec. However, a special unit for power has been named in honor of a famous mechanical engineer, James Watt (1736–1819). The unit of power in the MKS system is the watt W, and

one watt of power (the rate of doing work) is defined as one joule of work accomplished in one second,

In equation form

$$P\ (\text{W}) = \frac{W(\text{J})}{t\ (\text{sec})} \tag{3.6}$$

• • •

Example 3.8 Calculate the power when the box in Fig. 3.5 is moved in: (*a*) 30 sec; (*b*) 10 sec; (*c*) 1 sec.

Solution The total work has been determined as 300 ft-lb. Using the conversion factor from Table 3.1, the work in J is

$$W\ (\text{J}) = 300\ \text{ft-lb} \times 1.356\ \text{J/ft-lb} = 407\ \text{J}$$

Now, using Eq. 3.6, the power is:

$$(a)\quad P\ (\text{W}) = \frac{407\ \text{J}}{30\ \text{sec}} = 13.6\ \text{watts} = 13.6\ \text{W}$$

$$(b)\quad P\ (\text{W}) = \frac{407\ \text{J}}{10\ \text{sec}} = 40.7\ \text{W}$$

$$(c)\quad P\ (\text{W}) = \frac{407\ \text{J}}{1\ \text{sec}} = 407\ \text{W}$$

• • •

From Example 3.8 it is quite clear that as a given amount of work is performed in shorter periods of time, the power increases rapidly.

It is interesting to note that even though the watt is usually considered to be a unit of electric power, it is also a unit of mechanical power; it was in fact originally defined as a unit of mechanical power. Since we are primarily interested in the electric power in a circuit, it would be desirable to be able to calculate the watts of power in terms of other electrical quantities. This can be easily accomplished by making use of Eqs. 2.3, 2.4, and 3.6, which are shown here for convenience.

$$t = \frac{Q}{I} \tag{2.3}$$

$$V = \frac{W}{Q} \tag{2.4}$$

Eq. 2.4 can be rearranged as

$$W = V \times Q$$

If we now substitute these two equations into the equation for power, Eq. 3.6, the result is

$$P = \frac{W}{t} = \frac{V \times Q}{Q/I} = V \times Q \times \frac{I}{Q} = V \times I \tag{3.7}$$

Thus the power in W, in an electric circuit, is the product of the voltage in V and the current in A.

Using Ohm's Law, Eq. 3.7 can also be modified to give the power in terms of other circuit quantities. Substituting V (V) = I (A) × R (Ω) into Eq. 3.7 yields

$$P = V \times I = [I \times R] \times I = I^2 \times R \tag{3.8}$$

Thus the power in a resistance can be found as the product of the square of the current through the resistor and the resistance.

Substituting $I = V/R$ into Eq. 3.7 yields

$$P = V \times I = V \times [V/R] = \frac{V^2}{R} \tag{3.9}$$

Using Eq. 3.9, the power in a resistance can also be found as the square of the voltage across the resistor divided by the resistance.

• • •

Example 3.9 At what rate is energy being delivered to the resistor in Fig. 3.6 if the current $I = 0.25$ A?

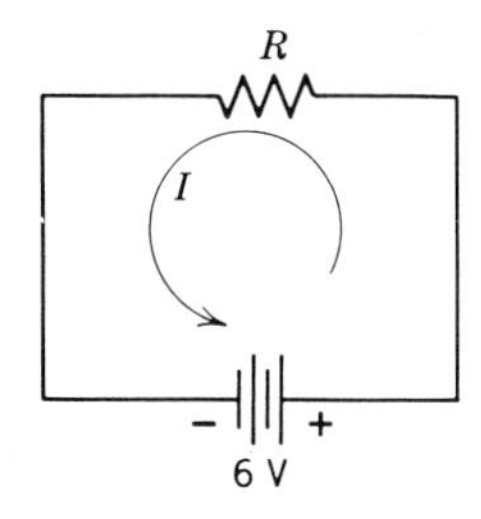

Figure 3.6

Solution Using Eq. 3.7

$$P = V \times I = 6 \text{ V} \times 0.25 \text{ A} = 1.50 \text{ W}$$

Energy is being delivered at a rate of 1.50 W or 1.50 J/sec.

• • •

Example 3.10 What must be the minimum power rating of the resistor in Fig. 3.7?

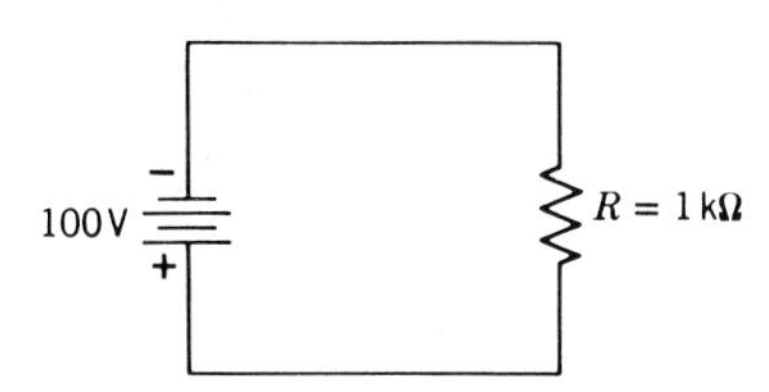

Figure 3.7

Solution Using Eq. 3.9,

$$P = \frac{V^2}{R} = \frac{(100 \text{ V})^2}{1 \text{ k}\Omega} = 10 \text{ W}$$

• • •

Example 3.11 What is the maximum allowable current which can be forced through the resistor shown in Fig. 3.8 without exceeding its rating?

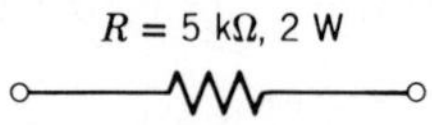

Figure 3.8

Solution Eq. 3.8 can be rearranged as

$$I^2 = \frac{P}{R}, \text{ and thus } I = \sqrt{\frac{P}{R}}$$

The maximum current is then

$$I = \sqrt{\frac{2\text{ W}}{5\text{ k}\Omega}} = \sqrt{4 \times 10^{-4}} = 20\text{ mA}$$

• • •

3.4 HORSEPOWER—HP

In the late 1700's when James Watt was perfecting the steam engine, he needed a unit of power to determine the rate at which the steam engine could do work. In those days the horse was widely used as the *beast of burden* and Watt quite naturally wanted to compare the performance of his engine with that of a good horse. After studying the work done by a number of different horses, Watt determined that the average horse could lift 550 lb a distance of 1 ft in 1 sec. Furthermore, the horse was capable of performing work at this rate over a sustained period of time, and Watt therefore defined this rate of doing work as one *horsepower* (*hp*). Thus

one hp is the rate of doing work (power) when 550 ft-lb of work is done in a period of 1 sec.

If work is being done at the rate of 550 ft-lb/sec (1 hp) then it is clear that the work done in one min is

$$550\text{ ft-lb/sec} \times 60\text{ sec/min} = 33{,}000\text{ ft-lb/min.}$$

Therefore one horsepower can also be defined as 33,000 ft-lb/min.

• • •

Example 3.12 How many hp are required in each of the three cases discussed in Example 3.8?

Solution The total work required is 300 ft-lb.

$$(a)\ P = \frac{W}{t} = \frac{300\text{ ft-lb}}{30\text{ sec}} = 10\text{ ft-lb/sec}$$

Then

$$P\text{ (hp)} = \frac{P\text{ (ft-lb/sec)}}{550} = \frac{10\text{ ft-lb/sec}}{550} = 0.0182\text{ hp}$$

$$(b)\ P\text{ (hp)} = \frac{300\text{ ft-lb}}{550 \times 10\text{ sec}} = 0.0545\text{ hp}$$

$$(c)\ P\ (\text{hp}) = \frac{300\ \text{ft-lb}}{550 \times 1\ \text{sec}} = 0.545\ \text{hp}$$

• • •

The definition of hp in ft-lb/sec is in units of mechanical hp. We can easily express hp in electrical quantities by using the conversion factor in Table 3.1 to change ft-lb to J. Thus

$$1\ \text{hp} = 550\ \text{ft-lb/sec} \times 1.356\ \text{J/ft-lb} = 746\ \text{J/sec}$$

But 1 J/sec is the definition of 1 W. Therefore

one horsepower equals 746 watts; or 1 hp = 746 W

• • •

Example 3.13 What hp is required for an electric motor which must lift a 2000-lb elevator a distance of 10 ft in 10 sec?

Solution

$$P = \frac{10\ \text{ft} \times 2000\ \text{lb}}{10\ \text{sec}} = 2000\ \text{ft-lb/sec}$$

Thus,

$$P\ (\text{hp}) = \frac{2000\ \text{ft-lb/sec}}{550} = 3.64\ \text{hp}$$

• • •

Example 3.14 What current must be supplied to the motor in Example 3.13 if the motor operates on a voltage of 230 V?

Solution Since 1 hp = 746 W, there are 746 W/hp. Therefore,

$$P = 3.64\ \text{hp} \times 746\ \text{W/hp} = 2.72\ \text{kW}$$

Using Eq. 3.7, $P = V \times I$. Thus

$$I = \frac{P}{V} = \frac{2720\ \text{W}}{230\ \text{V}} = 11.8\ \text{A}$$

• • •

.5 KILOWATTHOUR—kWh

The definition of the W given in Eq. 3.6 can be rearranged as

$$W(\text{J}) = P(\text{W}) \times t\ (\text{sec}) = W\ (\text{W-sec}) \qquad (3.10)$$

Thus,

one J is equal to one W-sec.

Note that this is the *product* of W and sec, and not W per sec. It is then possible to measure electric energy or work in W-sec as well as in J. Since the electric motor discussed in Example 3.14 uses electric energy at the rate of 2.72 kW, the total energy required in 10 sec is

$$W\ (\text{W-sec}) = P\ (\text{W}) \times t\ (\text{sec}) = 2.72\ \text{kW} \times 10\ \text{sec} = 2.72 \times 10^4\ \text{W-sec}$$

The W-sec is quite a satisfactory unit for the measurement of energy in small systems, but is seen to be quite small in large systems, such as the elevator motor in Example 3.14. For this reason, electric power companies measure electric energy in a larger unit, the *kilowatthour* (*kWh*). The electric energy in kWh is found by taking the product of the power in kW and the time in hours. Thus Eq. 3.10 can be used to find the total energy in kWh by simply changing units as follows:

$$W\ (\text{kWh}) = P\ (\text{kW}) \times t\ (\text{h}) \tag{3.11}$$

The kWh *meter* is used to measure the total *quantity* of electric energy used and power companies make use of such meters in billing their customers.

Again a comment concerning the improper use of words. It is *not* correct to speak of the *amount of power* used, since it is *electric energy* that is used and power expresses the *rate* at which the energy is used.

• • •

Example 3.15 What is the power rating of the lamp shown in Fig. 3.9 if the current in the circuit is 0.50 A? How much electric energy is used by the lamp in 4 hours?

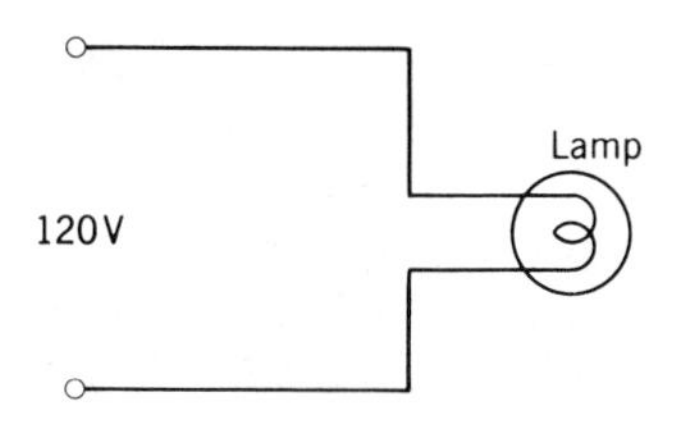

Figure 3.9

Solution $$P = V \times I = 120\ \text{V} \times 0.50\ \text{A} = 60\ \text{W}$$

The electric energy is found using Eq. 3.10

$$W = P \times t = 60\ \text{W} \times 4\ \text{h} \times 3600\ \text{sec/h}$$
$$= 8.64 \times 10^5\ \text{W-sec}$$

Or, using Eq. 3.11

$$W = P \times t = 0.060\ \text{kW} \times 4\ \text{h} = 0.24\ \text{kWh}$$

From Example 3.15 it is easily seen that

$$W\ (\text{kWh}) = P\ (\text{kW}) \times 1000\ \text{W/kW} \times t\ (\text{h}) \times 3600\ \text{sec/h}$$
$$= 3.6 \times 10^6\ \text{W-sec}$$

Thus

one kWh is equal to 3.6×10^6 W-sec.

• • •

Example 3.16 Information on the back of a color television set gives the power rating of the set as 420 W. What is the cost of the electric energy to operate this set for 24 h if electric energy costs $3\frac{1}{2}$ cents per kWh?

Solution The total electric energy needed is found using Eq. 3.11,

$$W = P \times t\ (\text{h}) = 0.420\ \text{kW} \times 24\ \text{h} = 10.1\ \text{kWh}$$

The total cost is then

$$\text{Cost} = 10.1\ \text{kWh} \times 3.5\ \text{cents/kWh} = 35.4\ \text{cents}$$

• • •

.6 EFFICIENCY

In our discussion of work using Fig. 3.1, it was pointed out that all of the work done in raising the block was returned when the block was lowered. On the other hand, the work done in moving the block in Fig. 3.2 was not recoverable. Most systems are not as ideal as the one shown in Fig. 3.1 and it is rare when *all* of the work put into a system can be recovered. Most systems can be considered to be a combination of these two cases. Any system can be represented in diagram form as shown in Fig. 3.10. In this case, the law of conservation of energy can be stated as

$$W_{in} = W_{out} + W_{loss} \tag{3.12}$$

Thus the total energy supplied W_{in} is equal to the sum of the energy lost (e.g., heat losses) W_{loss} and the useful output energy W_{out}. The only time that all of the energy put into a system can be completely recovered is when there are no losses, or when W_{loss} = zero. This is the *ideal* case in Fig. 3.1. The other extreme occurs when all of the energy put into a system is lost and there is no available energy remaining to do work; thus W_{out} = zero and this is the situation shown in Fig. 3.2.

Most systems operate somewhere between the two extremes discussed above. It would of course be desirable to always have an ideal system. A measure of how close a system is to the ideal is given by the *efficiency* η (the Greek letter eta). Efficiency is defined as the ratio of output energy to input energy. Thus

$$\text{efficiency} = \eta = \frac{W_{out}}{W_{in}} \tag{3.13}$$

In the ideal case, $W_{out} = W_{in}$ and the efficiency is seen to be *unity*. However, as the losses increase, W_{out} becomes less than W_{in} and the efficiency is some number less than 1.0. In the worst case, all of the input energy is lost, W_{out} = zero, and the efficiency is zero. Thus the efficiency of a system is always some number *between* 1.00 and 0.00 (it may equal one or zero in the two extreme cases). The closer it is to unity (1.0) the better and more efficient the system.

Efficiency can also be expressed as the ratio of output power to input power, since $W = P \times t$. Thus

$$\eta = \frac{W_{out}}{W_{in}} = \frac{P_{out} \times t}{P_{in} \times t} = \frac{P_{out}}{P_{in}} \tag{3.14}$$

Occasionally efficiency is expressed in percent. *Percent efficiency*, or the efficiency in percent, can be found by multiplying Eq. 3.13 by 100. Therefore

$$\text{percent efficiency} = \eta\ (\%) = 100\ \eta \tag{3.15}$$

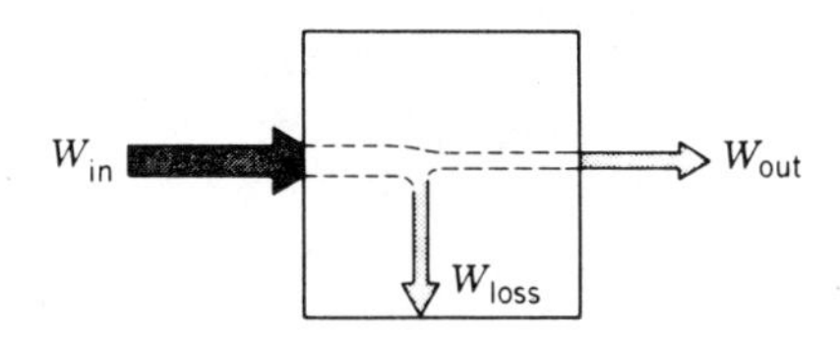

Figure 3.10 A general system diagram

• • •

Example 3.17 If the efficiency of the elevator system in Example 3.13 is 0.8 (80%), what hp motor is required?

Solution The actual work which must be performed is clearly 2000 ft-lb. Since this must be done in 1 sec, the actual output power P_{out} is 2000 ft-lb/sec, or 3.64 hp. If the system efficiency is 0.8, the required input power P_{in} can be found using Eq. 3.14. Thus

$$\eta = \frac{P_{out}}{P_{in}}; \text{ rearranging, } P_{in} = \frac{P_{out}}{\eta} = \frac{3.64 \text{ hp}}{0.8} = 4.55 \text{ hp}$$

• • •

Example 3.18 At what rate is energy lost in the example above?

Solution Since 80% of the input power is used in the output, 20% of the input power must be lost in order to satisfy the conservation of energy given in Eq. 3.12. Therefore

$$P_{loss} = 20\% \times P_{in} = 0.20 \times 4.55 \text{ hp} = 0.91 \text{ hp}$$

Or

$$P_{loss} = 0.91 \text{ hp} \times 550 = 500 \text{ ft-lb/sec}$$

• • •

Example 3.19 If a $1\frac{1}{2}$ hp motor is required to pump water at the rate of 240 gal/min as shown in Fig. 3.11, what is the system efficiency? Water weighs 8.35 lb/gal.

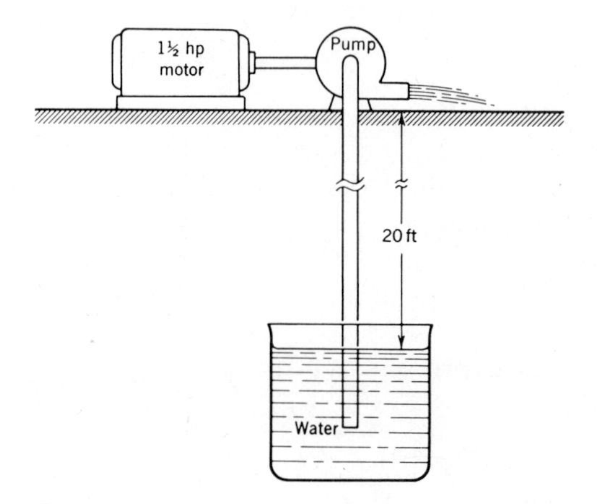

Figure 3.11

Solution The power input to the system is $P_{in} = 1\frac{1}{2}$ hp. The rate at which water is raised the 20-ft height is the power out of the system and is

$$\begin{aligned} P_{out} \text{ (ft-lb/sec)} &= 20 \text{ ft} \times 240 \text{ gal/min} \times 8.35 \text{ lb/gal} \times 1/60 \text{ min/sec} \\ &= 668 \text{ ft-lb/sec} \end{aligned}$$

Or

$$P_{out} \text{ (hp)} = \frac{668 \text{ ft-lb/sec}}{550} = 1.21 \text{ hp}$$

The efficiency is then

$$\eta = \frac{P_{out}}{P_{in}} = \frac{1.21 \text{ hp}}{1.50 \text{ hp}} = 0.81 \qquad \eta\,(\%) = 81\%$$

• • •

SUMMARY

There are many different sources of natural energy available. Energy is defined as the capacity to do work, and it can be changed from one form to another to do useful work. Work is defined as the product of an applied force multiplied by a distance and can be calculated or measured accurately. Power is the rate at which work is done and can be expressed in mechanical or electrical units. The power in an electric circuit can be determined entirely in terms of the electrical quantities describing the circuit. The ratio of output power to input power gives the efficiency of a system which can never exceed unity since the principle of conservation of energy must be observed.

GLOSSARY

Btu. *British Thermal Unit; the amount of heat required to raise the temperature of 1 lb of water by 1°F.*
Centigrade *(Celsius). The unit of temperature in the metric system.*
Efficiency. *A measure of how nearly a system approaches the ideal.*
Energy. *The capacity to do work.*
eV *(electron-volt). The work required to move one electron through a potential difference of one volt.*
ft-lb *(foot-pound). A unit of mechanical work.*
hp *(horsepower). 550 ft-lb/sec or 746* W.
Joule. *The unit of work in the metric system. One newton-meter.*
kWh *(kilowatt-hour). A unit of electric energy.*
MKS. *Meter kilogram second system of units.*
Power. *The rate of doing work.*
Watt. *A unit of power.* 1 *J/sec.*
Work. *The product of applied force and distance moved.*

IMPORTANT RELATIONSHIPS

	Definition		Alternative forms
• Work	$W = F \times d$	Eq. 3.1	$W = P \times t = V \times Q$
• Power	$P = W/t$	Eq. 3.5	$P = V \times I = V^2/R = I^2 \times R$
• Current	$I = Q/t$	Eq. 2.1	$I = V/R = P/V = \sqrt{P/R}$
• Voltage	$V = W/Q$	Eq. 2.4	$V = I \times R = P/I = \sqrt{P \times R}$
• Resistance	$R = V/I$	Eq. 2.5	$R = V^2/P = P/I^2$

$$°F = \frac{9}{5}°C + 32 \qquad °C = \frac{5}{9}(°F - 32)$$

$$W_{in} = W_{out} + W_{loss} \qquad \eta = \frac{W_{out}}{W_{in}} = \frac{P_{out}}{P_{in}}$$

REVIEW QUESTIONS

1. Work is found by taking the product of ____ and ____.
2. Energy is defined as the ____.
3. The principle of conservation of energy states that energy can neither be created nor destroyed. (TF)
4. The amount of work done when a 25 lb weight is raised a vertical distance of 5 ft is: (*a*) 5 ft-lb, (*b*) 125 ft-lb, (*c*) 125 joules, (*d*) 30 ft-lb.
5. One eV is the amount of energy required to move one C of charge through a potential difference of one V. (TF)
6. Electric energy is transformed into heat energy in a resistor. The proper units for measuring heat energy are Btu, and the conversion factor for changing J to Btu is ____.
7. One Btu is the amount of heat energy needed to raise the temperature of one lb of water by one °C. (TF)
8. The basic units of measurement in the MKS system are ____, ____, and ____.
9. The basic unit of work in the MKS system is the ____.
10. The rate of doing work is defined as ____.
11. A power of one W is equivalent to one ____ per ____.
12. When a given amount of work is done in shorter and shorter periods of time, the power (increases, decreases) rapidly.
13. One hp is equivalent to ____ ft-lb/sec, or ____ W.
14. The power in a resistor can be found by taking the product of the current in A and the voltage in V. (TF)
15. Two hp is equivalent to 66,000 ft-lb/min. (TF)
16. The energy used by a 100 W lamp in a period of 5 h is ____ kWh.
17. It is possible to achieve an efficiency of 1.50. (TF)
18. If a certain electric motor is found to have an efficiency of 0.8, the useful output power is 80% of the input power. What happens to the other 20%?
19. A one hp motor with an efficiency of 90% will more than likely be (larger, smaller) than a one hp motor with an efficiency of 70%.
20. Fans and blowers are frequently installed in cabinets containing electronic equipment. What might be the result of operating such equipment without the fans and blowers?

PROBLEMS

1. How many ft-lb of work are required to raise a 5-ton elevator a vertical distance of 25 ft?
2. If 4000 ft-lb of work are required to raise a certain block a vertical distance of 50 ft, what must be the weight of the block?
3. How many eV of work are required to move an electron through a potential difference of 5000 V?
4. How many Btu of energy are required to raise the temperature of 1 quart of water by 30°F? How many J of energy does this represent?
5. The water in Problem 4 is heated on an electric stove which operates on 120 V at a current of 2.0 A. If half of the energy supplied to the stove is used to heat the water, and the other half is lost, how long will it take to heat the water?

6. Make the following conversions: (*a*) 21.5 cm to in.; (*b*) 34 lb to kg; (*c*) 0.762 mils to mm; (*d*) 4.67 g to lb.
7. Make the following conversions: (*a*) 40°C to °F; (*b*) 40°F to °C.
8. What power (in ft-lb/sec) is required to raise a 1-ton elevator a distance of 12 ft in 10 sec? How many hp does this represent?
9. A 35 W soldering iron is operated on 120 V. What is the current in the heating element of the iron?
10. What is the maximum voltage which can be safely applied across the terminals of a 4.7 kΩ, $\frac{1}{2}$ W resistor?
11. What is the power dissipation in a 330 Ω resistor if the current through it is 25.0 mA?
12. What must be the current through a 6.3 V pilot lamp if it is rated at 2 W?
13. What is the maximum allowable current in a 2 MΩ, 1 W resistor?
14. What is the power rating of a 28 V indicator lamp if it carries a current of 25 mA?
15. Is it safe to apply 100 V across the terminals of a 10 kΩ, $\frac{1}{2}$ W resistor? Why, or why not?
16. A 12 V electric motor is operated with a current of 0.50 A. How many hp does this represent?
17. What is the cost of operating a 100 W lamp for 24 h if electric energy costs $3\frac{1}{2}$ cents per kWh?
18. If a 5-hp motor is used to operate the elevator in Problem 8, what is the system efficiency?
19. What is the cost of operating a 150 W soldering iron for 8 h if electric energy costs $3\frac{1}{2}$ cents per kWh?
20. A certain 9 V transistor radio battery will supply a current of 50 mA for 50 h. How long will the battery operate if the radio consumes electric energy at the rate of 450 mW?

4 SOURCES OF ELECTRIC ENERGY

The operation of *electric energy sources* depends on the transformation of some form of energy into electric energy. In voltaic cells, chemical energy is continuously converted into electric energy when the chemically active electrolyte reacts with the metallic electrodes, and such cells are called *chemical cells.** Chemical cells are used in many portable electric devices ranging from flashlights and transistor radios to complex missile and satellite systems. Some other sources of electric energy which are becoming increasingly important are *thermocouples* which convert heat energy into electric energy, *solar cells* which convert light energy into electric energy, *fuel cells* and other chemical cells which convert chemical energy into electric energy at very high efficiencies.

The primary source of electric energy in the world today is the *dynamo* or *electric generator.* This is a rotating machine in which mechanical energy is changed into electric energy. The mechanical energy needed to turn the generator is usually obtained from a fluid turbine operated by the falling water from a large dam, a steam turbine operated by the steam from a nuclear reactor, or an internal combustion engine. It is important to have a good understanding of the operating characteristics and limitations of electric energy sources, and we will devote this chapter to the study of a number of different sources.

*It is interesting to note that a voltaic cell can be formed by sticking an iron nail and a copper wire (two dissimilar metals) into an orange (the orange juice is the electrolyte).

4.1 THE VOLTAIC CELL

The EMF between the electrodes of a voltaic cell is due to the chemical activity between the electrodes and the electrolyte. Some metals react with an electrolyte more readily than others and metals can therefore be listed in a series according to their activities (tendencies to accept electrons). Such a listing is called the *electromotive-force series* and is given in Table 4.1. Hydrogen, though it is not a metal, is used as the reference. All of the metals in the table listed above hydrogen are more active than those listed below hydrogen. In fact, any metal in the table will react more vigorously, or replace, any other metal listed below it.

Table 4.1. The Electromotive-Force Series.

Metal	Symbol	Electrode potential (volts)
Magnesium	Mg	−2.34
Aluminum	Al	−1.67
Manganese	Mn	−1.05
Zinc	Zn	−0.76
Chromium	Cr	−0.71
Iron	Fe	−0.44
Cadmium	Cd	−0.40
Cobalt	Co	−0.28
Nickel	Ni	−0.25
Tin	Sn	−0.14
Lead	Pb	−0.13
Hydrogen	H	0.00
Copper	Cu	+0.34
Mercury	Hg	+0.80
Silver	Ag	+0.80
Platinum	Pt	+1.20
Gold	Au	+1.42

For example, if a piece of zinc (Zn) is immersed in a copper sulphate $CuSO_4$ solution, the zinc will displace the copper ions in the solution since zinc is more active than copper (the copper sulphate solution contains positive copper ions Cu^{++} and negative sulphate ions SO_4^{--}). The chemical reaction which occurs is shown in Fig. 4.1. When a neutral zinc atom leaves the zinc electrode to go into solution, it must leave behind two electrons since it must enter the solution as a positive zinc ion Zn^{++}. A copper ion is then attracted from the solution to the zinc electrode and it combines with the two excess electrons to produce a neutral copper atom. Thus as the zinc goes into solution, copper is deposited on the zinc electrode. No net electric energy is produced since the two excess electrons from each zinc atom are used by a copper atom; instead, the chemical energy of the cell is changed into heat energy.

In the cell shown in Fig. 4.2, a zinc electrode and a copper electrode are immersed in a solution of water and hydrochloric acid, HCl, (the electrolyte). The HCl ionizes in the water forming positive hydrogen ions H^+ and negative

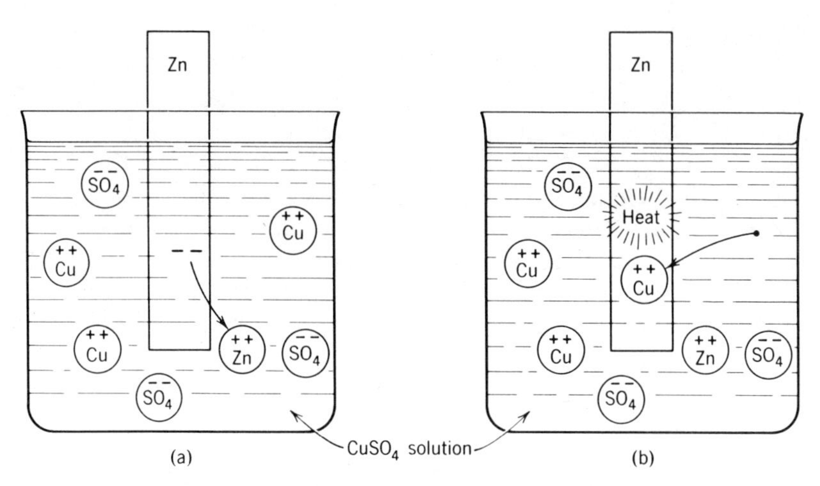

Figure 4.1 Displacement of copper by zinc (*a*) Zn atom goes into solution as Zn^{++} leaving 2 electrons on Zn electrode (*b*) Cu^{++} from solution is deposited on Zn electrode as Cu after taking 2 excess electrons

chlorine ions Cl^-. Since zinc is more active than hydrogen, zinc atoms will leave the zinc electrode and go into solution as zinc ions Zn^{++}. For each Zn^{++} formed, an excess of 2 electrons is left on the zinc electrode, and 2 hydrogen ions $2H^+$ are displaced from the solution. Hydrogen is more active than copper, and therefore the 2 hydrogen ions will travel to the copper electrode, take 2 electrons from the copper, and form a molecule of hydrogen gas H_2. Thus a negative charge accumulates on the zinc electrode and a positive charge forms on the copper electrode resulting in a potential difference between the two electrodes. The chemical reaction between the electrodes and the electrolyte will continue until the accumulated charge is great enough to establish an equilibrium condition.

The magnitude of the potential difference, or the EMF of the cell, is then found by taking the *difference* between the electrode potentials given in Table 4.1. In this case, the copper has an electrode potential of +0.34 V with respect to hydrogen while the zinc has an electrode potential of −0.76 V with respect

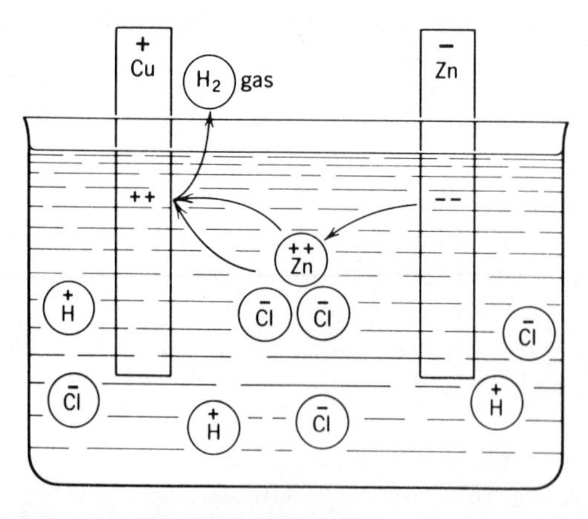

Figure 4.2 A simple Cu–Zn chemical cell

to hydrogen. Thus the EMF of the cell is

$$+0.34-(-0.76)=0.34+0.76=1.10\ \text{V}$$

The copper electrode is positive while the zinc electrode is negative.

The EMF of any simple chemical cell, such as the Zn–Cu cell, can be found in a similar manner; thus the magnitude of the EMF is the difference between the two electrode potentials given in Table 4.1 and the positive electrode is the *more positive* element as listed in the table.

• • •

Example 4.1 What would be the EMF of the cell shown in Fig. 4.2 if the electrodes were Ni and Zn? Which would be the positive electrode?

Solution The EMF is found by taking the difference between the electrode potentials given in Table 4.1. Thus

$$\text{EMF}=-0.25-(-0.76)=-0.25+0.76=0.51\ \text{V}$$

Nickel is the more positive of the two elements listed in the table and will therefore be the positive electrode.

• • •

In the copper–zinc cell the hydrogen gas liberated at the copper electrode will usually form a thin layer of bubbles which completely covers the copper bar. This layer of bubbles is undesirable since it will effectively insulate the copper from the electrolyte. As a result, the EMF of the cell is lowered since the hydrogen begins to act as the electrode in place of the copper. This effect is called *polarization.* The two main remedies for eliminating or reducing polarization are (*a*) the addition to the electrolyte of an agent that will remove the hydrogen gas and (*b*) the use of an electrolyte that does not cause the formation of a gas.

In the Daniell cell shown in Fig. 4.3, no gas is formed at either electrode and thus polarization is avoided. The cell operates just as the copper–zinc cell previously discussed except that the positive hydrogen ions which pass through the porous cup on their way to the copper electrode displace copper ions in the copper sulphate $CuSO_4$ solution (hydrogen is more active than copper). Thus pure copper from the solution is collected at the copper

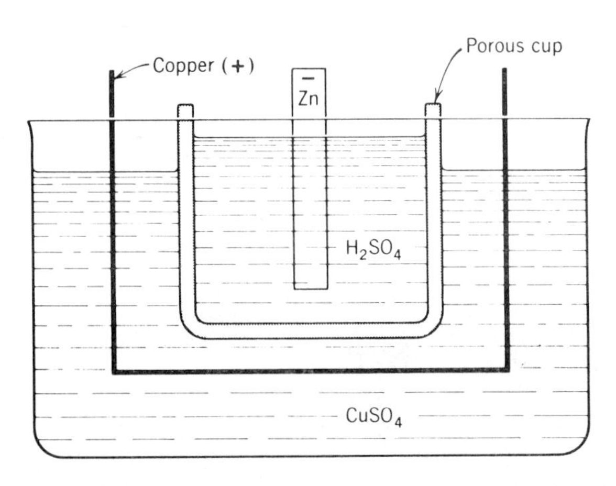

Figure 4.3 Daniell cell $CuSO_4$—copper sulphate H_2SO_4—sulphuric acid

electrode instead of hydrogen gas. The Daniell cell therefore provides a fairly stable voltage source.

There are certain chemical cells that will provide a very precise EMF when constructed from very pure materials. One such cell is the *Weston Standard* cell shown in Fig. 4.4. This cell is not too difficult to construct and will provide an EMF of 1.0188 V at a temperature of 20°C. The electrodes are mercury and cadmium amalgam and the electrolyte is a cadmium sulphate $CdSO_4$ solution which also contains cadmium and mercury sulphates. It is sealed in glass to prevent evaporation. The Weston cell is used as a laboratory standard, but it must be used with care since it will deteriorate rapidly if too much current is drawn from it.

Occasionally a chemical cell is created quite by accident and the results may be destructive. Consider for example the results of connecting copper lead-in wires to an aluminum antenna. If the antenna is used in an atmosphere having a high salt content, such as near the ocean, or in an atmosphere containing a high percentage of pollutants, the moisture in the atmosphere (dew or rain) may lead to the formation of a chemical cell between the copper and the aluminum. As a result, there may be considerable corrosion at the junction of the two metals. In another situation, the saliva of the mouth may act as an electrolyte between gold and silver amalgam fillings; if the fillings touch one another, an electric current may cause the eventual decomposition of the fillings.

Most of the practical chemical cells in use today are somewhat more complicated than the simple cells discussed here. The chemical reactions are usually more complicated because of additives in the electrolyte which reduce polarization, lead to greater efficiencies, and provide longer cell life. Nevertheless, the basic principles are the same and the chemical energy of the cell is used to create a potential difference. Thus when the cell is used in a circuit to produce an electric current, the chemical energy of the cell is transformed into electric energy.

As chemical energy is changed into electric energy in the Cu–Zn cell, or the

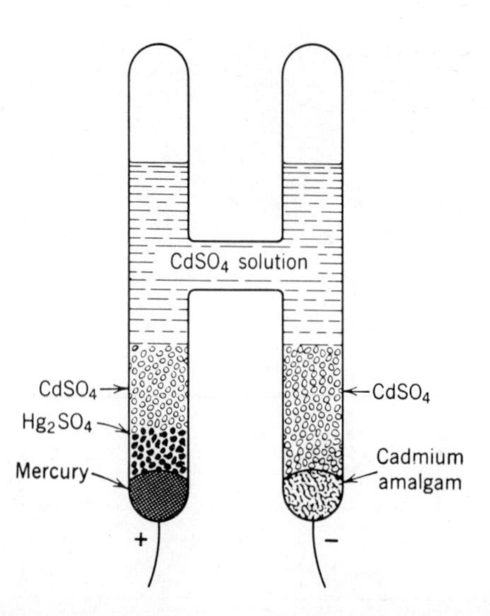

Figure 4.4 Weston Standard cell

Daniell cell, the zinc electrode is dissolved by the electrolyte. Eventually all of the zinc will be used up and the cell will no longer be able to maintain a constant EMF. At this point the cell can be *recharged* by forcing a current through the cell in the opposite direction. This can be accomplished by connecting a source having a higher EMF than the Cu–Zn cell as shown in Fig. 4.5. The charging current forces copper atoms from the copper electrode to go back into solution as copper ions, and it causes the zinc ions from the electrolyte to be deposited on the zinc electrode, thus restoring the cell to its original condition. During the recharging process, electric energy is stored in the cell in the form of chemical energy with some energy loss due to heat. Even though the Cu–Zn cell can be recharged, it is not usually done since it would cost more to recharge it than to simply replace it with a new cell. Cells which are not rechargeable or which are replaced when exhausted are called *primary* cells.

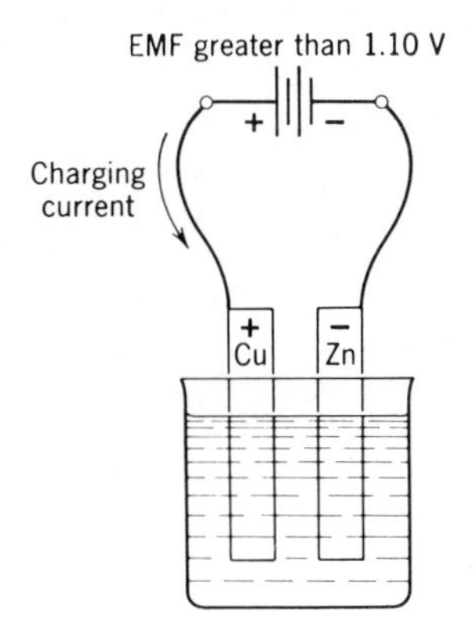

Figure 4.5 Recharging a Cu–Zn cell

4.2 DRY CELL

The chemical cells discussed in the previous section are often referred to as *wet* cells since they have liquid electrolytes. One of the obvious disadvantages of such a cell is the danger of spilling the electrolyte. The *dry* cell was developed to overcome this difficulty, and a number of different dry cells are shown in Fig. 4.6. These cells are used as portable sources of electric energy in such applications as flashlights, hearing aids, and transistor radios.

The construction of a typical dry cell is shown in Fig. 4.7. The zinc container serves as the negative electrode and the carbon rod in the center is the positive electrode. The term dry cell is somewhat misleading since the electrolyte is not really dry, but is rather a moist paste containing ammonium chloride, zinc chloride, black manganese dioxide, and carbon. The chemical action in the dry cell is not quite as simple as that in the Cu–Zn cell, but there are similarities. The zinc container is dissolved by the electrolyte, and in the process electrons are stored on the container making it the negative electrode. The hydrogen ions present in the electrolyte rob the carbon rod of electrons making it the positive electrode. The hydrogen ions form hydrogen gas near the carbon electrode and polarization would occur without the presence of a *depolarizing agent.* The depolarizer in the dry cell is the manganese dioxide

Figure 4.6 (*a*) 1.5 V "D" cell (*b*) 1.5 V "C" cell (*c*) 1.5 V "AA" cell (*d*) 1.5 V "No. 6" cell (*e*) Four "6" cells (*f*) 9 V battery (*g*) 9 V battery (*h*) 9 V battery (Courtesy of Union Carbide Corp., Common Products Division, New York, N.Y.)

which reacts with the hydrogen gas to form water. Since the gas formed is eliminated by the depolarizing agent, the cell can be sealed without any danger of bursting due to the generation of gases. The chemical reactions in the cell result in the separation of charges and the formation of an EMF of 1.4 to 1.6 V. Typically, the carbon–zinc dry cell is said to have an EMF of 1.5 V.

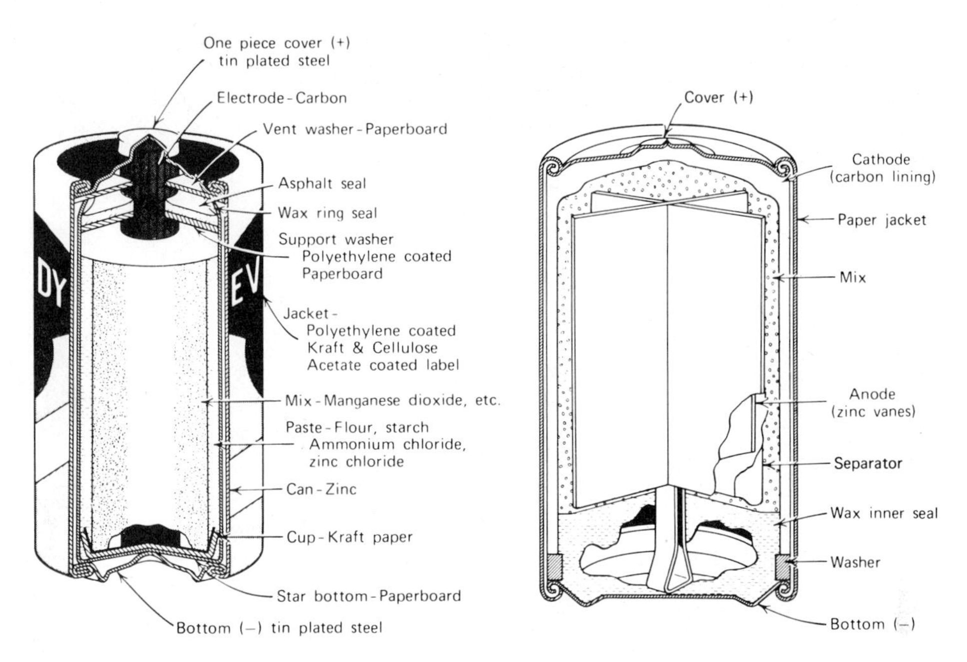

Figure 4.7 Dry cell construction (*a*) Cross section of standard round cell (*b*) Cross section of external cathode or "inside-out" round cell

The chemical reactions in carbon-zinc dry cells are somewhat more complicated than those in Cu–Zn cells. As a result dry cells cannot be effectively recharged and they are usually destroyed when their chemical energy is expended. Carbon–zinc dry cells are therefore primary cells.

There are a number of inherent drawbacks in the carbon–zinc dry cell. First of all, if there are impurities in the zinc, small local voltaic cells may be created which cause the energy of the cell to be dissipated even though the cell is not in use. This undesirable effect is called *local chemical action.* Secondly, the paste electrolyte may dry out, resulting in a reduction of the cell terminal voltage under load. As a result of these two deteriorating characteristics, dry cells can be stored for only a limited period of time. Most manufacturers stamp a date on their cells, and if the cells are not used within a prescribed period of time they may become defective. The period of time which a cell can be stored before using it is called the *shelf life.* The shelf life of a cell depends on its size and may vary from a few months for small penlite cells up to a year for larger cells.

4.3 LEAD–ACID CELL

Another very important chemical cell is the lead–acid cell shown in Fig. 4.8. The positive electrode is made of lead peroxide, PbO_2, the negative electrode is spongy lead, Pb, and the electrolyte is sulphuric acid, H_2SO_4, in water. The sulphuric acid ionizes in the water to form two positive hydrogen ions $2H^+$ and one negative sulphate ion SO_4^{--}. According to Table 4.1 lead is more active than hydrogen, and thus lead atoms from *both* electrodes will go into solution as lead ions Pb^{++}, and they will displace the hydrogen ions in the

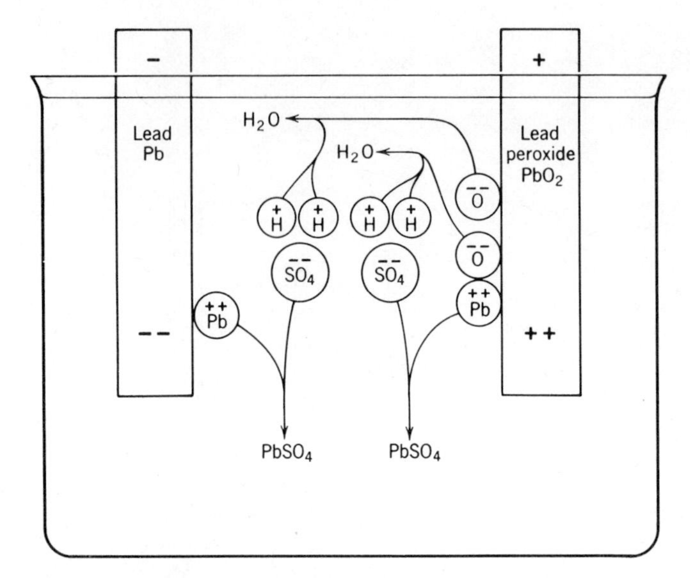

Figure 4.8 A lead–acid cell when discharging

electrolyte. There are therefore two separate chemical reactions to consider in the lead–acid cell.

When a lead atom leaves the spongy lead electrode to go into solution as a lead ion Pb^{++}, it must leave 2 electrons behind. The lead ion will combine immediately with a sulphate ion SO_4^{--} to form lead sulphate $PbSO_4$. Thus for every atom of lead which combines, there will be 2 excess electrons on the spongy lead electrode and 2 excess hydrogen ions in the electrolyte. The equation for this reaction is

$$Pb + SO_4^{--} \rightarrow PbSO_4 + 2 \text{ electrons}$$

It can be shown that the EMF developed between the electrolyte and the spongy lead electrode is 0.35 V with the electrode being more negative.

When a lead atom leaves the lead peroxide electrode to go into solution as a lead ion Pb^{++}, it also leaves behind 2 excess electrons, combines immediately with a sulphate ion SO_4^{--} to form lead sulphate $PbSO_4$, and displaces 2 hydrogen ions in the electrolyte. However, at this plate there will be 2 free oxygen atoms created for every lead atom which leaves. These 2 oxygen atoms will take 4 electrons from the lead peroxide plate and become oxygen ions O^{--} creating a *deficit* of 2 electrons on the plate. The 2 oxygen ions $2O^{--}$ will combine with 4 hydrogen ions $4\,H^+$ to form 2 molecules of water $2H_2O$. The equation for this reaction is

$$2 \text{ electrons} + PbO_2 + 4H^+ + SO_4^{--} \rightarrow PbSO_4 + 2H_2O$$

It can be shown that the EMF developed between the lead peroxide electrode and the electrolyte is 1.68 V with the electrode being more positive.

Since the EMF's developed between each electrode and the electrolyte have opposite polarities, the total EMF of the lead–acid cell is the sum of the two separate EMF's and is $0.35 + 1.68 = 2.03$ V. The lead peroxide is clearly the positive electrode since it has a deficit of electrons, while the spongy lead is the negative electrode since it has an excess of electrons. The operation of the lead–acid cell can be summarized as follows: (1) lead atoms leave both electrodes and combine with sulphate ions to form lead sulphate; (2) oxygen atoms from the lead peroxide combine with the excess hydrogen ions in the electrolyte to form water; (3) for each lead atom which leaves the spongy lead plate, an excess of 2 electrons is created on that plate; (4) for each lead peroxide molecule which leaves the lead peroxide plate, a deficit of 2 electrons is created on that plate.

It is clear that as the chemical energy of the cell is changed into electric energy, the lead from both plates is used up and the electrolyte is weakened. The energy of the cell is therefore limited, and its EMF may change since it is dependent on the strength of the electrolyte. An instrument called a *hydrometer* is used to check the *specific gravity* of the electrolyte. Specific gravity is the ratio of the weight of a volume of a substance to the weight of an equal volume of water. Concentrated sulphuric acid has a specific gravity of about 1.83 and pure water has a specific gravity of 1.00. Thus sulphuric acid is about 1.83 times as heavy as water. The specific gravity of the electrolyte gives an indication of the amount of sulphuric acid remaining and is thus a measure of the condition of a cell. A reading of 1.00 would mean that the electrolyte is pure water with no acid left. A typical value for the specific gravity of a fully charged cell is 1.28 (or 1280), and a fully discharged cell might have a specific gravity of around 1.15 (or 1150).

One of the most important aspects of the lead–acid cell is the fact that the hydrogen ions from the electrolyte combine with the oxygen ions from the lead peroxide to form water instead of hydrogen and oxygen gas. Thus no polarization occurs and the lead–acid cell is capable of delivering a large current for a short period of time without any reduction in cell EMF. Another important aspect of the lead–acid cell is that it can be recharged by connecting an external source of EMF to the cell as shown in Fig. 4.9. Any cell that can be

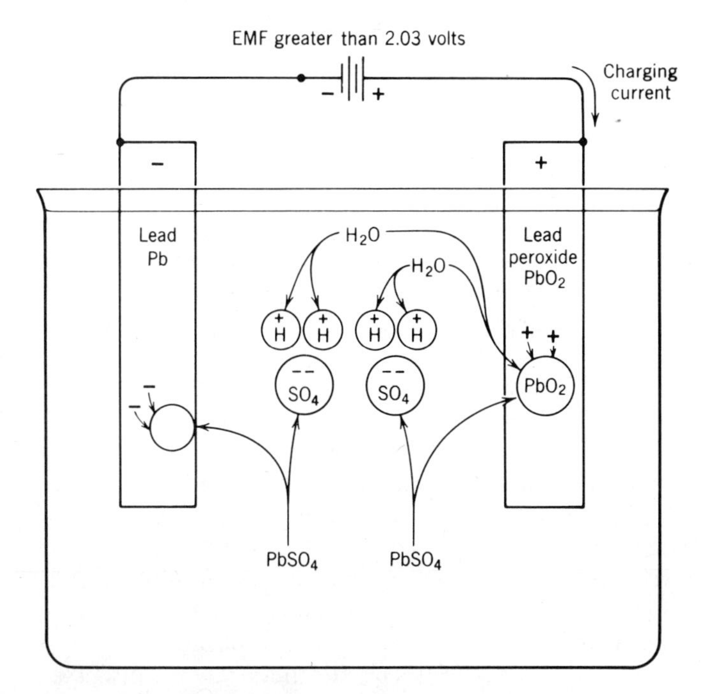

Figure 4.9 Recharging a lead–acid cell

recharged such as the lead–acid cell is called a *secondary* cell. During recharging, the equations describing the operation of the cell are simply reversed and the chemical energy of the cell is restored. Some heat is generated in the cell during recharging, and the recharging process must be slow enough so that the heat does not damage the cell. A discharged lead–acid cell must not be left standing too long without recharging since the lead sulphate formed may fall away from the plates and pile up in the bottom of the cell. If this occurs, the cell cannot be completely recharged. Furthermore, the lead sulphate may pile high enough to bridge the gap between oppositely charged plates resulting in permanent destruction of the cell.

4.4 OTHER CHEMICAL CELLS

One other wet cell which deserves mention because of its historical importance is the nickel–iron, or *Edison,* cell developed by T. A. Edison (1847–1931) in 1909. The Edison cell has a positive electrode of nickel oxide and a negative electrode of powdered iron immersed in an electrolyte of caustic potash (potassium hydroxide, KOH) dissolved in water. This cell produces a very constant EMF of 1.2 V since the specific gravity of the electrolyte does not change as the cell is discharged. It is a secondary cell since it is completely rechargeable, and it is not damaged if left in a fully discharged state for long periods of time. The Edison cell is more rugged than the lead–acid cell since it will stand more mechanical and electrical abuse, but it is not capable of supplying as large a current as the lead–acid cell.

Three important types of dry cells shown in Fig. 4.10 are the *alkaline, mercury,* and *nickel–cadmium* Ni–Cd cells. The manganese–alkaline cells have negative zinc electrodes, positive manganese dioxide electrodes, and electrolytes of potassium hydroxide. They produce an EMF of 1.5 V and some of them are rechargeable for a limited number of times. The manganese–alkaline cells can maintain a constant EMF during use for nearly twice as long as comparable carbon–zinc dry cells, but the alkaline cells are usually more expensive.

Mercury dry cells have positive zinc electrodes, negative mercuric oxide electrodes, and electrolytes of potassium hydroxide. They produce a very constant EMF of 1.3 V. During discharge, there are no gases formed in the chemical reactions, and there is therefore no need for additives to prevent polarization. As a result, very tiny mercury cells can be constructed and they are widely used in miniaturized equipment (hearing aids, transistor radios, heart pacers, and other medical electronic applications). Mercury cells are not rechargeable and are thus primary cells.

Nickel–cadmium cells have positive nickel electrodes, negative cadmium hydroxide electrodes, and electrolytes of potassium hydroxide. They produce a constant EMF of 1.25 V and are capable of maintaining their EMF's at very high load currents. They are rechargeable and thus are secondary cells. Furthermore, they can be stored indefinitely in either the charged or uncharged state without suffering any damage. These cells will of course lose their charge when stored, but they can be recharged very rapidly. Nickel–cadmium cells are widely used in cordless appliances such as shavers, flashlights, photographic equipment, and in aerospace applications.

A summary of the important characteristics of the chemical cells discussed so far is given in Table 4.2.

Figure 4.10 Dry cells (*a*) Alkaline (*b*) Mercuric oxide (*c*) Nickel–cadium (Courtesy of Union Carbide Corp., Consumer Products Division, New York, N.Y.)

Table 4.2. Chemical Cells.

Type	+ Electrode	− Electrode	Electrolyte	EMF (Volts)	Comments
Weston Standard	mercury	cadmium amalgam	cadmium mercury sulphate	1.0188	wet standard
Daniell	copper	zinc	sulphuric acid	1.10	wet secondary
Lead–acid	lead peroxide	lead	sulphuric acid	2.0	wet secondary automotive
Edison	nickel oxide	iron	potassium hydroxide	1.2	wet secondary rugged
Carbon–zinc	carbon	zinc	ammonium and zinc chloride	1.5	dry primary
Manganese–alkaline	manganese dioxide	zinc	potassium hydroxide	1.5	dry primary or secondary
Mercury	zinc	mercuric oxide	potassium hydroxide	1.3	dry primary small
Nickel–cadmium	nickel	cadmium hydroxide	potassium hydroxide	1.25	dry secondary small

4.5 BATTERIES

A battery is a group of cells connected together as defined in Chapter 2, and the proper symbols for a cell or a battery are given in Fig. 2.15. To provide a battery having a greater EMF than a single cell we connect cells in *series*; connecting cells in *parallel* will provide a battery having a greater current-carrying capacity.

A group of cells are said to be connected in series if the current is the same in each cell. The cells shown in Fig. 4.11*a* are connected in series since the

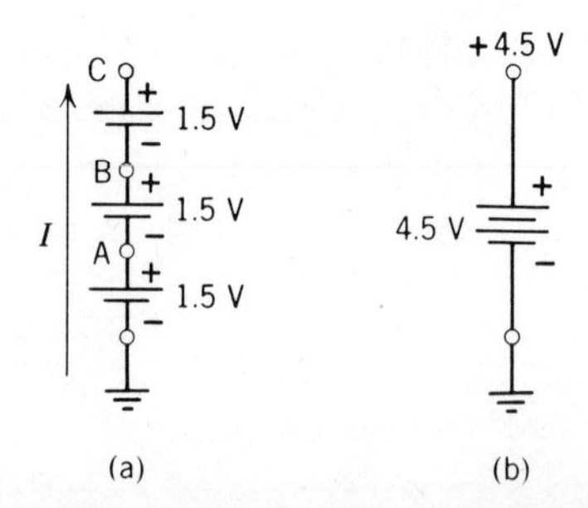

Figure 4.11 (*a*) Cells connected in series (*b*) 4.5 volt battery composed of three 1.5 V cells in series

current I must be the same in each of the three cells. The negative terminal of the lower cell is connected to ground which acts as the reference point for measuring voltage (ground is zero volts). Point A is clearly at a potential of +1.5 V with respect to ground. Point B has a potential of +1.5 V with respect to point A, or is at a voltage of +3.0 V with respect to ground. Similarly, point C is at a voltage of +4.5 V with respect to ground. These three cells connected in series are then the equivalent of the 4.5 V battery shown in Fig. 4.11*b*.

• • •

Example 4.2 How many nickel–cadmium cells would be required to construct a 5.0 V battery? Show how to connect these cells to provide a voltage source of −5.0 V with respect to ground.

Solution Since each cell has an EMF of 1.25 V, it will require 5.0 V/1.25 V = 4 cells to form a 5.0 V source. These cells should be connected as shown in Fig. 4.12 to form a −5.0 V source.

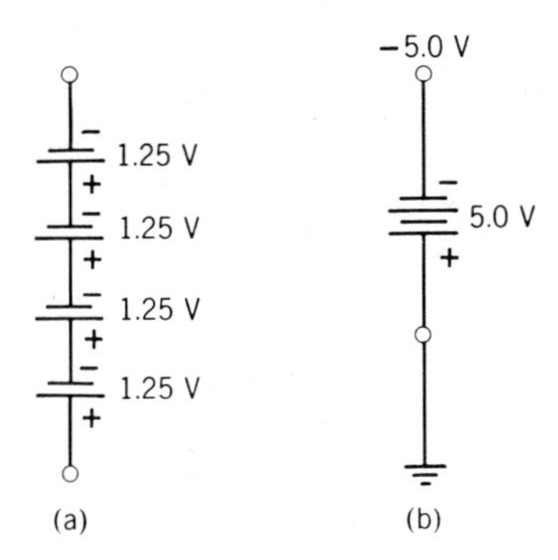

Figure 4.12 5.0 V Ni–Cd battery (*a*) Four Ni–Cd cells (*b*) −5.0 V source

• • •

Example 4.3 What is the voltage at point C in Fig. 4.13*a*? What would be the value of the current through a 1 kΩ resistor connected between points A and C or between A and B? Repeat for the battery shown in Fig. 4.13*b*.

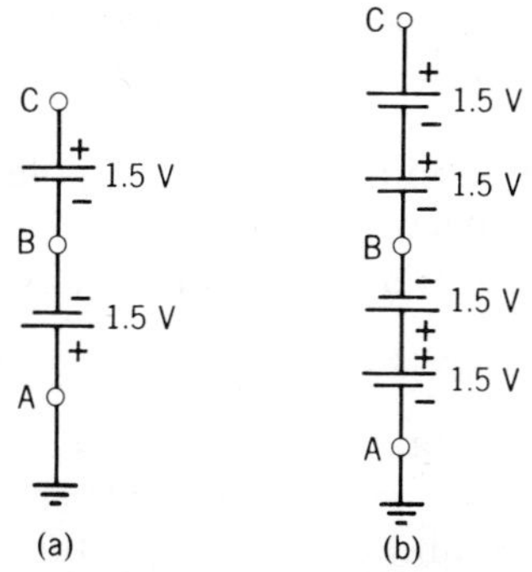

Figure 4.13 (*a*) Series cells in opposition (*b*) Series cells

Solution For Fig. 4.13*a*: In moving from point A to point B, we go down to −1.5 V. In moving from point B to point C, we go back up 1.5 V and thus point C is at 0.0 V with respect to ground. These two cells are connected in series but with opposite polarities, and as a result their EMF's cancel one another. If a 1 kΩ resistor is connected between points A and C, the voltage across the resistor will be 0.0 V and thus there will be no current. If a 1 kΩ resistor is connected between A and B, the voltage across the resistor will be 1.5 V and the current will be $I = 1.5\ \text{V}/1\ \text{k}\Omega = 1.5\ \text{mA}$.

For Fig. 4.13*b*: The two cells between points A and B are also connected in opposition and thus the voltage at point B is 0.0 V with respect to ground. These two cells can therefore be removed from the battery since they contribute nothing. The two cells between points B and C are connected in series with the correct polarities and thus the voltage at C is clearly +3.0 V. If a 1 kΩ resistor is connected between A and C, the current in the resistor is $I = 3.0\ \text{V}/1\ \text{k}\Omega = 3\ \text{mA}$. Since the potential between points A and B is zero, there would be no current through a resistor connected between these points.

• • •

Cells can also be connected in parallel to form batteries as shown in Fig. 4.14. Cells are connected in parallel only if the voltage across *every* cell is *exactly* the same (the currents through the cells may be different). If the three cells in Fig. 4.14 each have an EMF of 1.5 V, the battery EMF is 1.5 V. However, if the three cells are identical, the battery will be capable of delivering three times the current of a single cell. Thus cells are connected in parallel to increase the current capacity of the source.

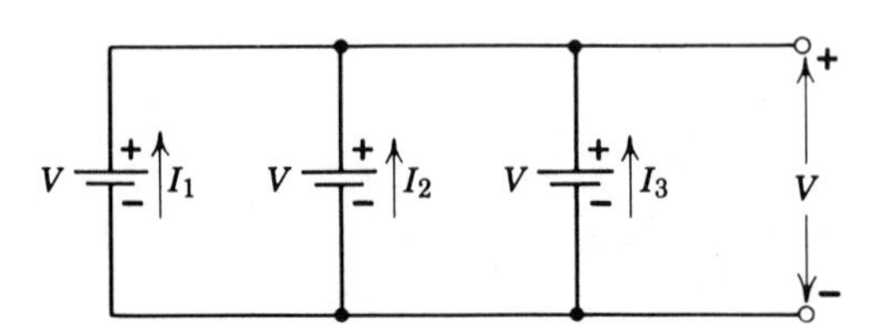

Figure 4.14 Cells connected in parallel

• • •

Example 4.4 A 10 Ω resistor is to be placed across a 1.5 V source. If the source must be constructed using 1.5 V cells capable of supplying 50 mA, how must the source be connected?

Solution The total current which must be supplied to the load resistor is $I = 1.5\ \text{V}/10\ \Omega = 0.15\ \text{A} = 150\ \text{mA}$. Since each cell can only supply 50 mA, there must be 150 mA/50 mA = 3 cells connected in parallel exactly as in Fig. 4.14.

• • •

A cell or battery converts stored chemical energy into electric energy when used in an electric circuit. The amount of energy that can be stored in a cell or battery depends on the size and type of cell. Thus every battery is given an energy rating. The energy could be measured in J, but is usually given in terms of W-sec (one J equals one W-sec). A measure of the quality of a battery or cell is often stated in terms of energy density, that is, the energy stored per pound, or in W-sec/lb. Several types of cells are compared with gasoline in Table 4.3.

Table 4.3.

Source	Energy Densities (W-sec/lb)
Lead–acid cell	3.6×10^4
Nickel–cadmium cell	5.0×10^4
Zinc–air battery	23.4×10^4
Fuel cells	72×10^4
Gasoline	180×10^4

The energy storage capacity of a battery can be stated in equation form as

$$\text{energy} = W = \text{watt-sec} = \text{volts} \times \text{amps} \times \text{sec} = V \times I \times t \tag{4.1}$$

If we assume that the EMF of the cells remains constant over the useful life of the battery, Eq. 4.1 can be rearranged as

$$\frac{W}{V} = I \times t = \text{constant} \tag{4.2}$$

This equation is not quite true since the EMF's of the cells (V in Eq. 4.2) do change; nevertheless, Eq. 4.2 provides a useful expression for the operating capacity of a battery in terms of the amount of current the battery is capable of supplying. The units of Eq. 4.2 are A-sec which are quite small and the units of ampere-hours (Ah) are more common. The Ah rating of a battery or cell is usually supplied by the manufacturer and is simply a statement of the length of time a source is capable of supplying a certain current without an appreciable drop in EMF.

• • •

Example 4.5 The Ah rating of a certain dry cell is given as 1.0 Ah. What will be the expected life of the cell if it supplies a load current of 100 mA?

Solution From Eq. 4.2, Ah = $I \times t$. Solving for t,

$$t = \frac{\text{Ah}}{I} = \frac{1.0\ \text{Ah}}{0.1\ \text{A}} = 10\ \text{h}$$

• • •

Example 4.6 What must be the Ah rating of the battery in Fig. 4.15 if the circuit must function unaided for 5000 h?

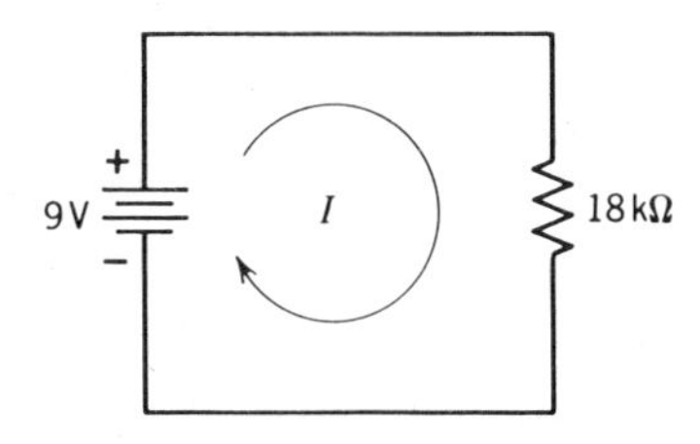

Figure 4.15

Solution The current in the circuit is

$$I = \frac{V}{R} = \frac{9\ \text{V}}{18\ \text{k}\Omega} = 0.5\ \text{mA}$$

The Ah rating is found using Eq. 4.2. Thus

$$\begin{aligned} \text{Ah} &= I \times t = 0.5 \times 10^{-3}\ \text{A} \times 5 \times 10^{3}\ \text{h} \\ &= 2.5\ \text{Ah} \end{aligned}$$

• • •

A much more accurate determination of cell performance can be obtained from the cell *voltage discharge curves.* Some typical discharge curves are shown in Fig. 4.16 where the terminal voltage of the cell is shown as a function of time for various values of load current. The *service life* of the cell can be determined by following the proper load current curve down to the minimum acceptable cell voltage.

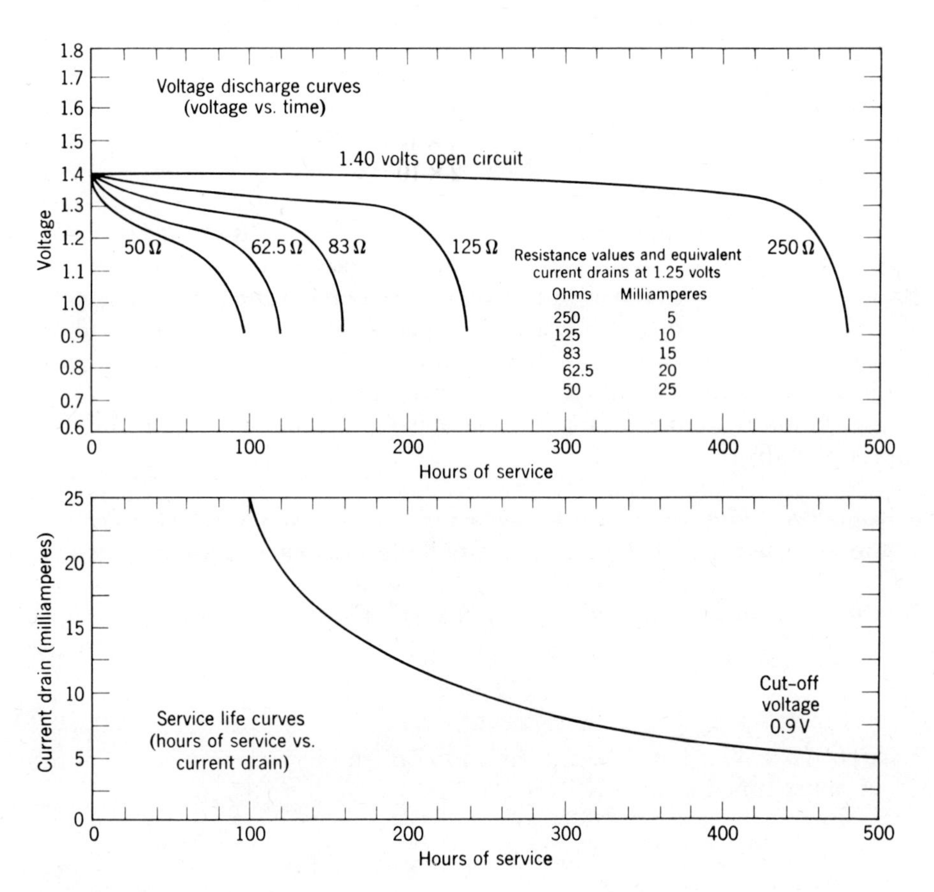

Figure 4.16 (Courtesy of Union Carbide Corp., Consumer Product Division, New York, N.Y.)

• • •

Example 4.7 Determine the service life of the cell shown in Fig. 4.16 if the load current is constant at 10 mA.

Solution If the cell is considered to be expended when the terminal voltage drops to 0.90 V, the service life read from the discharge curve is 240 h.

• • •

It must be emphasized that connecting cells in parallel as shown in Fig. 4.14 is permitted *only* when the cells have identical EMF's. If two cells are connected in parallel as shown in Fig. 4.17*a*, and the EMF of one of the cells is

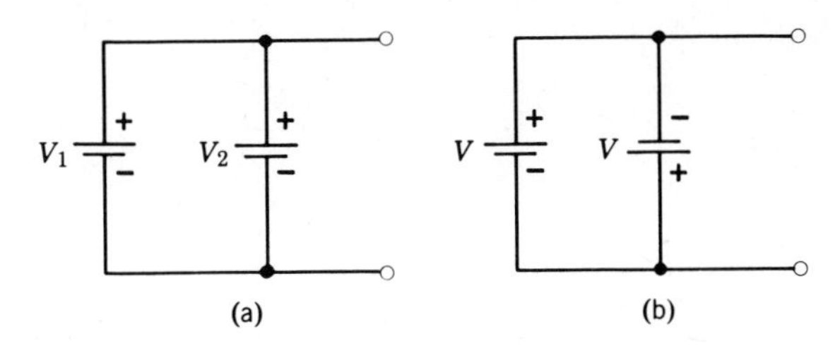

Figure 4.17 (*a*) Perfect cells must not be connected in parallel unless they have equal EMF's (*b*) Cells connected in parallel with incorrect polarity resulting in a short circuit

greater than the other, a very large current will flow in the circuit and the cells will very likely be destroyed. If the two cells are perfect (i.e., they have no resistance to current), there is no resistance in the circuit and the current approaches infinity. We will completely disallow the parallel connection of any perfect cells having different EMF's.

In the situation shown in Fig. 4.17*b*, the two cells are connected in parallel but the polarities are incorrect. If the two cells are perfect, there is no resistance and the current in the circuit will approach infinity resulting in destruction of the cells. Such an incorrect connection is again disallowed.

4.6 REAL AND IDEAL SOURCES

In the previous section, we stated that a perfect source is one which has no resistance to electric current. Such a source is shown in Fig. 4.18, and since this represents the ideal case it is called an *ideal source.* A graph of the terminal voltage V_t of the source versus current through the source (called a volt-ampere, V-A, characteristic) is also given in the figure, and it is clear that V_t is constant for any value of current. A negative value of current on the graph simply means that current is being forced through the source in a direction opposite to the normal direction shown in Fig. 4.18*a*. The reason that V_t is constant for any value of current is simply because an ideal source has no internal resistance.

There is no ideal source in practice since any source has some internal resistance. A model for a *real source* which takes into account the internal resistance is given in Fig. 4.19*a*. The real source model consists of an ideal

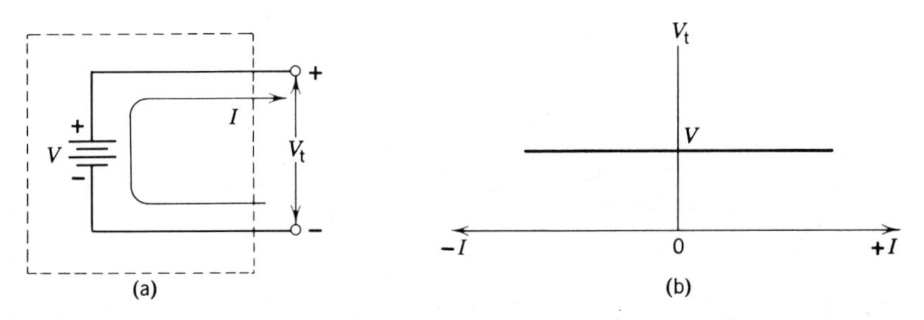

Figure 4.18 (*a*) Ideal voltage source (*b*) Volt-ampere (V-A) characteristic curve for an ideal voltage source

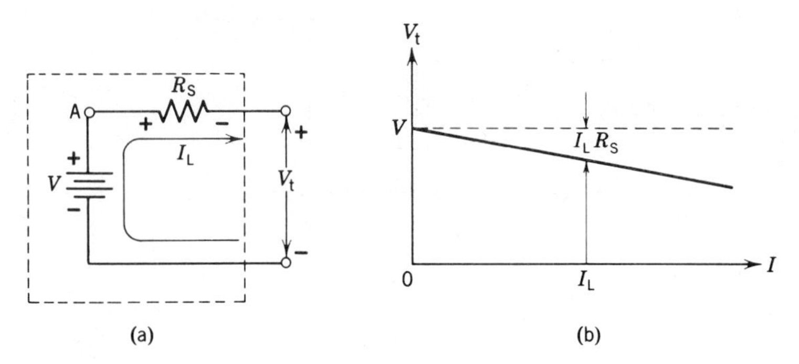

Figure 4.19 (*a*) A real voltage source having internal resistance (*b*) V-A characteristic for a real voltage source

EMF source V in series with a resistance R_S representing the internal resistance of the source. From the V-A characteristic for the real source given in Fig. 4.19*b*, it can be seen that the terminal voltage V_t is not constant and *falls off* as the current increases. This drop in terminal voltage is due entirely to the source resistance R_S since the internal EMF of the source V is constant.

When a load resistance is connected to the source, a load current I_L will be delivered. Since the load current must pass through the internal resistance, a voltage drop is developed across R_S as shown in Fig. 4.19*a*. The voltage at point A in Fig. 4.19*a* is $+V$ volts with respect to ground, but in moving to the terminals of the source, the voltage is reduced by the voltage drop across R_S. The voltage drop across R_S is $I_L \times R_S$ and thus the terminal voltage is

$$V_t = V - I_L R_S \tag{4.3}$$

If no load is connected to the source, the current is reduced to zero and the source is said to be *open circuited.* If there is no load current, the term $I_L R_S$ in Eq. 4.3 is zero and the *open circuit* terminal voltage of the source is equal to the EMF of the ideal source V.

• • •

Example 4.8 Calculate the terminal voltage of the real source shown in Fig. 4.20*a* for the following load currents: 0.00 A, 0.05 A, 0.10 A, 0.50 A, 1.00 A. Plot the results.

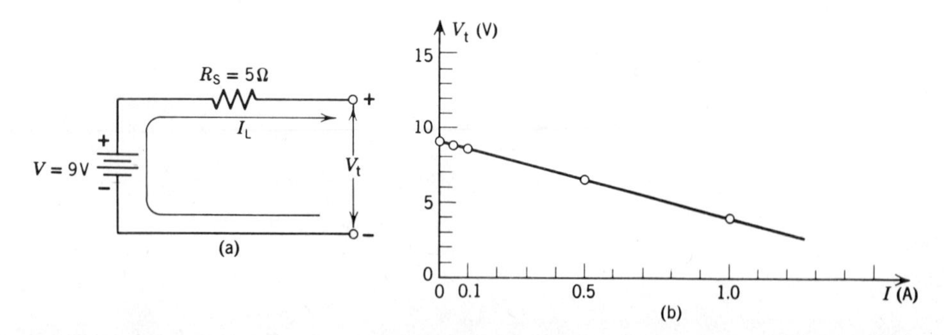

Figure 4.20 (*a*) A real voltage source (*b*) V_t versus I_L for a real voltage source

Solution The terminal voltage is found using Eq. 4.3.

For $I_L = 0.00$ A, $V_t = 9.00 \text{ V} - 0 = 9.00$ V

For $I_L = 0.05$ A, $V_t = 9.00 \text{ V} - 0.05 \text{ A} \times 5\ \Omega = 8.75$ V

For $I_L = 0.10$ A, $V_t = 8.50$ V

For $I_L = 0.50$ A, $V_t = 6.50$ V

For $I_L = 1.00$ A, $V_t = 4.00$ V

The results are plotted in Fig. 4.20*b*.

• • •

The internal resistance of a source can also be determined by measurement. Rearranging Eq. 4.3 yields

$$R_S = \frac{V - V_t}{I_L} \tag{4.4a}$$

We can check the validity of Eq. 4.4*a* for calculating internal resistance by using the data from Example 4.8 for $I_L = 1.00$ A. Thus

$$R_S = \frac{9.00 \text{ V} - 4.00 \text{ V}}{1.00 \text{ A}} = \frac{5.00 \text{ V}}{1.00 \text{ A}} = 5\ \Omega$$

The equation will work equally well for any other values of I_L in the example. Thus the internal resistance of a source can be found by making the following measurements and calculations:

1. Measure the open circuit terminal voltage V.
2. Connect a load resistance and measure the terminal voltage V_t.
3. Determine the load current I_L in part 2 above.
4. Calculate R_S using Eq. 4.4*a*.

• • •

Example 4.9 The open circuit terminal voltage of the source in Fig. 4.21 is 6.00 V. Determine the internal resistance of the source R_S if the terminal voltage drops to 5.00 V when a resistance R_L of 10 Ω is connected to the terminals.

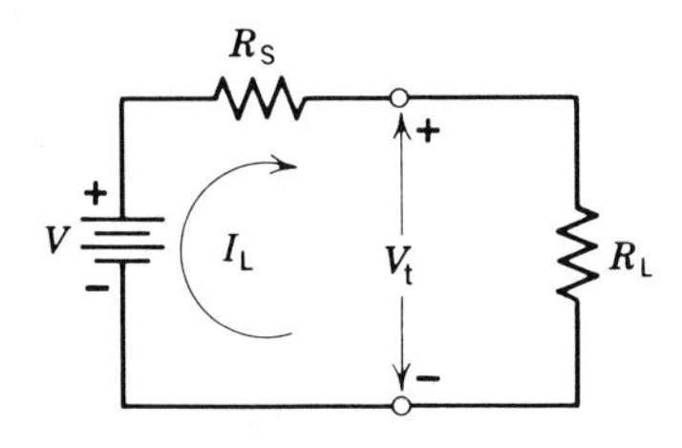

Figure 4.21

Solution The open circuit terminal voltage is equal to the EMF of the ideal source V and is 6.00 V. The terminal voltage with a 10 Ω load is $V_t = 5.00$ V. The load current I_L is the terminal voltage V_t divided by the load resistance R_L, or $I_L = 5.00 \text{ V}/10\ \Omega = 0.500$ A. Using Eq. 4.4*a*;

$$R_S = \frac{6.00 \text{ V} - 5.00 \text{ V}}{0.500 \text{ A}} = \frac{1.00 \text{ V}}{0.500 \text{ A}} = 2.00\ \Omega$$

• • •

Notice that Eq. 4.4*a* can be expressed in a slightly different form by making use of the relation $I_L = V_t/R_L$. Thus

$$R_S = \frac{V - V_t}{I_L} = \frac{V - V_t}{\dfrac{V_t}{R_L}} = \left(\frac{V - V_t}{V_t}\right) R_L = \left(\frac{V}{V_t} - 1\right) R_L \qquad (4.4b)$$

Example 4.9 can also be solved using Eq. 4.4*b*, and the result is of course the same. Thus

$$R_S = \left(\frac{V}{V_t} - 1\right) R_L = \left(\frac{6.00\ \text{V}}{5.00\ \text{V}} - 1\right) 10 = (1.20 - 1.00)10$$
$$= (0.20)10 = 2.00\ \Omega$$

The internal resistance of a source leads to two undesirable source characteristics: a drop in source terminal voltage as the load current is increased; and the power dissipated in the source resistance (I^2R_S) leads to heating the source. The source resistance of a real source should always be taken into account if accurate results are expected in circuit analysis.

4.7 OTHER ELECTRIC ENERGY SOURCES

All of the chemical cells discussed in the previous sections convert stored chemical energy into electric energy and they are important sources, but there are many other electric energy sources. As mentioned previously, most of the world's electric energy is produced by rotating machinery called generators. These generators act as voltage sources and some of them produce constant voltage (dc voltage), and some of them produce voltages that vary with time (ac voltage). The symbols for these two types of generators along with the types of voltages they produce are shown in Fig. 4.22. The resistance included in each figure represents the internal resistance of the generator. If we are concerned only with an ideal source, the internal resistance is zero and the

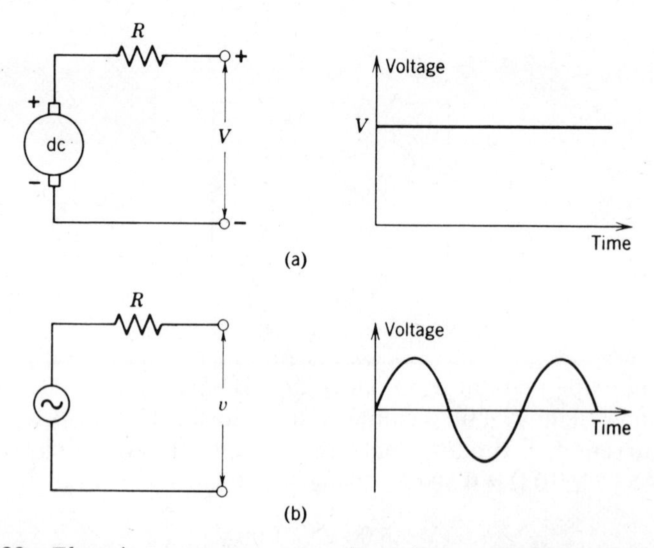

Figure 4.22 Electric generators (*a*) dc voltage generator (*b*) ac voltage generator

resistors are omitted. These generators convert mechanical energy into electric energy.

There are a number of different *electronic* sources that convert electric energy from one form to another. These sources are usually called *power supplies* and a number of them are shown diagramatically in Fig. 4.23. The dc power supply shown in Fig. 4.23*a* converts ac voltage into dc voltage. The dc to dc *converter* shown in Fig. 4.23*b* converts the dc voltage at the input to a dc voltage having a different value at the output. The output voltage may be either greater or smaller than the input voltage. The dc to ac *inverter* shown in Fig. 4.23*c* converts a dc input voltage to a varying or ac voltage at the output. The operation of most of these sources depends on the use of active elements such as vacuum tubes, transistors, or other devices.

Another interesting source of electric energy is the *thermocouple,* which is formed by making a pair of connections between two dissimilar metals as shown in Fig. 4.24. Any time two dissimilar metals are brought into contact, a number of electrons will travel from one metal to the other creating a potential difference across the junction between the two. Such a potential difference is called *contact potential* and is due to differences in the properties of the two metals (crystal structure, density, conductivity, etc.). In the thermocouple shown in Fig. 4.24, electrons will flow from the iron to the copper creating potential differences across both junctions X and Y. The two wires form a closed circuit, but since the two contact potentials are equal, there will be no current in the wires. The contact potentials are strongly temperature-dependent and if the temperature of one junction is different from the other, the two contact potentials will be unequal and a current will exist in the circuit. A typical EMF versus temperature curve for the copper–iron thermocouple is shown in Fig. 4.24*b*. It is interesting to note that the action of a thermocouple is reversible. Thus if a battery is placed in series with

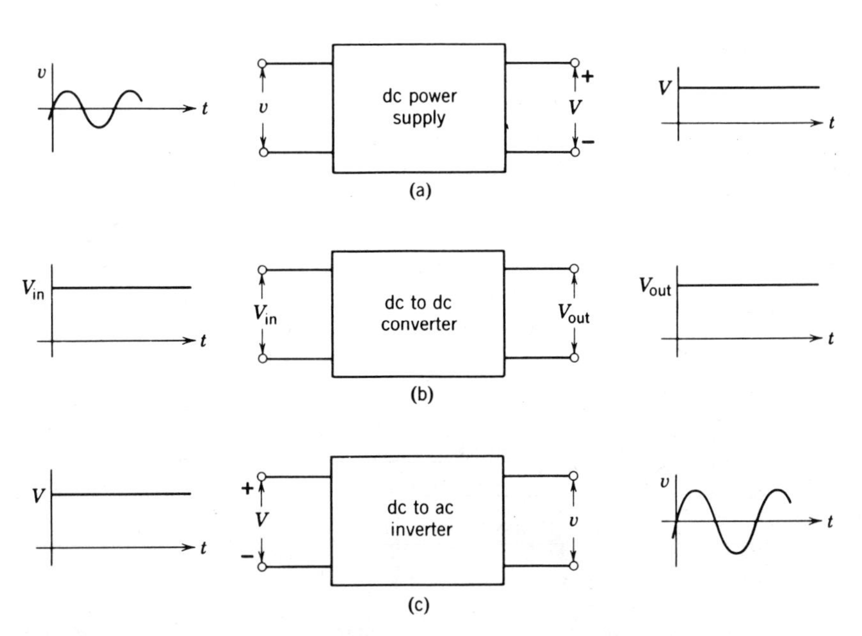

Figure 4.23 Electronic voltage sources (*a*) dc power supply (*b*) dc to dc converter (*c*) dc to ac inverter

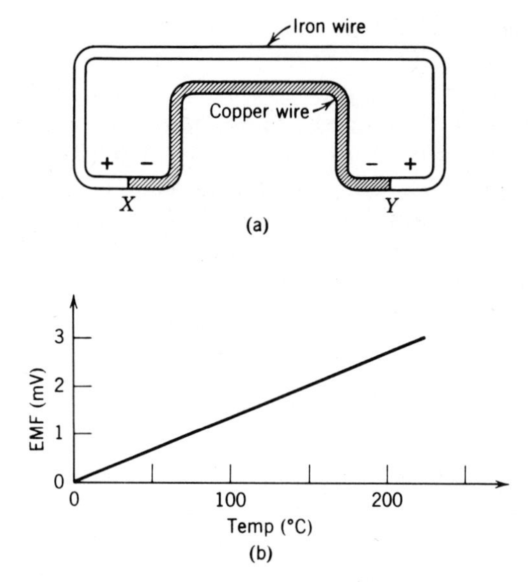

Figure 4.24 (*a*) A thermocouple (two junctions) (*b*) Thermocouple EMF as the temperature of Y is varied while the temperature of X is held at 0°C

the thermocouple circuit as shown in Fig. 4.25, junction X will get hot and junction Y will get cold. Furthermore, if the polarity of the battery is reversed, junction Y will get hot and junction X will get cold. These fundamental principles have been used to construct thermoelectric heaters and refrigerators. Thermocouples are also widely used in instrumentation systems for temperature measurement.

Another important source of electric energy is the *solar cell* in which light energy is converted directly into electric energy. A typical solar cell is made of silicon and might be about one inch square and 0.010 in. thick. Such a cell could produce an EMF of 0.5 V while supplying a current of 100 mA. These cells are wired in series to provide a solar battery having a higher EMF, and they are wired in parallel to increase the current capability of the battery. Solar batteries are widely used in missile and space applications.

Fuel cells form another very important class of electric energy sources. In each of these cells, a chemical reaction is carefully controlled and electric energy is extracted as a direct result of the chemical action. As an example, in the hydrogen–oxygen fuel cell the reaction between the hydrogen and oxygen gases forms pure water H_2O and electric energy is created directly as a

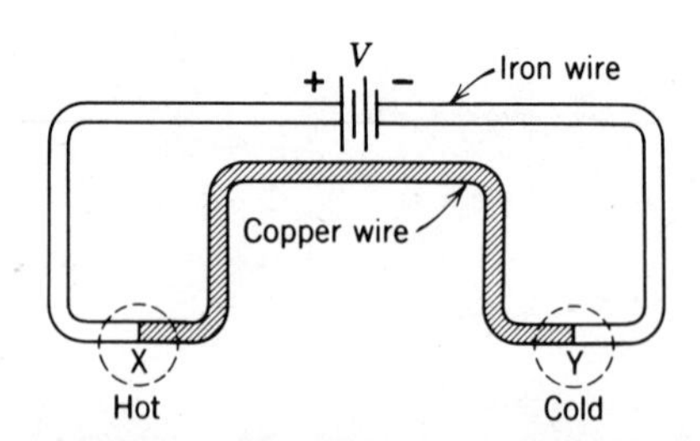

Figure 4.25 Thermoelectric heating and refrigeration

result of this chemical reaction. There is presently a considerable amount of research being conducted on fuel cells and this seems to be a very promising area for electric energy sources of the future.

SUMMARY

The operation of all voltage sources depends on the transformation of energy of some form into electric energy. Generators convert mechanical energy into electric energy and produce either a constant output voltage (dc) or a varying output voltage (ac). Chemical cells are widely used as voltage sources and their operation depends on chemical reactions in the cells to produce dc voltages. Fuel cells also depend on chemical reactions to produce constant dc voltages. Thermocouples transform heat energy directly into electric energy and solar cells transform light energy into electric energy.

Any voltage source is considered an ideal source if it has zero internal resistance; this means that the terminal voltage of the source does not change as the load current changes.

Chemical cells and batteries can only store a limited amount of energy and their energy capacities are given by their Ah ratings. The service life of a chemical cell or battery can be determined from its Ah rating or more accurately from its voltage-discharge curve.

GLOSSARY

Electromotive-force-series. *A listing of the metals according to their chemical activities.*

Fuel cell. *A cell in which electricity is produced as a direct result of a chemical reaction.*

Hydrometer. *An instrument for measuring the specific gravity of an electrolyte.*

Polarization. *The formation of an insulating layer around an electrode in a chemical cell (gas bubbles).*

Primary cell. *A chemical cell which cannot be recharged.*

Secondary cell. *A chemical cell which can be recharged by reversing the chemical process which occurs during discharging.*

Solar cell. *A device which transforms light energy into electric energy.*

Specific gravity. *The ratio of the weight of a substance to the weight of an equal volume of water.*

Thermocouple. *A device which transforms heat energy into electric energy.*

Weston cell. *A chemical cell having an EMF of 1.0188 V at 20°C which is used as a voltage standard.*

IMPORTANT RELATIONSHIPS

- energy = watt-sec = volts × amperes × seconds = $W = V \times I \times t$
- Ah = $I \times t$
- terminal voltage = $V_t = V - I_L R_S$
- source resistance = $R_S = \dfrac{V - V_t}{I_L}$

REVIEW QUESTIONS

1. The EMF's of simple chemical cells can be found from the ____ ____ given in Table 4.1.
2. The EMF's of more complicated cells depend on the ____ ____ between the electrodes and the electrolytes.
3. The nominal EMF of a carbon–zinc dry cell is ____ V.
4. A dry cell is a (primary, secondary) cell.
5. A secondary cell cannot be recharged. (TF)
6. Polarization can be caused by the formation of gases. (TF)
7. A lead–acid cell is a ____ cell.
8. No polarization occurs in a lead–acid cell because ____ ____.
9. Additives in an electrolyte are often used to prevent polarization. (TF)
10. The EMF of a lead–acid cell is dependent on the ____ ____ of the electrolyte.
11. A hydrometer is an instrument for measuring the specific gravity of an ____.
12. Three types of dry cells other than the carbon–zinc cell are ____, ____ and ____.
13. What is the difference between a battery and a cell?
14. Cells can be connected in ____ to form a higher EMF, or in ____ for an increased current capacity.
15. Ideal batteries having different EMF's should never be connected in parallel. (TF)
16. The capacity of a battery to supply a load current is measured in ____.
17. An ideal voltage source has zero internal ____.
18. In a real voltage source, the terminal voltage is dependent on the open circuit ____, the internal ____, and the load ____.
19. The use of iron nails on a galvanized tin roof could cause undesirable chemical cells. (TF)
20. The energy density of a source is measured in ____ ____/____.

PROBLEMS

1. What would be the EMF of the cell shown in Fig. 4.2 if the electrodes were aluminum and copper? Which would be the positive electrode?
2. If the EMF of a cell similar to that in Fig. 4.2 is 1.86 V and one of the electrodes is iron, what must the other electrode be?
3. Explain how polarization is avoided in the Daniell cell.
4. Why must a Weston Standard cell be used with care?
5. In construction handbooks there are tables which give permissible combinations of metals which can be used in metal structures. Can you explain the basis of these tables?
6. Make a sketch of a carbon–zinc dry cell showing the important components.
7. A reading of 1.10 is taken from a hydrometer while checking the electrolyte of a certain lead–acid battery. Is the battery usable, or should it be recharged?
8. What should be the hydrometer reading of a fully charged lead–acid cell?

9. A 470 Ω resistor is connected across the terminals of a mercury cell. What is the current through the resistor? Use the EMF given in Table 4.2.

10. What is the power dissipated in the resistor in Problem 9?

11. A car battery is formed by connecting 6 lead–acid cells in series. What is the EMF of the battery? What current must the battery be capable of delivering to a 10 Ω load?

12. How many carbon–zinc dry cells must be connected in series to construct a battery having an EMF of 13.5 V?

13. If the battery in Problem 11 is rated at 30 Ah, how long will it be capable of supplying current to the 10 Ω load?

14. An alkaline cell rated at 0.30 Ah is connected to a 1.5 kΩ load. What is the approximate service life of the cell?

15. A certain Ni–Cd cell is rated at 0.40 Ah. How many of these cells must be connected in parallel in order to supply current to a 10 Ω load for a period of at least 300 h?

16. A 9 V source must be constructed using 1.5 V carbon–zinc cells rated at 0.45 Ah. This source must supply current to a 200 Ω load for a period of 500 h. Make a drawing showing the proper connection of the cells.

17. The source resistance of a certain battery is 1.2 Ω. What will be the terminal voltage of the battery when supplying a load current of 100 mA if its open circuit EMF is 6.0 V?

18. A lead–acid car battery has an open circuit EMF of 12.0 V. If the terminal voltage drops to 11.8 V when supplying a current of 20 A, what is the internal resistance of the battery?

19. An electronic power supply has an open circuit EMF of 100 V. What is the source resistance of the supply if the terminal voltage drops to 95 V when a 470 Ω resistor is connected across its terminals?

20. An electronic power supply has an open circuit EMF of 24.0 V. The terminal voltage drops to 23.5 V when a 330 Ω resistor is connected across its terminals. What will the terminal voltage be when a current of 25.0 mA is drawn from the supply?

5

RESISTANCE

Electric *resistance* is a measure of the opposition which a material exhibits toward electric current. Nearly all materials exhibit resistance to electric current under ordinary conditions, although there are a few materials that have almost zero resistance when cooled to absolute zero. As seen in Chapter 2, the resistance of a copper wire is due to the collisions between drifting electrons and fixed copper ions, and the copper gains heat energy as a result of the collisions. A similar situation exists in other materials and thus they are heated when an electric current is forced through them. The transformation of electric energy into heat energy in a resistance is an important application of electrical principles, and resistive heating elements are designed specifically for this purpose. Some practical examples are toasters, electric ovens and ranges, and electric heaters. Resistive elements can also be used to limit the current in a circuit and *resistors* are constructed for this purpose.

5.1 CALCULATION OF RESISTANCE

The electrical unit for measuring resistance is the ohm as defined in Chapter 2. Equation 2.5 is repeated here for convenience and it states that the resistance is equal to the applied voltage divided by the resulting circuit current (Ohm's Law).

$$R \text{ (ohms)} = \frac{V \text{ (volts)}}{I \text{ (amperes)}} \tag{2.5}$$

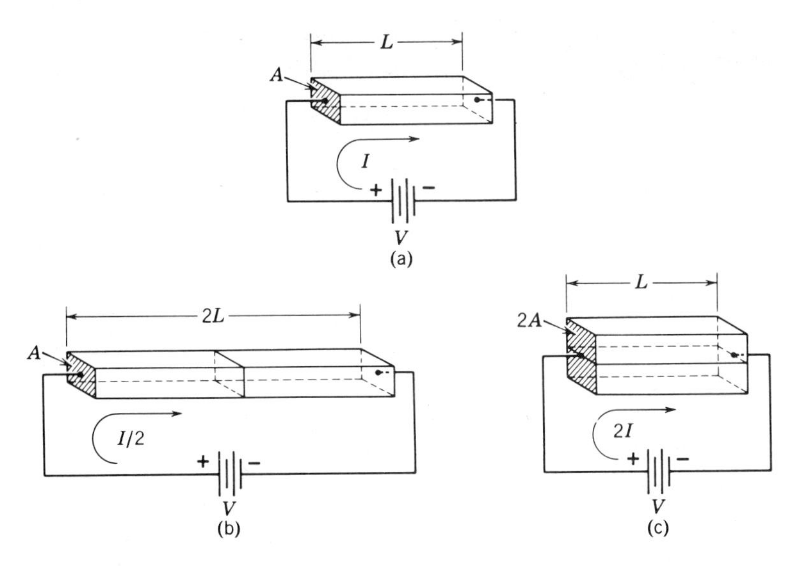

Figure 5.1 Resistance versus length and cross-sectional area

The resistance of a bar of test material can be determined by connecting the bar in a circuit as shown in Fig. 5.1*a* and calculating the ratio of the applied voltage V to the resulting current I. Suppose the resistance of the bar in Fig. 5.1*a* is found to be $R\ \Omega$. Now if the length of the bar is doubled (two bars connected in series) as shown in Fig. 5.1*b*, the current is reduced by a factor of 2. Since the applied voltage is constant, the resistance of the bar must be doubled to $2R\ \Omega$. If these measurements are repeated as the length of the bar is varied, it will be found that the resistance of the bar is *directly* proportional to its *length.*

The bar shown in Fig. 5.1*c* has the original length L, but the cross-sectional area is doubled (two bars connected in parallel), and as a result the current has increased to $2I$. Since the applied voltage is the same, the resistance of the bar must be cut in half to $R/2$. If these measurements are repeated as the cross-sectional area of the bar is varied, it will be found that the resistance of the bar is *inversely* proportional to its *cross-sectional* area.

These observations can be summarized by stating that the resistance of an object is directly proportional to its length and inversely proportional to its cross-sectional area. In the form of an equation

$$R = \rho \frac{L}{A} \tag{5.1}$$

where R is the resistance in ohms, L is the length, A is the cross-sectional area, and ρ (Greek letter rho) is a proportionality constant which depends on the material the bar is made from (e.g., copper, iron, germanium, silicon, mica, etc.). The proportionality constant ρ is called the *resistivity,* or *specific resistance,* and is different for different materials. Table 5.1 shows the approximate resistivities for a number of materials at 20°C.

The units of resistivity given in Table 5.1 are Ω-cm, and in order to keep the units consistent in Eq. 5.1 the dimensions of the bar must be in cm. Thus

$$R\ (\Omega) = \rho\ (\Omega\text{-cm}) \frac{L\ (\text{cm})}{A\ (\text{cm}^2)} \tag{5.2}$$

Table 5.1. Comparative Resistivity (at 20°C) of Some Materials.

Material	Resistivity, ρ, (Ω-cm)
Silver	1.5×10^{-6}
Copper	1.7×10^{-6}
Gold	2.4×10^{-6}
Aluminum	2.8×10^{-6}
Iron	1.0×10^{-5}
Nichrome	1.0×10^{-4}
Carbon	3.5×10^{-3}
Germanium	4.5×10^{1}
Silicon	2.3×10^{5}
Bakelite	1.0×10^{12}
Glass	1.0×10^{14}
Mica	1.0×10^{16}

Exact values depend on purity, temperature, and so forth.

Other units for measurement could be used, such as meters, feet, or inches, but the resistivity would have to be changed into the proper units also.

• • •

Example 5.1 Calculate the end-to-end resistance of the copper rod shown in Fig. 5.2.

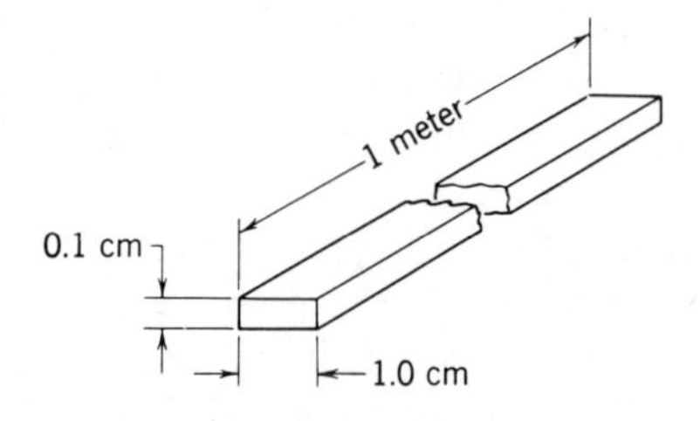

Figure 5.2

Solution Using Eq. 5.2

$$R = \rho \frac{L}{A} = 1.7 \times 10^{-6}\ \Omega\text{-cm} \frac{100\text{ cm}}{0.1\text{ cm} \times 1.0\text{ cm}} = 1.7 \times 10^{-3}\ \Omega$$

• • •

Example 5.2 Calculate the end-to-end resistance of a 100 ft length of iron wire having a diameter of 10 mils (0.01 in).

Solution The units of length and diameter must first be changed to cm. Thus

$$L = 100\text{ ft} \times 12 \frac{\text{in}}{\text{ft}} \times 2.54 \frac{\text{cm}}{\text{in}} = 3.05 \times 10^{3}\text{ cm}$$

The diameter of the wire d is 0.01 in which is

$$d = 0.01\text{ in} \times 2.54 \frac{\text{cm}}{\text{in}} = 2.54 \times 10^{-2}\text{ cm}$$

The cross-sectional area is then

$$A = \pi r^2 = \pi \frac{d^2}{4} = \pi \frac{(2.54 \times 10^{-2}\ \text{cm})^2}{4} = 5.07 \times 10^{-4}\ \text{cm}^2$$

Then, using Eq. 5.2,

$$R = \rho \frac{L}{A} = 1.0 \times 10^{-5}\ \Omega\text{-cm} \frac{3.05 \times 10^3\ \text{cm}}{5.07 \times 10^{-4}\ \text{cm}^2} = 60\ \Omega$$

• • •

5.2 RESISTIVITY

Resistivity ρ is an electrical property of a substance which depends on the purity of the substance as well as the temperature at which ρ is determined. The resistivities given in Table 5.1 are typical values and they can be measured or calculated. An examination of Table 5.1 shows that good conductors such as copper have very low values of resistivity, while good insulators such as mica have very high values of resistivity. Semiconductors such as germanium or silicon have resistivities that are midway between the two extremes. In fact, conductors, semiconductors, and insulators are often defined in terms of their resistivities.

In Chapter 1 it was pointed out that good conductors contain a large number of free electrons which can drift under the influence of an electric field. On the other hand, good insulators have very few free electrons. The number of free electrons is an important factor in determining the resistivity of a material. A second important factor in determining resistivity is the *ease* with which the current carriers drift through the material, and this is called the *mobility* μ (Greek letter mu). Using these two ideas, it can be shown that the resistivity of a material is given by

$$\rho\ (\Omega\text{-cm}) = \frac{1}{q\mu N} \tag{5.3}$$

where q = the charge of one electron, 1.6×10^{-19} C
μ = the current carrier mobility, cm^2/V-sec
N = the number of current carriers per cm^3

In metals such as copper, the *only* current carriers are electrons and the resistivity is calculated using *electron mobility* μ_n. Table 5.2 gives the current carrier mobility and the number of carriers per cm^3 for a number of important materials.

Table 5.2. Current Carrier Densities and Mobilities.

Material	Carriers/cm^3	Electron mobility μ_n (cm^2/V-sec)	Hole mobility μ_p (cm^2/V-sec)
Copper	8.5×10^{22}	43	XXXXX
Silver	9.0×10^{22}	46	XXXXX
Aluminum	5.1×10^{22}	44	XXXXX
Germanium	2.4×10^{13}	3900	1900
Silicon	1.5×10^{10}	1350	480

Approximate values at 20°C

In semiconductor materials such as germanium and silicon there are two current carriers due to the crystalline structure which these materials form (covalent bonding). The two carriers are the electron and a ficticious carrier called a *hole*. Thus Eq. 5.3 must be modified slightly in order to calculate the resistivity of a semiconductor. The proper equation is

$$\rho\ (\Omega\text{-cm}) = \frac{1}{q(\mu_n + \mu_p)N} \tag{5.4}$$

where μ_n = electron mobility, and μ_p = hole mobility.

• • •

Example 5.3 Calculate the resistivity of pure copper at 20°C.

Solution Using Eq. 5.3 and the data from Table 5.2,

$$\rho\ (\Omega\text{-cm}) = \frac{1}{q\mu_n N} = \frac{1}{1.6 \times 10^{-19} \times 43 \times 8.5 \times 10^{22}}$$
$$= 1.7 \times 10^{-6}\ \Omega\text{-cm}$$

• • •

Example 5.4 Calculate the resistivity of pure germanium at 20°C.

Solution Using Eq. 5.4 and the data from Table 5.2,

$$\rho\ (\Omega\text{-cm}) = \frac{1}{1.6 \times 10^{-19}(3900 + 1900)2.4 \times 10^{13}} = 45\ \Omega\text{-cm}$$

• • •

5.3 WIRE RESISTANCE

The resistance of a length of wire can always be calculated using Eq. 5.1. However, nearly all wire in common use has a circular cross-sectional area, and thus the cross-sectional area is found using the equation for the area of a circle ($A = \pi r^2 = \pi d^2/4$). Thus Eq. 5.1 can be modified for conductors having a circular cross-sectional area and can be written as

$$R = \rho \frac{L}{\pi d^2/4} = \frac{4\rho}{\pi} \times \frac{L}{d^2}$$

Furthermore, it is more convenient to measure the length of a wire L in feet, and to measure the diameter d in thousandths of an inch, mils, instead of cm. We can therefore include in the above equation the necessary conversion factors to convert the length from cm to ft and the diameter from cm to mils. The resulting equation is

$$R\ (\Omega) = \frac{4\rho \times 30.5\ \text{cm/ft}}{\pi(2.54 \times 10^{-3}\ \text{cm/mil})^2} \times \frac{L\ (\text{ft})}{[d\ (\text{mils})]^2}$$
$$= \rho' \frac{L\ (\text{ft})}{[d\ (\text{mils})]^2} \tag{5.5}$$

where ρ' is a new resistivity which is the product of the resistivity in Ω-cm and some constants. The units of ρ' are Ω-circular mils/ft, or Ω-cmil/ft, and the resistivities for a number of different materials in Ω-cmil/ft are given in Table 5.3. The resistance of any wire can then be easily calculated from Eq. 5.5 when the length and diameter of the wire are known and the resistivity is given in

Table 5.3. Approximate Resistivity (20°C).

Material	Resistivity, ρ', (Ω-cmil/ft)
Silver	9.9
Copper	10.4
Aluminum	17
Iron	58
Nichrome	660

Ω-cmil/ft. One circular mil (cmil) is defined as the area of a circle having a diameter of one mil. Thus, $A\ (\text{cmil}) = [d\ (\text{mils})]^2 = (4/\pi)A\ (\text{sq mils})$.

• • •

Example 5.6 Find the resistance of a 500-ft length of aluminum wire which has a diameter of 0.050 in.

Solution The diameter of the wire is 0.050 in which is 50 mils. Thus, using Eq. 5.5,

$$R\ (\Omega) = 17\ \Omega\text{-cmil/ft}\ \frac{500\ \text{ft}}{(50\ \text{mils})^2} = 3.4\ \Omega$$

• • •

Example 5.7 A length of iron wire having a diameter of 10 mils is found to have a resistance of 21 Ω. How long is the wire?

Solution Equation 5.5 can be solved for the length L as

$$L\ (\text{ft}) = \frac{R\ (\Omega) \times [d\,(\text{mils})]^2}{\rho'} = \frac{21\ \Omega \times (10\ \text{mils})^2}{58\ \Omega\text{-cmil/ft}} = 36.2\ \text{ft}$$

• • •

5.4 WIRE TABLES

In the interest of uniformity, most North American manufacturers produce copper wire in standard sizes. The different size wires are each given an American Wire Gage number (AWG no.) as shown in Table 5.4. The diameter, the cross-sectional area, and the resistance for each wire are also given in the table. Notice that the resistance of the wire increases as the AWG no. increases, since the cross-sectional area is smaller for larger AWG numbers.

The resistance of any length of a given size copper wire can be calculated from the data in Table 5.4 instead of using Eq. 5.5.

• • •

Example 5.8 Calculate the resistance of a 200 ft length of AWG no. 22 copper wire.

Solution From Table 5.4, the resistance of 1000 ft of AWG no. 22 wire is 16.5 Ω. Since resistance is directly proportional to length, and since 200 ft is 1/5 of 1000 ft, the resistance must be

$$R\,(\Omega) = \frac{16.5}{5} = 3.3\ \Omega$$

• • •

Table 5.4. Standard Annealed Copper Wire, Solid*. American Wire Gage (B.&S.). English Units.

Gage Number	Diameter, Mils	Cross section: Circular Mils	Cross section: Square Inches	Ω per 1000 ft: 25°C (= 77°F)	Ω per 1000 ft: 65°C (= 149°F)	Ω per Mile 25°C (= 77°F)	lb per 1000 ft
0000	460.0	212,000.0	0.166	0.0500	0.0577	0.264	641.0
000	410.0	168,000.0	0.132	0.0630	0.0727	0.333	508.0
00	365.0	133,000.0	0.105	0.0795	0.0917	0.420	403.0
0	325.0	106,000.0	0.0829	0.100	0.116	0.528	319.0
1	289.0	83,700.0	0.0657	0.126	0.146	0.665	253.0
2	258.0	66,400.0	0.0521	0.159	0.184	0.839	201.0
3	229.0	52,600.0	0.0413	0.201	0.232	1.061	159.0
4	204.0	41,700.0	0.0328	0.253	0.292	1.353	126.0
5	182.0	33,100.0	0.0260	0.319	0.369	1.685	100.0
6	162.0	26,300.0	0.0206	0.403	0.465	2.13	79.5
7	144.0	20,800.0	0.0164	0.508	0.586	2.68	63.0
8	128.0	16,500.0	0.0130	0.641	0.739	3.38	50.0
9	114.0	13,100.0	0.0103	0.808	0.932	4.27	39.6
10	102.0	10,400.0	0.00815	1.02	1.18	5.38	31.4
11	91.0	8,230.0	0.00647	1.28	1.48	6.75	24.9
12	81.0	6,530.0	0.00513	1.62	1.87	8.55	19.8
13	72.0	5,180.0	0.00407	2.04	2.36	10.77	15.7
14	64.0	4,110.0	0.00323	2.58	2.97	13.62	12.4
15	57.0	3,260.0	0.00256	3.25	3.75	17.16	9.86
16	51.0	2,580.0	0.00203	4.09	4.73	21.6	7.82
17	45.0	2,050.0	0.00161	5.16	5.96	27.2	6.20
18	40.0	1,620.0	0.00128	6.51	7.51	34.4	4.92
19	36.0	1,290.0	0.00101	8.21	9.48	9.48	3.90
20	32.0	1,020.0	0.000802	10.4	11.9	54.9	3.09
21	28.5	810.0	0.000636	13.1	15.1	69.1	2.45
22	25.3	642.0	0.000505	16.5	19.0	18.1	1.94
23	22.6	509.0	0.000400	20.8	24.0	109.8	1.54
24	20.1	404.0	0.000317	26.2	30.2	138.3	1.22
25	17.9	320.0	0.000252	33.0	38.1	174.1	0.970
26	15.9	254.0	0.000200	41.6	48.0	220.0	0.769
27	14.2	202.0	0.000158	52.5	60.6	277.0	0.61
28	12.6	160.0	0.000126	66.2	76.4	350.0	0.484
29	11.3	127.0	0.0000995	83.4	96.3	440.0	0.384
30	10.0	101.0	0.0000789	105.0	121.0	554.0	0.304
31	8.9	79.7	0.0000626	133.0	153.0	702.0	0.241
32	8.0	63.2	0.0000496	167.0	193.0	882.0	0.191
33	7.1	50.1	0.0000394	211.0	243.0	1,114.0	0.152
34	6.3	39.8	0.0000312	266.0	307.0	1,404.0	0.120
35	5.6	31.5	0.0000248	335.0	387.0	1,769.0	0.0954
36	5.0	25.0	0.0000196	423.0	488.0	2,230.0	0.0757
37	4.5	19.8	0.0000156	533.0	616.0	2,810.0	0.0600
38	4.0	15.7	0.0000123	673.0	776.0	3,550.0	0.0476
39	3.5	12.5	0.0000098	848.0	979.0	4,480.0	0.0377
40	3.1	9.9	0.0000078	1,070.0	1,230.0	5,650.0	0.0299

*From *Circ. 31,* U.S. Bureau of Standards.

Note 1. The *fundamental festivity* used in calculating the tables is the International Annealed Copper Standard, *viz.*, 0.15328 Ω (meter, gram) at 20°C. The *temperature coefficient* for this particular resistivity is $\alpha_{20} = 0.00393$, or $\alpha_0 = 0.00427$. The *density* is 8.89 g per cm^3.

Note 2. The values given in the table are only for annealed copper of the standard resistivity. The user of the table must apply the proper correction for copper of any other resistivity. Hard-drawn copper may be taken as about 2.7 per cent higher resistivity than annealed copper.

Note 3. Pounds per mile may be obtained by multiplying the respective values above by 5.28.

In setting up the AWG wire table, the conductor sizes were selected so that a decrease of one gage number results in a 25% increase in cross-sectional area. Furthermore, a decrease of 3 gage numbers results in an increase in cross-sectional area of $1.25 \times 1.25 \times 1.25 \cong 2$. Since resistance is inversely proportional to cross-sectional area, a decrease of 3 gage numbers results in doubling the area and halving the resistance. Alternatively, an increase of 3 gage numbers will double the resistance. If we note that the resistance of 1000 ft of AWG no. 10 wire is approximately 1 Ω, we can estimate the resistance of any wire without the use of the tables.

• • •

Example 5.9 Estimate the resistance of 1000 ft of AWG no. 16 copper wire.

Solution The resistance of 1000 ft of AWG no. 10 wire is about 1 Ω. Increasing 3 gages to AWG no. 13 would double the resistance to 2 Ω, and increasing 3 more gages to AWG no. 16 would double the resistance again to about 4 Ω. (Table 5.4 gives 4.09 Ω.)

• • •

It is also useful to note that a change of 10 gage numbers results in a 10:1 change in resistance.

• • •

Example 5.10 Estimate the resistance of 50 ft of AWG no. 30 copper wire.

Solution The resistance of 1000 ft of no. 10 wire is about 1 Ω. Going up 10 gages to no. 20 would give 10 Ω, and going up 10 more gages to no. 30 would give 100 Ω. Thus the resistance of 1000 ft of no. 30 wire is about 100 Ω, but, the 50 ft length is only 1/20 as long and thus the resistance of the 50 ft length must be $100/20 = 5$ Ω.

• • •

As a matter of safety, there is a maximum current that a conductor of a certain size is capable of handling. The maximum safe current-carrying capacity for various conductors is usually specified in a table of Electrical Safety Orders. The tables given in different publications are not in universal agreement, but some typical values are given in Table 5.5 for reference purposes. A number of different wire sizes and their typical uses are shown in Fig. 5.3.

Table 5.5. Maximum Safe Current-Carrying Capacity For a Single Insulated Copper Conductor (type THHN). National Electrical Code, 1975, pp. 70–133.

AWG no.	*I* (amperes)	AWG no.	*I* (amperes)
0000	385	8	70
000	330	10	40
00	285	12	25
0	245	14	20
1	210		
2	180		
4	135		
6	100		

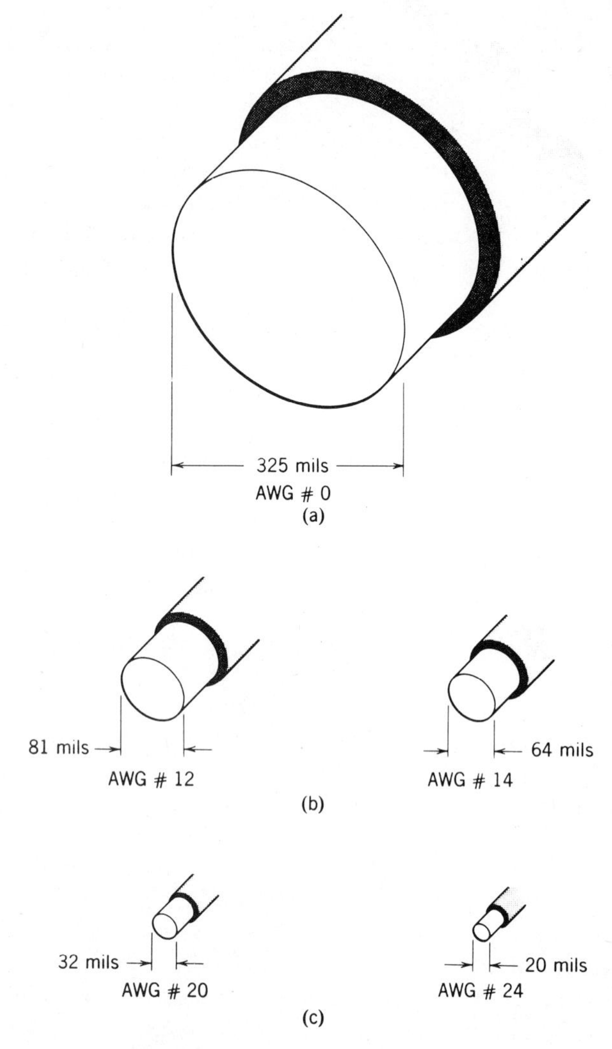

Figure 5.3 Copper conductors and their uses (*a*) Power distribution (*b*) Home and industrial wiring; appliance outlets (*c*) Electronics; radio and television

5.5 RESISTORS

Resistors are used in electric circuits primarily to limit the circuit current and there is a wide variety of resistors available from many different manufacturers. The *wire-wound* resistor shown in Fig. 5.4 is normally used in applications where considerable power is dissipated in the resistor. Wire-wound resistors are readily available in many different resistance sizes with wattage ratings from a few W up to a few hundred W.

Metal film resistors such as those shown in Fig. 5.5 are used primarily in low power applications when a very accurate amount of resistance is required. Metal film resistors are usually *precision* resistors and they are available in standard resistance sizes, or they can be ordered from manufacturers in special sizes.

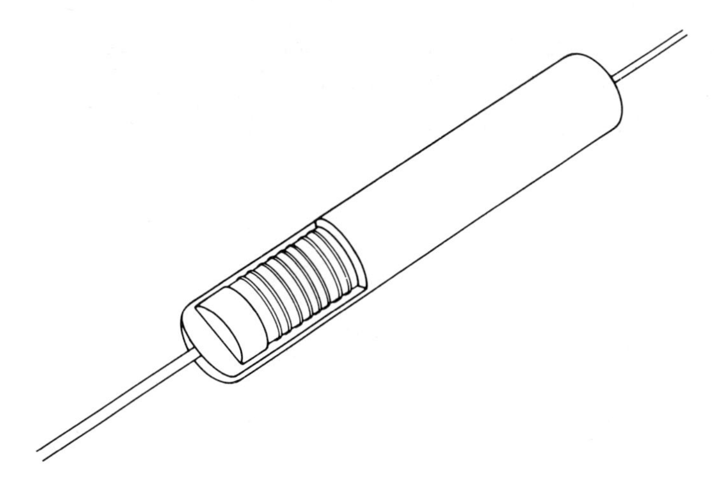

Figure 5.4 Wirewound resistors (power)

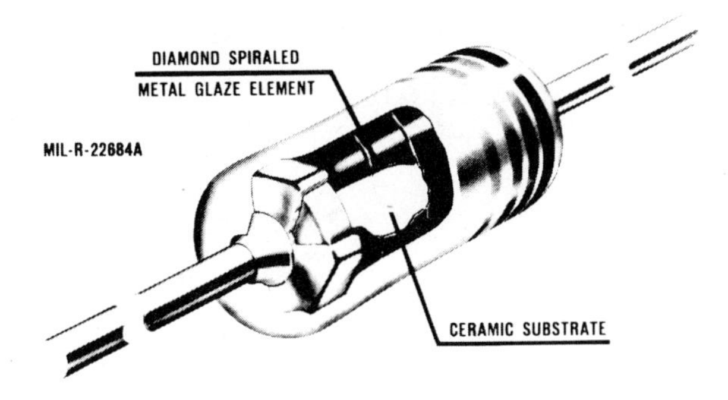

Figure 5.5 Metal film resistors (precision)

The most widely used general purpose resistors are the *carbon–composition* resistors such as those shown in Fig. 5.6. These resistors are widely available in resistance sizes from a few Ω up to a few megohms, and in wattage ratings from 1/8 W up to 2 W. Carbon–composition resistors are made in standard resistance sizes and their resistances are specified with *two* significant figures and a multiplier. For example, a 47,000 Ω resistor is specified using the two significant figures 47 and the multiplier 10^3; a 220 Ω resistor is specified using the two significant figures 22 and the multiplier 10. The standard resistance sizes for carbon–composition resistors are given in Table 5.6. Notice that the

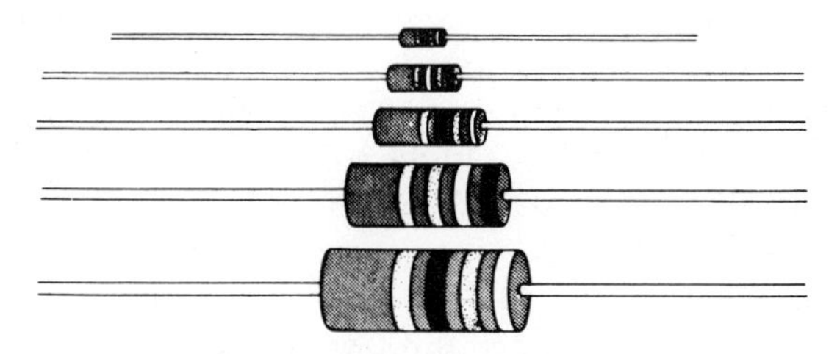

Figure 5.6 Carbon composition resistors (general purpose)

Table 5.6. List of Standard Carbon–Composition Resistors Available.

Resistance Tolerance 20%	10%	5%
1.0	1.0	1.0
		1.1
	1.2	1.2
		1.3
1.5	1.5	1.5
		1.6
	1.8	1.8
		2.0
2.2	2.2	2.2
		2.4
	2.7	2.7
		3.0
3.3	3.3	3.3
		3.6
	3.9	3.9
		4.3
4.7	4.7	4.7
		5.1
	5.6	5.6
		6.2
6.8	6.8	6.8
		7.5
	8.2	8.2
		9.1

available sizes depend on the tolerance. Only the two significant figures are given in Table 5.6, and the actual resistor size is found by multiplying the significant figures by any multiplier between 10^{-1} and 10^{7}. The smallest resistor is $2.4 \times 10^{-1} = 0.24\ \Omega$, and the largest is $22 \times 10^{7} = 22\ \text{M}\Omega$. (Note: these are not absolute minimum and maximum values, but are typical).

If we were to measure the resistance of a number of different 1000 Ω resistors, we would find that virtually *none* of them would have a resistance of *exactly* 1000 Ω. This is simply a matter of economics, since it is easy to manufacture a large quantity of resistors having resistances *close* to 1000 Ω, but it is difficult to manufacture resistors having a resistance of *exactly* 1000 Ω. The *tolerance* of a resistor tells how close the resistance is to the stated value. A tolerance of 20% means that the resistance may be as much as 20% higher or as much as 20% lower than the stated value. Thus a 20% tolerance means *within* 20%, or ±20%. For example, a 1000 Ω resistor having a 20% tolerance may be up to $0.20 \times 1000\ \Omega = 200\ \Omega$ higher, or as much as 200 Ω lower than the stated value of 1000 Ω. Thus a 1000 Ω, 20% resistor has an actual resistance value of $1000\ \Omega \pm 200\ \Omega$, or somewhere between 800 Ω and 1200 Ω.

• • •

Example 5.11 What are the upper and lower limits of resistance for a 4.7 kΩ, 5% resistor?

Solution We first compute the tolerance; 5% of $4700\ \Omega = 0.05 \times 4700\ \Omega = 235\ \Omega$. The lower limit of resistance is $4700\ \Omega - 235\ \Omega = 4465\ \Omega$, and the upper limit is $4700\ \Omega + 235\ \Omega = 4935\ \Omega$.

• • •

Occasionally the resistance values are stamped on the bodies of the resistors, but more often than not the resistance values are given by means of colored bands on the bodies of the resistors. The color coding of resistance values by means of bands is shown in Fig. 5.7. Notice that the bands are grouped closer to the left end of the resistor. When located this way, the left band is the *first* significant figure, the next band is the *second* significant figure, and the third band is the *multiplier*. If there is a *fourth* band, it is the tolerance; the *absence* of a fourth band implies a 20% tolerance. The numbers corresponding to the colors of the bands are given in Table 5.7.

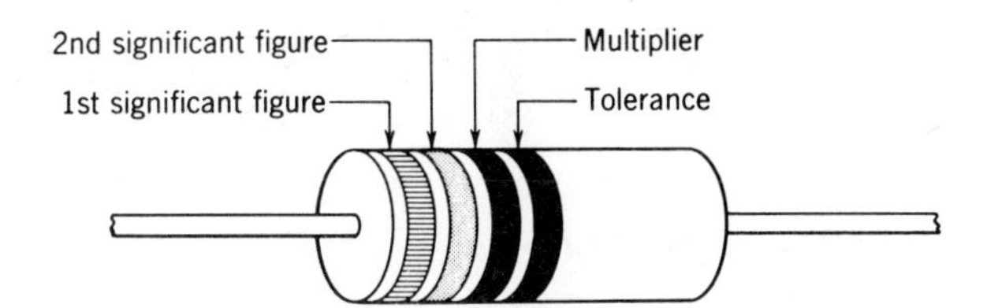

Figure 5.7 Color code bands on resistors

Table 5.7. Resistor Color Code.

Color	Significant figure	Multiplier	Tolerance
Silver	—	10^{-2}	10%
Gold	—	10^{-1}	5%
Black	0	1	—
Brown	1	10	—
Red	2	10^{2}	—
Orange	3	10^{3}	—
Yellow	4	10^{4}	—
Green	5	10^{5}	—
Blue	6	10^{6}	—
Violet	7	—	—
Gray	8	—	—
White	9	—	—
No color	—	—	20%

• • •

Example 5.12 What is the value of the resistor given in Fig. 5.8?

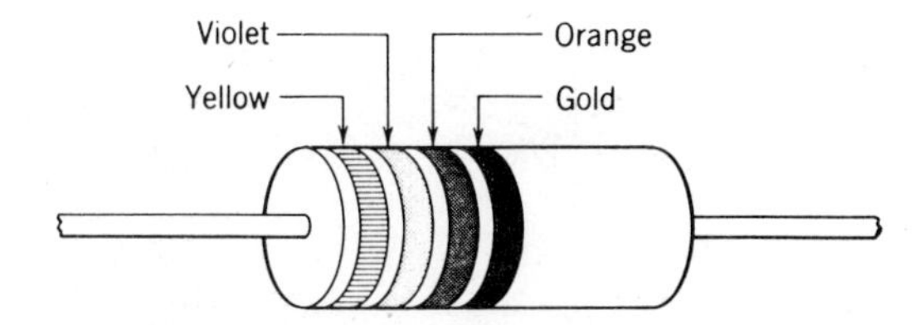

Figure 5.8

Solution The first band is yellow giving the first significant figure as 4. The second band is violet giving the second significant figure as 7. The multiplier band is orange giving a multiplier of 10^3. Thus the resistance value is $47 \times 10^3 = 47\ \mathrm{k}\Omega$. The tolerance band is gold giving a tolerance of $\pm 5\%$.

• • •

Example 5.13 What must be the color of the bands on a 5.1 kΩ, 20% resistor?

Solution The first band must be green (5); the second band must be brown (1); the multiplier band must be red (10^2); there is no band for a 20% tolerance. The resistor is shown in Fig. 5.9.

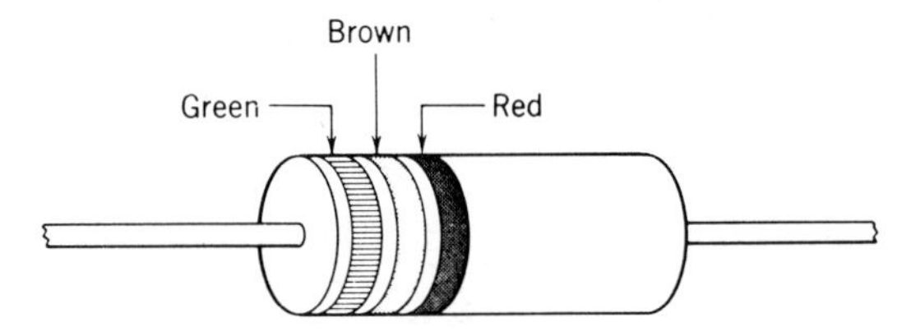

Figure 5.9 A 5.1 kΩ, 20% resistor

• • •

5.6 RESISTANCE VERSUS TEMPERATURE

When Georg Simon Ohm conducted his experiments showing the relationship between current and voltage (Ohm's Law, $R = V/I$), it was necessary to maintain the temperature of the wire at a constant value. The reason for this is that the resistance of a piece of wire changes as its temperature changes. In fact, the resistance of most materials changes as the temperature is varied.

A typical resistance versus temperature curve is shown in Fig. 5.10, and it is clear that the curve is essentially a straight line over a wide range of temperatures. At very low temperatures (the dashed portion of the curve) the curve may bend up or down, depending on the material involved. In any case, we will not be concerned with such low temperatures and we will simply extend the curve as a straight line until the resistance drops to zero Ω. The temperature at which the resistance becomes zero T_0 is different for different materials as shown in Table 5.8.

If we consider the curve in Fig. 5.10 to be a straight line all the way down to T_0, we can write an equation for the resistance as a function of temperature by

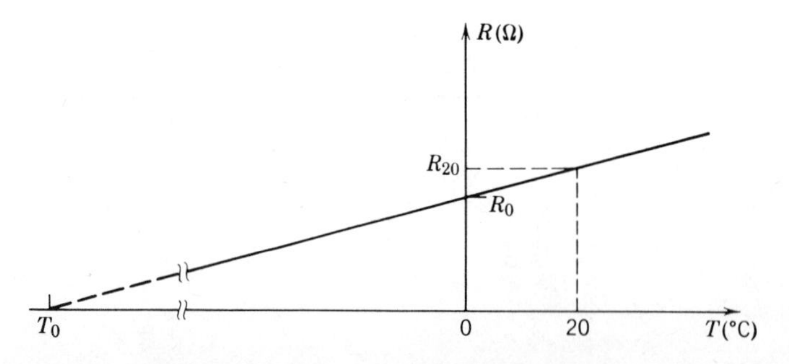

Figure 5.10 Resistance versus temperature

Table 5.8. Apparent Temperature at Which the Resistance Becomes Zero in Fig. 5.10.

Material	T_0 (°C)
Nickel	−147
Iron	−180
Tungsten	−202
Copper	−235
Aluminum	−236
Silver	−243
Nichrome II	−6250
Constantan	-125×10^3

simply using the standard equation for a straight line. It can be shown that the general equation for a straight line is $y = \mathrm{m}x + \mathrm{b}$, where m is the slope of the line and b is the y-intercept. In the case of Fig. 5.10, the equation is

$$R = \mathrm{m}T + R_0$$

where m is the slope of the line and R_0 is the R-intercept (the value of R when $T = 0$). The slope of the line is *rise/run*, and can be expressed as R_0/T_0. Thus

$$R = \frac{R_0}{T_0}T + R_0 = R_0\left(\frac{T}{T_0} + 1\right) = R_0\left(\frac{T + T_0}{T_0}\right) \tag{5.6}$$

Equation 5.6 can be used to find the resistance R at any temperature T when R_0 and T_0 are known. This is not a very convenient equation however, since we do not generally know R_0; it is more common to know the resistance at room temperature R_{20} (room temperature is approximately 20°C). We would then like to rearrange Eq. 5.6 in terms of R_{20}.

The resistance at 20°C can be found from Eq. 5.6 by setting $T = 20$. Thus

$$R_{20} = R_0\left(\frac{T_0 + 20}{T_0}\right)$$

Solving this equation for R_0 yields

$$R_0 = R_{20}\left(\frac{T_0}{T_0 + 20}\right)$$

If this value of R_0 is now substituted into Eq. 5.6, the result is

$$R = R_{20}\left(\frac{T_0}{T_0 + 20}\right)\left(\frac{T + T_0}{T_0}\right) = R_{20}\left(\frac{T + T_0}{T_0 + 20}\right)$$

Finally, this equation can be changed into a more convenient form as follows:

$$R = R_{20}\left[\frac{(T_0 + 20) + (T - 20)}{T_0 + 20}\right] = R_{20}\left(1 + \frac{T - 20}{T_0 + 20}\right)$$

Or,

$$R = R_{20}[1 + \alpha(T - 20)] \tag{5.7}$$

where $\alpha = 1/(T_0 + 20)$ is defined as the *temperature coefficient of resistance.* Equation 5.7 can be used to find the resistance R of a material at any temperature T by using the temperature coefficient of resistance α given in Table 5.9.

Table 5.9. Temperature Coefficient of Resistance at 20°C.

Material	α
Nickel	0.006
Iron	0.0055
Tungsten	0.0045
Copper	0.00393
Aluminum	0.0039
Silver	0.0038
Nichrome II	0.00016
Constantan	0.000008
Carbon	−0.0005

• • •

Example 5.14 The resistance of a certain piece of copper wire is 1 Ω at 20°C. What is its resistance at 100°C?

Solution Using Eq. 5.7,

$$R = R_{20}[1 + \alpha(T - 20)] = 1[1 + 0.00393(100 - 20)]$$
$$1(1 + 0.316) = 1.316\ \Omega$$

• • •

Example 5.15 The resistance of a certain piece of constantan wire (55% copper, 45% nickel) is 1 Ω at 20°C. What is its resistance at 100°C?

Solution Using Eq. 5.7,

$$R = 1[1 + 0.000008(100 - 20)] = 1(1 + 0.00064)$$
$$= 1.00064\ \Omega$$

• • •

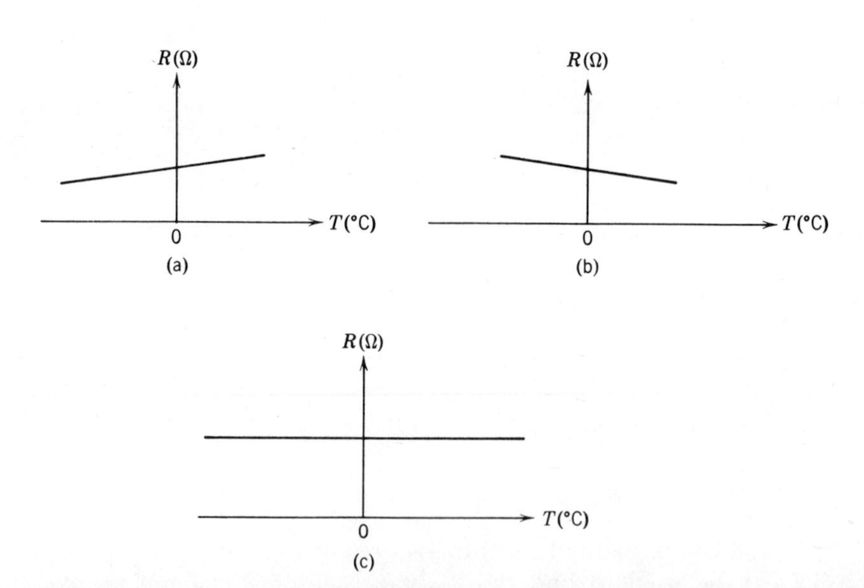

Figure 5.11 Temperature coefficients (*a*) Positive (*b*) Negative (*c*) Temperature coefficient equals zero

From the two previous examples, it is clear that some materials (e.g., copper) exhibit a considerable change in resistance as the temperature is varied, while the resistance of other materials (e.g., constantan) changes very little. It is also clear in the two examples above that the resistance *increases* as the temperature *increases.* Such materials have *positive* temperature coefficients as given in Table 5.9. There are materials however, such as carbon, which have *negative* temperature coefficients. The resistance of such materials *decreases* as the temperature increases. The effects of a positive, zero, or negative temperature coefficient are shown graphically in Fig. 5.11.

• • •

Example 5.16 A certain carbon resistor has a resistance of 1 kΩ at 20°C. What is its resistance at 100°C?

Solution Using Eq. 5.7,

$$R = 1\ \text{k}\Omega[1 + (-0.0005)(100 - 20)] = 960\ \Omega$$

• • •

Example 5.17 What is the resistance of the carbon resistor in Example 5.16 if the temperature is −100°C?

Solution Using Eq. 5.7,

$$R = 1\ \text{k}\Omega[1 + (-0.0005)(-100 - 20)] = 1.06\ \text{k}\Omega$$

• • •

5.7 CONDUCTANCE

We now know that the resistance of a circuit or an element is a measurement of its opposition to electric current. The exact *opposite* of resistance is a measurement of how well a circuit or element will conduct electric current, and this is called *conductance.* Conductance G is found by taking the *reciprocal* of resistance and it is measured in the unit mho (ohm spelled backwards). The symbol for conductance is the greek letter omega turned upside down ℧. Thus,

$$G(\mho) = \frac{1}{R(\Omega)} \tag{5.8}$$

• • •

Example 5.18 Find the conductance of the following resistors: (*a*) 10 Ω, (*b*) 2 kΩ, (*c*) 47 kΩ.

Solution Using Eq. 5.8,

(*a*) $G(\mho) = 1/R(\Omega) = 1/(10\ \Omega) = 0.1\ \mho$
(*b*) $G(\mho) = 1/(2 \times 10^3\ \Omega) = 500\ \mu\mho$
(*c*) $G(\mho) = 1/(47 \times 10^3\ \Omega) = 21.2\ \mu\mho$

• • •

Ohm's Law can be easily arranged to use conductance G instead of resistance R by simply making the substitution $R = 1/G$. Thus

$$V = I \times R = \frac{I}{G} \tag{5.9}$$

or,

$$I = \frac{V}{R} = G \times V \tag{5.10}$$

or,

$$R = \frac{V}{I} = \frac{1}{G} \quad \text{and} \quad G = \frac{I}{V} \tag{5.11}$$

• • •

Example 5.19 Calculate the battery voltage in the circuit shown in Fig. 5.12. What is the value of the resistor?

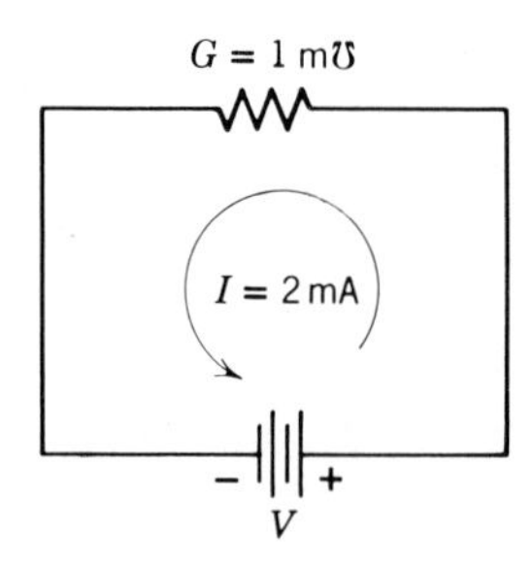

Figure 5.12

Solution Using Eq. 5.9,

$$V = \frac{I}{G} = \frac{2\text{ mA}}{1\text{ m}\mho} = 2\text{ V}$$

The resistor has a resistance of

$$R = \frac{1}{G} = \frac{1}{1\text{ m}\mho} = 1\text{ k}\Omega$$

• • •

The resistivity of a substance as discussed in Sections 5.1 and 5.2 can also be expressed in reciprocal form. *Conductivity* σ (the greek letter sigma) is found by taking the *reciprocal* of resistivity and is measured in units of ℧ per centimeter ℧/cm (sometimes expressed as ℧-cm^{-1}). Conductivity is frequently used instead of resistivity when discussing the properties of semiconductor materials. Thus

$$\sigma(\mho/\text{cm}) = \frac{1}{\rho(\Omega\text{-cm})} \tag{5.12}$$

Equation 5.4 can also be rearranged as

$$\sigma(\mho/\text{cm}) = \frac{1}{\rho(\mho\text{-cm})} = q(\mu_n + \mu_p)N \tag{5.13}$$

• • •

Example 5.20 Using the results of Example 5.4, calculate the conductivity of pure germanium at 20°C.

Solution Using Eq. 5.12,

$$\sigma(\mho/\text{cm}) = \frac{1}{45\ \Omega\text{-cm}} = 2.22 \times 10^{-2}\ \mho/\text{cm}$$

• • •

.8 NONLINEAR RESISTANCE

A resistor is usually considered to be a *linear* element since Ohm's Law in the form $V = I \times R$ is the equation of a straight line as shown in Fig. 5.13 (this is demonstrated in Example 2.9). The slope of the line is R and it is considered constant for any value of current or voltage. Notice in this figure that the current and voltage can both have *negative* values as well as positive values. This shows that current can flow through the resistor with equal ease in *either direction* and resistors are therefore called *bilateral* elements.

Ohm's Law can also be plotted in the form $I = V/R$ and this still results in a straight line as shown in Fig. 5.14. In this case, however, the slope of the line is $1/R = G$ (conductance).

There are a number of different elements whose characteristic curves (V-I plots) are not straight lines and they are therefore not linear elements; they are called *nonlinear* elements since their resistances are not constant. A *varistor* is a resistor made of carborundum crystals held in a suitable binder, and its resistance changes with applied voltage as shown in Fig. 5.15. When the voltage across the varistor increases beyond a certain value (100 V in this case), the current through the element increases very rapidly; this implies a rapid decrease in resistance. Varistors are used in circuits to provide overvoltage protection. Semiconductor and vacuum tube diodes also have nonlinear characteristic curves as shown in Fig. 5.16. These devices are used as switches in logic applications or as rectifiers to change ac voltages to dc voltages (diodes are discussed in detail in basic electronics). Notice that diodes are *not* bilateral devices.

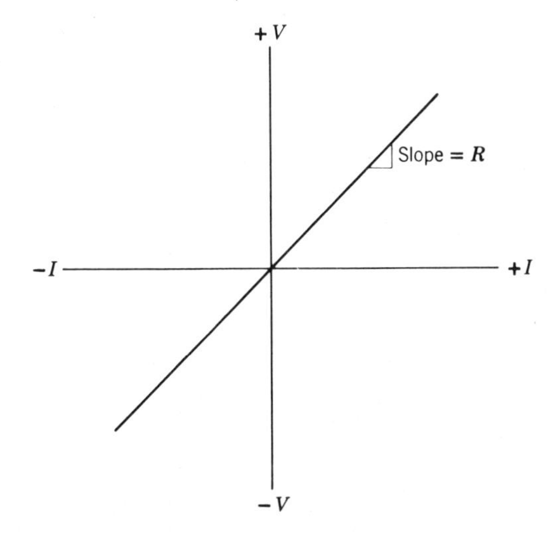

Figure 5.13 A linear resistance

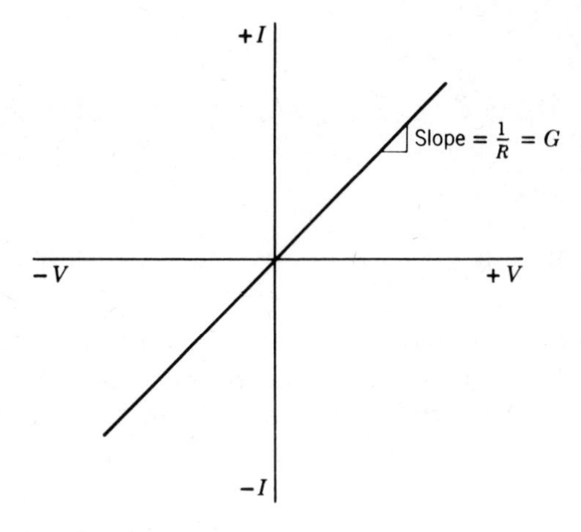

Figure 5.14 A linear conductance

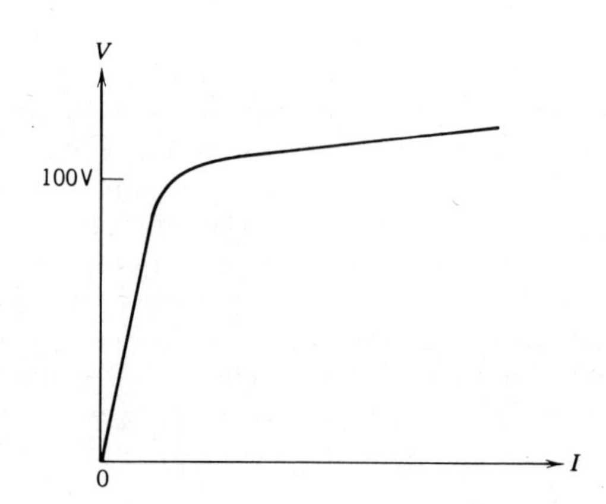

Figure 5.15 A varistor curve

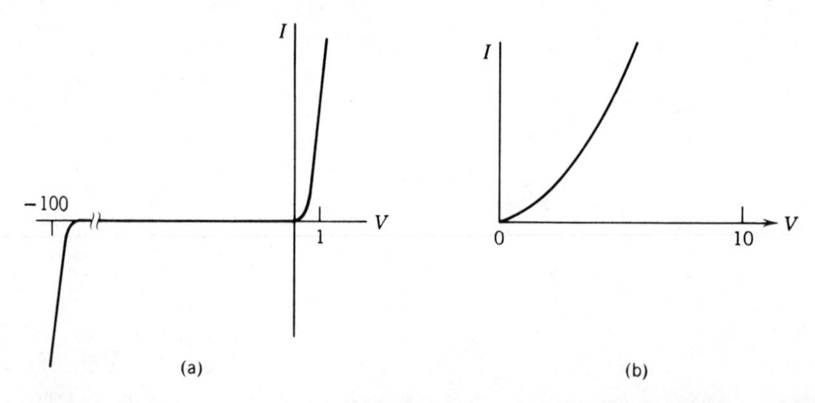

Figure 5.16 (*a*) Typical semiconductor diode (*b*) Typical vacuum tube diode

Resistors are also considered nonlinear if their resistances change very rapidly with temperature. A *thermistor* is a small resistor made of special metallic oxides in a suitable binder, and it has a large negative temperature coefficient of resistance. Thermistors have resistance versus temperature curves similar to that shown in Fig. 5.11*b*. The resistance of a typical thermistor might be reduced by a factor of 2 for only a 20°C increase in temperature. Thermistors are often used in electronic circuits to stabilize against temperature changes.

A *ballast* resistor is a resistor having a large positive temperature coefficient and its resistance increases rapidly with temperature as shown in Fig. 5.11*a*. Ballast resistors are often used as current limiters in circuits. The ordinary incandescent lamp behaves as a ballast resistor since the resistance of its filament changes from a very low value when it is cold to a much higher value when it is hot. As an example, the filament resistance of a typical 40 W lamp is around 30 Ω when it is cold, but it increases to 360 Ω when hot ($R = V^2/P = 120^2/40 = 360\ \Omega$).

SUMMARY

Resistivity (or conductivity) is a basic electrical property of a material. The resistance of any element is directly proportional to its length, inversely proportional to its cross-sectional area, and the resistance can be calculated using the length, area and resistivity. The resistances of copper wires are given in tables according to AWG wire sizes. Copper wire resistance can, however, be estimated by remembering that the resistance of 1000 ft of AWG no. 10 copper wire is about 1 Ω. Resistors are used as circuit elements primarily to limit current flow. Power resistors are usually wirewound, precision resistors are often metal film types, and carbon-composition resistors are used as general purpose resistors. Resistors are readily available only in standard sizes, depending on the tolerances, and the resistance and tolerance are often stated using colored bands on the bodies of the resistors. Nonlinear resistors have resistances which are not constant and they are used for a number of different special purposes.

GLOSSARY

AWG. *American Wire Gauge.*

Circular mil. *The area in circular mils of a circle having a diameter of one mil, cmil.*

Color code. *A system of colored bands used to designate resistance values and tolerances (also used on other elements).*

Conductance. *The reciprocal of resistance, G, measured in ℧.*

Conductivity. *The reciprocal of resistivity, σ.*

Linear resistance. *A resistance which is constant.*

Mobility. *A measure of the ease with which carriers move through a material, μ.*

Nonlinear resistance. *A resistance which varies with temperature, voltage, or current.*

Resistance. *A measure of the opposition to electric current, R.*

Resistivity. *A basic electrical property of a material, ρ.*

Temperature coefficient of resistance. *A constant which shows how the resistance of a material changes with temperature, α.*

IMPORTANT RELATIONSHIPS

- $R = \rho \dfrac{L}{A}$ $\qquad R(\Omega) = \rho'(\Omega\text{-cmil/ft}) \dfrac{(\text{ft})}{[d(\text{mils})]^2}$
- $R = R_{20}[1 + \alpha(T - 20)]$ $\qquad \rho(\Omega\text{-cm}) = \dfrac{1}{q(\mu_n + \mu_p)N}$
- $G(\mho) = 1/R(\Omega)$
- 1000 ft of AWG no. 10 copper wire $\cong 1\ \Omega$
- Resistor color code—Table 5.7

REVIEW QUESTIONS

1. Resistance is ____ proportional to length.
2. Resistance is ____ proportional to cross-sectional area.
3. Silicon and germanium are classed as semiconductors. (TF)
4. A measure of the ease with which carriers drift through a material is: (*a*) mobility, (*b*) resistivity, (*c*) conductivity.
5. Copper wire is produced in standard sizes according to an ____ ____ ____ wire number.
6. The maximum safe current in AWG no. 14 copper wire is ____ A.
7. Wire-wound resistors are usually used in high ____ applications.
8. Precision resistors are often made from ____ ____.
9. The four color bands on a resistor are ____ significant figure, ____ significant figure, ____, and ____.
10. Carbon-composition resistors are readily available only in standard sizes. (TF)
11. The color bands on a 3.9 kΩ, 20% resistor are ____, ____, and ____.
12. The bands on a resistor are red, red, yellow, and gold. The resistance value is ____ Ω, ± ____ %.
13. A positive temperature coefficient means that resistance ____ with temperature.
14. A negative temperature coefficient means that resistance ____ with temperature.
15. A zero temperature coefficient means that resistance does not change with temperature. (TF)
16. A thermistor has a ____ temperature coefficient.
17. The resistance of a ballast resistor varies with applied voltage. (TF)
18. The reciprocal of resistance is ____.
19. The units of conductance are ____.

PROBLEMS

1. Calculate the end-to-end resistance of a 100-ft aluminum rod having a cross-sectional area of $1\ \text{cm}^2$.
2. Calculate the resistance of a 2 m length of Nichrome wire having a cross-sectional area of $10^{-2}\ \text{cm}^2$.
3. Calculate the resistance between opposite faces of a 1-cm cube of glass.

4. Calculate the end-to-end resistance of a bar of pure silicon 1 cm square and 4 cm long.
5. Calculate the resistivity of pure silicon at 20°C.
6. Calculate the resistance of a 200 ft length of iron wire having a diameter of 0.01 in.
7. Calculate the resistance of a 150 ft length of aluminum wire having a diameter of 0.01 in.
8. The resistance of a 10 ft length of wire is 264 Ω, and its diameter is 5 mils. Of what material must the wire be made?
9. What length of Nichrome wire having a diameter of 10 mils must be used to obtain a resistance of 33 Ω?
10. Calculate the resistance of 1000 ft of AWG no. 20 copper wire.
11. Calculate the resistance of 100 ft of AWG no. 13 copper wire.
12. If a 250 ft length of copper wire must have a resistance less than 0.1 Ω, what is the smallest wire that can be used?
13. What is the minimum size copper wire that is safe to connect a 3 Ω load to a 120 V source?
14. What are the upper and lower limits of a 1.5 kΩ, 10% resistor?
15. The resistance needed for a circuit is calculated as 520 Ω. What is the closest standard resistor size if the tolerance is: (*a*) 20%, (*b*) 10%, (*c*) 5%?
16. What is the resistance of 1000 ft of AWG no. 10 copper wire at 0°C?
17. What must be the temperature coefficient of a thermistor whose resistance doubles when the temperature is decreased from 20°C to 0°C?
18. What is the conductance G of the following resistors? (*a*) 500 Ω, (*b*) 1 MΩ, (*c*) 2.2 kΩ.
19. What voltage must be applied across a resistor having a conductance of 4 m℧ in order to cause a current of 16 mA? What is the resistance of the resistor?

6

KIRCHHOFF'S LAWS

All circuits we have thus far encountered are easily solved using Ohm's Law, but we will have to use Kirchhoff's Laws to find solutions for more complex electric circuits. Gustav Robert Kirchhoff (1842–1887) formulated two laws which govern the behavior of voltages around a closed circuit, and currents at a junction (a connection between two or more elements). These laws are called Kirchhoff's Voltage Law (KVL) and Kirchhoff's Current Law (KCL) and they form the basis for all network analysis. We will devote this entire chapter to the study of KVL and KCL, and these laws will be applied to general network analysis in Chapter 10.

6.1 SERIES CIRCUITS

When two or more elements are connected end to end to form an electric circuit, as shown in Fig. 6.1, the elements are said to be connected in *series* and they form a *series circuit.* From our previous work, we know that the current must be the same in every element in a series circuit since it is impossible to *lose* any electrons at any point around the circuit, and it is also impossible to *gain* any electrons at any point. Thus we can say,

if two or more elements are connected in series, the current is the same in each element.

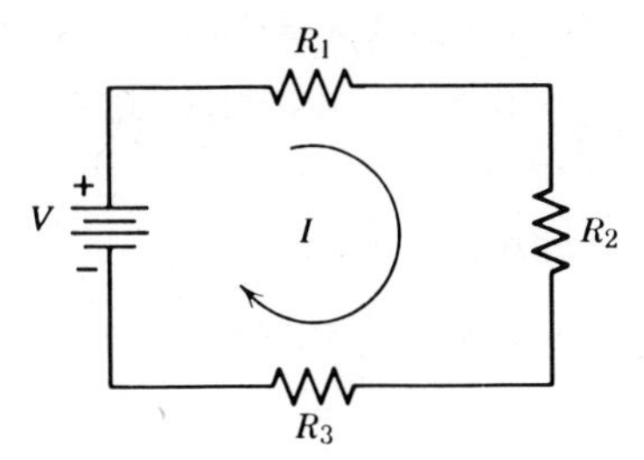

Figure 6.1 Series circuit

In Chapter 5 it was shown that the resistance of a wire is directly proportional to its length. Thus, if the resistance of 1000 ft of AWG no. 10 copper wire is 1 Ω, then the resistance of 2000 ft of the same wire is 2 Ω. However, this is the same as connecting two 1000 ft sections of the wire end to end (in series), and the total resistance is simply the *sum* of the two individual resistances. Similarly, a 2000 ft length of wire could be considered the same as four 500 ft sections connected in series; the total resistance would be the sum of the individual sections, or $0.5 + 0.5 + 0.5 + 0.5 = 2\ \Omega$. These ideas are summarized in Fig. 6.2.

A logical extension of these ideas can be used to determine the total resistance of a number of resistors connected in series. Thus in Fig. 6.3, the total resistance of the 3 resistors R_1, R_2, and R_3 connected in series is simply the sum of the 3 individual resistances, or $R_T = R_1 + R_2 + R_3$. This is true for

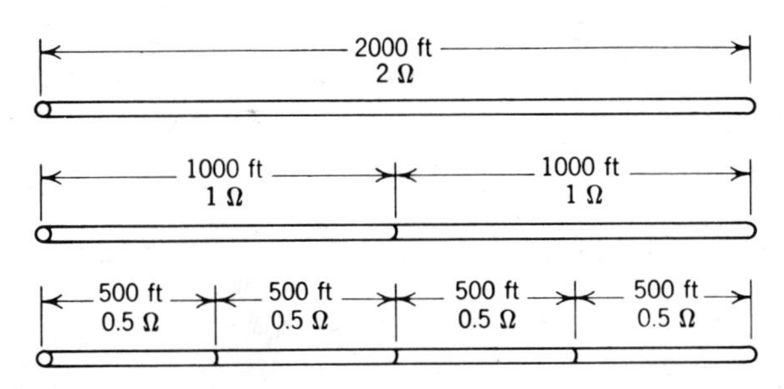

Figure 6.2 The total resistance of 2000 ft of AWG #10 copper wire is 2 Ω. The total series resistance is equal to the sum of the individual resistances

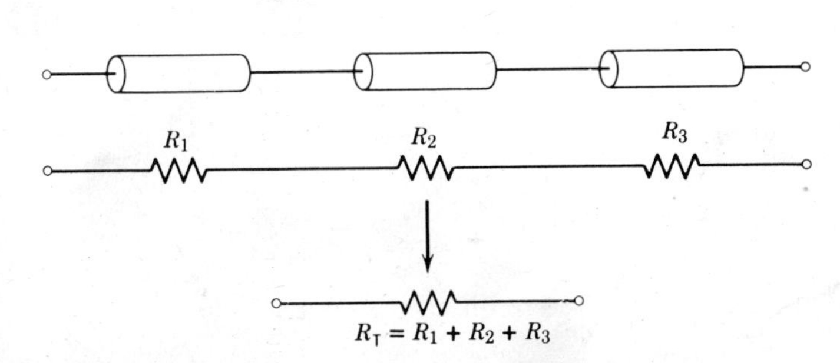

Figure 6.3

any number of resistors connected in series, and thus,

the total resistance of any number of resistors connected in series is simply the sum of the individual resistances.

Expressed in the form of an equation,

$$R_T = R_1 + R_2 + R_3 + R_4 + \cdots + R_N \tag{6.1}$$

where N equals total number of resistors.

• • •

Example 6.1 Find the total resistance of the resistors in Fig. 6.3 if $R_1 = 1\ \text{k}\Omega$, $R_2 = 4.7\ \text{k}\Omega$, and $R_3 = 6.8\ \text{k}\Omega$.

Solution The total resistance is given by Eq. 6.1, and thus,

$$R_T = 1\ \text{k}\Omega + 4.7\ \text{k}\Omega + 6.8\ \text{k}\Omega = 12.5\ \text{k}\Omega$$

• • •

The method of finding the total resistance in a series circuit given in Eq. 6.1 is very useful in solving for the current in a series circuit. As an example, the 3 resistors in the circuit shown in Fig. 6.1 can be replaced with a single resistor having a value of $R_T = R_1 + R_2 + R_3$ as shown in Fig. 6.4. Since R_T has a value equal to the sum of the 3 resistors in Fig. 6.1, it is said to be an *equivalent resistance*, and the circuit in Fig. 6.4 is said to be the *equivalent circuit* of Fig. 6.1. Thus the resistances in a series circuit can be combined into an equivalent resistance to form a very simple equivalent circuit, and Ohm's Law can then be used to solve for the circuit current.

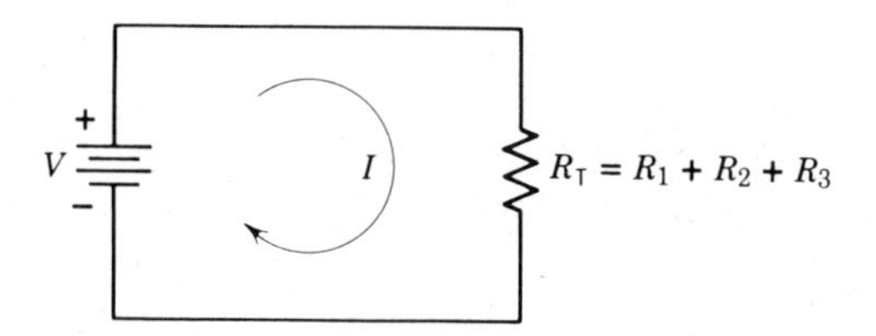

Figure 6.4 Equivalent circuit of Fig. 6.1

• • •

Example 6.2 Solve for the current in Fig. 6.1 if $V = 10\ \text{V}$, $R_1 = 500\ \Omega$, $R_2 = 1\ \text{k}\Omega$, and $R_3 = 3.5\ \text{k}\Omega$.

Solution The equivalent circuit in Fig. 6.4 can be determined by first finding the equivalent resistance R_T. Thus

$$R_T = 500\ \Omega + 1000\ \Omega + 3500\ \Omega = 5000\ \Omega = 5\ \text{k}\Omega$$

The current is then found using Ohm's Law, and

$$I = \frac{V}{R_T} = \frac{10\ \text{V}}{5\ \text{k}\Omega} = 2\ \text{mA}$$

• • •

Example 6.3 Find the current in the circuit shown in Fig. 6.5*a*.

Solution This is a series circuit since all of the elements are connected end to end. The current must be the same in every element, and the circuit can

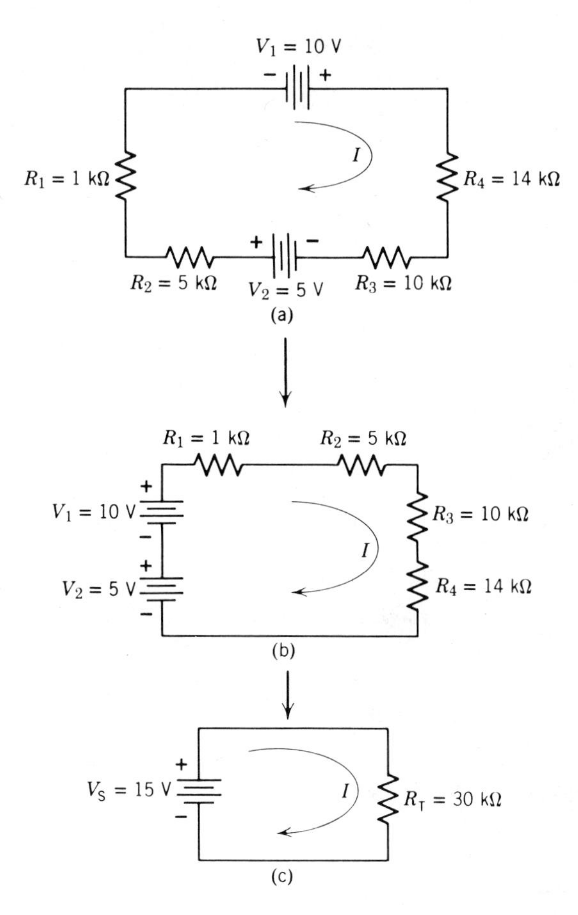

Figure 6.5 (*a*) Original circuit (*b*) Resistors and sources shifted (*c*) Equivalent circuit

therefore be redrawn by *shifting* the elements around the loop as shown in Fig. 6.5*b*. The circuit is unchanged as far as the circuit current is concerned since the total source voltage is the same and the total series resistance is the same; however, it is now easier to develop an equivalent circuit since the resistors are all grouped together and the sources are also grouped together. The total source voltage is clearly the sum of the two source voltages, and thus,

$$V_S = V_1 + V_2 = 10 \text{ V} + 5 \text{ V} = 15 \text{ V}$$

The total circuit resistance is the sum of the individual resistances and is

$$\begin{aligned} R_T &= R_1 + R_2 + R_3 + R_4 \\ &= 1 \text{ k}\Omega + 5 \text{ k}\Omega + 10 \text{ k}\Omega + 14 \text{ k}\Omega \\ &= 30 \text{ k}\Omega \end{aligned}$$

The final equivalent circuit is then given in Fig. 6.5*c* and the circuit current is found by using Ohm's Law.

$$I = \frac{V_S}{R_T} = \frac{15 \text{ V}}{30 \text{ k}\Omega} = 0.5 \text{ mA}$$

• • •

6.2 KIRCHHOFF'S VOLTAGE LAW—KVL

The equivalent circuit shown in Fig. 6.4 was found by adding the resistances in Fig. 6.1 according to the following relation

$$R_T = R_1 + R_2 + R_3$$

We can multiply *both* sides of this equation by the same quantity without changing its value, and if we multiply by the circuit current I, the result is

$$IR_T = I(R_1 + R_2 + R_3) = IR_1 + IR_2 + IR_3$$

If we now compare this equation with Fig. 6.4, it is clear that IR_T is equal to the source voltage V. Similarly, in comparing this equation with Fig. 6.1, it is clear that IR_1 is equal to the voltage V_{R_1} across the resistor R_1, IR_2 is equal to the voltage V_{R_2} across the resistor R_2, and IR_3 is equal to the voltage V_{R_3} across resistor R_3. These quantities are shown in Fig. 6.6. By making the appropriate substitutions, this equation can then be rewritten as

$$V = V_{R_1} + V_{R_2} + V_{R_3}$$

Thus, in the series circuit shown in Fig. 6.6, the total applied voltage must be equal to the sum of the voltage drops across the resistors.

This idea can of course be extended to a series circuit containing any number of resistors and sources and therefore,

in a series circuit, the algebraic sum of all the source voltages must be equal to the algebraic sum of all the voltage drops.

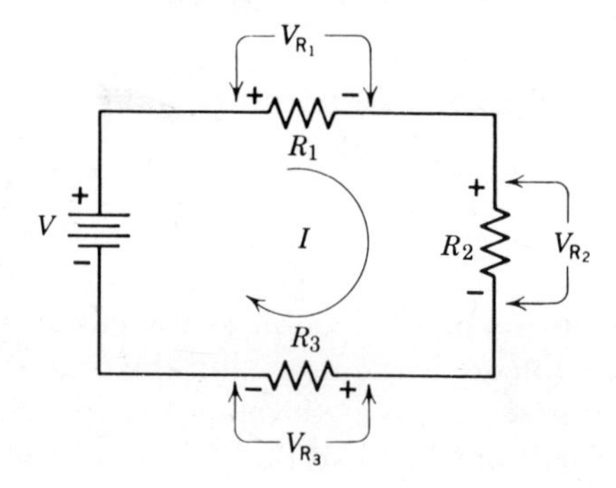

Figure 6.6

This result was noted by Kirchhoff and is known as Kirchhoff's Voltage Law (KVL). KVL can be stated in equation form as

$$\overbrace{V_{S_1} + V_{S_2} + V_{S_3} + \cdots + V_{S_M}}^{\text{Source voltages}} = \overbrace{V_{R_1} + V_{R_2} + V_{R_3} + \cdots + V_{R_N}}^{\text{Voltage drops}} \qquad (6.2)$$

where M is the number of voltage sources and N is the number of resistors.

• • •

Example 6.4 Check the validity of KVL for the circuit in Example 6.2.

Solution The complete circuit is shown in Fig. 6.7. The voltage across the

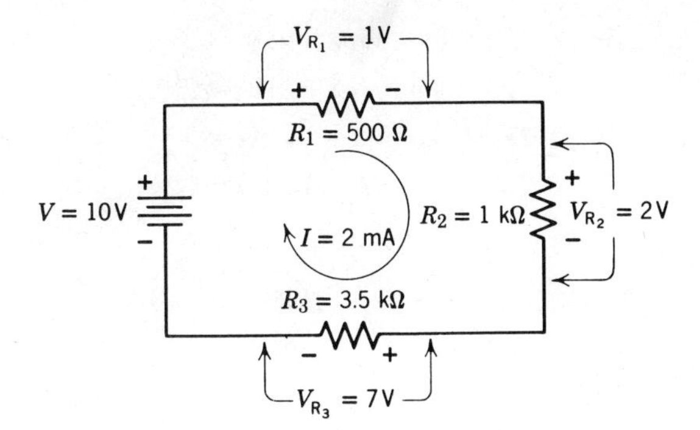

Figure 6.7

resistor R_1 is

$$V_{R_1} = IR_1 = 2 \text{ mA} \times 500 \, \Omega = 1 \text{ V}$$

The voltage across resistor R_2 is

$$V_{R_2} = IR_2 = 2 \text{ mA} \times 1 \text{ k}\Omega = 2 \text{ V}$$

The voltage across resistor R_3 is

$$V_{R_3} = IR_3 = 2 \text{ mA} \times 3.5 \text{ k}\Omega = 7 \text{ V}$$

Using Eq. 6.2,

$$V = V_{R_1} + V_{R_2} + V_{R_3}$$

$$10 \text{ V} = 1 \text{ V} + 2 \text{ V} + 7 \text{ V}$$

Thus Eq. 6.2 checks and KVL is satisfied.

• • •

You will notice that KVL provides an excellent method for checking the results of a problem, for if some error has been made in the solution of a problem, it will surely show up when the results are checked using KVL.

• • •

Example 6.5 Check the results of Example 6.3 using KVL.

Solution The complete circuit is shown in Fig. 6.8.

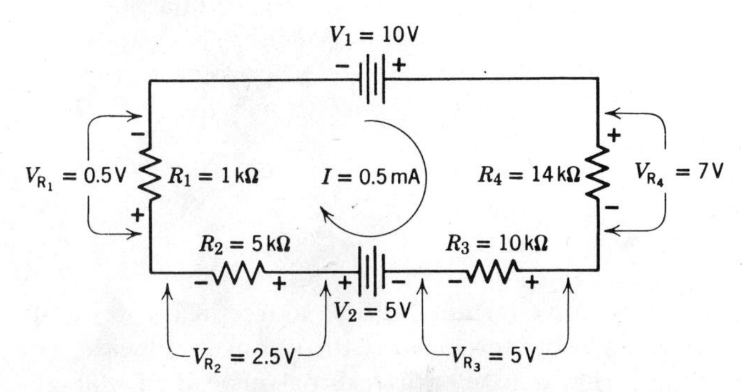

Figure 6.8

$$V_{R_1} = IR_1 = 0.5\text{ mA} \times 1\text{ k}\Omega = 0.5\text{ V}$$
$$V_{R_2} = IR_2 = 0.5\text{ mA} \times 5\text{ k}\Omega = 2.5\text{ V}$$
$$V_{R_3} = IR_3 = 0.5\text{ mA} \times 10\text{ k}\Omega = 5\text{ V}$$
$$V_{R_4} = IR_4 = 0.5\text{ mA} \times 14\text{ k}\Omega = 7\text{ V}$$

Using Eq. 6.2,

$$10\text{ V} + 5\text{ V} = 0.5\text{ V} + 2.5\text{ V} + 5\text{ V} + 7\text{ V}$$

and KVL is seen to be satisfied.

• • •

In applying KVL around a closed circuit, it is necessary to pay careful attention to the *polarities* of the voltage drops and the sources. We can apply KVL to a series circuit by beginning at any point on the circuit, tracing a path completely around the circuit, and recording the voltage drops across the resistors and the voltage rises across the sources according to Eq. 6.2. In tracing the closed path, if we pass through a resistor in the direction shown in Fig. 6.9*a* we record the voltage drop as positive since it is truly a voltage drop (we pass from the positive side to the negative side); otherwise we record it as a negative value. If we pass through a source in the direction shown in Fig. 6.9*b*, we record the source voltage as a positive value since this is the normal direction of current from a source; if we pass through a source in the other direction, the source voltage would be entered in Eq. 6.2 as a negative value.

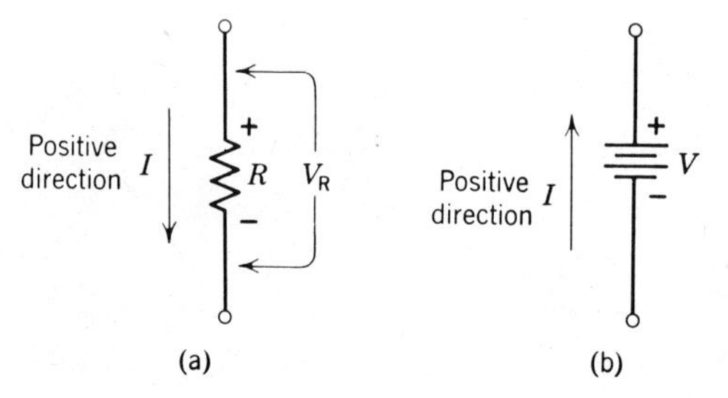

Figure 6.9 (*a*) Voltage drop across a resistor (*b*) Voltage rise across a voltage source

We can apply these criteria to the circuit shown in Fig. 6.7 by beginning at the negative terminal of the voltage source and tracing a path in a clockwise direction around the circuit. We will travel through the source in the positive direction, and we will also travel through a voltage drop across each resistor; thus all of the entries in the KVL equation are positive and the result is

$$V = V_{R_1} + V_{R_2} + V_{R_3}$$
$$10\text{ V} = 1\text{ V} + 2\text{ V} + 7\text{ V}$$

and the KVL equation is satisfied.

We could also begin at the negative terminal of the source in Fig. 6.7 and trace a path around the circuit in a counterclockwise direction. In this case, we would travel through voltage *rises* across each resistor (instead of voltage drops), and we would pass through the source in the negative direction. Thus all of the entries in the KVL equation would be negative, and the result would be

$$-V = -V_{R_1} - V_{R_2} - V_{R_3}$$
$$-10\text{ V} = -1\text{ V} - 2\text{ V} - 7\text{ V}$$

and the KVL equation is still satisfied since $-10\text{ V} = -10\text{ V}$.

• • •

Example 6.6 Check the voltages shown in the circuit in Fig. 6.10 using KVL.

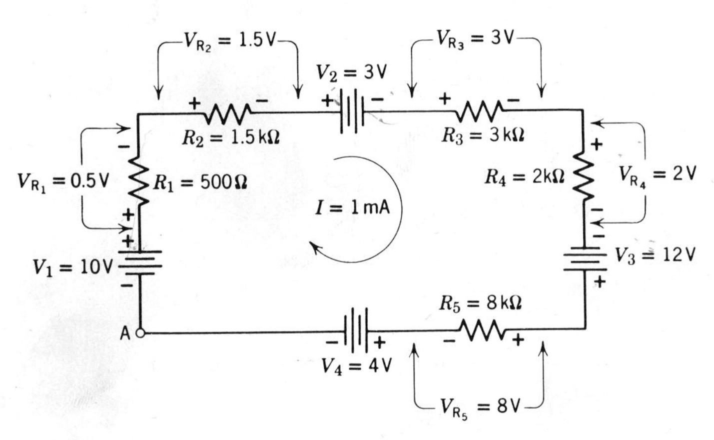

Figure 6.10

Solution We begin at the point labeled A in the circuit and trace a path around the circuit in a clockwise direction. It is clear from the circuit that we will travel through a voltage drop across each resistor, and thus all of the voltage drops in the KVL equation, Eq. 6.2, will have positive values. In considering the voltage sources, we note the following: we travel through V_1 in a *positive* direction; we travel through V_2 in a *negative* direction; we travel through V_3 in a *positive* direction; we travel through V_4 in a *negative* direction. The proper KVL equation is then

$$V_1 - V_2 + V_3 - V_4 = V_{R_1} + V_{R_2} + V_{R_3} + V_{R_4}$$
$$10\text{ V} - 3\text{ V} + 12\text{ V} - 4\text{ V} = 0.5\text{ V} + 1.5\text{ V} + 3\text{ V} + 2\text{ V} + 8\text{ V}$$
$$15\text{ V} = 15\text{ V}$$

The reader should verify the KVL equation by tracing through the circuit in a counterclockwise direction.

• • •

POWER IN SERIES CIRCUITS

It is sometimes necessary to calculate the power dissipated in each of the resistors in a series circuit. The power dissipated in each of the resistors in the circuit shown in Fig. 6.11 can be found in a number of ways. The power in a resistor can be found using any of the relations $P = IV = I^2R = V^2/R$, and the relationship used depends on the quantities given in the problem. In Fig. 6.11,

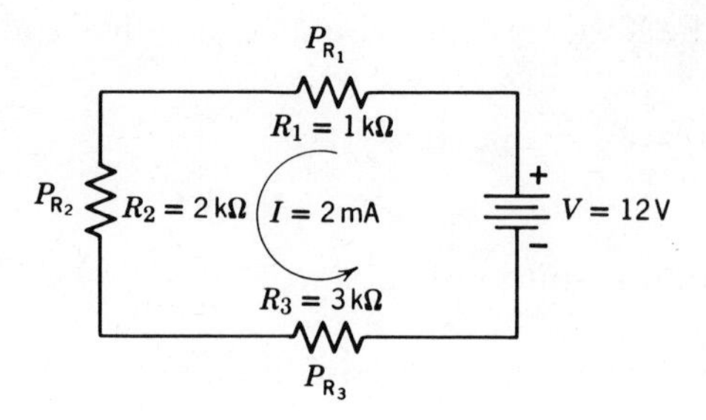

Figure 6.11

the resistance values and the source voltage are given and it is a simple matter to calculate the circuit current I. Thus,

$$I = \frac{V}{R_T} = \frac{12\text{ V}}{1\text{ k}\Omega + 2\text{ k}\Omega + 3\text{ k}\Omega} = \frac{12\text{ V}}{6\text{ k}\Omega} = 2\text{ mA}$$

The power in each resistor can then be found using the relationship $P = I^2R$. Thus the power in each resistor is given as

$$P_{R_1} = I^2R_1 \qquad P_{R_2} = I^2R_2 \qquad P_{R_3} = I^2R_3$$

and

$$P_{R_1} = (2\text{ mA})^2 \times 1\text{ k}\Omega = 4\text{ mW}$$
$$P_{R_2} = (2\text{ mA})^2 \times 2\text{ k}\Omega = 8\text{ mW}$$
$$P_{R_3} = (2\text{ mA})^2 \times 3\text{ k}\Omega = 12\text{ mW}$$

The total resistance for this circuit is given by

$$R_T = R_1 + R_2 + R_3$$

If we multiply both sides of this equation by I^2, the result is

$$I^2R_T = I^2(R_1 + R_2 + R_3) = I^2R_1 + I^2R_2 + I^2R_3$$

The term I^2R_T is the *total* power delivered to the circuit by the source, and thus it is clear the total power delivered by the source is equal to the sum of the individual powers dissipated in each resistor. This is really nothing more than a confirmation of the conservation of energy, since power is energy per unit time, and we would not expect to *gain* or *lose* any power. This idea can be extended to a series circuit containing any number of resistances and sources and it follows that

the total power delivered to a circuit by a source (or sources) must be equal to the total power dissipated in the resistances.

This can be stated in equation form as

$$\overbrace{P_{S_1} + P_{S_2} + \cdots + P_{S_M}}^{\text{Power delivered by sources}} = \overbrace{P_{R_1} + P_{R_2} + \cdots + P_{R_N}}^{\text{Power dissipated in resistances}} \tag{6.3}$$

where M is equal to the number of sources in the circuit and N is equal to the number of resistances.

In the circuit shown in Fig. 6.11, the total power delivered by the voltage source is given by

$$P_S = I^2 R_T = (2\text{ mA})^2 \times 6\text{ k}\Omega = 24\text{ mW}$$

If we now apply Eq. 6.3 to the powers calculated in Fig. 6.11, the results are

$$P_S = P_{R_1} + P_{R_2} + P_{R_3}$$
$$24\text{ mW} = 4\text{ mW} + 8\text{ mW} + 12\text{ mW}$$

and Eq. 6.3 is clearly satisfied.

6.4 PARALLEL CIRCUITS; KIRCHHOFF'S CURRENT LAW—KCL

When two or more elements have their terminals or end points connected to the same two points as shown in Fig. 6.12, the elements are said to be connected in *parallel*. You will notice that the circuit in Fig. 6.12 could be

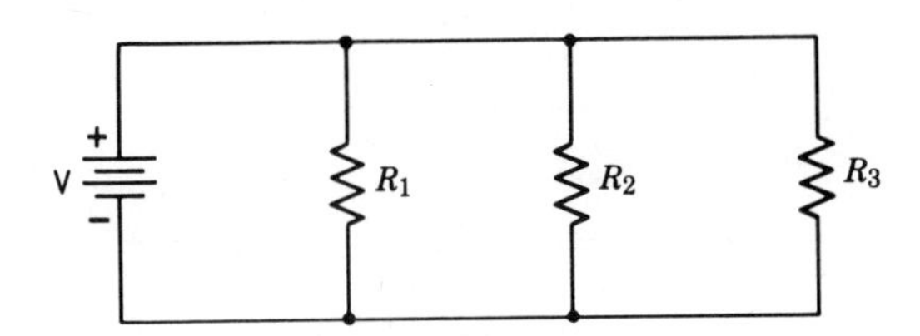

Figure 6.12 Elements connected in parallel

connected (or drawn) as shown in Fig. 6.13 without any change in the electrical quantities of the circuit. All of the circuit arrangements shown in these two figures are equivalent and it is quite clear that all the elements are connected between two common points and they are therefore connected in parallel. A careful examination of these circuits shows that the voltage is the same across every element, and thus we can say,

if two or more elements are connected in parallel, the voltage must be exactly the same across every element.

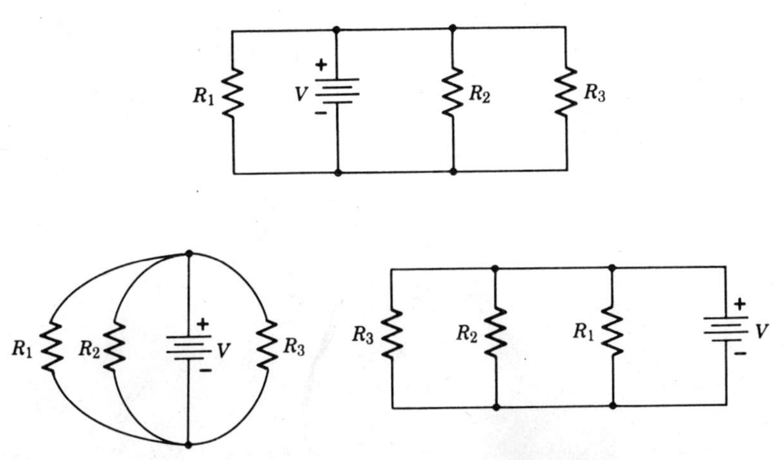

Figure 6.13 Different ways of drawing the circuit of Fig. 6.12

A common point of connection between two or more elements is called a *node*, and the circuit shown in Fig. 6.14 has two nodes labeled A and B. The total current from the source I_T leaves the positive terminal of the source, and enters node A. At node A, I_T splits into two components. The component of I_T which goes through R_1 is I_{R_1}, and the component which goes through R_2 is I_{R_2}. Since we can neither *gain* nor *lose* current at any point along a circuit, the total current entering node A must be equal to the total current leaving node A. Therefore,

$$I_T = I_{R_1} + I_{R_2}$$

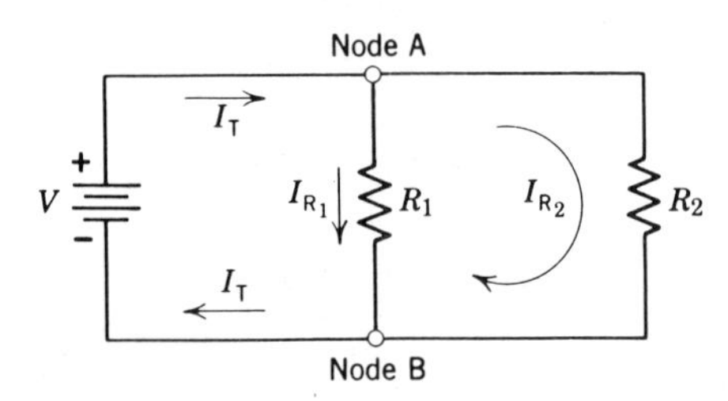

Figure 6.14

This equation also holds true at node B since the total current entering this node is $I_{R_1} + I_{R_2}$, and the total current leaving is I_T. Since current is the rate of flow of charge, this equation is really a statement showing the conservation of charge at nodes A and B (i.e., we can neither gain nor lose charge at any point in a circuit), and this idea is more generally known as Kirchhoff's Current Law (KCL). In general terms, Kirchhoff's Current Law states

the algebraic sum of all the currents entering a node must be equal to the algebraic sum of all the currents leaving that node.

Applying KCL to the more general node shown in Fig. 6.15 yields the equation,

$$\overbrace{I_1 + I_2 + I_3 + I_4}^{\text{Currents entering}} = \overbrace{I_5 + I_6}^{\text{Currents leaving}}$$

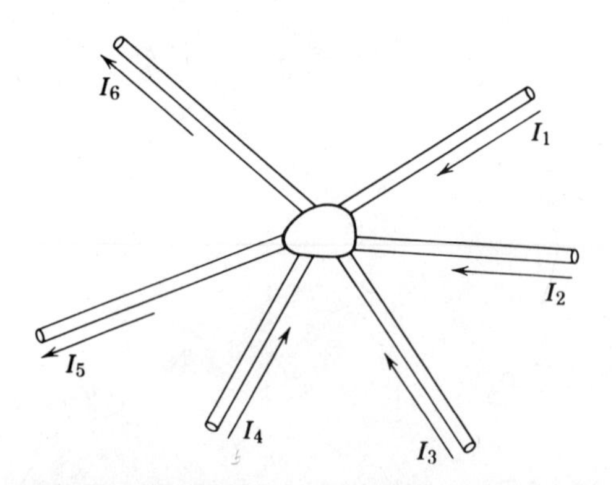

Figure 6.15 A generalized node

• • •

Example 6.7 Apply KCL to check the currents given in the circuit shown in Fig. 6.16.

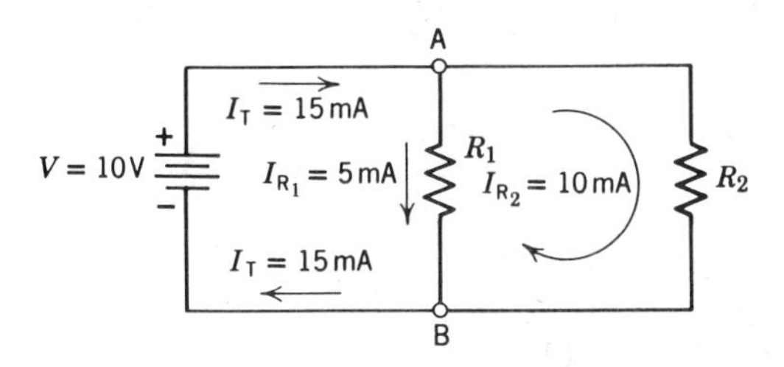

Figure 6.16

Solution At node A,

$$I_T = I_{R_1} + I_{R_2}$$
$$15 \text{ mA} = 5 \text{ mA} + 10 \text{ mA}$$

At node B,

$$I_{R_1} + I_{R_2} = I_T$$
$$5 \text{ mA} + 10 \text{ mA} = 15 \text{ mA}$$

• • •

Example 6.8 Find the values of the resistors R_1 and R_2 in Fig. 6.16.

Solution Since this is a parallel circuit, the voltage across the two resistors is the same. Thus,

$$R_1 = \frac{V}{I_{R_1}} = \frac{10 \text{ V}}{5 \text{ mA}} = 2 \text{ k}\Omega$$

and

$$R_2 = \frac{V}{I_{R_2}} = \frac{10 \text{ V}}{10 \text{ mA}} = 1 \text{ k}\Omega$$

• • •

Example 6.9 In the circuit shown in Fig. 6.14, $V = 20$ V, $R_1 = 470\ \Omega$, and $R_2 = 1.5$ kΩ. Find the values of the currents I_T, I_{R_1}, and I_{R_2}.

Solution The resistors are connected in parallel and the voltage is therefore the same across both resistors. The resistor currents can then be found using Ohm's Law, and thus

$$I_{R_1} = \frac{V}{R_1} = \frac{20 \text{ V}}{470\ \Omega} = 42.6 \text{ mA}$$

$$I_{R_2} = \frac{V}{R_2} = \frac{20 \text{ V}}{1.5 \text{ k}\Omega} = 13.3 \text{ mA}$$

Using KCL, the total current delivered by the source is

$$I_T = I_{R_1} + I_{R_2} = 42.6 \text{ mA} + 13.3 \text{ mA} = 55.9 \text{ mA}$$

• • •

6.5 RESISTANCES IN PARALLEL

The application of KCL to the circuit in Fig. 6.16 yields the equation

$$I_T = I_{R_1} + I_{R_2}$$

But, it is shown in Example 6.9 that the voltage across each resistor is the same and therefore,

$$I_{R_1} = \frac{V}{R_1} \qquad \text{and} \qquad I_{R_2} = \frac{V}{R_2}$$

If we substitute these two relationships into the KCL equation above, the result is

$$I_T = I_{R_1} + I_{R_2} = \frac{V}{R_1} + \frac{V}{R_2} = V\left(\frac{1}{R_1} + \frac{1}{R_2}\right)$$

We can make an equivalent circuit of the circuit shown in Fig. 6.16 by combining the two resistors R_1 and R_2 to form an equivalent resistance R_T as shown in Fig. 6.17. From this equivalent circuit it is clear that $I_T = V/R_T$, and if we substitute this relationship into the above equation, the result is

$$I_T = \frac{V}{R_T} = V\left(\frac{1}{R_1} + \frac{1}{R_2}\right)$$

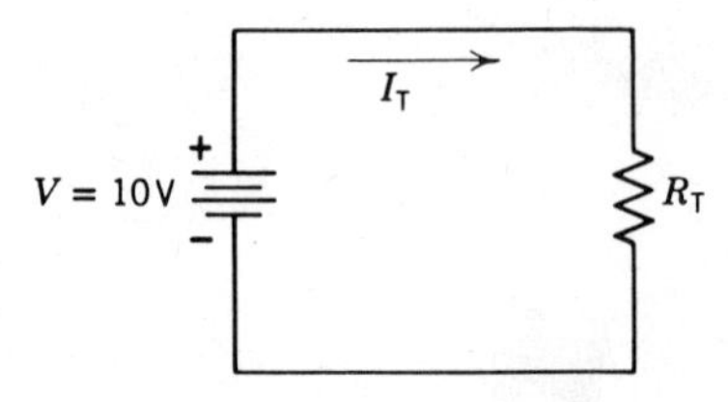

Figure 6.17 Equivalent circuit of Fig. 6.16

Finally, dividing both sides of this equation by V yields

$$\frac{1}{R_T} = \frac{1}{R_1} + \frac{1}{R_2} \tag{6.4}$$

The equivalent resistance of two resistors connected in parallel is then found by taking the reciprocal of Eq. 6.4, and this is

$$R_T = \frac{1}{\dfrac{1}{R_1} + \dfrac{1}{R_2}} = \frac{R_1 R_2}{R_1 + R_2} \tag{6.5}$$

Thus the equivalent resistance of two resistors connected in parallel is found by taking the *product* of the two resistances and dividing it by the *sum* of the two resistances.

• • •

Example 6.10 Find the equivalent resistance of the resistors in Fig. 6.18 when: (*a*) $R_1 = 100\ \Omega$, $R_2 = 200\ \Omega$; (*b*) $R_1 = 3\ \text{k}\Omega$, $R_2 = 6\ \text{k}\Omega$.

Solution The equivalent resistance is found by taking the product divided

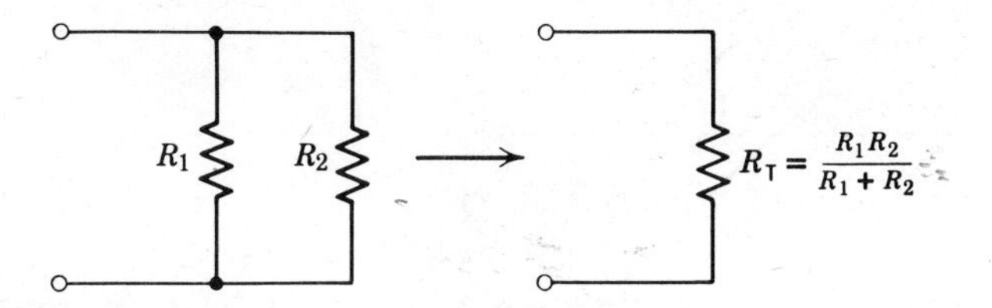

Figure 6.18 Finding equivalent parallel resistance

by the sum. Thus,

$$(a)\ R_T = \frac{R_1 R_2}{R_1 + R_2} = \frac{100\ \Omega \times 200\ \Omega}{100\ \Omega + 200\ \Omega} = 66.7\ \Omega$$

$$(b)\ R_T = \frac{3\ k\Omega \times 6\ k\Omega}{3\ k\Omega + 6\ k\Omega} = 2\ k\Omega$$

• • •

Example 6.11 Check the total current I_T in Fig. 6.16 by finding the total resistance R_T and using Ohm's Law.

Solution The total resistance R_T is given by

$$R_T = \frac{R_1 R_2}{R_1 + R_2} = \frac{2\ k\Omega \times 1\ k\Omega}{2\ k\Omega + 1\ k\Omega} = 2/3\ k\Omega$$

The total current I_T is then found using Ohm's Law as

$$I_T = \frac{V}{R_T} = \frac{10\ V}{2/3\ k\Omega} = 15\ mA$$

• • •

Equation 6.4 can be easily extended to any number of resistors connected in parallel and the result is

$$\frac{1}{R_T} = \frac{1}{R_1} + \frac{1}{R_2} + \cdots + \frac{1}{R_N} \tag{6.6}$$

The equivalent resistance of any number of resistors connected in parallel is found by taking the reciprocal of Eq. 6.6 and is

$$R_T = \frac{1}{\frac{1}{R_1} + \frac{1}{R_2} + \cdots + \frac{1}{R_N}} \tag{6.7}$$

where N is the number of resistors.

• • •

Example 6.12 Find the equivalent resistance of the four resistors in Fig. 6.19.

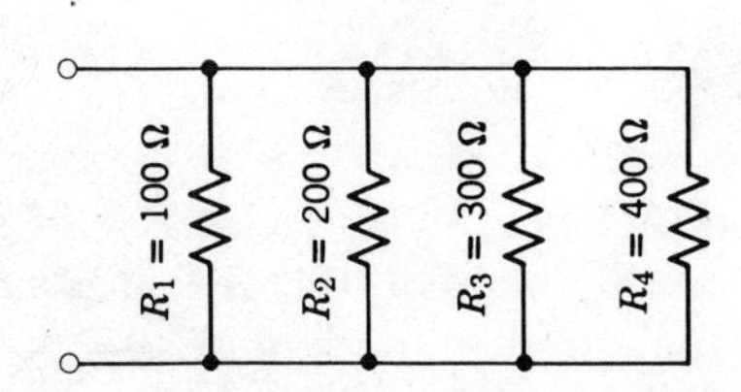

Figure 6.19

Solution Using Eq. 6.7,

$$R_T = \frac{1}{\frac{1}{100\ \Omega}+\frac{1}{200\ \Omega}+\frac{1}{300\ \Omega}+\frac{1}{400\ \Omega}} = \frac{1}{\frac{12+6+4+3}{1200}}\ \Omega = \frac{1200\ \Omega}{25} = 48\ \Omega$$

• • •

If a number of resistors N are connected in parallel, and the resistors *all* have the *same* resistance, then Eq. 6.7 reduces to

$$R_T = \frac{1}{\frac{1}{R}+\frac{1}{R}+\cdots+\frac{1}{R_N}} = \frac{1}{N\left(\frac{1}{R}\right)} = \frac{R}{N} \tag{6.8}$$

For example, the equivalent resistance of five 1 kΩ resistors connected in parallel is $R_T = 1\ \mathrm{k}\Omega/5 = 200\ \Omega$. Similarly, the equivalent resistance of two 10 kΩ resistors connected in parallel is $R_T = 10\ \mathrm{k}\Omega/2 = 5\ \mathrm{k}\Omega$.

It is sometimes easier to find the equivalent resistance of a number of parallel resistances by taking the resistances *two at a time* (using Eq. 6.5) as shown in the following example.

• • •

Example 6.13 Find the equivalent resistance of the four resistors shown in Fig. 6.20*a*.

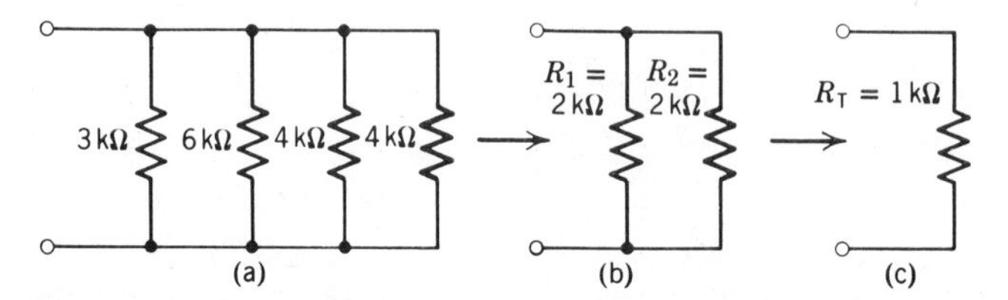

Figure 6.20

Solution Rather than use the general equation Eq. 6.7, it is easier to combine these resistors two at a time. Combining the 3 kΩ and the 6 kΩ using Eq. 6.5, forms a resistance of

$$R_1 = \frac{3\ \mathrm{k}\Omega \times 6\ \mathrm{k}\Omega}{3\ \mathrm{k}\Omega + 6\ \mathrm{k}\Omega} = 2\ \mathrm{k}\Omega$$

Combining the two 4 kΩ resistors according to Eq. 6.8 gives a resistance of

$$R_2 = \frac{R}{N} = \frac{4\ \mathrm{k}\Omega}{2} = 2\ \mathrm{k}\Omega$$

These two equivalent resistances are shown in Fig. 6.20*b*. Finally, combining R_1 and R_2 to find R_T as in Fig. 6.20*c* yields

$$R_T = \frac{R_1}{N} = \frac{R_2}{N} = \frac{2\ \mathrm{k}\Omega}{2} = 1\ \mathrm{k}\Omega$$

• • •

If we recall that conductance is the reciprocal of resistance, or $G = 1/R$, then it is easy to see from Eq. 6.6 that the total circuit conductance is equal to the sum of the individual conductances or

$$G_T = G_1 + G_2 + \cdots + G_N \tag{6.9}$$

Thus the equivalent conductance, or resistance, of a number of resistances connected in parallel can be found using Eq. 6.9.

• • •

Example 6.14 Find the equivalent conductance and resistance of the circuit in Fig. 6.19.

Solution The individual conductances are:

$$G_1 = 1/R_1 = 1/100\ \Omega = 0.010\ \mho = 10\ \text{m}\mho$$
$$G_2 = 1/R_2 = 1/200\ \Omega = 0.005\ \mho = 5.0\ \text{m}\mho$$
$$G_3 = 1/R_3 = 1/300\ \Omega = 0.0033\ \mho = 3.3\ \text{m}\mho$$
$$G_4 = 1/R_4 = 1/400\ \Omega = 0.0025\ \mho = 2.5\ \text{m}\mho$$

The total conductance is then

$$G_T = G_1 + G_2 + G_3 + G_4 = 10\ \text{m}\mho + 5.0\ \text{m}\mho + 3.3\ \text{m}\mho + 2.5\ \text{m}\mho = 20.8\ \text{m}\mho$$

The equivalent resistance is the reciprocal of the equivalent conductance and it is given by

$$R_T = \frac{1}{G_T} = \frac{1}{20.8\ \text{m}\mho} = \frac{1}{20.8 \times 10^{-3}\ \mho} = 48\ \Omega$$

• • •

6.6 POWER IN PARALLEL CIRCUITS

It is sometimes necessary to calculate the power dissipated in each resistor in a parallel circuit. The power in each resistor shown in Fig. 6.16 can most easily be found using the equation $P = V^2/R$ since all of the resistors are connected in parallel and therefore they all have the same voltage across them. Thus the power in each of the resistors in Fig. 6.16 is,

$$P_{R_1} = \frac{V^2}{R_1} = \frac{(10\ \text{V})^2}{2\ \text{k}\Omega} = 50\ \text{mW}$$

$$P_{R_2} = \frac{V^2}{R_2} = \frac{(10\ \text{V})^2}{1\ \text{k}\Omega} = 100\ \text{mW}$$

The total power delivered by the source in this circuit is $P_S = I_T V$ and is clearly,

$$P_S = I_T V = 15\ \text{mA} \times 10\ \text{V} = 150\ \text{mW}$$

It is then quite clear that the total power delivered by the source is equal to the sum of the power dissipated in the individual resistances. Thus

$$P_S = P_{R_1} + P_{R_2}$$

which is exactly the same result obtained for the series circuit.

If we apply Eq. 6.6 to this circuit, it appears as

$$\frac{1}{R_T} = \frac{1}{R_1} + \frac{1}{R_2}$$

We can now multiply both sides of this equation by V^2 to obtain

$$\frac{V^2}{R_T} = \frac{V^2}{R_1} + \frac{V^2}{R_2}$$

But, V^2/R_T is clearly the power delivered by the source, V^2/R_1 is the power in resistor R_1, and V^2/R_2 is the power in resistor R_2. Thus,

$$P_S = P_{R_1} + P_{R_2}$$

This result can of course be extended to any number of resistors connected in parallel, and the result is exactly the same as that for resistors connected in series, namely Eq. 6.3.

6.7 PARALLEL RESISTANCES CONNECTED IN SERIES

Many of the more complicated electric circuits can be easily solved by treating them as interconnected groups of series and parallel circuits. The rules for combining series and parallel resistances can be used to find equivalent resistances, and thus the complicated circuit can be reduced to a simple series or parallel equivalent circuit. A *complete solution* for the circuit can then be found using Ohm's Law. A complete solution means finding the *current through* and the *voltage across every element* in the circuit. In finding a complete solution for a circuit, the importance of labeling all of the voltages and currents in the circuit cannot be overemphasized—slovenly notation will surely be disastrous!

The circuit shown in Fig. 6.21 is easily seen to be three groups of parallel resistors connected in series. A complete solution for the circuit can be found by first combining the parallel resistances to form the simple series circuit shown in Fig. 6.22*a*. This is accomplished by finding the parallel equivalent of R_1 and R_2, R_3 and R_4, and, R_5 and R_6. These parallel equivalent resistances can then be added (since they are in series) to form the final equivalent circuit shown in Fig. 6.22*b*. This equivalent circuit can then be used to find the

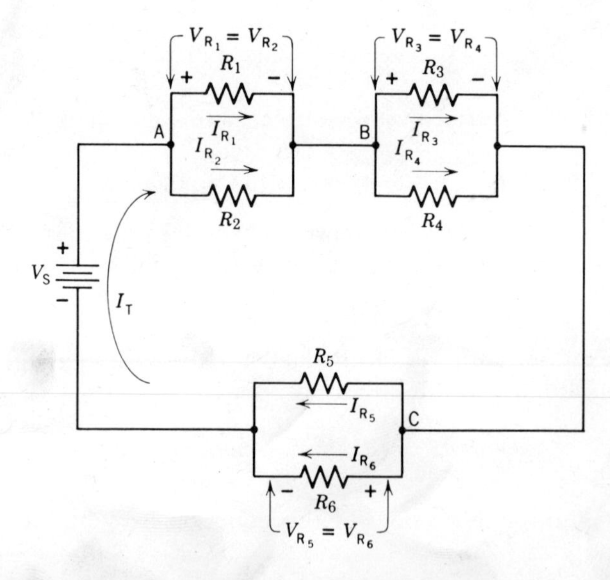

Figure 6.21 Parallel resistors connected in series

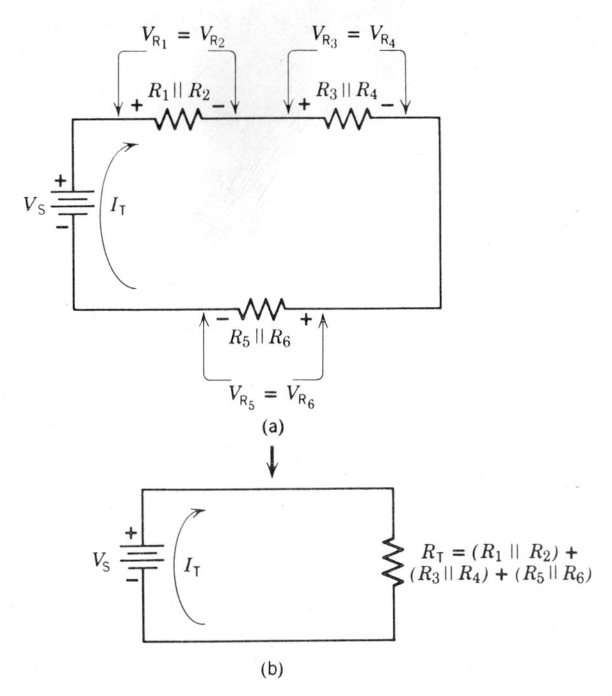

Figure 6.22 (*a*) First equivalent circuit (*b*) Final equivalent circuit

current delivered by the source I_T. Thus

$$I_T = \frac{V_S}{R_T} = \frac{V_S}{(R_1 \| R_2) + (R_3 \| R_4) + (R_5 \| R_6)}$$

where the lines $\|$ mean "in parallel with."

Referring to Fig. 6.21, it is easily seen that the voltage across R_1, V_{R_1} is equal to the voltage across R_2, V_{R_2}, since R_1 and R_2 are connected in parallel, and this is the voltage across the equivalent resistance $R_1 \| R_2$ in Fig. 6.22*a*. A similar situation exists for V_{R_3} and V_{R_4}, and V_{R_5} and V_{R_6}. Thus the voltages across the 6 resistors are found from Fig. 6.22*a* and I_T as,

$$V_{R_1} = V_{R_2} = I_T(R_1 \| R_2)$$
$$V_{R_3} = V_{R_4} = I_T(R_3 \| R_4)$$
$$V_{R_5} = V_{R_6} = I_T(R_5 \| R_6)$$

Now that the voltages across each of the resistors are known, we can return to Fig. 6.21 and use Ohm's Law to find the currents through each of the resistors. Thus,

$$I_{R_1} = \frac{V_{R_1}}{R_1} \qquad I_{R_2} = \frac{V_{R_2}}{R_2} \qquad I_{R_3} = \frac{V_{R_3}}{R_3}$$

$$I_{R_4} = \frac{V_{R_4}}{R_4} \qquad I_{R_5} = \frac{V_{R_5}}{R_5} \qquad I_{R_6} = \frac{V_{R_6}}{R_6}$$

We have now obtained a complete solution for the circuit, and the solution can be checked by applying KVL around the circuit, and KCL at the nodes.

• • •

Example 6.15 Find the complete solution for the circuit in Fig. 6.21 for the following circuit values: $V_S = 12\text{ V}$, $R_1 = 6\text{ k}\Omega$, $R_2 = 3\text{ k}\Omega$, $R_3 = R_4 = 2\text{ k}\Omega$, $R_5 = 12\text{ k}\Omega$, $R_6 = 4\text{ k}\Omega$.

Solution The first step is to find the parallel equivalent resistances and form the equivalent circuit in Fig. 6.22*a*.

$$R_1\|R_2 = \frac{R_1R_2}{R_1+R_2} = \frac{6\text{ k}\Omega \times 3\text{ k}\Omega}{6\text{ k}\Omega + 3\text{ k}\Omega} = 2\text{ k}\Omega$$

$$R_3\|R_4 = \frac{R_3}{\text{N}} = \frac{R_4}{\text{N}} = \frac{2\text{ k}\Omega}{2} = 1\text{ k}\Omega$$

$$R_5\|R_6 = \frac{R_5R_6}{R_5+R_6} = \frac{12\text{ k}\Omega \times 4\text{ k}\Omega}{12\text{ k}\Omega + 4\text{ k}\Omega} = 3\text{ k}\Omega$$

The total circuit resistance R_T in Fig. 6.22*b* is the sum of these equivalent resistances and is given by

$$R_T = (R_1\|R_2) + (R_3\|R_4) + (R_5\|R_6) = 2\text{ k}\Omega + 1\text{ k}\Omega + 3\text{ k}\Omega = 6\text{ k}\Omega$$

The current from the source I_T is then found as

$$I_T = \frac{V_S}{R_T} = \frac{12\text{ V}}{6\text{ k}\Omega} = 2\text{ mA}$$

The resistor voltages are then,

$$V_{R_1} = V_{R_2} = I_T(R_1\|R_2) = 2\text{ mA} \times 2\text{ k}\Omega = 4\text{ V}$$
$$V_{R_3} = V_{R_4} = I_T(R_3\|R_4) = 2\text{ mA} \times 1\text{ k}\Omega = 2\text{ V}$$
$$V_{R_5} = V_{R_6} = I_T(R_5\|R_6) = 2\text{ mA} \times 3\text{ k}\Omega = 6\text{ V}$$

These results are immediately checked by using KVL since

$$V_S = V_{R_1} + V_{R_3} + V_{R_5} = V_{R_2} + V_{R_4} + V_{R_6}$$

Thus,

$$12\text{ V} = 4\text{ V} + 2\text{ V} + 6\text{ V}$$

The individual resistor currents can now be found using Ohm's Law,

$$I_{R_1} = \frac{V_{R_1}}{R_1} = \frac{4\text{ V}}{6\text{ k}\Omega} = 2/3\text{ mA} \qquad I_{R_2} = \frac{V_{R_2}}{R_2} = \frac{4\text{ V}}{3\text{ k}\Omega} = 4/3\text{ mA}$$

$$I_{R_3} = \frac{V_{R_3}}{R_3} = \frac{2\text{ V}}{2\text{ k}\Omega} = 1\text{ mA} \qquad I_{R_4} = \frac{V_{R_4}}{R_4} = \frac{2\text{ V}}{2\text{ k}\Omega} = 1\text{ mA}$$

$$I_{R_5} = \frac{V_{R_5}}{R_5} = \frac{6\text{ V}}{12\text{ k}\Omega} = 1/2\text{ mA} \qquad I_{R_6} = \frac{V_{R_6}}{R_6} = \frac{6\text{ V}}{4\text{ k}\Omega} = 3/2\text{ mA}$$

The results can be immediately checked by applying KCL to each of the nodes A, B, and C in the circuit. Thus

$$\text{A: } I_T = I_{R_1} + I_{R_2},\ 2\text{ mA} = 2/3\text{ mA} + 4/3\text{ mA}$$
$$\text{B: } I_T = I_{R_3} + I_{R_4},\ 2\text{ mA} = 1\text{ mA} + 1\text{ mA}$$
$$\text{C: } I_T = I_{R_5} + I_{R_6},\ 2\text{ mA} = 1/2\text{ mA} + 3/2\text{ mA}$$

• • •

6.8 SERIES RESISTANCES CONNECTED IN PARALLEL

The circuit shown in Fig. 6.23*a* consists of two groups of series resistors connected in parallel. This circuit can be simplified by first combining the series resistors and drawing the parallel equivalent circuit shown in Fig. 6.23*b*. From this simplified equivalent circuit, it is easy to find the currents in the parallel branches, and they are,

$$I_1 = \frac{V_S}{R_1 + R_2 + R_3} \qquad I_2 = \frac{V_S}{R_4 + R_5 + R_6}$$

The total current from the source I_T can then be found by applying KCL at node A in the equivalent circuit, and

$$I_T = I_1 + I_2$$

We can now return to the original circuit shown in Fig. 6.23*a* and solve for the individual resistor voltages using Ohm's Law. The current I_1 flows through all three resistors R_1, R_2, and R_3 (since they are connected in series), and the voltage drops across these resistors are then,

$$V_{R_1} = I_1 R_1 \qquad V_{R_2} = I_1 R_2 \qquad V_{R_3} = I_1 R_3$$

Similarly, the current I_2 is the same in all three resistors R_4, R_5, and R_6, and the voltage drops across these resistors are found using Ohm's Law:

$$V_{R_4} = I_2 R_4 \qquad V_{R_5} = I_2 R_5 \qquad V_{R_6} = I_2 R_6$$

We have now obtained a complete solution for the circuit and the results can be checked by applying KVL around the circuit.

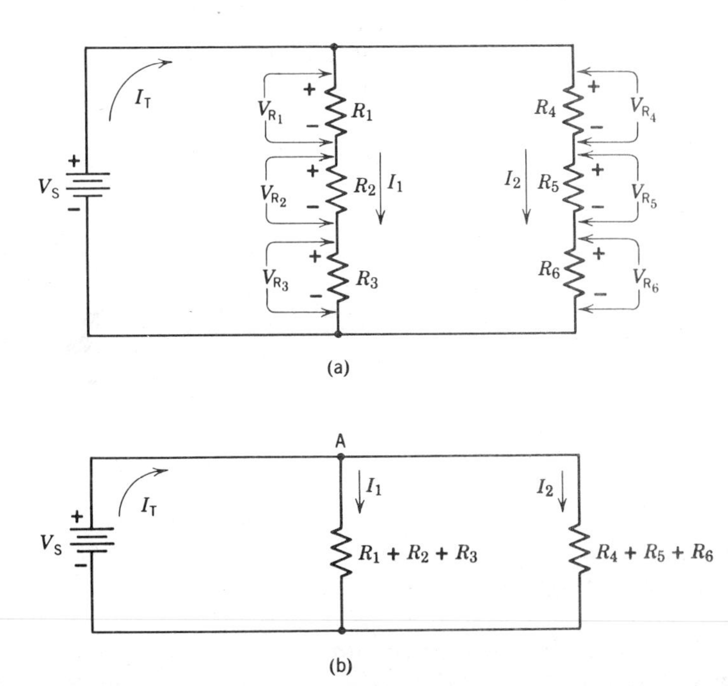

Figure 6.23 (*a*) Series resistors connected in parallel (*b*) Equivalent circuit

• • •

Example 6.16 Find a complete solution for the circuit in Fig. 6.23 for the following component values: $V_S = 24$ V, $R_1 = 200\ \Omega$, $R_2 = 6.8$ kΩ, $R_3 = 3$ kΩ, $R_4 = 4.7$ kΩ, $R_5 = 3.3$ kΩ, $R_6 = 4$ kΩ.

Solution The first step in solving this circuit is to find the simplified parallel circuit shown in Fig. 6.23*b* by combining resistances. Thus

$$R_1 + R_2 + R_3 = 200\ \Omega + 6.8\ \text{k}\Omega + 3\ \text{k}\Omega = 10\ \text{k}\Omega$$

$$R_4 + R_5 + R_6 = 4.7\ \text{k}\Omega + 3.3\ \text{k}\Omega + 4\ \text{k}\Omega = 12\ \text{k}\Omega$$

The two branch currents I_1 and I_2 are then found to be

$$I_1 = \frac{V_S}{R_1 + R_2 + R_3} = \frac{24\ \text{V}}{10\ \text{k}\Omega} = 2.4\ \text{mA}$$

$$I_2 = \frac{V_S}{R_4 + R_5 + R_6} = \frac{24\ \text{V}}{12\ \text{k}\Omega} = 2.0\ \text{mA}$$

The total current from the source I_T is now found using KCL at node A and,

$$I_T = I_1 + I_2 = 2.4\ \text{mA} + 2.0\ \text{mA} = 4.4\ \text{mA}$$

The individual resistor voltages can now be found from the original circuit using Ohm's Law and the calculated values of I_1 and I_2. Thus,

$$V_{R_1} = I_1R_1 = 2.4\ \text{mA} \times 200\ \Omega = 0.48\ \text{V}$$

$$V_{R_2} = I_1R_2 = 2.4\ \text{mA} \times 6.8\ \text{k}\Omega = 16.3\ \text{V}$$

$$V_{R_3} = I_1R_3 = 2.4\ \text{mA} \times 3\ \text{k}\Omega = 7.2\ \text{V}$$

These resistor voltages can be immediately checked by applying KVL around the circuit, and

$$V_S = V_{R_1} + V_{R_2} + V_{R_3}$$

Thus,

$$24\ \text{V} = 0.48\ \text{V} + 16.3\ \text{V} + 7.2\ \text{V} = 23.98\ \text{V}$$

and the results check.

The remaining resistor voltages are found and checked by KVL in a similar fashion. Thus,

$$V_{R_4} = I_2R_4 = 2.0\ \text{mA} \times 4.7\ \text{k}\Omega = 9.4\ \text{V}$$

$$V_{R_5} = I_2R_5 = 2.0\ \text{mA} \times 3.3\ \text{k}\Omega = 6.6\ \text{V}$$

$$V_{R_6} = I_2R_6 = 2.0\ \text{mA} \times 4\ \text{k}\Omega = 8.0\ \text{V}$$

By KVL

$$V_S = V_{R_4} + V_{R_5} + V_{R_6} = 9.4\ \text{V} + 6.6\ \text{V} + 8.0\ \text{V} = 24\ \text{V}$$

and the calculations are verified.

• • •

6.9 SERIES-PARALLEL CIRCUITS

There are many circuits which consist of numerous resistors connected in series and parallel combinations and the techniques of the previous two sections can often be used to find complete solutions for these circuits. The technique is to combine series and parallel resistances to form equivalent resistances, and thus form simplified equivalent circuits. The techniques are best illustrated by examples, and the reader will gain in understanding and confidence by studying the examples in detail.

• • •

Example 6.17 Find a complete solution for the circuit shown in Fig. 6.24*a*. Let $V_S = 12$ V.

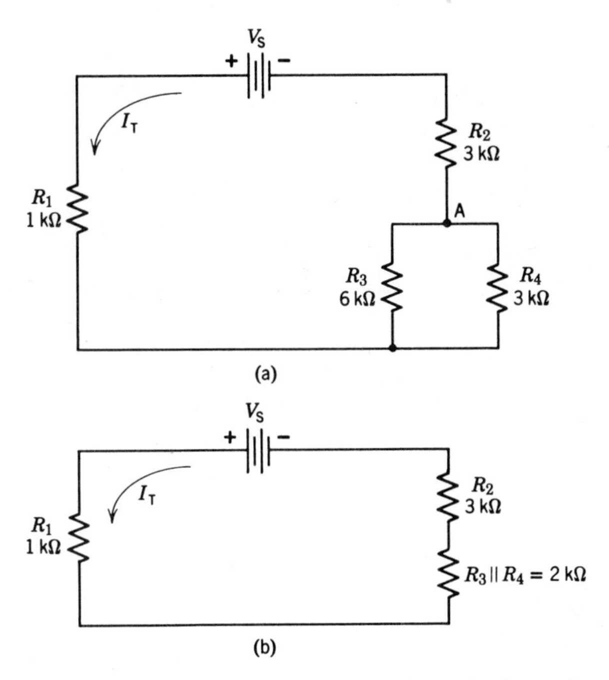

Figure 6.24 (*a*) Original circuit (*b*) Simplified equivalent circuit

Solution The simplified equivalent circuit shown in Fig. 6.24*b* is found by combining the two parallel resistors R_3 and R_4. Thus,

$$R_3 \| R_4 = \frac{R_3 R_4}{R_3 + R_4} = \frac{6\ \text{k}\Omega \times 3\ \text{k}\Omega}{6\ \text{k}\Omega + 3\ \text{k}\Omega} = 2\ \text{k}\Omega$$

The total source current I_T is now found from the simplified circuit as,

$$I_T = \frac{V_S}{R_1 + R_2 + (R_3 \| R_4)} = \frac{12\ \text{V}}{1\ \text{k}\Omega + 3\ \text{k}\Omega + 2\ \text{k}\Omega} = 2.0\ \text{mA}$$

The voltages across the resistors are now found using Ohm's Law,

$$V_{R_1} = I_T R_1 = 2.0\ \text{mA} \times 1\ \text{k}\Omega = 2.0\ \text{V}$$
$$V_{R_2} = I_T R_2 = 2.0\ \text{mA} \times 3\ \text{k}\Omega = 6.0\ \text{V}$$
$$V_{R_3} = V_{R_4} = I_T(R_3 \| R_4) = 2.0\ \text{mA} \times 2\ \text{k}\Omega = 4.0\ \text{V}$$

The results are immediately checked using KVL; thus,

$$V_S = V_{R_1} + V_{R_2} + V_{R_3}$$
$$12\ \text{V} = 2.0\ \text{V} + 6.0\ \text{V} + 4.0\ \text{V}$$

The currents through R_3 and R_4 are now found as,

$$I_{R_3} = \frac{V_{R_3}}{R_3} = \frac{4.0\text{ V}}{6\text{ k}\Omega} = 2/3\text{ m/A}$$

$$I_{R_4} = \frac{V_{R_4}}{R_4} = \frac{4.0\text{ V}}{3\text{ k}\Omega} = 4/3\text{ mA}$$

These results are checked using KCL at node A in Fig. 6.24. Thus

$$I_T = I_{R_3} + I_{R_4} \qquad 2.0\text{ mA} = 2/3\text{ mA} + 4/3\text{ mA}$$

• • •

Quite often it is easier to see how to simplify a circuit if it is redrawn in a different form. It is perhaps not quite so easy to see how to simplify the circuit in Fig. 6.25. However, if this circuit is redrawn as shown in Fig. 6.26*a*, it is much easier to see how to simplify the circuit. The circuit is simplified to that shown in Fig. 6.26*b* by adding the series resistors R_1, R_2, and R_3 to form an equivalent resistance R_X. Finally, the simple series circuit in Fig. 6.26*c* is found by combining the two parallel resistors R_X and R_4. Notice that throughout the simplification process, the node A is carefully labeled.

The total source current I_T is now found from Fig. 6.26*c*, and is

$$I_T = \frac{V_S}{R_5 + (R_X \| R_4)}$$

The voltage drop across R_5 is then

$$V_{R_5} = I_T R_5$$

The voltage at node A, V_A, is equal to the voltage across R_4, that is, V_{R_4}, and this is the same as the voltage across R_X. Thus,

$$V_A = V_{R_4} = V_{R_X} = I_T(R_X \| R_4) = V_S - V_{R_5}$$

The current through R_4, I_{R_4}, and the current through R_X, I_X, are now easily found as

$$I_{R_4} = \frac{V_{R_4}}{R_4} \qquad I_X = \frac{V_{R_X}}{R_X}$$

The current in each of the resistors R_1, R_2, and R_3 is equal to I_X, and the

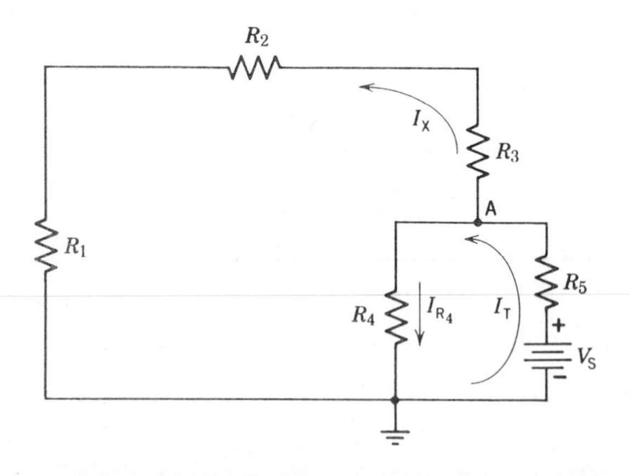

Figure 6.25

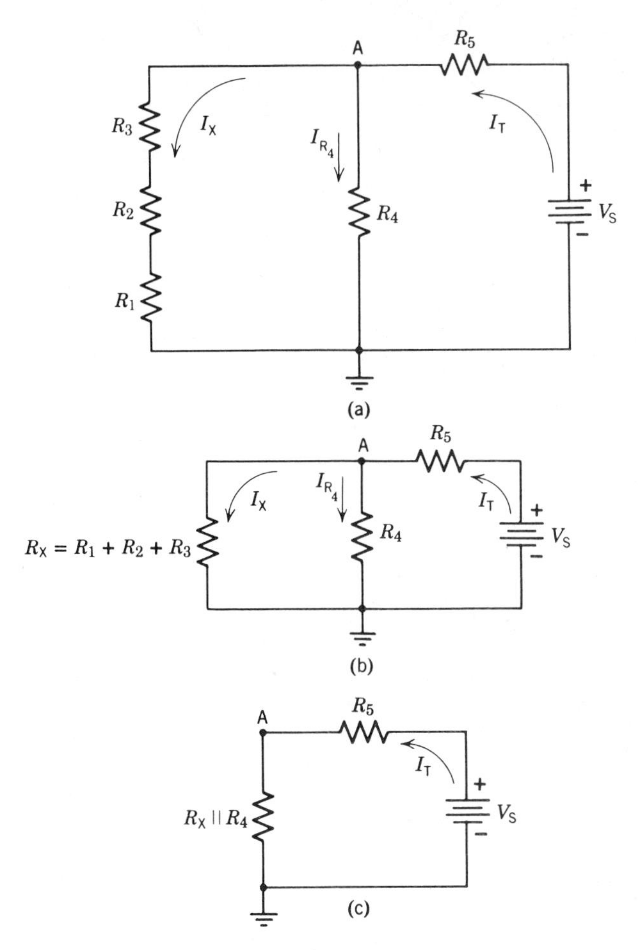

Figure 6.26 (*a*) Figure 6.25 redrawn (*b*) First equivalent circuit (*c*) Final equivalent circuit

voltage drops across these resistors are found using Ohm's Law. Thus,

$$V_{R_1} = I_X R_1 \qquad V_{R_2} = I_X R_2 \qquad V_{R_3} = I_X R_3$$

The results are now easily checked using KVL and KCL.

• • •

Example 6.18 Find a complete solution for the circuit in Fig. 6.25 for the following component values: $V_S = 10$ V, $R_1 = 3.3$ kΩ, $R_2 = 4.7$ kΩ, $R_3 = 8$ kΩ, $R_4 = 16$ kΩ, $R_5 = 2$ kΩ.

Solution Referring to Fig. 6.26,

$$R_X = R_1 + R_2 + R_3 = 3.3\text{ k}\Omega + 4.7\text{ k}\Omega + 8\text{ k}\Omega = 16\text{ k}\Omega$$

then,

$$R_X \| R_4 = 16\text{ k}\Omega \| 16\text{ k}\Omega = \frac{16\text{ k}\Omega}{2} = 8\text{ k}\Omega$$

The source current is then

$$I_T = \frac{V_S}{R_5 + (R_X \| R_4)} = \frac{10\text{ V}}{2\text{ k}\Omega + 8\text{ k}\Omega} = 1\text{ mA}$$

then,

$$V_{R_5} = I_T R_5 = 1\text{ mA} \times 2\text{ k}\Omega = 2.0\text{ V}$$

and

$$V_A = V_{R_4} = V_{R_X} = I_T(R_X \| R_4) = 1\text{ mA} \times 8\text{ k}\Omega = 8.0\text{ V}$$

Also

$$V_{R_4} = V_S - V_{R_5} = 10\text{ V} - 2.0\text{ V} = 8.0\text{ V}$$

The currents in Fig. 6.26*b* are,

$$I_{R_4} = \frac{V_{R_4}}{R_4} = \frac{8.0\text{ V}}{16\text{ k}\Omega} = 1/2\text{ m/A}$$

$$I_X = \frac{V_{R_X}}{R_X} = \frac{8.0\text{ V}}{16\text{ k}\Omega} = 1/2\text{ m/A}$$

The other resistor voltages are,

$$V_{R_1} = I_X R_1 = 1/2\text{ mA} \times 3.3\text{ k}\Omega = 1.65\text{ V}$$
$$V_{R_2} = I_X R_2 = 1/2\text{ mA} \times 4.7\text{ k}\Omega = 2.35\text{ V}$$
$$V_{R_3} = I_X R_3 = 1/2\text{ mA} \times 8\text{ k}\Omega = 4.0\text{ V}$$

The solution is easily checked using KVL, since

$$V_A = V_{R_1} + V_{R_2} + V_{R_3}, \qquad 8.0\text{ V} = 1.65\text{ V} + 2.35\text{ V} + 4.0\text{ V}$$

A second check is to apply KCL at node A in Fig. 6.25. Thus,

$$I_T = I_X + I_{R_4}, \qquad 1\text{ mA} = 1/2\text{ mA} + 1/2\text{ mA}$$

• • •

Another type of circuit often encountered is shown in Fig. 6.27. This circuit may appear more complicated at first glance, but after a little thought, it is seen that the voltage at node A is *always* equal to the source voltage V_S, and thus the resistors to the *right* of the source are all connected in parallel with all the resistors to the *left* of the source. This means that the circuit can actually be broken into two parts as shown in Fig. 6.28. This clearly demonstrates that the total source current I_T is equal to the sum of the two currents I_1 and I_2, and that these two components of current are independent of one another. The solution for this circuit will be demonstrated by an example.

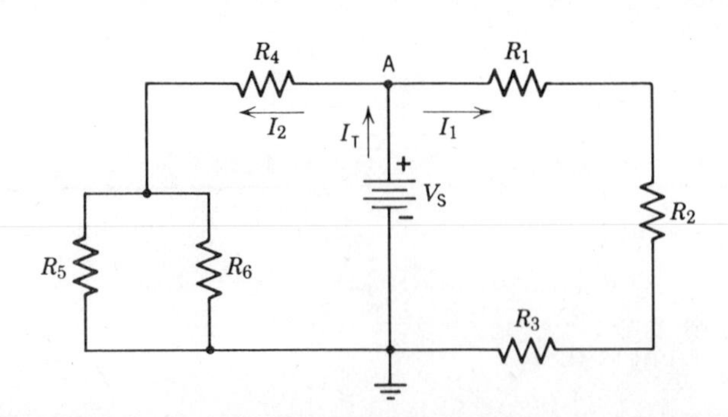

Figure 6.27

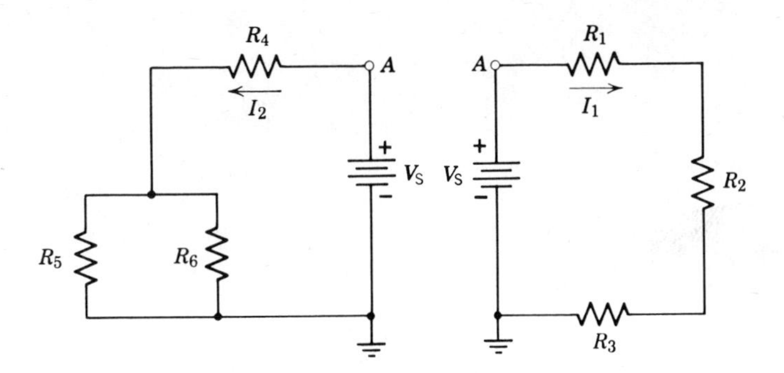

Figure 6.28

• • •

Example 6.19 Find a complete solution for the circuit in Fig. 6.27 for the following component values: $V_S = 100$ V, $R_1 = 47$ kΩ, $R_2 = 33$ kΩ, $R_3 = 20$ kΩ, $R_4 = 22$ kΩ, $R_5 = 12$ kΩ, $R_6 = 4$ kΩ.

Solution The circuit is first broken into two parts as shown in Fig. 6.28. The current I_1 is then,

$$I_1 = \frac{V_S}{R_1 + R_2 + R_3} = \frac{100\text{ V}}{47\text{ k}\Omega + 33\text{ k}\Omega + 20\text{ k}\Omega} = 1\text{ mA}$$

The voltages across these three series resistors are,

$$V_{R_1} = I_1R_1 = 1\text{ mA} \times 47\text{ k}\Omega = 47\text{ V}$$
$$V_{R_2} = I_1R_2 = 1\text{ mA} \times 33\text{ k}\Omega = 33\text{ V}$$
$$V_{R_3} = I_1R_3 = 1\text{ mA} \times 20\text{ k}\Omega = 20\text{ V}$$

The solution is checked using KVL,

$$V_S = V_{R_1} + V_{R_2} + V_{R_3}, \qquad 100\text{ V} = 47\text{ V} + 33\text{ V} + 20\text{ V}$$

The current I_2 in Fig. 6.28 is determined by finding the total equivalent resistance. Thus,

$$I_2 = \frac{V_S}{R_4 + (R_5 \| R_6)} = \frac{100\text{ V}}{22\text{ k}\Omega + (12\text{ k}\Omega \| 4\text{ k}\Omega)} = \frac{100\text{ V}}{22\text{ k}\Omega + 3\text{ k}\Omega} = 4\text{ mA}$$

The voltage drop across R_4 is then

$$V_{R_4} = I_2R_4 = 4\text{ mA} \times 22\text{ k}\Omega = 88\text{ V}$$

then using KVL, the voltage across R_5 and R_6 is

$$V_{R_5} = V_{R_6} = V_S - V_{R_4} = 100\text{ V} - 88\text{ V} = 12\text{ V}$$

The currents through R_5 and R_6 are then found as,

$$I_{R_5} = \frac{V_{R_5}}{R_5} = \frac{12\text{ V}}{12\text{ k}\Omega} = 1\text{ mA}$$

$$I_{R_6} = \frac{V_{R_6}}{R_6} = \frac{12\text{ V}}{4\text{ k}\Omega} = 3\text{ mA}$$

A check of the resulting currents is immediately made by applying KCL at the node connecting R_4, R_5, and R_6. Thus,

$$I_2 = I_{R_5} + I_{R_6}, \qquad 4\text{ mA} = 1\text{ mA} + 3\text{ mA}$$

The application of KCL at node A yields the total source current I_T.

$$I_T = I_1 + I_2 = 1\text{ mA} + 4\text{ mA} = 5\text{ mA}$$

• • •

SUMMARY

Kirchhoff's Voltage Law (KVL) states that the algebraic sum of the source voltages must be equal to the algebraic sum of the voltage drops around a closed circuit. KVL provides a powerful method for analyzing a series circuit, or for checking calculations. Kirchhoff's Current Law (KCL) states that the algebraic sum of the currents entering a node must be equal to the algebraic sum of the currents leaving that node. KCL provides a powerful method for analyzing parallel circuits, or for checking calculations.

The total resistance of a number of resistances connected in series is simply the sum of the individual resistances. The equivalent resistance of a number of resistances connected in parallel can always be found by taking the product divided by the sum (taking two resistances at a time).

The total power delivered to a circuit by a source (or sources) is always equal to the sum of the powers dissipated in the individual resistors.

GLOSSARY

Equivalent resistance. *The total resistance which results when a number of resistors are combined in series or parallel.*
KCL. *Kirchhoff's Current Law*
KVL. *Kirchhoff's Voltage Law*
Parallel circuit. *Elements connected between two common points.*
Series circuit. *Elements connected end to end.*

IMPORTANT RELATIONSHIPS

- KVL $V_{S_1} + V_{S_2} + V_{S_3} + \cdots + V_{S_M} = V_{R_1} + V_{R_2} + V_{R_3} + \cdots + V_{R_N}$
- KCL $I_{in} = I_{out}$
- Resistances in series $R_T = R_1 + R_2 + \cdots + R_N$
- Resistances in parallel $R_T = \dfrac{1}{\dfrac{1}{R_1} + \dfrac{1}{R_2} + \cdots + \dfrac{1}{R_N}}$
- Power in a circuit (series or parallel):

$$\overbrace{P_{S_1} + P_{S_2} + \cdots + P_{S_M}}^{\text{Power delivered by sources}} = \overbrace{P_{R_1} + P_{R_2} + \cdots + P_{R_N}}^{\text{Power dissipated in resistances}}$$

EVIEW QUESTIONS

1. The equivalent resistance of resistors connected in series is found by taking the ____ of the individual resistances.
2. According to KVL, the algebraic sum of the voltage drops around a circuit must be equal to the algebraic sum of the source voltages. (TF)
3. Total resistance in a (series, parallel) circuit is found by taking the sum of all resistances in the circuit.
4. The electrical quantity which is common to all the elements in a series circuit is ____.
5. In applying KVL to a series circuit, a path is traced around the circuit and if the path travels through a voltage source from the negative to the positive terminal, the voltage of that source is entered in the KVL equation as a (positive, negative) quantity.
6. The power dissipated in the resistances in a series circuit could be greater or less than the total power delivered by the source. (TF)
7. The power dissipated in a resistor is proportional to the square of the current in the resistor. (TF)
8. If two equal resistors are connected in series with a source voltage, the voltage across either resistor is ____% of the source voltage.
9. The total resistance of two resistors connected in parallel is (larger, smaller) than either individual resistance.
10. The electrical quantity which is common to a number of elements connected in parallel is ____.
11. KCL states that the algebraic sum of the currents entering a node must be equal to the algebraic sum of the currents leaving that node. (TF)
12. If five 10-kΩ resistors are connected in parallel, the equivalent resistance is ____.
13. An easy way to find the equivalent resistance of a number of different resistors connected in parallel is to take them (two, three, four) at a time.
14. When resistors are connected in series, their resistances can be added to find the total resistance, but when they are connected in parallel, the quantities which must be added are ____.
15. The equation $I_T = I_{R_1} + I_{R_2} + I_{R_3}$ is KCL applied to a (series, parallel) circuit.
16. The total power dissipated in a number of resistors connected in parallel must be equal to the power delivered by the source. (TF)
17. A complete solution means finding the current through and the voltage across every element. (TF)

OBLEMS

1. Solve for the current in Fig. 6.1 if $V = 25$ V, $R_1 = 220\ \Omega$, $R_2 = 1.5\ \text{k}\Omega$, and $R_3 = 560\ \Omega$.
2. Solve for the current in Fig. 6.1 if $V = 150$ V, $R_1 = 15\ \text{k}\Omega$, $R_2 = 22\ \text{k}\Omega$, and $R_3 = 47\ \text{k}\Omega$.
3. Find R_T in Fig. 6.3 if $R_1 = 390\ \Omega$, $R_2 = 1.8\ \text{k}\Omega$, and $R_3 = 470\ \Omega$.
4. Find R_T in Fig. 6.3 if $R_1 = 560\ \Omega$, $R_2 = 22\ \text{k}\Omega$, and $R_3 = 39\ \text{k}\Omega$.
5. Find the current I in Fig. 6.29. Draw the circuit and show the current direction.

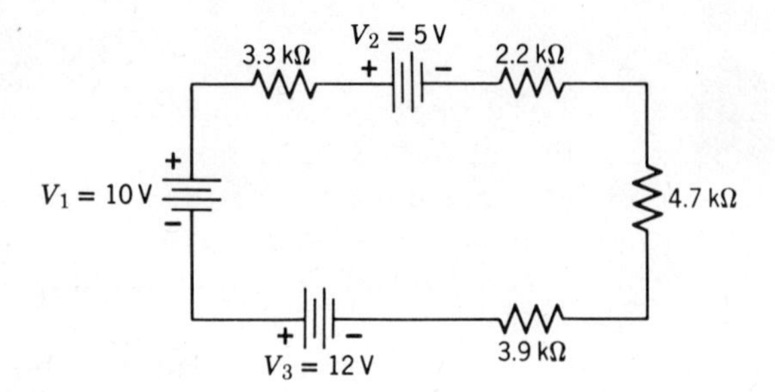

Figure 6.29

6. Find the current in Fig. 6.29 if V_2 is changed to 39 V. Show the current direction.
7. Find the current I in Fig. 6.30. What must be the EMF of the voltage source V?
8. Find the current I in Fig. 6.30 if the voltage across the 5.6 kΩ resistor is 23 V. What must be the EMF of the voltage source V?
9. Check Problem 1 using KVL.
10. Check Problem 2 using KVL.
11. Check Problem 5 using KVL.
12. Check Problem 6 using KVL.
13. Find the voltage across the 1.5 kΩ resistor in Fig. 6.31. Check using any method.
14. If the 4.7 kΩ resistor in Fig. 6.31 is *shorted*, find the voltage across the 1.5 kΩ resistor. Check using any method.
15. Find the power dissipated in each of the resistors in Problem 1. Check the results by finding the total power delivered by the source.

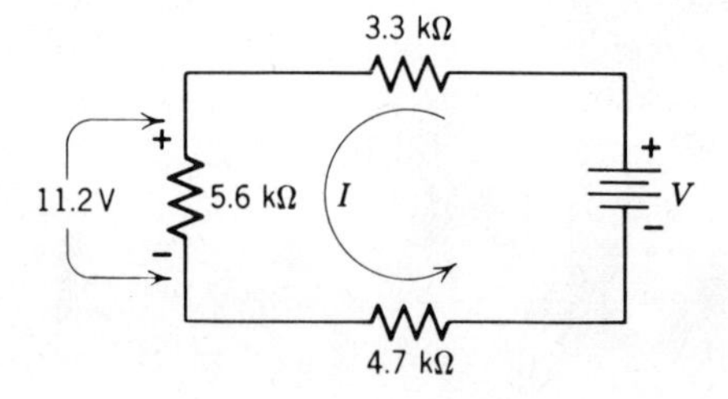

Figure 6.30

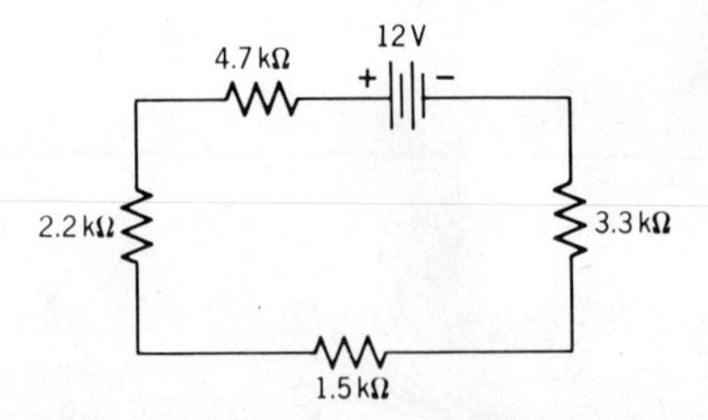

Figure 6.31

16. Find the power dissipated in each of the resistors in Problem 2. Check the results by finding the total power delivered by the source.
17. What must be the value of R_L for $V_T = V_S/2$ in Fig. 6.32 if $V_S = 26$ V and $R_S = 120\ \Omega$? What power is delivered to the load R_L under this condition?

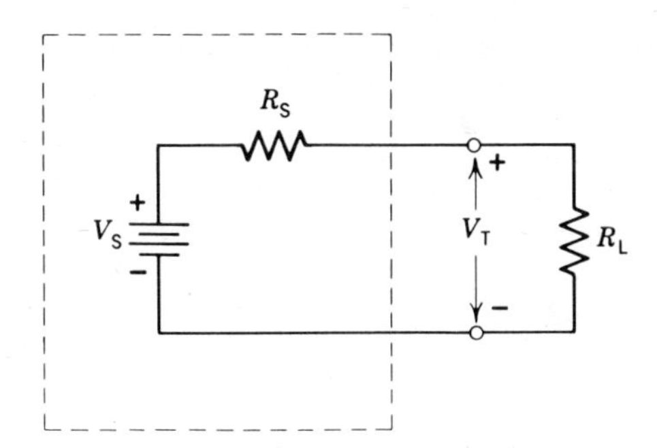

Figure 6.32

18. What is V_T in Fig. 6.32 if $V_S = 85$ V, $R_S = 220\ \Omega$, and $R_L = 430\ \Omega$? What power is dissipated in R_L and R_S?
19. What is the equivalent resistance of a 4.7 kΩ and a 2.2 kΩ resistor connected in parallel?
20. Find the equivalent resistance of four 22 kΩ resistors connected in parallel.
21. Draw an equivalent circuit of Fig. 6.12 if $R_1 = 15$ kΩ, $R_2 = 22$ kΩ, $R_3 = 47$ kΩ, and $V = 6$ V.
22. What resistor must be connected in parallel with a 10 kΩ resistor to form (*a*) 5 kΩ, (*b*) 1.8 kΩ, (*c*) 15 kΩ.
23. Find the current in each resistor in Problem 21. Find the source current using KCL. Check by any method.
24. Three 3.3 kΩ resistors are connected across a 22 V source. Find the current in each resistor and the total current supplied by the source.
25. Solve Problem 23 using conductances.
26. Find the power dissipated in each resistor in Problem 24.
27. Find the power dissipated in each resistor in Problem 21. Check by finding the total power delivered by the source.
28. What power is dissipated in each resistor in Fig. 6.33? What is the power delivered by the source? Check by any method.
29. What is the current in each resistor in Fig. 6.33? What is the source current? Check by any method.

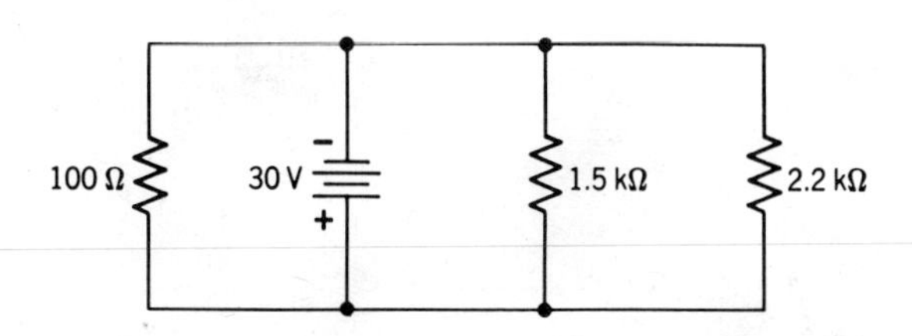

Figure 6.33

30. What voltage must be applied to the circuit in Fig. 6.20 in order to dissipate 150 mW of power in the 3 kΩ resistor? What power will be dissipated in the other resistors? What will be the source current and power?

31. Find a complete solution for the circuit in Fig. 6.21 when: $R_1 = 6.8\ \text{k}\Omega$, $R_2 = 3.9\ \text{k}\Omega$, $R_3 = R_4 = 10\ \text{k}\Omega$, $R_5 = 1\ \text{k}\Omega$, $R_6 = 2.2\ \text{k}\Omega$, $V_S = 22\ \text{V}$.

32. Find a complete solution for the circuit in Fig. 6.21 when: $R_1 = 12\ \text{k}\Omega$, $R_2 = 4\ \text{k}\Omega$, $R_3 = 15\ \text{k}\Omega$, $R_4 = 9.1\ \text{k}\Omega$, $R_5 = 470\ \text{k}\Omega$, $R_6 = 470\ \text{k}\Omega$, $V_S = 100\ \text{V}$.

33. Find a complete solution for the circuit in Fig. 6.34. Draw the circuit carefully. Check using KVL.

34. Find a complete solution for the circuit in Fig. 6.34 if V_{S_2} is changed to 6 V. Check using KVL.

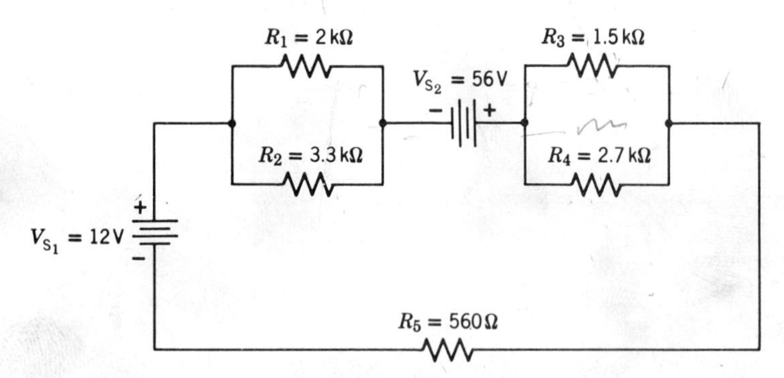

Figure 6.34

35. Find a complete solution for the circuit in Fig. 6.23 when: $R_1 = 1.5\ \text{k}\Omega$, $R_2 = 3.3\ \text{k}\Omega$, $R_3 = 2.2\ \text{k}\Omega$, $R_4 = 1.0\ \text{k}\Omega$, $R_5 = 3.6\ \text{k}\Omega$, $R_6 = 5.6\ \text{k}\Omega$, $V_S = 30\ \text{V}$.

36. Find a complete solution for the circuit in Fig. 6.23 when: $R_1 = 150\ \Omega$, $R_2 = 330\ \Omega$, $R_3 = 220\ \Omega$, $R_4 = 100\ \Omega$, $R_5 = 360\ \Omega$, $R_6 = 560\ \Omega$, $V_S = 3\ \text{V}$.

37. Find a complete solution for the circuit in Fig. 6.35. What is the voltage at the positive terminal of V_{S_2}?

38. Find a complete solution for the circuit in Fig. 6.35 if V_{S_2} is changed to 3.0 V.

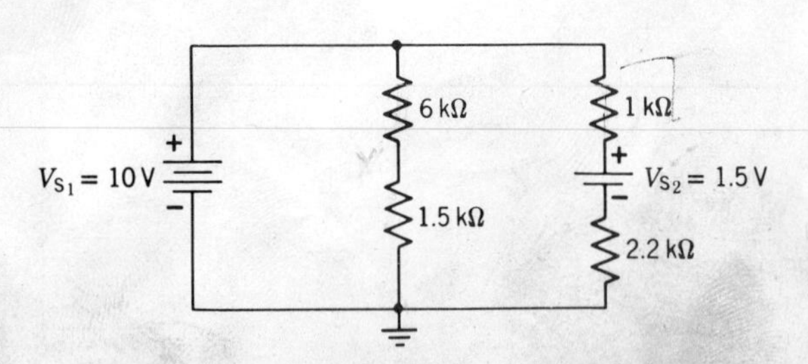

Figure 6.35

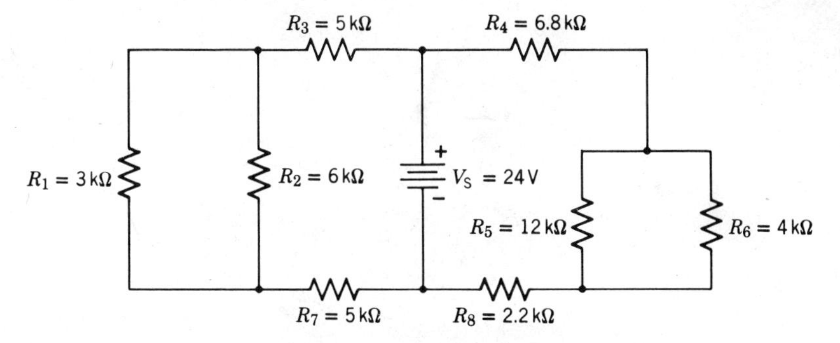

Figure 6.36

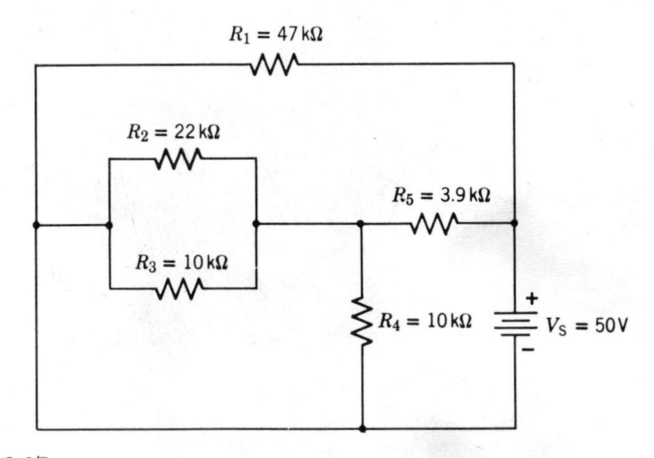

Figure 6.37

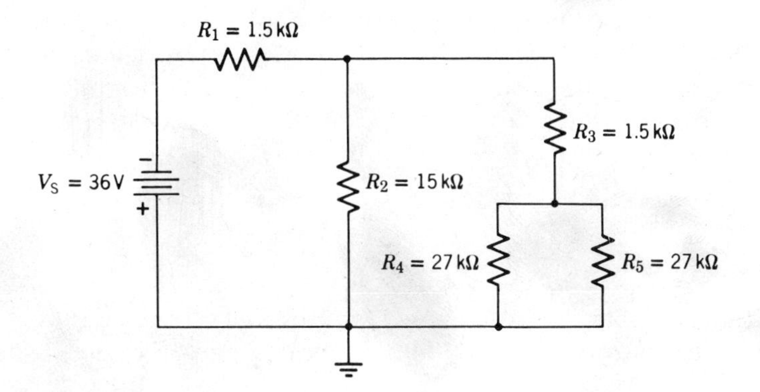

Figure 6.38

39. Find a complete solution for the circuit in Fig. 6.36.
40. Find a complete solution for the circuit in Fig. 6.36 if V_S is changed to 120 V.
41. Find a complete solution for the circuit in Fig. 6.37.
42. Find a complete solution for the circuit in Fig. 6.37 if R_1 is changed to 4.7 kΩ.
43. Find a complete solution for the circuit in Fig. 6.38.

USING KVL AND KCL

The application of Ohm's Law and the rules for combining resistances in series and parallel can always be used to find the element currents and voltages in a circuit. The labor required to obtain a solution can frequently be reduced by using the *voltage divider* theorem with series resistances, and the *current divider* theorem with parallel resistances. In addition, there is a special technique referred to as the *ladder method* which can be used to obtain solutions when dealing with ladder type circuits.

7.1 VOLTAGE DIVIDER THEOREM

Quite often it is desirable to find the voltage across one of the resistors in a series circuit such as shown in Fig. 7.1. This can be easily accomplished by first

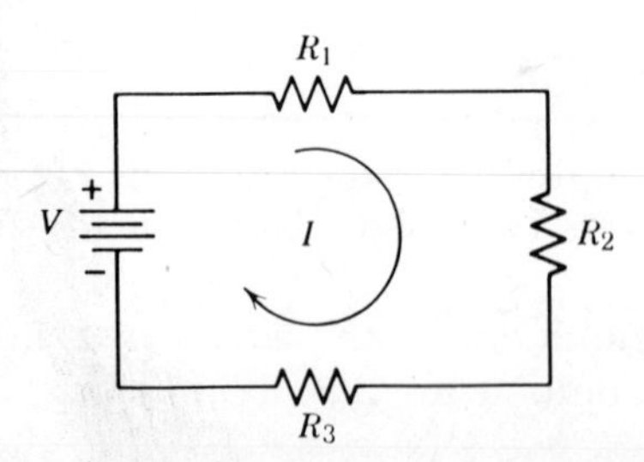

Figure 7.1 Series circuit

finding the circuit current I (combining all the series resistances to form a total resistance R_T), and then multiplying the current by the value of the resistance. The circuit current for Fig. 7.1 is given by

$$I = \frac{V}{R_1 + R_2 + R_3} = \frac{V}{R_T}$$

The voltage across each resistor is then given by

$$V_{R_1} = IR_1 = V\frac{R_1}{R_T}$$

$$V_{R_2} = IR_2 = V\frac{R_2}{R_T}$$

$$V_{R_3} = IR_3 = V\frac{R_3}{R_T}$$

After a little thought, one can recognize a pattern in these three resistor voltage equations, and it is seen that the voltage across any resistor is equal to the product of the applied voltage and the resistor in question, all divided by the total resistance. This is the *voltage divider theorem,* and it can be extended to a series circuit containing any number of resistors. In equation form, the voltage divider theorem is

$$V_X = V\frac{R_X}{R_T} \tag{7.1}$$

where V_X is the voltage across resistor R_X, V is the total applied EMF, and R_T is the total series circuit resistance.

• • •

Example 7.1 Apply the voltage divider theorem to find the voltage across each of the resistors in Fig. 7.2.

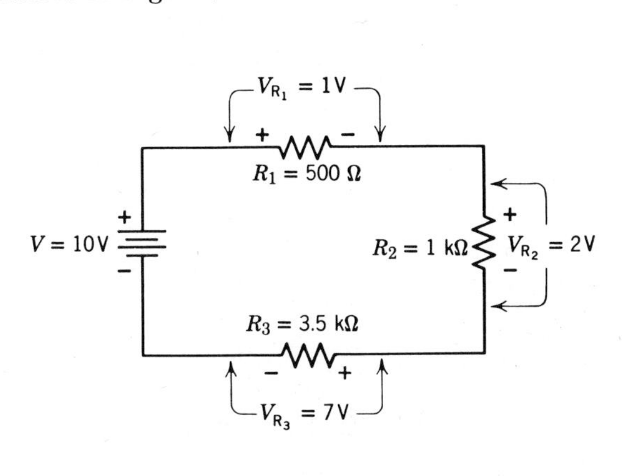

Figure 7.2

Solution Using the voltage divider theorem, Eq. 7.1:
across R_1

$$V_{R_1} = V\frac{R_1}{R_T} = 10\text{ V}\frac{500\ \Omega}{500\ \Omega + 1\text{ k}\Omega + 3.5\text{ k}\Omega} = 1\text{ V}$$

across R_2

$$V_{R_2} = V\frac{R_2}{R_T} = 10\text{ V}\frac{1\text{ k}\Omega}{5\text{ k}\Omega} = 2\text{ V}$$

across R_3

$$V_{R_3} = V\frac{R_3}{R_T} = 10\text{ V}\frac{3.5\text{ k}\Omega}{5\text{ k}\Omega} = 7\text{ V}$$

• • •

The voltage divider theorem can be applied to find the terminal voltage of a real voltage source (having internal resistance) for various values of load resistance.

• • •

Example 7.2 Find the terminal voltage V_T of the voltage source shown in Fig. 7.3.

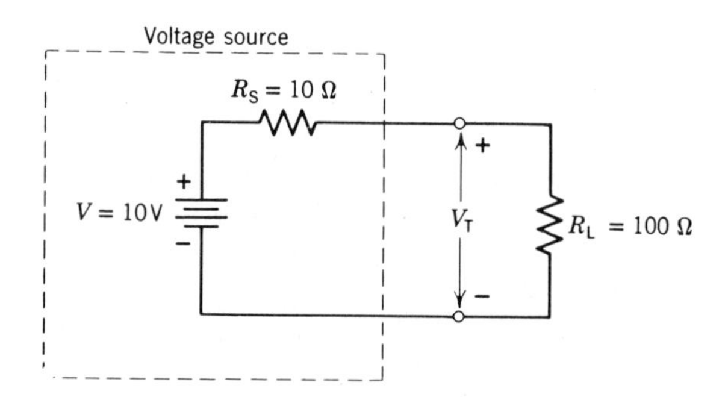

Figure 7.3

Solution The terminal voltage is easily found by a simple application of the voltage divider theorem. Thus

$$V_T = V\frac{R_L}{R_T} = V\frac{R_L}{R_S + R_L}$$

This equation will work for any value of R_L, but for the values given in the circuit,

$$V_T = 10\text{ V}\frac{100\ \Omega}{10\ \Omega + 100\ \Omega} = 9.09\text{ V}$$

• • •

7.2 USING THE VOLTAGE DIVIDER THEOREM

The voltage divider theorem can often be used to great advantage in solving series-parallel circuits. Since the voltage divider theorem is used with circuits having resistances in series, we should consider using it whenever we encounter a circuit having resistors connected in series (either the original circuit or a simplified equivalent circuit).

Consider the application of the voltage divider theorem to the circuit shown in Fig. 7.4*a*. The simplified equivalent circuit in Fig. 7.4*b* can be easily formed by combining the two parallel resistors R_2 and R_3. Since this equivalent circuit consists of three series resistors, the voltage divider theorem can be used to

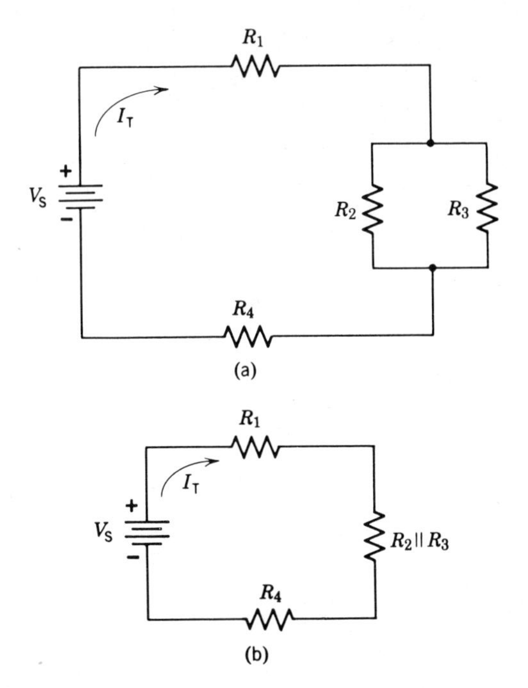

Figure 7.4 (*a*) Original circuit (*b*) Equivalent circuit

find the voltages across each of the resistors. Thus,

$$V_{R_1} = V_S \frac{R_1}{R_1 + (R_2 \| R_3) + R_4} = V_S \frac{R_1}{R_T}$$

$$V_{R_2} = V_{R_3} = V_S \frac{(R_2 \| R_3)}{R_1 + (R_2 \| R_3) + R_4} = V_S \frac{(R_2 \| R_3)}{R_T}$$

$$V_{R_4} = V_S \frac{R_4}{R_1 + (R_2 \| R_3) + R_4} = V_S \frac{R_4}{R_T}$$

The total series resistance is

$$R_T = R_1 + (R_2 \| R_3) + R_4.$$

The currents through R_2 and R_3 are now found as

$$I_{R_2} = \frac{V_{R_2}}{R_2} \qquad I_{R_3} = \frac{V_{R_3}}{R_3}$$

The total current from the source I_T can be found as

$$I_T = \frac{V_{R_1}}{R_1} = \frac{V_{R_4}}{R_4} = \frac{V_{R_2}}{(R_2 \| R_3)} = \frac{V_{R_3}}{(R_2 \| R_3)} = I_{R_1} + I_{R_2}$$

• • •

Example 7.3 Find a complete solution for the circuit in Fig. 7.4 if all the resistors are 10 kΩ, and $V_S = 50$ V.

Solution The total series circuit resistance R_T is

$$R_T = R_1 + (R_2 \| R_3) + R_4 = 10\ \text{k}\Omega + (10\ \text{k}\Omega \| 10\ \text{k}\Omega) + 10\ \text{k}\Omega = 25\ \text{k}\Omega$$

The resistor voltages are then

$$V_{R_1} = V_S \frac{R_1}{R_T} = 50\text{ V} \times \frac{10\text{ k}\Omega}{25\text{ k}\Omega} = 20\text{ V}$$

$$V_{R_2} = V_{R_3} = V_S \frac{(R_2\|R_3)}{R_T} = 50\text{ V} \times \frac{5\text{ k}\Omega}{25\text{ k}\Omega} = 10\text{ V}$$

$$V_{R_4} = V_S \frac{R_4}{R_T} = 50\text{ V} \times \frac{10\text{ k}\Omega}{25\text{ k}\Omega} = 20\text{ V}$$

These results are immediately checked using KVL, and

$$V_S = V_{R_1} + V_{R_2} + V_{R_4}, \qquad 50\text{ V} = 20\text{ V} + 10\text{ V} + 20\text{ V}$$

The currents through R_2 and R_3 are calculated as

$$I_{R_2} = \frac{V_{R_2}}{R_2} = \frac{10\text{ V}}{10\text{ k}\Omega} = 1\text{ mA}$$

$$I_{R_3} = \frac{V_{R_3}}{R_3} = \frac{10\text{ V}}{10\text{ k}\Omega} = 1\text{ mA}$$

That $I_{R_2} = I_{R_3}$ should be immediately obvious without calculation since $R_2 = R_3$. The total current from the source can be found in a number of ways;

$$I_T = \frac{V_{R_1}}{R_1} = \frac{20\text{ V}}{10\text{ k}\Omega} = 2\text{ mA}$$

$$I_T = \frac{V_{R_4}}{R_4} = \frac{20\text{ V}}{10\text{ k}\Omega} = 2\text{ mA}$$

$$I_T = I_{R_1} + I_{R_2} = 1\text{ mA} + 1\text{ mA} = 2\text{ mA}$$

$$I_T = \frac{V_{R_2}}{R_2\|R_3} = \frac{V_{R_3}}{R_2\|R_3} = \frac{10\text{ V}}{5\text{ k}\Omega} = 2\text{ mA}$$

• • •

The use of the voltage divider theorem in obtaining other than a complete solution should be very clear. For example, if a complete solution to the circuit in Fig. 7.4 is not desired but, instead, it is only desired to find the voltage across one of the resistors, a simple application of the voltage divider theorem will provide the solution.

7.3 CURRENT DIVIDER THEOREM

We may at times encounter a circuit such as that shown in Fig. 7.5 where the input current I_T and the resistances are known, and it is desired to find the individual resistor currents. Since the resistors are connected in parallel, we can assume a voltage V across them. If the two resistors are combined to form

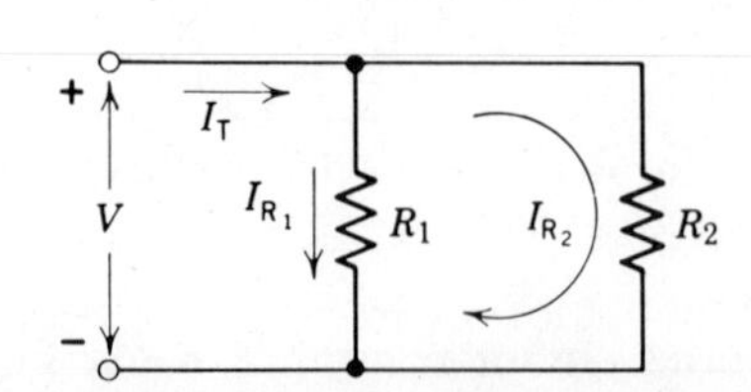

Figure 7.5

an equivalent resistance R_T, Ohm's Law must be satisfied for the circuit, and thus

$$V = I_T R_T = I_T \frac{R_1 R_2}{R_1 + R_2}$$

The individual resistor currents are also found using Ohm's Law and they are

$$I_{R_1} = \frac{V}{R_1}; \qquad I_{R_2} = \frac{V}{R_2}$$

If we substitute the equation above for the voltage V into these two relationships, the result is

$$I_{R_1} = \frac{V}{R_1} = \frac{I_T \dfrac{R_1 R_2}{R_1 + R_2}}{R_1} = I_T \frac{R_2}{R_1 + R_2}$$

$$I_{R_2} = \frac{V}{R_2} = \frac{I_T \dfrac{R_1 R_2}{R_1 + R_2}}{R_2} = I_T \frac{R_1}{R_1 + R_2}$$

There is a very definite pattern in these two equations (similar to the voltage divider equations), and after careful examination we can see that the current through either resistor is equal to the incoming current multiplied by the *opposite* resistance and divided by the sum of the two resistances. This is known as the *current divider theorem*, and can be stated for two resistors in parallel as,

the current through one of two resistances connected in parallel is equal to the product of the incoming current and the opposite resistance, all divided by the sum of the resistances.

The current divider theorem can be stated in equation form as

$$I_{R_X} = I_T \frac{R_{OPP}}{R_{OPP} + R_X} \tag{7.2}$$

• • •

Example 7.4 Find I_{R_1} and I_{R_2} in Fig. 7.5 when $R_1 = 6\ \text{k}\Omega$, $R_2 = 14\ \text{k}\Omega$, and $I_T = 10$ mA. What is the voltage V across the resistors?

Solution Using the current divider theorem Eq. 7.2,

$$I_{R_1} = I_T \frac{R_2}{R_1 + R_2} = 10\ \text{mA} \times \frac{14\ \text{k}\Omega}{6\ \text{k}\Omega + 14\ \text{k}\Omega} = 7\ \text{mA}$$

$$I_{R_2} = I_T \frac{R_1}{R_1 + R_2} = 10\ \text{mA} \times \frac{6\ \text{k}\Omega}{6\ \text{k}\Omega + 14\ \text{k}\Omega} = 3\ \text{mA}$$

The validity of the solution can be checked by applying KCL at the top node which yields

$$I_T = I_{R_1} + I_{R_2}$$
$$10\ \text{mA} = 7\ \text{mA} + 3\ \text{mA}$$

The voltage across the resistors can be found from

$$\begin{aligned} V &= I_{R_1} R_1 = I_{R_2} R_2 \\ &= 7\ \text{mA} \times 6\ \text{k}\Omega = 42\ \text{V} \\ &= 3\ \text{mA} \times 14\ \text{k}\Omega = 42\ \text{V} \end{aligned}$$

• • •

7.4 USING THE CURRENT DIVIDER THEOREM

The current divider theorem can also at times be used to great advantage in solving series-parallel circuits. Since this theorem is used with parallel resistances when the total current is known, we should consider its use whenever we encounter parallel resistors.

Consider the application of the current divider theorem to help in solving the circuit shown in Fig. 7.6*a*. The circuit is first reduced to the simple series circuit shown in Fig. 7.6*b* by combining parallel resistors. The total source current I_T is then found as

$$I_T = \frac{V_S}{R_T} = \frac{V_S}{(R_1 \| R_2) + (R_3 \| R_4)}$$

The individual resistor currents can then be found using the current divider theorem. Thus,

$$I_{R_1} = I_T \frac{R_2}{R_1 + R_2} \qquad I_{R_2} = I_T \frac{R_1}{R_1 + R_2}$$

$$I_{R_3} = I_T \frac{R_4}{R_3 + R_4} \qquad I_{R_4} = I_T \frac{R_3}{R_3 + R_4}$$

The resistor voltage drops are then found using Ohm's Law.

$$V_{R_1} = I_{R_1} R_1 \qquad V_{R_2} = I_{R_2} R_2$$

$$V_{R_3} = I_{R_3} R_3 \qquad V_{R_4} = I_{R_4} R_4$$

Note that these voltages could also have been found by applying the voltage

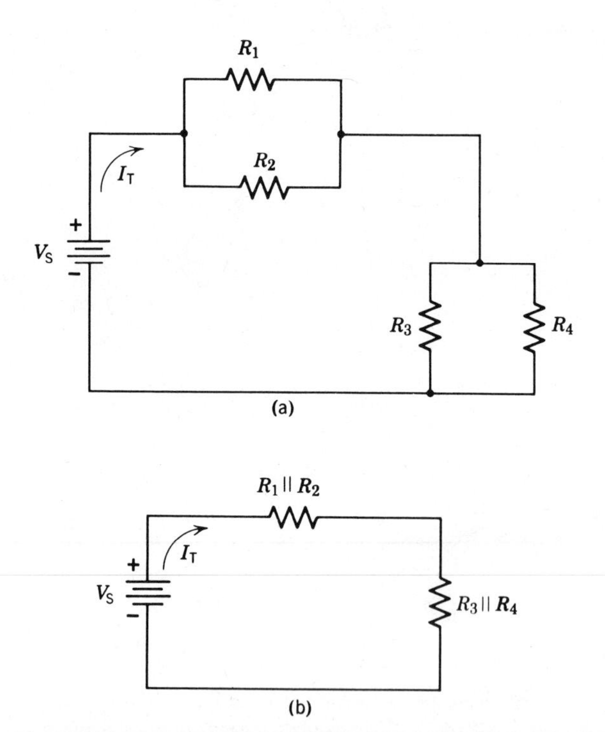

Figure 7.6 (*a*) Original circuit (*b*) Equivalent circuit

divider theorem to the equivalent circuit in Fig. 7.6*b*. This technique is now demonstrated with an example.

• • •

Example 7.5 Find a complete solution for the circuit in Fig. 7.6*a* making use of the current divider theorem. $V_S = 35$ V, $R_1 = R_2 = 10$ kΩ, $R_3 = 3$ kΩ, $R_4 = 6$ kΩ.

Solution The total circuit resistance as shown in Fig. 7.6*b* is found by combining the parallel resistors. Thus,

$$R_T = (R_1 \| R_2) + (R_3 \| R_4) = \frac{10\text{ k}\Omega}{2} + \frac{6\text{ k}\Omega \times 3\text{ k}\Omega}{6\text{ k}\Omega + 3\text{ k}\Omega} = 5\text{ k}\Omega + 2\text{ k}\Omega = 7\text{ k}\Omega$$

The source current is then seen to be,

$$I_T = \frac{V_S}{R_T} = \frac{35\text{ V}}{7\text{ k}\Omega} = 5\text{ mA}$$

The individual resistor currents are then found as

$$I_{R_1} = I_T \frac{R_2}{R_1 + R_2} = 5\text{ mA} \times \frac{10\text{ k}\Omega}{10\text{ k}\Omega + 10\text{ k}\Omega} = 2.5\text{ mA}$$

$$I_{R_2} = I_T \frac{R_1}{R_1 + R_2} = 5\text{ mA} \times \frac{10\text{ k}\Omega}{10\text{ k}\Omega + 10\text{ k}\Omega} = 2.5\text{ mA}$$

$$I_{R_3} = I_T \frac{R_4}{R_3 + R_4} = 5\text{ mA} \times \frac{6\text{ k}\Omega}{3\text{ k}\Omega + 6\text{ k}\Omega} = 10/3\text{ mA}$$

$$I_{R_4} = I_T \frac{R_3}{R_3 + R_4} = 5\text{ mA} \times \frac{3\text{ k}\Omega}{3\text{ k}\Omega + 6\text{ k}\Omega} = 5/3\text{ mA}$$

The resistor voltage drops are then found as

$$V_{R_1} = I_{R_1} R_1 = 2.5\text{ mA} \times 10\text{ k}\Omega = 25\text{ V}$$
$$V_{R_2} = V_{R_1} \text{ (since they are in parallel)} = 25\text{ V}$$

As a check,

$$V_{R_2} = I_{R_2} R_2 = 2.5\text{ mA} \times 10\text{ k}\Omega = 25\text{ V}$$
$$V_{R_3} = I_{R_3} R_3 = 10/3\text{ mA} \times 3\text{ k}\Omega = 10\text{ V}$$
$$V_{R_4} = V_{R_3} \text{ (since they are in parallel)} = 10\text{ V}$$

As a check,

$$V_{R_4} = I_{R_4} R_4 = \frac{5}{3}\text{ mA} \times 6\text{ k}\Omega = 10\text{ V}$$

• • •

The current divider theorem will be found more useful in circuits containing *constant current sources* (the dual of a constant voltage source, which we will discuss in the next chapter). Constant current sources are used in equivalent circuits of vacuum tubes and transistors as well as other active circuits.

THE LADDER METHOD

There are a number of circuits which can be solved by a special technique called the *ladder method.* The circuits to which this method can be easily

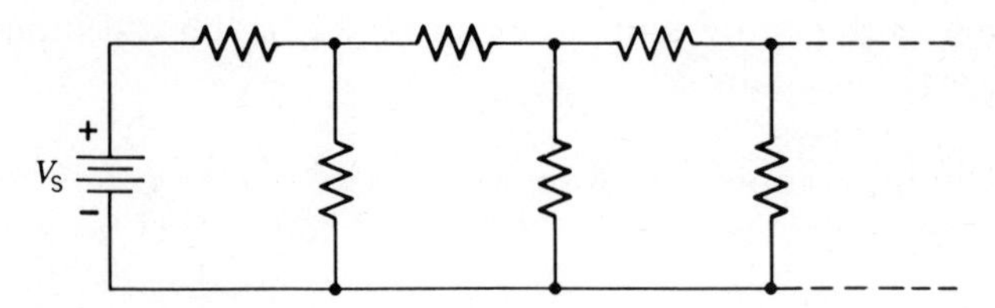

Figure 7.7 A ladder circuit

applied are sometimes called *ladder circuits* since their construction resembles a ladder as shown in Fig. 7.7. The technique here is to begin with the element *furthest* from the source, and *assume* a value for the current through or the voltage across that element. The other currents and voltages are then expressed in terms of the assumed value, until an expression containing the source voltage is obtained. If the source voltage is known, the assumed current or voltage can then be calculated.

Consider the application of the ladder method to the circuit in Fig. 7.8. The

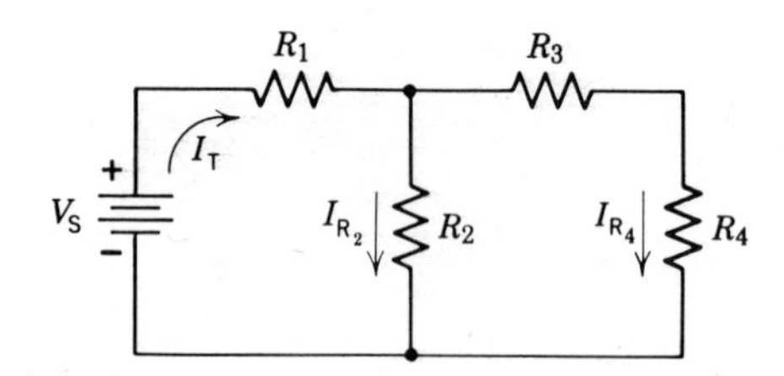

Figure 7.8

first step is to *assume* a value for the current I_{R_4}. We then begin to work our way back through the circuit assuming that we know I_{R_4}. Thus,

$$V_{R_4} = I_{R_4}R_4$$

$$V_{R_3} = I_{R_4}R_3$$

and

$$V_{R_2} = V_{R_3} + V_{R_4}$$

Then,

$$I_{R_2} = \frac{V_{R_2}}{R_2}$$

and

$$I_T = I_{R_2} + I_{R_4}$$

Now,

$$V_{R_1} = I_T R_1$$

and

$$V_S = V_{R_1} + V_{R_2}$$

If all the resistances in the circuit are known, this final expression for V_S will have only I_{R_4} in it as an unknown, and we can then solve for I_{R_4}. A specific example will best illustrate the method.

• • •

Example 7.6 Solve the circuit in Fig. 7.8 using the ladder method if, $V_S = 30$ V, $R_1 = 6$ kΩ, $R_2 = 8$ kΩ, $R_3 = 3.3$ kΩ, $R_4 = 4.7$ kΩ.

Solution The first step is to assume a value for I_{R_4}. Then

$$V_{R_4} = I_{R_4}R_4 = I_{R_4} \times 4.7\ \text{k}\Omega$$

and,

$$V_{R_3} = I_{R_4}R_3 = I_{R_4} \times 3.3\ \text{k}\Omega$$

Now,

$$V_{R_2} = V_{R_3} + V_{R_4} = 3.3\ \text{k}\Omega\ I_{R_4} + 4.7\ \text{k}\Omega\ I_{R_4} = 8.0\ \text{k}\Omega\ I_{R_4}$$

Then,

$$I_{R_2} = \frac{V_{R_2}}{R_2} = \frac{8.0\ \text{k}\Omega\ I_{R_4}}{8\ \text{k}\Omega} = I_{R_4}$$

and

$$I_T = I_{R_2} + I_{R_4} = I_{R_4} + I_{R_4} = 2\ I_{R_4}$$

Now,

$$V_{R_1} = I_T R_1 = 2\ I_{R_4} \times 6\ \text{k}\Omega = 12\ \text{k}\Omega\ I_{R_4}$$

and

$$V_S = V_{R_1} + V_{R_2} = 12\ \text{k}\Omega\ I_{R_4} + 8.0\ \text{k}\Omega\ I_{R_4} = 20\ \text{k}\Omega\ I_{R_4}$$

But, $V_S = 30$ V, and thus

$$30\ \text{V} = 20\ \text{k}\Omega\ I_{R_4}$$

So,

$$I_{R_4} = \frac{30\ \text{V}}{20\ \text{k}\Omega} = 1.5\ \text{mA}$$

Now, having found I_{R_4}, we can return to the original equations to find all of the other circuit currents and voltages.

• • •

Even though the ladder method may seem lengthy, the ideas are simple (KCL and KVL) and the numerical calculations are easy, all of which leads to a relatively simple solution. The ladder method is again quite valuable when a complete solution is not desired.

One final example will be used to illustrate the method.

• • •

Example 7.7 Use the ladder method to find the current I_{R_4} in Fig. 7.9.

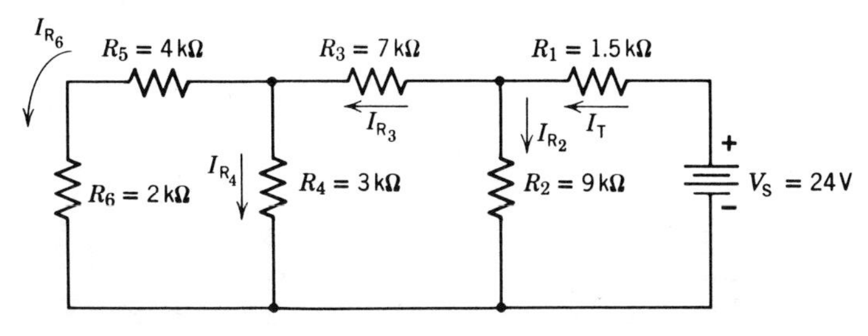

Figure 7.9

Solution We begin by assuming a current I_{R_6} through R_6. We then work our way back through the circuit towards V_S. Thus,

$$V_{R_4} = I_{R_6}(R_6 + R_5) = I_{R_6}(2\ \mathrm{k\Omega} + 4\ \mathrm{k\Omega}) = 6\ \mathrm{k\Omega}\ I_{R_6}$$

$$I_{R_4} = \frac{V_{R_4}}{R_4} = \frac{6\ \mathrm{k\Omega}\ I_{R_6}}{3\ \mathrm{k\Omega}} = 2\ I_{R_6}$$

$$I_{R_3} = I_{R_6} + I_{R_4} = I_{R_6} + 2\ I_{R_6} = 3\ I_{R_6}$$

$$V_{R_3} = I_{R_3}R_3 = 3\ I_{R_6} \times 7\ \mathrm{k\Omega} = 21\ \mathrm{k\Omega}\ I_{R_6}$$

$$V_{R_2} = V_{R_4} + V_{R_3} = 6\ \mathrm{k\Omega}\ I_{R_6} + 21\ \mathrm{k\Omega}\ I_{R_6} = 27\ \mathrm{k\Omega}\ I_{R_6}$$

$$I_{R_2} = \frac{V_{R_2}}{R_2} = \frac{27\ \mathrm{k\Omega}\ I_{R_6}}{9\ \mathrm{k\Omega}} = 3\ I_{R_6}$$

$$I_T = I_{R_2} + I_{R_3} = 3\ I_{R_6} + 3\ I_{R_6} = 6\ I_{R_6}$$

$$V_{R_1} = I_T R_1 = 6\ I_{R_6} \times 1.5\ \mathrm{k\Omega} = 9\ \mathrm{k\Omega}\ I_{R_6}$$

$$V_S = V_{R_2} + V_{R_1} = 27\ \mathrm{k\Omega}\ I_{R_6} + 9\ \mathrm{k\Omega}\ I_{R_6} = 36\ \mathrm{k\Omega}\ I_{R_6}$$

But $V_S = 24$ V; thus,

$$24\ \mathrm{V} = 36\ \mathrm{k\Omega}\ I_{R_6}$$

and

$$I_{R_6} = \frac{24\ \mathrm{V}}{36\ \mathrm{k\Omega}} = 2/3\ \mathrm{mA}$$

Now, returning to the second equation,

$$I_{R_4} = 2\ I_{R_6} = 2 \times 2/3\ \mathrm{mA} = 4/3\ \mathrm{mA}$$

The ladder method is clearly seen to provide a powerful technique for finding a *specific* current or voltage in a series-parallel circuit.

7.6 SERIES-PARALLEL CIRCUIT PROBLEMS

All of the problems discussed so far in this chapter have been concerned with finding the currents and voltages in a circuit when the source voltage and the resistance values are given. Occasionally some of the voltages and currents are known in a circuit, and it is desired to find the existing resistance values. This simply requires the application of the same basic principles (finding equivalent resistances and applying KCL and KVL), although the order of application is not necessarily the same. The solution of these types of problems with ease comes only after some experience, and with this in mind, the remainder of this section will simply be devoted to the solution of a number of problems. The reader is urged to study the solutions carefully, and indeed to look for other methods of solution. There is almost always more than one way to solve a given problem, and the process of solving problems in a number of different ways is an extremely valuable learning technique.

• • •

Example 7.8 A battery charger having an open-circuit voltage of 12 V is used to recharge an old lead–acid battery having an open circuit voltage of 11 V. The internal resistance of the battery charger is known to be 0.1 Ω, and the internal resistance of the old battery is 0.4 Ω (assume the internal resistances are constant). What is the initial charging current? What must be the open circuit voltage of the old battery when the charging current has dropped to 100 mA?

Solution The first step in the solution is to draw a complete circuit as shown in Fig. 7.10. It is seen from the figure that the battery and charger form a simple series circuit. The equation for the charging current is then

$$I_{CH} = \frac{V_{S_1} - V_{S_2}}{R_{S_1} + R_{S_2}} = \frac{12 - V_{S_2}}{0.1\,\Omega + 0.4\,\Omega} = 2(12 - V_{S_2})\text{A}$$

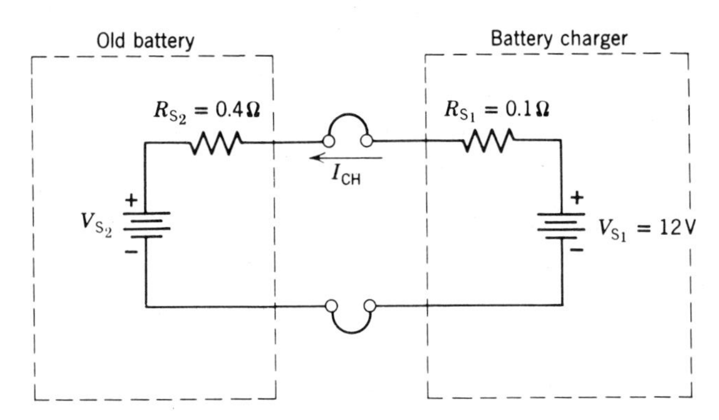

Figure 7.10

The initial voltage of the old battery V_{S_2} is 11 V, and thus the initial charging current is

$$I_{CH} = 2(12 - 11)\text{A} = 2\text{ A}$$

When the charging current has dropped to 100 mA, or 0.1 A, the old battery voltage can be found using the same equation. Thus

$$I_{CH} = 2(12 - V_{S_2}) = 24 - 2\,V_{S_2}$$

and

$$V_{S_2} = \frac{24 - I_{CH}}{2} = 12 - \frac{I_{CH}}{2} = 12 - \frac{(0.1\text{ A})}{2} = 12 - 0.05 = 11.95\text{ V}$$

• • •

Example 7.9 What is the value of the current I_T in Fig. 7.11*a*? What must be the resistance of R_3?

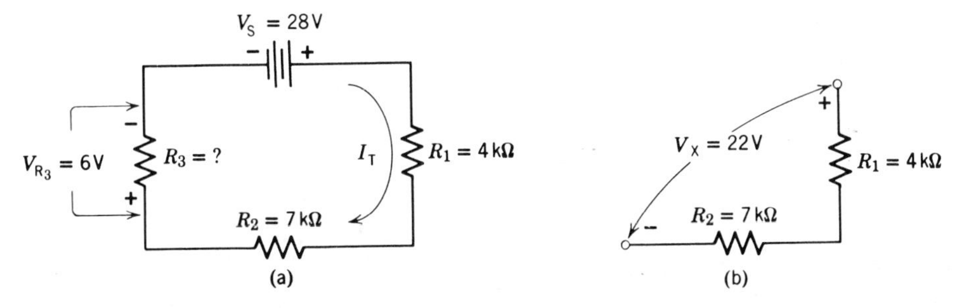

Figure 7.11 (*a*) Original circuit (*b*) Partial equivalent circuit

Solution The first step in solving this problem is to realize a partial equivalent circuit as shown in Fig. 7.11*b*; the voltage across these two resistors is found using KVL around the loop. Thus,

$$V_S = V_x + V_{R_3}$$

and

$$V_x = V_S - V_{R_3} = 28\text{ V} - 6\text{ V} = 22\text{ V}$$

From this equivalent circuit, we can easily calculate the circuit current I_T, and

$$I_T = \frac{V_x}{R_1 + R_2} = \frac{22\text{ V}}{4\text{ k}\Omega + 7\text{ k}\Omega} = 2\text{ mA}$$

We can now return to the original circuit and find R_3. Thus,

$$R_3 = \frac{V_{R_3}}{I_T} = \frac{6\text{ V}}{2\text{ mA}} = 3\text{ k}\Omega$$

• • •

Example 7.10 Find the source current I_T and the resistance of R_5 in Fig. 7.12.

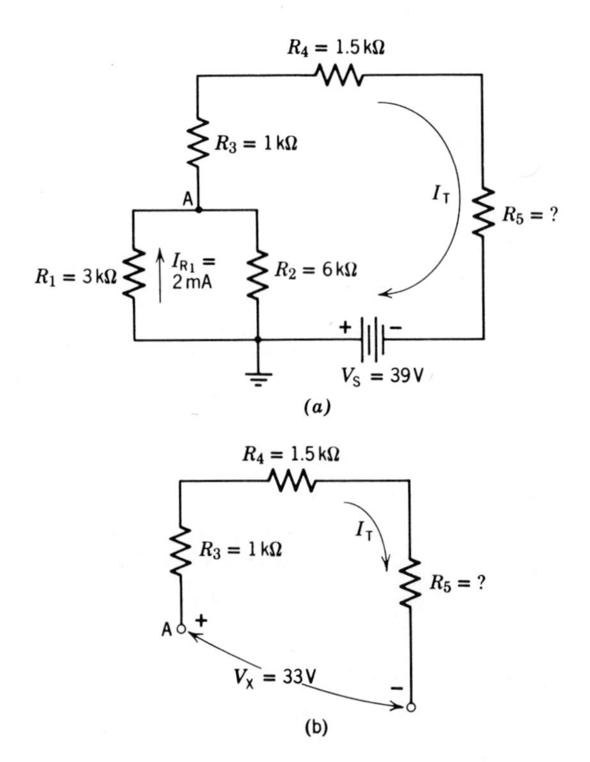

Figure 7.12 (*a*) Original circuit (*b*) Partial equivalent circuit

Solution Since the current I_{R_1} and the resistance of R_1 are both known, the voltage across R_1 is easily found as

$$V_{R_1} = I_{R_1} R_1 = 2\text{ mA} \times 3\text{ k}\Omega = 6\text{ V}$$

The current through R_2 is then found as,

$$I_{R_2} = \frac{V_{R_1}}{R_2} = \frac{6\text{ V}}{6\text{ k}\Omega} = 1\text{ mA}$$

Applying KCL at node A then yields the source current I_T.

$$I_T = I_{R_1} + I_{R_2} = 2\ \text{mA} + 1\ \text{mA} = 3\ \text{mA}$$

We can now apply KVL to form a partial equivalent circuit as shown in Fig. 7.12*b*. Since

$$V_S = V_{R_1} + V_x$$

Then,

$$V_x = V_S - V_{R_1} = 39\ \text{V} - 6\ \text{V} = 33\ \text{V}$$

The total series resistance of this circuit is then,

$$R_T = R_3 + R_4 + R_5 = \frac{V_x}{I_T} = \frac{33\ \text{V}}{3\ \text{mA}} = 11\ \text{k}\Omega$$

Thus,

$$R_5 = R_T - R_3 - R_4 = 11\ \text{k}\Omega - 1\ \text{k}\Omega - 1.5\ \text{k}\Omega = 8.5\ \text{k}\Omega$$

• • •

SUMMARY

In this chapter we have seen that many circuits can be considered as combinations of resistors connected in series and parallel, and these circuits can be solved by using the rules for combining series and parallel resistors to form simplified equivalent circuits. The current and voltage divider theorems along with KCL and KVL form the basis for finding a complete solution for these series-parallel circuits. Another method which is useful in finding a specific current or voltage in a network is the ladder method. KVL and KCL provide a quick and convenient means for checking any circuit solution.

GLOSSARY

Complete solution. *Finding the voltage across and the current through every element in a circuit.*

Ladder network. *A particular form of network whose structure resembles a ladder.*

Series-parallel circuit. *A circuit composed of elements connected in series and parallel combinations.*

IMPORTANT RELATIONSHIPS

- Combining series resistances $R_T = R_1 + R_2 + \cdots R_N$.
- Combining parallel resistances $R_T = \dfrac{1}{\dfrac{1}{R_1} + \dfrac{1}{R_2} + \cdots \dfrac{1}{R_N}}$
- KVL, KCL.
- Voltage divider theorem $V_X = V\dfrac{R_X}{R_T}$.
- Current divider theorem $I_X = I_T\dfrac{R_{OPP}}{R_X + R_{OPP}}$.

REVIEW QUESTIONS

1. The equation $V_S = V_{R_1} = V_{R_2} = V_{R_3}$ is KVL applied to a (series, parallel) circuit.
2. The equation $I_T = I_{R_1} = I_{R_2} = I_{R_3}$ is KCL applied to a (series, parallel) circuit.
3. A complete solution means finding the current through and the voltage across every element in a circuit. (TF)
4. Two convenient methods for checking the solutions for a circuit are ____ and ____.
5. The voltage divider theorem is used with resistors in parallel. (TF)
6. The ____ divider theorem is useful in finding the currents in parallel resistances.
7. The ladder method is useful in finding a specific current or voltage in a circuit. (TF)
8. In using the ladder method, one begins by assuming a current or voltage for the element (nearest, furthest) from the source.

PROBLEMS

1. Use the voltage divider theorem to find the voltage across the 1.5 kΩ resistor in Fig. 7.13. Check using any method.

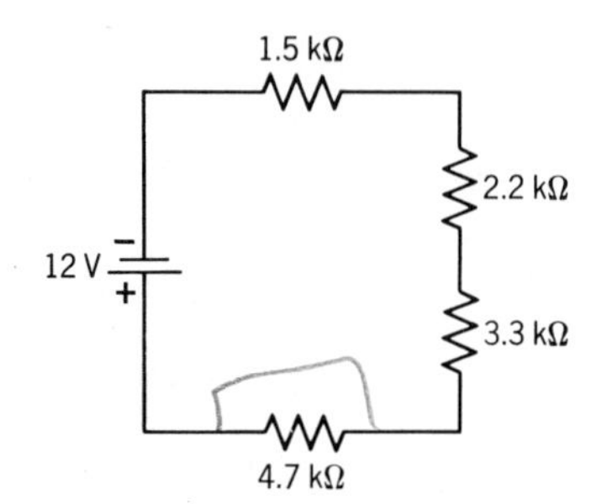

Figure 7.13

2. Use the voltage divider theorem to find the voltage across the 1.5 kΩ resistor in Fig. 7.13 if the 4.7 kΩ resistor is shorted. Check using any method.
3. What current would flow through a 1 kΩ resistor connected between nodes A and B in Fig. 7.14?

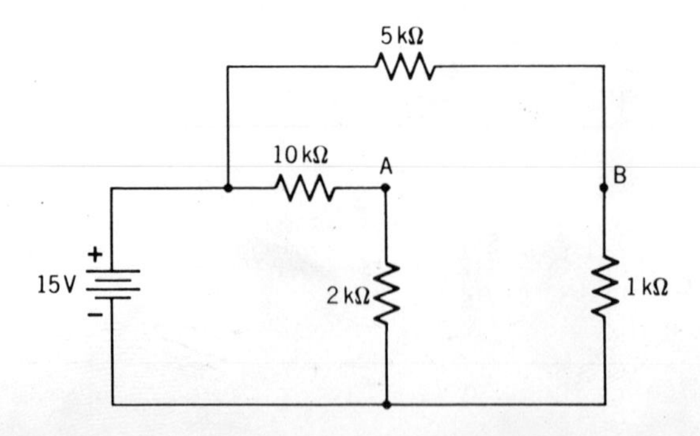

Figure 7.14

4. Find the voltage across R_4 in Fig. 7.15.

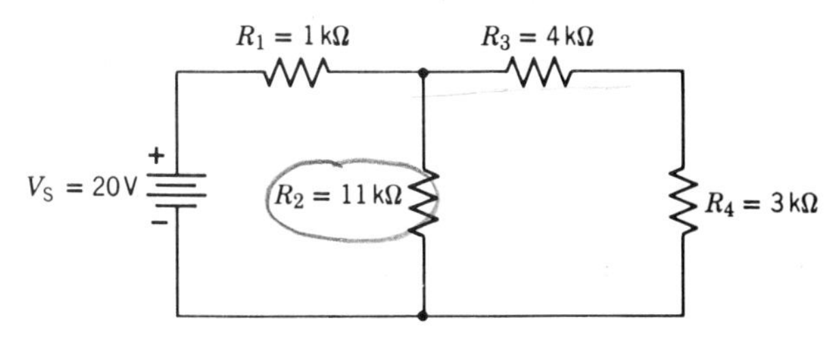

Figure 7.15

5. Find the voltage across R_4 in Fig. 7.15 if R_3 is shorted.
6. What is the current through R_2 in Problem 4?
7. What is the current through R_2 in Problem 5?
8. A voltage divider is shown in Fig. 7.16. How must the potentiometer be set to provide an output voltage of 30 V? (What is the resistance above and below the slider?)

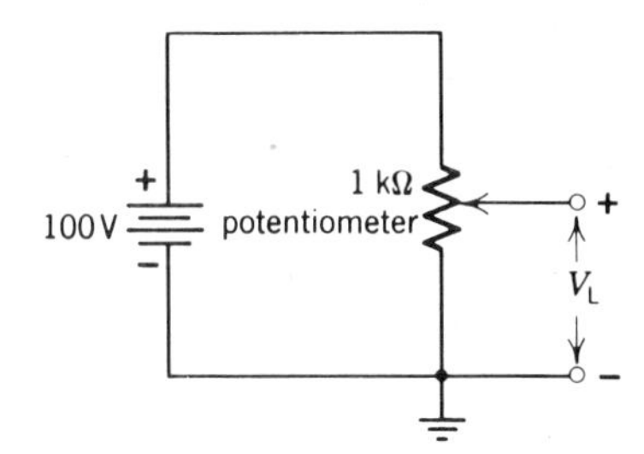

Figure 7.16

9. What will be the value of V_L in Problem 8 if a 300-Ω-load resistor is connected to the output? Where must the potentiometer be set in order to obtain 30 V across the load resistor?
10. A voltage divider to provide both positive and negative voltages is shown in Fig. 7.17. Calculate the output voltages V_A, V_B, V_C, and V_D.

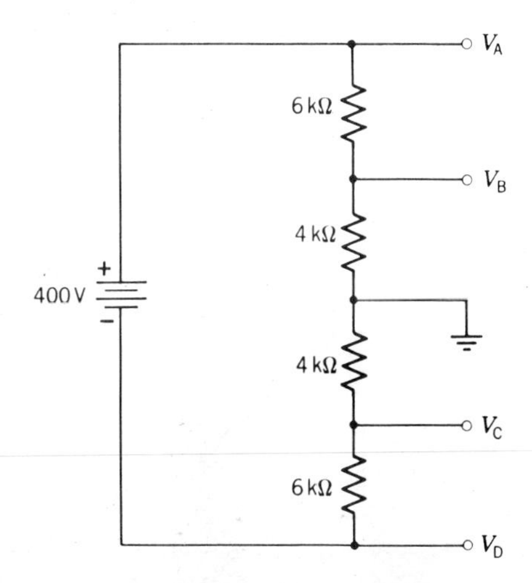

Figure 7.17

11. A 10 kΩ resistor is connected between V_B and ground in Fig. 7.17, all other taps are left unloaded. What is the voltage at V_B?

12. All of the resistors in the voltage divider in Fig. 7.17 are reduced by a factor of 10 to 600 Ω, 400 Ω, 400 Ω, and 600 Ω. What are the output voltages?

13. What is the voltage at V_B in Problem 12 when a 10 kΩ load resistor is connected from V_B to ground?

14. Use the current divider theorem to find I_{R_1} in Fig. 7.5 if $I_T = 200$ mA, $R_1 = 100\ \Omega$, and $R_2 = 200\ \Omega$. What must be the value of V across the resistors?

15. In Fig. 7.5, $R_1 = R_2 = 7.5$ kΩ. What must be the value of I_T if I_{R_1} is to be 15 mA?

16. Use the current divider theorem to find the value of I_{R_2} in Fig. 7.6*a* if $I_T = 15$ mA, $R_1 = 1$ kΩ, $R_2 = R_3 = 2.7$ kΩ, and $R_4 = 3.3$ kΩ.

17. What is the value of I_{R_3} for the circuit in Problem 16?

18. Use the current divider theorem to find the current in the 2 kΩ resistor in Fig. 7.14.

19. Find a complete solution for the circuit in Fig. 7.6*a* if $V_S = 35$ V, $R_1 = R_3 = 4.7$ kΩ, and $R_2 = R_4 = 3.3$ kΩ.

20. Use the results of Example 7.6 to find a complete solution for the circuit in Fig. 7.8. Check the results using KVL or KCL.

21. Use the ladder method to find I_{R_4} in Fig. 7.8 if the value of R_4 is changed to 10 kΩ.

22. Use the ladder method to find I_{R_6} in Fig. 7.9 if R_5 is shorted.

23. Use the results of Example 7.7 to find I_T in Fig. 7.9.

24. What must be the voltage across R_2 in Fig. 7.9? Use the results from Example 7.7 to obtain your answer.

25. Find a complete solution for the circuit in Fig. 7.9. Check results using KVL or KCL.

26. Find a complete solution for the circuit in Fig. 7.15. Check using KVL or KCL.

27. Find a complete solution for the circuit in Example 7.3 if the value of V_S is changed to 15 V.

28. Find a complete solution for the circuit in Fig. 7.12*a* if $R_5 = 4.7$ kΩ, and I_{R_1} is unknown.

8 NETWORK THEOREMS

There are four very important theorems that are widely used in solving electric circuits: Thévenin's theorem, Norton's theorem, the Superposition theorem, and Millman's theorem. They provide four very powerful techniques for finding complete solutions for electric circuits, and perhaps more importantly, they provide a means for finding a partial solution without obtaining a complete solution. Thus, these theorems can be used to find the voltage across, or the current through, a particular element in a circuit without having to find a complete solution—obviously saving time and labor!

8.1 IDEAL SOURCES

Before beginning a discussion of network theorems, we must expand our concepts of ideal sources to include *current sources* along with *voltage sources.* You will recall from our previous work that an ideal voltage source is one that has zero internal resistance and maintains a constant EMF no matter what current passes through the source. An ideal voltage source along with its V-I curve is shown in Fig. 8.1.

Since an ideal voltage source provides a constant voltage, an ideal current source must provide a constant current. An ideal current source along with its V-I curve is shown in Fig. 8.2. Notice that no matter what the terminal voltage is, the source provides a current which is maintained at a constant value I_S.

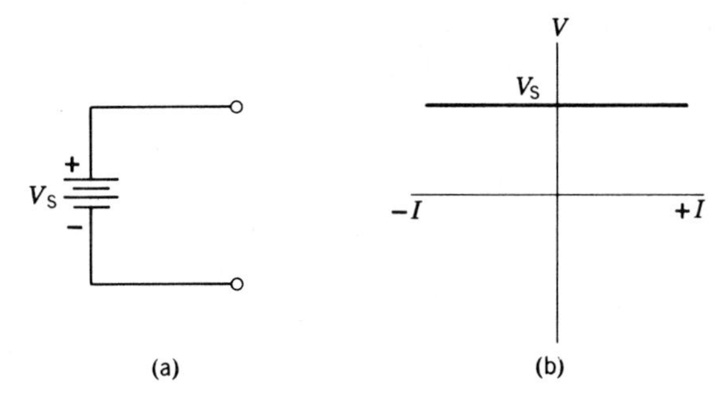

Figure 8.1 Ideal voltage source (*a*) Symbol (*b*) *V-I* curve

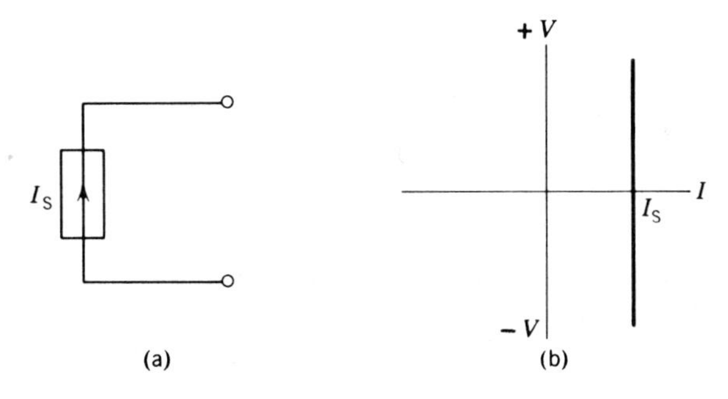

Figure 8.2 Ideal current source (*a*) Symbol (*b*) *V-I* curve

• • •

Example 8.1 A constant current source with a load resistor R_L is shown in Fig. 8.3*a*. Calculate the terminal voltage V_T of the current source as R_L is changed from 1 kΩ to 2 kΩ to 5 kΩ and finally to 10 kΩ. Plot the resulting *V-I* curve.

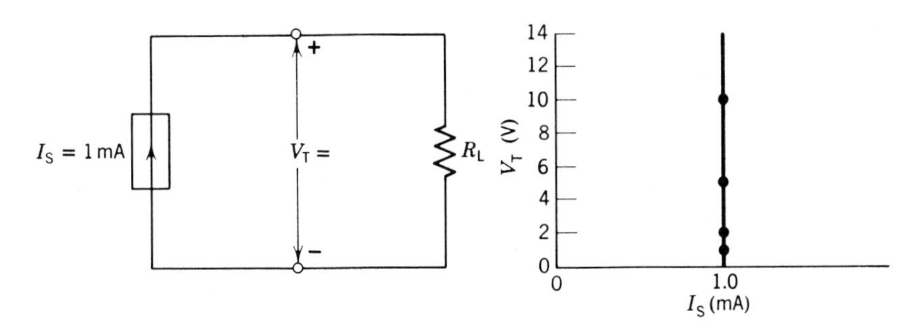

Figure 8.3 (*a*) Ideal current source with load R_L (*b*) *V-I* curve for (*a*)

Solution The current through the load resistor R_L is constant at $I_S = 1$ mA. The terminal voltage is then simply $V_T = I_S R_L$. Thus,

$$R_L = 1\ \text{k}\Omega;\quad V_T = I_S R_L = 1\ \text{mA} \times 1\ \text{k}\Omega = 1\ \text{V}$$
$$R_L = 2\ \text{k}\Omega;\quad V_T = I_S R_L = 1\ \text{mA} \times 2\ \text{k}\Omega = 2\ \text{V}$$
$$R_L = 5\ \text{k}\Omega;\quad V_T = I_S R_L = 1\ \text{mA} \times 5\ \text{k}\Omega = 5\ \text{V}$$
$$R_L = 10\ \text{k}\Omega;\quad V_T = I_S R_L = 1\ \text{mA} \times 10\ \text{k}\Omega = 10\ \text{V}$$

The V-I curve (V_T versus $I_S = I_L$) is shown in Fig. 8.3*b*, and it is seen that the terminal voltage V_T varies with R_L, but I_S is constant.

• • •

Just as there are restrictions on the use of a voltage source, there are also restrictions on the use of a current source. You will recall that an ideal voltage source must never be *shorted,* since the source current would then have to be infinite. Similarly, an ideal current source can never be left *open circuited,* since its terminal voltage would have to approach infinity in order to maintain I_S constant. These two forbidden situations are shown in Fig. 8.4*a* and 8.4*b*.

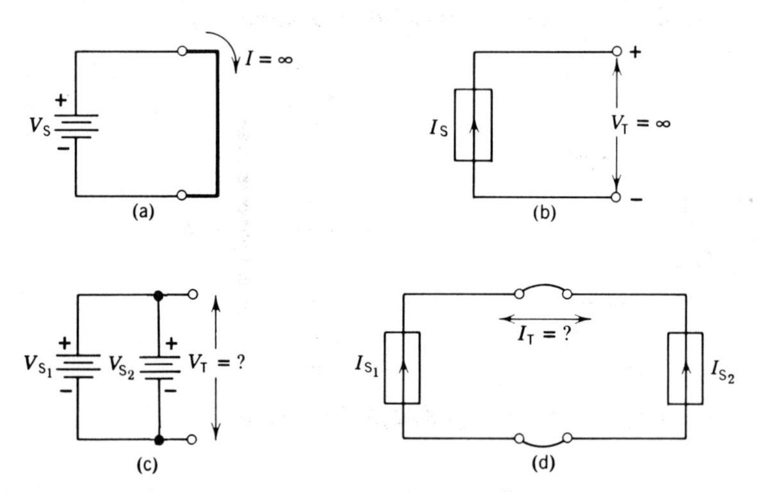

Figure 8.4 Forbidden connections of ideal sources (*a*) Shorted voltage source (*b*) Open current source (*c*) Unequal voltage sources in parallel (*d*) Unequal current sources in series

You will recall that it is also forbidden to connect two ideal voltage sources in parallel as shown in Fig. 8.4*c*, since the terminal voltage V_T is not realizable. Similarly, it is forbidden to connect two ideal current sources in series as shown in Fig. 8.4*d*, since the circuit current I_T is not realizable (i.e., $I_{S_1} \neq I_{S_2}$).

The internal resistance of a source can be found from its V-I curve by taking the ratio of $R_{\text{int}} = \Delta V/\Delta I$, where ΔV is a change in terminal voltage due to a change in source current ΔI. It is clear from Fig. 8.1 that ΔV is zero no matter what change is made in ΔI, and thus the internal resistance of an ideal voltage source is

$$R_{\text{int}} = \frac{\Delta V}{\Delta I} = \frac{0}{\Delta I} = 0\ \Omega$$

An examination of Fig. 8.2 shows that for any value of ΔV, ΔI is zero, and thus the internal resistance of an ideal current source is

$$R_{\text{int}} = \frac{\Delta V}{\Delta I} = \frac{\Delta V}{0} = ? \rightarrow \infty$$

Division by zero is undefined, but as ΔI approaches zero (becomes very small), then R_{int} approaches infinity (becomes very large). These two very important results can be summarized as,

the internal resistance of an ideal voltage source is zero, while the internal resistance of an ideal current source is infinite.

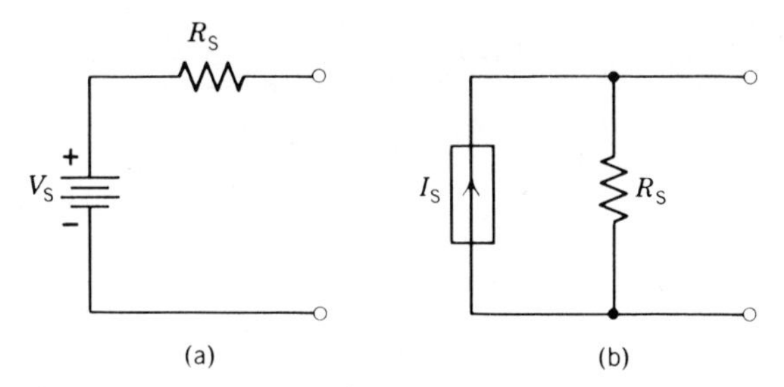

Figure 8.5 (*a*) Voltage (*b*) Current

A real voltage source does have some source resistance R_S which is connected in series with V_S as shown in Fig. 8.5*a*. Similarly, a *real* current source does have some source resistance R_S, and it is connected in *parallel* with I_S as shown in Fig. 8.5*b*.

Finally it should be noted that current sources are not simply dreamed up as a matter of convenience. They are in fact constructed using vacuum tube or transistor circuits and they are used to generate triangular or sawtooth wave forms (Chapter 1) to provide known currents in digital-to-analog or analog-to-digital converters and to perform many other useful functions; they are widely used in integrated circuits.

• • •

Example 8.2 A real current source is shown in Fig. 8.6*a*. Calculate the terminal voltage V_T as the load R_L is changed from 1 kΩ to 2 kΩ, to 5 kΩ and finally to 10 kΩ. Plot the resulting *V-I* curve and compare with Example 8.1.

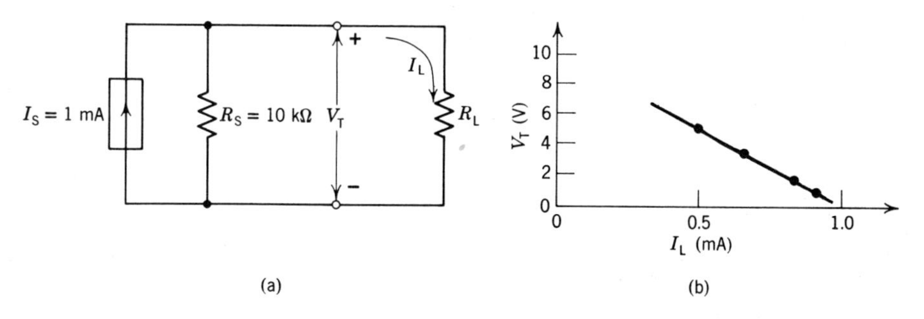

Figure 8.6 (*a*) A real current source (*b*) *V-I* curve

Solution The current through the load resistor R_L is *not* constant in this case because of the presence of R_S. We can find I_L using the current divider theorem, and then $V_T = I_L R_L$. Thus,

$$R_L = 1\text{ k}\Omega; \quad I_L = I_S \frac{R_S}{R_S + R_L} = 1\text{ mA} \times \frac{10\text{ k}\Omega}{10\text{ k}\Omega + 1\text{ k}\Omega} = 0.91\text{ mA}$$

$$V_T = I_L R_L = 0.91\text{ mA} \times 1\text{ k}\Omega = 0.91\text{ V}$$

$$R_L = 2\text{ k}\Omega; \quad I_L = 1\text{ mA} \times \frac{10\text{ k}\Omega}{10\text{ k}\Omega + 2\text{ k}\Omega} = 0.833\text{ mA}$$

$$V_T = 0.833\text{ mA} \times 2\text{ k}\Omega = 1.67\text{ V}$$

$$R_L = 5\ \text{k}\Omega; \quad I_L = 1\ \text{mA} \times \frac{10\ \text{k}\Omega}{10\ \text{k}\Omega + 5\ \text{k}\Omega} = 0.667\ \text{mA}$$

$$V_T = 0.667\ \text{mA} \times 5\ \text{k}\Omega = 3.33\ \text{V}$$

$$R_L = 10\ \text{k}\Omega; \quad I_L = 1\ \text{mA} \times \frac{10\ \text{k}\Omega}{10\ \text{k}\Omega + 10\ \text{k}\Omega} = 0.50\ \text{mA}$$

$$V_T = 0.50\ \text{mA} \times 10\ \text{k}\Omega = 5.0\ \text{V}$$

The *V-I* curve is shown in Fig. 8.6*b*, and it is clear that the load current I_L is *not* constant due to the presence of R_S. Thus the source resistance R_S is seen to *reduce* the current delivered to the load in a fashion similar to the reduction in terminal voltage in a real voltage source.

• • •

8.2 THÉVENIN'S THEOREM

Thévenin's theorem provides an extremely valuable means for reducing a complex circuit to a simple circuit containing an ideal voltage source in series with an equivalent resistance. Thus the complex circuit in Fig. 8.7*a* can always be reduced to the Thévenin equivalent shown in Fig. 8.7*b*.

Thévenin's theorem states that any circuit connected between two terminals can be replaced with a Thévenin voltage V_{TH} in series with a Thévenin resistance R_{TH} connected between the same two terminals.

The Thévenin voltage V_{TH} is equal to the voltage appearing between the two terminals. The Thévenin resistance is equal to the resistance between the two terminals when *all* sources are replaced with their *internal resistances* (replace ideal voltage sources with a *short,* and replace ideal current sources with an *open circuit*).* The procedure for finding the Thévenin equivalent of a circuit can be summarized in three steps:

1. Draw the Thévenin circuit (Fig. 8.7*b*).
2. V_{TH}. Determine the voltage appearing between terminals A and B in Fig. 8.7*a*. This is equal to V_{TH}.
3. R_{TH}. Replace all *voltage* sources with *short* circuits, and all *current* sources with *open* circuits in Fig. 8.7*a*. The resistance measured between terminals A and B is then R_{TH}.

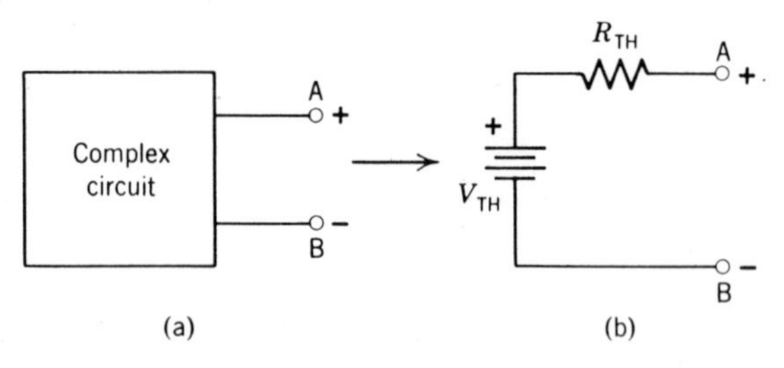

Figure 8.7 Thévenin's theorem (*a*) Original circuit (*b*) Thévenin equivalent

*We will not be concerned with *dependent* sources at this time.

• • •

Example 8.3 Determine the Thévenin equivalent of the circuit shown in Fig. 8.8*a*.

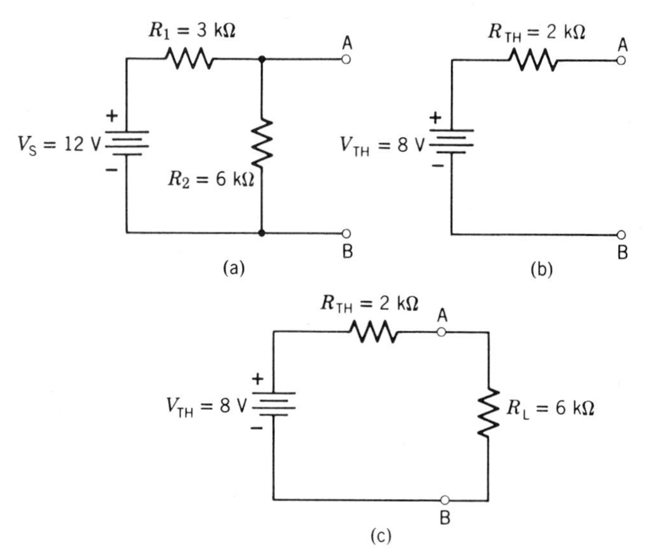

Figure 8.8 (*a*) A circuit (*b*) Thévenin equivalent (*c*) Thévenin equivalent with a 6 kΩ load resistor

Solution We determine the Thévenin equivalent following the three steps given above.

1. The Thévenin circuit is shown in Fig. 8.8*b*.
2. V_{AB} is equal to V_{TH} in Fig. 8.8*b*. Using the voltage divider theorem,

$$V_{AB} = V_{TH} = V_S \frac{R_2}{R_1 + R_2} = 12\text{ V} \times \frac{6\text{ k}\Omega}{3\text{ k}\Omega + 6\text{ k}\Omega} = 8\text{ V}$$

3. R_{TH} is found by shorting V_S in Fig. 8.8*a*. With V_S shorted, the total resistance connected between A and B is clearly $R_1 \| R_2$. Thus,

$$R_{TH} = R_1 \| R_2 = 3\text{ k}\Omega \| 6\text{ k}\Omega = 2\text{ k}\Omega$$

• • •

Example 8.4 Verify that the Thévenin equivalent in Fig. 8.8*b* is indeed equivalent to the circuit in Fig. 8.8*a* by connecting a load resistor $R_L = 6\text{ k}\Omega$ between terminals A and B finding I_{R_L} and V_{R_L}.

Solution If a 6-kΩ resistor is connected between terminals A and B as shown in Fig. 8.8*c*, the load current is clearly

$$I_L = \frac{V_{TH}}{R_{TH} + R_L} = \frac{8\text{ V}}{2\text{ k}\Omega + 6\text{ k}\Omega} = 1\text{ mA}$$

The load voltage is then

$$V_L = I_L R_L = 1\text{ mA} \times 6\text{ k}\Omega = 6\text{ V}$$

If the 6-kΩ load is connected between terminals A and B in Fig. 8.8*a*, the total source current is

$$I_S = \frac{V_S}{R_1 + (R_2 \| R_L)} = \frac{12\text{ V}}{3\text{ k}\Omega + (6\text{ k}\Omega \| 6\text{ k}\Omega)} = 2\text{ mA}$$

The source current will divide evenly between R_2 and R_L, since $R_2 = R_L$, and thus,

$$I_L = 1\text{ mA}$$

The load voltage is then

$$V_L = 1\text{ mA} \times 6\text{ k}\Omega = 6\text{ V}$$

A similar situation will exist for any value of R_L and the two circuits are seen to be equivalent.

• • •

Thévenin's theorem is extremely valuable in finding the current through or the voltage across any resistor in a circuit. The process is to "cut" the circuit, and it is best demonstrated by examples.

• • •

Example 8.5 Use Thévenin's theorem to find the current through R_4 in Fig. 8.9*a*.

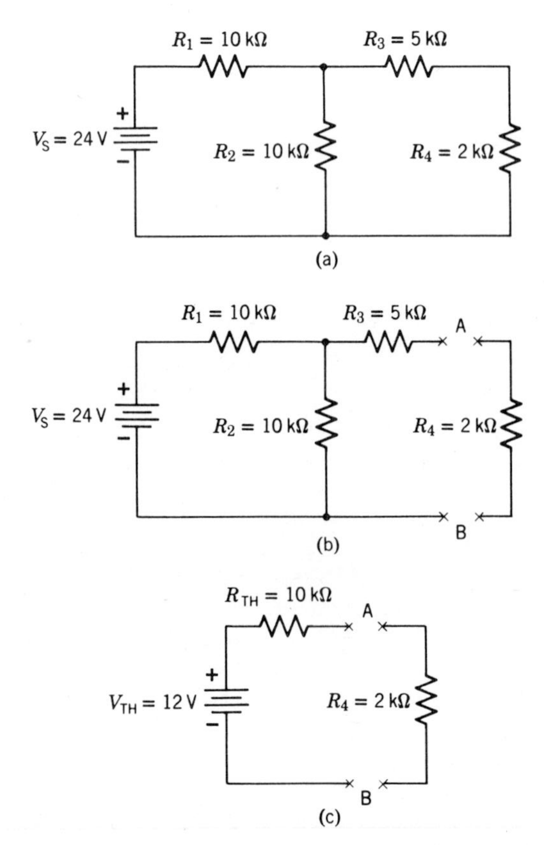

Figure 8.9 (*a*) Original circuit (*b*) "Cutting" out R_4 (*c*) Thévenin equivalent

Solution Since we are interested in finding the current through R_4, we begin by "cutting" R_4 out of the circuit as shown in Fig. 8.9*b*. We then find the Thévenin equivalent of the circuit to the *left* of the "cut."

1. The Thévenin equivalent is drawn in Fig. 8.9*c*.

2. The Thévenin voltage is found using the voltage divider theorem. (Note that there is *no* current in R_3 and therefore no voltage drop across it. Thus $V_{AB} = V_{R_2}$!)

$$V_{TH} = V_{AB} = V_S \frac{R_2}{R_1 + R_2} = 24\text{ V} \times \frac{10\text{ k}\Omega}{10\text{ k}\Omega + 10\text{ k}\Omega} = 12\text{ V}$$

3. R_{TH} is found from Fig. 8.9*b* by shorting V_S. Thus

$$R_{TH} = R_3 + (R_1 \| R_2) = 5\text{ k}\Omega + (10\text{ k}\Omega \| 10\text{ k}\Omega) = 10\text{ k}\Omega$$

The current through R_4 is now found using the Thévenin equivalent in Fig. 8.9*c*, and

$$I_{R_4} = \frac{V_{TH}}{R_{TH} + R_4} = \frac{12\text{ V}}{10\text{ k}\Omega + 2\text{ k}\Omega} = 1\text{ mA}$$

• • •

The reader should verify the solution by applying some other technique to the original circuit (Fig. 8.9*a*).

Thévenin's theorem can be applied equally well to solve circuits containing current sources as shown by the following example.

• • •

Example 8.6 Use Thévenin's theorem to find the current through R_3 in Fig. 8.10*a*.

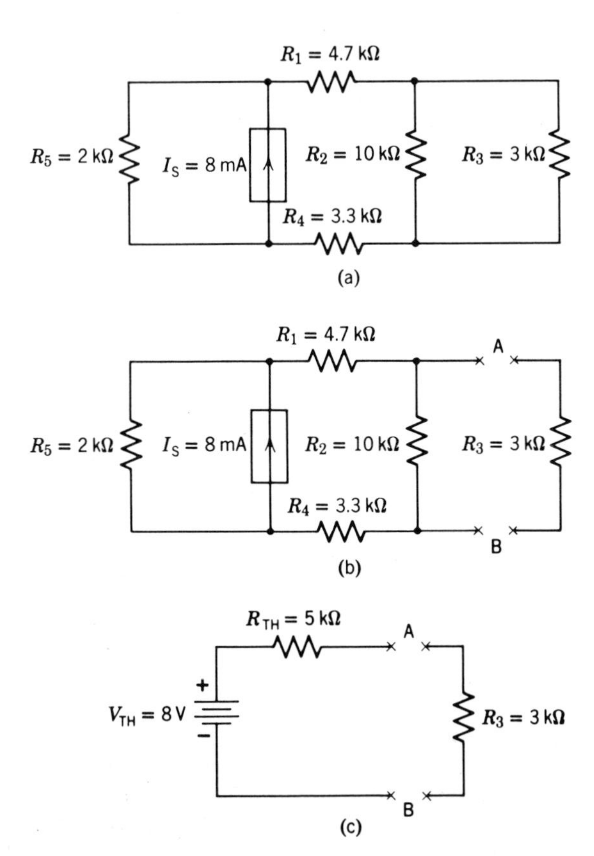

Figure 8.10 (*a*) Original circuit (*b*) "Cutting" out R_3 (*c*) Thévenin equivalent

Solution We begin by "cutting" R_3 out of the circuit as shown in Fig. 8.10*b*. We then find the Thévenin equivalent of the circuit to the *left* of the "cut."

1. The Thévenin equivalent is shown in Fig. 8.10*c*.
2. V_{TH} is the voltage across R_2 in Fig. 8.10*b*. Using the current divider theorem,

$$I_{R_2} = I_S \frac{R_5}{R_5 + (R_1 + R_2 + R_4)}$$

$$= 8\text{ mA} \times \frac{2\text{ k}\Omega}{2\text{ k}\Omega + (4.7\text{ k}\Omega + 10\text{ k}\Omega + 3.3\text{ k}\Omega)} = 4/5\text{ mA}$$

Then,

$$V_{R_2} = V_{AB} = V_{TH} = I_{R_2}R_2 = 4/5\text{ mA} \times 10\text{ k}\Omega = 8\text{ V}.$$

3. R_{TH} is found from Fig. 8.10*b* with I_S *removed* (replaced with an *open* circuit). Thus,

$$R_{TH} = R_2 \| (R_1 + R_4 + R_5) = 10\text{ k}\Omega \| (4.7\text{ k}\Omega + 3.3\text{ k}\Omega + 2\text{ k}\Omega) = 5\text{ k}\Omega$$

The current through R_3 is now found using the Thévenin equivalent in Fig. 8.10*c*. Thus,

$$I_{R_3} = \frac{V_{TH}}{R_{TH} + R_3} = \frac{8\text{ V}}{5\text{ k}\Omega + 3\text{ k}\Omega} = 1\text{ mA}$$

The reader should verify this solution by applying some other technique to the original circuit (Fig. 8.10*a*).

• • •

8.3 NORTON'S THEOREM

Norton's theorem is very similar to Thévenin's theorem and provides an equally powerful method for simplifying a complex circuit. Norton's theorem is used to construct an equivalent circuit containing an ideal current source in parallel with an equivalent resistance as shown in Fig. 8.11.

Norton's theorem states that any circuit connected between two terminals can be replaced with a Norton current source I_N in parallel with a Norton resistance R_N connected between the same two terminals.

The Norton current I_N is equal to the current which exists in a *short* circuit connected between the two terminals. The Norton resistance R_N is equal to the resistance measured between the two terminals when *all* sources are

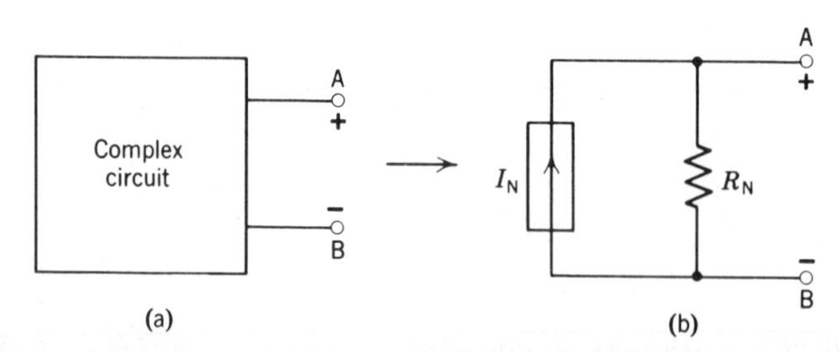

Figure 8.11 (*a*) Original circuit (*b*) Norton equivalent circuit

replaced with their *internal* resistances. Thus R_N is seen to be *exactly the same* as the Thévenin resistance R_{TH}.

The procedure for finding the Norton equivalent of a circuit can be summarized in three steps:

1. Draw the Norton circuit (Fig. 8.11).
2. I_N. Determine the current which flows through a *short* placed across terminals A and B in Fig. 8.11*a*. This is I_N.
3. R_N. Replace all *voltage* sources with *short* circuits, and all *current* sources with open circuits in Fig. 8.11*a*. The resistance measured between terminals A and B is then R_N (the same as R_{TH}).

• • •

Example 8.7 Determine the Norton equivalent of the circuit in Fig. 8.8*a*.

Solution We determine the Norton equivalent following the three steps above.

1. The Norton circuit is shown in Fig. 8.12.

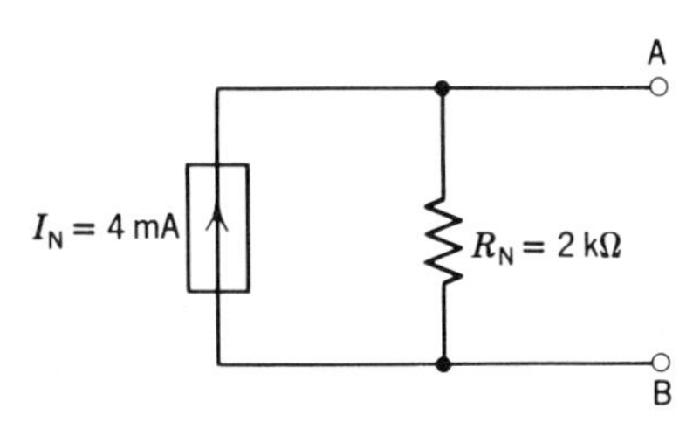

Figure 8.12 Norton equivalent of Figure 8.8

2. Placing a short across terminals A and B in Fig. 8.8*a* results in a current through the short of

$$I_{sh} = I_N = \frac{V_S}{R_1} = \frac{12\text{ V}}{3\text{ k}\Omega} = 4\text{ mA}$$

3. R_N is found by shorting V_S in Fig. 8.8*a*, and

$$R_N = R_1 \| R_2 = 3\text{ k}\Omega \| 6\text{ k}\Omega = 2\text{ k}\Omega$$

Note that this is the same result obtained in Example 8.3.

• • •

Example 8.8 Verify the validity of the Norton equivalent in Fig. 8.12 by connecting a 6 kΩ load resistor R_L and finding I_{R_L} and V_{R_L}, and comparing the results with Example 8.4.

Solution The current through a load $R_L = 6$ kΩ in Fig. 8.12 is found using the current divider theorem. Thus,

$$I_{R_L} = I_N \times \frac{R_N}{R_N + R_L} = 4\text{ mA} \times \frac{2\text{ k}\Omega}{2\text{ k}\Omega + 6\text{ k}\Omega} = 1\text{ mA}$$

This is clearly the same result obtained in Example 8.4 and the Norton is indeed equivalent.

• • •

The Norton equivalent in Fig. 8.12 could easily have been obtained using the Thévenin equivalent in Fig. 8.8*b*. The important relationships are

$$R_N = R_{TH} \tag{8.1}$$

$$I_N = \frac{V_{TH}}{R_{TH}} \tag{8.2}$$

Equation 8.2 can be solved for V_{TH} as

$$V_{TH} = I_N R_{TH} = I_N R_N \tag{8.3}$$

Equations 8.1 through 8.3 provide the means for changing a Thévenin equivalent to a Norton equivalent, or vice versa.

• • •

Example 8.9 Find the Norton equivalent of the circuit in Fig. 8.9.

Solution The Norton equivalent is most easily obtained from the Thévenin equivalent in Fig. 8.9*c*. The Norton circuit appears as shown in Fig. 8.13.

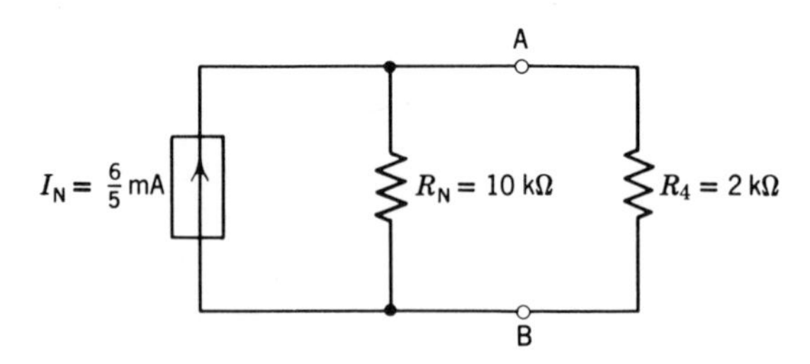

Figure 8.13

Using Eqs. 8.1 and 8.2,

$$R_N = R_{TH} = 10 \text{ k}\Omega$$

$$I_N = \frac{V_{TH}}{R_{TH}} = \frac{12 \text{ V}}{10 \text{ k}\Omega} = \frac{6}{5} \text{ mA}$$

The reader should verify the solution by calculating the current through $R_4 = 2$ kΩ using the Norton equivalent just constructed.

• • •

Occasionally, the use of Thévenin's or Norton's theorem to simplify a circuit will reveal some of the less obvious characteristics inherent in the circuit. These theorems then lead to a greater understanding of the operation of a circuit, as illustrated in the following example.

• • •

Example 8.10 Use Norton's theorem to solve for the current through R_6 in the circuit shown in Fig. 8.14*a*.

Solution

1. We begin by "cutting" R_6 from the circuit and finding the Norton equivalent between the "cut" points A and B. The Norton equivalent is shown in Fig. 8.14*b* (step 1).

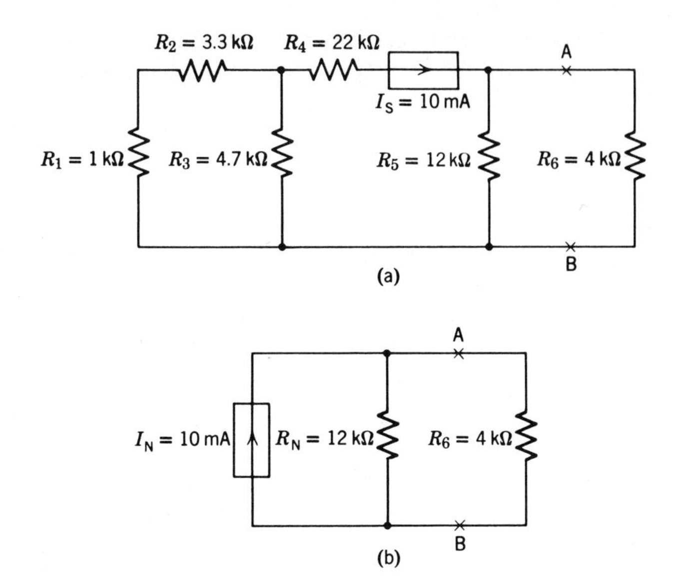

Figure 8.14 (*a*) Original circuit (*b*) Norton equivalent

2. If a *short* is placed across points A and B in Fig. 8.14*a*, the current through the short is equal to the source current I_S. Thus,

$$I_N = I_S = 10 \text{ mA}$$

3. R_N is found by removing I_S and measuring the resistance between points A and B, and it is

$$R_N = R_5 = 12 \text{ k}\Omega$$

The current through R_6 is now found by applying the current divider theorem to the Norton equivalent in Fig. 8.14*b*. Thus,

$$I_{R_6} = I_N \frac{R_N}{R_N + R_6} = 10 \text{ mA} \times \frac{12 \text{ k}\Omega}{12 \text{ k}\Omega + 4 \text{ k}\Omega} = 7.5 \text{ mA}$$

Notice that the resistors R_1, R_2, R_3, and R_4 did not appear at any time in the solution for I_{R_6}! These resistors could in fact have any value (could be changed at will) and the current I_{R_6} would be unaffected.

• • •

.4 THE SUPERPOSITION THEOREM

The superposition theorem is very useful in finding solutions for circuits containing *more* than a single source.

> ***The superposition theorem states that the voltage across (or the current through) any element in a circuit can be found by taking the algebraic sum of the voltages across (or the currents through) that element due to each individual source acting alone.***

Application of this theorem means finding the response in an element due to each source (all other sources are replaced with their internal resistances), and then adding these individual responses to obtain the overall response. The method is best illustrated by example.

• • •

Example 8.11 Use superposition to find a complete solution for the circuit in Fig. 8.15*a*.

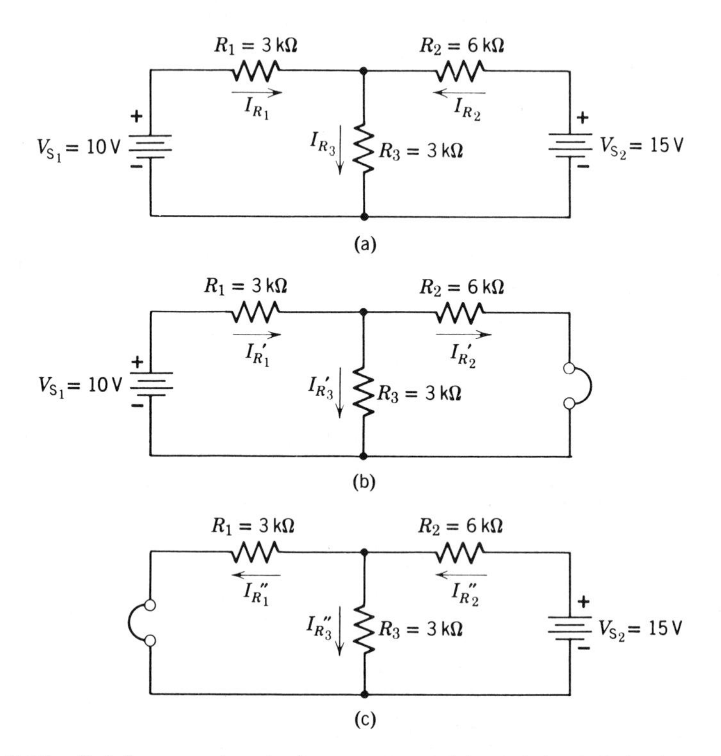

Figure 8.15 Solving a circuit by superposition (*a*) Original circuit (*b*) Currents due to V_{S_1}, V_{S_2} shorted (*c*) Currents due to V_{S_2}, V_{S_1} shorted

Solution The circuit in Fig. 8.15*b* is used to find the response due to the source V_{S_1} acting alone; V_{S_2} is replaced with its internal resistance—a short. The currents due to V_{S_1} are shown in the figure, and are,

$$I'_{R_1} = \frac{V_{S_1}}{R_1 + (R_2 \| R_3)} = \frac{10\ \text{V}}{3\ \text{k}\Omega + (6\ \text{k}\Omega \| 3\ \text{k}\Omega)} = 2\ \text{mA}$$

Applying the current divider theorem,

$$I'_{R_2} = I'_{R_1} \frac{R_3}{R_2 + R_3} = 2\ \text{mA} \times \frac{3\ \text{k}\Omega}{6\ \text{k}\Omega + 3\ \text{k}\Omega} = 2/3\ \text{mA}$$

$$I'_{R_3} = I'_{R_1} \frac{R_2}{R_2 + R_3} = 2\ \text{mA} \times \frac{6\ \text{k}\Omega}{6\ \text{k}\Omega + 3\ \text{k}\Omega} = 4/3\ \text{mA}$$

We must now find the currents due to V_{S_2} acting alone; this is done using Fig. 8.15*c* where V_{S_1} is now replaced with its internal resistance—a short. Thus,

$$I''_{R_2} = \frac{V_{S_2}}{R_2 + (R_1 \| R_3)} = \frac{15\ \text{V}}{6\ \text{k}\Omega + (3\ \text{k}\Omega \| 3\ \text{k}\Omega)} = 2\ \text{mA}$$

Using the current divider theorem,

$$I''_{R_3} = I''_{R_1} = I''_{R_2} \frac{R_3}{R_1 + R_3} = 2\text{ mA} \times \frac{3\text{ k}\Omega}{3\text{ k}\Omega + 3\text{ k}\Omega} = 1\text{ mA}$$

The overall response in the circuit due to both sources acting simultaneously is now found by adding algebraically the individual currents. Thus,

$$I_{R_1} = I'_{R_1} - I''_{R_1} = 2\text{ mA} - 1\text{ mA} = 1\text{ mA}$$
$$I_{R_2} = I''_{R_2} - I'_{R_2} = 2\text{ mA} - 2/3\text{ mA} = 4/3\text{ mA}$$
$$I_{R_3} = I'_{R_3} + I''_{R_3} = 4/3\text{ mA} + 1\text{ mA} = 7/3\text{ mA}$$

The importance of labeling the currents in *all* circuits cannot be overemphasized! The reader should check these results using KVL or KCL.

• • •

The superposition theorem can be applied equally well to circuits containing both voltage and current sources as the following example shows.

• • •

Example 8.12 Use superposition to find the current through R_2 in Fig. 8.16*a*.

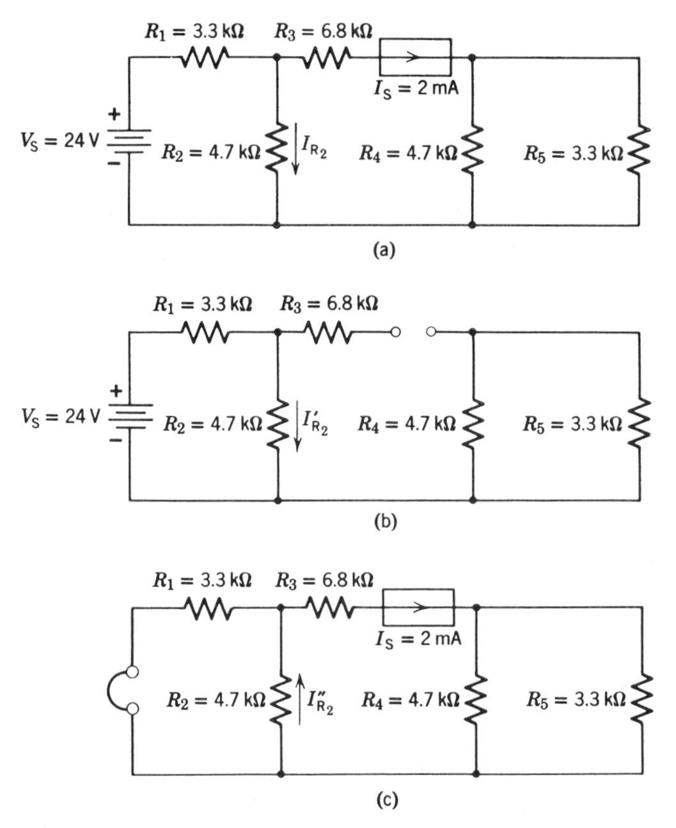

Figure 8.16 Solving a circuit by superposition (*a*) Original circuit (*b*) Current in R_2 due to V_S, I_S open (*c*) Current in R_2 due to I_S, V_S shorted

Solution The current in R_2 due to V_S acting alone I'_{R_2} is found using the circuit in Fig. 8.16*b*; the current source is replaced with its internal resistance—an *open* circuit. Thus,

$$I'_{R_2} = \frac{V_S}{R_1 + R_2} = \frac{24 \text{ V}}{3.3 \text{ k}\Omega + 4.7 \text{ k}\Omega} = 3 \text{ mA}$$

The current in R_2 due to the current source alone I''_{R_2} is found using the circuit in Fig. 8.16*c*; the voltage source is replaced with its internal resistance—a *short*. Applying a current divider theorem,

$$I''_{R_2} = I_S \frac{R_1}{R_1 + R_2} = 2 \text{ mA} \times \frac{3.3 \text{ k}\Omega}{3.3 \text{ k}\Omega + 4.7 \text{ k}\Omega} = 0.825 \text{ mA}$$

The total current through R_2 in Fig. 8.16*a* is the sum of these two components. Thus,

$$I_{R_2} = I'_{R_2} - I''_{R_2} = 3 \text{ mA} - 0.825 \text{ mA} = 2.175 \text{ mA}$$

• • •

Again, the importance of clearly labeling the circuit currents is clearly evident.

It must be emphasized that the superposition theorem is valid for adding currents and voltages in an element, but the power dissipated in a resistance *cannot* be found as the sum of the individual powers. This can be demonstrated by examining the power dissipated in R_2 in the previous example (Example 8.12). The power due to V_S alone would *appear to be*

$$P'_{R_2} = (I'_{R_2})^2 R_2 = (3 \text{ mA})^2 \times 4.7 \text{ k}\Omega = 42.3 \text{ mW}$$

The power due to I_S alone would *appear to be*

$$P''_{R_2} = (I''_{R_2})^2 R_2 = (0.825 \text{ mA})^2 \times 4.7 \text{ k}\Omega = 3.2 \text{ mW}$$

If we then *add* these two components, the total power would *appear to be*

$$P_{R_2} = P'_{R_2} + P''_{R_2} = 42.3 \text{ mW} + 3.2 \text{ mW} = 45.5 \text{ mW}$$

This is certainly *not* the correct result however, since the power in R_2 *is* clearly

$$P_{R_2} = I^2_{R_2} R_2 = (2.175 \text{ mA})^2 \times 4.7 \text{ k}\Omega = 22.2 \text{ mW}$$

It might at first seem that we could subtract the individual powers, since the currents are in opposite directions; it is easy to see that this does not work either, however, and we must always remember that superposition *cannot* be used to find total power by adding individual powers.

8.5 MILLMAN'S THEOREM

Another theorem which is sometimes useful in solving for the voltage in a circuit consisting of parallel resistances and sources is Millman's theorem. Millman's theorem really amounts to the repeated application of Norton's theorem in order to arrive at a general formula. The circuit in Fig. 8.17 consists of three voltage sources, each in series with a resistance, all connected in parallel. Since each of the sources in series with a resistor is identical to a Thévenin equivalent, they can easily be changed to Norton equivalents as shown in Fig. 8.18*a*. Since all of the elements in Fig. 8.18*a* are connected in parallel, we can combine them to form the simple equivalent circuit in Fig. 8.18*b*. Using KCL,

$$I_S = I_{S_1} + I_{S_2} + I_{S_3}$$

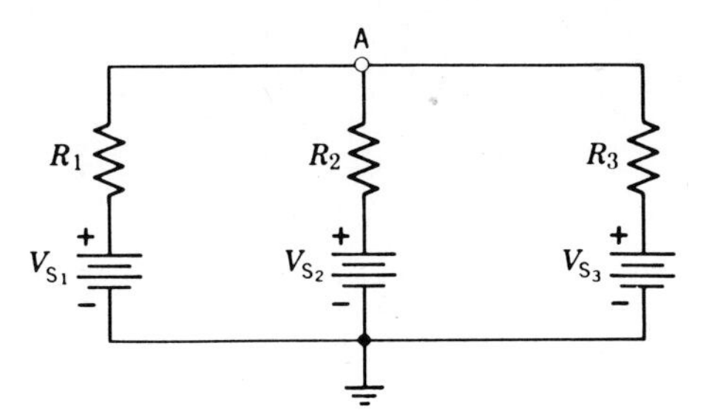

Figure 8.17

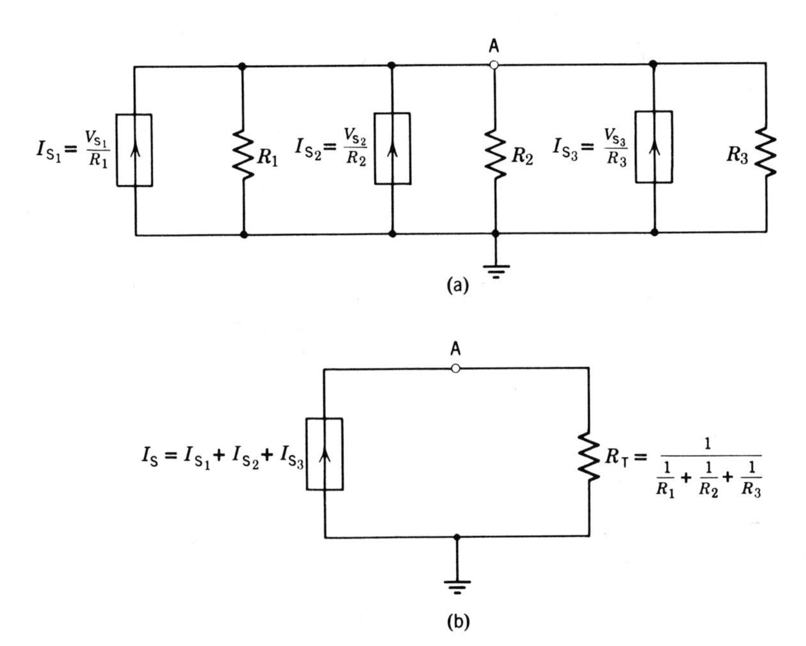

Figure 8.18 (*a*) Norton equivalent of Figure 8.17 (*b*) Combining sources and resistances

and

$$R_T = \frac{1}{\frac{1}{R_1} + \frac{1}{R_2} + \frac{1}{R_3}}$$

The voltage at node A is now found from this simple equivalent circuit as

$$V_A = I_S R_T = \frac{I_{S_1} + I_{S_2} + I_{S_3}}{\frac{1}{R_1} + \frac{1}{R_2} + \frac{1}{R_3}} = \frac{V_{S_1}/R_1 + V_{S_2}/R_2 + V_{S_3}/R_3}{\frac{1}{R_1} + \frac{1}{R_2} + \frac{1}{R_3}}$$

This equation is known as Millman's theorem (for this three-branch circuit), and it is clear that it can be extended to a circuit having any number of branches as,

$$V_A = \frac{V_{S_1}/R_1 + V_{S_2}/R_2 + V_{S_3}/R_3 + \cdots + V_{S_N}/R_N}{\frac{1}{R_1} + \frac{1}{R_2} + \frac{1}{R_3} \cdots + \frac{1}{R_N}} \tag{8.4}$$

• • •

Example 8.13 Find V_A in Fig. 8.17 if $V_{S_1} = 10$ V, $V_{S_2} = 6$ V, $V_{S_3} = 4$ V, $R_1 = 10$ kΩ, $R_2 = 2$ kΩ, and $R_3 = 1$ kΩ.

Solution Using Millman's theorem, Eq. 8.4,

$$V_A = \frac{10\text{ V}/10\text{ k}\Omega + 6\text{ V}/2\text{ k}\Omega + 4\text{ V}/1\text{ k}\Omega}{\dfrac{1}{10\text{ k}\Omega} + \dfrac{1}{2\text{ k}\Omega} + \dfrac{1}{1\text{ k}\Omega}} = \frac{1 \times 10\text{ V} + 5 \times 6\text{ V} + 10 \times 4\text{ V}}{1 + 5 + 10} = 5\text{ V}$$

• • •

The reader is urged to check this result using perhaps superposition.

One more example will be used to illustrate the utility of Millman's Theorem.

• • •

Example 8.14 Find V_A in Fig. 8.19 using Millman's theorem.

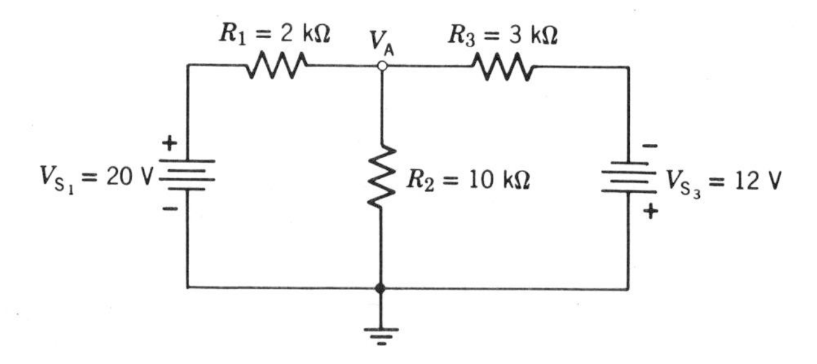

Figure 8.19

Solution Using Eq. 8.4,

$$V_A = \frac{20\text{ V}/2\text{ k}\Omega + 0\text{ V}/10\text{ k}\Omega - 12\text{ V}/3\text{ k}\Omega}{\dfrac{1}{2\text{ k}\Omega} + \dfrac{1}{10\text{ k}\Omega} + \dfrac{1}{3\text{ k}\Omega}} = 6.43\text{ V}$$

Notice that R_2 has no voltage source in series with it, and thus the numerator term is $V_{S_2}/R_2 = 0$ V/10 kΩ. Also notice that V_{S_3} is -12 V!

• • •

Again the reader is urged to check the solution by some other method.

8.6 MAXIMUM POWER TRANSFER THEOREM

We can discover some very important aspects of voltage sources by examining in detail the voltage source shown in Fig. 8.20 as the load resistance R_L is varied. First of all, the terminal voltage V_T can be calculated for various values of load resistance R_L by using the voltage divider theorem. The formula to be used is

$$V_T = V \frac{R_L}{R_S + R_L}$$

and the terminal voltages have been calculated for a number of different values of load resistances and they are recorded in Table 8.1. The load

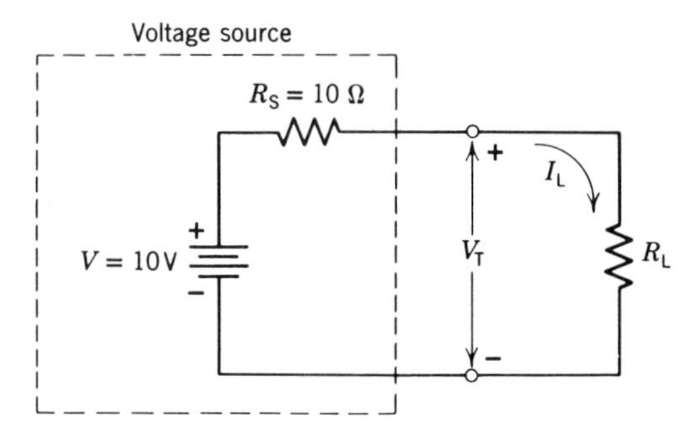

Figure 8.20

Table 8.1. Voltage Source Characteristics (Fig. 8.20).

$R_L(\Omega)$	V_T(V)	I_L(A)	P_{R_L}(W)	P_{R_S}(W)
0	0.0	1.00	0.0	10.0
2	1.67	0.834	1.39	6.95
4	2.86	0.715	2.04	5.11
6	3.75	0.625	2.34	3.91
8	4.45	0.555	2.47	3.08
10	5.00	0.500	2.50	2.50
12	5.45	0.455	2.48	2.07
14	5.83	0.416	2.43	1.73
20	6.67	0.333	2.22	1.11
40	8.00	0.200	1.60	0.40
100	9.10	0.091	0.831	0.082

current I_L is calculated from the equation

$$I_L = \frac{V}{R_S + R_L}$$

and the load currents for different values of load resistances are also given in Table 8.1. Finally, the power in the load P_{R_L} is found by taking the product of the load current and the terminal voltage, and the power dissipated in the source resistance P_{R_S} is found by multiplying the square of the load current by the source resistance. Thus,

$$P_{R_L} = I_L V_T \qquad P_{R_S} = I_L^2 R_S$$

and these values of power are also given in Table 8.1. These four quantities are plotted in Fig. 8.21 as a function of R_L.

Examination of the terminal voltage curve shows that the terminal voltage ranges from zero, when $R_L = 0$ (a short circuit) to 10 V when $R_L = \infty$ (an open circuit). Furthermore, the load resistance must be considerably greater than the source resistance if we desire a terminal voltage near the open circuit voltage of the source.

The load current curve shows that the load current ranges from a maximum of 1.0 A when the output is shorted (this maximum is determined by the source resistance and the open circuit voltage) to a minimum of zero when the output is open circuited.

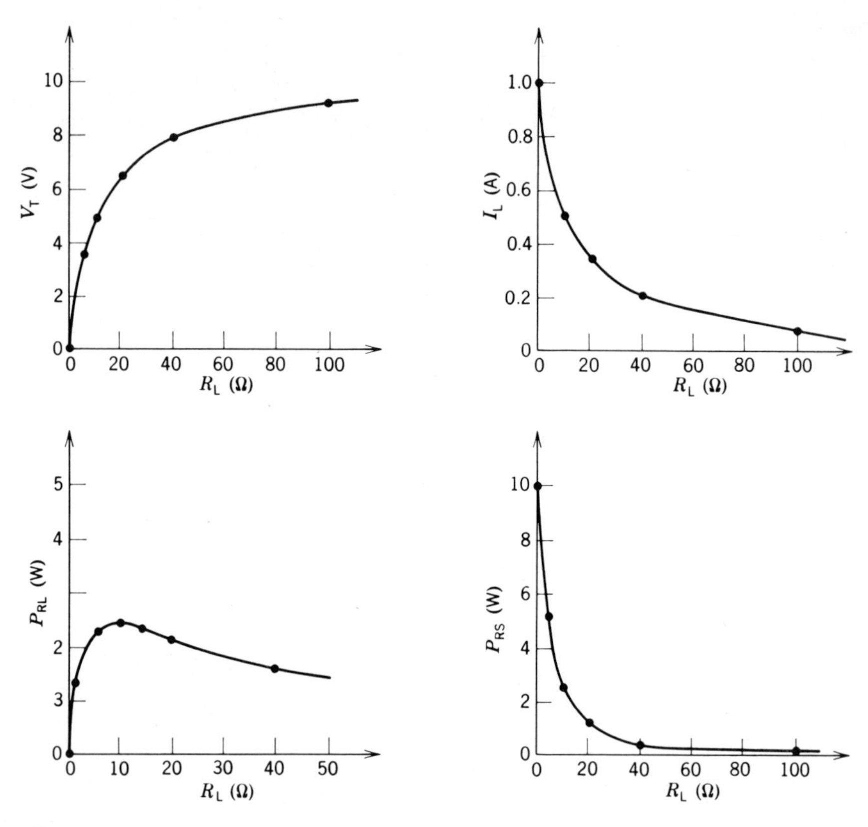

Figure 8.21

The curve showing the power in the load is very interesting since it has a *maximum* value. The curve shows clearly that the *maximum* power is delivered to the load when the load resistance is *equal* to the source resistance. This is the *maximum power transfer theorem.* When the load resistance is equal to the source resistance, maximum power is being *delivered* to the load, and the terminal voltage is exactly *one-half* the open circuit terminal voltage. The maximum power transfer theorem becomes important when considering the power delivered by large power installations, or in electronic power amplifier circuits.

The curve showing the power dissipated in the source is a maximum when the output is shorted, and is very high for low values of load resistances. Under these conditions, a great deal of heat may be generated inside the voltage source resulting in deterioration or damage to the source. This is referred to as an *overload* condition and is to be avoided.

• • •

Example 8.15 What should be the value of the load resistor R_L in Fig. 8.22 to obtain maximum power transfer to the load? What is the load voltage under this condition?

Solution For maximum power transfer to the load, the load resistance R_L must be equal to the source resistance R_S. Thus,

$$R_L = R_S = 50\ \Omega$$

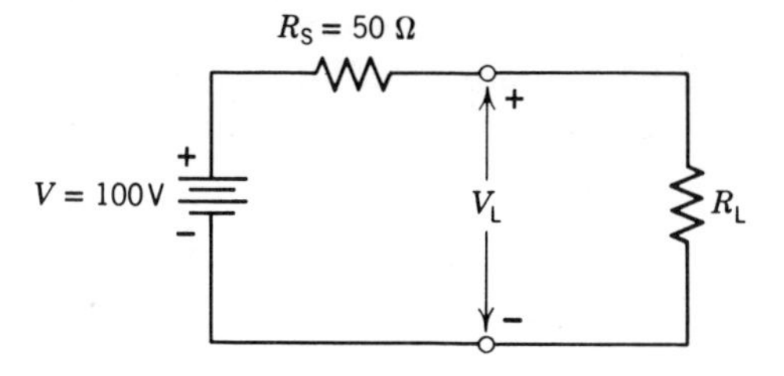

Figure 8.22

Under conditions of maximum power transfer, the load voltage is exactly one-half the open circuit voltage. Thus,

$$V_L = V/2 = 100\text{ V}/2 = 50\text{ V}$$

The load voltage can be checked by using the voltage divider theorem. Thus,

$$V_L = V\frac{R_L}{R_L + R_S} = 100\text{ V}\frac{50\ \Omega}{50\ \Omega + 50\ \Omega} = 50\text{ V}$$

• • •

7 USING THE THEOREMS

All theorems discussed in this chapter provide definite advantages when used in solving different circuits. In any particular circuit, one theorem may be more useful than the others, and the skill in choosing the proper theorem and applying it comes only after considerable practice. With this in mind, this section will be devoted to solving a number of different problems using all of the theorems so that a comparison between methods of solution can be made. The reader is urged to follow the solutions carefully, and to perhaps solve the circuits by the simplification techniques given in Chapter 7 in order to appreciate the utility of the theorems.

• • •

Example 8.16 A very common circuit in electronics is the *bridge* circuit shown in Fig. 8.23. When solving this circuit, it is usually desired to find the current through the resistor R_X. This circuit can be solved with very little effort by finding the Thévenin or Norton equivalent circuit with R_X "cut" from the circuit. The superposition theorem obviously does not offer any advantage since the bridge only has *one* voltage source, and superposition is useful with *multiple* source circuits. Millman's theorem likewise does not offer any advantages. We therefore want to solve for I_{RX} using Thévenin's and Norton's theorems.

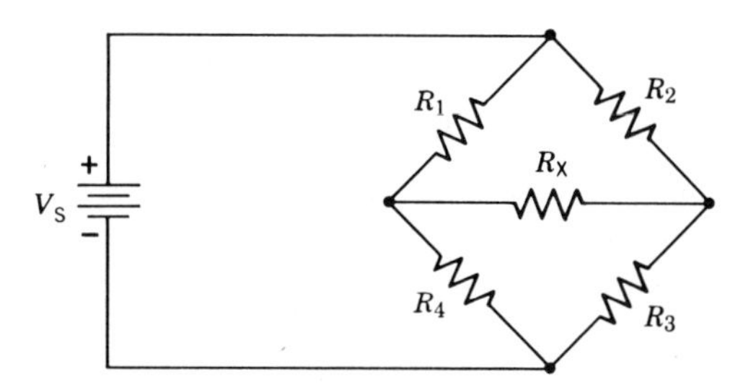

Figure 8.23 Bridge circuit

Solution Thévenin: We begin by removing R_X as shown in Fig. 8.24*a*. The desired Thévenin equivalent is shown in Fig. 8.24*b*. The voltage V_{AB} is equal to V_{TH} and is found by applying the voltage divider theorem twice. Thus,

$$V_A = V_S \frac{R_4}{R_1 + R_4} \qquad V_B = V_S \frac{R_3}{R_2 + R_3}$$

V_{TH} is then

$$V_{TH} = V_{AB} = V_A - V_B = V_S \left(\frac{R_4}{R_1 + R_4} - \frac{R_3}{R_2 + R_3} \right)$$

R_{TH} is found by replacing V_S with a *short*, and the circuit in Fig. 8.24*c* results.

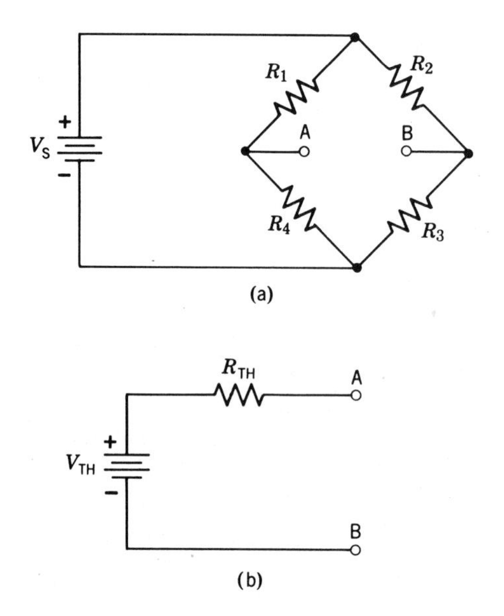

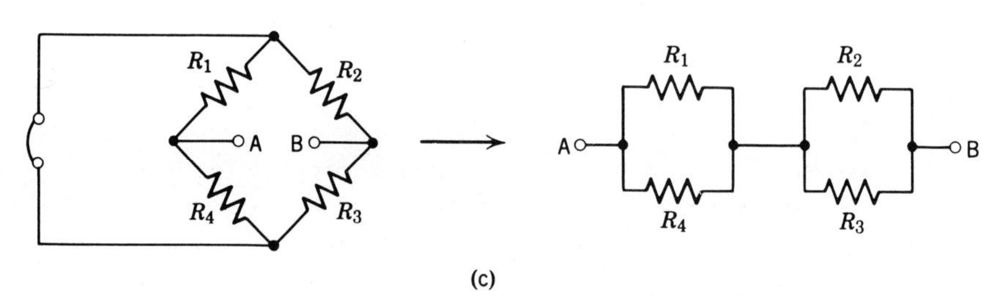

Figure 8.24 (*a*) "Cutting" the resistor R_x (*b*) Thévenin circuit (*c*) Finding R_{TH}

From this circuit it is clear that

$$R_{TH} = (R_1 \| R_4) + (R_2 \| R_3)$$

Using numerical values; $V_S = 9\ \text{V}$, $R_1 = R_3 = 1\ \text{k}\Omega$, $R_2 = R_4 = 2\ \text{k}\Omega$. Then

$$V_{TH} = 9\ \text{V} \left(\frac{2\ \text{k}\Omega}{1\ \text{k}\Omega + 2\ \text{k}\Omega} - \frac{1\ \text{k}\Omega}{2\ \text{k}\Omega + 1\ \text{k}\Omega} \right) = 3\ \text{V}$$

$$R_{TH} = (1\ \text{k}\Omega \| 2\ \text{k}\Omega) + (2\ \text{k}\Omega \| 1\ \text{k}\Omega) = \frac{4}{3}\ \text{k}\Omega$$

I_{R_X} is then found as

$$I_{R_X} = \frac{V_{TH}}{R_{TH} + R_X} = \frac{3\ \text{V}}{4/3\ \text{k}\Omega + R_X}$$

The Norton equivalent is most easily found from the Thévenin equivalent and it is shown in Fig. 8.25,

$$R_N = R_{TH} = \frac{4}{3}\ \text{k}\Omega$$

$$I_N = \frac{V_{TH}}{R_{TH}} = \frac{3\ \text{V}}{4/3\ \text{k}\Omega} = \frac{9}{4}\ \text{mA}$$

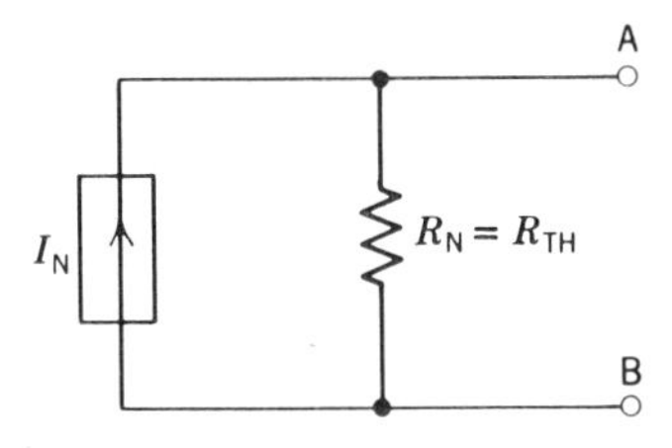

Figure 8.25

• • •

Example 8.17 The voltage V_A was found in Example 8.14 using Millman's theorem. Solve this problem using Thévenin's theorem, Norton's theorem, and Superposition.

Solution Thévenin:
The circuit can be simplified by "cutting" it as shown in Fig. 8.26*a* and finding

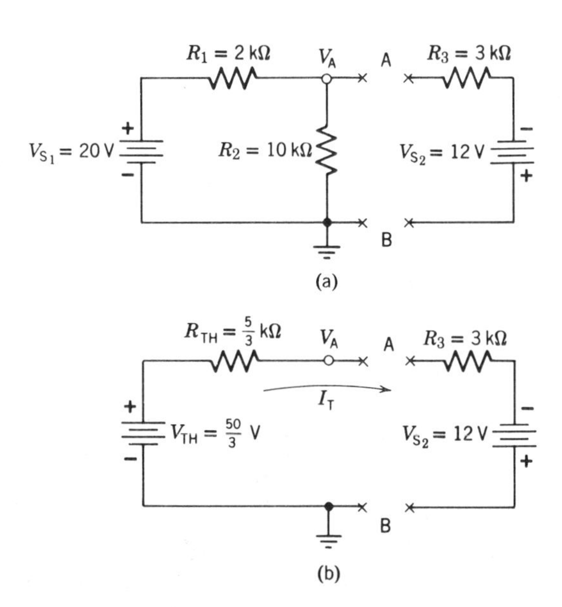

Figure 8.26 Finding a Thévenin equivalent (*a*) "Cutting" the circuit (*b*) Thévenin equivalent

the Thévenin equivalent to the *left* of the "cut." For this circuit,

$$R_{TH} = R_1 \| R_2 = 2\text{ k}\Omega \| 10\text{ k}\Omega = \frac{5}{3}\text{ k}\Omega$$

$$V_{TH} = V_{S_1}\frac{R_2}{R_1 + R_2} = 20\text{ V} \times \frac{10\text{ k}\Omega}{2\text{ k}\Omega + 10\text{ k}\Omega} = \frac{50}{3}\text{ V}$$

The current in the equivalent circuit is then

$$I_T = \frac{V_{TH} + V_{S_2}}{R_{TH} + R_3} = \frac{50/3\text{ V} + 12\text{ V}}{5/3\text{ k}\Omega + 3\text{ k}\Omega} = 43/7\text{ mA}$$

The voltage drop across R_{TH} is

$$V_{R_{TH}} = I_T R_{TH} = \frac{43}{7}\text{ mA} \times \frac{5}{3}\text{ k}\Omega = \frac{215}{21}\text{ V}$$

Finally,

$$V_A = V_{TH} - V_{R_{TH}} = \frac{50}{3}\text{ V} - \frac{215}{21}\text{ V} = 6.43\text{ V}.$$

Norton:

The circuit can be simplified by "cutting" it on both sides of R_2 and finding two Norton equivalents as shown in Fig. 8.27. On the *left* of the "cut,"

$$R_{N_1} = R_1 = 2\text{ k}\Omega$$

$$I_{N_1} = \frac{V_{S_1}}{R_1} = \frac{20\text{ V}}{2\text{ k}\Omega} = 10\text{ mA}$$

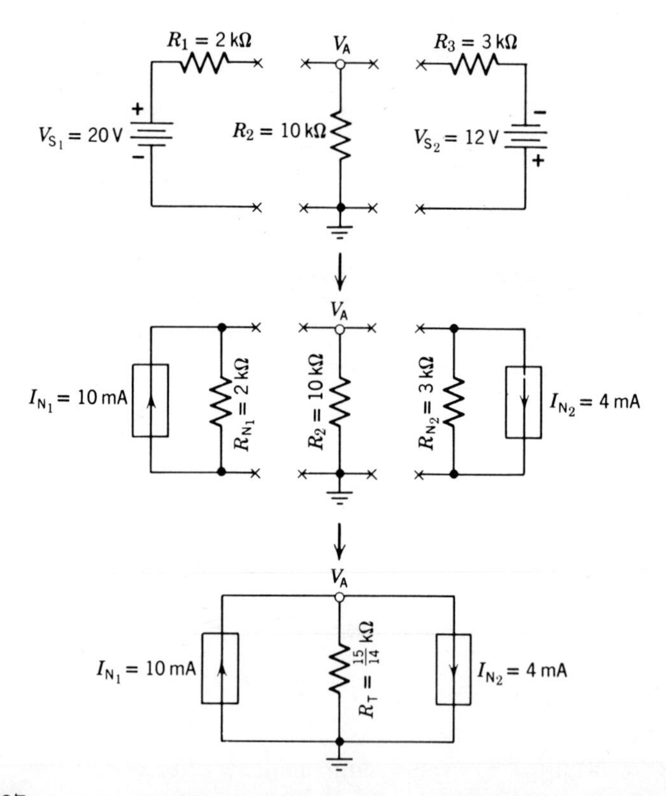

Figure 8.27

On the *right* of the "cut,"

$$R_{N_2} = R_3 = 3\text{ k}\Omega$$

$$I_{N_2} = \frac{V_{S_2}}{R_3} = \frac{12\text{ V}}{3\text{ k}\Omega} = 4\text{ mA}$$

The three parallel resistors can now be combined to form R_T,

$$R_T = R_{N_1} \| R_2 \| R_{N_2} = 2\text{ k}\Omega \| 10\text{ k}\Omega \| 3\text{ k}\Omega = \frac{15}{14}\text{ k}\Omega$$

The total current through R_T is clearly $I_{N_1} - I_{N_2} = 10\text{ mA} - 4\text{ mA} = 6\text{ mA}$, and it is in a direction from V_A toward ground. Thus,

$$V_A = (I_{N_1} - I_{N_2})R_T = 6\text{ mA} \times \frac{15}{14}\text{ k}\Omega = 6.43\text{ V}$$

Superposition:

The circuit is divided into two parts as shown in Fig. 8.28. From Fig. 8.28*a*, application of a voltage divider theorem gives,

$$V_A' = V_{S_1} \times \frac{(R_2 \| R_3)}{R_1 + (R_2 \| R_3)} = 20\text{ V} \times \frac{(10\text{ k}\Omega \| 3\text{ k}\Omega)}{2\text{ k}\Omega + (10\text{ k}\Omega \| 3\text{ k}\Omega)} = \frac{75}{7}\text{ V}$$

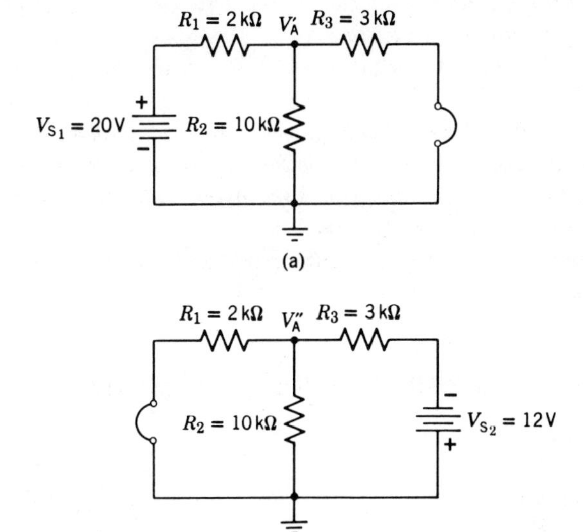

Figure 8.28 Solving by superposition (*a*) V_A' due to $V_{S_1} \cdot V_{S_2}$ shorted (*b*) V_A'' due to $V_{S_2} \cdot V_{S_1}$ shorted

From Fig. 8.28*b*, application of a voltage divider theorem gives,

$$V_A'' = V_{S_2} \times \frac{(R_1 \| R_2)}{R_3 + (R_1 \| R_2)} = -12\text{ V} \times \frac{(2\text{ k}\Omega \| 10\text{ k}\Omega)}{3\text{ k}\Omega + (2\text{ k}\Omega \| 10\text{ k}\Omega)} = -\frac{30}{7}\text{ V}$$

Finally, the voltage V_A is given as

$$V_A = V_A' + V_A'' = \frac{75}{7}\text{ V} - \frac{30}{7}\text{ V} = 6.63\text{ V}$$

• • •

SUMMARY

The use of network theorems has been shown to provide a powerful tool for the analysis of electric networks. Thévenin's and Norton's theorems can be used to reduce a complicated circuit to a very simple equivalent circuit, making the solution for the current through or the voltage across a single element a simple task. The Superposition theorem provides a very useful technique for evaluating currents and voltages in a circuit containing multiple sources. Millman's theorem is really an extension of Norton's theorem applied to elements connected in parallel, and provides a very simple and convenient means for finding node voltages. Each of these theorems provides definite advantages in different circuits, and the skill in choosing and applying the best theorem in any particular case will increase with practice.

GLOSSARY

Ideal current source. *Provides a constant output current for any finite load.*
Ideal voltage source. *Provides a constant terminal voltage for any load other than zero Ω.*
Internal resistance. *The resistance of a real source.*
Millman's theorem. *An extension of Norton's theorem used to find node voltages.*
Norton's theorem. *Any circuit can be replaced with an ideal current source in parallel with an equivalent resistance.*
Real current source. *An ideal current source in parallel with an internal resistance.*
Real voltage source. *An ideal voltage source in series with an internal resistance.*
Superposition theorem. *The total voltage across or current through an element can be found by taking the algebraic sum of the voltages or currents due to each individual source taken one at a time.*
Thévenin's theorem. *Any circuit can be replaced with an ideal voltage source in series with an equivalent resistance.*

IMPORTANT RELATIONSHIPS

- Thévenin Equivalent:
  1. Draw the Thévenin circuit.
  2. Determine V_{TH}.
  3. Determine R_{TH}.
- Norton Equivalent:
  1. Draw the Norton circuit.
  2. Determine I_N.
  3. Determine R_N.

$$R_{TH} = R_N \qquad I_N = \frac{V_{TH}}{R_{TH}} \qquad V_{TH} = I_N R_N$$

- Superposition:
  1. Draw a separate circuit for each source acting alone. Replace all other sources with their internal resistances.
  2. Determine each individual current or voltage in each of the circuits in (1) above.

3. Determine the total current or voltage by adding algebraically the components in (2) above.

- Millman's theorem:

$$V_A = \frac{V_{S_1}/R_1 + V_{S_2}/R_2 + \cdots + V_{S_N}/R_N}{\dfrac{1}{R_1} + \dfrac{1}{R_2} + \cdots + \dfrac{1}{R_N}}$$

- Maximum power transfer theorem: Maximum power to load when $R_L = R_S$.

EVIEW QUESTIONS

1. The internal resistance of an ideal voltage source is _____.
2. The internal resistance of an ideal current source is _____.
3. An ideal voltage source must *never* be shorted, and an ideal current source must *never* be left open circuited. (TF)
4. Two ideal current sources having different currents must never be connected in (series, parallel).
5. A *real* current source is represented by the parallel connection of an ideal _____ source and a _____.
6. The fact that the current from a *real* current source is not constant for various loads is due to the _____ _____.
7. In finding R_{TH} (Thévenin) all sources in the circuit are replaced by their internal resistances. (TF)
8. V_{TH} is equal to the circuit terminal voltage (with, without) the load.
9. The Norton resistance R_N is found exactly as the Thévenin in resistance R_{TH} and the two are exactly equal. (TF)
10. I_N (Norton) is found by (shorting, opening) the terminals of a circuit.
11. Thévenin's theorem is used to reduce a circuit to a simple (series, parallel) circuit, while Norton's theorem is used to reduce a circuit to a simple (series, parallel) circuit.
12. The voltage and current divider theorems are very useful in finding Thévenin and Norton equivalent circuits. (TF)
13. The Superposition theorem is useful in circuits containing (one, more than one) source(s).
14. Careful notation is extremely important when using the Superposition theorem. (TF)
15. The Superposition theorem is based on adding algebraically _____ and _____.
16. The total power in a resistance can be found by adding individual powers using the Superposition theorem. (TF)
17. Millman's theorem is an extension of _____ theorem.
18. Millman's theorem is useful in finding (voltages, currents).
19. The Superposition theorem would be useful in solving the bridge circuit in Fig. 8.20. (TF)
20. Circuit solutions using the theorems in this chapter can and always should be checked using KVL or KCL. (TF)
21. If a load resistance is exactly equal to a source resistance, the source is delivering maximum power to the load. (TF)

22. Under conditions of maximum power transfer, the terminal voltage of a source is ______% of the open circuit voltage.
23. A voltage source is said to be overloaded when the load resistance is much (larger, smaller) than the source resistance.

PROBLEMS

1. What must be the value of I_S in Fig. 8.29?

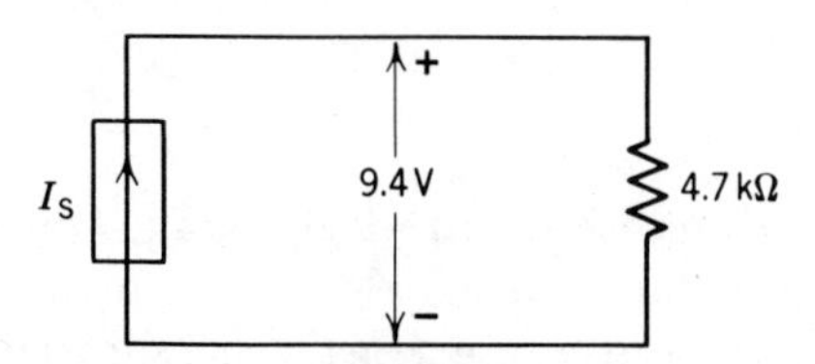

Figure 8.29

2. What is the value of I_L in Fig. 8.6 if R_L is: (*a*) 3.3 kΩ, (*b*) 330 Ω? Make a plot of I_L versus R_L for the values of R_L in this problem and in Example 8.2. Include the point for $R_L = 0\ \Omega$.
3. A 2.7 kΩ load is connected to the current source in Fig. 8.5*b*. What must be the internal resistance of the source R_S if $I_S = 10$ mA and $I_{R_L} =$ 9.7 mA?
4. What is the *largest* resistor R_L that can be connected to the current source in Fig. 8.6 without I_L falling *below* 0.95 mA?
5. Find the Thévenin equivalent of the circuit in Fig. 8.8*a* if R_1 is changed to 6 kΩ.
6. Find the Thévenin equivalent of the circuit in Fig. 8.8*a* if V_S is changed to 100 V.
7. A voltage source is constructed by connecting three 12 V batteries in parallel. Find the Thévenin equivalent of the source if the battery internal resistances are 0.10 Ω, 0.15 Ω, and 0.20 Ω.
8. Solve for the voltage across R_3 in Fig. 8.9*a* using Thévenin's theorem. Check by any method.
9. Solve for the voltage across R_2 in Fig. 8.9*a* using Thévenin's theorem. Check by any method.
10. An electronic power supply is capable of delivering 500 mA to a 50 Ω load. Draw the Thévenin equivalent of the power supply if it has an open circuit voltage of 28 V. Draw the Norton equivalent.
11. Find the Norton equivalent in Problem 5.
12. Find the Norton equivalent in Problem 6.
13. A current source can be simulated by connecting a voltage source in series with a resistance. Draw the Norton equivalent of a 500 V power supply connected in series with a 100 kΩ resistor.
14. A 10 mA current source is constructed by connecting a 100 V power supply in series with a 10 kΩ resistor. What is the largest permissible load resistor that can be used with this source if the load current must always be within 10% of the desired 10 mA? What is the smallest load resistor?

15. A small heating element is designed for a current between 90 mA and 100 mA. The element resistance is only 10 Ω when cold, but goes up to 100 Ω when hot. Use Norton's theorem to design a current source which will guarantee that the heater current will never exceed 100 mA and will never drop below 90 mA.

16. Find the current in R_3 in Fig. 8.10 using Norton's theorem.

17. Use Superposition to solve for V_A in Fig. 8.19 when the source V_{S_3} is reversed in polarity. Check by any method.

18. Use Superposition to find the voltage across R_3 in Fig. 8.15*a* when $V_{S_2} = 10$ V.

19. Use Superposition to solve for the current in R_1 in Fig. 8.16*a* when R_5 is removed.

20. Solve Problem 33 in Chapter 6 using Superposition.

21. Solve Problem 37 in Chapter 6 using Superposition.

22. Solve Problem 18 using Millman's theorem.

23. Solve Problem 17 using Millman's theorem.

24. Find the current through R_3 in Fig. 8.30. Make full use of theorems. Check your solution by working the problem a different way.

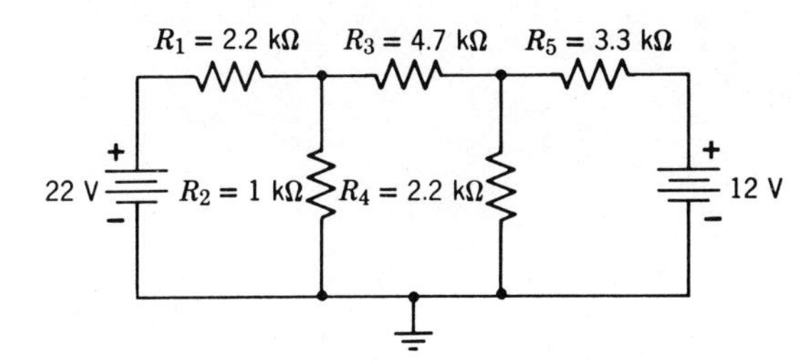

Figure 8.30

25. What must be the value of R_L for maximum power transfer in Fig. 8.20 if $V = 26$ V, and $R_S = 120$ Ω? What is V_T for this value of R_L? What power is delivered to the load R_L under this condition?

26. What is V_T in Fig. 8.20 if $V = 85$ V, $R_S = 220$ Ω, and $R_L = 430$ Ω? What power is dissipated in R_L and R_S? What must R_L be for maximum power transfer?

dc METERS

One very practical application of all the material we have discussed is the construction of dc meters. In this chapter we will use these network laws and theorems to study the operation of *ammeters, voltmeters,* and *ohmmeters.* We will then develop the techniques to design ammeters for current measurement, voltmeters for voltage measurement, and ohmmeters for resistance measurement. All of these dc meters are constructed using a basic meter movement called a D'Arsonval movement. The D'Arsonval meter is basically a current-measuring meter, and we will add circuitry to this basic meter to achieve the desired results.

9.1 D'ARSONVAL METER

A basic D'Arsonval or *moving-coil* meter is shown in Fig. 9.1 and consists of a coil of wire wound on an iron cylinder mounted between the pole faces of a permanent magnet. The cylinder is mounted on pivots at its ends and therefore is free to rotate in either direction (i.e., clockwise or counterclockwise). A spiral spring is used to hold the coil in a fixed position, and the tension on the spring can be adjusted to place the pointer in any desired position. When a current passes through the coil, a magnetic field is created around the coil. This magnetic field combines with the magnetic field of the permanent magnet and develops a torque which causes the cylinder to rotate

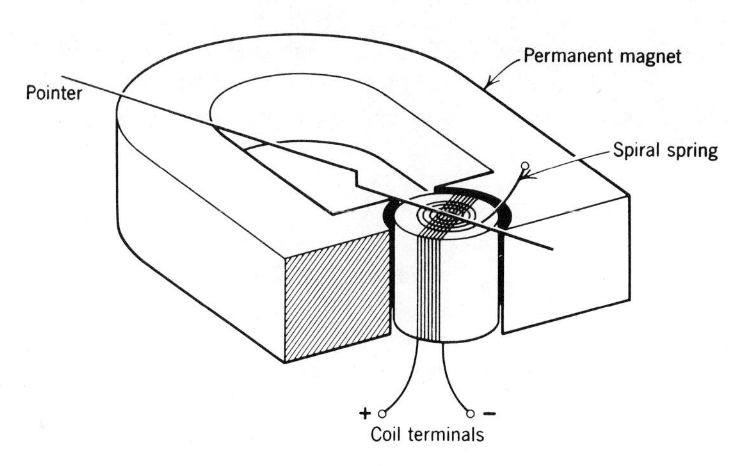

Figure 9.1 Basic D'Arsonval meter movement

(we will discuss this action in detail in Chapter 11). The cylinder will then rotate from its rest position (no current in the coil) until the counter-torque developed by the spiral spring is exactly equal to the magnetic torque. The pointer will then be displaced from its *rest* position by an amount depending on the current in the coil. In fact, the displacement of the pointer is very nearly proportional to the current in the coil. Thus, a scale can be placed under the pointer tip which specifies the current in the coil.

The meter in Fig. 9.2 is adjusted so that with no current in the coil the pointer is over 0 μA. When the current is passed through the coil in the direction shown in the figure, the pointer will move in a clockwise direction. When the meter current is 50 μA, the pointer will be over 50 on the scale, and when the meter current is 100 μA the pointer will be over 100. This meter has a full scale deflection of 100 μA, and if a larger current is forced through it, the meter may be permanently damaged. Therefore, great care must be taken *never* to exceed the full scale reading of the meter. Since 100 μA is *full scale*, 50 μA is *one-half* scale, and 33 μA is *one-third* scale. The meter in Fig. 9.2 is shown at just under one-third scale, or 30 μA.

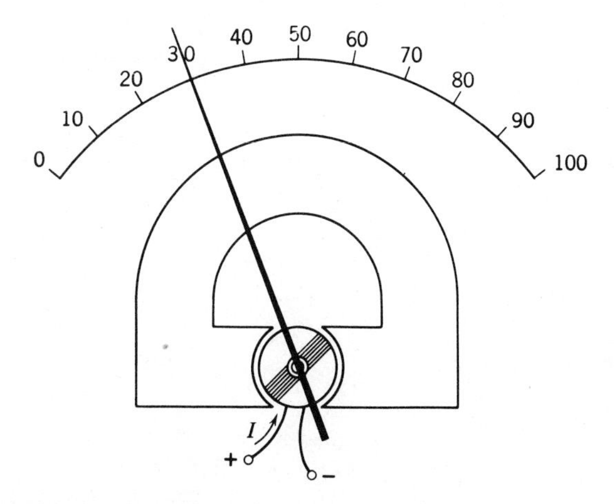

Figure 9.2 A 100 μA D'Arsonval meter

If the current direction in the meter in Fig. 9.2 is reversed, the pointer will try to rotate in a *counterclockwise* direction. This could cause damage to this particular meter since most meters have *stops* which prevent the pointer from moving off the face of the scale. Thus it is very important when using a meter to observe the correct meter polarity. All meters have their terminals marked + and −, and they *must* be connected such that current *enters* the + terminal and *exits* the − terminal.

Since the direction of rotation of the pointer depends on the current direction in the coil, it is possible to construct meters capable of measuring currents in either direction. Such meters are called *zero-center* scale meters, and a typical zero-center scale meter is shown in Fig. 9.3.

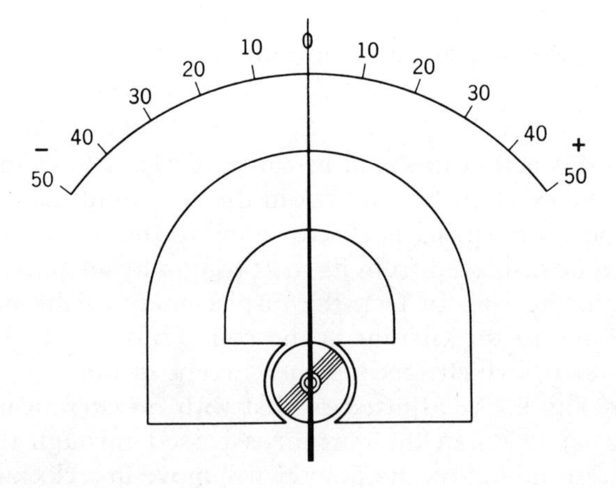

Figure 9.3 A zero center scale meter

The basic D'Arsonval meter was patented in 1881 and was later improved by Edward Weston (1850–1936) into the form of most meters used today. For this reason, these types of meters are often called Weston meters. A *galvanometer* is a very sensitive meter of this type which may be capable of measuring currents as small as 100 picoamperes.

An *ideal* D'Arsonval meter would have *zero* resistance in its windings. However, it is not possible to construct a meter of this type which has zero resistance, since the coil wire must have some resistance. Therefore, these meters are constructed to have the *smallest* convenient value of resistance possible, such as 10 Ω, 25 Ω, or 50 Ω, in order to simplify their use. We will then use the symbol in Fig. 9.4 to represent a *real* meter. R_M is the total meter resistance and I_{FS} is the *full scale* deflection current of the meter.

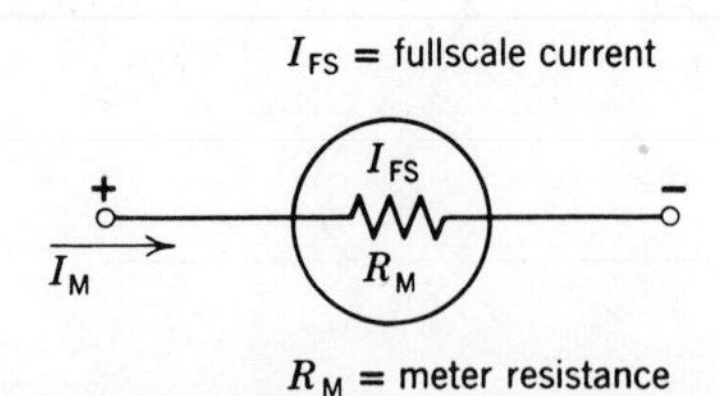

Figure 9.4 A real D'Arsonval meter

We conclude our discussion of D'Arsonval meter movements with the following summary:

1. A current greater than full scale must never be forced through a meter.
2. The correct current polarity must be observed when using a meter.
3. These meters are not ideal since they have some internal resistance R_M. The meters are designed such that R_M is the smallest convenient value.

• • •

Example 9.1 The meter shown in Fig. 9.4 has a resistance of $R_M = 25\ \Omega$, and is a 1 mA full scale movement. What is the voltage across the meter terminals when the current through the meter is, (*a*) full scale, (*b*) half scale?

Solution The terminal voltage of the meter V_M is found using Ohm's Law. Thus,

(*a*) $V_M = I_M R_M = 1\ \text{mA} \times 25\ \Omega = 25\ \text{mV}$
(*b*) $V_M = I_M R_M = 0.5\ \text{mA} \times 25\ \Omega = 12.5\ \text{mV}$

• • •

From this example it is quite clear that a voltage source must *never* be placed across the terminals of a D'Arsonval meter. For example, suppose a 1.5 V battery is connected across the terminals of the meter in Example 9.1. The meter current would be

$$I_M = \frac{1.5\ \text{V}}{R_M} = \frac{1.5\ \text{V}}{25\ \Omega} = 0.060\ \text{A} = 60\ \text{mA}$$

The meter would surely be damaged!

• • •

Example 9.2 The terminal voltage of the meter in Fig. 9.4 is found to be 5 mV when a current of 100 μA is passing through the meter. What is the meter resistance R_M?

Solution The meter resistance R_M is found using Ohm's Law. Thus,

$$R_M = \frac{V_M}{I_M} = \frac{5\ \text{mV}}{100\ \mu\text{A}} = 50\ \Omega$$

• • •

2 AMMETERS

An ammeter is a basic instrument which is used to measure current and the symbol for an ammeter is shown in Fig. 9.5. The current through any element can be measured with an ammeter by placing the meter in *series* with the element as shown in Fig. 9.6. That the ammeter must be placed in series with the element is clear, since the current in all parts of a series circuit is the same, and we are interested in measuring current. (Why must the ammeter *not* be

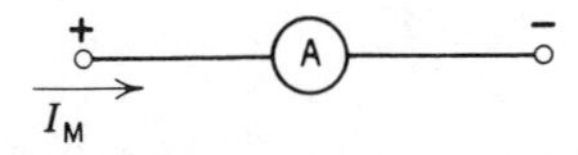

Figure 9.5 Ammeter symbol

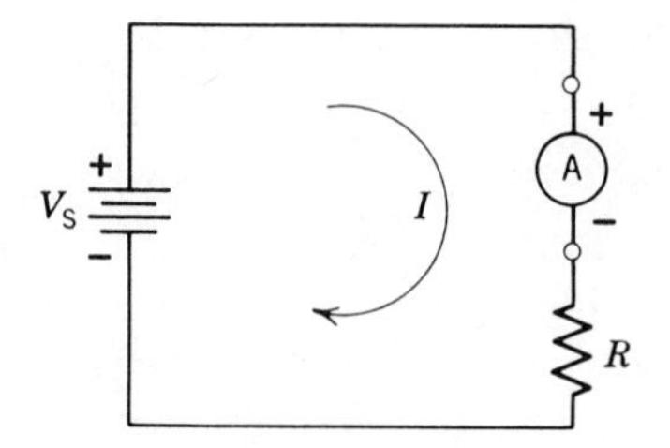

Figure 9.6 Measuring current with an ammeter

placed in parallel with R in Fig. 9.6?) Notice in the figure that the correct ammeter polarity is used since the current passes through the ammeter *from* the + terminal *to* the − terminal.

• • •

Example 9.3 Show how to place ammeters in Fig. 9.7*a* in order to measure each of the resistor currents.

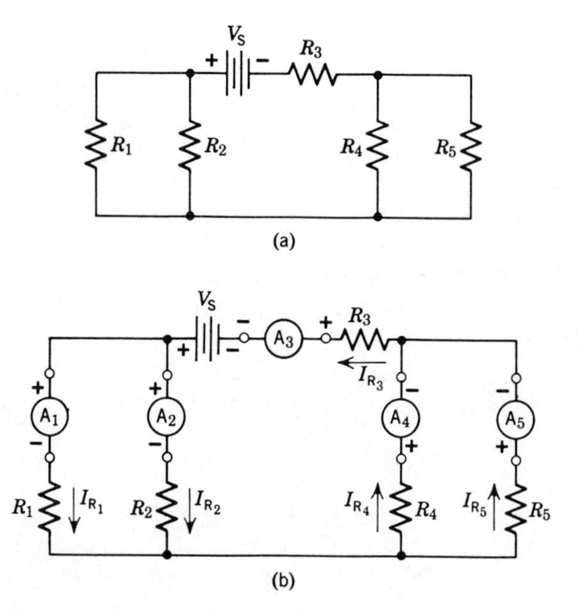

Figure 9.7 (*a*) Original circuit (*b*) Placement of ammeters to measure resistor currents

Solution An ammeter must be placed in *series* with a resistor in order to measure the current through that resistor. The correct polarities must be observed, and the ammeters and the currents they measure are (see Fig. 9.7*b*):

$$A_1 \rightarrow I_{R_1} \qquad A_2 \rightarrow I_{R_2}$$

$$A_3 \rightarrow I_{R_3} \qquad A_4 \rightarrow I_{R_4}$$

$$A_5 \rightarrow I_{R5}$$

• • •

The basic D'Arsonval meter shown in Fig. 9.4 can be used *as is* as an ammeter. The full scale of the ammeter would then be the full scale current of the D'Arsonval meter. For example, if the D'Arsonval meter has a full scale current of 100 μA, we immediately have an ammeter which we can use to measure currents *up to* 100 μA.

Now suppose that we want to use the D'Arsonval meter in Fig. 9.4 to construct an ammeter capable of measuring currents *greater* than the full scale current I_{FS} of the D'Arsonval meter. This can be accomplished quite easily, by simply placing an external resistor in *parallel* with the D'Arsonval meter as shown in Fig. 9.8. This additional resistor is called a *shunt* resistor

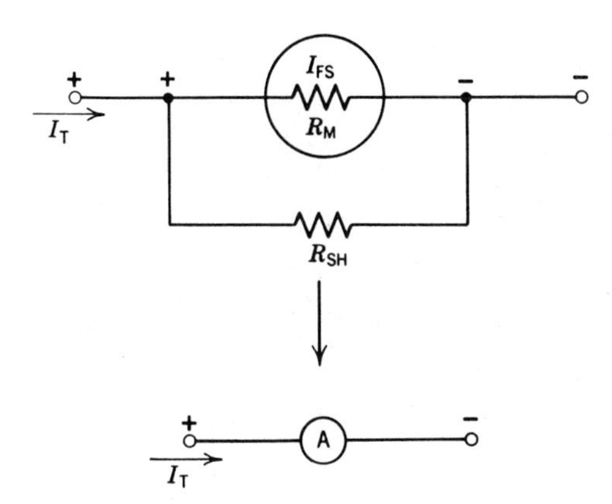

Figure 9.8 Constructing an ammeter by using a shunt resistor and a basic D'Arsonval meter movement

R_{SH} since a portion of the total ammeter current I_T is *shunted* through this resistor. When the current through the D'Arsonval meter is full scale I_{FS}, then the voltage across the D'Arsonval meter (and also across R_{SH}) is,

$$V_M = V_{SH} = I_{FS}R_M \tag{9.1}$$

Thus, the current through the shunt resistor is

$$I_{SH} = \frac{V_M}{R_{SH}} \tag{9.2}$$

KCL provides that the sum of the current through the D'Arsonval meter and the current through the shunt must equal the total current. Thus,

$$I_T = I_{FS} + I_{SH} \tag{9.3}$$

If we rearrange Eqs. 9.1–9.3 to find

$$V_M = I_{FS}R_M; \; R_{SH} = \frac{V_M}{I_{SH}}; \; I_{SH} = I_T - I_{FS}$$

We then solve these equations for the shunt resistance R_{SH}, and

$$R_{SH} = \frac{I_{FS}R_M}{I_T - I_{FS}} \tag{9.4}$$

Equation 9.4 can then be used to find the proper shunt resistance for any basic meter movement (I_{FS} and R_M) to provide an ammeter for any desired full scale current I_T.

• • •

Example 9.4 Find the proper shunt resistor R_{SH} to construct a 5.0 mA full scale ammeter using a D'Arsonval meter having $I_{FS} = 100\ \mu A$ and $R_M = 50\ \Omega$.

Solution The ammeter will be of the form shown in Fig. 9.8, and the complete ammeter is shown in Fig. 9.9. The correct shunt resistor R_{SH} is found using Eq. 9.4. Thus

$$R_{SH} = \frac{I_{FS}R_M}{I_T - I_{FS}} = \frac{100\ \mu A \times 50\ \Omega}{5.0\ mA - 100\ \mu A} = 1.02\ \Omega$$

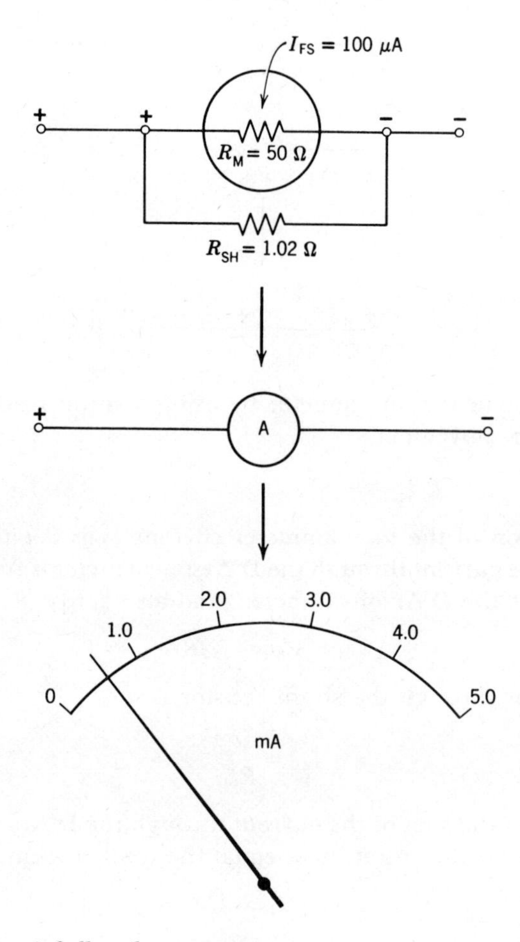

Figure 9.9 A 5 mA full scale ammeter

• • •

Example 9.5 What is the total ammeter resistance R_T of the meter constructed in Example 9.4, and what is its terminal voltage V_T at full scale current?

Solution The following relations must hold for the ammeter at full scale current (using Ohm's Law):

$$V_T = I_{FS}R_M = I_{SH}R_{SH} = I_TR_T$$

Thus,

$$V_T = I_{FS}R_M = 100\ \mu A \times 50\ \Omega = 5\ mV$$

Then,

$$V_T = I_T R_T,$$

and

$$R_T = \frac{V_T}{I_T} = \frac{5\ mV}{5\ mA} = 1.0\ \Omega$$

• • •

The technique of using a shunt resistor to construct an ammeter having any full scale current can be extended to construct a *multirange* ammeter as shown in Fig. 9.10. The shunt resistors are calculated using Eq. 9.4 as:

$$R_{SH_1} = \frac{1\ mA \times 50\ \Omega}{10\ mA - 1\ mA} = 5.56\ \Omega$$

$$R_{SH_2} = \frac{1\ mA \times 50\ \Omega}{100\ mA - 1\ mA} = 0.505\ \Omega$$

$$R_{SH_3} = \frac{1\ mA \times 50\ \Omega}{500\ mA - 1\ mA} = 0.1002\ \Omega$$

Notice that no shunt resistor is needed for the 1 mA position since this is the full scale current of the basic meter movement. Notice also that the total meter resistance is reduced as the range switch is moved to higher current positions.

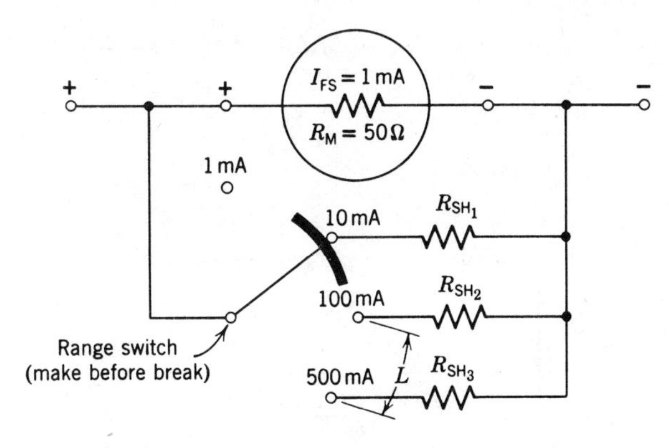

Figure 9.10 A multirange ammeter

Great care must be taken to use a "make before break" switch when using this configuration, since the basic meter movement must be protected while switching. A "make before break" switch means that the wiper arm of the switch will make contact with the next switch position *before* it loses contact with the previous position. Thus, the basic meter movement will *always* have a shunt across it when the range switch is moved. If a "break before make" switch is used (the opposite of a "make before break"), the wiper arm of the switch would be *open* when the switch is moved from one position to another. This means that *no* shunt would be across the basic meter movement for a short period of time during switching, and the meter movement could be damaged by high currents.

The disadvantage of having to use a "make before break" switch in the ammeter in Fig. 9.10 can be overcome by using a different shunt as shown in Fig. 9.11*a*. This arrangement is called an *Ayrton* or *universal* shunt. The basic meter movement is always protected since the shunt remains across it no matter what is done with the switch. This particular ammeter has three ranges and the correct shunt resistances can be found using the circuits in Fig. 9.11*b*, *c*, and *d*. From Fig. 9.11*b*,

$$V_T = I_{FS}R_M = I_{SH}(R_1 + R_2 + R_3)$$

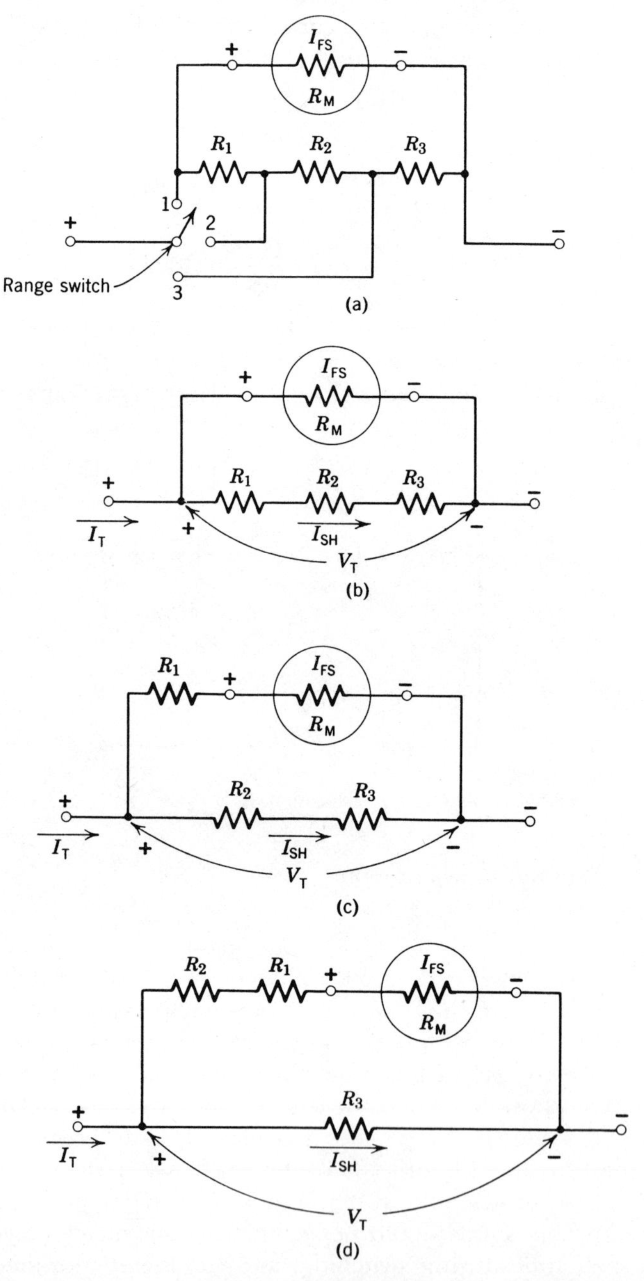

Figure 9.11 Multirange ammeter using an Ayrton or Universal shunt (*a*) Ayrton shunt type ammeter (*b*) Range 1 (*c*) Range 2 (*d*) Range 3

But $I_{SH} = I_T - I_{FS}$ by KCL. Therefore,

$$I_{FS}R_M = (I_T - I_{FS})(R_1 + R_2 + R_3) \tag{9.5}$$

From Fig. 9.11*c*,

$$V_T = I_{FS}(R_1 + R_M) = I_{SH}(R_2 + R_3)$$

So,

$$I_{FS}(R_1 + R_M) = (I_T - I_{FS})(R_2 + R_3) \tag{9.6}$$

From Fig. 9.11*d*,

$$V_T = I_{FS}(R_1 + R_2 + R_M) = I_{SH}R_3$$

So,

$$I_{FS}(R_1 + R_2 + R_M) = (I_T - I_{FS})R_3 \tag{9.7}$$

In solving for the resistances of an Ayrton shunt, we must solve all three equations 9.5, 9.6, and 9.7 simultaneously.

• • •

Example 9.6 Calculate the resistances needed for the ammeter in Fig. 9.11 if the basic meter movement has $I_{FS} = 100\ \mu\text{A}$ and $R_M = 100\ \Omega$, and the three desired ammeter current ranges are: 10 mA, 100 mA, 500 mA.

Solution We must solve the three equations 9.5–9.7 simultaneously for the three variables R_1, R_2, and R_3. The equations are:

$$I_{FS}R_M = (I_T - I_{FS})(R_1 + R_2 + R_3)$$
$$I_{FS}(R_1 + R_M) = (I_T - I_{FS})(R_2 + R_3)$$
$$I_{FS}(R_1 + R_2 + R_M) = (I_T - I_{FS})R_3$$

substituting numerical values,

$$100\ \mu\text{A} \times 100\ \Omega = (10\ \text{mA} - 100\ \mu\text{A})(R_1 + R_2 + R_3)$$
$$100\ \mu\text{A}(R_1 + 100\ \Omega) = (100\ mA - 100\ \mu\text{A})(R_2 + R_3)$$
$$100\ \mu\text{A}(R_1 + R_2 + 100\ \Omega) = (500\ \text{mA} - 100\ \mu\text{A})R_3$$

Performing the indicated arithmetic and rearranging leads to the equations

(1) $99R_1 + 99R_2 + 99R_3 = 100$

(2) $-R_1 + 999R_2 + 999R_3 = 100$

(3) $-R_1 - R_2 + 4999R_3 = 100$

Solving these three equations simultaneously yields:

subtracting (2) from (3),

$$-1000R_2 + 4000R_3 = 0$$
$$R_2 = 4R_3$$

substituting into (1) and (2)

$$99R_1 + 99(4R_3) + 99R_3 = 100$$
$$-R_1 + 999(4R_3) + 999R_3 = 100$$

simplifying,

$$99R_1 + 495R_3 = 100$$
$$-R_1 + 4995R_3 = 100$$

By elimination,

$$R_3 = 0.0202\ \Omega$$

then,

$$R_2 = 0.0808\ \Omega$$

and,

$$R_1 = 0.909\ \Omega$$

• • •

9.3 VOLTMETERS

The basic D'Arsonval meter movement can also be used to construct a voltmeter by placing a resistor R_V in *series* with the basic meter to limit the current. Consider the voltmeter shown in Fig. 9.12. The symbol for a voltmeter is shown in Fig. 9.12*a*, and the actual construction of the meter is shown in Fig. 9.12*b*.

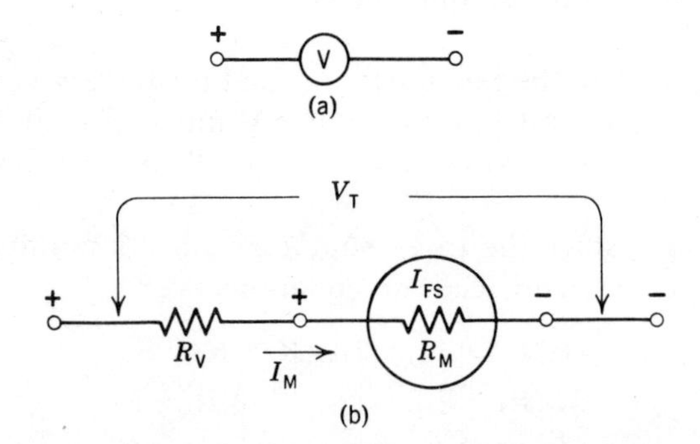

Figure 9.12 A basic voltmeter (*a*) Symbol (*b*) Construction of a voltmeter

Since a voltmeter is used to measure the voltage *across* an element, it must be connected in *parallel* with the element. This makes sense since the voltage across *parallel* elements is the same. The proper method for connecting a voltmeter to measure the voltage *across* a resistance R is shown in Fig. 9.13. Since the voltmeter has a large resistance R_V in series with a basic meter movement, the voltmeter must *never* be connected in *series* with other elements. Notice also that the correct meter polarity must be observed when measuring voltage.

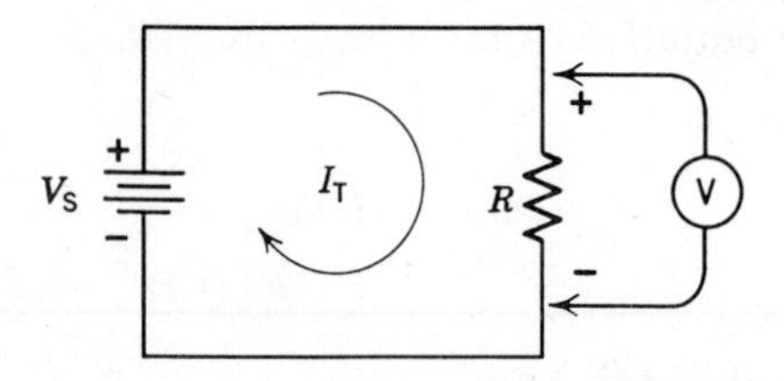

Figure 9.13 The proper connection of a voltmeter

• • •

Example 9.7 Show the proper connection of voltmeters to measure the voltages across all of the resistors in Fig. 9.7*a*.

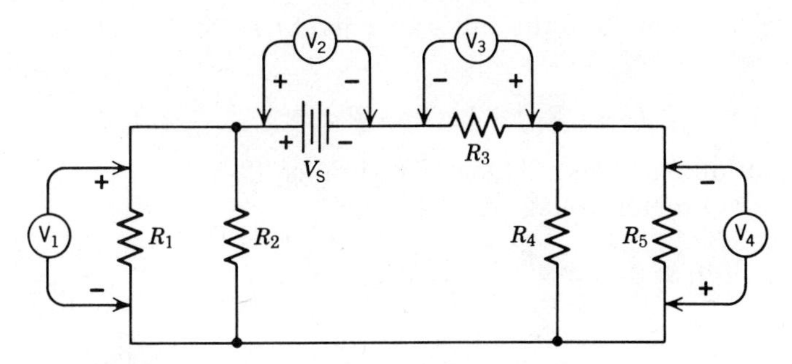

Figure 9.14 Proper voltmeter connections

Solution The proper connections are shown in Fig. 9.14. Note the correct polarities. V_2 measures the source voltage V_S, and the other meters measure the resistor voltages as follows:

$$\mathrm{V}_1 \rightarrow V_{R_1} = V_{R_2}$$

$$\mathrm{V}_3 \rightarrow V_{R_3}$$

$$\mathrm{V}_4 \rightarrow V_{R_4} = V_{R_5}$$

• • •

A voltmeter can also be used to measure the voltage at any point in a circuit with respect to ground as shown in Fig. 9.15. In Fig. 9.15*a*, all voltages in the circuit are more *positive* than ground, and thus the negative terminal of the meter is *grounded*, and the positive terminal is used as a *probe*. In Fig. 9.15*b*, all circuit voltages are more *negative* than ground, and thus the *negative* terminal of the meter is used as a probe.

Now, consider the basic voltmeter in Fig. 9.12. The series resistor R_V is used

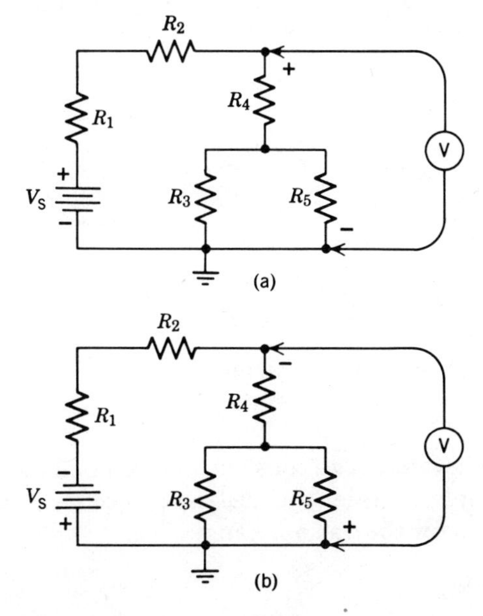

Figure 9.15 Measuring voltages with respect to ground (*a*) Positive voltages (*b*) Negative voltages

to limit the total meter current I_M to the full scale current I_{FS} at the full scale terminal voltage V_T. Thus,

$$V_T = I_M(R_V + R_M) = I_{FS}(R_V + R_M)$$

We can solve this equation for the series meter resistance R_V in terms of I_{FS}, R_M, and the full scale voltmeter voltage V_T. Thus,

$$R_V = \frac{V_T - I_{FS}R_M}{I_{FS}} = \frac{V_T}{I_{FS}} - R_M \tag{9.8}$$

• • •

Example 9.8 Determine the series resistor R_V to construct a 100 V full scale voltmeter using a basic meter movement having $I_{FS} = 200\ \mu\text{A}$ and $R_M = 100\ \Omega$.

Solution The meter is constructed as in Fig. 9.12, and using Eq. 9.8,

$$R_V = \frac{100\text{ V}}{200\ \mu\text{A}} - 100\ \Omega = 499.9\text{ k}\Omega$$

• • •

Example 9.9 What must be the value of R_V in Example 9.8 to provide a full scale voltage of 10 V?

Solution Using Eq. 9.8

$$R_V = \frac{10\text{ V}}{200\ \mu\text{A}} - 100\ \Omega = 49.99\text{ k}\Omega$$

• • •

The base design of a voltmeter can obviously be extended to construct a multiple range voltmeter using a range switch. Using the results of the previous two examples, and adding two more resistances, the multiple range voltmeter in Fig. 9.16 is easily constructed. The reader should verify the values of the other two resistors. Building a voltmeter in this way would not be very practical, however, since the resistors are all precision, and they are all of values which may be difficult to obtain.

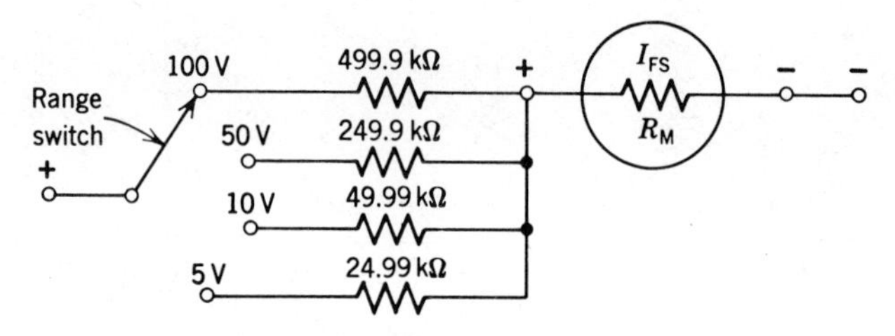

Figure 9.16 A multiple range voltmeter

A better approach would be to construct the voltmeter as shown in Fig. 9.17. In this case, we simply consider the *total* resistance R_T in the circuit for each voltage range. Thus on the various ranges,

1. $V_T = I_{FS}(R_1 + R_M)$ (9.9)

   and $R_1 = \dfrac{V_T}{I_{FS}} - R_M$

2. $V_T = I_{FS}(R_1 + R_2 + R_M)$

$$\text{and } R_2 = \frac{V_T}{I_{FS}} - R_1 - R_M$$

3. $V_T = I_{FS}(R_1 + R_2 + R_3 + R_M)$

$$\text{and } R_3 = \frac{V_T}{I_{FS}} - R_1 - R_2 - R_M$$

4. $V_T = I_{FS}(R_1 + R_2 + R_3 + R_4 + R_M)$

$$\text{and } R_4 = \frac{V_T}{I_{FS}} - R_1 - R_2 - R_3 - R_M$$

These equations can be continued for any number of ranges, and in general the N*th* resistor is,

$$R_N = \frac{V_T}{I_{FS}} - R_1 - R_2 - \cdots - R_{N-1} - R_M$$

• • •

Example 9.10 Using a basic meter movement having $I_{FS} = 200\ \mu A$ and $R_M = 100\ \Omega$, design a multiple range voltmeter as in Fig. 9.17 having 5 V, 10 V, 50 V, and 100 V ranges.

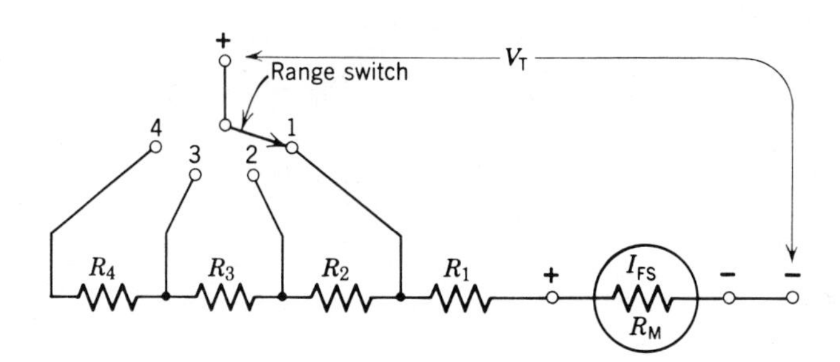

Figure 9.17 A practical multiple range voltmeter

Solution Using Equations 9.9,

5 V range:

$$R_1 = \frac{V_T}{I_{FS}} - R_M = \frac{5\text{ V}}{200\ \mu\text{A}} - 100\ \Omega = 24.9\text{ k}\Omega$$

10 V range:

$$R_2 = \frac{V_T}{I_{FS}} - R_1 - R_M = \frac{10\text{ V}}{200\ \mu\text{A}} - 24.9\text{ k}\Omega - 0.1\text{ k}\Omega = 25\text{ k}\Omega$$

50 V range:

$$R_3 = \frac{V_T}{I_{FS}} - R_1 - R_2 - R_M = \frac{50\text{ V}}{200\ \mu\text{A}} - 24.9\text{ k}\Omega - 25\text{ k}\Omega - 0.1\text{ k}\Omega = 200\text{ k}\Omega$$

100 V range:

$$R_4 = \frac{V_T}{I_{FS}} - R_1 - R_2 - R_3 - R_M$$

$$= \frac{100\text{ V}}{200\ \mu\text{A}} - 24.9\text{ k}\Omega - 25\text{ k}\Omega - 200\text{ k}\Omega - 0.1\text{ k}\Omega = 250\text{ k}\Omega$$

• • •

The voltmeter in this example is obviously more practical to build. Even though the resistors are precision, all but one are of a standard size, and they are therefore easy to obtain.

The input resistance of a voltmeter is a very important quantity because we can use it to determine how much a voltmeter *disturbs* or *loads* a circuit when connected for measuring voltage (we will discuss meter loading in Section 9.5). The resistance of an *ideal* voltmeter would be infinite, and it would not disturb a circuit at all when connected for measuring voltage. A real voltmeter does however have some finite resistance, and it will "rob" some current from any circuit to which it is connected; it will therefore *disturb* or *load* the circuit. The total resistance of the voltmeter R_{VM} on any range can be found as the ratio of the full scale voltage V_{FS} divided by the full scale current I_{FS} (Ohm's Law). Thus

$$R_{VM} = \frac{V_{FS}}{I_{FS}} \tag{9.10}$$

Thus if we know the full scale current of the meter movement used in a voltmeter we can always calculate its input resistance using this equation.

• • •

Example 9.11 Calculate the voltmeter resistances for each range in Example 9.10.

Solution Using Eq. 9.10,

5 V range:

$$R_{VM} = \frac{V_{FS}}{I_{FS}} = \frac{5\text{ V}}{200\ \mu A} = 25\text{ k}\Omega$$

10 V range:

$$R_{VM} = \frac{V_{FS}}{I_{FS}} = \frac{10\text{ V}}{200\ \mu\text{A}} = 50\text{ k}\Omega$$

50 V range:

$$R_{VM} = \frac{V_{FS}}{I_{FS}} = \frac{50\text{ V}}{200\ \mu\text{A}} = 250\text{ k}\Omega$$

100 V range:

$$R_{VM} = \frac{V_{FS}}{I_{FS}} = \frac{100\text{ V}}{200\ \mu\text{A}} = 500\text{ k}\Omega$$

• • •

Notice that the voltmeter resistance increases on the higher voltage ranges. In fact, the voltmeter resistance is seen to be *directly proportional* to the full scale voltage. That is,

$$R_{VM} = V_{FS} \times \frac{1}{I_{FS}}$$

In this equation, the quantity $1/I_{FS}$ is very important, and it is defined as the voltmeter *sensitivity*. Thus

$$\text{sensitivity} \triangleq \mathcal{S} \triangleq \frac{1}{I_{FS}} \tag{9.11}$$

The symbol $\triangleq$ means "is defined as." The sensitivity $\mathcal{S}$ is a constant for any given meter movement, and it is usually printed on the face of the meter. The units of sensitivity are,

$$\mathcal{S} = \frac{1}{I_{FS}} = \frac{1}{V_{FS}/R_M} = \frac{1}{\text{V}/\Omega} = \frac{\Omega}{\text{V}}$$

Thus meter sensitivity is expressed as *Ohms per volt*, Ω/V. If the meter sensitivity is known, it can be used to find the full scale meter current, and to find the total voltmeter resistance on any voltage scale.

• • •

Example 9.12 What is the sensitivity of a basic meter movement having $I_{FS} = 100\ \mu A$ and $R_M = 50\ \Omega$?

Solution Using Eq. 9.11,

$$\mathcal{S} = \frac{1}{I_{FS}} = \frac{1}{100\ \mu A} = 10\ k\Omega/V$$

• • •

Example 9.13 What is the full scale current of a meter having a sensitivity of 20 kΩ/V?

Solution Eq. 9.11 can be rearranged as,

$$I_{FS} = \frac{1}{\mathcal{S}}$$

Thus,

$$I_{FS} = \frac{1}{\mathcal{S}} = \frac{1}{20\ k\Omega/V} = 50\mu\ A$$

• • •

Example 9.14 Find the input resistances of the voltmeter in Example 9.10 using the meter sensitivity. Compare with Example 9.11.

Solution The meter sensitivity is

$$\mathcal{S} = \frac{1}{I_{FS}} = \frac{1}{200\ \mu A} = 5\ k\Omega/V$$

The meter resistance on each range is then found as,

$$R_{VM} = V_{FS}\mathcal{S}$$

$$\text{5 V range:}\quad R_{VM} = 5\ V \times 5\ k\Omega/V = 25\ k\Omega$$

$$\text{10 V range:}\quad R_{VM} = 10\ V \times 5\ k\Omega/V = 50\ k\Omega$$

$$\text{50 V range:}\quad R_{VM} = 50\ V \times 5\ k\Omega/V = 250\ k\Omega$$

$$\text{100 V range:}\quad R_{VM} = 100\ V \times 5\ k\Omega/V = 500\ k\Omega$$

• • •

Another interesting and important use of meter sensitivity can be found by examining the voltmeter in Fig. 9.18. The meter sensitivity is seen to be

$$\mathcal{S} = \frac{1}{I_{FS}} = \frac{1}{100\ \mu A} = 10\ k\Omega/V$$

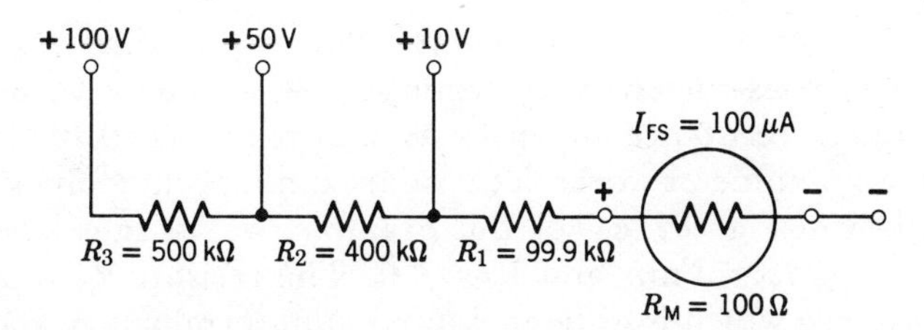

Figure 9.18 A multiple range voltmeter

If the meter is now used to measure 100 V, the current in the meter is $I_{FS} = 100\ \mu A$. The voltage drop across R_3 is then

$$V_{R_3} = I_{FS}R_3 = 100\ \mu A \times 500\ k\Omega = 50\ V$$

Notice that the voltage V_{R_3} is exactly equal to the *difference* between the +100 V and +50 V positions. Similarly, the voltage across R_2 is

$$V_{R_2} = I_{FS}R_2 = 100\ \mu A \times 400\ k\Omega = 40\ V$$

Again, this is exactly equal to the *difference* between the +50 V and +10 V positions. We can now rearrange these equations to show

$$R_3 = \frac{V_{R_3}}{I_{FS}} = \mathcal{S} V_{R_3} = \mathcal{S}(\Delta V)$$

where ΔV is the difference between the two voltage positions on either side of R_3. Similarly,

$$R_2 = \frac{V_{R_2}}{I_{FS}} = \mathcal{S} V_{R_3} = \mathcal{S}(\Delta V)$$

This result can of course be extended to any resistor in a voltmeter of this type, and thus

$$R_N = \mathcal{S}\Delta V \tag{9.12}$$

• • •

Example 9.15 The resistor R_2 is "burned out' in the voltmeter shown in Fig. 9.19. What value of resistance must be used to replace it?

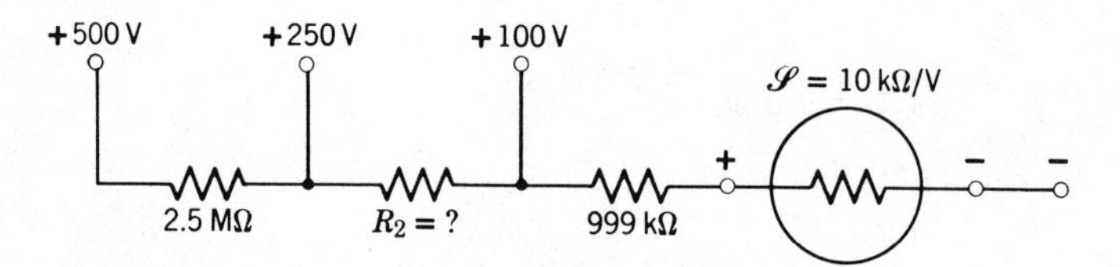

Figure 9.19 A damaged voltmeter

Solution Using equation 9.12, the ΔV across R_2 is,

$$\Delta V = 250\ V - 100\ V = 150\ V$$

then,

$$R_2 = \mathcal{S}\Delta V = 10\ k\Omega/V \times 150\ V = 1.5\ m\Omega$$

9.4 OHMMETERS

An *ohmmeter* is a basic dc instrument used to measure resistance. An ohmmeter can be constructed by connecting a voltage source V_S, a basic meter movement, and a resistance in series as shown in Fig. 9.20. In order to discover how an ohmmeter works, let's assign some specific values to the basic circuit as shown in Fig. 9.21*a*. For this example, we assume a *perfect* meter movement having $I_{FS} = 1$ mA and $R_M = 0\ \Omega$. The resistor R_X represents the *unknown* resistance which is to be measured, and terminals A and B are the ohmmeter terminals.

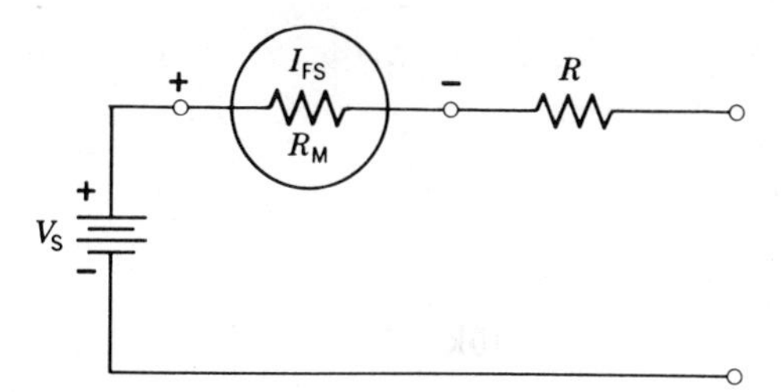

Figure 9.20 A basic ohmmeter

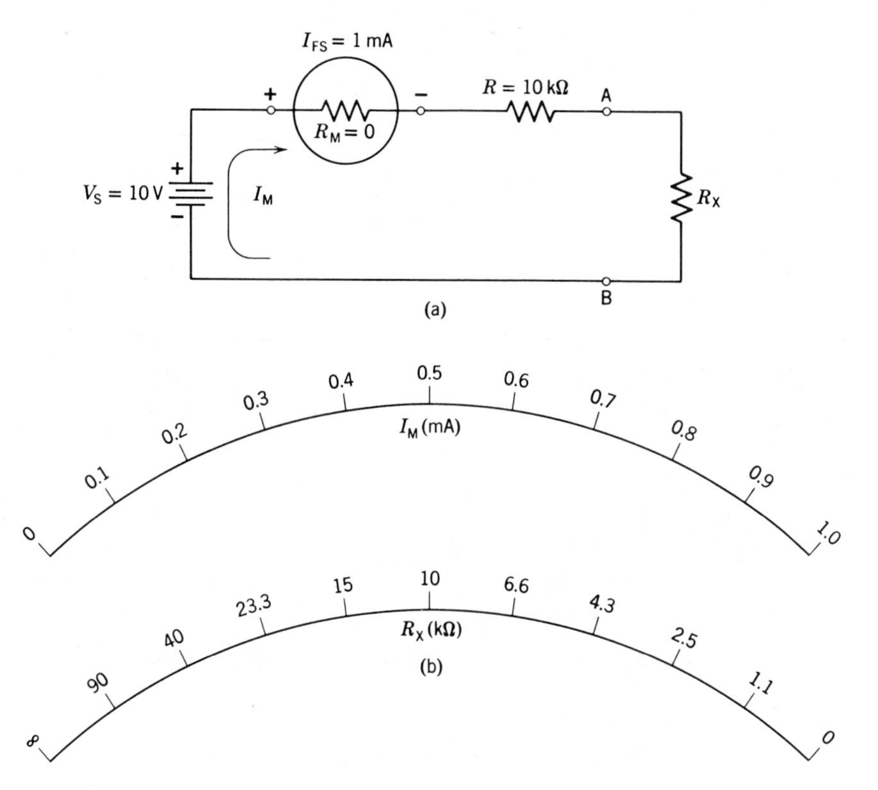

Figure 9.21 (*a*) An ohmmeter (*b*) The ohms scale compared with the meter current scale

Now, suppose the ohmmeter terminals are *shorted*, corresponding to a resistance $R_X = 0\ \Omega$. Then,

$$I_M = \frac{V_S}{R} = \frac{10\ \text{V}}{10\ \text{k}\Omega} = 1\ \text{mA}$$

Under this condition, I_M is equal to the full scale current of the basic meter movement, that is, $I_M = I_{FS} = 1$ mA. The pointer will then be over 1.0 on the *current* scale, and since this is obtained for $R_X = 0$, we can mark this position as zero on the *ohms* scale. This is done in Fig. 9.21*b*.

If we now remove R_X from the ohmmeter, the terminals are left *open* circuited, and this corresponds to $R_X = \infty\ \Omega$. The meter current I_M will obviously be zero, and we can therefore mark the zero position on the *current* scale as ∞ on the *ohms* scale (see Fig. 9.21*b*).

Now, for any value of R_X between the extremes, zero and infinity, the circuit current I_M is given by

$$I_M = \frac{V_S}{R + R_X} \tag{9.13a}$$

This equation can be solved for R_X as

$$R_X = \frac{V_S}{I_M} - R \tag{9.13b}$$

Let us now make a table of values of R_X for various values of I_M using Eq. 9.13b. The results are given in Table 9.1, and the reader is urged to verify the entries.

Table 9.1. R_X versus I_M for meter in Fig. 9.21.

I_M(mA)	0.1	0.2	0.3	0.4	0.5	0.6	0.7	0.8	0.9	1.0
R_X(Ω)	90k	40k	23.3k	15k	10k	6.6k	4.3k	2.5k	1.1k	0

These values are also marked on the scale shown in Fig. 9.21*b*. From these scales we make four important observations:

1. Full scale on the meter corresponds to zero Ω.
2. Zero on the meter ($I_M = 0$) corresponds to infinite Ω.
3. One-half scale on the meter is exactly equal to the total meter resistance; R in this case, and $(R + R_M)$ in general.
4. The ohms scale is not linear (the number of Ω between equidistant marks is *not* the same over the entire scale), and thus it is inconvenient to mark the resistance in Ω on the meter marks for current.

This last observation would suggest marking the *ohms* scale by calculating the current I_M for various desired values of R_X using Eq. 9.13a. This is in fact more convenient, and some appropriate calculations are given in Table 9.2 (the reader should use Eq. 9.13a to verify the entries). A better *ohms* scale for this ohmmeter is marked in Fig. 9.22 using the calculations from Table 9.2.

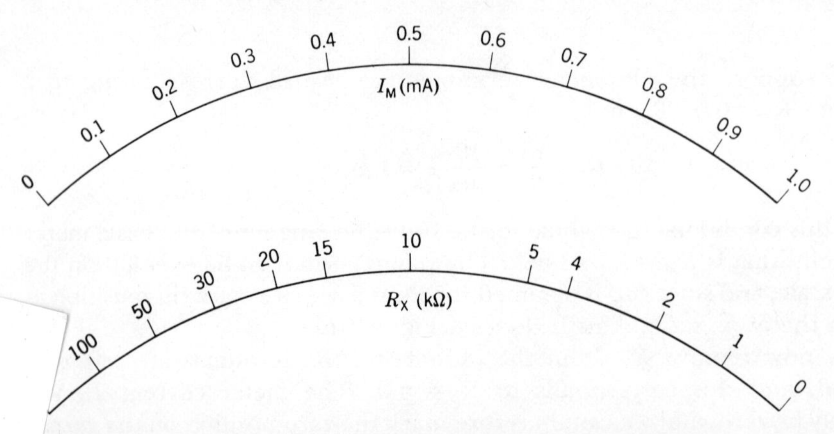

igure 9.22 An ohms scale for the ohmmeter in Fig. 9.21

Table 9.2. I_M versus R_X for meter in Fig. 9.21.

$R_X(\Omega)$	1k	2k	4k	5k	10k	15k	20k	30k	50k	100k
I_M(mA)	0.91	0.833	0.715	0.667	0.5	0.4	0.333	0.25	0.167	0.091

• • •

Example 9.16 Where should marks be placed on the ohms scale in Fig. 9.22 for $R_X = 3\ \text{k}\Omega$, and $R_X = 500\ \text{k}\Omega$?

Solution Using Eq. 9.13a,

for $R_X = 3\ \text{k}\Omega$,

$$I_M = \frac{V_S}{R + R_X} = \frac{10\ \text{V}}{10\ \text{k}\Omega + 3\ \text{k}\Omega} = 0.77\ \text{mA}$$

for $R_X = 500\ \text{k}\Omega$,

$$I_M = \frac{10\ \text{V}}{10\ \text{k}\Omega + 500\ \text{k}\Omega} = 0.0196\ \text{mA}$$

We can use the results of the previous discussion to establish a set of rules for designing an ohmmeter having any range. This is useful since it would be very difficult to measure a resistance of 93 Ω or 680 kΩ with any accuracy using the ohms scale in Fig. 9.22. (You might check this by calculating I_M using Eq. 9.13a and trying to locate the proper points on the meter face!) We note first of all that the center position (half-scale) is exactly equal to the total ohmmeter resistance. This is the key in determining the range of values of resistance which can be measured. If for example we wanted an ohmmeter capable of measuring 93 Ω, we might make the *half-scale* point 100 Ω. If we wanted to measure 680 kΩ, we might make the *half-scale* point 500 kΩ, and so on.

Determining the *half-scale* resistance fixes the value of R (Fig. 9.20), and we then choose a value of V_S which will cause full scale current in the meter movement when the terminals A-B are shorted.

Thus, having determined R and V_S, we then use Eq. 9.13a to lay out the ohmmeter scale. In summary, an ohmmeter such as that shown in Fig. 9.20 can be designed by:

1. Determine the center point (half-scale) on the ohmmeter face. This is exactly equal to $(R + R_M)$ in the ohmmeter.
2. For the meter movement used (I_{FS} and R_M given), determine V_S from

$$V_S = I_{FS}(R + R_M)$$

3. Lay out the ohmmeter scale using Eq. 9.13a.
4. A multiple scale ohmmeter is simply a number of these basic ohmmeters connected by a switch (Fig. 9.23). The potentiometer is a part of R (Fig. 9.20) and it eliminates the need for using precision resistors (it also accounts for changes in V_S due to aging). The potentiometer must be adjusted as the ohmmeter is switched from scale to scale, and this is accomplished by setting the pointer to read zero Ω when the ohmmeter terminals are *shorted* (i.e., zeroing the meter).*

*When switching from scale to scale, either V_S must be changed or a shunt resistor must be placed (switched) in parallel with R_M.

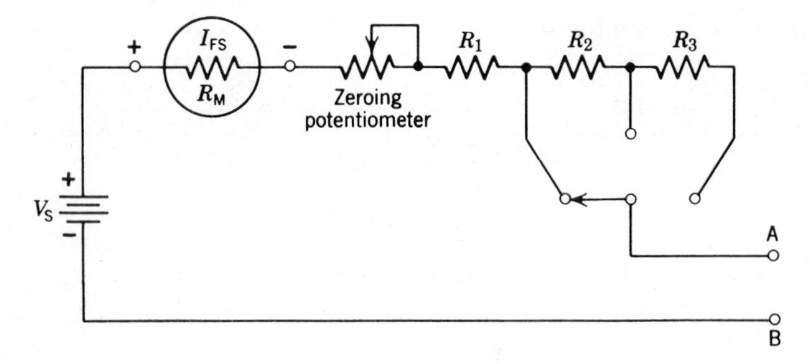

Figure 9.23 A multiple scale ohmmeter

• • •

Example 9.17 Using a basic meter movement having $I_{FS} = 2$ mA and $R_M = 100\ \Omega$, design a simple ohmmeter having 1 kΩ center scale.

Solution We proceed according to the design steps given above:

1. The meter circuit is shown in Fig. 9.24. Since center scale must be 1 kΩ, then $(R + R_M) = 1$ kΩ. Therefore, make $R_1 = 470\ \Omega$ and $R_P = 1$ kΩ. This will ensure that the total meter resistance can be adjusted to 1 kΩ. The series resistor R_1 is used with the potentiometer R_P to ensure that the meter movement is protected at all times. Why?

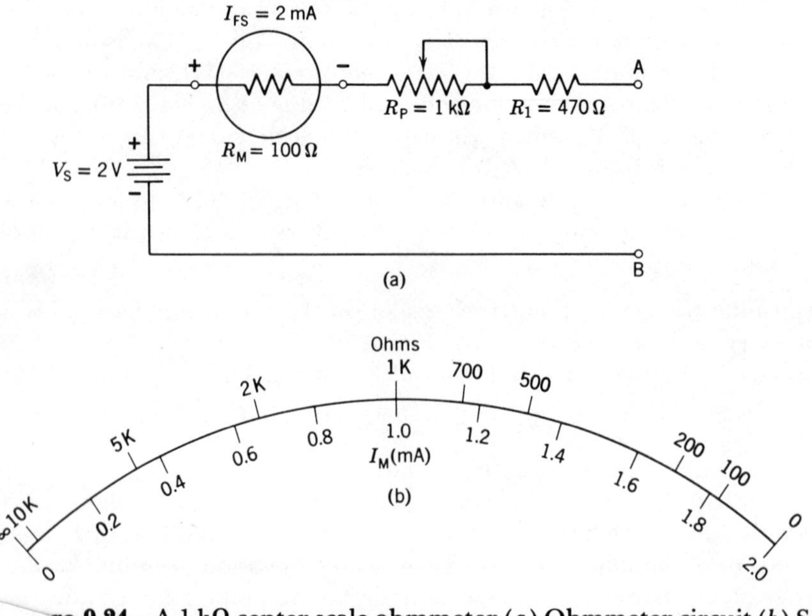

Figure 9.24 A 1 kΩ center scale ohmmeter (*a*) Ohmmeter circuit (*b*) Scale

V_S is now found as,

$$V_S = I_{FS}(R + R_M) = 2\text{ mA }(1\text{ k}\Omega) = 2\text{ V}$$

3. We use Eq. 9.13a to calculate the following values of I_M versus R_X. The scale is shown in Fig. 9.24*b*.

R_X (Ω)	100	200	500	700	1 k	2 k	5 k	10 k
I_M (mA)	1.82	1.67	1.33	1.18	1.0	0.667	0.333	0.0167

4. This step is not necessary since this is not a multiple range meter.

• • •

9.5 APPLICATIONS

In using an ammeter to measure current we have seen that the meter must be connected in series in the circuit. An *ideal* ammeter has zero resistance and therefore does not *add* any additional resistance to the circuit. The use of an ideal ammeter will not disturb the original circuit in any way, and therefore will give an exact reading of the current in the circuit. A *real* ammeter does however have some small resistance, and it is equal to $R_M \| R_{SH}$ as shown in Fig. 9.8. The total resistance of the ammeter is almost always negligibly small when compared with the total circuit resistance of the circuit. This is a very fortunate state of affairs, since it means that real ammeters will usually not disturb a circuit and thus they will give accurate readings.

• • •

Example 9.18 A 1 mA full scale ammeter is used to measure the current in the circuit of Fig. 9.25. The ammeter has a total resistance of $(R_M \| R_{SH}) = 0.2\ \Omega$. Compare the total circuit resistance with and without the ammeter. How does the use of the ammeter affect the circuit current?

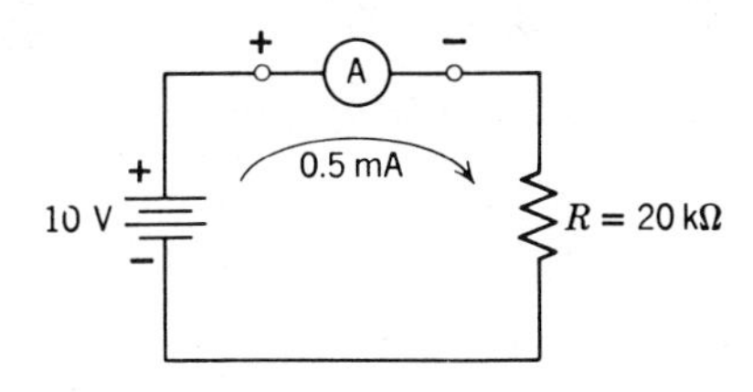

Figure 9.25

Solution Assuming an ideal source, the total circuit resistance without the ammeter is 20 kΩ, or 20,000 Ω. With the ammeter included, the total circuit resistance is $R + (R_M \| R_{SH}) = 20\ \text{k}\Omega + 0.2\ \Omega = 20{,}000.2\ \Omega$. The ammeter obviously does not affect the total circuit resistance *or* the circuit current to any appreciable extent.

• • •

A voltmeter must be connected in parallel (across an element in order to measure voltage. When using a voltmeter, some care must be taken since the voltmeter may very well disturb the circuit. The loading effect of a voltmeter is best illustrated by considering some examples.

• • •

Example 9.19 Calculate the actual voltage across the resistor R in Fig. 9.26, and the voltage measured by the voltmeter. Consider first a voltmeter having

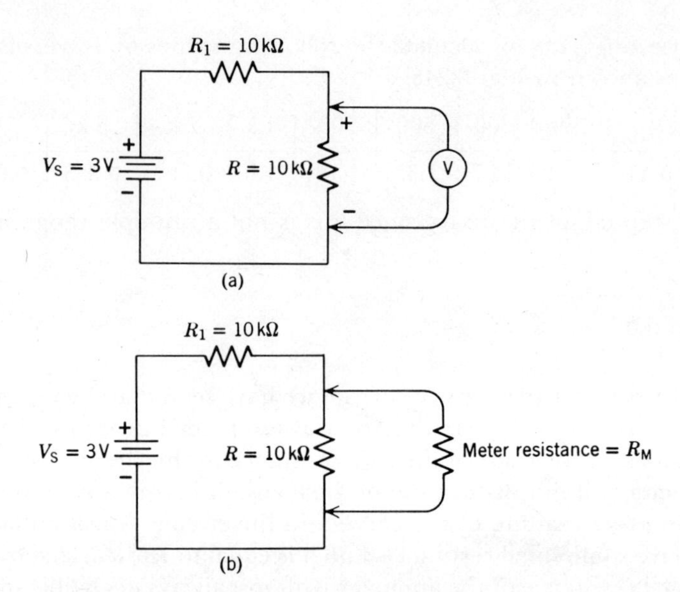

Figure 9.26 Voltmeter loading error (*a*) Original circuit (*b*) Accounting for voltmeter loading

a sensitivity of 2 kΩ/V used on the 5 V scale, and then a voltmeter having a sensitivity of 20 kΩ/V used on the 5 V scale.

Solution Using the 2 kΩ/V meter:
On the 5 V scale, the voltmeter has a total resistance of

$$R_M = V \times \mathcal{S} = 5\text{ V} \times 2\text{ k}\Omega/\text{V} = 10\text{ k}\Omega$$

Thus the voltage across the voltmeter terminals is

$$V_R = V_{RM} = V_S \frac{(R_M \| R)}{R_1 + (R_M \| R)} = 3\text{ V} \frac{(10\text{ k}\Omega \| 10\text{ k}\Omega)}{10\text{ k}\Omega + (10\text{ k}\Omega \| 10\text{ k}\Omega)} = 1\text{ V}$$

The actual circuit voltage without the voltmeter connected is

$$V_R = V_S \frac{R}{R_1 + R} = 3\text{ V} \times \frac{10\text{ k}\Omega}{10\text{ k}\Omega + 10\text{ k}\Omega} = 1.5\text{ V}$$

Thus the 2 kΩ/V voltmeter has introduced considerable error in this case.
Using the 20 kΩ/V voltmeter:
On the 5 V scale, the voltmeter has a total resistance of

$$R_M = V \times \mathcal{S} = 5\text{ V} \times 20\text{ k}\Omega/\text{V} = 100\text{ k}\Omega$$

The voltage across the voltmeter terminals is then

$$V_{RM} = V_S \frac{(R_M \| R)}{R_1 + (R_M \| R)} = 3\text{ V} \times \frac{(10\text{ k}\Omega \| 100\text{ k}\Omega)}{10\text{ k}\Omega + (10\text{ k}\Omega \| 100\text{ k}\Omega)} = 1.43\text{ V}$$

e actual circuit voltage is 1.50 V, and the meter is seen to introduce *ding* error.

• • •

the previous two examples, it is clear that voltmeters can indeed circuits to which they are connected, and we should always be aware of the *loading* error introduced by voltmeters. In all fairness, it must be pointed out that these two examples were chosen specifically to demonstrate voltmeter loading, and the loading error is not usually so large in a typical case.

Another important fact to be noted from these two examples is that voltmeter sensitivity (Ω/V) is a measure of the *quality* of the meter, that is, the *higher* the sensitivity in Ω/V, the *better* the voltmeter.

In using ohmmeters, the accuracy is primarily a function of the accuracy of the basic meter movement used. One word of caution, however: an ohmmeter must *never* be connected to a circuit containing sources (voltage or current) since the resistance measurement will surely be incorrect, and the ohmmeter may even be damaged. Why? (Draw an equivalent circuit using a battery in series with R_X in Fig. 9.21).

9.6 WHEATSTONE BRIDGE

The ohmmeter discussed in the previous sections will provide quite accurate resistance measurements so long as the terminal voltage of the voltage source remains constant. However, as the source ages, the terminal voltage will vary and the resistance measurements will soon reflect an unacceptable error. Thus, one of the disadvantages of this type of ohmmeter is the fact that the measurements depend on the terminal voltage of the source.

A circuit which overcomes this disadvantage is the *wheatstone bridge* shown in Fig. 9.27. This circuit can be used to measure an unknown resistance R_X in terms of the other three resistances, R_1, R_2, and R_3. A galvanometer is placed between nodes A and B, and R_3 is adjusted until the galvanometer shows zero deflection. This of course means there is no current through the galvanometer, and under this condition the bridge is said to be *balanced.*

When the bridge is balanced, the voltages at nodes A and B must be exactly the same since there is no current through the galvanometer. The voltages at nodes A and B can be found using the voltage divider theorem. Thus,

$$V_A = V_S \frac{R_2}{R_1 + R_2} \qquad V_B = V_S \frac{R_3}{R_X + R_3}$$

Since $V_A = V_B$, these two expressions can be equated to obtain

$$V_S \frac{R_2}{R_1 + R_2} = V_S \frac{R_3}{R_X + R_3}$$

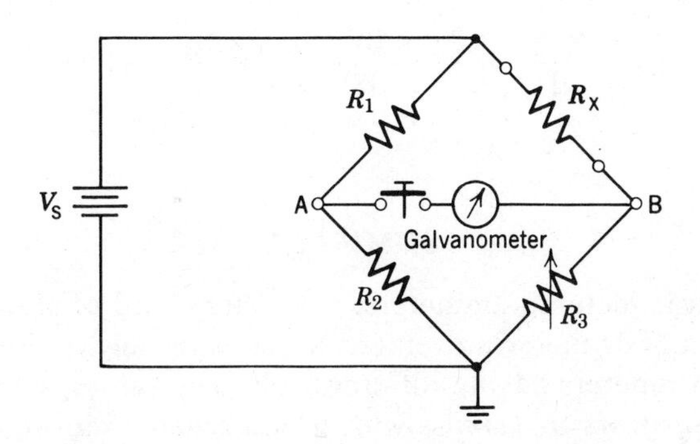

Figure 9.27

V_S can then be eliminated from both sides to obtain

$$\frac{R_2}{R_1 + R_2} = \frac{R_3}{R_X + R_3}$$

Finally, this expression can be simplified to

$$R_2(R_X + R_3) = R_3(R_1 + R_2)$$
$$R_2R_X + R_2R_3 = R_1R_3 + R_2R_3$$
$$R_2R_X = R_1R_3 \tag{9.14}$$

Equation 9.14 is known as the *bridge balance equation,* and it can be used to determine whether or not a bridge is balanced, or it can be used to determine an unknown resistance in terms of the other three known resistances. For example, Eq. 9.14 can be solved for the unknown resistance R_X as

$$R_X = \frac{R_1R_3}{R_2} \tag{9.15}$$

Notice that Eq. 9.15 does not depend on the value of the bridge source voltage V_S. This is a distinct advantage of the wheatstone bridge over the basic ohmmeter discussed previously. If resistors R_1, R_2, and R_3 are precision resistors, the value of R_X can be determined with a high degree of accuracy. Thus the wheatstone bridge is used as a *precision* resistance-measuring device.

• • •

Example 9.20 The resistors in the wheatstone bridge in Fig. 9.27 are: $R_1 = 10.0\ \text{k}\Omega$, $R_2 = 7.00\ \text{k}\Omega$, $R_3 = 3.92\ \text{k}\Omega$, $R_X = 5.60\ \text{k}\Omega$. Is the bridge balanced?

Solution From the bridge balance equation, Eq. 9.14,

$$R_2R_X = R_1R_3$$
$$7.00 \times 10^3 \times 5.60 \times 10^3 = 10.0 \times 10^3 \times 3.92 \times 10^3$$
$$39.2 \times 10^6 = 39.2 \times 10^6$$

Therefore, the bridge is balanced.

• • •

Example 9.21 In the wheatstone bridge in Fig. 9.27, $R_1 = 50.0\ \text{k}\Omega$, and $R_2 = 100\ \text{k}\Omega$. What must be the value of the unknown resistor R_X if R_3 must be adjusted to 68.4 kΩ in order to balance the bridge?

Solution Using Eq. 9.15,

$$R_X = \frac{R_1R_3}{R_2} = \frac{50.0 \times 10^3 \times 68.4 \times 10^3}{100 \times 10^3} = 34.2\ \text{k}\Omega$$

• • •

SUMMARY

The three basic meters—ammeters, voltmeters, and ohmmeters—are constructed using a basic meter movement which is usually of the D'Arsonval or Weston type. Ammeters having different full scale values, are constructed by placing *shunt* resistors in *parallel* with a basic meter movement. Ammeters must always be connected in *series* to measure current, and they introduce very little loading error.

Voltmeters having different full scale voltage ranges are constructed by connecting resistors in *series* with a basic meter movement. Voltmeters are always connected in *parallel* with an element in order to measure the existing voltage, and they may introduce considerable loading error.

Ohmmeters are constructed by connecting a basic meter movement in series with a resistance and a source. Center scale on the ohms scale is exactly equal to the total ohmmeter resistance. The ohmmeter scale is nonlinear and it must be calculated (using Eq. 9.13a). Ohmmeters must *never* be connected to a circuit containing a voltage or current source.

The wheatstone bridge is used for precision resistance measurements and has the advantage of being independent of source voltage.

GLOSSARY

Ammeter. *A basic meter used to measure current.*

Ayrton shunt. *A special shunt used in ammeters which protects the basic meter movement. Also called a Universal shunt.*

D'Arsonval meter. *A basic meter movement based on a rotating coil mounted in the magnetic field of a permanent magnet. Also called a Weston meter.*

Galvanometer. *A very sensitive D'Arsonval meter movement.*

Loading error. *The change in a circuit current or voltage caused by the connection of an ammeter or voltmeter.*

Ohmmeter. *A basic meter used to measure resistance.*

Sensitivity. *A measure of the quality of a voltmeter. Expressed in Ω/V.*

Voltmeter. *A basic meter used to measure voltage.*

Wheatstone bridge. *A precision resistance-measuring circuit.*

IMPORTANT RELATIONSHIPS

- Resistance of a basic meter movement $R_M = \dfrac{V_M}{I_M}$
- Ammeter shunt resistance $R_{SH} = \dfrac{I_{FS}R_M}{I_T - I_{FS}}$ (9.4)
- Voltmeter resistor $R_V = \dfrac{V_T}{I_{FS}} - R_M$ (9.8)
- Voltmeter sensitivity $\mathcal{S} \triangleq \dfrac{1}{I_{FS}}$ (9.11)
- Ohmmeter scale $I_M = \dfrac{V_S}{R + R_X}$ (9.13a)
- Bridge balance equation $R_2R_X = R_1R_3$ (9.14)

EVIEW QUESTIONS

1. The torque in a D'Arsonval meter is due to the interaction between two magnetic fields. (TF)
2. The pointer on a basic meter movement will move in opposite directions if the polarity of the coil current is _____.

3. A very sensitive D'Arsonval meter is called a ____.
4. An ideal meter movement has ____ resistance, while an actual meter does have some resistance.
5. Forcing a current greater than full scale through a basic meter movement may permanently damage the meter. (TF)
6. There is no such thing as meter polarity. (TF)
7. The resistance of a basic meter movement can be found by dividing the terminal (voltage, current) of the meter by the meter (voltage, current).
8. An ammeter must always be connected in a circuit in ____ in order to measure current.
9. The full scale current of an ammeter can be increased by placing a resistor in (series, parallel) with the meter.
10. An Ayrton shunt protects the basic meter movement of an ammeter since there is always a resistance in parallel with the basic movement. (TF)
11. An Ayrton shunt eliminates the need for using a ____ ____ ____ type switch with a multirange ammeter.
12. An ammeter will almost surely be damaged if it is connected across the terminals of a voltage source. (TF)
13. A voltmeter must always be connected in (series, parallel) to measure the voltage across an element.
14. The full scale voltage of a voltmeter can be increased by placing a resistor in (series, parallel) with the meter.
15. A voltmeter may be damaged if it is connected in series in a low resistance circuit. (TF)
16. The sensitivity of a voltmeter is defined as the reciprocal of the ____ ____ ____ of the basic meter movement.
17. The sensitivity of a voltmeter can be used to determine the total meter resistance. (TF)
18. The total resistance in the circuit of an ohmmeter is equal to the scale on the ohms scale.
19. The ohms scale on an ohmmeter (is, is not) linear.
20. An ohmmeter is *zeroed* by (shorting, opening) its terminals.
21. If an ohmmeter is connected to a circuit containing a voltage source, the resistance measurement will probably be inaccurate. (TF)
22. The resistance of an ammeter is usually small enough to be neglected when measuring current. (TF)
23. When a voltmeter loads a circuit, the voltage reading will be (smaller, larger) than the actual circuit voltage.
24. A high quality voltmeter has a (high, low) sensitivity in Ω/V.
25. One of the advantages of using a wheatstone bridge for resistance measurement is the fact that a bridge is independent of ____ ____.

PROBLEMS

1. The D'Arsonval meter movement shown in Fig. 9.4 has an R_M of 50 Ω and $I_{FS} = 100\ \mu A$. What is the terminal voltage of the meter at 100 μA?
2. What is the terminal voltage of a D'Arsonval meter having an R_M of 100 Ω and $I_{FS} = 500\ \mu A$?

3. If the voltage across a D'Arsonval meter is 20 mV when the current through it is 400 μA, what must be the meter resistance?
4. What must be the full scale current of a D'Arsonval meter having an R_M of 25 Ω if the terminal voltage at full scale current is 25 mV?
5. Find the proper shunt resistor to construct a 10 mA full scale ammeter using a D'Arsonval meter having $I_{FS} = 0.5$ mA and $R_M = 100\ \Omega$.
6. Find the proper shunt resistor to construct a 2 mA full scale ammeter using the meter movement in Problem 5.
7. What is the total ammeter resistance of the meter constructed in Problem 5, and what is its terminal voltage at full scale current?
8. What is the total ammeter resistance of the meter constructed in Problem 6, and what is its terminal voltage?
9. What will be the Ayrton shunt resistor values in Example 9.6 if a 1 mA meter movement is used, and all other values are the same?
10. Determine the series resistor needed to construct a 10 V full scale voltmeter using a basic meter movement having $I_{FS} = 0.5$ mA and $R_M = 200\ \Omega$.
11. Determine the series resistor needed to construct a 500 V full scale voltmeter using a basic meter movement having $I_{FS} = 1.0$ mA and $R_M = 200\ \Omega$.
12. Find the resistor values to construct a multiple range voltmeter as shown in Fig. 9.17 having ranges of 1 V, 10 V, 50 V, and 100 V. $I_{FS} = 0.5$ mA and and $R_M = 50\ \Omega$.
13. What is the total resistance of the voltmeter in Problem 10 and Problem 11?
14. Calculate the total resistance on each range for the voltmeter constructed in Problem 12.
15. What are the sensitivities of the voltmeters in Problems 10, 11, and 12? Which is the better voltmeter?
16. What is the full scale current of a voltmeter having a sensitivity of 50 kΩ/V?
17. A 10 V full scale voltmeter has a sensitivity of 20 kΩ/V. What resistance could be connected in series with it to construct a voltmeter having a full scale voltage of 50 V?
18. What series resistance could be used to change the original voltmeter in Problem 17 into a 100 V full scale voltmeter?
19. If the sensitivity of the voltmeter in Fig. 9.19 were 20 kΩ/V, what would the three resistor values have to be? Assume $R_M = 1$ kΩ.
20. What would the resistor values have to be in Fig. 9.19 if a meter having a sensitivity of 100 kΩ/V were used? $R_M \approx 0\ \Omega$.
21. What is the meter current in the ohmmeter used in Example 9.16 if $R_X = 8$ kΩ, and 200 kΩ?
22. Using a basic meter movement having $I_{FS} = 1$ mA and $R_M = 50\ \Omega$, design a simple ohmmeter having 1 kΩ center scale.
23. Design a second ohms scale for the ohmmeter in Problem 22 having a 10 kΩ center scale.
24. What would the ohmmeter in Fig. 9.24 read if it were accidentally connected to a 1 kΩ resistor in series with a 2 V battery?
25. Is it possible to alter the ohmmeter in Fig. 9.24 to have a 100 Ω center scale by simply changing resistor values? Why, or why not?

26. Alter the ohmmeter in Fig. 9.24 to have 500 Ω center scale. Change the resistance values (R_P and R_1) if necessary, and shunt the meter movement. Leave $V_S = 2.0$ V.

27. Calculate the actual and measured voltages across R_2 in Fig. 9.28 if the voltmeter sensitivity is 20 kΩ/V. The voltmeter is on the 20 V scale.

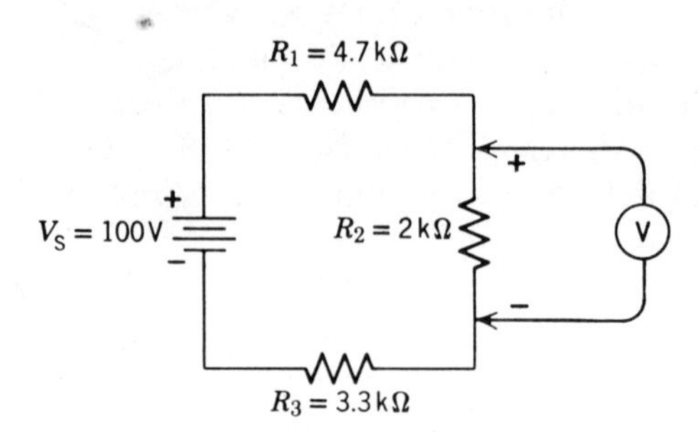

Figure 9.28

28. What are the actual and measured voltages in Problem 27 if the voltmeter has a sensitivity of 100 kΩ/V?

29. What would be the voltmeter reading in Problem 27 if the voltmeter were used on the 50 V scale?

30. The resistances in Fig. 9.27 have values of: $R_1 = 16$ kΩ, $R_2 = 21$ kΩ, $R_3 = 17$ kΩ, and $R_X = 18$ kΩ. Is the bridge balanced?

31. In Fig. 9.27, R_3 is adjusted to 281 Ω in order to balance the bridge. What must be the value of the unknown resistor R_X if $R_1 = 1000$ Ω and $R_2 = 500$ Ω?

32. What must be the value of R_X in Fig. 9.27 if $R_1 = 1.00$ MΩ, $R_2 = 1.00$ kΩ, and R_3 must be set at 915 Ω for balance?

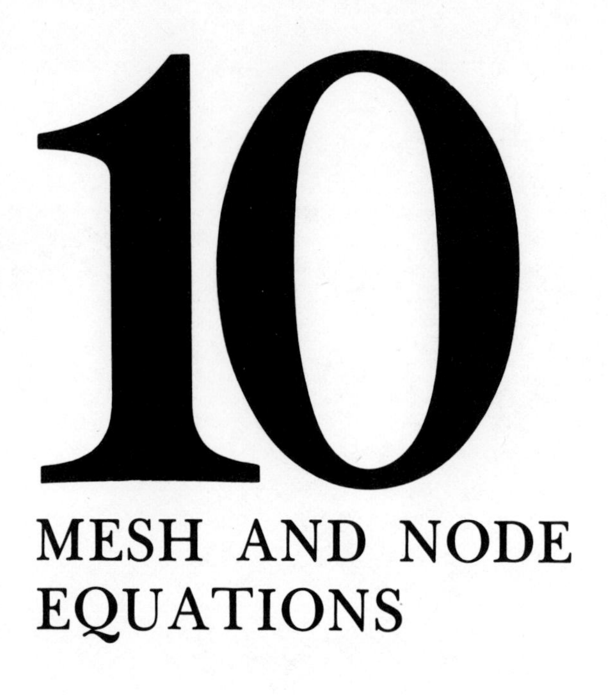

MESH AND NODE EQUATIONS

Network theorems and simplification techniques provide the most direct methods for solving simple circuits, but a more general, systematic approach is needed for more complicated circuits. The two most widely used methods for finding a complete solution for a circuit are based on KVL and KCL, and they are called *loop* (*mesh*) equations and *node* equations, respectively.* Mesh equations and node equations are really nothing new, since they are based on the already familiar ideas from KVL and KCL; they do however represent a very orderly and systematic method for finding solutions for networks. The great importance of mesh and node equations can be emphasized by making the following observations:

1. An orderly, systematic approach is desirable, since once mastered, it becomes an "old friend" which can be applied to any circuit with a high degree of *confidence.*
2. An orderly, systematic approach will greatly reduce the chance of error in finding a solution for a circuit.
3. Both mesh and node equations can be applied with equal ease to both dc and ac circuits, and a mastery of the techniques here will greatly simplify the analysis of ac circuits in the latter portion of this book.
4. The solution for a circuit found using mesh equations can be checked by solving the circuit using node equations, and vice versa.

*As we shall see, mesh equations are a special case of the more general method of loop equations.

We will devote this entire chapter to the *writing* of mesh and node equations, the *solution* of mesh and node equations, and the *use* of mesh and node equations to find solutions for electric circuits.

10.1 BASIC CONCEPTS

Suppose it is desired to find a complete solution for the circuit in Fig. 10.1. The circuit could be solved by repeatedly "cutting" it and finding Thévenin equivalents until it is reduced to a simple series circuit. This would lead to a solution for the voltage across and the current through *one* resistor. We could then return to the original circuit and solve for the remaining voltages and currents. This would however be a somewhat tedious process, particularly if *arithmetic* were involved (i.e., for resistor values such as 4.7 kΩ, 5.1 kΩ, 6.8 kΩ, 9.1 kΩ). On the other hand, the principle of Superposition could be used, but this would require the solution of *three* separate circuits; again, not a very pleasant prospect. Furthermore, neither Norton's nor Millman's theorem offers any better solution for the problem. There are however two very simple methods for finding a complete solution for the circuit. The first is to write three *mesh equations* and solve them, and the second is to write two *node equations* and solve them.

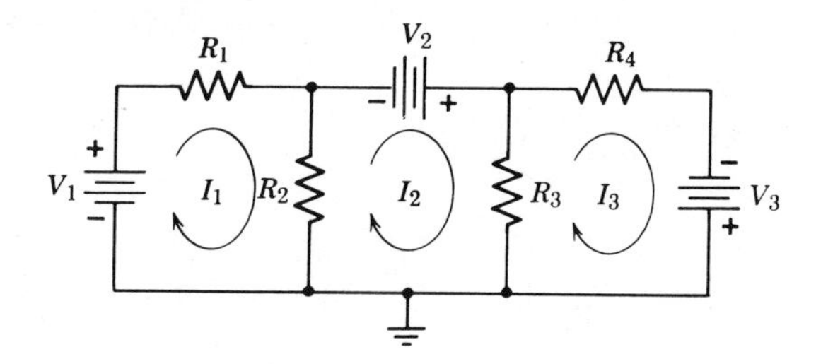

Figure 10.1

Before we can successfully write mesh or node equations, there are a few simple definitions which must be made clear. First of all, an electric circuit consists of a number of *elements* connected together. An element is any circuit component such as a voltage source, a current source, a resistor, and so forth. The circuit in Fig. 10.1 clearly has 7 elements.

A connection between two or more elements is called a *node*. The circuit in Fig. 10.1 has 5 nodes. Note carefully that the entire bottom line in the circuit is really only one node, since the circuit can be redrawn as shown in Fig. 10.2.

The circuit in Fig. 10.1 is also a *planar* circuit, since it can be drawn in a plane (on a flat surface such as a chalk board) *without* any elements *crossing over* any other elements. The circuit in Fig. 10.3*a* is *not* planar since R_4 crosses over other elements; however, it can be redrawn in planar form as shown in Fig. 10.3*b* by simply moving R_4. On the other hand, the circuit in Fig. 10.3*c* *cannot* be redrawn in planar form (you might try it). We are interested in planar circuits since the method of writing mesh equations which we will use is applicable only to planar circuits.

Any electric circuit contains one or more *closed paths* around which current can flow. In general, such a closed path is called a *loop*. Any loop which

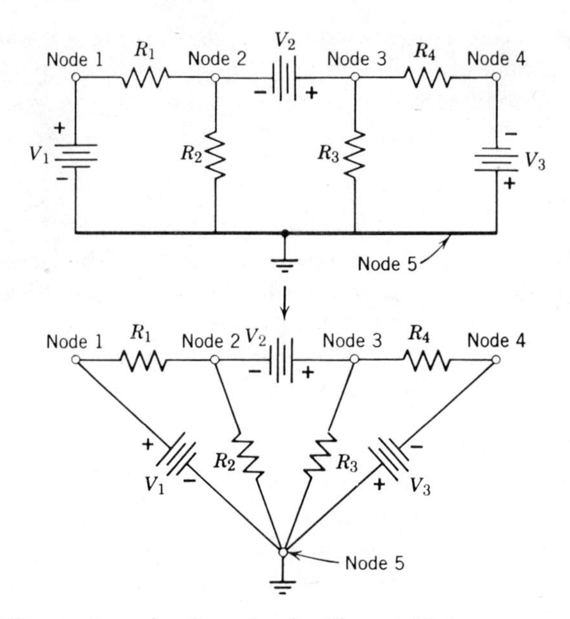

Figure 10.2 Illustrating the 5 nodes in Figure 10.1

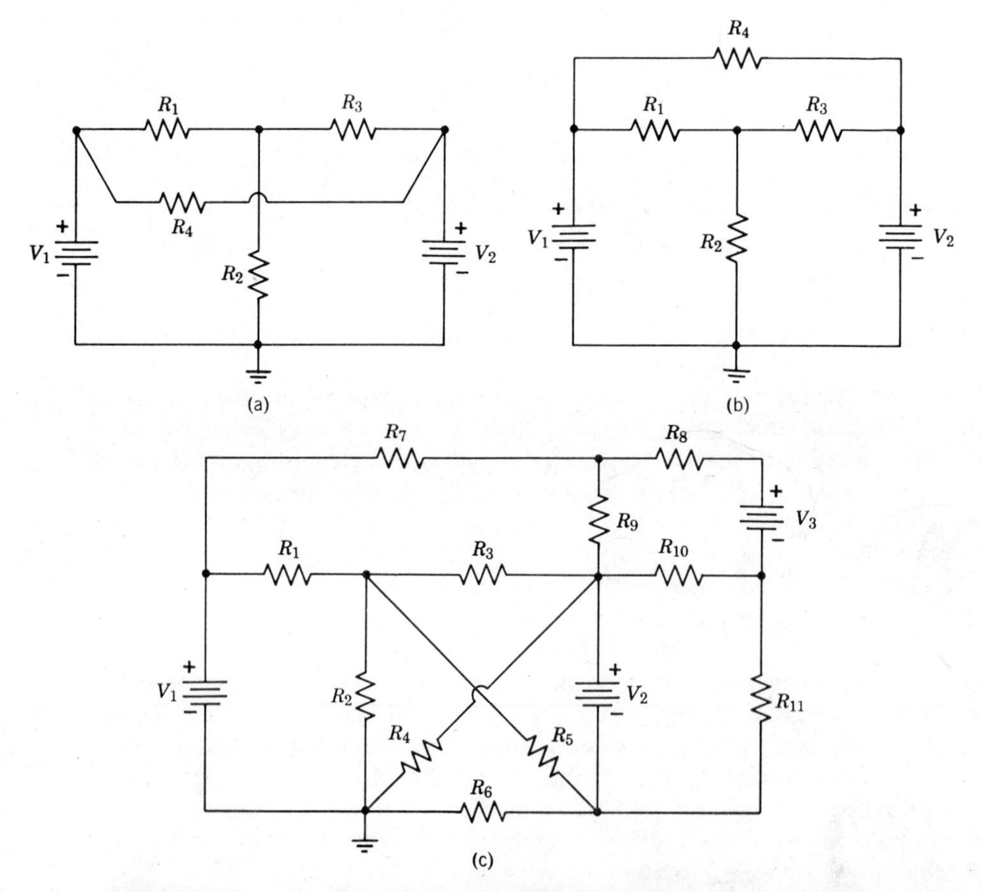

Figure 10.3 (*a*) Nonplanar circuit (*b*) Redrawing (*a*) as a planar circuit (*c*) A nonplanar circuit which cannot be redrawn as a planar circuit

contains no other paths within it is defined as a *mesh.* The meshes in a planar circuit are easily found by simply considering the circuit as a "window" and drawing a mesh around each "window pane." For example, a *graphical* representation of the circuit in Fig. 10.1 is shown in Fig. 10.4 (all of the elements have simply been replaced with straight lines). This *graph* clearly resembles a "window" containing three "window panes." The meshes in the circuit are simply the closed paths around each "pane," and there are exactly 3 in this circuit. In general, any *planar* circuit can be considered as a "*window,*" and the meshes in the circuit are then assigned around the "window panes." Notice that the path around the outside of the circuit is a loop, but it is not a mesh, since it contains three smaller paths within it.

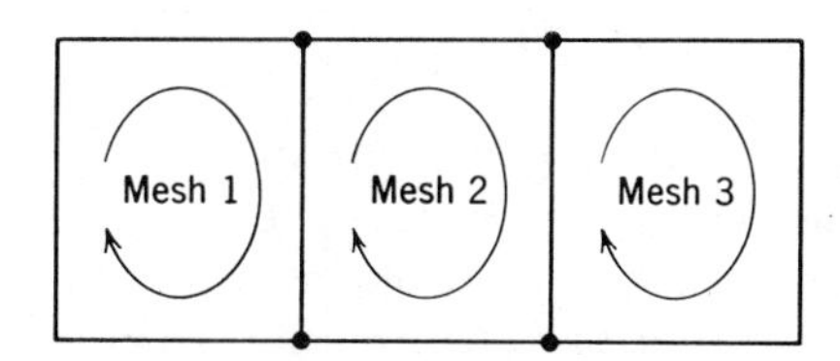

Figure 10.4 A graphical representation of the circuit in Figure 10.1

• • •

Example 10.1 How many elements, nodes, and meshes are there in the circuit in Fig. 10.3*b*?

Solution There are 6 elements; 4 resistors, and 2 voltage sources. There are 4 nodes. There are 3 meshes, since there are 3 "panes" in the "window."

• • •

.2 KVL

We now direct our attention to the application of KVL to a number of circuits. Consider first the circuit in Fig. 10.5. It is assumed that we know the values of V_1, R_1, R_2, and R_3, and a complete solution for the circuit is desired. The

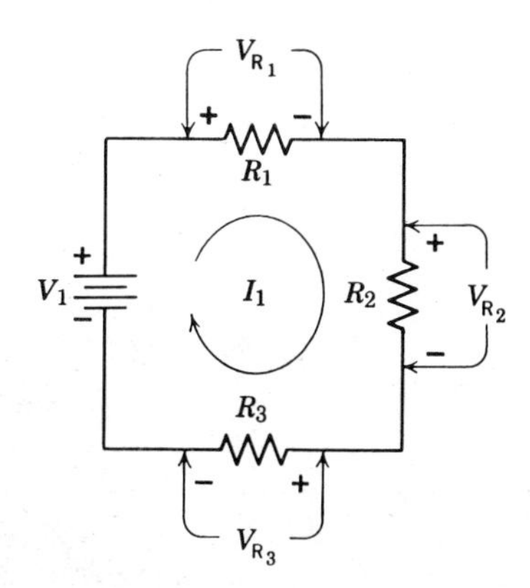

Figure 10.5 A simple 1-mesh circuit

unknowns are then I, V_{R_1}, V_{R_2}, and V_{R_3}. First of all, we note that the circuit is planar and has only one mesh. The first step is then to assign the mesh current I (it is carefully drawn and labeled on the circuit). The application of KVL around the mesh yields,

$$V_1 = V_{R_1} + V_{R_2} + V_{R_3}$$

But, since this is a series circuit, it is clear that

$$V_{R_1} = IR_1, \qquad V_{R_2} = IR_2, \qquad V_{R_3} = IR_3$$

Substituting these three equations into the first equation yields

$$V_1 = IR_1 + IR_2 + IR_3$$

We can now simplify this equation by factoring; thus,

$$V_1 = (R_1 + R_2 + R_3)I \tag{10.1}$$

This is then the *mesh equation* for this circuit.

Consider now the application of KVL to the circuit in Fig. 10.6. We assume that the resistor values and the source voltages are known. Note first of all that the circuit is planar and there are two meshes (two "window panes"). Two unknown mesh currents I_1 and I_2 are then assigned and labeled clearly on the circuit. The application of KVL around mesh 1 yields

$$V_1 = V_{R_1} + V_{R_2} \tag{10.2}$$

The application of KVL around mesh 2 yields

$$V_2 = -V_{R_2} + V_{R_3} \tag{10.3}$$

Note very carefully the use of the minus sign on V_{R_2}!

Now, inspection of mesh 1 reveals that

$$V_{R_1} = I_1R_1, \text{ and } V_{R_2} = (I_1 - I_2)R_2$$

It is important to note that the total current through R_2 is $(I_1 - I_2)$, and *not* $(I_2 - I_1)$, in order to give V_{R_2} the proper polarity (why?). If these two relationships are substituted into Eq. 10.2, the result is the first mesh equation,

$$\begin{aligned} V_1 &= V_{R_1} + V_{R_2} = I_1R_1 + (I_1 - I_2)R_2 \\ &= I_1R_1 + I_1R_2 - I_2R_2 = (R_1 + R_2)I_1 - R_2I_2 \end{aligned}$$

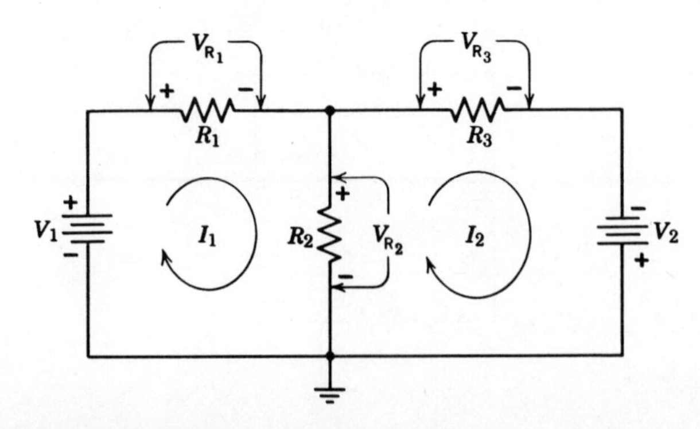

Figure 10.6 A 2-mesh circuit

Inspection of mesh 2 reveals

$$V_{R_2} = (I_1 - I_2)R_2, \qquad \text{and} \qquad V_{R_3} = I_2 R_3$$

Substituting these relationships into Eq. 10.3 yields the second mesh equation,

$$V_2 = -V_{R_2} + V_{R_3} = -(I_1 - I_2)R_2 + I_2 R_3$$
$$= -I_1 R_2 + I_2 R_2 + I_2 R_3 = -R_2 I_1 + (R_2 + R_3) I_2$$

Thus the two mesh equations for this circuit are

$$V_1 = (R_1 + R_2)I_1 - R_2 I_2 \tag{10.4a}$$

$$V_2 = -R_2 I_1 + (R_2 + R_3) I_2 \tag{10.4b}$$

This is a set of two equations in two unknowns, and they must be solved simultaneously. The two unknowns are the mesh currents I_1 and I_2, since the other quantities in Eqs. 10.4 are all known.

• • •

Example 10.2 Using the two mesh equations just developed, solve for the two mesh currents in Fig. 10.6 if: $V_1 = 10\text{ V}$, $V_2 = 20\text{ V}$, $R_1 = 10\ \Omega$, $R_2 = 20\ \Omega$, and $R_3 = 30\ \Omega$.

Solution Substituting the known quantities into Eqs. 10.4 yields,

$$V_1 = (R_1 + R_2)I_1 - R_2 I_2 \qquad\qquad V_2 = -R_2 I_1 + (R_2 + R_3) I_2$$
$$10 = (10\ \Omega + 20\ \Omega)I_1 - (20\ \Omega)I_2 \qquad 20 = -(20\ \Omega)I_1 + (20\ \Omega + 30\ \Omega)I_2$$

Simplifying,

$$10 = 30I_1 - 20I_2$$
$$20 = -20I_1 + 50I_2$$

Solving simultaneously (multiply first equation by 2; multiply second equation by 3; add),

$$\begin{aligned} 20 &= 60I_1 - 40I_2 \\ 60 &= -60I_1 + 150I_2 \ (+) \\ \hline 80 &= 0 + 110I_2 \end{aligned}$$

thus

$$I_2 = 8/11\text{ A}$$

By substitution,

$$I_1 = 9/11\text{ A}$$

The solution is easily checked by finding V_{R_1}, V_{R_2}, and V_{R_3} and using KVL. Thus,

$$V_{R_1} = I_1 R_1 = \tfrac{9}{11} \times 10 = \tfrac{90}{11}\text{ V}$$
$$V_{R_2} = (I_1 - I_2)R_2 = (\tfrac{9}{11} - \tfrac{8}{11})20 = \tfrac{20}{11}\text{ V}$$
$$V_{R_3} = I_2 R_3 = \tfrac{8}{11} \times 30 = \tfrac{240}{11}$$

Now, by KVL,

$$V_1 = V_{R_1} + V_{R_2}; \qquad 10 = \tfrac{90}{11} + \tfrac{20}{11} = \tfrac{110}{11} = 10$$
$$V_2 = -V_{R_2} + V_{R_3}; \qquad 20 = -\tfrac{20}{11} + \tfrac{240}{11} = \tfrac{220}{11} = 20$$

and the solution is seen to check.

• • •

We conclude this section by noting that the mesh equations for a planar circuit can always be found by applying KVL to each mesh in the circuit and then simplifying the equations. There is one mesh equation for each mesh in the circuit and the unknowns are the mesh currents. Thus the number of equations and the number of unknowns is exactly equal to the number of meshes in a planar circuit. The reader is urged to apply KVL to the circuit in Fig. 10.1 in order to verify that the mesh equations for that circuit are

$$V_1 = (R_1 + R_2)I_1 - R_2I_2 + 0$$
$$V_2 = -R_2I_1 + (R_2 + R_3)I_2 - R_3I_3$$
$$V_3 = 0 - R_3I_2 + (R_3 + R_4)I_3$$

10.3 MESH EQUATIONS

The process of applying KVL to each mesh in a planar circuit as outlined in the previous section can *always* be used to find the mesh equations for the circuit. This process is however somewhat tedious, and there is in fact a very simple method of writing these equations by a simple *inspection* of the circuit. We will devote this section to developing this technique.

A careful examination of the mesh equations developed in the previous section reveals that they can all be written in a *standard form.* It is this standard *format* which will allow us to write mesh equations by a simple inspection of the circuit. The standard format for mesh equations is

$$\begin{aligned}
&\text{Mesh 1} \quad && \Sigma V_{11} = R_{11}I_1 - R_{12}I_2 - R_{13}I_3 - \cdots - R_{1N}I_N \\
&\text{Mesh 2} \quad && \Sigma V_{22} = -R_{21}I_1 + R_{22}I_2 - R_{23}I_3 - \cdots - R_{2N}I_N \\
&\quad\cdot && \cdot \\
&\quad\cdot && \cdot \\
&\quad\cdot && \cdot \\
&\text{Mesh N} \quad && \Sigma V_{NN} = -R_{N1}I_1 - R_{N2}I_2 - R_{N3}I_3 - \cdots + R_{NN}I_N
\end{aligned} \tag{10.5}$$

These equations may seem rather formidable at first, but if we carefully examine them we will find that each component has a very simple meaning. First of all, there are N equations in N unknowns (i.e., N mesh currents). If N = 1, there is only one mesh, and we therefore need only one equation. It is

$$\Sigma V_{11} = R_{11}I_1$$

This corresponds to the 1-mesh circuit in Fig. 10.5 and the mesh equation for this circuit (Eq. 10.1). Comparing these two equations,

$$\Sigma V_{11} = R_{11}I_1$$
$$V_1 = (R_1 + R_2 + R_3)I_1$$

The mathematical symbol Σ means "the summation of," and the double 1 subscript on V_{11} means "the voltage sources in mesh 1." Thus ΣV_{11} simply means to take the *algebraic sum* of all the voltage sources around mesh 1. Similarly, the double 1 subscript on R_{11} means "the total resistance around mesh 1." If we compare these two equations, and the circuit in Fig. 10.5, it is clear that ΣV_{11} is indeed just V_1, and R_{11} is $(R_1 + R_2 + R_3)$.

Let us now consider the case when N = 2. This means a 2-mesh circuit; there must then be two mesh equations and two unknown mesh currents to solve for. The proper standard form is then

$$\Sigma V_{11} = R_{11}I_1 - R_{12}I_2$$
$$\Sigma V_{22} = -R_{21}I_1 + R_{22}I_2$$

These two equations correspond exactly to the 2-mesh circuit in Fig. 10.6, and its two mesh equations from the previous section are

$$V_1 = (R_1 + R_2)I_1 - R_2 I_2$$
$$V_2 = -R_2 I_1 + (R_2 + R_3)I_2$$

We can compare these equations more easily by writing them one above the other

$$\Sigma V_{11} = R_{11}I_1 - R_{12}I_2 \qquad \Sigma V_{22} = -R_{21}I_1 + R_{22}I_2$$
$$V_1 = (R_1 + R_2)I_1 - R_2 I_2 \qquad V_2 = -R_2 I_1 + (R_2 + R_3)I_2$$

ΣV_{11} is the algebraic sum of all the voltage sources around mesh 1 and is clearly V_1. Similarly, ΣV_{22} is the algebraic sum of all the voltage sources around mesh 2, and it is clearly V_2. It is very important to note at this point that V_1 and V_2 are both *positive* since I_1 and I_2 pass through the voltage sources in the *positive* direction. If, for example, the voltage source V_1 were reversed, V_1 would appear as a *negative* quantity in the equation since I_1 would be passing through it in the *reverse* direction.

Now, notice that R_{11} is the sum of all the resistances around mesh 1 and is simply $(R_1 + R_2)$. Similarly, R_{22} is the sum of all the resistances around mesh 2, and is seen to be $(R_2 + R_3)$.

R_{12} and R_{21} are the resistances which have *both* currents I_1 and I_2 passing through them. They are said to be the mutual *resistances* between meshes 1 and 2. From the circuit, it is clear that $R_{12} = R_{21} = R_2$.

At this point, it is very important to note that the *subscripts* are the *keys* to all of the components in the equations. Notice that when the subscripts are *alike* (11, 22, 33, etc.) we are looking for a *mesh* quantity; the algebraic sum of the voltages *around* a mesh, or the sum of the resistances *around* a mesh. When the subscripts are *not alike* (12, 21, 13, 23, etc.) we are looking for a resistance which is *common* between *two meshes*, and the subscripts define which meshes.

We can summarize these ideas by writing the standard equations for a 3-mesh circuit and adding the descriptive notation as follows:

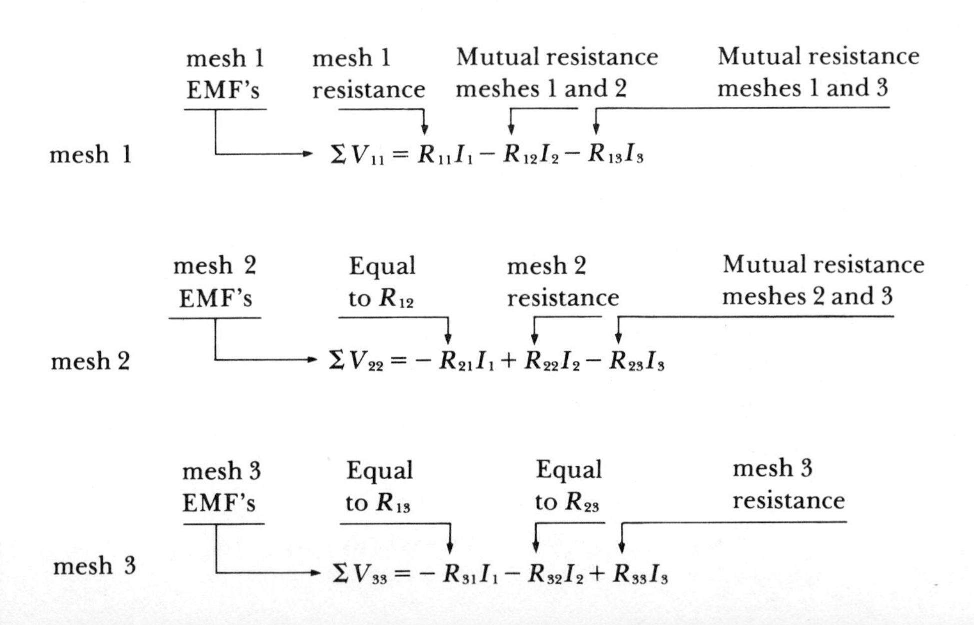

We now note that a set of three mesh equations can be written in the form,

$$V_{11} = R_{11}I_1 - R_{12}I_2 - R_{13}I_3$$
$$V_{22} = -R_{21}I_1 + R_{22}I_2 - R_{23}I_3$$
$$V_{33} = -R_{31}I_1 - R_{32}I_2 + R_{33}I_3$$
(+)

Notice that *all* of the resistances along the main diagonal line (the light line drawn through the equations) have *positive* signs, and *all others* have *negative* signs. This is a consequence of *always* assigning *all* of the mesh currents in a *clockwise* direction. The directions of the mesh currents are of course arbitrary, but we will *always* assign them *clockwise* so that the *sign symmetry* will always appear in the mesh equations. This will relieve us of the burden of having to determine the sign of each and every term, since they will *always* be as shown above. This represents great savings in labor and the elimination of a source of errors.

We now state a systematic method for writing mesh equations:

1. Draw the circuit in planar form.
2. Assign and carefully label mesh currents in a clockwise direction.
3. Write the *blank form* of the mesh equations as

$$(\Sigma\ V_S') = (\qquad)I_1 - (\qquad)I_2 - (\qquad)I_3 \ldots$$

4. Fill in the empty blanks in the mesh equations by inspecting the circuit.

• • •

Example 10.3 Write the mesh equations for the circuit in Fig. 10.7.

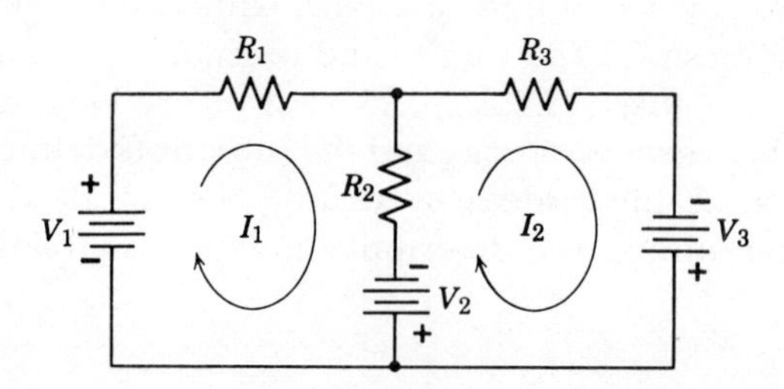

Figure 10.7

Solution We follow the four steps above:

1. The circuit is already planar.
2. The mesh currents are assigned and drawn in clockwise.
3. This is a 2-mesh circuit. We need two mesh equations, and in *blank form* they are:

$$\text{mesh 1}\quad (\qquad) = (\qquad)I_1 - (\qquad)I_2$$
$$\text{mesh 2}\quad (\qquad) = -(\qquad)I_1 + (\qquad)I_2$$

4. The *blank forms* are filled in by inspecting the circuit:

$$\text{mesh 1}\quad (V_1 + V_2) = (R_1 + R_2)I_1 - (R_2)I_2$$
$$\text{mesh 2}\quad (V_3 - V_2) = -(R_2)I_1 + (R_2 + R_3)I_2$$

Notice carefully that the sum of the voltage sources around mesh 2, ΣV_{22}, is $(V_3 - V_2)$ since I_2 passes through V_3 in the positive direction, but it passes through V_2 in the *negative* direction.

• • •

Example 10.4 Write the mesh equations for the circuit in Fig. 10.1.

Solution Again we follow the four steps.

1. The circuit is planar.
2. The mesh currents are assigned.
3. The three *blank form* equations are:

$$\begin{aligned}
&\text{mesh 1} \quad (\qquad) = (\qquad)I_1 - (\qquad)I_2 - (\qquad)I_3 \\
&\text{mesh 2} \quad (\qquad) = -(\qquad)I_1 + (\qquad)I_2 - (\qquad)I_3 \\
&\text{mesh 3} \quad (\qquad) = -(\qquad)I_1 - (\qquad)I_2 + (\qquad)I_3
\end{aligned}$$

4. The *blank forms* are filled in by inspecting the circuit.

$$\begin{aligned}
&\text{mesh 1} \quad V_1 = (R_1 + R_2)I_1 - (R_2)I_2 - (0)I_3 \\
&\text{mesh 2} \quad V_2 = -(R_2)I_1 + (R_2 + R_3)I_2 - (R_3)I_3 \\
&\text{mesh 3} \quad V_3 = -(0)I_1 - (R_3)I_2 + (R_3 + R_4)I_3
\end{aligned}$$

Notice that R_{13} (in mesh 1) and R_{31} (in mesh 3) are both zero since there is *no* resistance through which *both* I_1 and I_3 flow. Compare these equations with those given at the end of the previous section.

• • •

Example 10.5 Write the mesh equations for the circuit in Fig. 10.8.

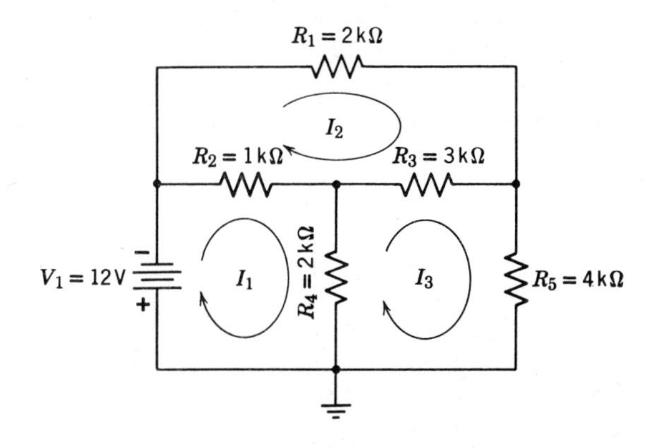

Figure 10.8

Solution

1. The circuit in planar form.
2. The mesh currents are assigned.
3. The *blank form* equations are:

$$\begin{aligned}
&\text{mesh 1} \quad (\qquad) = (\qquad)I_1 - (\qquad)I_2 - (\qquad)I_3 \\
&\text{mesh 2} \quad (\qquad) = -(\qquad)I_1 + (\qquad)I_2 - (\qquad)I_3 \\
&\text{mesh 3} \quad (\qquad) = -(\qquad)I_1 - (\qquad)I_2 + (\qquad)I_3
\end{aligned}$$

4. The *blank forms* are filled in by inspecting the circuit:

$$\begin{aligned}
&\text{mesh 1} \quad & -V_1 &= (R_2 + R_4)I_1 - (R_2)I_2 - (R_4)I_3 \\
&\text{mesh 2} \quad & 0 &= -(R_2)I_1 + (R_1 + R_2 + R_3)I_2 - (R_3)I_3 \\
&\text{mesh 3} \quad & 0 &= -(R_4)I_1 - (R_3)I_2 + (R_3 + R_4 + R_5)I_3
\end{aligned}$$

Notice that there are no sources in meshes 1 and 3, and the voltage source terms are thus zero.
We now fill in the proper circuit values:

$$\begin{aligned}
&\text{mesh 1} \quad & -12 &= (1\text{ k}\Omega + 2\text{ k}\Omega)I_1 - (1\text{ k}\Omega)I_2 - (2\text{ k}\Omega)I_3 \\
&\text{mesh 2} \quad & 0 &= -(1\text{ k}\Omega)I_1 + (2\text{ k}\Omega + 1\text{ k}\Omega + 3\text{ k}\Omega)I_2 - (3\text{ k}\Omega)I_3 \\
&\text{mesh 3} \quad & 0 &= -(2\text{ k}\Omega)I_1 - (3\text{ k}\Omega)I_2 - (3\text{ k}\Omega + 2\text{ k}\Omega + 4\text{ k}\Omega)I_3
\end{aligned}$$

Simplifying:

$$\begin{aligned}
&\text{mesh 1} \quad & -12 &= 3\text{ k}\Omega I_1 - 1\text{ k}\Omega I_2 - 2\text{ k}\Omega I_3 \\
&\text{mesh 2} \quad & 0 &= -1\text{ k}\Omega I_1 + 6\text{ k}\Omega I_2 - 3\text{ k}\Omega I_3 \\
&\text{mesh 3} \quad & 0 &= -2\text{ k}\Omega I_1 - 3\text{ k}\Omega I_2 + 9\text{ k}\Omega I_3
\end{aligned}$$

This process may appear lengthy here, but in the examples all of the details have been written out for clarity. The reader will find that after some practice, it will be possible to omit the intermediate steps and write the mesh equations directly.

10.4 USING MESH EQUATIONS

Confidence in writing mesh equations and using them to solve circuits come only with practice, and this entire section will therefore be devoted to *writing* and *solving* mesh equations.

We begin by solving for the mesh currents I_1 and I_2 in Fig. 10.9.

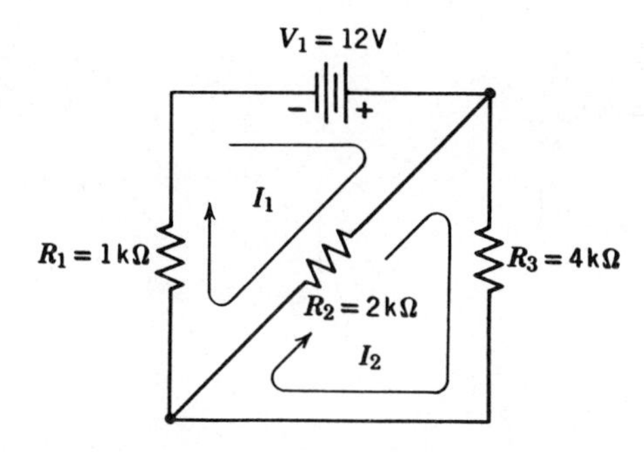

Figure 10.9 A 2-mesh circuit

1. The circuit is planar.
2. The mesh currents are assigned.
3. The two *blank forms* are:

$$\begin{aligned}
&\text{mesh 1} \quad & (\qquad) &= (\qquad)I_1 - (\qquad)I_2 \\
&\text{mesh 2} \quad & (\qquad) &= -(\qquad)I_1 + (\qquad)I_2
\end{aligned}$$

4. The final equations are found by inspection:

$$\text{mesh 1} \quad 12 = (2\ \text{k}\Omega + 1\ \text{k}\Omega)I_1 - (2\ \text{k}\Omega)I_2$$
$$\text{mesh 2} \quad 0 = -(2\ \text{k}\Omega)I_1 + (2\ \text{k}\Omega + 4\ \text{k}\Omega)I_2$$

Simplifying,

$$\text{mesh 1} \quad 12 = 3\ \text{k}\Omega I_1 - 2\ \text{k}\Omega I_2$$
$$\text{mesh 2} \quad 0 = -2\ \text{k}\Omega I_1 + 6\ \text{k}\Omega I_2$$

We now solve by first dividing both equations by 1 kΩ. Thus,

$$\text{mesh 1} \quad 12 \times 10^{-3} = 3I_1 - 2I_2$$
$$\text{mesh 2} \quad 0 = -2I_1 + 6I_2$$

Then, multiplying the first equation by 3 and adding to the second, we have

$$36 \times 10^{-3} = 7I_1$$

Therefore,

$$I_1 = \frac{36}{7} \times 10^{-3}\ \text{A} = \frac{36}{7}\ \text{mA}$$

By substitution,

$$I_2 = \frac{12}{7}\ \text{mA}$$

The solution is easily checked by redrawing the circuit to show that it is a simple series-parallel circuit.

The circuit in Fig. 10.10*a* is drawn in nonplanar form. The first step in solving this circuit is to redraw it in planar form as in Fig. 10.10*b*. The three mesh currents are assigned, and the mesh equations are:

$$\text{mesh 1} \quad -10 = 4\ \text{k}\Omega I_1 - 3\ \text{k}\Omega I_2 - 1\ \text{k}\Omega I_3$$
$$\text{mesh 2} \quad -20 = -3\ \text{k}\Omega I_1 + 7\ \text{k}\Omega I_2 - 0$$
$$\text{mesh 3} \quad 20 = -1\ \text{k}\Omega I_1 - 0 + 3\ \text{k}\Omega I_3$$

The reader should carefully verify the accuracy of *every* term in these three equations.

Dividing all three equations by 1 kΩ;

$$\text{mesh 1} \quad -10 \times 10^{-3} = 4I_1 - 3I_2 - I_3$$
$$\text{mesh 2} \quad -20 \times 10^{-3} = -3I_1 + 7I_2 - 0$$
$$\text{mesh 3} \quad 20 \times 10^{-3} = -I_1 - 0 + 3I_3$$

These three equations can be solved by substitution or elimination, but it is much easier to make use of determinants.* Thus, using Cramer's rule

$$\Delta = \begin{vmatrix} 4 & -3 & -1 \\ -3 & 7 & 0 \\ -1 & 0 & 3 \end{vmatrix} = 50$$

$$I_1 = \frac{\begin{vmatrix} -10 \times 10^{-3} & -3 & -1 \\ -20 \times 10^{-3} & 7 & 0 \\ 20 \times 10^{-3} & 0 & 3 \end{vmatrix}}{\Delta} = \frac{-25 \times 10^{-2}}{50} = -5\ \text{mA}$$

*The reader who is not familiar with determinants is urged to study the presentation in the appendix. Determinants provide an extremely powerful tool for solving simultaneous equations.

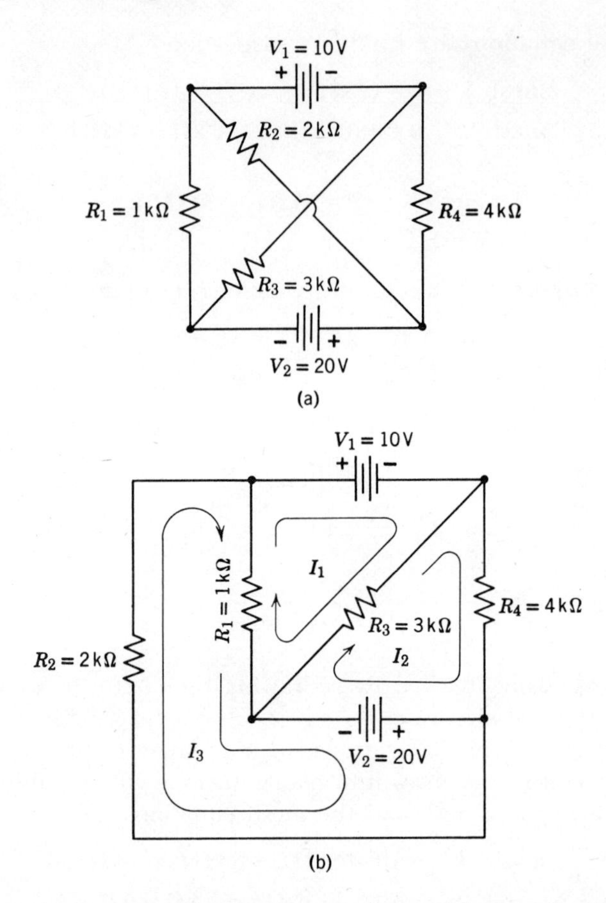

Figure 10.10 (*a*) A circuit in nonplanar form (*b*) the circuit redrawn in planar form

$$I_2 = \frac{\begin{vmatrix} 4 & -10\times 10^{-3} & -1 \\ -3 & -20\times 10^{-3} & 0 \\ -1 & 20\times 10^{-3} & 3 \end{vmatrix}}{\Delta} = \frac{-25\times 10^{-2}}{50} = -5 \text{ mA}$$

$$I_3 = \frac{\begin{vmatrix} 4 & -3 & -10\times 10^{-3} \\ -3 & 7 & -20\times 10^{-3} \\ -1 & 0 & 20\times 10^{-3} \end{vmatrix}}{\Delta} = \frac{25\times 10^{-2}}{50} = 5 \text{ mA}$$

The reader is urged to check the solution using KVL. Notice carefully that the *negative* sign in the solution for I_1 and I_2 simply means that these currents *actually* flow in a *counterclockwise* direction, or just the opposite of the two assumed mesh currents.

• • •

Example 10.6 Write the mesh equations for the circuit in Fig. 10.11.

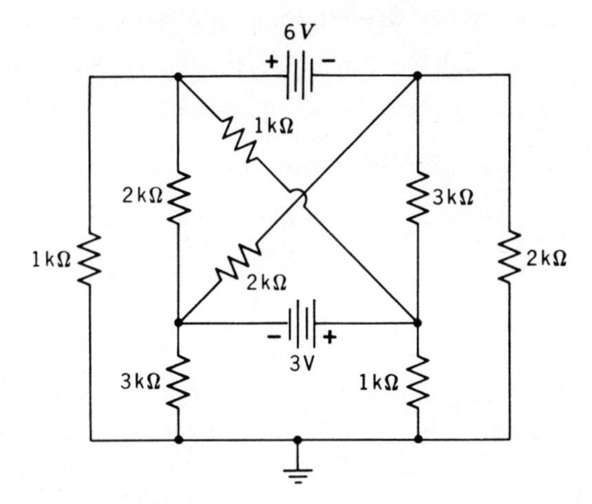

Figure 10.11 A nonplanar circuit

Solution This circuit is nonplanar, and it *cannot* be redrawn in planar form. We cannot use our technique of writing mesh equations to solve this circuit. A more general method of loop equations could be used, or node equations could be used.

• • •

Example 10.7 Solve for the currents in the circuit in Fig. 10.12.

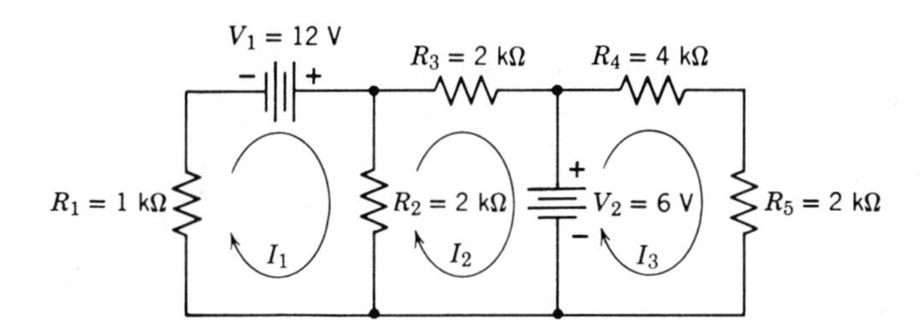

Figure 10.12

Solution The circuit is planar and the mesh currents are assigned. The three mesh equations are:

$$\text{mesh 1} \quad 12 = 3\text{ k}\Omega I_1 - 2\text{ k}\Omega I_2 - 0$$

$$\text{mesh 2} \quad -6 = -2\text{ k}\Omega I_1 + 4\text{ k}\Omega I_2 - 0$$

$$\text{mesh 3} \quad 6 = 0 - 0 + 6\text{ k}\Omega I_3$$

The equation for mesh 3 is interesting because it has no I_1 or I_2 terms. This shows that I_3 is completely *independent* of I_1 and I_2. Similarly, the first two mesh equations have no I_3 terms, which simply verifies that I_1 and I_2 depend in no way on I_3. This is easy to see from the circuit, since we could clearly remove R_4 and R_5 (making $I_3 = 0$) and I_1 and I_2 would in no way be affected. The voltage source V_2 "splits" the circuit into two parallel circuits. From the third mesh equation,

$$I_3 = 1\text{ mA}$$

We then solve the first two mesh equations simultaneously for I_1 and I_2. Thus,

$$\text{mesh 1} \qquad 12 = 3\ \text{k}\Omega I_1 - 2\ \text{k}\Omega I_2$$

$$\text{mesh 2} \qquad -6 = -2\ \text{k}\Omega I_1 + 4\ \text{k}\Omega I_2$$

Multiplying the first equation by 2 and adding to the second

$$18 = 4\ \text{k}\Omega I_1$$

$$I_1 = 4.5\ \text{mA}$$

By substitution

$$I_2 = 0.75\ \text{mA}$$

• • •

We conclude this section by solving the circuit in Fig. 10.13 using mesh equations. This is an important example since it shows a method of dealing with a current source. The circuit is planar, and the three mesh currents have been assigned. The first and third mesh equations are straightforward, and the only apparent difficulty is in mesh 2. The difficulty is the current source I_S; it must have a voltage across it, but we *do not* know the value of this voltage. However, notice that the mesh current I_2 must be exactly the same as the current source current I_S. Thus we immediately know I_2

$$I_2 = I_S = 1\ \text{mA}$$

Since I_2 is known, we need only write an equation for mesh 1 and for mesh 3 as follows

$$\text{mesh 1} \quad V_{S_1} = (R_1 + R_2)I_1 - R_2 I_2 - 0$$

$$\text{mesh 3} \quad V_{S_2} = 0 - R_2 I_2 + (R_3 + R_4)I_3$$

Substituting values

$$\text{mesh 1} \quad 12 = 3\ \text{k}\Omega I_1 - 2\ \text{k}\Omega I_2 + 0$$

$$\text{mesh 3} \quad 6 = 0 - 3\ \text{k}\Omega I_2 + 7\ \text{k}\Omega I_3$$

We can now substitute $I_2 = 1$ mA into mesh 1 equation to find I_1

$$12 = 3\ \text{k}\Omega I_1 - 2\ \text{k}\Omega(1\ \text{mA})$$

$$I_1 = \frac{14}{3}\ \text{mA}$$

We can also substitute the value of I_2 into the mesh 3 equation to find I_3

$$6 = -3\ \text{k}\Omega(1\ \text{mA}) + 7\ \text{k}\Omega I_3$$

$$I_3 = \frac{9}{7}\ \text{mA}$$

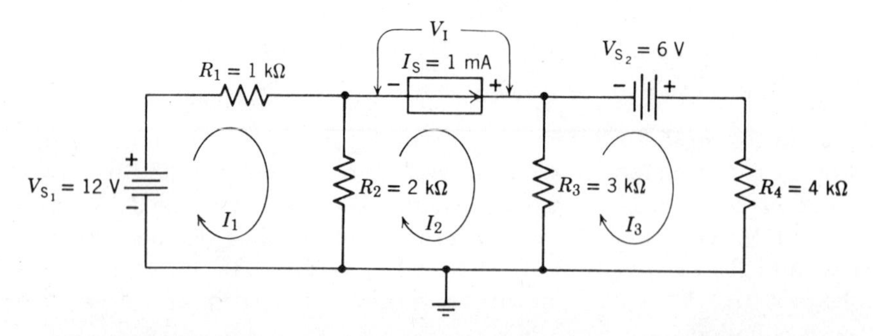

Figure 10.13 A 3-mesh circuit containing a current source

Knowing the three mesh currents, it is now a simple matter to determine all the resistor voltages and currents, and the reader is urged to verify the solution by doing so.

10.5 KCL

In the previous sections we have seen that the application of KVL to a circuit leads to a set of mesh equations; we will now consider the *dual* which is the application of KCL, and this will lead to a set of *node* equations. We will then solve these equations for the *node voltages.*

An examination of the circuit in Fig. 10.14 reveals that there are 3 elements, 2 meshes, and 2 nodes. Since the voltage at a node must be measured with respect to some reference, we will *always* connect one node to ground, and this is referred to as the *reference node.* The reference node voltage is then simply zero volts. The circuit in Fig. 10.14 has only one node voltage which is unknown, and we clearly label it as V_A.

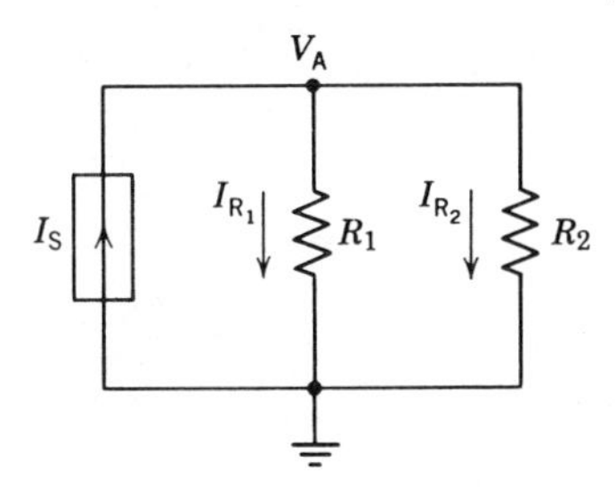

Figure 10.14 A circuit requiring one node equation

The application of KCL at node A in Fig. 10.14 yields

$$I_S = I_{R_1} + I_{R_2}$$

Since R_1 and R_2 are in parallel, the voltage across both of them is V_A, and thus

$$I_{R_1} = \frac{V_A}{R_1} \qquad I_{R_2} = \frac{V_A}{R_2}$$

Substituting these two relationships into the KCL equation yields

$$I_S = I_{R_1} + I_{R_2} = \frac{V_A}{R_1} + \frac{V_A}{R_2}$$

Factoring,

$$I_S = \left(\frac{1}{R_1} + \frac{1}{R_2}\right) V_A \tag{10.6}$$

This is the node equation for Fig. 10.14. The known quantities are the circuit element values R_1, R_2, and I_S, and the unknown is the node voltage V_A.

• • •

Example 10.8 Using the node equation just developed, solve for V_A in Fig. 10.14 if $I_S = 6$ mA, $R_1 = 3$ kΩ, and $R_2 = 6$ kΩ.

Solution Using Eq. 10.6,

$$I_S = \left(\frac{1}{R_1} + \frac{1}{R_2}\right) V_A$$

Substituting values,

$$6\text{ mA} = \left(\frac{1}{3\text{ k}\Omega} + \frac{1}{6\text{ k}\Omega}\right) V_A$$

Multiply both sides of the equation by 6 kΩ.

$$6\text{ k}\Omega \times 6\text{ mA} = 6\text{ k}\Omega\left(\frac{1}{3\text{ k}\Omega} + \frac{1}{6\text{ k}\Omega}\right) V_A$$

$$36 = (2 + 1)V_A = 3V_A$$

Therefore,

$$V_A = 12\text{ V}$$

The solution is easily verified by solving for I_{R_1} and I_{R_2} and applying KCL. Thus,

$$I_{R_1} = \frac{V_A}{R_1} = \frac{12\text{ V}}{\text{k}\Omega} = 4\text{ mA} \qquad I_{R_2} = \frac{V_A}{R_2} = \frac{12\text{ V}}{6\text{ k}\Omega} = 2\text{ mA}$$

By KCL,

$$I_S = I_{R_1} + I_{R_2}, \qquad 6\text{ mA} = 4\text{ mA} + 2\text{ mA}$$

and the solution is seen to be correct.

• • •

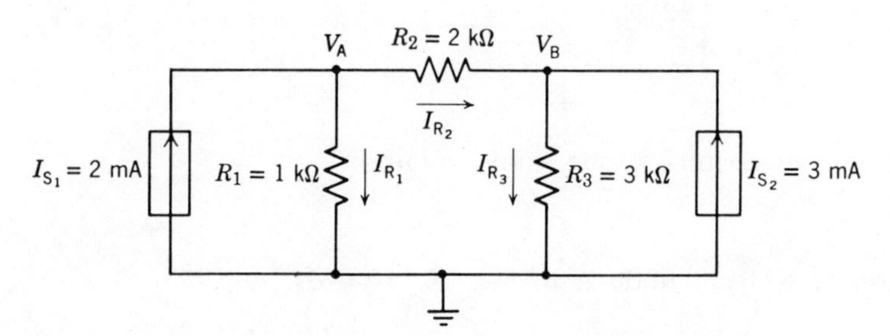

Figure 10.15 A circuit requiring two node equations

We now turn our attention to the circuit in Fig. 10.15. This circuit clearly has 3 nodes, one of which is the reference node (grounded), and the other 2 labeled V_A and V_B. The resistor currents are also drawn in and labeled I_{R_1}, I_{R_2}, and I_{R_3}. The application of KCL at node A yields,

$$I_{S_1} = I_{R_1} + I_{R_2}$$

and at node B,

$$I_{S_2} = -I_{R_2} + I_{R_3}$$

Note carefully the *minus* sign in this equation. We can now examine the circuit to show that

$$I_{R_1} = \frac{V_A}{R_1} \qquad I_{R_2} = \frac{(V_A - V_B)}{R_2} \qquad I_{R_3} = \frac{V_B}{R_3}$$

Notice that we must use $(V_A - V_B)$ in the expression for I_{R_2} in order that I_{R_2} is in the direction shown (Why?). If we now substitute these relationships into the two KCL equations written at the two nodes, the results are

$$I_{S_1} = I_{R_1} + I_{R_2} = \frac{V_A}{R_1} + \frac{(V_A - V_B)}{R_2} = \frac{V_A}{R_1} + \frac{V_A}{R_2} - \frac{V_B}{R_2}$$

$$I_{S_2} = -I_{R_2} + I_{R_3} = \frac{-(V_A - V_B)}{R_2} + \frac{V_B}{R_3} = \frac{-V_A}{R_2} + \frac{V_B}{R_2} + \frac{V_B}{R_3}$$

We now factor these two equations to get

$$\text{Node A} \quad I_{S_1} = \left(\frac{1}{R_1} + \frac{1}{R_2}\right) V_A - \frac{1}{R_2} V_B \tag{10.7a}$$

$$\text{Node B} \quad I_{S_2} = -\frac{1}{R_2} V_A + \left(\frac{1}{R_2} + \frac{1}{R_3}\right) V_B \tag{10.7b}$$

These are the two node equations for this circuit.

• • •

Example 10.9 Solve for the node voltages in Fig. 10.15 using node equations.

Solution Using Eq. 10.7,

$$\text{Node A} \quad 2\text{ mA} = \left(\frac{1}{1\text{ k}\Omega} + \frac{1}{2\text{ k}\Omega}\right) V_A - \frac{1}{2\text{ k}\Omega} V_B$$

$$\text{Node B} \quad 3\text{ mA} = -\frac{1}{2\text{ k}\Omega} V_A + \left(\frac{1}{2\text{ k}\Omega} + \frac{1}{3\text{ k}\Omega}\right) V_B$$

Multiplying the first equation by 2 kΩ, and the second equation by 6 kΩ,

$$\text{Node A} \quad 2\text{ k}\Omega \times 2\text{ mA} = 2\text{ k}\Omega \left(\frac{1}{1\text{ k}\Omega} + \frac{1}{2\text{ k}\Omega}\right) V_A - \frac{2\text{ k}\Omega}{2\text{ k}\Omega} V_B$$

$$\text{Node B} \quad 6\text{ k}\Omega \times 3\text{ mA} = -\frac{6\text{ k}\Omega}{2\text{ k}\Omega} V_A + 6\text{ k}\Omega \left(\frac{1}{2\text{ k}\Omega} + \frac{1}{3\text{ k}\Omega}\right) V_B$$

Simplifying,

$$\text{Node A} \quad 4 = (2 + 1) V_A - V_B = 3V_A - V_B$$

$$\text{Node B} \quad 18 = -3V_A + (3 + 2) V_B = -3V_A + 5V_B$$

Adding the two equations,

$$22 = 4\, V_B$$

Therefore,

$$V_B = \frac{11}{2}\text{ V}$$

By substitution,

$$V_A = \frac{19}{6}\text{ V}$$

The solutions are easily verified by solving for the resistor currents and applying KCL at each node (the reader is urged to do so).

• • •

From the two preceding examples it is clear that it is always possible to write the node equations for any circuit by: (1) identifying all the nodes in the circuit, (2) grounding one of the nodes as a reference, and (3) applying KCL at the remaining nodes. Notice that in any circuit containing v nodes, it will require exactly $(v - 1)$ node equations, since one of the nodes is always a reference.

10.6 NODE EQUATIONS

Even though the node equations for a circuit can always be written by applying KCL to each node as outlined in the previous section, this is a somewhat tedious process. As might be expected, there is a set of standard node equations, just as there is a set of standard mesh equations; this should not be surprising since KCL and KVL are duals. We will make use of the *standard form* node equations to write the node equations for a circuit by a simple inspection of the circuit. The standard form node equations are:

$$
\begin{aligned}
&\text{Node A} \quad \Sigma I_{AA} = G_{AA}V_A - G_{AB}V_B - \cdots - G_{AN}V_N \\
&\text{Node B} \quad \Sigma I_{BB} = -G_{BA}V_A + G_{BB}V_B - \cdots - G_{BN}V_N \\
&\quad \vdots \qquad\quad \vdots \\
&\text{Node N} \quad \Sigma I_{NN} = -G_{NA}V_A - G_{NB}V_B - \cdots + G_{NN}V_N \qquad (10.8)
\end{aligned}
$$

These equations are dual to the standard mesh equations, Eq. 10.5, since we substitute currents for voltages, voltages for currents, and G's for R's. You will recall that G is a conductance and is equal to $1/R$. Note first of all that there are N equations (for a circuit having N nodes), and N unknowns (the unknowns are the node voltages V_A, $V_B \ldots V_N$).

The left-hand terms are the algebraic sum of the current sources *connected* to a node. Thus, ΣI_{AA} is the algebraic sum of all the current sources connected to node A, ΣI_{BB} is the algebraic sum of all the current sources connected to node B, and so on. Currents *entering* the node from current sources are *positive*, and currents *leaving* the node to current sources are *negative.*

The G's having *like* subscripts (G_{AA}, G_{BB}, G_{CC}, etc.) are the sums of the conductors connected to nodes. Thus G_{AA} is the *total* conductance *connected* to node A ($1/R_1 + 1/R_2 + \cdots$), G_{BB} is the total conductance *connected to* node B ($1/R_3 + 1/R_4 + \cdots$), and so on.

The G's with *unlike* subscripts (G_{AB}, G_{BC}, G_{AC}, etc.) represent the conductance connected *between* two nodes. Thus, G_{AB} is the conductance connected *between* nodes A and B, G_{BC} is the conductance connected between nodes B and C, and so forth. Notice that the *subscripts* tell which two nodes the conductance is connected between.

If we now take the case where N = 1, we have a circuit requiring only one node equation, and Eq. 10.8 provides

$$\Sigma I_{AA} = G_{AA}V_A$$

We can compare this to the circuit in Fig. 10.14 and its node equation, Eq. 10.6

$$I_S = \left(\frac{1}{R_1} + \frac{1}{R_2}\right) V_A$$

By inspection of the circuit, we see that the algebraic sum of the currents (due to current sources) entering node A is $\Sigma I_{AA} = I_S$, which corresponds exactly to the node equation developed. Note here that if I_S were reversed in Fig. 10.14, it would be entered as $-I_S$ in the node equation. The term G_{AA} in the node equation is the sum of all the conductances connected to node A, and this is indeed ($1/R_1 + 1/R_2$).

If we now take the case when N = 2, we have a circuit requiring two node equations. The two standard form node equations are then

$$\Sigma I_{AA} = G_{AA}V_A - G_{AB}V_B$$
$$\Sigma I_{BB} = -G_{BA}V_A - G_{BB}V_B$$

These standard forms are now compared with the two node equations developed for the circuit in Fig. 10.15 (Eq. 10.7).

$$\text{Node A} \quad \Sigma I_{AA} = G_{AA}V_A - G_{AB}V_B$$
$$I_{S_1} = \left(\frac{1}{R_1} + \frac{1}{R_2}\right)V_A - \frac{1}{R_2}V_B$$
$$\text{Node B} \quad \Sigma I_{BB} = -G_{BA}V_A + G_{BB}V_B$$
$$I_{S_2} = -\frac{1}{R_2}V_A + \left(\frac{1}{R_2} + \frac{1}{R_3}\right)V_B$$

Inspection of the circuit verifies that the sum of the current sources entering node A, ΣI_{AA}, is simply I_{S_1}; similarly, ΣI_{BB} at node B is I_{S_2}. The total conductance connected to node A, G_{AA}, is clearly $(1/R_1 + 1/R_2)$, and the total conductance connected to node B, G_{BB}, is $(1/R_2 + 1/R_3)$. The terms G_{AB} and G_{BA} represent the conductance connected between nodes A and B. They are clearly equal since they represent the same resistor R_2, and the conductance is $1/R_2$.

As an aid in writing node equations, we write the three standard form node equations with the descriptive notation:

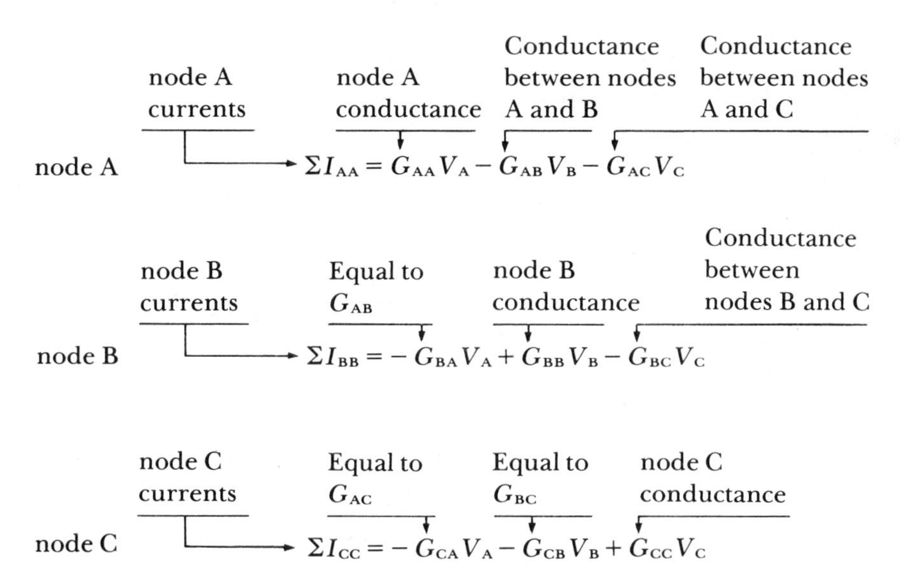

Notice that the same sign symmetry exists in node equations as in mesh equations; that is, all G's on the *main diagonal* are *positive*, and all other G's are negative. Notice also that node equations provide a more powerful technique than the mesh equations we have discussed, since node equations apply to *any* circuit, no matter whether it is planar or not!

We now state a systematic procedure for writing node equations:

1. Identify all the nodes in the circuit. Designate a ground (reference node and label the other nodes clearly.

2. Write the blank form of the standard node equations (one equation for each node) as

$$(\Sigma I\text{'s}) = (\qquad)V_A - (\qquad)V_B - (\qquad)V_C \ldots$$

3. Fill in the empty blanks in the node equations by inspecting the circuit.

• • •

Example 10.10 Write the node equation for the circuit in Fig. 10.16.

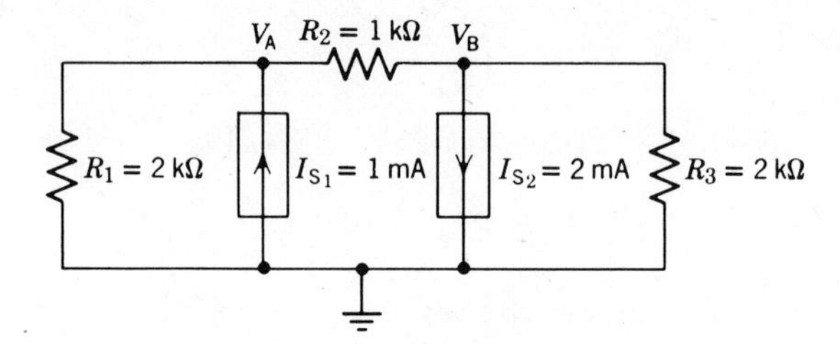

Figure 10.16

Solution We follow the above procedure.

1. There are three nodes in the circuit; the bottom node is ground, and the other two nodes are labeled V_A and V_B.
2. Since there are two unknown node voltages, we need two node equations, and the *blank forms* are:

$$\text{node A} \quad (\qquad) = (\qquad)V_A - (\qquad)V_B$$
$$\text{node B} \quad (\qquad) = -(\qquad)V_A + (\qquad)V_B$$

3. We complete the node equations by inspecting the circuit to fill in the "blanks." Thus

$$\text{node A} \qquad I_{S_1} = \left(\frac{1}{R_1} + \frac{1}{R_2}\right) V_A - \left(\frac{1}{R_2}\right) V_B$$

$$\text{node B} \qquad -I_{S_2} = -\left(\frac{1}{R_2}\right) V_A + \left(\frac{1}{R_2} + \frac{1}{R_3}\right) V_B$$

Notice the minus sign of I_{S_2} in the second equation. Filling in the proper circuit values,

$$\text{node A} \qquad 1\text{ mA} = \left(\frac{1}{2\text{ k}\Omega} + \frac{1}{1\text{ k}\Omega}\right) V_A - \left(\frac{1}{1\text{ k}\Omega}\right) V_B$$

$$\text{node B} \qquad -2\text{ mA} = -\left(\frac{1}{1\text{ k}\Omega}\right) V_A + \left(\frac{1}{1\text{ k}\Omega} + \frac{1}{2\text{ k}\Omega}\right) V_B$$

• • •

Example 10.11 Write the node equations for the circuit in Fig. 10.17.

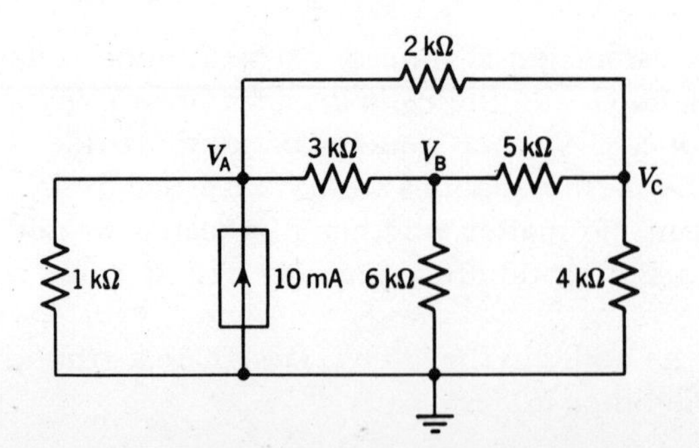

Figure 10.17

Solution

1. There are 4 nodes in the circuit; one is grounded and the other 3 are labeled.
2. The three *blank form* node equations are:

$$\begin{aligned}&\text{node A} \quad (\qquad) = (\qquad)V_A - (\qquad)V_B - (\qquad)V_C\\&\text{node B} \quad (\qquad) = -(\qquad)V_A + (\qquad)V_B - (\qquad)V_C\\&\text{node C} \quad (\qquad) = -(\qquad)V_A - (\qquad)V_B + (\qquad)V_C\end{aligned}$$

3. The node equations are completed by inspecting the circuit.

$$\text{node A} \quad 10\text{ mA} = \left(\frac{1}{1\text{ k}\Omega} + \frac{1}{2\text{ k}\Omega} + \frac{1}{3\text{ k}\Omega}\right)V_A - \left(\frac{1}{3\text{ k}\Omega}\right)V_B - \left(\frac{1}{2\text{ k}\Omega}\right)V_C$$

$$\text{node B} \quad 0 = -\left(\frac{1}{3\text{ k}\Omega}\right)V_A + \left(\frac{1}{3\text{ k}\Omega} + \frac{1}{5\text{ k}\Omega} + \frac{1}{6\text{ k}\Omega}\right)V_B - \left(\frac{1}{5\text{ k}\Omega}\right)V_C$$

$$\text{node C} \quad 0 = -\left(\frac{1}{2\text{ k}\Omega}\right)V_A - \left(\frac{1}{5\text{ k}\Omega}\right)V_B + \left(\frac{1}{2\text{ k}\Omega} + \frac{1}{4\text{ k}\Omega} + \frac{1}{5\text{ k}\Omega}\right)V_C$$

The equations can be simplified as follows: (1) multiply the first equation by 6 kΩ; (2) multiply the second equation by 30 kΩ; (3) multiply the third equation by 20 kΩ. The results are:

$$\begin{aligned}&\text{node A} \quad 60 = 11V_A - 2V_B - 3V_C\\&\text{node B} \quad 0 = -10V_A + 21V_B - 6V_C\\&\text{node C} \quad 0 = -10V_A - 4V_B + 19V_C\end{aligned}$$

Note that this circuit would require the simultaneous solution of *four* mesh equations!

• • •

7 USING NODE EQUATIONS

Again the confidence in writing node equations and using them to solve circuits comes only with practice. We will therefore devote this section to the solution of a number of circuits using node equations. We begin by solving for the node voltages in Fig. 10.16 using the node equations from Example 10.10. The equations are,

$$\text{node A} \quad 1\text{ mA} = \left(\frac{1}{2\text{ k}\Omega} + \frac{1}{1\text{ k}\Omega}\right)V_A - \left(\frac{1}{1\text{ k}\Omega}\right)V_B$$

$$\text{node B} \quad -2\text{ mA} = -\left(\frac{1}{1\text{ k}\Omega}\right)V_A + \left(\frac{1}{1\text{ k}\Omega} + \frac{1}{2\text{ k}\Omega}\right)V_B$$

We first simplify the equations by multiplying both of them by 2 kΩ. The results are,

$$\begin{aligned}&\text{node A} \quad 2 = 3V_A - 2V_B\\&\text{node B} \quad -4 = -2V_A + 3V_B\end{aligned}$$

We then use determinants to solve for V_A and V_B, and by Cramer's rule

$$V_A = \frac{\begin{vmatrix} 2 & -2 \\ -4 & 3 \end{vmatrix}}{\begin{vmatrix} 3 & -2 \\ -2 & 3 \end{vmatrix}} = \frac{6-8}{9-4} = -\frac{2}{5}\text{ V}$$

$$V_B = \frac{\begin{vmatrix} 3 & 2 \\ -2 & -4 \end{vmatrix}}{\begin{vmatrix} 3 & -2 \\ -2 & 3 \end{vmatrix}} = \frac{-12+4}{5} = -\frac{8}{5}\text{ V}$$

The solutions are easily checked by finding the individual resistor currents and applying KCL at each node.

Consider now the circuit in Fig. 10.18. At first glance it would seem to have four nodes, but after more careful examination it is seen to have only three

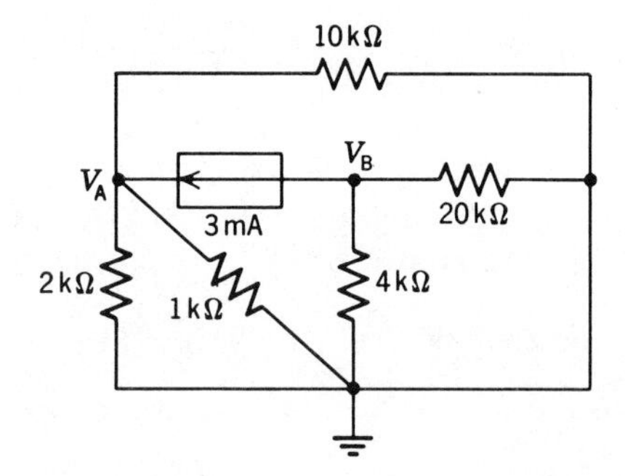

Figure 10.18

nodes; one node is ground, and the other two are labeled V_A and V_B. Note that it would require four mesh equations to solve the circuit as it stands. The two required node equations are:

$$\text{node A} \quad 3\text{ mA} = \left(\frac{1}{1\text{ k}\Omega} + \frac{1}{2\text{ k}\Omega} + \frac{1}{10\text{ k}\Omega}\right) V_A - (0)\, V_B$$

$$\text{node B} \quad -3\text{ mA} = -(0)\, V_A + \left(\frac{1}{4\text{ k}\Omega} + \frac{1}{20\text{ k}\Omega}\right) V_B$$

The first equation is easily solved for V_A by multiplying by 10 kΩ,

$$\text{node A} \quad 30 = 16 V_A$$

Therefore,

$$V_A = \frac{15}{8}\text{ V}$$

The second equation is then solved for V_B by multiplying by 20 kΩ. Thus,

$$\text{node B} \quad -60 = 6 V_B$$

Therefore,

$$V_B = -10\text{ V}$$

Note carefully that V_B is 10 volts *negative* with respect to ground. As a check, we will find all resistor currents and show them in Fig. 10.19. Using Ohm's Law,

$$I_{2k} = \frac{V_A}{2\text{ k}\Omega} = \frac{15/8}{2\text{ k}\Omega} = \frac{15}{16}\text{ mA}$$

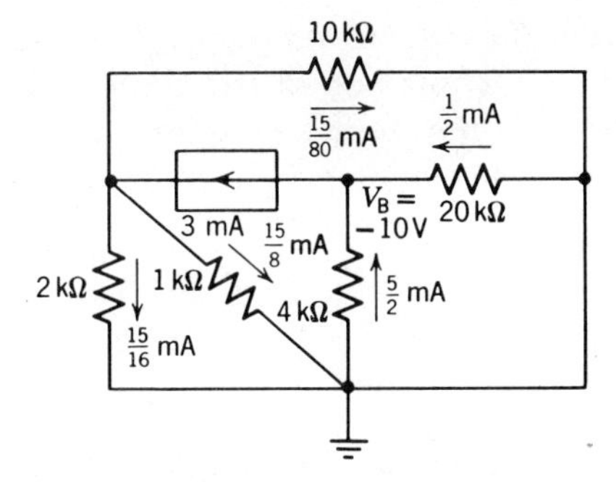

Figure 10.19 Solution for the circuit in Figure 10.18

$$I_{1k} = \frac{V_A}{1\ \text{k}\Omega} = \frac{15/8}{1\ \text{k}\Omega} = \frac{15}{8}\ \text{mA}$$

$$I_{10k} = \frac{V_A}{10\ \text{k}\Omega} = \frac{15/8}{10\ \text{k}\Omega} = \frac{15}{80}\ \text{mA}$$

$$I_{4k} = \frac{V_B}{4\ \text{k}\Omega} = \frac{10}{4\ \text{k}\Omega} = \frac{5}{2}\ \text{mA}$$

$$I_{20k} = \frac{V_B}{20\ \text{k}\Omega} = \frac{10}{20\ \text{k}\Omega} = \frac{1}{2}\ \text{mA}$$

We now check by applying KCL at each node.

$$\text{node A} \quad 3\ \text{mA} = I_{1k} + I_{2k} + I_{10k} = \frac{15}{8}\ \text{mA} + \frac{15}{16}\ \text{mA} + \frac{15}{80}\ \text{mA}$$

$$= \frac{150 + 75 + 15}{80}\ \text{mA} = \frac{240}{80}\ \text{mA} = 3\ \text{mA}$$

$$\text{node B} \quad 3\ \text{mA} = I_{4k} + I_{20k} = \frac{5}{2}\ \text{mA} + \frac{1}{2}\ \text{mA} = \frac{6}{2}\ \text{mA} = 3\ \text{mA}$$

Let us now consider solving a circuit using node equations, when the circuit contains *voltage* sources. We will find that voltage sources *may* reduce the number of equations which need to be solved. Consider the circuit in Fig. 10.20. The circuit has four nodes, one of which is grounded, and the other three are labeled. The three required blank node equations are

$$\text{node A} \quad (\qquad) = (\qquad)V_A - (\qquad)V_B - (\qquad)V_C$$

$$\text{node B} \quad (\qquad) = -(\qquad)V_A + (\qquad)V_B - (\qquad)V_C$$

$$\text{node C} \quad (\qquad) = -(\qquad)V_A - (\qquad)V_B + (\qquad)V_C$$

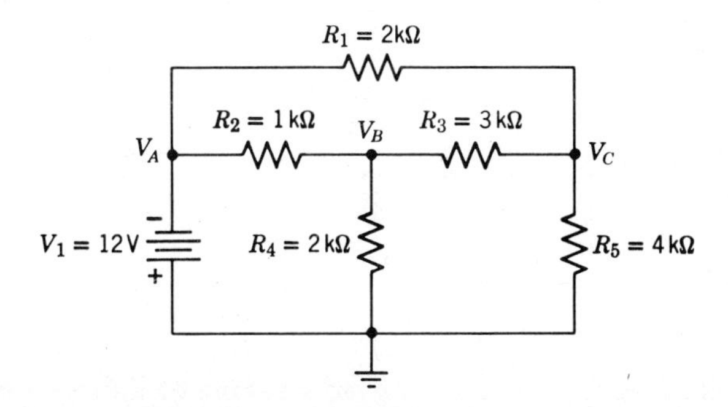

Figure 10.20

At this point, notice that the voltage at node A, V_A, is simply $-V_1$ or -12 V. We therefore immediately know one of the three unknown node voltages. Thus we actually need only *two* node equations, the ones at nodes B and C, and we simply eliminate node equation A. The two required equations are:

$$\text{node B} \quad 0 = -\left(\frac{1}{1\text{ k}\Omega}\right)V_A + \left(\frac{1}{1\text{ k}\Omega} + \frac{1}{2\text{ k}\Omega} + \frac{1}{3\text{ k}\Omega}\right)V_B - \left(\frac{1}{3\text{ k}\Omega}\right)V_C$$

$$\text{node C} \quad 0 = -\left(\frac{1}{2\text{ k}\Omega}\right)V_A - \left(\frac{1}{3\text{ k}\Omega}\right)V_B + \left(\frac{1}{2\text{ k}\Omega} + \frac{1}{3\text{ k}\Omega} + \frac{1}{4\text{ k}\Omega}\right)V_C$$

Notice carefully that we *must include* the V_A terms in these equations. We now substitute the known value for V_A and transpose to get

$$\text{node B} \quad \frac{-12}{1\text{ k}\Omega} = \left(\frac{1}{1\text{ k}\Omega} + \frac{1}{2\text{ k}\Omega} + \frac{1}{3\text{ k}\Omega}\right)V_B - \left(\frac{1}{3\text{ k}\Omega}\right)V_C$$

$$\text{node C} \quad \frac{-12}{2\text{ k}\Omega} = -\left(\frac{1}{3\text{ k}\Omega}\right)V_B + \left(\frac{1}{2\text{ k}\Omega} + \frac{1}{3\text{ k}\Omega} + \frac{1}{4\text{ k}\Omega}\right)V_C$$

We now simplify these equations by multiplying the first by 6 kΩ, and the second by 12 kΩ. Thus,

$$\text{node B} \quad -72 = 11V_B - 2V_C$$

$$\text{node C} \quad -72 = -4V_B + 13V_C$$

Subtracting these two equations,

$$0 = 15V_B - 15V_C$$

or

$$V_B = V_C$$

Then by substitution,

$$V_B = V_C = -8\text{ V}$$

Thus we see that voltage sources in a circuit can actually be an aid when using node equations, since their presence may reduce the number of equations needed.

We now conclude this section with another example.

• • •

Example 10.12 Find the node voltages in the circuit in Fig. 10.21.

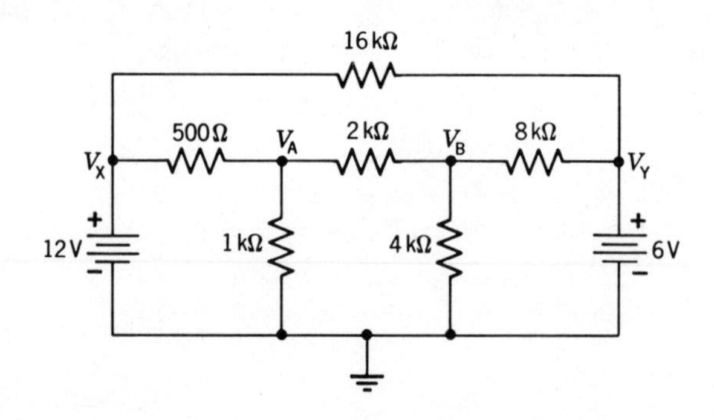

Figure 10.21

Solution There are five nodes in the circuit. One of them is grounded, and the other four are labeled V_A, V_B, V_X, and V_Y. The node voltages V_X and V_Y are immediately seen to be

$$V_X = 12\text{ V} \qquad V_Y = 6\text{ V}$$

We therefore need only two node equations written at nodes A and B. The proper forms are (notice carefully that V_X and V_Y are used!)

$$\text{node A} \quad (\quad) = (\quad)V_A - (\quad)V_B - (\quad)V_X - (\quad)V_Y$$

$$\text{node B} \quad (\quad) = -(\quad)V_A + (\quad)V_B - (\quad)V_X - (\quad)V_Y$$

By inspection of the circuit,

$$\text{node A} \quad 0 = \left(\frac{1}{500\ \Omega} + \frac{1}{1\text{ k}\Omega} + \frac{1}{2\text{ k}\Omega}\right) V_A - \left(\frac{1}{2\text{ k}\Omega}\right) V_B - \left(\frac{1}{500\ \Omega}\right) V_X - (0)\,V_Y$$

$$\text{node B} \quad 0 = -\left(\frac{1}{2\text{ k}\Omega}\right) V_A + \left(\frac{1}{2\text{ k}\Omega} + \frac{1}{4\text{ k}\Omega} + \frac{1}{8\text{ k}\Omega}\right) V_B - (0)\,V_X - \left(\frac{1}{8\text{ k}\Omega}\right) V_Y$$

Substituting in values for V_X and V_Y and transposing,

$$\text{node A} \quad \left(\frac{12}{0.5\text{ k}\Omega}\right) = \left(\frac{1}{0.5\text{ k}\Omega} + \frac{1}{1\text{ k}\Omega} + \frac{1}{2\text{ k}\Omega}\right) V_A - \frac{1}{2\text{ k}\Omega} V_B$$

$$\text{node B} \quad \frac{6}{8\text{ k}\Omega} = -\left(\frac{1}{2\text{ k}\Omega}\right) V_A + \left(\frac{1}{2\text{ k}\Omega} + \frac{1}{4\text{ k}\Omega} + \frac{1}{8\text{ k}\Omega}\right) V_B$$

We now multiply the first equation by 2 kΩ and the bottom equation by 8 kΩ to obtain,

$$\text{node A} \quad 48 = 7V_A - V_B$$

$$\text{node B} \quad 6 = -4V_A + 7V_B$$

Multiplying the top equation by 7 and adding to the bottom equation yields,

$$342 = 45V_A$$

Therefore, $$V_A = 7.6\text{ V}$$

Then by substitution,

$$V_B = 5.2\text{ V}$$

• • •

The reader is urged to check the solution by finding the resistor currents and applying KCL at each node. Does the 16 kΩ resistor have any effect on the solution? Why? The reader might also attempt a solution using mesh equations.

.8 CHOOSING BETWEEN MESH AND NODE EQUATIONS

We have now studied two completely general methods for solving electric circuits: mesh equations and node equations. For any one particular circuit one of the methods *may* be better than the other, and we would like to find a simple test to determine which method to use.

Since both methods lead to a set of simultaneous equations which must be solved, we would obviously choose the method requiring the fewest number of equations. For example, the circuit in Fig. 10.21 requires only *two* node equations, but would require *four* mesh equations. Most any one would agree that node equations will be the easier of the two methods in this case. On the other hand, the circuit in Fig. 10.22 would only require *two* mesh equations, but it would require *nine* node equations! Obviously, mesh equations would provide the best method.

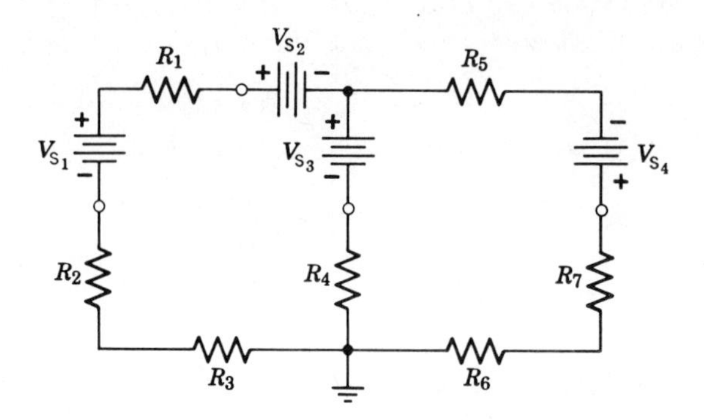

Figure 10.22

From our previous work, we know that a circuit having v nodes will in general require (v − 1) node equations, since one node is always referenced to ground. Thus, we have the formula,

$$\text{number of node equations} = \text{v} - 1$$
$$\text{where v is the number of nodes in the circuit.} \quad (10.9)$$

The number of required mesh equations can be found by simply counting the number of "window panes" if the circuit is drawn in planar form. It can be shown that the number of mesh equations required is equal to (e − v + 1), where e is the number of elements and v is the number of nodes. This equation is always true and it can be used whether or not the circuit is planar. We then have the formula,

$$\text{number of mesh equations} = (\text{e} - \text{v} + 1)$$
$$\text{where e is the number of elements,}$$
$$\text{and v is the number of nodes.} \quad (10.10)$$

• • •

Example 10.13 What is the best method (mesh or node equations) for solving (1) the circuit in Fig. 10.20, (2) the circuit in Fig. 10.13, (3) the circuit in Fig. 10.23?

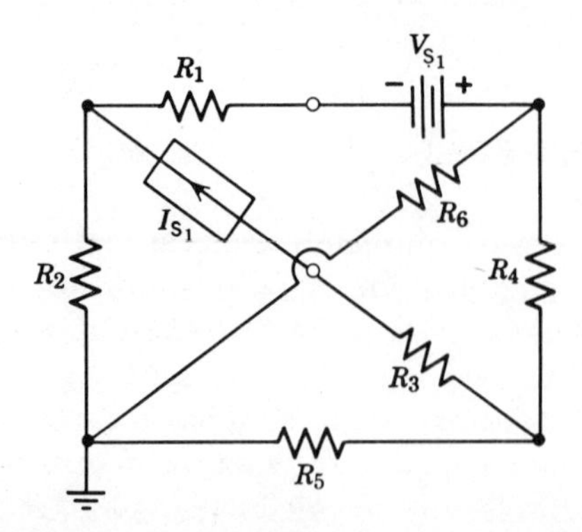

Figure 10.23

Solution

1. The number of elements, e = 6. The number of nodes v = 4. Thus,

$$\text{node equations} = v - 1 = 4 - 1 = 3$$
$$\text{mesh equations} = e - v + 1 = 6 - 4 + 1 = 3$$

It would seem to require 3 equations in either case; however, since V_A is immediately known, it would require only 2 node equations.

2. The number of elements, e = 7. The number of nodes, v = 5. Thus,

$$\text{node equations} = v - 1 = 5 - 1 = 4$$

But the node voltage at the positive side of V_{S_1} is known, so it only requires 3 node equations.

$$\text{mesh equations} = e - v + 1 = 7 - 5 + 1 = 3$$

However, $I_2 = I_S = 1$ mA, and thus only two mesh equations are required.

3. The number of elements, e = 8. The number of nodes, V = 6. Thus

$$\text{node equations} = v - 1 = 6 - 1 = 5$$
$$\text{mesh equations} = e - v + 1 = 8 - 6 + 1 = 3$$

Obviously, mesh equations would be the easier method. We now close this chapter with a word of warning. The solution for any circuit should be obtained in the *easiest* way possible. Therefore, one should always give consideration to the simplification techniques and the circuit theorems first. If these methods fail to produce results, then either mesh or node equations can always be used to provide a correct solution.

SUMMARY

Mesh and node equations provide two powerful methods for the solution of circuits. Mesh equations are based on the application of KVL around the meshes in the circuit, and they represent a system of equations which must be solved for the assigned mesh currents. Node equations are based on the application of KCL at all but the reference node in a circuit, and they represent a system of equations which must be solved for the assigned node voltages. The method of writing mesh equations described here is applicable only to planar circuits, but the node equations will always work for any circuit. One usually makes a choice between mesh and node equations based on the least number of required equations.

GLOSSARY

Element. *A component in a circuit; resistor, source, and so forth.*
Mesh. *A closed path which contains no other closed paths.*
Mesh equations. *The standard form given in Eq. 10.5.*
Loop. *A closed path which may or may not contain another closed path. Therefore, any closed path.*
Mutual resistance. *A resistor through which two different mesh currents flow (also called transfer resistance).*

Node. *A connection between two or more elements.*
Node equations. *The standard form given in Eq. 10.8.*
Planar circuit. *A circuit which can be drawn in a plane without any element crossing over any other element.*
Reference node. *The single ground (reference) node in any circuit.*

IMPORTANT RELATIONSHIPS

- Number of mesh equations = e − v + 1
- Mesh equations

$$\text{(1)}\quad \Sigma V_{11} = R_{11}I_1 - R_{12}I_2 - R_{13}I_3 \cdots$$
$$\text{(2)}\quad \Sigma V_{22} = -R_{21}I_1 + R_{22}I_2 - R_{23}I_3 \cdots$$
$$\text{(3)}\quad \Sigma V_{33} = -R_{31}I_1 - R_{32}I_2 + R_{33}I_3 \cdots$$
$$\vdots$$

- Number of node equations = v − 1
- Node equations

$$\text{(A)}\quad \Sigma I_{AA} = G_{AA}V_A - G_{AB}V_B - G_{AC}V_C \cdots$$
$$\text{(B)}\quad \Sigma I_{BB} = -G_{BA}V_A + G_{BB}V_B - G_{BC}V_C \cdots$$
$$\text{(C)}\quad \Sigma I_{CC} = -G_{CA}V_A - G_{CB}V_B + G_{CC}V_C \cdots$$
$$\vdots$$

REVIEW QUESTIONS

1. Two general methods for solving circuits are ____ equations and ____ equations.
2. Mesh and node equations can be used to solve ac circuits as well as dc circuits. (TF)
3. A general term which describes any component in a circuit is ____.
4. A connection between two or more elements is called a ____.
5. A nonplanar circuit (can, cannot) be drawn in a plane such that no element crosses over any other element.
6. A mesh is *any* closed path in a circuit. (TF)
7. A mesh must not contain any other ____ ____ within it.
8. If a planar circuit is considered as a "window," the meshes correspond to the "____ ____."
9. The application of KVL to a circuit leads to a system of (mesh, node) equations.
10. The unknowns which must be solved for in mesh equations are the ____ ____.
11. The number of mesh equations needed in a planar circuit is exactly equal to the number of "window panes." (TF)
12. The mathematical symbol Σ means "____ ____ ____."
13. Refer to the standard mesh equations given in Eqs. 10.5. The R's with (like, unlike) subscripts refer to total mesh resistances, and the R's with (like, unlike) subscripts refer to mutual resistances.
14. The mutual resistance R_{23} is common to mesh currents ____ and ____.

15. The *sign symmetry* in mesh equations means that all the R terms on the main diagonal are ____ and all other R terms are ____.
16. The *sign symmetry* in mesh equations is a result of assigning all of the mesh currents in a ____ direction.
17. List the four systematic steps for writing mesh equations.
18. In solving a set of mesh equations, one of the mesh currents is found to have a *negative* value. This means that the actual mesh current flows in the opposite direction from the assigned mesh current. (TF)
19. The application of KCL to a circuit leads to a system of (mesh, node) equations.
20. The unknowns which must be solved for in node equations, are the ____ ____.
21. The number of node equations needed for any circuit is exactly equal to the number of nodes, minus one. (TF)
22. Node equations can be used with any circuit, whether it is planar or not. (TF)
23. Refer to the standard node equations given in Eq. 10.8. The G's with (like, unlike) subscripts refer to the total conductance connected to a node, and the G's with (like, unlike) subscripts refer to conductances connected between two nodes.
24. The conductance G_{AB} is connected between nodes ____ and ____, and it is equal to G ____.
25. Currents from current sources connected to a node are entered as positive values in the node equation if they (enter, leave) that node.
26. List the three systematic steps for writing node equations.
27. The number of mesh equations required for the solution of a circuit is given by the formula ____, and the number of node equations is given by the formula ____.

OBLEMS

1. How many elements, nodes, and meshes are there in the circuit in Fig. 10.10?
2. How many elements, nodes, and meshes are there in the circuit in Fig. 10.12?
3. Apply KVL to the circuit in Fig. 10.24 to develop the mesh equation. Now solve the mesh equation and check the solution by any method.
4. Apply KVL to the circuit in Fig. 10.25 to develop the mesh equation. Now solve the mesh equation and check the solution by any method.
5. Apply KVL to the circuit in Fig. 10.26 to develop the two mesh equations.
6. Solve for the mesh currents in Fig. 10.26 if $R_1 = 6.8\ \text{k}\Omega$, $R_2 = 2.2\ \text{k}\Omega$, $R_3 = 1\ \text{k}\Omega$, $R_4 = 3.3\ \text{k}\Omega$, $R_5 = 4.7\ \text{k}\Omega$, and $V_S = 27\ \text{V}$. Check by any method.
7. Solve for the mesh currents in Fig. 10.26 if $R_1 = 600\ \Omega$, $R_2 = 500\ \Omega$, $R_3 = 100\ \Omega$, $R_4 = 200\ \Omega$, $R_5 = 300\ \Omega$, and $V_S = 12\ \text{V}$.
8. Write the mesh equation for the circuit in Fig. 10.25 using the four steps. Solve for the mesh current if V_{S_2} is changed to 40 V, and check by any method.

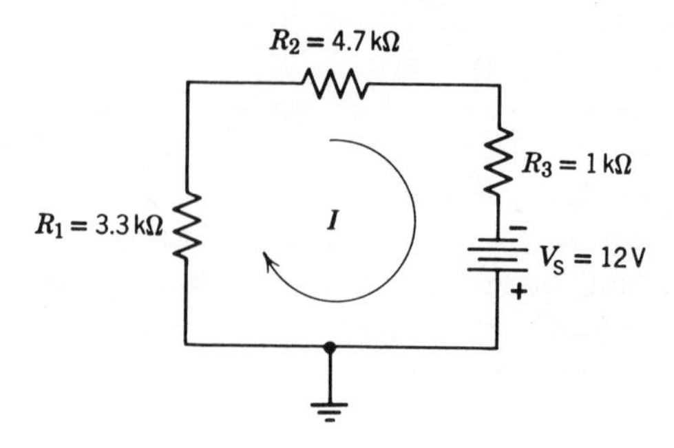

Figure 10.24

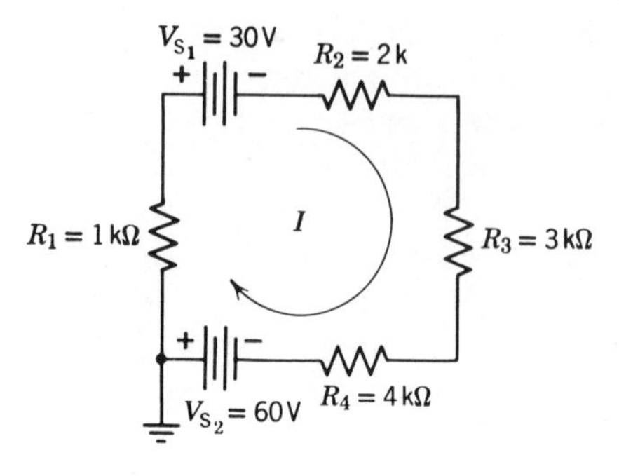

Figure 10.25

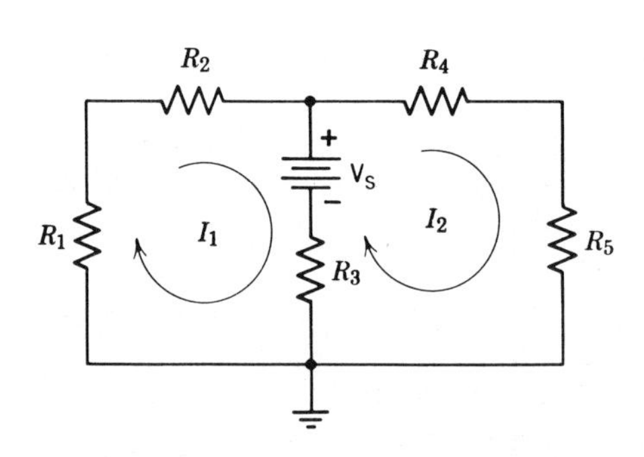

Figure 10.26

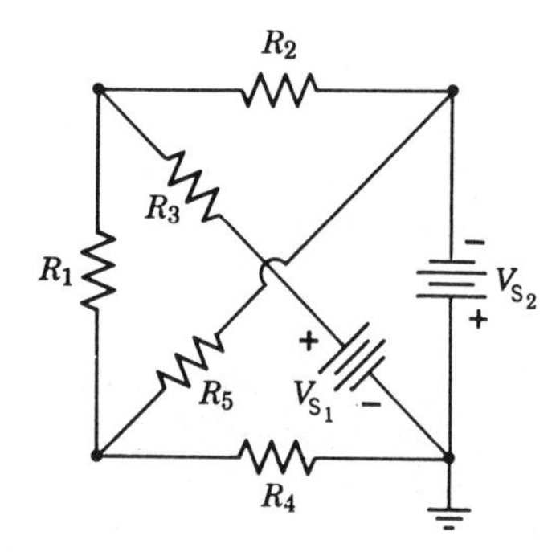

Figure 10.27

9. Use the four steps to write the mesh equations for the circuit in Fig. 10.26. Solve for the mesh currents if $R_1 = 10\ \text{k}\Omega$, $R_2 = 68\ \text{k}\Omega$, $R_3 = 22\ \text{k}\Omega$, $R_4 = 68\ \text{k}\Omega$, $R_5 = 110\ \text{k}\Omega$, and $V_S = 100\ \text{V}$. Check by any method.
10. Write the mesh equations for the circuit in Fig. 10.27.
11. Solve for the mesh currents in Fig. 10.27 if $R_1 = 100\ \Omega$, $R_2 = 100\ \Omega$, $R_3 = 200\ \Omega$, $R_4 = 300\ \Omega$, $R_5 = 400\ \Omega$, $V_{S_1} = 5\ \text{V}$, and $V_{S_2} = 10\ \text{V}$. Check by any method.
12. Use mesh equations to solve for the mesh currents in Problem 11 if V_{S_2} is changed to 5 V.
13. Write the mesh equations for the circuit in Fig. 10.20. Solve for the three mesh currents.
14. Solve the circuit in Fig. 10.15 using mesh equations.
15. Use mesh equations to find the current through the 2 kΩ resistor in Fig. 10.28. Check by any method.
16. Apply KCL to the circuit in Fig. 10.29 to develop a node equation. Now solve for the node voltage and check by any method.
17. Apply KCL to the circuit in Fig. 10.30 to develop two node equations.
18. Solve for the node voltages in Fig. 10.30 if $R_1 = 1\ \text{k}\Omega$, $R_2 = 2\ \text{k}\Omega$, $R_3 = 3\ \text{k}\Omega$, $I_{S_1} = 5\ \text{mA}$, and $I_{S_2} = 10\ \text{mA}$ (use the results of Problem 17). Check by any method.
19. Solve Problem 11 using node equations.
20. Solve the circuit in Fig. 10.12 using node equations. Use the bottom line of the circuit as the reference node and replace R_1 and V_1 with a Norton equivalent. Check by any method.

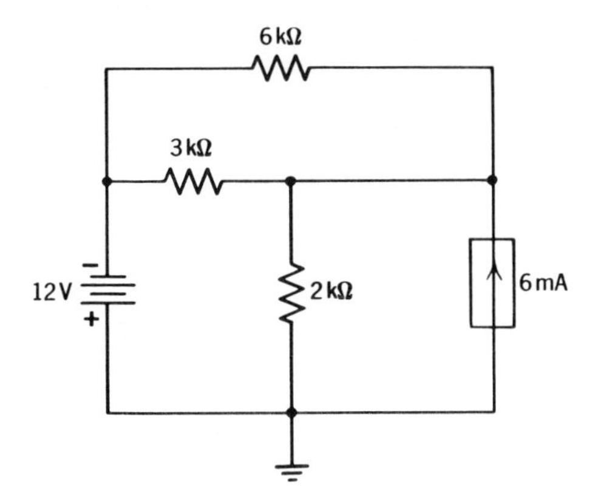

Figure 10.28

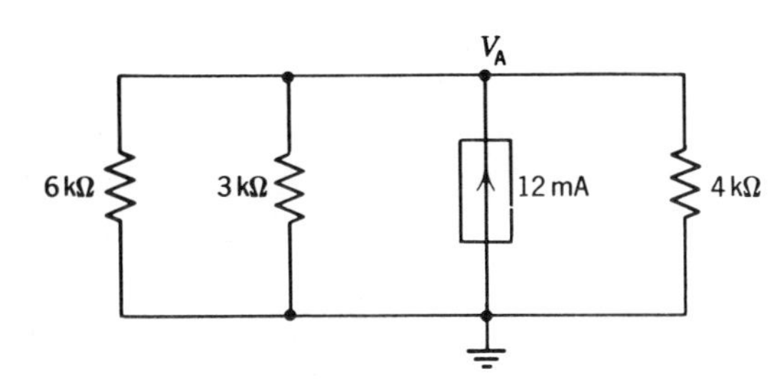

Figure 10.29

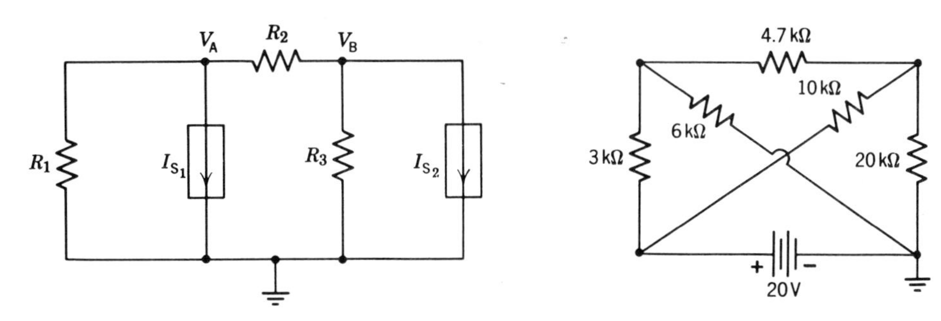

Figure 10.30

Figure 10.31

21. Solve for the node voltages in Fig. 10.28 using node equations. Check the results with Problem 15.

22. Use node equations to find the node voltages in Fig. 10.31. Check by finding all of the resistor currents.

23. Use node equations to solve for the node voltage V_X in Fig. 10.32. Check by any method.

24. Solve for V_X in Fig. 10.32 using mesh equations. Consider replacing the 0.001 A current source in parallel with the 10 kΩ (a Norton) with its Thévenin equivalent first.

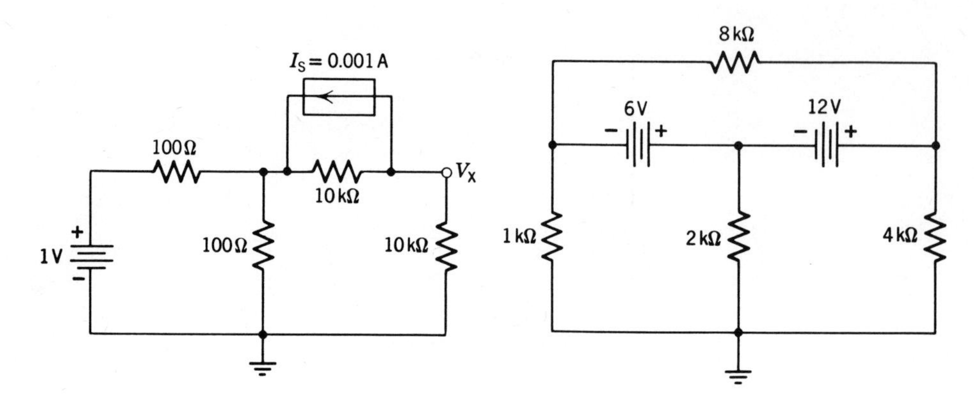

Figure 10.32

Figure 10.33

25. Solve the circuit in Fig. 10.33 using mesh equations. Can you check the results by solving using node equations? What are the difficulties encountered?

26. Solve for the node voltages in Fig. 10.34 using node equations. Check your solution using mesh equations. Which method is easier? Why?

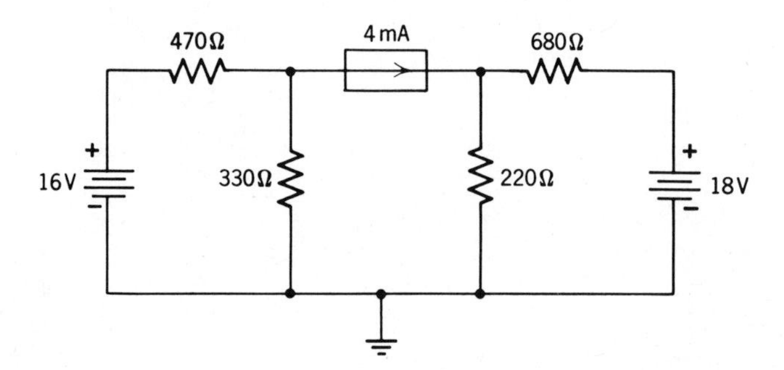

Figure 10.34

MAGNETICS

A magnetic field is fundamental in nearly all applications of electricity. Magnetic fields are associated with permanent magnets, electric currents, and electromagnetic radiation (radio and television signals). The forces exerted by magnetic fields account for the operation of many electrical devices—motors, generators, TV sets, D'Arsonval meters, to name but a few. The study of magnetism is therefore essential for understanding basic electricity.

Magnetism has been known to man since the first century A.D. A black stone, called *lodestone,* was found to attract iron; later lodestone was shaped into the form of a needle and used as a compass. One story attributes the discovery of lodestone to a Greek shepherd named Magnus; another story states that it was first discovered in Magnesia, a province of Asia Minor. In any case, we know lodestone as a naturally occurring iron oxide called *magnetite.*

Following the discovery of lodestone, a great deal of fear and superstition gathered in its wake. It was thought to contain magical powers over life and health and was often used as medicine. Sailors would not eat onions or garlic fearing that *demons* in the onion and garlic juice would destroy the magnetism in the compass and they would become lost at sea.* Sir Thomas Browne (1605–1682) dispelled these fears by soaking a compass in union juice until it was rusty thus demonstrating there was no loss of magnetism. The fear and

*A modern example of ignorance leading to superstition and fear is the advertisement by *Quacks* of *radioactive* materials and machines which will heal nearly any ailment.

superstition associated with magnetism prevailed until approximately the year 1600. In that year, Dr. William Gilbert (1540–1603), a physician to Queen Elizabeth, published his famous work, *De Magnete.* This book, containing all of the accumulated knowledge of electricity, marked the beginning of the written history of electricity. Thereafter, the scientific method began to replace and reduce ignorance and fear by experiments such as those of Sir Thomas Browne.

1.1 PERMANENT MAGNETS

Nearly everyone has observed the forces of attraction and repulsion between two magnets. These forces can be accounted for by assigning *magnetic poles* as shown on the *bar* magnet in Fig. 11.1. The end of the magnet marked N is the *north* pole; the other end is the *south* pole. A bar magnet always has a north and a south pole. If the magnet is broken into a number of smaller pieces, each becomes a bar magnet having two poles as shown in Fig. 11.2.

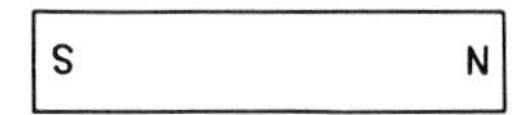

Figure 11.1 A bar magnet

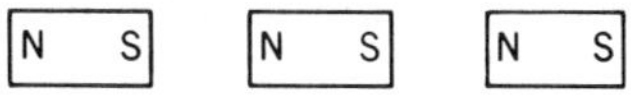

N S N S N S

Figure 11.2 A bar magnet broken into three pieces forms three smaller bar magnets

If two bar magnets are placed with a south pole near a north pole, as shown in Fig. 11.3*a*, there will be a force of *attraction* between the two magnets. If the two magnets are allowed to move, they will come together and "stick" to one another; the force of attraction between the *unlike* poles will hold them together. *Thus there is a force of attraction between unlike poles.*

On the other hand, if two bar magnets are placed with their south poles (or north poles) near to one another, the magnets will *repel* each other. If the magnets are allowed to move, they will move apart. *Thus there is a force of repulsion between like poles* (Fig. 11.3*b*).

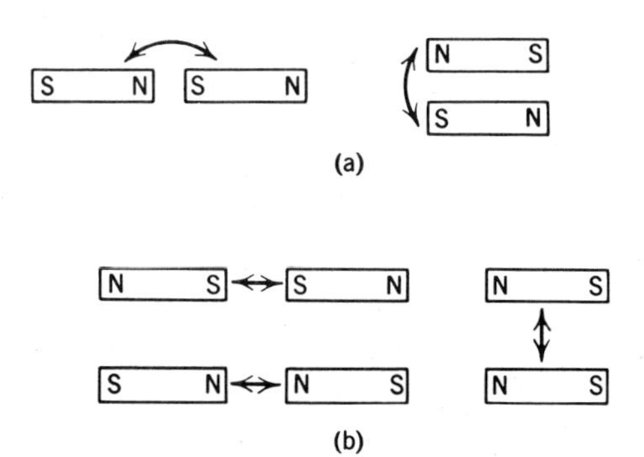

Figure 11.3 Forces between magnetic poles (*a*) Unlike poles attract (*b*) Like poles repel

The forces between magnetic poles are governed by Coulomb's law which states

the force between two magnetic poles is directly proportional to the product of the pole strengths, and inversely proportional to the square of the distance between them.

In equation form,

$$F = \frac{M_1 M_2}{\mu d^2} \tag{11.1}$$

where, F is the force, M_1 and M_2 are the magnetic pole strengths, μ is the permeability, and d is the distance between the poles. The permeability μ is a constant which is different for different materials. The pole strengths M_1 and M_2 are idealized values, since it is *not* possible to isolate a pole; electric charge of both kinds can be separated, but breaking a magnet into smaller pieces results in smaller magnets, all of which still have both N and S poles. An isolated pole can be simulated by using a *long thin* magnet where the poles are near the ends. The reader should notice the similarity between this equation and those for gravitational and electrostatic forces in Chapter 1.

The fact that magnets exert forces on one another even when they are not touching shows the *force at a distance* relationship. This is explained by the existence of a *magnetic field* around a magnet. A magnetic field is said to exist in any space in which a force is exerted on a magnetic pole; forces are also exerted on electrons *moving* through a magnetic field.

The magnetic field surrounding a bar magnet can be demonstrated using a compass as shown in Fig. 11.4. Wherever the compass is placed in the magnetic field, it will line itself up with the *field lines.* The field lines can be traced out in this manner. Notice that the "arrow" end of the needle is a *north* pole, and it therefore points along the field lines toward the *south* pole of the bar magnet.

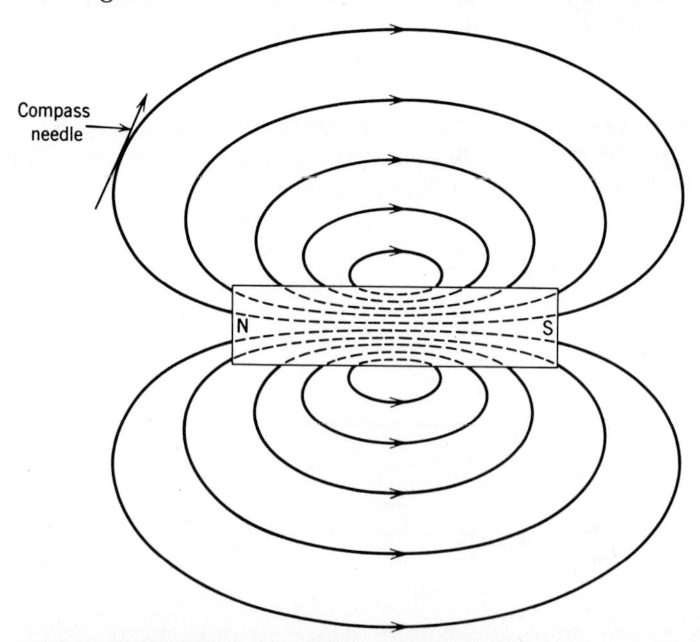

Figure 11.4 Magnetic field around a bar magnet

We make the following important observations regarding a magnetic field:

1. The field completely surrounds the magnet and occupies a space similar to a "fat donut."
2. The field lines in Fig. 11.4 are representative of the field; the field occupies *all* of the space around the magnet, and there are no gaps or holes.
3. The field lines are *continuous* loops; they do not begin or end at some point.
4. The field lines have an orientation: They emerge from the north end of the magnet and enter the south end.
5. The field lines tend to repel one another; they tend to diverge as one moves outward from the magnet. Therefore, the field lines *never* cross one another. At the same time, the field lines tend to shorten themselves like stretched rubber bands.

When two magnets are brought close together, there is an interaction between the two fields. The *resulting field* can be used to demonstrate the forces of attraction and repulsion. In Fig. 11.5*a*, the fields from the two magnets are in the *same* direction. They therefore tend to *reinforce* one another, and the resulting field is stronger. This leads to a force of attraction. The field tends to confine itself to the space directly between the ends of the magnets, and the "bulging" of the field is called *fringing*.

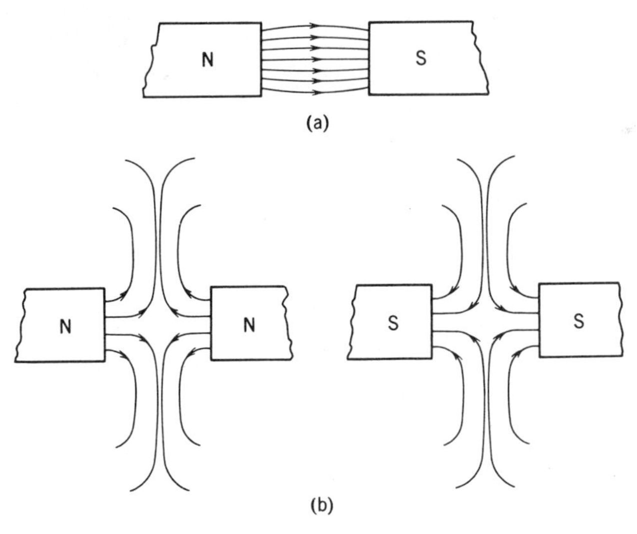

Figure 11.5 (*a*) Reinforcing fields (*b*) Opposing fields

The fields from the two magnets in Fig. 11.5*b* are in opposite directions. They therefore tend to *oppose* one another, and the field lines seem to be bent away from one another, resulting in a force of repulsion.

• • •

Example 11.1 Show the magnetic field around a ring magnet *without* a gap, and then with a gap.

Solution A ring magnet is shown in Fig. 11.6*a*. With no gap, the field is almost completely confined to the interior of the magnet itself. The field lines

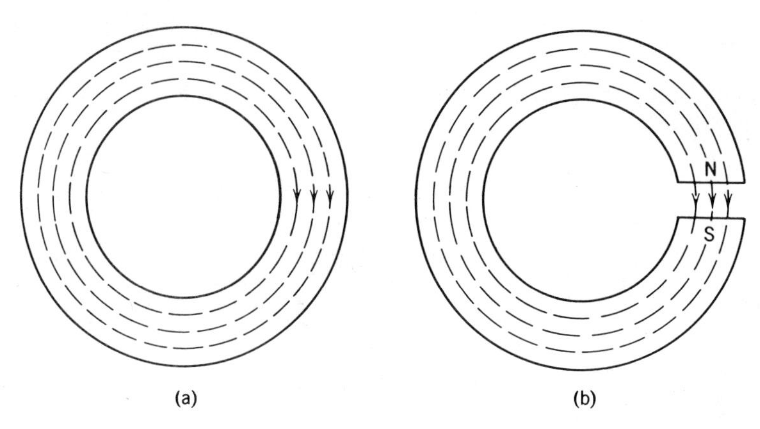

Figure 11.6 Example 11.1 (*a*) Ring magnet (*b*) Ring magnet with a gap

are continuous (no beginning or ending), and there are therefore no north or south poles. A ring magnet with a gap is shown in Fig. 11.6*b*. The field is still largely confined to the interior of the magnet. A north and a south pole are now formed at the gap. The field across the gap is a reinforcing field and is largely confined to the cross-sectional area of the gap.

• • •

In *De Magnete,* Dr. Gilbert suggests that the earth itself acts as a huge magnet. The earth is indeed surrounded by a magnetic field, as evidenced by the use of the compass. The earth might be considered as a huge bar magnet as shown in Fig. 11.7. The north pole of the earth must actually be the south pole of our large bar magnet, since the pointing end of the compass needle (or the north-seeking pole) is traditionally called a north pole.

The magnetic poles of the earth are not aligned with the geographical poles. The geographical north pole is used to designate *true north,* and the north magnetic pole is used to designate *magnetic north* as measured with a compass. The difference between the two varies depending on the latitude of the observer. The variation is called the *magnetic declination.*

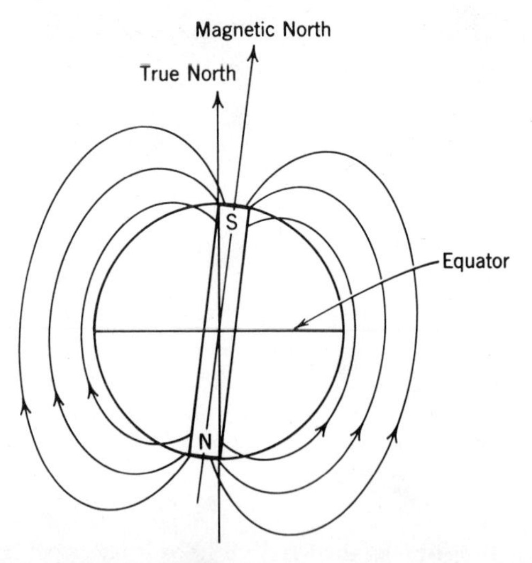

Figure 11.7 The earth as a magnet

11.2 MAGNETIC FLUX Φ

The magnetic field lines in Fig. 11.4 are only representations of the magnetic field surrounding the magnet. Actually there are an infinite number of such lines, and they are collectively referred to as the *magnetic flux* surrounding the magnet. The Greek letter phi **Φ** is the symbol for magnetic flux (pronounced fee or fí).

In early works, magnetic fields of different strengths were represented by drawing them with different numbers of field lines. Thus the *line* was used as the unit of magnetic flux; the magnet in Fig. 11.4, for example, has a flux of 10 lines since there are 10 lines representing the field. A stronger field having a flux of 20 lines would be drawn with exactly 20 field lines. Thus the terms *flux* and *lines* are synonymous. A line of flux is called a *maxwell* in honor of the Scottish physicist James Clerk Maxwell (1831–1879). The maxwell is the unit of flux in the centimeter-gram-second (cgs) system of measurement, but we will use the MKS unit of flux, the *weber*. One weber equals 10^8 lines (or maxwells).

You will notice that the symbol for flux **Φ** is printed in **bold face** type. This means that magnetic flux is a *vector* quantity, that is, it has *both magnitude* and *direction*. The magnitude is the *amount* of flux measured in webers; the *direction* corresponds to the direction of the field lines.

• • •

Example 11.2 Draw the representation of magnetic flux of $\mathbf{\Phi} = 5 \times 10^{-8}$ webers between the poles of a horseshoe magnet.

Solution The magnet and the field are shown in Fig. 11.8. Notice that 5×10^{-8} webers = 5 lines. There are thus 5 lines, and they are directed from the N pole to the S pole.

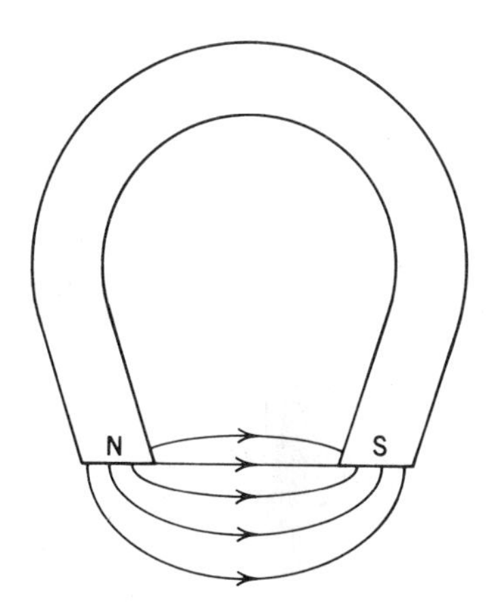

Figure 11.8

• • •

A very useful quantity in dealing with magnetic fields is magnetic *flux density*. The symbol for flux density is **B**. Magnetic flux density is simply the amount of magnetic flux passing through a plane at right angles to the magnetic field. For example, in Fig. 11.9 the flux density **B** is 2 lines (maxwells) per square centimeter. The basic unit of flux density in the cgs system is one maxwell per square centimeter, and this is defined as one gauss (after Karl Gauss, Germany, 1777–1855). In the MKS system, the unit of flux is the weber,

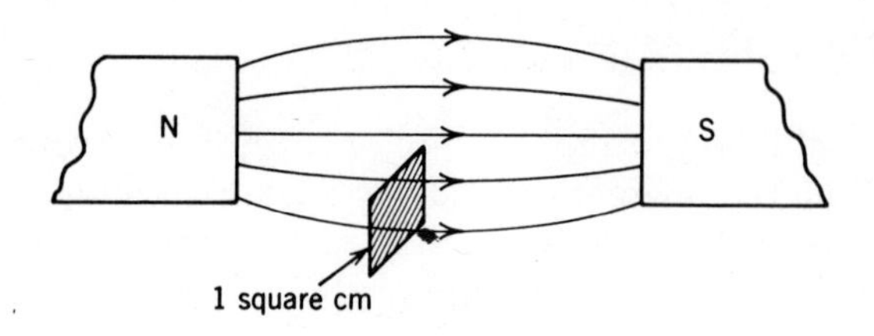

Figure 11.9

the unit of length is the meter, and thus the unit of flux density is webers per square meter. In equation form,

$$\mathbf{B}(\mathrm{Wb/m^2}) = \frac{\mathbf{\Phi}(\mathrm{Wb})}{A\,(\mathrm{m^2})} \tag{11.2}$$

where Wb is the abbreviation for webers.

• • •

Example 11.3 What is the flux density next to the pole face of the bar magnet in Fig. 11.10 if the total flux is 10^{-5} Wb?

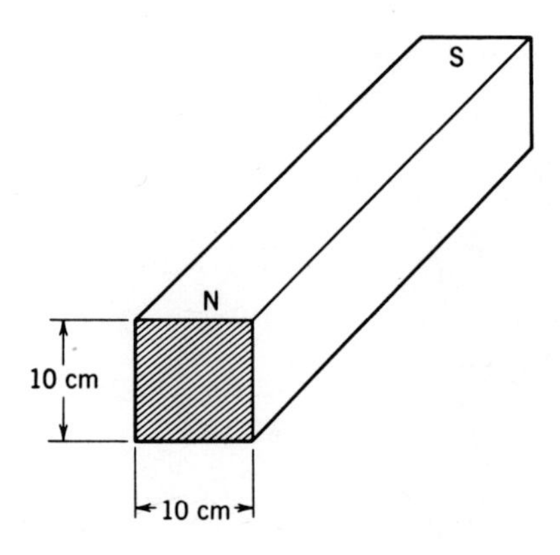

Figure 11.10

Solution We assume all of the flux passes through the end (pole face) of the magnet. The pole face area is

$$A = 10\ \mathrm{cm} \times 10\ \mathrm{cm} = 10^{-2}\ \mathrm{m^2}$$

Then using Eq. 11.2,

$$\mathbf{B} = \frac{\mathbf{\Phi}}{A} = \frac{10^{-5}\ \mathrm{Wb}}{10^{-2}\ \mathrm{m^2}} = 10^{-3}\ \mathrm{Wb/m^2}$$

• • •

Example 11.4 What is the total flux passing through a square 10.0 cm on a side if the flux density is 5000 μWb/m^2?

Solution Equation 11.2 can be rearranged as

$$\mathbf{\Phi} = \mathbf{B}A$$

$$\mathbf{\Phi} = 5000 \times 10^{-6}\,\frac{\mathrm{Wb}}{\mathrm{m^2}} \times 10^{-2}\ \mathrm{m^2} = 50\ \mu\mathrm{Wb}$$

• • •

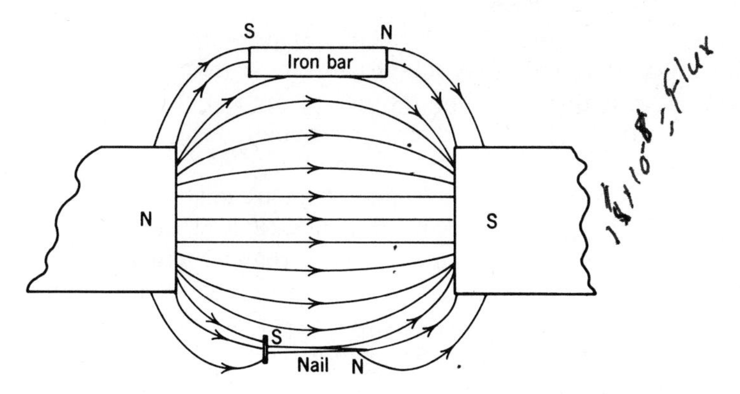

Figure 11.11 Forming a permanent magnet by induction

When a magnetic material such as iron is placed in a magnetic field, the field lines are distorted as shown in Fig. 11.11. The lines all tend to bend toward and travel through the iron bar (or nail). Since there are flux lines passing through the iron, it must have a north and a south pole, and it in fact behaves as a magnet. If the bar or the nail is then removed from the field, it will remain a magnet, and it is said to be *magnetized.* This is called magnetizing by *induction* and is one method to form permanent magnets.

The magnetic field surrounding a bar magnet can be shown by placing a paper or glass plate over the magnet and sprinkling *iron filings* on the paper. Each of the little pieces of iron becomes a magnet by induction, and they arrange themselves in the form N—SN—SN—S along the magnetic field lines. The pattern formed by the filings is then representative of the magnetic field lines around the magnet as shown in Fig. 11.12.

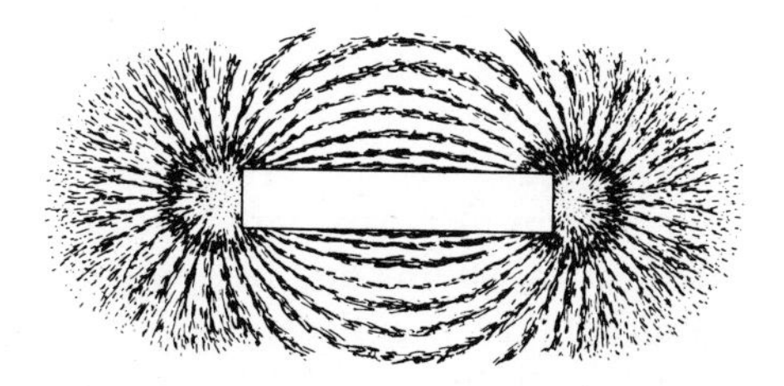

Figure 11.12 Iron filings show magnetic field line pattern around a bar magnet

MAGNETIC MATERIALS

Not all materials are magnetic, as evidenced by the fact that not all materials are attracted to a permanent magnet. There are in general three classes of materials, *diamagnetic, paramagnetic,* and *ferromagnetic,* and the magnetic behavior of these materials can be explained on the basis of their atomic structure.

The electrons in an atom are assumed to be traveling around a nucleus in fixed orbits (the Bohr model). Since these electrons are in motion they constitute electric currents and they therefore have magnetic fields associated with them.* Furthermore, each electron *spins* on its own axis as it travels in an orbit (similar to the earth's rotation) and there is an additional magnetic field associated with this spinning. These magnetic fields are very small for an individual atom, but can produce a strong field if the atoms are arranged such that all of the fields reinforce one another.

In one class of materials, the *spin* magnetic fields and the *orbital* magnetic fields are in opposite directions and they tend to cancel one another. Such materials are said to be *diamagnetic.* When diamagnetic materials are placed in a magnetic field, the magnetic fields associated with each atom tend to *oppose* the external field. However, the net magnetic flux *inside* the diamagnetic material is the *same* as *outside* the material. Wood and paraffin are examples of two diamagnetic materials.

In a second class of materials, the orbital and spin magnetic fields do not quite cancel, and as a result each atom has a slight magnetic field associated with it. The atoms are all arranged randomly in the material, and thus it does not exhibit the properties of a permanent magnet. If such a material is placed in a magnetic field the atoms align themselves with the applied field, and thus the field *inside* the material is *slightly greater* than the applied field. Such materials are said to be *paramagnetic.* Two such materials are aluminum and platinum.

The third class of materials, known as *ferromagnetic* materials, are very important in magnetics. In ferromagnetic materials, each atom has a relatively large magnetic field associated with it since the orbital and spin fields reinforce one another. When a ferromagnetic material is placed in an external magnetic field, the atoms align themselves with the applied field. As a result, the net magnetic field *inside* the material is *much greater* than the applied field. Specifically, the flux per unit area (flux density) within the material may be many times greater than the flux density of the applied field. Examples of ferromagnetic materials are iron, nickel, cobalt, and their alloys. In Fig. 11.11 it is clear that the flux density in the iron bar or nail is much greater than the flux density of the applied field. Ferromagnetic materials have the important characteristic that the atoms tend to remain aligned after the removal of the applied field. Thus ferromagnetic materials can be magnetized and they will remain as permanent magnets.

It is interesting to note that the atoms in a permanent magnet will tend to rearrange themselves randomly if the magnet is subjected to physical shock. Thus permanent magnets can be destroyed by vibration or "pounding."

It is also interesting to note that the atoms do not change orientation individually but seem to "switch" together in groups called *domains.* The "switching" of these domains can be observed under a microscope by shining polarized light on the polished surface of a magnet held in a variable magnetic field.

If the temperature of a ferromagnetic material is raised above a certain value called the *Curie temperature* (about 770°C for iron), the ferromagnetic properties disappear and the material behaves paramagnetically. It is believed *thermal agitation* becomes so great above the Curie temperature that atoms "switch" individually and domains disappear. Permanent magnets can be easily made by heating above the Curie temperature and then cooling in an

*The magnetic field associated with an electric current is discussed in the next section.

applied magnetic field. Alternatively, magnets can be destroyed by subjecting them to high temperatures.

11.4 ELECTROMAGNETISM

A relationship between electricity (electric current) and magnetism was discovered in 1819 by Hans Christian Oersted (1770–1851). Oersted noticed that a compass needle could be deflected when brought near a current-carrying conductor. In fact, the needle would always set itself at right angles to the direction of the current as shown in Fig. 11.13. Based on these observations, André Ampère explained that the magnetic field, associated with the current, must encircle the conductor. The magnetic field lines would then form rings (cylinders) around the conductor as shown in Fig. 11.13.

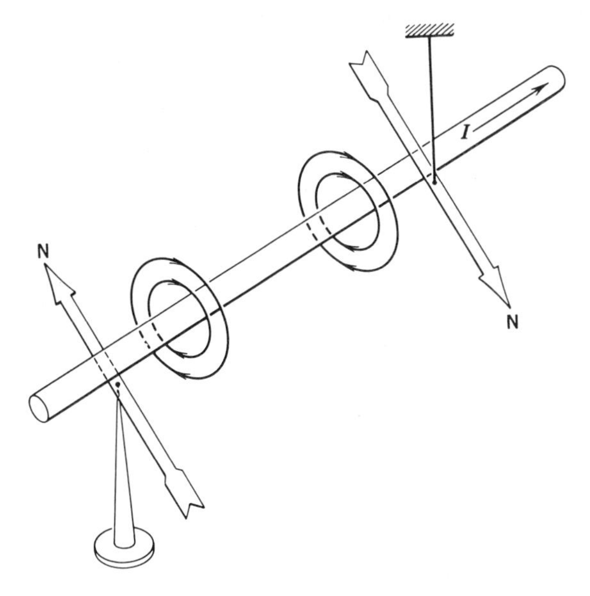

Figure 11.13 Deflection of a compass needle placed near a current-carrying conductor

The magnetic field lines around a current-carrying conductor can be demonstrated by passing the conductor through a plane (sheet of cardboard or glass) which is held at right angles to the conductor. The magnetic field lines can then be drawn on the surface of the plane using a compass; alternatively, iron filings sprinkled on the surface will align themselves with the magnetic field lines, and they will form concentric rings as shown in Fig. 11.14.

A careful inspection of the magnetic field lines around a current-carrying conductor leads to a very important result known as the *right-hand-rule.*

The right-hand-rule: If the fingers of the right hand encircle a conductor such that the fingers point in the direction of the magnetic field, the thumb will then point in the direction of the conductor current.

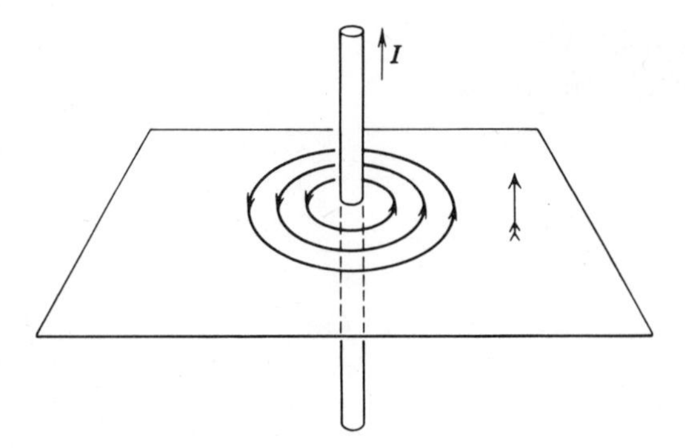

Figure 11.14 Indicating magnetic field lines around a current-carrying conductor using a compass, or iron filings

The right-hand-rule can be used to determine the magnetic field direction if the current is known, or to determine the current direction if the field direction is known (Fig. 11.15).

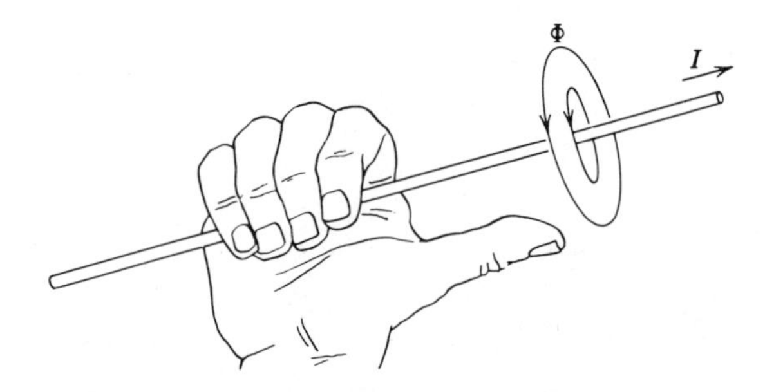

Figure 11.15 The right-hand-rule

• • •

Example 11.5 Use the right-hand-rule to determine the direction of the magnetic field in Fig. 11.16*a*. Determine the direction of the current in Fig. 11.16*b*.

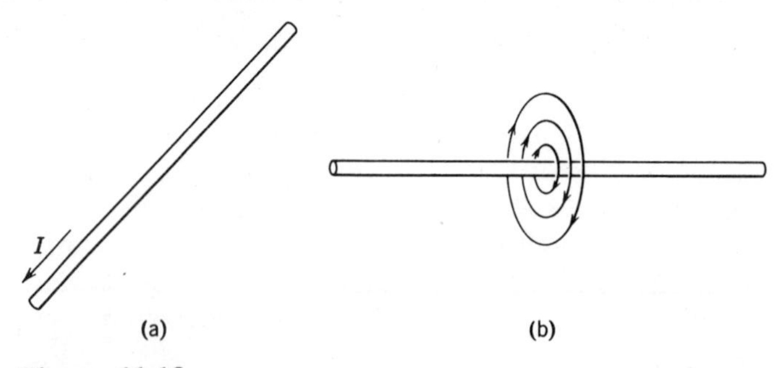

Figure 11.16

Solution If the conductor in Fig. 11.16*a* is grasped with the thumb pointing down and to the left in the direction of *I*, the fingers of the right hand encircle the wire in a counterclockwise direction. Thus the magnetic field is counterclockwise around this conductor. If the fingers of the right hand encircle the conductor in Fig. 11.16*b* in the direction of the magnetic field as shown, the thumb must point to the right. Thus the current direction must be from left to right.

• • •

We now turn our attention to a coil of wire formed by winding a length of wire around a cylindrical form as shown in Fig. 11.17*a*. A coil of wire such as this is often called a *solenoid*. If a current is forced through the coil, there will be a magnetic field set up around the coil as shown in the figure. You will notice that the magnetic field around the coil is very similar to the field around a permanent magnet, that is, the flux lines are continuous loops, and they emerge from the left end of the coil and enter the right end of the coil.

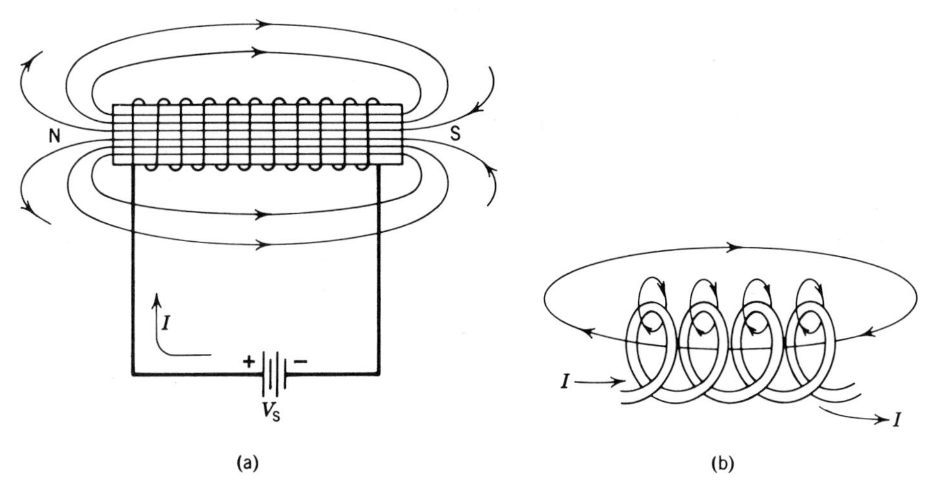

Figure 11.17 (*a*) The magnetic field due to current in a solenoid (*b*) Composite magnetic field due to field around each loop of wire

Thus, a coil such as this can be used as a magnet; the left end of this coil is the *north pole* and the right end is the *south pole*. If the coil is wound on a soft iron core, the iron is magnetized by the field, and we have an *electromagnet*.

There are three important points to be noted. First, if the current in the coil is reduced to zero, there is no field. Thus an electromagnet can be "turned off," whereas a permanent magnet cannot. Second, if the current direction is reversed, the magnetic field lines will also reverse, resulting in a change of polarity of the poles. Thus the poles of an electromagnet can be changed by simply changing the current direction. Finally, the strength of the magnetic field, or the flux in the field, can be varied by simply changing the magnitude of the current: the greater the current in the coil, the greater the magnetic flux.

The magnetic field around a solenoid can be justified by applying the right-hand-rule to each wire as shown in Fig. 11.17*b*. It is easily seen that the field lines between adjacent wires cancel one another. Also, the fingers always point from right to left *inside* the coil, as the hand travels along the wire. Thus the field lines inside the coil reinforce one another resulting in a composite field going from right to left. As the hand travels around the loops of the coil, the fingers always point from left to right *outside* the coil. Thus the field lines outside the coil also reinforce one another, resulting in a composite field from left to right. We can incorporate these results to form the *right-hand-rule for solenoids.*

> ***Right-hand-rule for solenoids: If the solenoid is grasped with the right hand such that the fingers point in the direction of current in the wire, the thumb will point in the direction of magnetic flux (toward the north pole).***

• • •

Example 11.6 Which end of the electromagnet in Fig. 11.18 is the north pole?

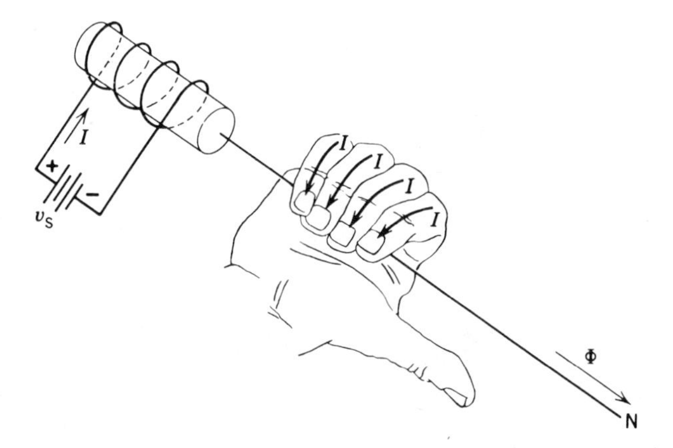

Figure 11.18

Solution Applying the *right-hand-rule for solenoids*, it is clear that the right end is the north pole.

• • •

11.5 MAGNETIC CIRCUIT QUANTITIES

In the previous section we saw that a current in a solenoid produces a magnetic field. The total flux $\mathbf{\Phi}$ in the field is directly proportional to I, the current in the coil, and to N, the number of turns of wire on the solenoid. Since the current produces the magnetic flux, we can think of the product of the current and the number of turns as the *magnetizing force* or magnetomotive force. This is analogous to an electromotive force (emf) producing a current in an electric circuit. Thus in a magnetic circuit

$$\text{magnetizing force} = \text{magnetomotive force} = \text{mmf} = IN$$

The unit for measuring current is the ampere, and the number of turns is simply a number. Thus the unit for measuring *magnetomotive force,* mmf, is also the ampere. Frequently the mmf is expressed in ampere-turns, At, to emphasize that it is the product of the current and the number of turns in a multiturn coil. The quantity At still carries the dimension of amperes however since the number of turns is a dimensionless quantity. We will then express mmf in At to emphasize its origin. The script letter $\mathscr{F}$ is used as the symbol for mmf, and thus

$$\text{mmf} = \mathcal{F}\ (\text{At}) = I\ (\text{A}) \times N \tag{11.3}$$

• • •

Example 11.7 What is the mmf in a 200 turn solenoid if the current is 100 mA?

Solution Using Eq. 11.3,

$$\begin{aligned}\text{mmf} = \mathcal{F} &= I \times N \\ &= 0.1\ \text{A} \times 200 = 20\ \text{At}\end{aligned}$$

• • •

Let us explore further the analogy between an electric circuit and a magnetic circuit. Recall that Ohm's Law for an electric circuit can be expressed as

$$I = \frac{\text{emf}}{R} = \frac{V}{R}$$

In a magnetic circuit, the flux is directly proportional to the mmf; thus for the magnetic circuit

$$\mathbf{\Phi} = \frac{\text{mmf}}{\text{constant}} = \frac{\mathcal{F}}{\text{constant}}$$

If we compare this equation with Ohm's Law, we see that flux $\mathbf{\Phi}$ is analogous to current I, mmf is analogous to emf, and thus the *constant* must be analogous to resistance R. The resistance of a magnetic circuit is defined as *reluctance,* and the script letter $\mathcal{R}$ is used as the symbol. Thus

$$\mathbf{\Phi} = \frac{\text{mmf}}{\mathcal{R}} = \frac{\mathcal{F}}{\mathcal{R}}$$

This equation is not quite correct since $\mathbf{\Phi}$ is a *vector* quantity having both magnitude and direction, while $\mathcal{F}$ and $\mathcal{R}$ are *scalar* quantities; a scalar quantity has *only* magnitude, no direction. Thus we should use the symbol Φ for flux, instead of $\mathbf{\Phi}$, in this equation. This symbol shows that the equation gives only the *magnitude* of the flux; the *direction* must then be obtained using the *right-hand-rule.* The resulting equation is called Ohm's Law for magnetic circuits, and it is

$$\Phi = \frac{\mathcal{F}}{\mathcal{R}} \tag{11.4}$$

In the MKS system, the unit for flux is the weber and the unit for mmf is the ampere-turn. Equation 11.4 can be rewritten as

$$\mathcal{R} = \frac{\mathcal{F}}{\Phi} = \frac{\text{At}}{\text{Wb}}$$

and thus the units for reluctance $\mathcal{R}$ are At/Wb. Hence Ohm's Law for magnetic circuits in the MKS system is

$$\Phi\ (\text{Wb}) = \frac{\mathcal{F}\ (\text{At})}{\mathcal{R}\ (\text{At/Wb})} \tag{11.5}$$

• • •

Example 11.8 What is the flux in the ring solenoid in Fig. 11.19 if the total reluctance $\mathcal{R}$ is 10^7 At/Wb? What is the direction of the flux?

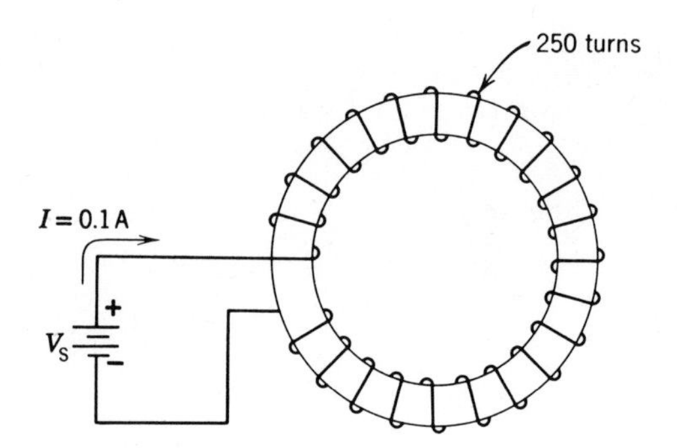

Figure 11.19

Solution Using Eq. 11.5,

$$\Phi = \frac{\mathcal{F}}{\mathcal{R}} = \frac{10^{-1}\ \text{A} \times 250\ \text{(turns)}}{10^{7}\ \text{At/Wb}} = 2.5\ \mu\text{Wb}$$

The direction of the flux is found to be clockwise around the ring (use the *right-hand-rule*).

• • •

Example 11.9 If the total resistance of the wire in Fig. 11.19 is 5 Ω, what must be the battery voltage V_S in order to produce a flux Φ of 10 μWb?

Solution We must first find the total mmf required. Rearranging Eq. 11.5,

$$\mathcal{F} = \Phi \times \mathcal{R} = 10 \times 10^{-6}\ \text{Wb} \times 10^{7}\ \text{At/Wb} = 100\ \text{At}$$

Now since $\mathcal{F} = IN$, the current required is

$$I = \frac{\mathcal{F}}{N} = \frac{100\ \text{A}}{250} = 0.4\ \text{A}$$

The battery voltage is then

$$V_S = IR = 0.4\ \text{A} \times 5\ \Omega = 2\ \text{V}$$

• • •

At this point we can easily find flux Φ and mmf $\mathcal{F}$ but we have not yet discussed reluctance $\mathcal{R}$. We can again make use of the analogy between electric and magnetic circuits. You will recall that the resistance of a conductor can be calculated from

$$R = \rho\frac{l}{A} = \frac{l}{\sigma A}$$

In a similar fashion, the reluctance of a magnetic circuit can be found from

$$\mathcal{R} = \frac{l}{\mu A} \qquad (11.6)$$

where l = the length of the circuit through which the flux travels
A = the cross-sectional area of the circuit through which the flux travels
μ = the permeability, given in reference tables or on curves

In MKS units, the proper units are

$$\mathcal{R}\ \text{(At/weber)} = \frac{l\ \text{(meter)}}{\mu A\ \text{(square meters)}} \qquad (11.7)$$

The permeability μ is actually the product of two terms, the *relative permeability* μ_r, and the *permeability of free space* μ_0. Thus

$$\mu = \mu_r \mu_0 \tag{11.8}$$

In the MKS system, $\mu_0 = 4\pi \times 10^{-7}$, and the value of μ_r is very nearly 1 (unity) for any material other than ferromagnetic materials such as iron, nickel, cobalt, and alloys of these materials.*

• • •

Example 11.10 Calculate the reluctance $\mathcal{R}$ of the air-core toroid shown in Fig. 11.20. What would be the reluctance if the toroid had a brass core?

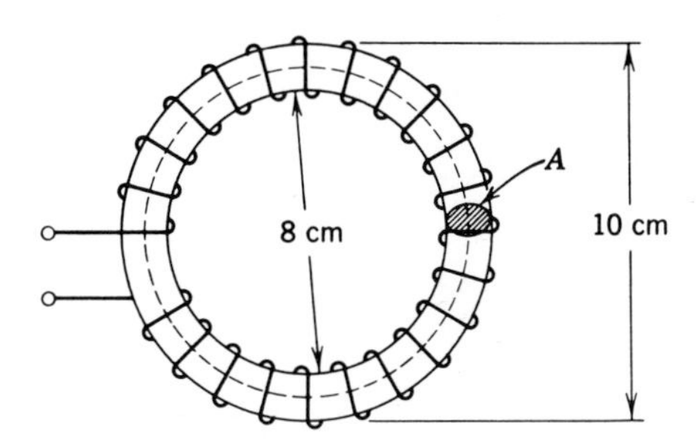

Figure 11.20

Solution The flux in the coil will travel around a closed path inside the core of the toroid. The *average* length of this path is shown by the dotted line, and is seen to be

$$l = \pi d = \pi \times 9 \text{ cm} = 9\pi \times 10^{-2} \text{ m}$$

The cross-sectional area of the core is circular, and has a diameter of 1 cm, or 0.01 m. Thus

$$A = \frac{\pi d^2}{4} = \frac{\pi(10^{-2} \text{ m})^2}{4} = \frac{\pi}{4} \times 10^{-4} \text{ m}^2$$

The permeability of air is essentially that of free space, $4\pi \times 10^{-7}$. Thus the reluctance is found using Eq. 11.7,

$$\mathcal{R} = \frac{l}{\mu A} = \frac{9\pi \times 10^{-2} \text{ m}}{\pi \times 10^{-7} \times \frac{\pi}{4} \times 10^{-4} \text{ m}^2} = 2.86 \times 10^9 \text{ At/Wb}$$

If the core were made of brass, the reluctance would be essentially the same since brass is nonmagnetic, that is, $\mu = \mu_r \mu_0$, and μ_r for both air and brass is very nearly 1.0.

• • •

Example 11.11 Calculate the reluctance of the magnetic circuit shown in Fig. 11.21. The cross-sectional area of the core A is 1 cm^2, or 10^{-4} m^2. The core is made of iron, and $\mu_r = 4000$. There is an air gap 1 cm long in the core.

*For example, μ_r for paraffin is 0.99999942.

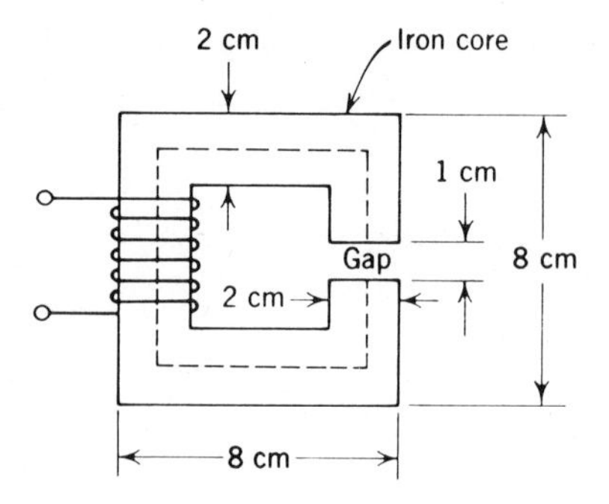

Figure 11.21

Solution Since the core is iron, nearly all of the flux will be confined to the core and the 1-cm air gap. The average path length is shown by the dotted line, and is clearly 24 cm, or 0.24 m. The total reluctance of the circuit is equal to the reluctance of the iron core $\mathcal{R}_C$ added to the reluctance of the airgap $\mathcal{R}_G$. The length of the gap is 1 cm, or 0.01 m, its cross-sectional area is 10^{-4} m^2, and thus

$$\mathcal{R}_G = \frac{l}{\mu A} = \frac{0.01 \text{ m}}{4\pi \times 10^{-7} \times 10^{-4} \text{ m}^2} = 7.95 \times 10^7 \text{ At/Wb}$$

The length of the path in the iron is 23 cm = 0.23 m, and the cross-sectional area is 10^{-4} m^2. However, the permeability in the iron is

$$\mu = \mu_r \mu_0 = 4000 \times 4\pi \times 10^{-7} = 1.6\,\pi \times 10^{-3}$$

Thus the reluctance of the iron is

$$\mathcal{R}_C = \frac{l}{\mu A} = \frac{0.23 \text{ m}}{1.6\pi \times 10^{-3} \times 10^{-4} \text{ m}^2} = 4.56 \times 10^5 \text{ At/Wb}$$

The total reluctance is then

$$\mathcal{R} = \mathcal{R}_G + \mathcal{R}_C = 795 \times 10^5 + 4.56 \times 10^5 = 799.6 \times 10^5 \text{ At/Wb}$$

• • •

From the above examples it is clear that the reluctance of a circuit containg ferromagnetic materials is much smaller than that of circuits containing paramagnetic or diamagnetic materials. Thus, iron (or other ferromagnetic materials) are used in magnetic circuits to reduce the reluctance. An examination of the equation $\Phi = \mathcal{F}/\mathcal{R}$ shows that for a given mmf, a greater flux can be created by reducing the reluctance, and a ferromagnetic material will do just that.

11.6 PERMEABILITY—μ

In the two magnetic circuit equations

$$\Phi = \frac{\mathcal{F}}{\mathcal{R}} = \frac{NI}{\mathcal{R}} \qquad \mathcal{R} = \frac{l}{\mu A}$$

all of the quantities except the permeability μ are fixed by the physical dimensions of the circuit and the applied current I. The permeability μ will however require some special attention.

For diamagnetic and paramagnetic materials, the relative permeability μ_r is very nearly unity (1.00). Thus μ is *constant* and is given by

$$\mu = \mu_r\mu_0 \approx \mu_0 = 4\pi \times 10^{-7} \text{ MKS units}$$

for nonmagnetic materials. Since μ is constant, these are said to be *linear* magnetic circuits.

For ferromagnetic materials, μ_r is *not* unity, and furthermore it is *not* a constant: μ changes for different values of mmf or flux Φ. Thus, magnetic circuits containing ferromagnetic materials are said to be *nonlinear* magnetic circuits.

We now examine the two magnetic circuit equations given in the beginning of this section in order to develop a relationship for μ. Notice first of all that we can obtain the following equation by substitution.

$$\Phi = \frac{NI}{\mathcal{R}} = \frac{NI}{\dfrac{l}{\mu A}} = \frac{NI}{l}\mu A$$

Dividing both sides by A yields

$$\frac{\Phi}{A} = \mu\frac{NI}{l} \tag{11.9}$$

The term Φ/A has been previously defined as the flux density B. Thus

$$B = \frac{\Phi}{A} \quad \text{or} \quad \mathbf{B} = \frac{\mathbf{\Phi}}{A}$$

The term NI/l is the mmf per unit length, and this is defined as the magnetic *field intensity* **H**.* Thus

$$\text{magnetic field intensity} = \mathbf{H} \triangleq \frac{NI}{l} \tag{11.10}$$

We can now substitute these relationships into Eq. 11.9 to get

$$\mathbf{B} = \mu\mathbf{H} \tag{11.11}$$

Thus, the permeability μ can be expressed as

$$\mu = \frac{\mathbf{B}}{\mathbf{H}} \tag{11.12}$$

If we plot Eq. 11.11 for a diamagnetic or a paramagnetic material, the result will be a straight line as shown in Fig. 11.22*a*. This shows that μ is a constant (the equation of a straight line) and we have a linear magnetic circuit. However, if we plot Eq. 11.11 for a ferromagnetic material, the result appears as shown in Fig. 11.22*b*. This is *not* a straight line, and thus μ is *not* a constant. This is then a nonlinear magnetic circuit. This plot of **B** *versus* **H** is called a *magnetization curve*, or simply a **B-H** curve. A typical magnetization curve is shown in Fig. 11.23. Notice that MKS units are used.

*Sometimes called magnetic field strength.

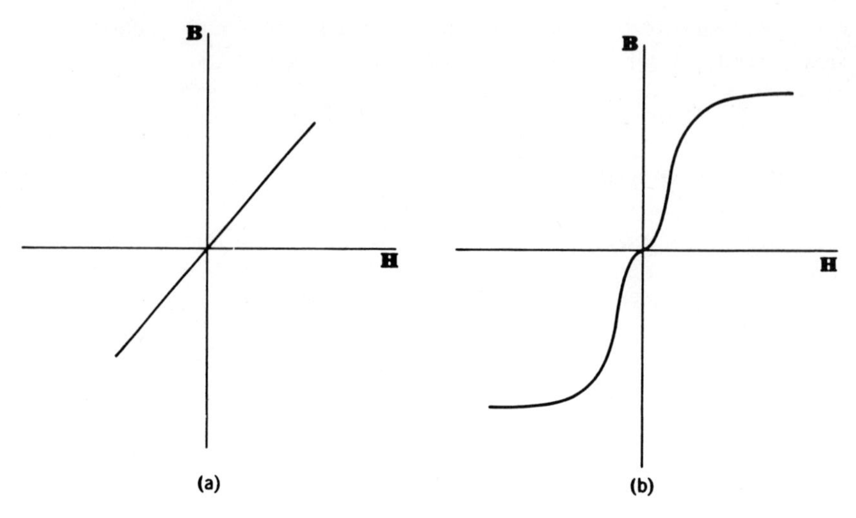

Figure 11.22 Magnetization curves (*a*) Paramagnetic or diamagnetic material (*b*) Ferromagnetic material

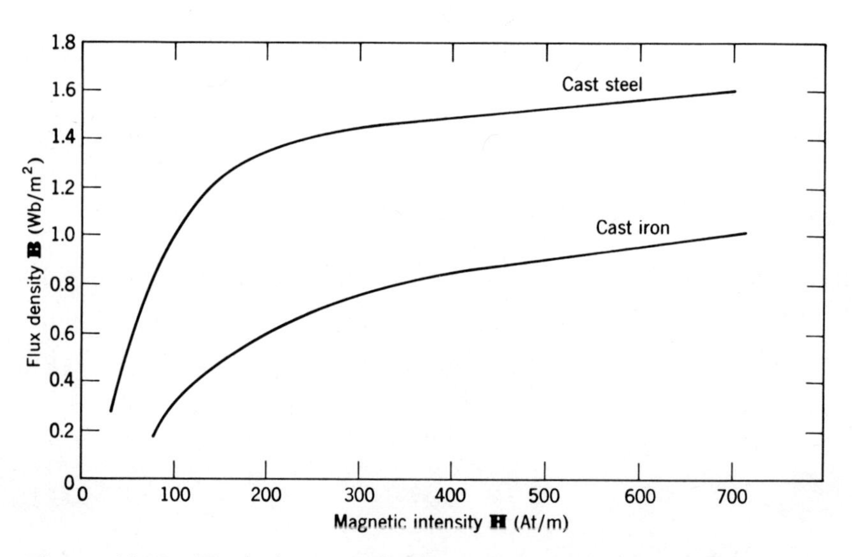

Figure 11.23 Typical magnetization curves

• • •

Example 11.12 If the coil in Fig. 11.20 is wound on a cast iron rod (core), what is the permeability if the current I is 0.10 A? $N = 566$.

Solution From Eq. 11.10, the magnetic intensity **H** is

$$\mathbf{H} = \frac{NI}{l} = \frac{566 \times 0.10\ A}{0.283\ \text{m}} = 200\,\frac{\text{At}}{\text{m}}$$

Using the curve for cast iron in Fig. 11.23, a magnetic intensity **H** of 200 At/m corresponds to a flux density **B** of 0.6 Wb/m^2. Then using Eq. 11.12,

$$\mu = \frac{\mathbf{B}}{\mathbf{H}} = \frac{0.6\ \text{Wb/m}^2}{200\ \text{At/m}} = 3 \times 10^{-3}\ \text{MKS units}$$

• • •

Example 11.13 If the core in Fig. 11.24 is made of cast steel, what current I is required to produce a flux density of 1.5 Wb/m^2?

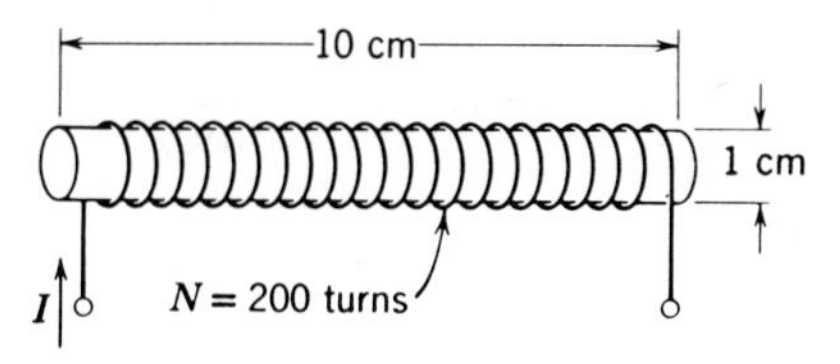

Figure 11.24

Solution Using the curve for cast steel in Fig. 11.23, a flux density **B** of 1.5 Wb/m^2 corresponds to a magnetic intensity **H** of 400 At/m. Using Eq. (11.10),

$$\mathrm{H} = \frac{NI}{l},$$

Thus

$$I = \frac{\mathrm{H}l}{N} = \frac{400 \text{ At/m} \times 0.1 \text{ m}}{200} = 0.2 \text{ A} = 200 \text{ mA}$$

• • •

Magnetization curves are very often given in *English units* rather than MKS units as shown in Fig. 11.25. Thus it is sometimes easier to work magnetic

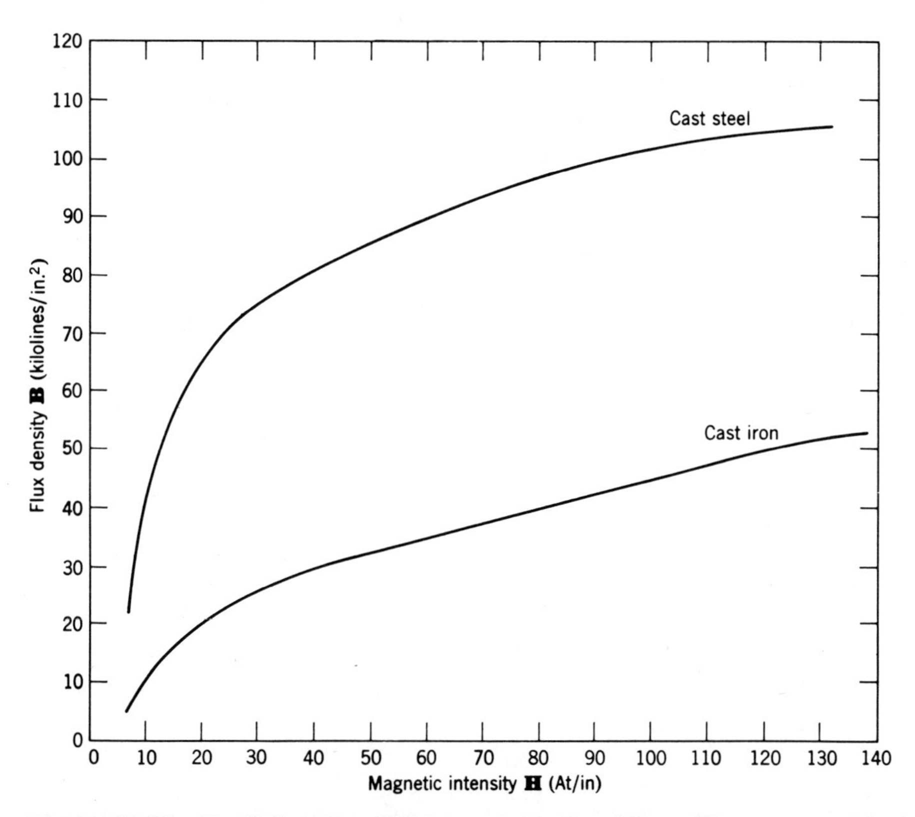

Figure 11.25 Typical magnetization curves (English units)

problems using English units. All of the formulas developed are still valid, but it is necessary to be consistent and use English units throughout. The formulas with the proper English units are given as:

For English units

$$\mu_0 = 3.2 \text{ English units}$$

$$\Phi\ (\text{lines}) = \frac{\mathcal{F}\ (\text{At})}{\mathcal{R}\ (\text{At/line})} \tag{11.13}$$

$$\mathcal{R}\ (\text{At/line}) = \frac{l\ (\text{in.})}{\mu A\ (\text{in.}^2)} \tag{11.14}$$

$$\mathbf{B}\ (\text{lines/in.}^2) = \frac{\Phi\ (\text{lines})}{A\ (\text{in.}^2)} \tag{11.15}$$

$$\mathbf{H}\ (\text{At/in.}) = \frac{IN\ (\text{At})}{l\ (\text{in.})} \tag{11.16}$$

These formulas are exactly the same as those using MKS units, except we use the *line* instead of the weber as the unit of flux (10^8 lines = 1 weber), and we use the *inch* instead of the meter as the unit of length.

• • •

Example 11.14 If the coil in Fig. 11.26 is wound on a cast iron core, what is the permeability μ if the current I is 0.60 A?

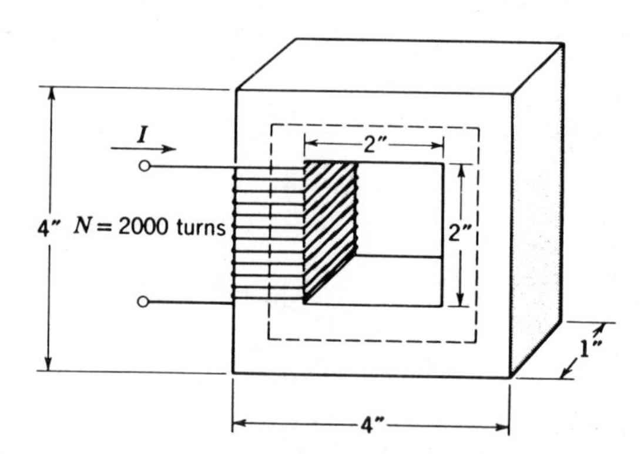

Figure 11.26

Solution From Eq. 11.16, the magnetic intensity **H** is

$$\mathbf{H}\ (\text{At/in.}) = \frac{I(A)N}{l\ (\text{in.})} = \frac{0.60\ \text{A} \times 2000\ (\text{turns})}{12\ \text{in.}} = 100\ \text{At/in.}$$

Notice that the average path length of the flux in Fig. 11.26 is 12 in. Using the curve for cast iron in Fig. 11.25, a magnetic intensity **H** of 100 At/in. corresponds to a flux density **B** of 45 kilolines/in.^{2}, or 45,000 lines/in.^{2} Then using Eq. 11.12 (with English units),

$$\mu = \frac{\mathbf{B}}{\mathbf{H}} = \frac{45 \times 10^3\ \text{lines/in.}^2}{100\ \text{At/in.}} = 450 \text{ English units}$$

• • •

Example 11.15 What is the reluctance $\mathcal{R}$ of the magnetic circuit in Fig. 11.26?

Solution Using the results of the previous example and Eq. 11.14,

$$\mathcal{R}\ (\text{At/line}) = \frac{l\ (\text{in.})}{\mu A\ (\text{in.}^2)} = \frac{12\ \text{in.}}{450 \times 1\ \text{in.}^2} = 0.0267\ \text{At/line}$$

• • •

If we were to extend the magnetization curves in Figs. 11.23 and 11.25 to have both positive and negative values of **H** (this simply means reversing the current direction since $\mathbf{H} = IN/l$), the curves would look something like those shown in Fig. 11.27*a*. Such a curve is a magnetization curve, but it is sometimes called a *hysteresis curve*, or *hysteresis loop*. This curve shows that a flux remains in the core ($\mathbf{\Phi} = \mathbf{B}A$), even after the magnetizing current I is removed. That is, the core can be magnetized by applying a current to the coil for a short period of time. Thus if a positive current I is applied to the winding in Fig. 11.27*b* and then removed, the *core* will be magnetized in one direction (point *a* on the hysteresis curve). If a current in the *opposite* direction is applied and then removed, the *core* will be magnetized in the opposite direction (point *b* on the hysteresis curve).* This is the basic principle used in *magnetic core memories* in digital computers.

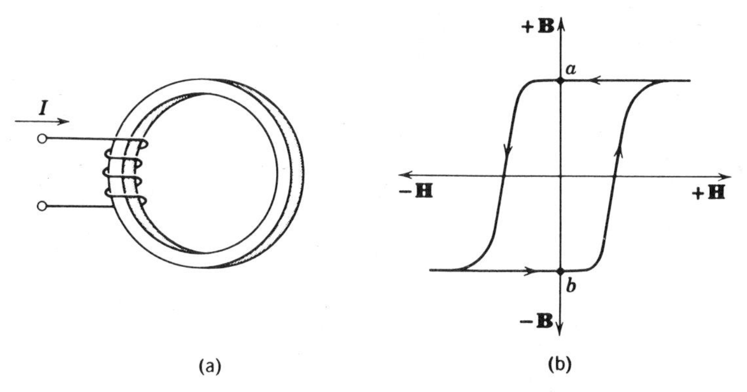

Figure 11.27 (*a*) A magnetic core (*b*) Hysteresis loop for a magnetic core

.7 FORCES BETWEEN MAGNETIC FIELDS

In our discussion of permanent magnets, we have seen that the forces of attraction and repulsion can be explained on the basis of magnetic fields aiding or cancelling one another. Since a current-carrying conductor is surrounded by a magnetic field, there must be a force exerted on a current-carrying conductor placed in a magnetic field. Similarly, there must be forces exerted on two (or more) current-carrying conductors placed close together.

Consider first the force exerted on a current-carrying conductor placed in a magnetic field. The magnetic field around a conductor is shown in Fig. 11.28*a*, and the magnetic field of a permanent magnet is shown in Fig. 11.28*b*. If this conductor is placed between the poles of the magnet, the *resulting* (composite) field will appear as shown in Fig. 11.28*c*. Notice that the two fields tend to *cancel* one another *above* the conductor, and they tend to *reinforce* one

*Check the direction of flux in the core using the right-hand-rule.

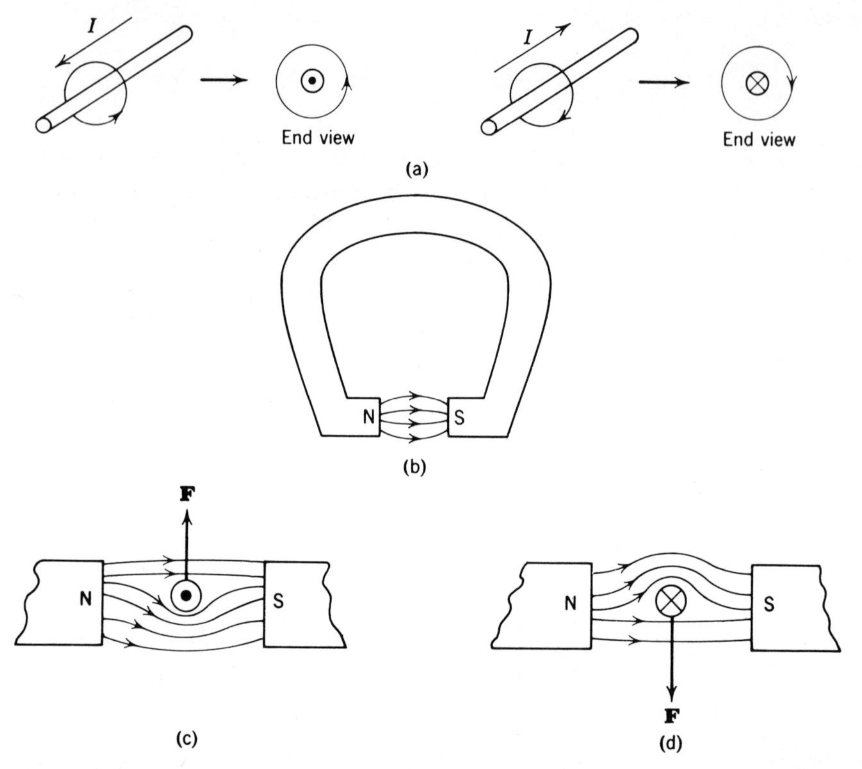

Figure 11.28 Force on a current carrying conductor in a magnetic field (*a*) Magnetic field around a current-carrying conductor (*b*) The magnetic field of a permanent magnet (*c*) Upward force **F** (*d*) Downward force **F**

another *below* the conductor. Since the field lines tend to straighten themselves (like stretched rubber bands), there will clearly be an *upward* force **F** exerted on the conductor. This force is directly proportional to the flux density of the permanent magnet field **B**, the current in the conductor *I*, and the length of the conductor *l*. Thus, in MKS units,

$$\mathbf{F}\ (\text{newtons}) = \mathbf{B}\ (\text{Wb/m}^2) \times I\ (\text{A}) \times l\ (\text{m}) \tag{11.17}$$

In Fig. 11.28*d*, the current direction has been reversed (it is going *into* the page) and the force on the conductor **F** has changed direction.

• • •

Example 11.16 Calculate the force on the conductor in Fig. 11.29 if the flux density of the permanent magnet field is $\mathbf{B} = 0.2\ \text{Wb/m}^2$.

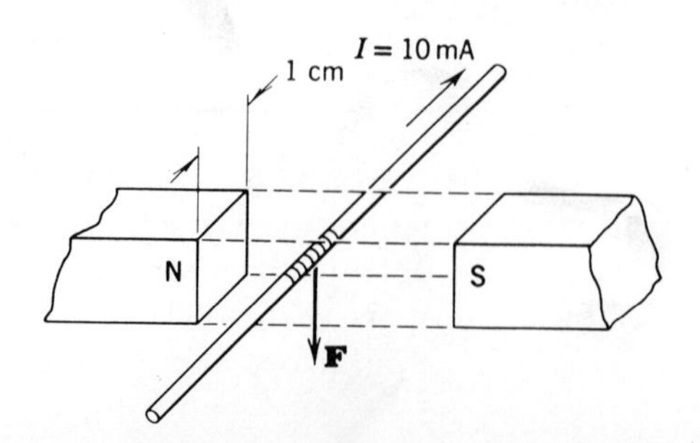

Figure 11.29

Solution The length of the conductor *in the field* is approximately 1 cm. Thus using Eq. 11.17,

$$\mathbf{F} = \mathbf{B} \times I \times l$$
$$= 0.2 \text{ Wb/m}^2 \times 0.01 \text{ A} \times 0.01 \text{ m} = 2 \times 10^{-5} \text{ newtons*}$$

• • •

Example 11.17 Show the composite magnetic field and describe the forces associated with two parallel conductors when (*a*) the currents are in the same direction, (*b*) the currents are in opposite directions.

Solution (*a*) When the currents are in the *same* direction, the field appears as shown in Fig. 11.30*a*. The conductors are clearly forced *together.* (*b*) When the currents are in *opposite* directions the field appears as shown in Fig. 11.30*b*. The conductors are clearly forced apart.

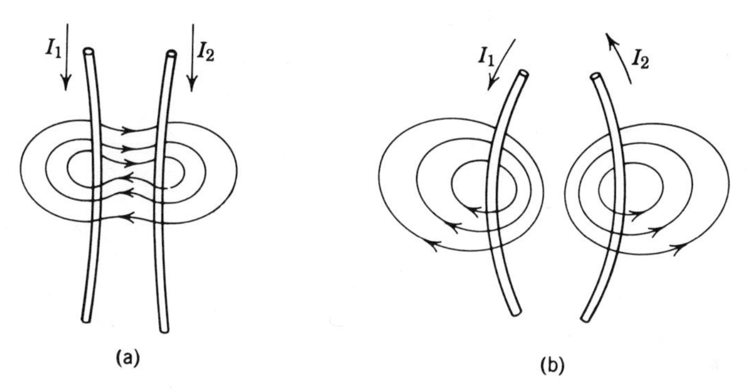

Figure 11.30 Example 11.7 (*a*) Conductors forced together (*b*) Conductors forced apart

Consider now the effect of placing a *loop* of wire in a magnetic field as shown in Fig. 11.31*a*. The loop is mounted on pivots and is free to rotate about its center line. Examination of Fig. 11.31*b* shows that there is a downward force on the right side of the loop, and an upward force on the left side. These two forces produce a *torque* **T** which causes the loop to rotate in a clockwise direction. Torque is defined as the product of a force and a lever arm. The torque on each side of the loop is given as

$$\mathbf{T}_{\text{right}} = \mathbf{T}_{\text{left}} = \mathbf{F}r$$

where r is the *radius* of the loop. The total torque tending to rotate the loop is equal to the sum of the torques on each side of the loop, and thus

$$\mathbf{T} = \mathbf{T}_{\text{right}} + \mathbf{T}_{\text{left}} = 2\mathbf{F}r$$

But $2r$ is simply the *diameter* of the loop d; thus

$$\mathbf{T} = \mathbf{F}d$$

We also know that the force on a conductor **F** is given by $\mathbf{B}Il$. Thus, by substitution,

$$\mathbf{T} = \mathbf{B}Ild$$

*One pound of force lbf is the pull of the earth on a one pound mass, and one newton is equal to 0.225 lbf.

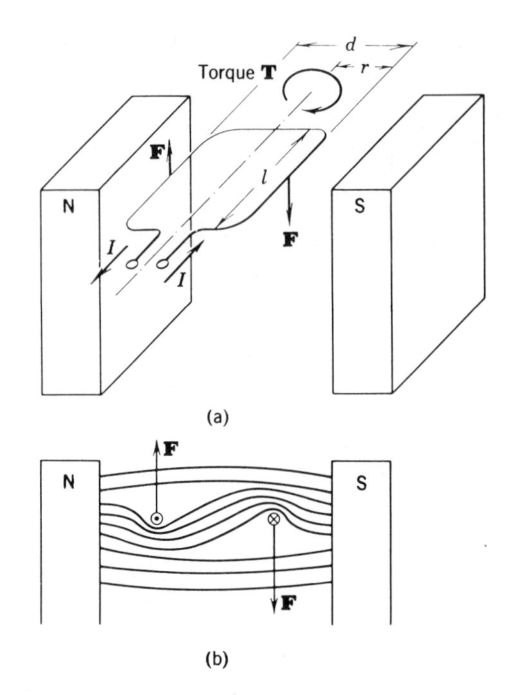

Figure 11.31 (*a*) Current-carrying loop in a magnetic field (*b*) Composite magnetic field

If there is more than one loop (i.e., a coil of N turns), we simply multiply by the number of turns N. Thus

$$\mathbf{T} = \mathbf{B} l I N d \tag{11.18}$$

In MKS units, $\mathbf{T}$ = torque in newton-m
$\mathbf{B}$ = flux density in Wb/m^2
l = length of coil in m
I = coil current in A
N = number of turns in the coil
d = diameter of the coil in m

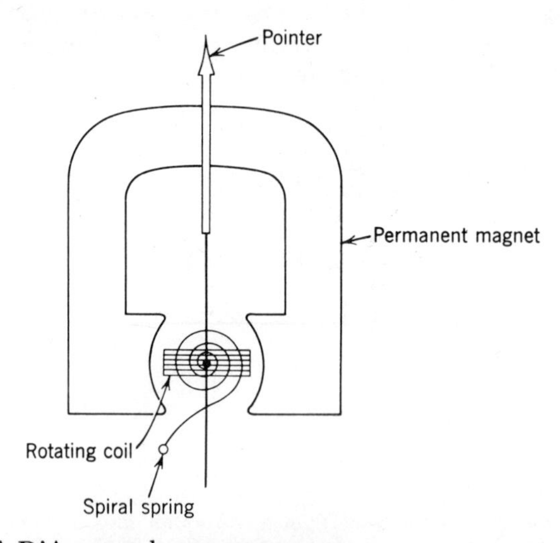

Figure 11.32 A D'Arsonval meter movement

If the movement of the loop in Fig. 11.31 is restrained by a spiral spring as shown in Fig. 11.32, the result is a D'Arsonval meter movement. Notice that a current of $+I$ will rotate the coil in a clockwise direction, and a current of $-I$ will rotate it in a counterclockwise direction.

• • •

Example 11.18 What is the torque on the D'Arsonval meter movement in Fig. 11.32 if $\mathbf{B} = 0.2\ \text{Wb/m}^2$, $l = 1$ cm, $I = 200$ mA, $N = 100$, and $d = 1$ cm?

Solution Using Eq. 11.18,

$$T = 0.2\frac{\text{Wb}}{\text{m}^2} \times 0.01\ \text{m} \times 0.2\ \text{A} \times 100 \times 0.01\ \text{m} = 4 \times 10^{-4}\ \text{newton-m}$$

This is a very small torque and illustrates the delicate nature of D'Arsonval meter movements.

• • •

Equation 11.18 is known as the *basic electric motor equation,* and it is clear that the loop in Fig. 11.31 will rotate if no counter torque spring is used to restrict its motion. However, when the loop moves to the position shown in Fig. 11.33*a,* it will stop. The forces on the conductors in the loop are equal and they are in opposite directions. Thus the net torque on the coil is zero. If the

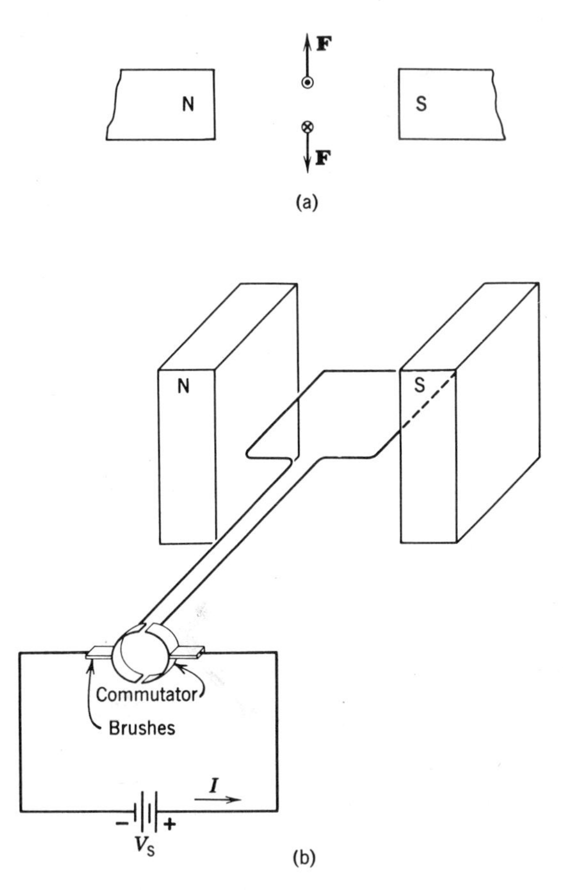

Figure 11.33 (*a*) Coil stops rotating in this position (*b*) A basic dc motor

current were reversed in polarity, the forces would also change direction, and the coil would then rotate another 180° and stop. A method for changing current direction every 180° of rotation is provided by means of the *commutator* shown in Fig. 11.33*b*. This arrangement will cause the coil to rotate continuously, and this is the basis of a dc motor. The motor would of course rotate in the opposite direction if the battery polarity in Fig. 11.33*b* were reversed. There must necessarily be enough *inertia* associated with the coil so that it will always move past the *vertical* position, and it will therefore not get "stuck."

11.8 ELECTROMAGNETIC INDUCTION

Since passing a current through a conductor in a magnetic field can be used to cause the conductor to move, it would seem reasonable to expect that the movement of a conductor through a magnetic field could be used to cause a current in the conductor. Such an effect does indeed occur, and it can be demonstrated experimentally as shown in Fig. 11.34. If the conductor is moved *vertically* (up or down) in the field, a deflection is noted on the galvanometer. This indicates the presence of an *induced current* in the conductor. If the conductor is moved horizontally (right or left), there is no galvanometer deflection and thus no induced current. If the conductor is held motionless and the magnet is moved (therefore the field lines move), the same results are noted. Thus we make the important observations:

If the flux lines of a magnetic field "cut across" (move across) a conductor there is a current induced in the conductor; but if the direction of motion is parallel to the field lines, there is no induced current.

We also can make another important observation:

The faster the movement of the lines across the conductor, the greater the induced current.

The effects of this experiment can be increased by using a coil of wire as shown in Fig. 11.35. The effects are the same, and it is only necessary to move

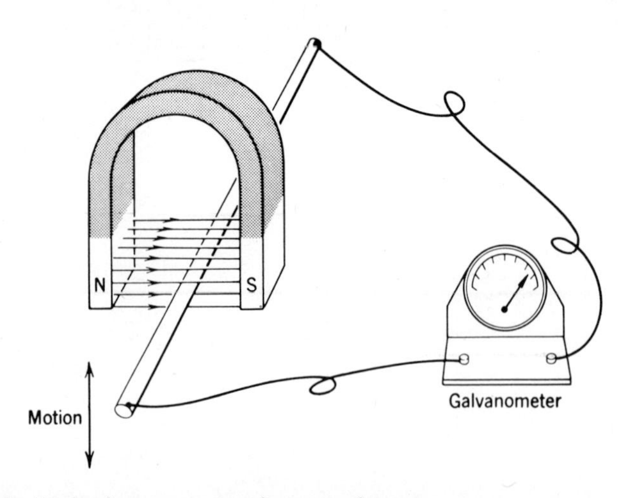

Figure 11.34 Inducing a current in a conductor

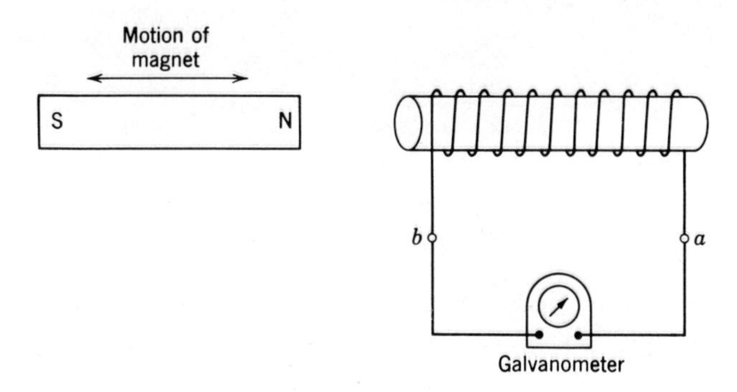

Figure 11.35

the magnet (or the coil) in such a way that the flux lines from the magnet "cut across" the turns in the coil.

If the meters are omitted in Figs. 11.34 and 11.35, no current can flow since there is not a complete circuit. However, the electrons in the conductors will move to one end of the conductor (a current) resulting in a separation of charge. This we recognize as the requirement for producing a voltage. Thus a voltage is induced in a wire *moving* across the lines of a magnetic field. These effects were all noticed by Faraday, and the results are summed up in *Faraday's Law of induced voltage*:

$$\text{induced voltage} = v\ (\text{V}) = N\frac{\Delta\phi\ (\text{Wb})}{\Delta t\ (\text{sec})} = N\frac{d\phi\ (\text{Wb})}{dt\ (\text{sec})} \tag{11.19}$$

Since the movement of the field across the turns of the coil may change with time (depending on how fast, how often, and in what direction the field moves), the voltage may also change with time. Thus the induced voltage is not necessarily a constant, and we therefore use the *lower case letter* v for the induced voltage. Similarly, the magnetic flux may change with time, and we therefore use the symbol ϕ for flux instead of Φ, since Φ denotes a *constant* value of flux. The symbol $\Delta\phi/\Delta t$ or $d\phi/dt$ specifies *how fast* the flux ϕ cuts across the turns. N is of course the number of turns involved.

• • •

Example 11.19 Suppose that the conductor in Fig. 11.34 (without the meter) is moved through the field such that 10^8 lines cut across the conductor each second. What is the induced voltage?

Solution We must use Eq. 11.19. $d\phi/dt$ tells how many webers (lines) cut across the conductor each second. Since 1 weber = 10^8 lines,

$$\frac{d\phi}{dt} = \frac{1\ \text{Wb}}{1\ \text{sec}} = 1\ \text{Wb/sec}$$

there is only one conductor, and thus $N = 1$. Therefore,

$$v = N\frac{d\phi}{dt} = 1 \times 1\ \text{Wb/sec} = 1\ \text{volt}$$

• • •

Example 11.20 The coil in Fig. 11.35 has 500 turns. If the meter is removed, what will be the induced voltage across the coil terminals if the magnet is moved such that 0.2 weber cuts across the coil in 4 seconds?

Solution

$$\frac{d\phi}{dt} = \frac{\text{Wb}}{\text{sec}} = \frac{0.2\text{ Wb}}{4\text{ sec}} = 0.05\text{ Wb/sec}$$

Thus

$$v\text{ (V)} = N\frac{d\phi}{dt} = 500 \times 0.05\text{ Wb/sec} = 25\text{ V}$$

• • •

Faraday's law can be used to establish the *magnitude* of an induced voltage, but the polarity must still be determined. Lenz's law can be used to determine polarity.

Lenz's law: The induced voltage must have a polarity such that any current produced by it will set up a flux in opposition to any change in the original flux.

Let us now apply Lenz's law to the coil in Fig. 11.35. First suppose the magnet is moving to the right. In order to *oppose this motion,* the left end of the coil must be an N pole. Using the *right-hand-rule,* the induced current must then *enter* terminal *b*, and exit from terminal *a*. Since the coil is acting as a *source,* terminal *a* must be positive and terminal *b* must be negative. These results are summarized in Fig. 11.36*a*.

Now suppose the magnet is moved to the left in Fig. 11.35. In order to *oppose* this motion, the left end of the coil must be an S pole. Using the *right-hand-rule* shows that the induced current must enter terminal *a*, and thus terminal *b* must now be positive while terminal *a* must be negative. These results are summarized in Fig. 11.36*b*.

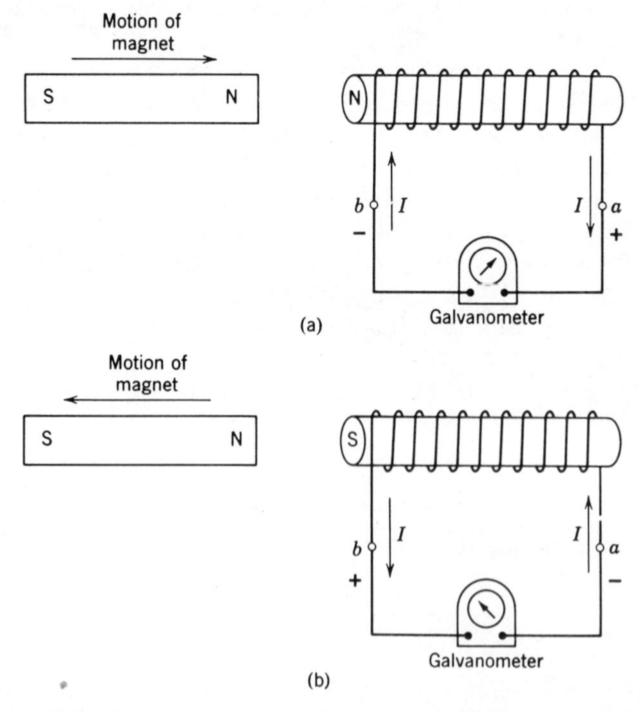

Figure 11.36 Applying Lenz's law (*a*) Induced current and voltage (*b*) Induced current and voltage

If the loop (coil) in Fig. 11.31 is rotated by some external torque, an electric current will be induced in the loop (or a voltage across the loop terminals). This forms the basis of an *electric generator,* and it can be used to produce ac currents (voltages) or dc currents (voltages) depending on whether slip rings or a commutator is used to connect the loop to an external circuit. The proper arrangements for ac and dc generators are shown in Fig. 11.37, and the reader is encouraged to apply Lenz's law to verify the currents (voltages) generated.

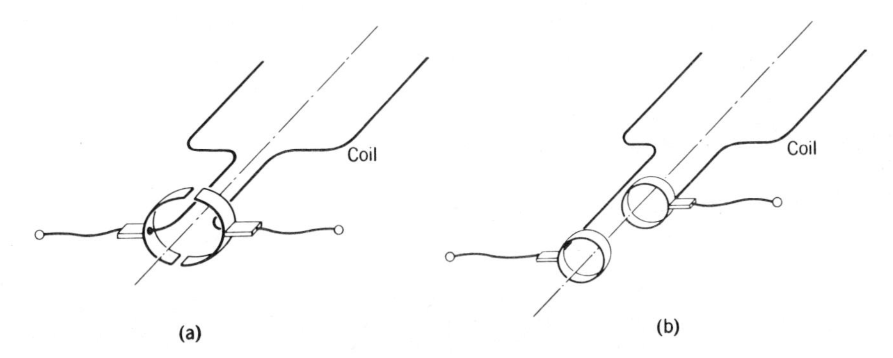

Figure 11.37 Generators (*a*) dc generator using a commutator (*b*) ac generator using slip rings

.9 ELECTROMAGNETIC DEVICES

One of the simplest applications of the principles of electromagnetism is the formation of a *lifting magnet* as shown in Fig. 11.38. The maximum lifting force for such a magnet is

$$\mathbf{F} = 1.4 \times 10^{-8}\,\mathbf{B}^2 A \tag{11.20}$$

where $\mathbf{F}$ = the lifting force in pounds
$\mathbf{B}$ = the flux density in lines/in.^{2}
A = surface area of the magnet face in in.^{2}

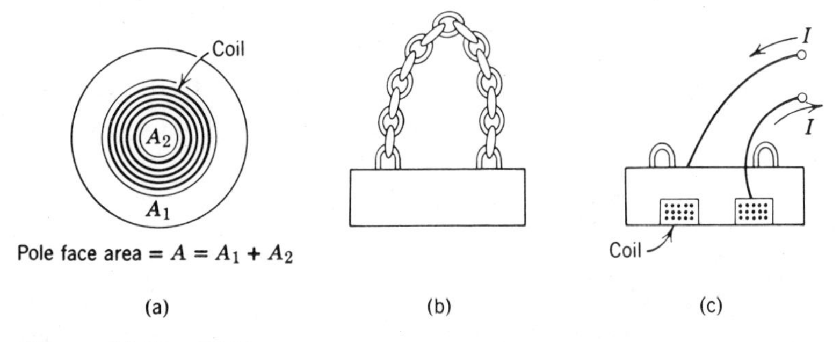

Figure 11.38 Lifting magnet (*a*) Bottom view (*b*) Side view (*c*) Sectional view

• • •

Example 11.21 Calculate the maximum lifting force of the electromagnet in Fig. 11.38 if $A_1 = 75$ in.^{2}, $A_2 = 125$ in.^{2}, and the flux density is 50 kilolines/in.^{2}

Solution Using Eq. 11.20,

$$\mathbf{F}\text{ (lb)} = 1.4 \times 10^{-8} \times (50 \times 10^{3}\text{ lines/in.})^{2} \times (75\text{ in.}^{2} + 125\text{ in.}^{2}) = 7000\text{ lb}$$

• • •

A very common electrical device, the relay, consists of a solenoid (electromagnet) in conjunction with a movable arm held in place by a spring. A typical relay is shown in Fig. 11.39*a*. When a current is applied to the relay coil, the iron arm is attracted to the pole face of the relay core, and when the current is interrupted, the arm is returned to its original position by the spring. The arm can then be used to close or open a set of contacts. The three standard arrangements for relay contacts are shown using the proper symbols in Fig. 11.39*b*.

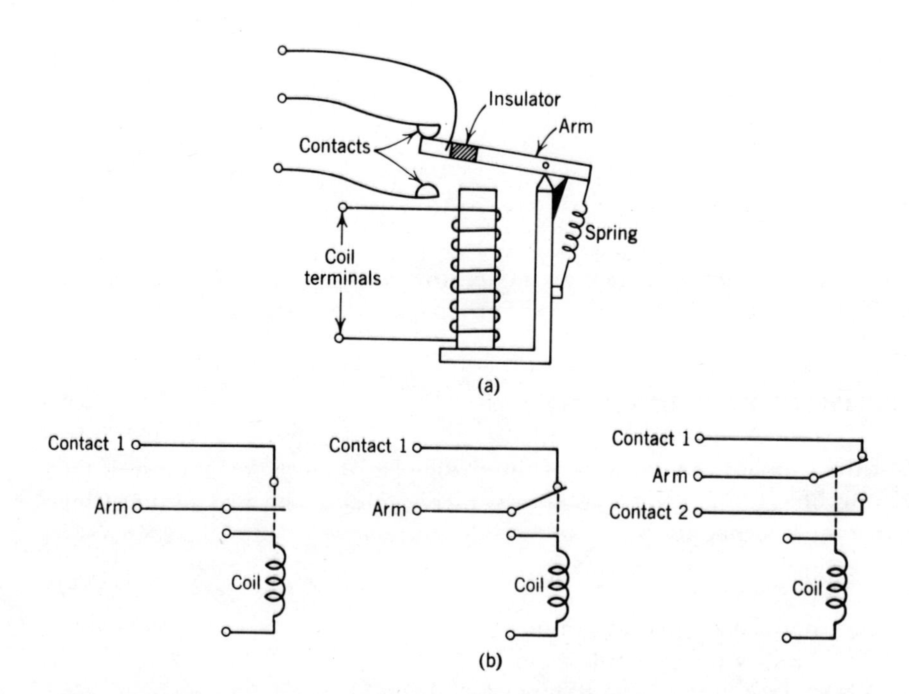

Figure 11.39 An electromagnetic relay (*a*) Relay circuit (shown with form C contacts) (*b*) Relay contacts

Another very common electromagnetic device is the *speaker* in a radio or television set. The construction of a typical *permanent magnet* (PM) speaker is shown in Fig. 11.40*a*. The voice coil terminals are connected to the output of an *audio amplifier* and the speaker will then translate the electric current variations into vibrations of the air (sound).

Consider the case when the current in the voice coil has the direction shown in Fig. 11.40*b*. An inspection of the composite magnetic field shows that the voice coil will move to the left. Reversing the voice coil current will of course cause the coil to move to the right. Thus the coil will move back and forth in response to the variations in voice coil current. Since the diaphragm is connected to the voice coil, it will also move in response to the coil current; thus the *audio* signal currents are translated into audio sounds by the movement of the coil and diaphragm.

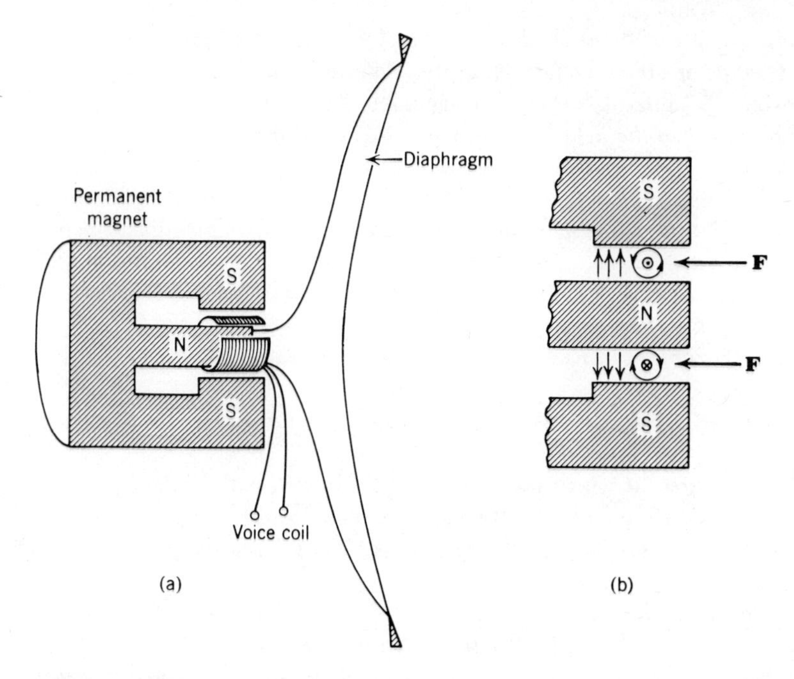

Figure 11.40 A permanent magnet speaker (*a*) Cross-sectional view (*b*) Illustrating the magnetic fields

SUMMARY

A magnetic field exists in any space in which a force is exerted on a magnetic pole. Magnetic fields exist around permanent magnets and current-carrying conductors, and they can be represented by magnetic field lines (flux). The forces associated with magnetic fields can be accounted for by examining areas of reinforcement and cancellation between two (or more) fields. The various quantities in a magnetic circuit can be solved using *Ohm's law for magnetic circuits* and noting the following analogies: (*a*) flux $\mathbf{\Phi}$ is analogous to electric current, (*b*) magnetomotive force (mmf) $\mathcal{F}$ is analogous to emf (voltage), and (*c*) reluctance $\mathcal{R}$ is analogous to resistance. The permeability used in finding the reluctance of a magnetic circuit is constant for paramagnetic and diamagnetic materials; however, it is not constant for ferromagnetic materials, and the proper values can be found from magnetization curves (**B-H** curves). The force exerted in a current-carrying conductor in a magnetic field is the basis for D'Arsonval meter movements as well as electric motors. Alternatively, a current (or voltage) is induced in a conductor whenever magnetic lines of flux "cut across" it. The induced current (or voltage) is directly proportional to the rate at which the lines cut across the conductor. Electromagnetic induction is the basis for electric generators.

GLOSSARY

Commutator. *An arrangement for connecting conductors to the moving coil of a dc motor or generator.*

Curie temperature. *The temperature above which ferromagnetic materials behave as paramagnetic materials.*

Diamagnetic. *A material in which there is a slight opposition to magnetic flux.*
Domains. *Groups of atoms in ferromagnetic materials that "switch" together.*
Ferromagnetic. *A material in which a magnetic flux is greatly aided.*
Flux Φ. *Equivalent to the field lines drawn to represent a magnetic field.*
Flux density B. *The magnetic flux per unit area.*
Fringing. *The bulging of magnetic field lines between two adjacent unlike poles.*
Magnetic field. *Exists in any space in which a force is exerted on a magnetic pole.*
Magnetic field intensity. H *The magnetomotive force per unit length, mmf/l.*
Magnetization curve. *A plot of magnetic flux density versus magnetic field intensity,* **B-H** *curve, hysteresis curve.*
Magnetomotive force. $\mathcal{F}$ *The product of applied current and the number of turns, IN, mmf.*
Nonlinear magnetic circuit. *A magnetic circuit containing a ferromagnetic material.*
North pole. *The area at which the magnetic field lines leave a magnet.*
Paramagnetic. *A material in which a magnetic flux is aided slightly.*
Permeability. μ *The product of relative permeability* μ_r *and the permeability of free space* μ_0.
PM. *Permanent magnet.*
Reluctance. $\mathcal{R}$ *The resistance to flux in a magnetic circuit.*
Right-hand-rule. *The relationship describing the direction of a magnetic field in terms of the direction of the current producing the field.*
Slip rings. *An arrangement for connecting to the moving coil of an ac generator or motor.*
Solenoid. *A coil of wire wound on a form.*
South pole. *The area at which the magnetic field lines enter a magnet.*
Vector quantity. *A quantity having both magnitude and direction.*

IMPORTANT RELATIONSHIPS

- Ohm's law for magnetic circuits $\Phi = \dfrac{\mathcal{F}}{\mathcal{R}}$

$$\Phi = \mathbf{B}A \qquad \text{mmf} = \mathcal{F} = IN \qquad \mathcal{R} = \frac{l}{\mu A}$$

$$\mu = \mu_r \mu_0 = \frac{\mathbf{B}}{\mathbf{H}} \qquad \mathbf{H} = \frac{NI}{l}.$$

- Force on a conductor $\mathbf{F} = \mathbf{B}Il.$
- Basic motor equation $\text{torque} = \mathbf{T} = \mathbf{B}lINd.$
- Faraday's law of induced voltage $v = N\dfrac{d\phi}{dt}$

Term	flux	flux density	mmf	magnetic intensity	reluctance	permeability	rate of change of flux
Symbol	Φ	$\mathbf{B}$	$\mathcal{F}$	$\mathbf{H}$	$\mathcal{R}$	μ	$\dfrac{d\phi}{dt}$
MKS units	weber, Wb	Wb/m^2	At	At/m	At/Wb	$4\pi \times 10^{-7}\ \mu_r$	Wb/sec
English units	line	lines/in.^{2}	At	At/in.	At/line	$3.2\mu_r$	lines/sec

1 Weber = 10^8 lines

REVIEW QUESTIONS

1. There is a force of (attraction, repulsion) between unlike poles of two PM's.
2. There is a force of repulsion between two ____ magnetic poles.
3. A force is exerted on an electron traveling through a magnetic field. (TF)
4. A magnetic field exerts a force on an electron only if the electron is (in motion, at rest) with respect to the field.
5. Magnetic field lines form continuous loops. (TF)
6. Magnetic field lines emerge from the South pole of a permanent magnet. (TF)
7. Magnetic field lines in the ____ direction reinforce one another, while field lines in the ____ direction cancel one another.
8. The reason that iron filings arrange themselves along magnetic field lines is that the filings are ____ ____ ____.
9. The flux density inside a ferromagnetic material held in a magnetic field is (greater, less than) the flux density of the applied field.
10. Vibration can be harmful to permanent magnets. (TF)
11. The magnetic characteristics of a ferromagnetic material (are, are not) affected by temperature.
12. A magnetic field exists around a current carrying conductor. (TF)
13. The direction of the magnetic field around a current carrying conductor can be found using the ____ ____ ____ if the current direction is known.
14. A solenoid wound on a soft iron core is called a (an) ____.
15. The North and South poles of an electromagnet can be reversed by simply reversing the ____.
16. In using the right-hand-rule for solenoids, the thumb points towards the ____ pole.
17. State Ohm's law for magnetic circuits.
18. In Ohm's law for magnetic circuits, the flux is analogous to voltage in an electric circuit. (TF)
19. The permeability of ____ and ____ materials is very nearly equal to that of free space; but the permeability of ferromagnetic materials is much (smaller, larger).
20. The **B-H** curve for ferromagnetic materials is a straight line. (TF)
21. The **B-H** curve for a ferromagnetic core is sometimes called a ____.
22. The force exerted on a current carrying conductor in a magnetic field is due to the interaction between two magnetic fields. (TF)
23. The torque on a loop in a magnetic field is directly proportional to the current in the loop. (TF)
24. The spiral spring in a D'Arsonval meter movement counter balances the ____ set up by the current in the coil.
25. The ____ in a simple one coil dc motor changes the current direction in the rotating coil every 180° of rotation.
26. If a conductor is moved through a magnetic field along a line parallel to the field lines, there (will, will not) be a current induced in the conductor.
27. Faraday's law is used to determine the magnitude of an induced (voltage, current).
28. The polarity of an induced current or voltage can be determined using ____ law.

29. The magnitude of an induced voltage is directly proportional to the rate at which the flux lines cut across the conductor(s). (TF)
30. An electromagnet in conjunction with a movable iron arm can be used to construct a ______.

PROBLEMS

1. Sketch the magnetic field lines around a bar magnet which is $\frac{1}{4}'' \times \frac{1}{4}'' \times 2''$.
2. Sketch the magnetic field lines around a U shaped permanent magnet.
3. Two permanent magnets similar to the one in Problem 1 are held placed end to end with like poles touching. Sketch the resulting magnetic field.
4. A nail is placed on a wooden block floating in a tank of water. Describe the movements of the nail as a bar magnet is moved about in the vicinity of the nail.
5. If we used the system in Example 11.2 to represent a magnetic flux of 5×10^{-3} Wb, we would have to draw 500,000 lines! Describe how we could still use this system by *scaling* the lines on the drawing.
6. What is the flux density next to the pole face of the bar magnet in Fig. 11.10 if the total flux is 3×10^{-4} Wb?
7. The cross-sectional area of the ring magnet in Fig. 11.6*b* is 5×10^{-2} m^2. What is the flux entering the North pole if the flux density at the South pole face is 400 μWb/m^2?
8. There is a magnetic flux of 20 μWb in the ring magnet in Fig. 11.6*a*. What must be the cross-sectional area of the magnet if the flux density is 6 mWb/m^2?
9. An aluminum bar is placed in a uniform magnetic field having a flux density of 25 MWb/m^2. What is the flux density *inside* the aluminum bar?
10. Describe a step-by-step procedure for making a permanent magnet from an iron bar using a furnace and a magnetic field.
11. Sketch the direction of the magnetic field in Fig. 11.41.

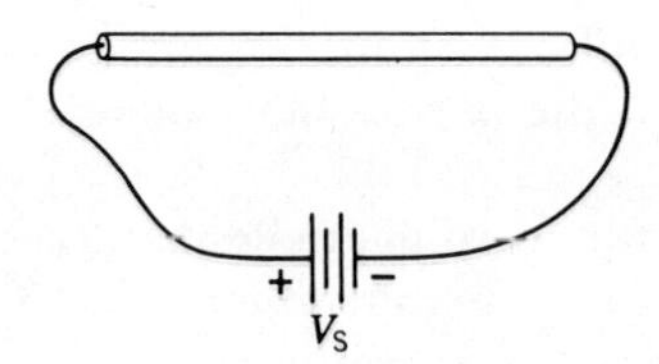

Figure 11.41

12. Show the proper battery polarity to create the magnetic field in Fig. 11.42.
13. Label the poles and sketch the magnetic field around the solenoid in Fig. 11.43.
14. Show the proper battery polarity to produce the poles shown in Fig. 11.44.
15. What is the mmf in a 750 turn solenoid if the current is 0.2 A?
16. How many turns are required to produce an mmf of 50 A in a solenoid if the coil current is 25 mA?
17. The solenoid in Fig. 11.43 has 500 turns, and the total coil resistance is 25 Ω. What must be the value of V_S to produce an mmf of 40 A?

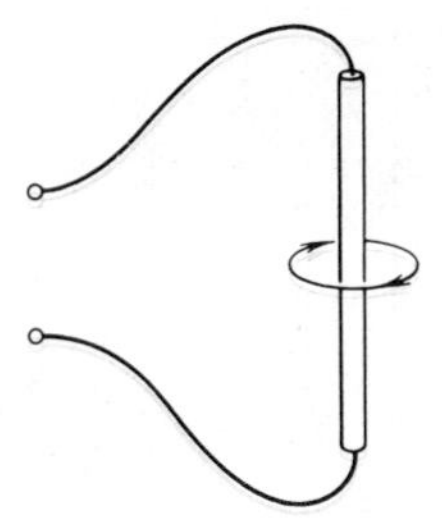

Figure 11.42

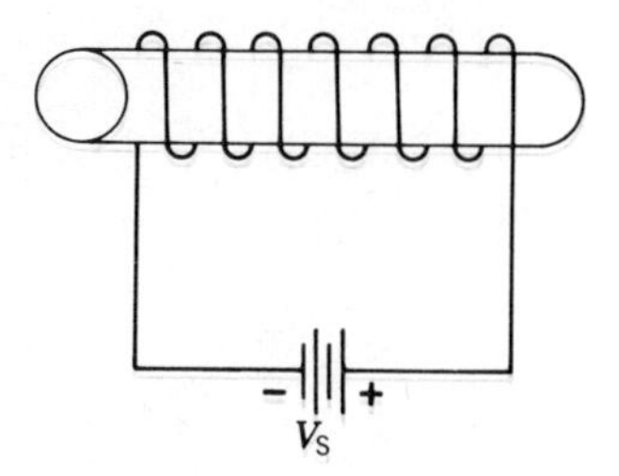

Figure 11.43

18. The solenoid in Fig. 11.44 has 350 turns and the total coil resistance is 12 Ω. What is the mmf if $V_S = 3$ V?
19. What is the total flux in the solenoid in Problem 16 if the reluctance is 2.0×10^6 At/Wb?
20. What must be the reluctance of the circuit in Problem 17 if the total flux is 0.1 mWb?
21. The solenoid core in Fig. 11.43 has a cross-sectional area of 5×10^{-3} m^2. If the flux density in the core is 4×10^{-2} Wb/m^2, and the circuit reluctance is 3×10^4 At/Wb, what must be the mmf?
22. What must be the coil current in Problem 21 if the coil has 200 turns?
23. What is the reluctance of a 2-cm square aluminum bar which is 9 cm long?
24. What is the reluctance of the ring in Fig. 11.45 if it is half aluminum and half brass?
25. What is the reluctance of the ring in Fig. 11.45 if it is half aluminum and half iron? Assume μ_r for the iron is 4000.
26. The core in Fig. 11.46 is made of cast steel and there are 400 turns in the winding. Find the magnetic intensity **H** if the coil current I is 280 mA, and: $a = 2$ cm, $b = c = 16$ cm, $d = e = 12$ cm. What is the flux density, and the total flux in the core? What is the permeability at this operating point? What is the reluctance of the core?

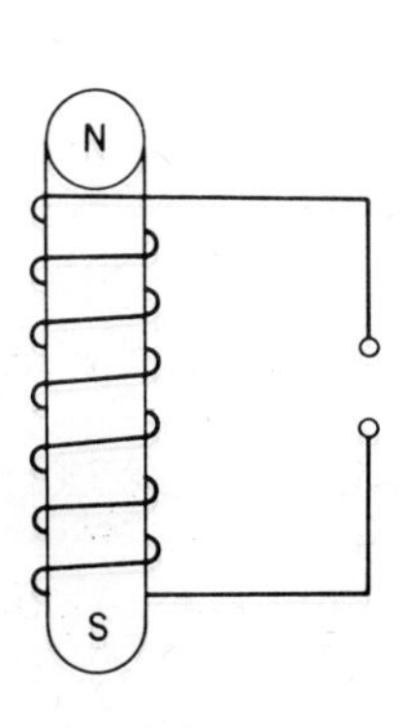

Figure 11.44

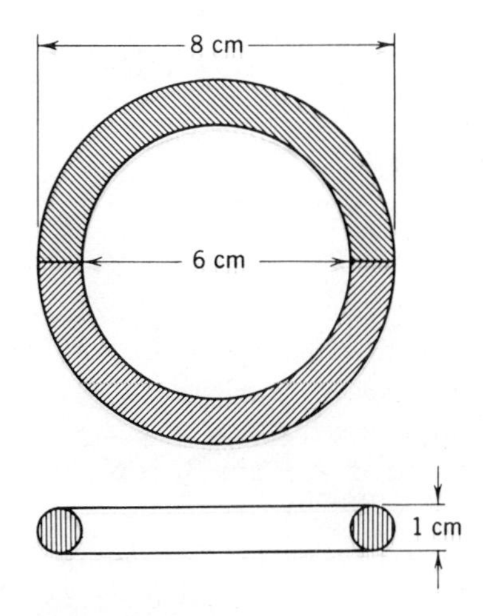

Figure 11.45

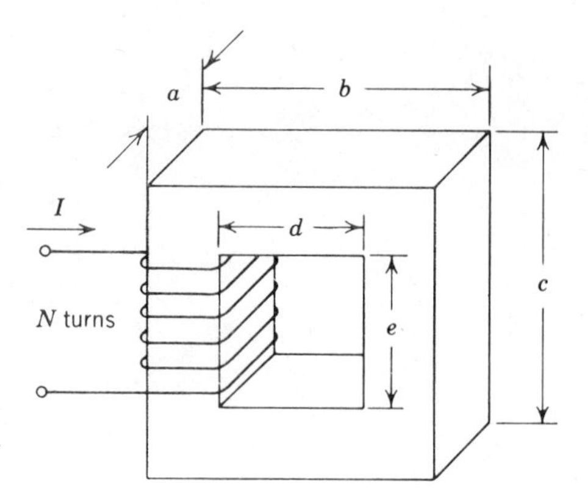

Figure 11.46

27. Repeat Problem 26 if the core is made of cast iron.
28. The core in Fig. 11.46 is made of cast steel and there are 5000 turns in the winding. Find the magnetic intensity **H** if the coil current I is 0.4 A, and: $a = 1''$, $b = c = 6''$, $d = e = 4''$. What is the flux density, and the total flux in the core? What is the permeability at this operating point? What is the reluctance of the core?
29. Repeat Problem 28 if the core is made of cast iron, and $I = 0.2$ A.
30. What is the torque on a 2000 turn, 1 cm × 1 cm D'Arsonval meter coil if the magnetic flux density is 0.07 Wb/m^2 and the coil current is 1 mA?
31. The torque on the coil in Problem 30 is 1.4×10^{-4} newton-meters. What must be the coil current?
32. What is the voltage across the terminals of a 700 turn coil if flux lines are "cutting across" the coil at the rate of 10^{-2} Wb/sec?
33. At what rate must flux lines cut across a 250 turn coil in order to induce a terminal voltage of 100 V?
34. Show the polarity of the induced voltage in Fig. 11.47 as the magnet moves *into* the coil.

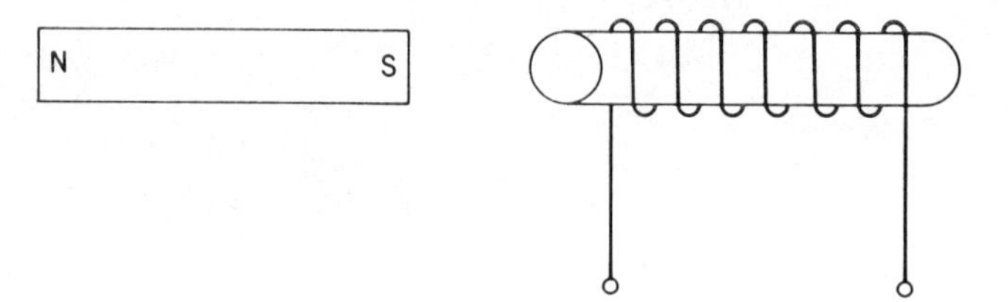

Figure 11.47

35. Show the polarity of the induced voltage in Fig. 11.47 as the magnet moves *away* from the coil.
36. Draw the schematic of a circuit using relays which will simultaneously disconnect A from B, connect A to C, and connect D to E.

12 INDUCTANCE

Faraday's law provides the means for determining the voltage induced in a conductor as a result of *changing* magnetic flux; but the magnetic flux surrounding a conductor can be produced by an electric current in that conductor, and thus we would like to develop a more convenient relationship to express *induced voltage* in terms of *changing current.* The proportionality constant between induced voltage and changing current is defined as *inductance L.* Thus we can say that inductance is a measure of the ability of a conductor to produce an induced voltage in response to a changing current. An *inductor* is an electric circuit element which has a known value of *inductance,* just as a resistor has a known value of resistance. The study of current (and voltage) changes in a circuit when switches are being opened or closed is called *transient* circuit analysis, and we will investigate transient circuits containing inductors.

12.1 INDUCTANCE

The basic unit of inductance is the *henry* (H), named in honor of the American physicist Joseph Henry (1797–1878). One henry can be defined as follows:

An element is said to have an inductance of one henry when a current changing at the rate of one ampere per second induces a terminal voltage of one volt.

Recall from the last chapter that the magnetization curve (**B-H**) for any nonmagnetic material (air, brass, etc.) is a straight line as shown in Fig. 12.1*a*. We can change the vertical scale of this curve from **B** to **Φ** simply by multiplying it by the cross-sectional area A, since $\mathbf{\Phi} = \mathbf{B}A$; similarly we can change the horizontal scale from **H** to I by multiplying by l/N, since $\mathbf{H} = NI/l$. The resulting curve is still a straight line, but the variables are now **Φ** and I as shown in Fig. 12.1*b*. The **B-H** curve for a magnetic material is shown in Fig. 12.1*c*, and it can also be drawn with **Φ** and I as the variables as shown in Fig. 12.1*d*.

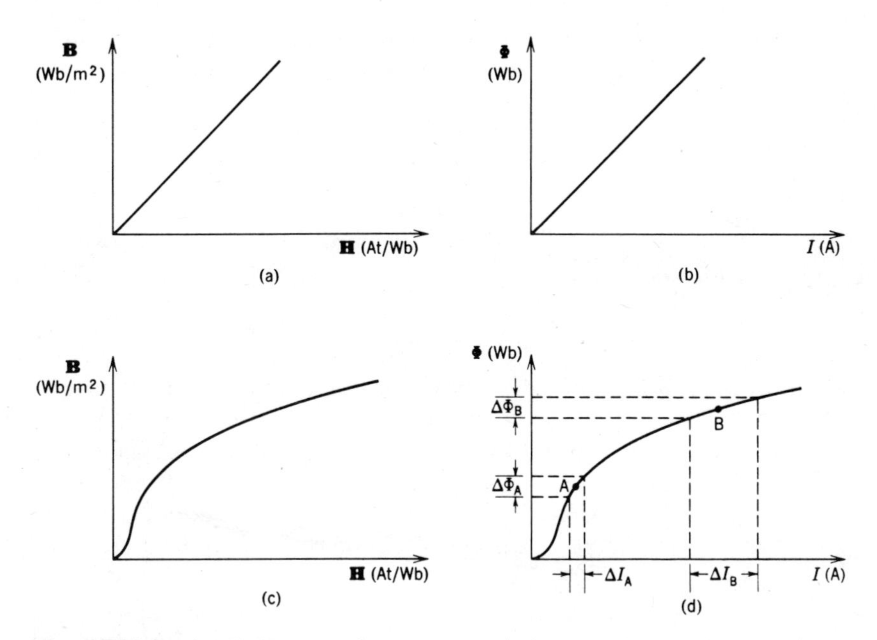

Figure 12.1 (*a*) **B-H** curve for nonmagnetic materials (*b*) Flux versus current for nonmagnetic materials (*c*) **B-H** curve for a magnetic material (*d*) Flux versus current for a magnetic material

Inductance* is sometimes defined in terms of the *slopes* of the curves in Fig. 12.1*b* and 12.1*d*. Since we know that slope is *rise over run*, the *rise* in these curves is a change in flux, or $\Delta\Phi$, and the *run* is a change in current, or ΔI. Thus for these curves, the slope is $\Delta\Phi/\Delta I$. Inductance is then defined as

$$L \triangleq N\frac{\Delta\Phi}{\Delta I} = N\frac{d\phi}{di} \tag{12.1}$$

where N is the number of turns, Φ is in webers, I is in amperes, and L is in henries.† For nonmagnetic materials, the curve in Fig. 12.1*b* is a straight line and thus $\Delta\Phi/\Delta I$, or $d\phi/di$ (the slope) is a constant. However, for magnetic materials the curve in Fig. 12.1*d* is *not* a straight line and thus the slope $\Delta\Phi/\Delta I$ changes depending on the current in the conductor. Inspection of Fig. 12.1*d*

*The terms inductance and self-inductance are synonymous.

†The lower case ϕ and i are used in $d\phi/di$, since in general ϕ and i may be time-varying quantities.

clearly shows that the slope determined by taking a small ΔI around point A is different than that at point B. Thus when working with inductors having magnetic cores, we must use the *incremental inductance* taken about some operating point (such as A or B). If the curve in Fig. 12.1*d* is extended further to the right, it will become nearly horizontal. Thus the slope $\Delta\Phi/\Delta I$ becomes zero, and therefore the inductance also becomes zero. Such an inductor is said to be *saturated,* and it no longer behaves as an inductance.

Equation 12.1 can be written in a different form by multiplying both sides by ΔI. Thus

$$L\,\Delta I = N\,\Delta\Phi$$

If we now divide both sides by Δt (this means examining the behavior of both sides of the equation as time t is varied), the result is

$$L\frac{\Delta I}{\Delta t} = N\frac{\Delta\Phi}{\Delta t}$$

But, Faraday's equation is

$$v = N\frac{\Delta\Phi}{\Delta t}$$

and thus, by substitution,

$$v = L\frac{\Delta I}{\Delta t} = L\frac{di}{dt} \qquad (12.2)$$

where v is in volts, i is in amperes, t is in seconds, and L is in henries. Equation 12.2 is a very important relationship which relates the *inductance* of a *circuit element* to the current through and the voltage across the element.

The circuit symbols for inductors (coils) are shown in Fig. 12.2*a* and 12.2*b*. The symbol in Fig. 12.2*a* is for any coil wound on a nonmagnetic form, and the two *parallel lines* in Fig. 12.2*b* denote a coil wound on a magnetic core.

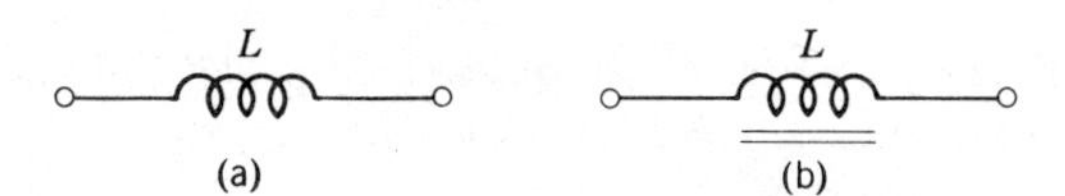

Figure 12.2 Inductance circuit symbols (*a*) Nonmagnetic core (*b*) Magnetic core

• • •

Example 12.1 The current in a 5 mH inductance changes steadily from 0.100 A to 0.900 A in a period of 4 msec. What is the terminal voltage across the inductor?

Solution Since the current changes from 0.100 A to 0.900 A, the *change* in current is

$$\Delta I = 0.900\text{ A} - 0.100\text{ A} = 0.800\text{ A}$$

The time period Δt is 4 msec, and thus

$$\frac{\Delta I}{\Delta t} = \frac{0.800\text{ A}}{4\text{ msec}} = 0.200 \times 10^3\text{ A/sec}$$

Then, using Eq. 12.2

$$v = L\frac{\Delta I}{\Delta t} = 5 \times 10^{-3}\text{ H} \times 0.200 \times 10^3\text{ A/sec} = 1.00\text{ V}$$

• • •

Example 12.2 What is the inductance of a coil if 35 V exists across the terminals when the current changes at the rate of 2 mA per μsec?

Solution The rate of change of current per unit time is

$$\frac{\Delta I}{\Delta t} = \frac{2 \times 10^{-3}\ \text{A}}{10^{-6}\ \text{sec}} = 2 \times 10^{3}\ \text{A/sec.}$$

Equation 12.2 is now rearranged as

$$L = \frac{v}{\frac{\Delta I}{\Delta t}} = \frac{35\ \text{V}}{2 \times 10^{3}\ \text{A/sec}} = 17.5 \times 10^{-3}\ \text{H} = 17.5\ \text{mH}$$

• • •

Example 12.3 At what rate must the current in a 1 H inductance change in order to induce a terminal voltage of 1.0 V?

Solution Rearranging Eq. 12.2,

$$\frac{\Delta I}{\Delta t} = \frac{v}{L} = \frac{1.0\ \text{V}}{1\ \text{H}} = 1.0\ \text{A/sec}$$

• • •

Example 12.4 A relay coil has an inductance of 20 mH. When the relay is energized, the coil conducts a current of 100 mA. When the relay is deenergized, the coil current goes from 100 mA to 0.0 mA in 5 μsec. What is the induced voltage across the coil terminals?

Solution Using Eq. 12.2

$$v = L\frac{\Delta I}{\Delta t} = 20 \times 10^{-3}\ \text{H} \times \frac{0.100\ \text{A}}{5 \times 10^{-6}\ \text{sec}} = 400\ \text{V}$$

• • •

The large voltage in Example 12.4 is known as *inductive kickback*, and it can be very damaging to neighboring circuit components unless it is suppressed. On the other hand, inductive kickback is exactly the mechanism used to provide a high voltage to the spark plugs in an internal combustion engine (can you draw the circuit and explain why?). The relay coil can be protected by placing a voltage variable resistor (thyrite) or a diode across the relay terminals.

.2 INDUCTORS

A general formula for the inductance of a coil can be found by using Ohm's law for magnetic circuits and the formula for reluctance. These two formulas are

$$\Phi = \frac{\mathcal{F}}{\mathcal{R}} = \frac{IN}{\mathcal{R}} \qquad \mathcal{R} = \frac{l}{\mu A}$$

By substituting the formula for reluctance in the first relationship, we get

$$\Phi = \frac{\mathcal{F}}{\mathcal{R}} = \frac{IN}{\frac{l}{\mu A}} = \frac{\mu INA}{l}$$

The terms N, A, and l are constant for any one coil, and if we take small changes (Δ's) in Φ and I, we get

$$\Delta\Phi = \Delta I \frac{\mu NA}{l}$$

Now, dividing both sides by ΔI,

$$\frac{\Delta\Phi}{\Delta I} = \frac{\mu NA}{l}$$

If we now substitute this relationship into Eq. 12.1, we obtain,

$$L = N\frac{\Delta\Phi}{\Delta I} = N\frac{\mu NA}{l} = \frac{\mu N^2 A}{l}$$

Thus a general formula for inductance is

$$L = \frac{\mu N^2 A}{l} \tag{12.3}$$

where L is in henries, μ is MKS units, A is in square meters, and l is in meters.

• • •

Example 12.5 Calculate the inductance of the coil in Fig. 12.3 if the coil form (core) is made of plastic.

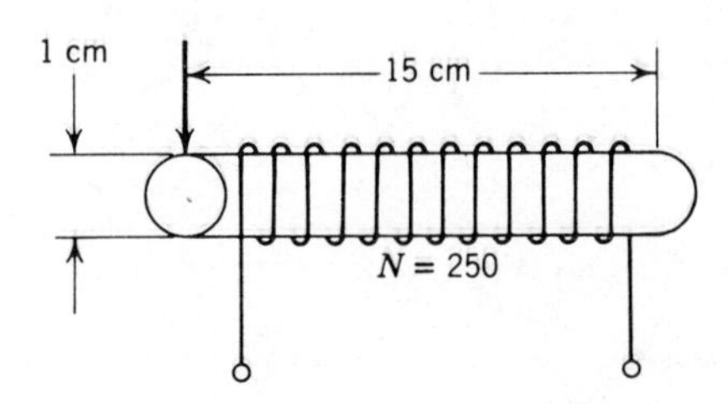

Figure 12.3

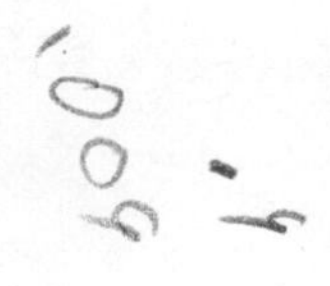

Solution We assume that the permeability of the plastic core is the same as that of air; $\mu = 4\pi \times 10^{-7}$ MKS units. Then using Eq. 12.3,

$$L = \frac{\mu N^2 A}{l} = \frac{4\pi \times 10^{-7} \times (250)^2 \times \dfrac{\pi(10^{-2})^2}{4}\ \text{m}^2}{0.15\ \text{m}} = 41.9\ \mu\text{H}$$

• • •

Equation 12.3 gives accurate results only for coils of the type shown in Fig. 12.3; that is, *single layer* solenoids having a length l several times greater than the diameter. More accurate formulas and design curves for coils of other geometries are given in a number of reference books.* Two coils which require special formulas are shown in Fig. 12.4. The formula for the ***pancake*** coil is

$$L(\mu\text{H}) = \frac{r^2 N^2}{8r + 11t} \tag{12.4}$$

*For example, *Reference Data for Radio Engineers*, International Telephone and Telegraph Corp., New York, 1956.

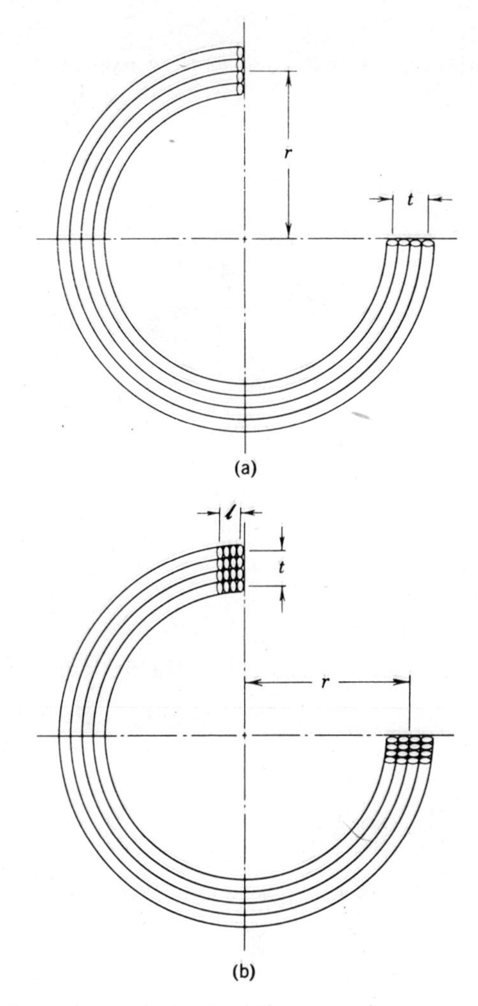

Figure 12.4 (*a*) Pancake coil (*b*) Multilayer coil

where r and t are in inches. The formula for the multilayer coil is

$$L\,(\mu\text{H}) = \frac{0.8r^2N^2}{6r + 9l + 10t} \tag{12.5}$$

where r, l, and t are in inches.

• • •

Example 12.6 Calculate the inductance of a pancake coil as shown in Fig. 12.4*a* if it consists of 100 turns of AWG no. 30 wire, and $r = 5$ in.

Solution From the standard wire tables (Table 5.4), the diameter of AWG no. 30 wire is 0.010 in. Thus the dimension t in Fig. 12.4*a* is equal to the thickness of 99 wires, and

$$t = 99 \times 0.010 = 0.99 \text{ in.}$$

Then, using Eq. 12.4

$$L\ (\mu\text{H}) = \frac{(5)^2 \times (100)^2}{8 \times 5 + 11 \times 0.99} = 4.92 \text{ mH}$$

• • •

Example 12.7 Calculate the inductance of the multilayer coil shown in Fig. 12.5.

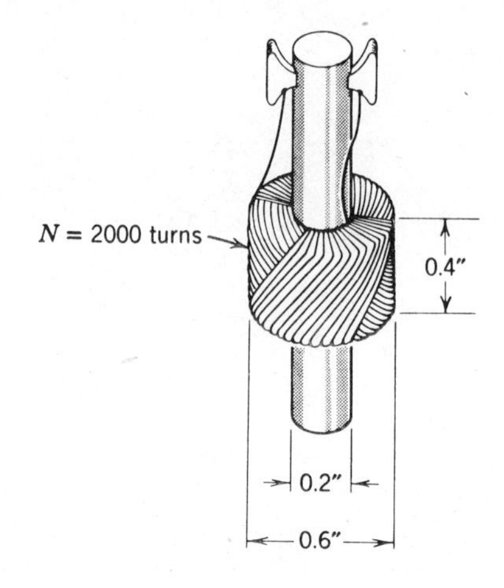

Figure 12.5

Solution Comparing the coil with Fig. 12.4*b*, the dimensions are seen to be

$r = 0.2''$ $\quad l = 0.4''$ $\quad t = 0.2''$ (we neglect the wire thickness)

Then, using Eq. 12.5

$$L\ (\mu\text{H}) = \frac{0.8r^2N^2}{6r + 9l + 10t} = \frac{0.8(0.2)^2(2000)^2}{6 \times 0.2 + 9 \times 0.4 + 10 \times 0.2} = 18.8\ \text{mH}$$

• • •

12.3 MUTUAL INDUCTANCE

When two inductors are placed close to one another as shown in Fig. 12.6, the flux from one coil will *link* the turns of the other coil. If the current in coil A is *changing with time* (not steady dc), the flux lines from coil A will *cut across* the

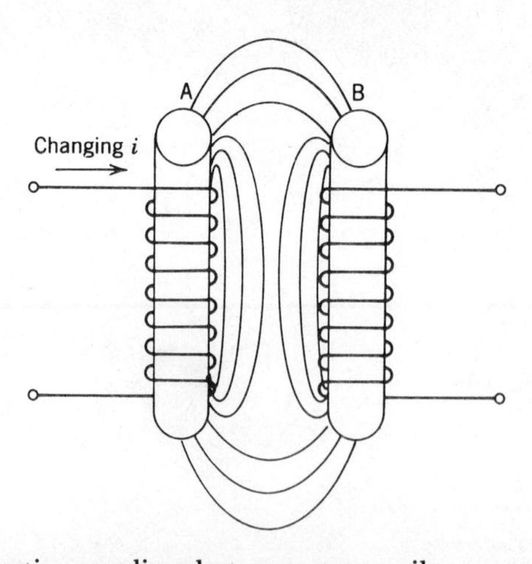

Figure 12.6 Magnetic coupling between two coils

turns in coil B and thus there will be a current induced in B as a result of changing current in A. At the same time, the induced current in B sets up a magnetic field around B, and some of these flux lines will *link* the turns in A. These two coils are then said to be *magnetically coupled.* The ratio of the flux linking both coils to the total flux in coil A is defined as the coefficient of coupling k. Thus

$$\text{coefficient of coupling} = k \triangleq \frac{\text{flux linkages between A and B}}{\text{flux produced by A}} \tag{12.6}$$

If *all* of the flux in A links the turns in B, then the coefficient of coupling k is equal to 1.0; if only half the flux produced by A links B, k is equal to 0.5, and so on. Thus the closer together the two coils are placed, the greater the value of k (the maximum value of k is 1.0, and the minimum value is 0.0). When k is close to 1.0, the coils are said to be *tightly* coupled, and when k is close to 0.0 the coils are said to be *loosely* coupled.

When it is desired to *minimize* the interaction between two inductors, they should be placed as far apart as possible with their core axes perpendicular to one another as shown in Fig. 12.7*a*. Such an arrangement will provide the minimum number of flux linkages between the two coils, and thus the coefficient of coupling k will be very nearly zero. In order to provide *tight* coupling between two coils, they can be wound on the same core as shown in Fig. 12.7*b*. If the core is nonmagnetic, this arrangement will provide a coefficient of coupling k of around 0.1 to 0.3. These arrangements are known as *air core transformers.* If the coils are wound on a common magnetic core as shown in Fig. 12.7*c*, virtually all of the flux from one coil links the other coil, and k is nearly equal to 1.0. This is the arrangement used to construct most *transformers* where the maximum possible coupling is desired. The symbols for transformers are shown in Fig. 12.8. One winding is usually called the *primary*, and the other winding is called the *secondary*.

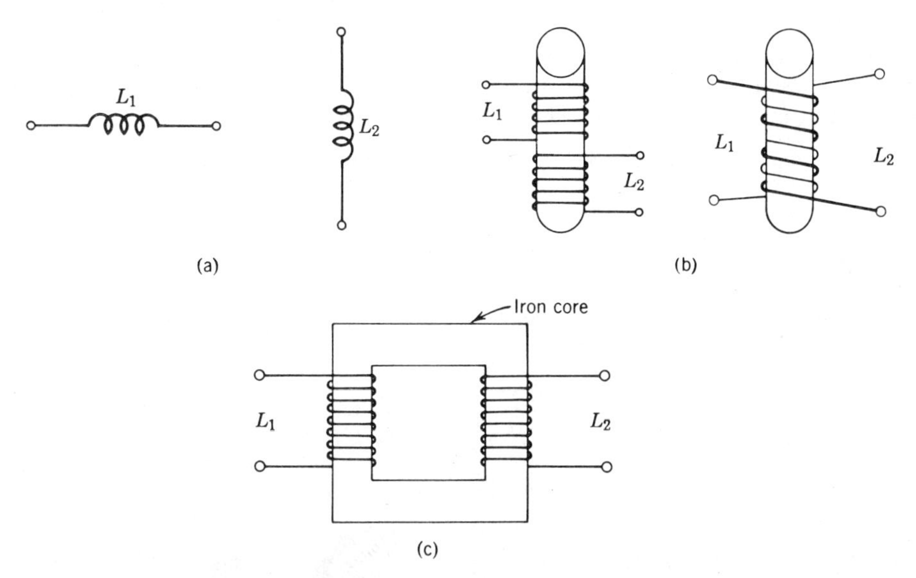

Figure 12.7 Coil arrangements (*a*) Minimum coupling. $k = 0.0$ (*b*) Air core transformers. $k = 0.1$ to 0.3 (*c*) Iron core transformers. $k = 1.0$

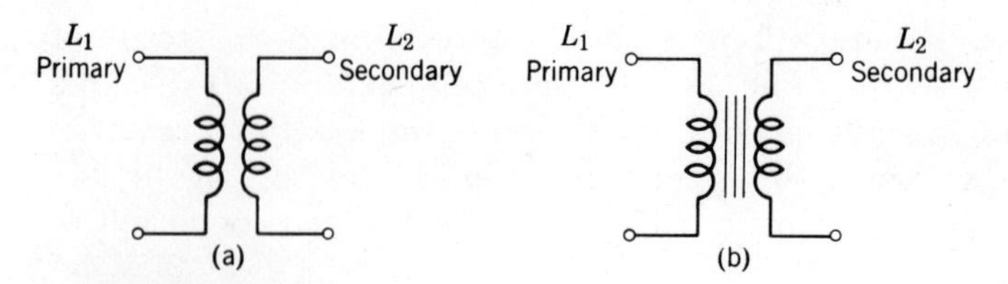

Figure 12.8 Transformer symbols (*a*) Air core transformer (*b*) Iron core transformer

It is important to note that a current will be induced in the secondary of a transformer only when the flux linking the two windings is *changing*; thus the current in the primary must also be changing. As a result, a steady dc current in the primary of a transformer will *not* induce a current in the secondary. On the other hand, since an ac current is always changing, an ac current in the primary will induce an ac current in the secondary. Thus transformers are very useful circuit elements in ac circuits.

• • •

Example 12.8 There are 150 μWb of changing flux in the primary of the transformer in Fig. 12.8*a*. If 30 μWb of this flux links the secondary, what is the coefficient of coupling k?

Solution Using Eq. 12.6

$$k = \frac{30\ \mu\text{Wb}}{150\ \mu\text{Wb}} = 0.2$$

• • •

When two coils have magnetic coupling between them, each coil is capable of inducing a current in the other coil, and thus there is a *mutual inductance* L_M between the two coils. The unit of mutual inductance is the henry, and the *maximum* possible theoretical value is

$$\text{maximum } L_M(\text{H}) = \sqrt{L_1 L_2}$$

where L_1 and L_2 are the inductances of the two coils in henries. This maximum value of L_M is based on a coefficient of coupling k of 1.0, and in general the mutual inductance L_M is given by

$$L_M = k\sqrt{L_1 L_2} \tag{12.7}$$

where L_M, L_1, and L_2 are in henries, and k is the coefficient of coupling.

• • •

Example 12.9 The primary of the transformer in Fig. 12.8*b* has an inductance of 30 mH, and the secondary has an inductance of 120 mH. What is the mutual inductance L_M?

Solution Since this is an iron core transformer, we assume the coefficient of coupling $k = 1.0$. Then using Eq. 12.7,

$$L_M = k\sqrt{L_1 L_2} = 1.0\sqrt{30 \times 10^{-3} \times 120 \times 10^{-3}} = 60\text{ mH}$$

• • •

Example 12.10 One of the air core coils in Fig. 12.7*b* has the following values: $L_1 = 100\ \mu\text{H}$, $L_2 = 400\ \mu\text{H}$, and $L_M = 20\ \mu\text{H}$. What is the coefficient of coupling k?

Solution Equation 12.7 can be rearranged to solve for k, and thus

$$k = \frac{L_M}{\sqrt{L_1 L_2}} = \frac{20 \times 10^{-6}\text{ H}}{\sqrt{100 \times 10^{-6}\text{ H} \times 400 \times 10^{-6}\text{ H}}} = 0.1$$

• • •

We will have occasion to use the concept of mutual inductance in the next section when we consider the series and parallel connections of inductors.

12.4 INDUCTORS IN SERIES AND PARALLEL

The two coils in Fig. 12.9*a* are connected in series (the current is the same in both) and there is no coupling between them. In order to find the *total inductance* of two such coils connected in series, we apply KVL to the loop. Notice that we must use lower case symbols since the currents and voltages are time-varying (not steady dc). Application of KVL yields

$$v_t = v_{L_1} + v_{L_2} \tag{12.8}$$

Now, the terminal voltage of the coil in terms of the coil current is given by Eq. 12.2; thus

$$v_t = L_T \frac{di}{dt} \qquad v_{L_1} = L_1 \frac{di}{dt} \qquad v_{L_2} = L_2 \frac{di}{dt}$$

where L_T is the *total* inductance of the circuit. Substituting these relationships into Eq. 12.8 yields

$$L_T \frac{di}{dt} = L_1 \frac{di}{dt} + L_2 \frac{di}{dt}$$

Dividing through by di/dt,

$$L_T = L_1 + L_2$$

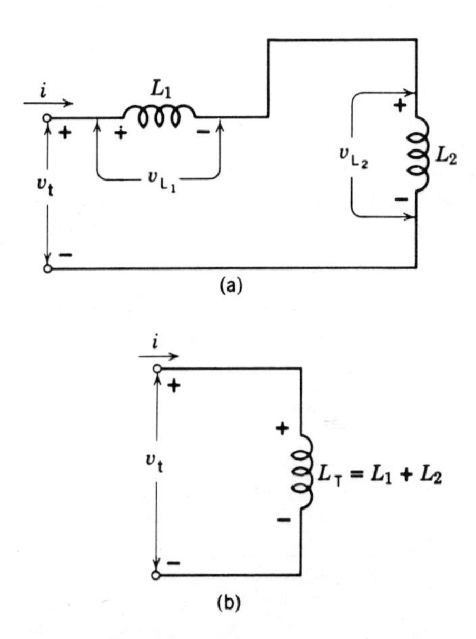

Figure 12.9 Two coils connected in series with no mutual magnetic coupling (*a*) Original circuit (*b*) Equivalent circuit

Thus the total inductance is simply the *sum* of the two inductances, and the circuit in Fig. 12.9*a* can be reduced to the equivalent circuit in Fig. 12.9*b*. This result can be extended to account for any number of inductances connected in series, and thus the total inductance of N coils connected in series, without any mutual inductance, is

$$L_T = L_1 + L_2 + \cdots + L_N \tag{12.9}$$

• • •

Example 12.11 The coils L_1 and L_2 in Fig. 12.9*a* have inductances of 1 mH and 450 μH respectively. What is the total inductance of the circuit?

Solution The total inductance is simply the sum; using Eq. 12.9,

$$L_T = L_1 + L_2 = 1\text{ mH} + 450\ \mu\text{H} = 1.45\text{ mH}$$

• • •

Example 12.12 What is the total inductance of the circuit in Fig. 12.10?

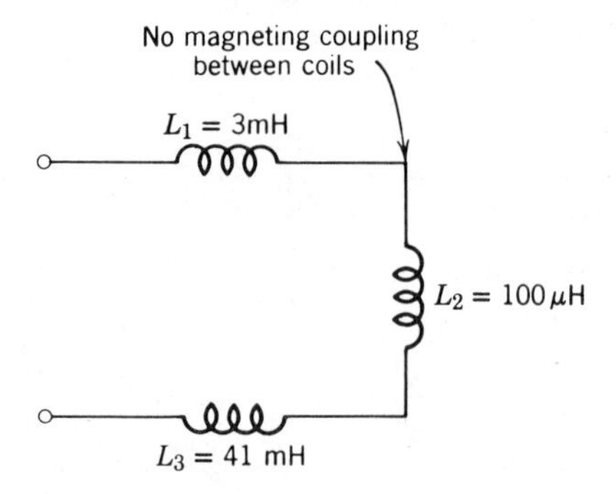

Figure 12.10

Solution The coils are connected in series, and thus we use Eq. 12.9.

$$L_T = L_1 + L_2 + L_3 = 3\text{ mH} + 100\ \mu\text{H} + 41\text{ mH} = 44.1\text{ mH}$$

• • •

Example 12.13 If the total inductance of the circuit in Fig. 12.11 is 640 μH, what must be the value of L_1?

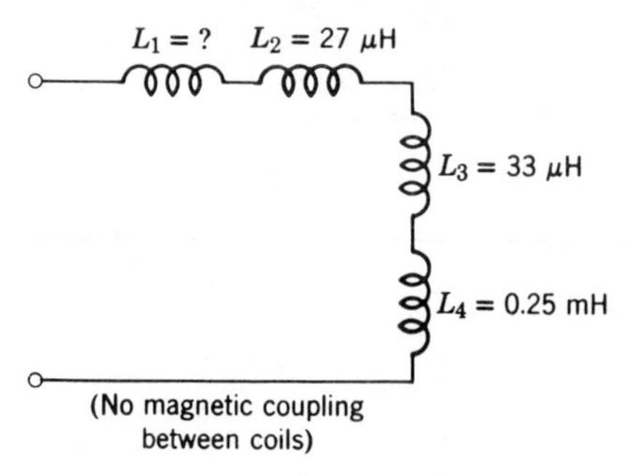

Figure 12.11

Solution Rearranging Eq. 12.9,

$$\begin{aligned} L_1 &= L_T - L_2 - L_3 - L_4 \\ &= 640\ \mu\text{H} - 27\ \mu\text{H} - 33\ \mu\text{H} - 250\ \mu\text{H} \\ &= 330\ \mu\text{H} \end{aligned}$$

• • •

The two coils in Fig. 12.12*a* are connected in parallel and there is no magnetic coupling between them. The voltage across both coils is clearly the same (since they are in parallel) but the current is different in each coil. Application of KCL to the circuit yields

$$i_t = i_{L_1} + i_{L_2}$$

Now, taking small changes in the currents,

$$\Delta i_t = \Delta i_{L_1} + \Delta i_{L_2}, \qquad \text{or} \qquad di_t = di_{L_1} + di_{L_2}$$

Dividing both sides of this equation by dt,

$$\frac{di_t}{dt} = \frac{di_{L_1}}{dt} + \frac{di_{L_2}}{dt} \tag{12.10}$$

But the application Eq. 12.2 to each coil in the circuit yields

$$v_t = L_T \frac{di_t}{dt} \qquad v_t = L_1 \frac{di_{L_1}}{dt} \qquad v_t = L_2 \frac{di_{L_2}}{dt}$$

where L_T is the total circuit inductance. Rearranging these relationships,

$$\frac{v_t}{L_T} = \frac{di_t}{dt} \qquad \frac{v_t}{L_1} = \frac{di_{L_1}}{dt} \qquad \frac{v_t}{L_2} = \frac{di_{L_2}}{dt}$$

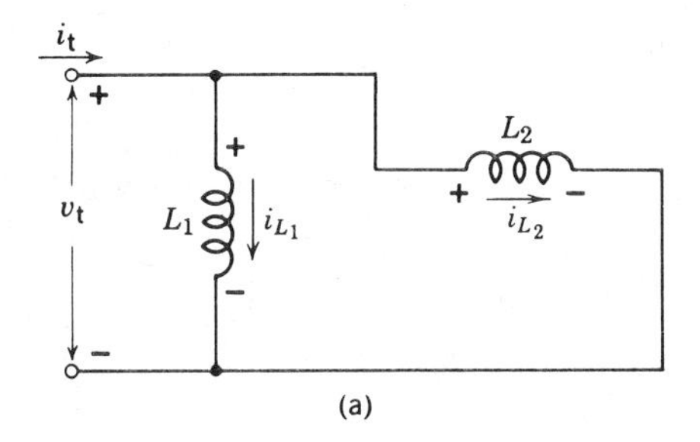

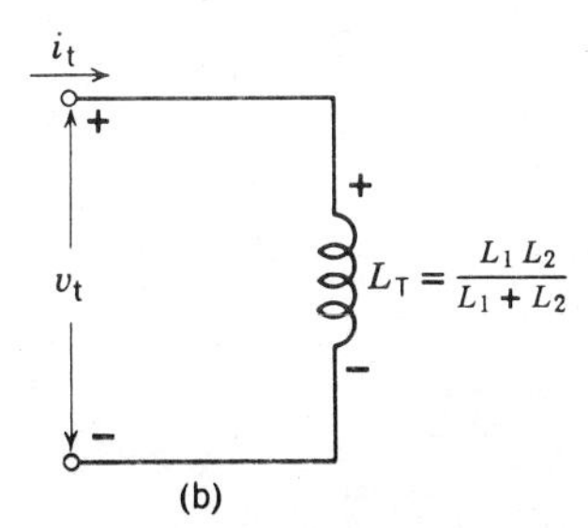

Figure 12.12 Two coils connected in parallel with no magnetic coupling (*a*) Original circuit (*b*) Equivalent circuit

Now, substituting these equations into Eq. 12.10,

$$\frac{v_t}{L_T} = \frac{v_t}{L_1} + \frac{v_t}{L_2}$$

and dividing by v_t,

$$\frac{1}{L_T} = \frac{1}{L_1} + \frac{1}{L_2}$$

Thus the total inductance of two coils connected in parallel without magnetic coupling is

$$L_T = \frac{1}{\frac{1}{L_1} + \frac{1}{L_2}} = \frac{L_1 L_2}{L_1 + L_2} \tag{12.11}$$

This equation can be extended to the general case of N coils connected in parallel without mutual coupling; thus

$$L_T = \frac{1}{\frac{1}{L_1} + \frac{1}{L_2} + \cdots + \frac{1}{L_N}} \tag{12.12}$$

From Eqs. 12.9, 12.11, and 12.12, we note that inductors combine in series and parallel in exactly the same manner that *resistors* combine.

• • •

Example 12.14 What is the total inductance of the circuit in Fig. 12.12 if $L_1 = 4$ mH and $L_2 = 6$ mH?

Solution Using Eq. 12.11

$$L_T = \frac{L_1 L_2}{L_1 + L_2} = \frac{4\text{ mH} \times 6\text{ mH}}{4\text{ mH} + 6\text{ mH}} = 2.4\text{ mH}$$

• • •

Example 12.15 What is the total inductance of the circuit in Fig. 12.13?

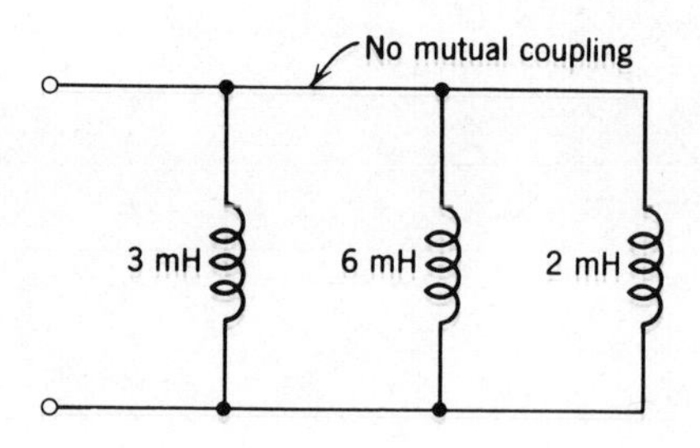

Figure 12.13

Solution Using Eq. 12.12

$$L_T = \frac{1}{\frac{1}{L_1} + \frac{1}{L_2} + \frac{1}{L_3}} = \frac{1}{\frac{1}{3\text{ mH}} + \frac{1}{6\text{ mH}} + \frac{1}{2\text{ mH}}} = 1\text{ mH}$$

• • •

Notice that Example 12.15 could have been solved by combining the coils two at a time; the reader should do this for verification.

• • •

Example 12.16 What is the value of L_1 in Fig. 12.12*a* if $L_T = 12\ \mu H$ and $L_2 = 30\ \mu H$?

Solution Equation 12.11 can be solved for L_1, and thus

$$L_1 = \frac{L_2 L_T}{L_2 - L_T} = \frac{30\ \mu H \times 12\ \mu H}{30\ \mu H - 12\ \mu H} = 20\ \mu H$$

• • •

The two coils in Fig. 12.14*a* are placed such that there is mutual coupling between them (i.e., $k \neq 0$), and there is therefore some mutual inductance L_M. Since these two coils are in *series*, and since the flux in one *aids* the flux in the other, the total inductance is given by

$$L_T = L_1 + L_2 + 2L_M \tag{12.13}$$

If the two coils are connected in series such that the flux in one "bucks" the flux in the other as shown in Fig. 12.14*b* (check with the right-hand-rule), the total inductance is given by*

$$L_T = L_1 + L_2 - 2L_M \tag{12.14}$$

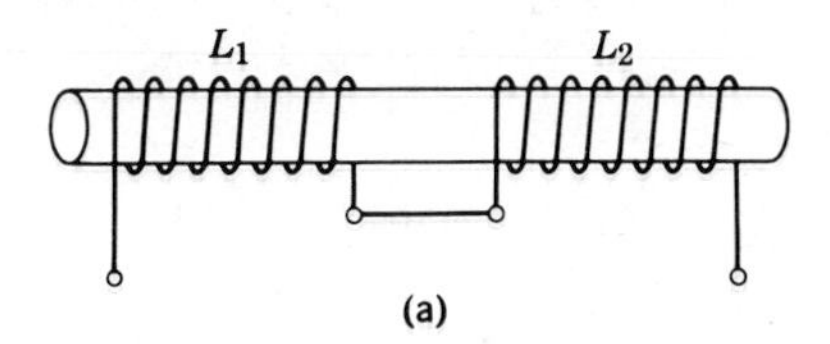

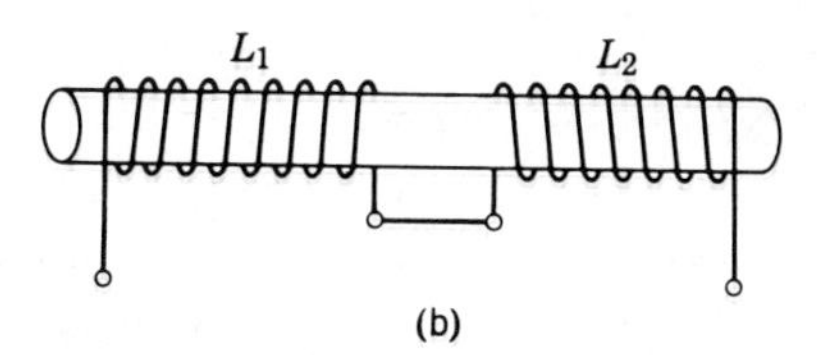

Figure 12.14 Series connected coils with mutual inductance (coupling) (*a*) Aiding flux (*b*) Cancelling flux

• • •

Example 12.17 What is the total inductance of the coils in Fig. 12.14*a* if

$$L_1 = 33\ \text{mH},\ L_2 = 25\ \text{mH, and } L_M = 2.5\ \text{mH?}$$

Solution Using Eq. 12.13,

$$L_T = L_1 + L_2 + 2L_M = 33\ \text{mH} + 25\ \text{mH} + 2(2.5\ \text{mH}) = 63\ \text{mH}$$

• • •

Example 12.18 If the total inductance of the coils in Fig. 12.14*b* is 15 mH, and if $L_1 = 8$ mH and $L_2 = 8$ mH, what is the mutual inductance?

*The total inductance of two coils connected in parallel with mutual coupling involves solving simultaneous, second order differential equations, and we will therefore not discuss it here.

Solution Equation 12.14 can be solved for L_M, and thus

$$L_M = \frac{L_1 + L_2 - L_T}{2} = \frac{8\text{ mH} + 8\text{ mH} - 15\text{ mH}}{2} = 0.5\text{ mH}$$

• • •

The mutual inductance of two coils connected in series can easily be found by measurement. The total inductance is first measured with an ac bridge with the coils connected as in Fig. 12.14*a*; call this inductance L_{aid}. The total inductance is then measured with the coils connected as in Fig. 12.14*b*; call this L_{buck}. Then using Eqs. 12.13 and 12.14,

$$L_{\text{aid}} = L_1 + L_2 + 2L_M$$

$$L_{\text{buck}} = L_1 + L_2 - 2L_M$$

Subtracting these two equations,

$$L_{\text{aid}} - L_{\text{buck}} = 4L_M$$

Thus

$$L_M = \frac{L_{\text{aid}} - L_{\text{buck}}}{4} \tag{12.15}$$

• • •

Example 12.19 Two measurements are made on the coils in Fig. 12.14 with the results, $L_{\text{aid}} = 590\ \mu\text{H}$ and $L_{\text{buck}} = 510\ \mu\text{H}$. What is the mutual inductance?

Solution Using Eq. 12.15,

$$L_M = \frac{L_{\text{aid}} - L_{\text{buck}}}{4} = \frac{590\ \mu\text{H} - 510\ \mu\text{H}}{4} = 20\ \mu\text{H}$$

• • •

12.5 CURRENT RISE IN AN INDUCTOR

Suppose an inductor is connected to a voltage source through a switch as shown in Fig. 12.15*a*. The switch is then closed at time $t = 0$, and we wish to examine the current in the circuit. Since this is a simple series circuit, the voltage across the inductor v_L must be equal to the source voltage V_S at all

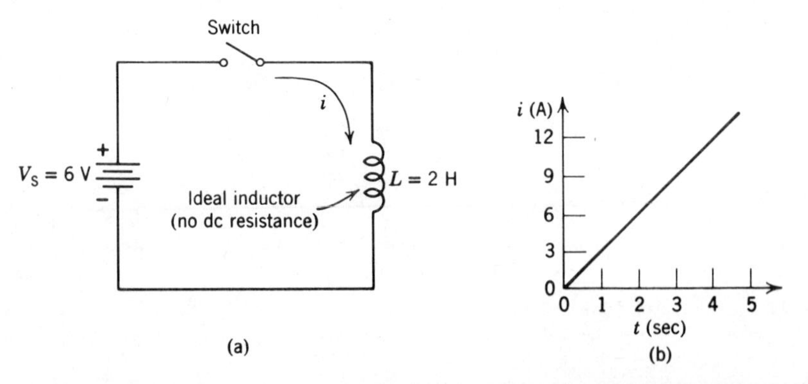

Figure 12.15 Current in an ideal inductance (*a*) Circuit (*b*) Current versus time

times; using Eq. 12.2,

$$v_L = V_S = L\frac{di}{dt}$$

If we assume that the inductance L is constant, then the current must increase at a constant rate since

$$\frac{V_S}{L} = \frac{6\text{ V}}{2\text{ H}} = 3\text{ A/sec} = \frac{di}{dt} = \frac{\Delta i}{\Delta t}$$

Since $\Delta i/\Delta t$ is a constant, we can remove the Δ's to obtain

$$3 = \frac{\Delta i}{\Delta t} = \frac{i}{t}, \qquad \text{or} \qquad i = 3t$$

Thus a plot of the current in the inductor as a function of time would appear as shown in Fig. 12.15*b*. When $t = 0$, $i = 0$; when $t = 1$ sec, $i = 3$ A; then $t = 2$ sec, $i = 6$ A, and so on. The current would then increase constantly without limit. Notice that we use the *lower case* letter i for current since the value of the current depends on time (it is *time-varying*); we also use the *upper case* letter V_S for the source voltage, since its value does not depend on time (i.e., it is constant).

The current in an *ideal inductor* would increase without limit as shown in Fig. 12.15, since an ideal inductor, like an *ideal source*, has no dc resistance. A *real inductor* does have some dc resistance, and it is exactly equal to the resistance of the wire used to wind the coil. As a result of the coil resistance R_L, the current will be limited to some maximum value V_S/R_L as shown in Figs. 12.16*a* and 12.16*b*. The behavior of the current in the circuit from the time the switch is closed until it reaches its final *steady state* value is called the *transient response*, and this involves finding an equation for the curve in Fig. 12.16*b*. Such an equation will allow us to determine the current in the circuit at any time, and thus the length of the transient period.

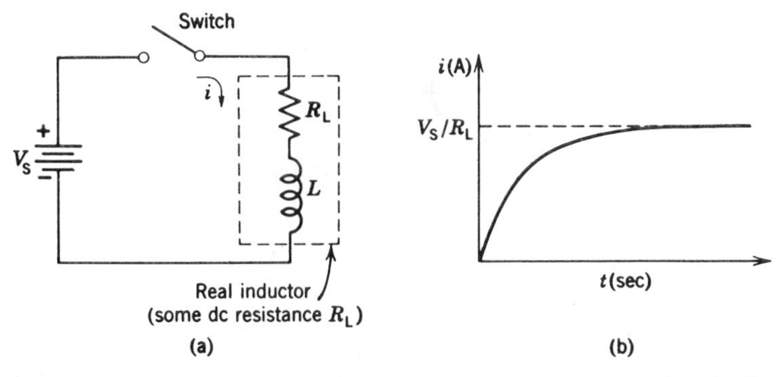

Figure 12.16 The current in a real inductance (*a*) Circuit (*b*) Current versus time

Consider the general R-L circuit in Fig. 12.17. The coil resistance R_L may be added to the resistance of the resistor R, but usually R_L will be much smaller than R. The switch is closed at time $t = 0$, and we want to find an expression for the circuit current i. Applying KVL around the loop,

$$V_S = v_R + v_L$$

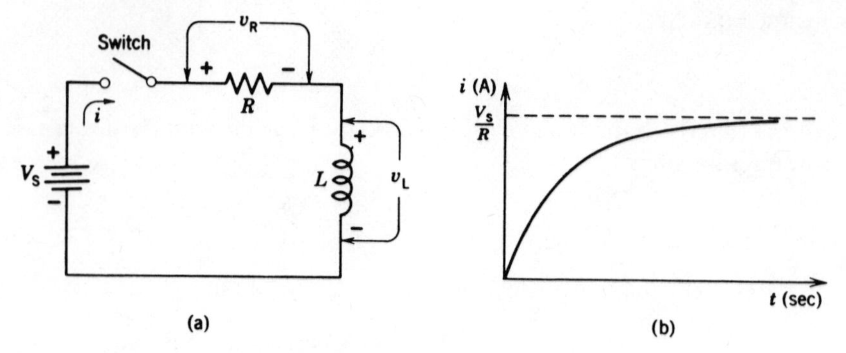

Figure 12.17 A general R-L circuit (*a*) Circuit (*b*) Current versus time

Using Ohm's law, and Eq. 12.2,

$$v_R = iR \qquad \text{and} \qquad v_L = L\frac{di}{dt}$$

By substitution,

$$V_S = iR + L\frac{di}{dt} \tag{12.16}$$

Equation 12.16 is a differential equation which can be easily solved for the current i using calculus. The result is

$$i = \frac{V_S}{R}[1 - e^{-(R/L)t}] \tag{12.17}$$

where $e = 2.718$ (the base of natural logarithms).

Equation 12.17 can be used to find the current in the circuit at any time t. As a check, let us assume the switch has just been closed and $t = 0$ sec. Substituting $t = 0$ in Eq. 12.17*

$$i = \frac{V_S}{R}[1 - e^{-(R/L)\times 0}] = \frac{V_S}{R}(1 - 1) = 0$$

since

$$e^{-0} = e^{0} = 1$$

and thus the equation gives $i = 0$ for $t = 0$.

Now let us assume the switch has been closed a long time and t is very large, for example, $t \to \infty$. Then

$$i = \frac{V_S}{R}[1 - e^{-(R/L)\times\infty}] = \frac{V_S}{R}(1 - 0) = \frac{V_S}{R}$$

since

$$e^{-\infty} = \frac{1}{e^{\infty}} = 0$$

and thus the equation gives the final value of i as V_S/R.

• • •

Example 12.20 Write the equation for the current in Fig. 12.17 if $V_S = 10$ V, $R = 100\ \Omega$, and $L = 20$ mH. What is the value of the current 0.2 msec after the switch is closed?

*The values of e^{-x} can be determined from the tables in the appendix; also, most engineering slide rules and electronic calculators can be used to determine e^{x} and e^{-x}.

Solution Using Eq. 12.17,

$$i = \frac{V_S}{R}[1 - e^{-(R/L)t}] = \frac{10\text{ V}}{100\ \Omega}[1 - e^{-[100/(20\times10^{-3})]t}]$$
$$= 0.1[1 - e^{-5\times10^3 t}]$$

when $t = 0.2$ msec,

$$i = 0.1[1 - e^{-1}]$$
$$= 0.1(1 - 0.368) = 63.2\text{ mA}$$

• • •

Example 12.21 What is the final value of the current in Example 12.20? How long will it take for the current to reach one-half the final value?

Solution The final value of the current is

$$I_t = \frac{V_S}{R} = \frac{10\text{ V}}{100\ \Omega} = 0.1\text{ A}$$

One-half the final value is then 50 mA. Using Eq. 12.17,

$$50\text{ mA} = 0.05\text{ A} = \frac{10\text{ V}}{100\ \Omega}[1 - e^{-[100/(20\times10^{-3})]t}]$$

or

$$0.05 = 0.1[1 - e^{-5\times10^3 t}]$$

We must solve for the time t. Thus, dividing both sides by 0.1,

$$\frac{0.05}{0.1} = [1 - e^{-5\times10^3 t}]$$

simplifying,

$$0.5 = e^{-5\times10^3 t}$$

From the table of exponentials,

$$e^{-0.69} = 0.5$$

Thus

$$5\times10^3 t = 0.69$$

and

$$t = \frac{0.69}{5\times10^3} = 0.14\text{ msec}$$

• • •

We now turn our attention to the voltages v_R and v_L in the circuit in Fig. 12.17. The source voltage V_S is constant and does not change with time as shown in Fig. 12.18. The voltage across the resistor v_R is the product of the

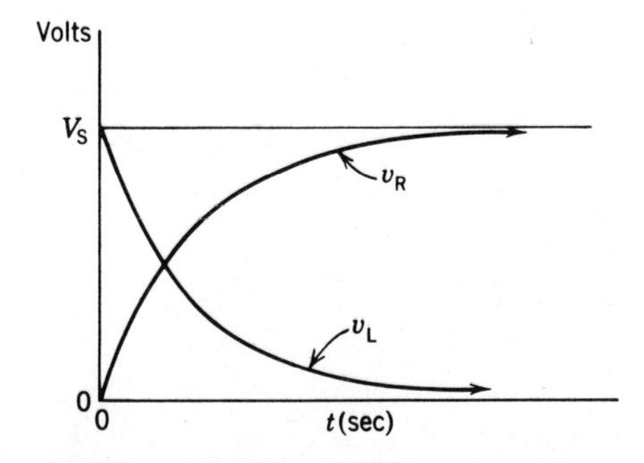

Figure 12.18 The voltages in Figure 12.17

current i and the resistance R, and thus

$$v_R = iR = \left[\frac{V_S}{R}(1 - e^{-(R/L)t})\right] R = V_S(1 - e^{-(R/L)t})$$

The resistor voltage is then a curve which has exactly the same shape as the current curve, but its final value is simply changed to V_S as shown in Fig. 12.18. Now, using the KVL equation for this circuit, we find that the voltage across the inductor v_L is simply the *difference* between the constant source voltage V_S and the resistor voltage v_R. Thus

$$v_L = V_S - v_R$$

By substitution,

$$\begin{aligned} v_L &= V_S - V_S[1 - e^{-(R/L)t}] = V_S - V_S + V_S e^{-(R/L)t} \\ &= V_S e^{-(R/L)t} \end{aligned}$$

This curve is simply an exponential as shown in Fig. 12.18.

• • •

Example 12.22 What is the voltage across the inductor in Fig. 12.17 100 μsec after the switch is closed if: $V_S = 6$ V, $R = 1$ $k\Omega$, and $L = 200$ mH?

Solution Using Eq. 12.19

$$\begin{aligned} v_L &= V_S e^{-(R/L)t} = 6e^{-[10^3\ \Omega/(200\times10^{-3}\ \mathrm{H})]\times10^{-4}\ \mathrm{sec}} \\ &= 6e^{-0.5} = 6 \times 0.606 = 3.64\ \mathrm{V} \end{aligned}$$

• • •

12.6 *R-L* TIME CONSTANT τ

We now know the shape of the current curve in an R-L circuit, and we also know that the *final value* of the current is V_S/R as shown in Fig. 12.19. Consider now how long it would take for the current to reach its final value if it could continue at its *initial rate* (i.e., no resistance in the circuit). Immediately after the switch is closed ($t = 0^+$), all of the voltage V_S is across the inductor. Thus

$$v_L = V_S = L\frac{di}{dt}$$

But di/dt is the initial rate of increase in current, and since we want this to be a

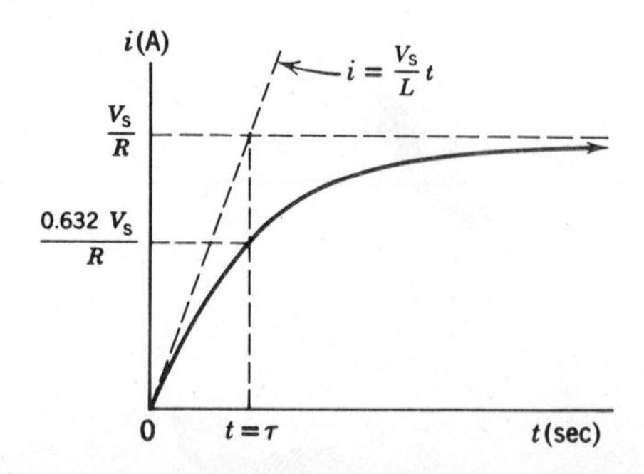

Figure 12.19 The time constant τ of an R-L circuit

constant, we can write

$$\frac{di}{dt} = \frac{\Delta i}{\Delta t} = \text{constant} = \frac{V_S}{L}$$

We can therefore drop the Δ's, and we then have the equation of a straight line (dotted line in Fig. 12.19)

$$i = \frac{V_S}{L} t$$

The final value of $i = V_S/R$ and thus the time required for the current to reach its final value at the initial rate is found from

$$i = \frac{V_S}{R} = \frac{V_S}{L} t$$

Thus

$$t = \frac{L}{R}$$

This length of time has a very special meaning, and it is defined as the *time constant* τ of the circuit.

The time constant τ of an R-L circuit is equal to L/R and is the time required for the current to reach its final value if it were to continue to increase at its initial rate.

Thus

$$\text{time constant} = \tau \triangleq \frac{L}{R} \text{ (sec)} \tag{12.18}$$

In one time constant, the current in an R-L circuit will increase from zero to 63.2 percent of its final value. We can show this by putting the time t equal to one time constant L/R into Eq. 12.17.

$$i = \frac{V_S}{R}[1 - e^{-(R/L)t}] = \frac{V_S}{R}[1 - e^{-(R/L)\times(L/R)}] = \frac{V_S}{R}(1 - e^{-1})$$

The value of e^{-1} is 0.368, and thus

$$i = \frac{V_S}{R}(1 - 0.368) = 0.632\frac{V_S}{R}$$

But the final value of i is V_S/R, and thus in *one time* constant the value of i is 63.2 percent of the final value.

• • •

Example 12.23 What is the time constant of the circuit in Example 12.22? What will be the value of the current after one time constant?

Solution The time constant τ is found from Eq. 12.18.

$$\tau = \frac{L}{R} = \frac{200 \times 10^{-3}\ H}{10^{3}\ \Omega} = 200 \times 10^{-6} \text{ sec}$$

The final value of the current is

$$i_f = \frac{V_S}{R} = \frac{6\text{ V}}{1\text{ k}\Omega} = 6\text{ mA}$$

and thus the current after one time constant is

$$i = 0.632\, i_f = 0.632 \times 6 \text{ mA} = 3.79 \text{ mA}$$

• • •

Example 12.24 What is the value of the current in an R-L circuit after a period of 5 time constants?

Solution Substituting $t = 5L/R$ in Eq. 12.17,

$$i = \frac{V_S}{R}[1 - e^{-(R/L)t}] = \frac{V_S}{R}[1 - e^{-(R/L)\times(5L/R)}] = \frac{V_S}{R}(1 - e^{-5})$$

But $e^{-5} = 0.0067$. Thus

$$i = \frac{V_S}{R}(1 - 0.0067) = 0.993\frac{V_S}{R}$$

For all practical purposes, the current has reached its final value.

• • •

From Example 12.24, we will make the following agreement:

After 5 time constants, 5τ, the current in an R-L circuit has reached its final value, and the transient period is ended.

• • •

Example 12.25 What is the length of the transient period for the circuit in Fig. 12.20?

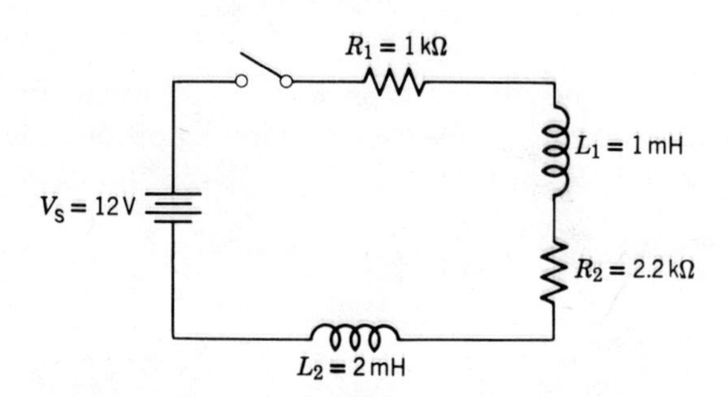

Figure 12.20

Solution The total circuit resistance is

$$R = R_1 + R_2 = 1 \text{ k}\Omega + 2.2 \text{ k}\Omega = 3.2 \text{ k}\Omega$$

The total circuit inductance is

$$L = L_1 + L_2 = 1 \text{ mH} + 2 \text{ mH} = 3 \text{ mH}$$

The time constant is then

$$\tau = \frac{L}{R} = \frac{3 \times 10^{-3} \text{ H}}{3.2 \times 10^{3}\ \Omega} = 0.938\ \mu\text{sec}$$

The transient period is over in 5 time constants, or

$$t = 5\tau = 5 \times 0.938\ \mu\text{sec} = 4.69\ \mu\text{sec}$$

• • •

12.7 CURRENT FALL IN AN INDUCTOR

We now consider the case of a *decreasing* current in an inductor. In the circuit in Fig. 12.21*a*, the switch has been in position A for a long time and the current in the inductor L is $I_0 = V_S/R_1$. When the switch is moved to position B, we have the equivalent circuit shown in Fig. 12.21*b*, and the current in the inductor will decay to zero. Applying KVL to this circuit,

$$0 = v_{R_T} - v_L$$

But

$$v_{R_T} = iR_T \qquad \text{and} \qquad v_L = L\frac{di}{dt} \qquad \text{where} \qquad R_T = R_1 + R_2$$

Notice the change in polarity of v_L, since the inductor is now behaving like a source to maintain i. This voltage is called *counter emf* cemf since it tends to maintain the existing coil current. Thus,

$$0 = iR_T - L\frac{di}{dt} \tag{12.19}$$

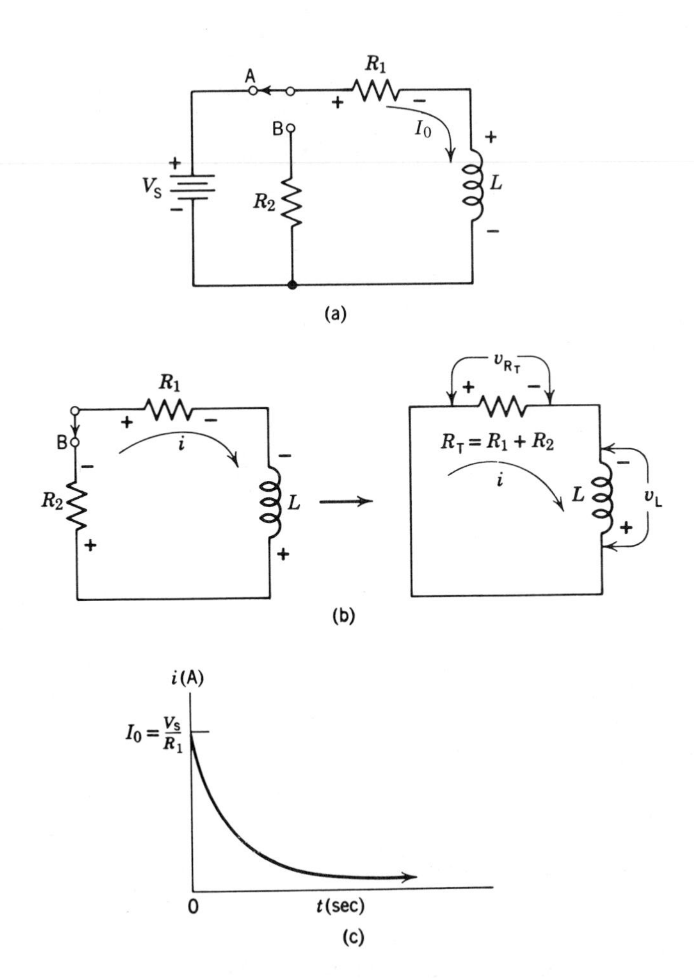

Figure 12.21 Current fall in an inductance (*a*) Establishing I_0. (*b*) Equivalent circuit (*c*) Decaying current

Solving this differential equation using calculus yields

$$i = \frac{V_S}{R_1} e^{-(R_T/L)t}$$

But the initial current in the inductance is $I_0 = V_S/R_1$, and thus

$$i = I_0 e^{-(R_T/L)t} \tag{12.20}$$

• • •

Example 12.26 What is the inductor current in Fig. 12.21. 100 μsec after the switch is moved to position B if: $V_S = 12$ V, $R_1 = 200\ \Omega$, $R_2 = 100\ \Omega$, and $L = 60$ mH?

Solution Using Eq. 12.20,

$$i = I_0 e^{-(R_T/L)t} = \frac{12\text{ V}}{200\ \Omega} e^{-[(200\Omega+100\Omega)/(60\times 10^{-3}\text{H})]\times 100\times 10^{-6}\text{sec}}$$

$$= 60 \times 10^{-3}\, e^{-0.5} = 60 \times 10^{-3} \times 0.606 = 36.4 \text{ mA}$$

• • •

The voltage across the total resistance v_{R_T} in Fig. 12.21*b* is simply

$$v_{R_T} = iR_T = I_0 R_T\, e^{-(R_T/L)t} \tag{12.21}$$

This curve has exactly the same shape as the current curve, but the initial value is different as shown in Fig. 12.22. The KVL equation for this circuit shows that v_L must be equal to v_{R_T}, and thus the curve in Fig. 12.22 is equal to both v_L and v_{R_T}.

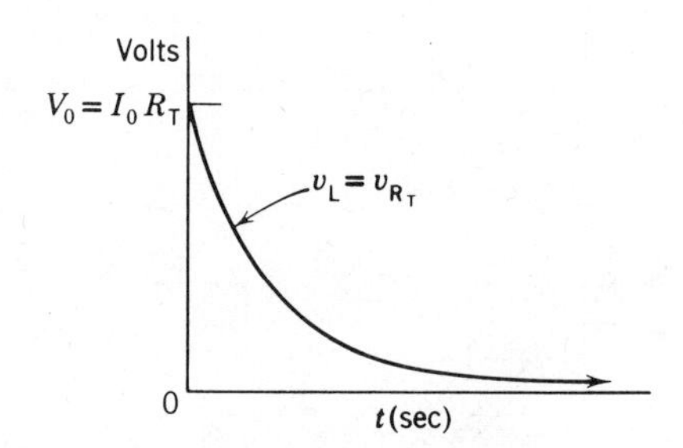

Figure 12.22 Element voltages in Figure 21.21*b*

The time constant for a decaying current in an *R-L* circuit is found in exactly the same way as for an increasing current in an *R-L* circuit. The time constant τ is equal to the *reciprocal* of the coefficient of t in Eq. 12.20. Thus for a decaying *R-L* circuit,

$$\text{time constant} = \tau = \frac{L}{R}$$

In one time constant, the current in a decaying *R-L* circuit will fall to 36.8 percent of its initial value, and it will be zero for all practical purposes after *five* time constants.

• • •

Example 12.27 What is the time constant for the circuit in Fig. 12.21*a* when: (*a*) switch in position A, (*b*) switch in position B? $V_S = 6$ V, $R_1 = 1$ k, $R_2 = 2$ k, $L = 400$ mH.

Solution (*a*) with the switch in position A, the time constant is

$$\tau = \frac{L}{R_1} = \frac{400 \text{ mH}}{1 \text{ k}\Omega} = 400 \ \mu\text{sec}$$

(*b*) with the switch in position B, the time constant is

$$\tau = \frac{L}{R_1 + R_2} = \frac{400 \text{ mH}}{1 \text{ k}\Omega + 2 \text{ k}\Omega} = 133 \ \mu\text{sec}$$

• • •

Example 12.28 Make a plot of the current in the inductor in Fig. 12.23*a* if the switch is moved from A to B and held for 1 msec, and then moved to position C.

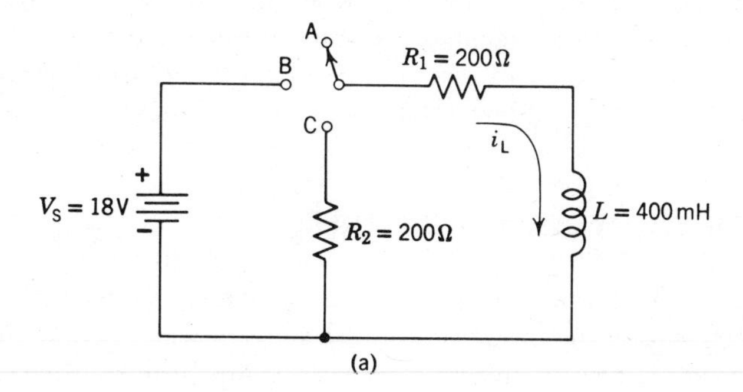

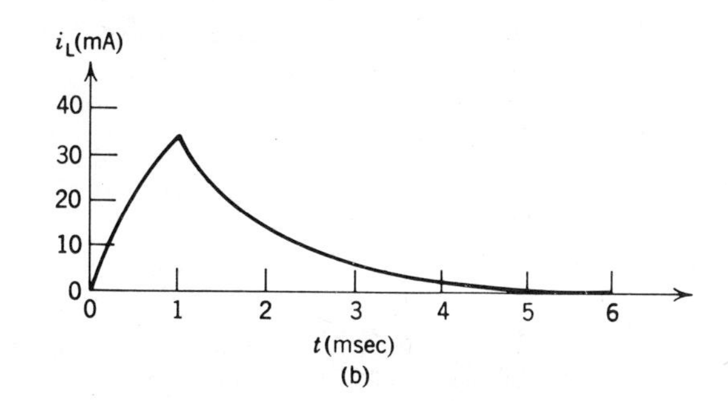

Figure 12.23 (*a*) *R-L* circuit (*b*) Inductor current

Solution When the switch is in position B, the current is increasing in the inductance. The current in L after 1 msec is found using Eq. 12.17,

$$i = \frac{V_S}{R_1}[1 - e^{-(R_1/L)t}] = \frac{18 \text{ V}}{200 \ \Omega}[1 - e^{-[200\,\Omega/(400\times10^{-3}\,\text{H})]\times10^{-3}\,\text{sec}}]$$
$$= 90 \times 10^{-3}(1 - e^{-0.5}) = 35.4 \text{ mA}$$

This becomes I_0 when the switch is moved to position C. In position C, the current in L is decaying and we use Eq. 12.20 to find

$$i = I_0 e^{-[(R_1+R_2)/L]t} = 35.4 \times 10^{-3} e^{-[(200\Omega+200\Omega)/(400\times10^{-3}\text{H})]t}$$
$$= 35.4 \times 10^{-3} e^{-10^3 t}$$

One decay time constant after the switch is moved to C, the current will have decayed to 36.8 percent of I_0. The decay time constant is

$$\tau = \frac{L}{R_1 + R_2} = \frac{400 \text{ mH}}{200\ \Omega + 200\ \Omega} = 10^{-3} \text{ sec}$$

Thus, 1 msec after moving the switch to C, the current is

$$i = 36.8 \text{ percent } I_0 = 0.368 \times 35.4 \text{ mA} = 13.0 \text{ mA}$$

After 5 decay time constants, the current will be essentially zero. Thus

$$5\tau = 5 \times 10^{-3} \text{ sec}$$

The current curve is shown in Fig. 12.23*b*.

• • •

From the examples in this and the previous section, we observe a very important fact: the *voltage* across an inductor *can* change instantly, but the *current* through an inductor *cannot* change instantly. The current must in fact change according to an *exponential* increase or decrease.

12.8 ENERGY STORED BY AN INDUCTOR

Whenever an inductor is connected to a source as shown in Fig. 12.24, a current flows through the inductor and the source must expend energy. If the inductor is *ideal*, it has no internal resistance and one might rightfully ask what happens to the energy delivered by the source, since energy can be neither created nor destroyed. The answer is, the energy is stored in the magnetic field surrounding the coil. A derivation for the energy stored requires calculus, but the result is simple, and

$$\text{energy stored by an inductance} = W \text{ (J)} = \tfrac{1}{2}LI_0^2 \qquad (12.22)$$

where W is in J, L is in H, and I_0 is in A. I_0 represents an increase or decrease in inductor current.

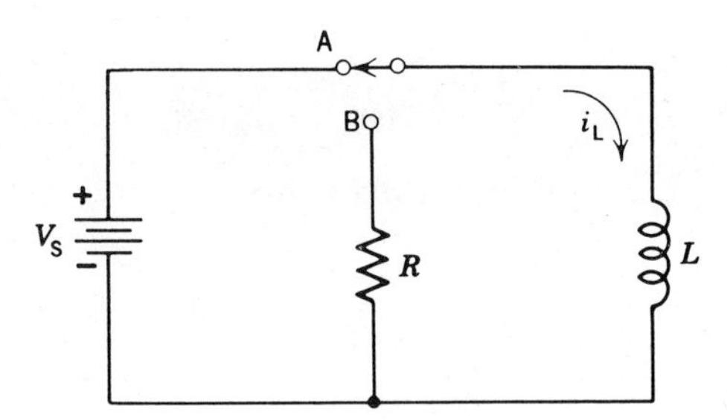

Figure 12.24

When the switch is moved to B in Fig. 12.24, the inductor voltage reverses polarity in an effort to maintain the current i_L. The inductance then acts as a *source*, and the energy stored in the magnetic field is used to maintain the current i_L (it decays exponentially of course). The energy of the magnetic field is then dissipated as heat in the resistor R. Thus an inductance can store energy in its magnetic field, and then return the energy to another circuit element.

• • •

Example 12.29 What is the energy stored in L in Fig. 12.24 if the switch is held in position A while the inductor current increases from zero to 100 mA? The switch is now moved to B; what is the energy delivered to the resistor R by the inductance? $L = 100$ mH and $R = 100\ \Omega$.

Solution The energy stored in L is found using Eq. 12.22.

$$W\ (\text{J}) = \tfrac{1}{2}LI_0^2 = \tfrac{1}{2} \times 100 \times 10^{-3}\ \text{H}(100 \times 10^{-3}\ \text{A})^2 = 0.5 \times 10^{-3}\ \text{J}$$

When the switch is moved to B, all of the energy stored in L is dissipated in R. Thus the total energy dissipated in R is 0.5×10^{-3} J.

• • •

SUMMARY

Inductance has been defined as the amount of flux set up around a coil per unit current in the coil. Using Faraday's law and the definition of inductance, we have shown that the terminal voltage of an inductor is equal to the product of the inductance and the *rate of change* of current in the inductor. Since there is no voltage induced in an *ideal* inductor unless the current is changing, an inductance has no effect in a dc circuit. The magnetic field lines around an inductor, may *link* the turns of another inductor, and thus we have defined mutual inductance to account for interaction between coils. Transformers are designed using the principles of mutual inductance and *flux linkages.* Inductors and transformers are useful circuit elements in ac circuits, and we can combine L's in series and parallel to find equivalent circuits just as we combine R's.

The current in an inductance *cannot* change instantaneously, and the study of changing currents when switches are opened and/or closed is called transient analysis. The currents in inductors change exponentially during transient periods. The time constant τ of an R-L circuit can be used to determine the length of a transient period, as well as the transient current at different times.

The magnetic field around an inductor stores energy during an increase in inductor current, and returns the stored energy to some circuit when the current is decreasing.

GLOSSARY

Coefficient of coupling. k. *Ratio of the flux linking two coils to the flux in one of the coils.*

Counter emf. cemf. *The induced voltage in an inductance which tends to maintain the existing coil current.*

Flux linkages. *Loops of flux which* link *two different coils.*

Incremental inductance. *The ratio* $\Delta\phi/\Delta i$ *taken around a point on a* **B-H** ($\phi - i$) *curve.*

Inductance. *The proportionality constant between induced voltage and* changing *current in a coil; the change in flux due to a change in current.*

Inductor. *A circuit element having the property of inductance.*

Inductive kickback. *The large counter emf generated across an inductance when an existing current is interrupted.*

Loose coupling. *Arranging two inductors to achieve a minimum of flux linkages between them. Thus a low value of k.*

Mutual inductance. *An additional inductance developed as a result of flux linkages between two coils.*

Tight coupling. *Arranging two inductors to achieve a maximum of flux linkages between them. Thus a high value of k.*

Time constant. τ. *A measure of how fast currents (voltages) change in a transient circuit.*

Transient circuit. *A circuit in which the currents are disturbed from their steady state values for a period of time.*

Transient period. 5τ. *The time required for currents (voltages) in a transient circuit to reach final steady-state values.*

Transient response. *The behavior of currents (voltages) in a transient circuit.*

IMPORTANT RELATIONSHIPS

- $L = N\dfrac{d\phi}{di} \qquad v_L = L\dfrac{di}{dt}.$
- coefficient of coupling $= k = \dfrac{\text{flux linking A and B}}{\text{flux in A}}$
- mutual inductance $= L_M = k\sqrt{L_1 L_2}$
- L's in series: $L_T = L_1 + L_2 + \cdots + L_N$.
- L's in parallel: $L_T = \dfrac{1}{\dfrac{1}{L_1} + \dfrac{1}{L_2} + \cdots + \dfrac{1}{L_N}}$.
- L's in series with L_M: $L_T = L_1 + L_2 \pm L_M$.
- $L_M = \dfrac{L_{aid} - L_{buck}}{4}$.
- current rise in series R-L circuit — current fall in series R-L circuit

$$i = \frac{V_S}{R}\left(1 - e^{-(R/L)t}\right) \qquad\qquad I = I_0 e^{-(R/L)t}$$

$$\text{time constant} = \tau = \frac{L}{R}$$

$$\text{stored energy} = W = \tfrac{1}{2}LI_0^2$$

REVIEW QUESTIONS

1. There is no induced voltage in an inductor unless the current is changing. (T, F)
2. The induced voltage in a coil is directly proportional to the rate of change of ____.
3. The induced voltage in an inductor is (inversely, directly) proportional to the inductance.
4. An iron core inductor will become *saturated* if the current is too (low, high).

5. Draw the symbol for an air core inductor.
6. Draw the symbol for an iron core inductor.
7. Two coils with a coefficient of coupling of 0.95 are said to be ______ coupled.
8. The maximum value for the coefficient of coupling k is ______.
9. If two coils have no mutual coupling, the value of k is ______.
10. In order to minimize coupling, two coils should be placed parallel to one another. (TF)
11. Draw the symbols for air core and iron core transformers.
12. Transformers and coils (are, are not) useful in dc circuits.
13. The maximum possible value of mutual inductance is obtained when $k =$ ______.
14. The coefficient of coupling for two coils wound on an iron core is essentially ______.
15. The equivalent inductance of a number of coils connected in series can be found by taking the sum of the individual inductances. (TF)
16. Inductors in parallel can be combined just like resistors in (series, parallel).
17. The mutual inductance between two coils can be found by connecting the coils in series with the flux ______ and then with the flux ______, and measuring the total inductance in each case.
18. Any practical inductance has some resistance due to the resistance of the ______.
19. The transient current in a series *RL* circuit has an exponential shape. (TF)
20. A resistor in series with a coil is suddenly connected across the terminals of a battery. The inductor voltage (rises, falls) exponentially.
21. The voltage across the resistor in Question 20 above has the same shape as the circuit ______.
22. The current in a series *R-L* circuit will increase ______ per cent of its final value in one time constant.
23. The current in a series *R-L* circuit will decay to ______ per cent of its initial value in one time constant.
24. The transient period in a series *R-L* circuit is essentially ended after ______ time constants.
25. The current in an inductance (can, cannot) change instantly, but the voltage (can, cannot).

PROBLEMS

1. The current in a 15 mH inductor changes from 0.0 mA to 25 mA in a period of 75 msec. What is the induced coil voltage?
2. What is the induced voltage in a 500 mH coil if the current changes from 1 A to 500 mA in 2 seconds?
3. What must be the inductance of a coil if a current changing at the rate of 950 A/sec induces a voltage of 85 mV?
4. At what rate must the current change in a 300 mH coil in order to induce a voltage of 100 V?
5. How fast must a 100 mA current be reduced to zero in order to induce 500 volts in a 250 mH coil?

6. A relay coil requires 125 mA to energize the relay. What is the inductive kickback voltage across the coil if the coil current goes to zero in 2 msec? $L_{coil} = 75$ mH.

7. What is the inductance of a 150 turn coil wound on a cylindrical plastic core 150 cm long and 0.9 cm in diameter?

8. What will be the inductance of the coil in Problem 7 if the number of turns is increased to 900?

9. Calculate the inductance of a pancake coil (Fig. 12.4*a*) if it consists of 50 turns of AWG no. 20 wire and $r = 12''$.

10. What would be the inductance of a coil equivalent to 50 of the pancake coils in Problem 9 placed side by side (wound as one continuous coil) as in Fig. 12.4*b*?

11. Two coils having inductances of 5 mH and 10 mH are placed close together and have a coefficient of coupling of 0.4. What is the mutual inductance between the coils?

12. What is the maximum possible L_M for the two coils in Problem 11?

13. What is the coefficient of coupling between two 20 μH coils if their mutual inductance is 3 μH?

14. The mutual inductance between 2 coils wound on an iron core is 58 μH. If the inductance of one coil is 75 μH, what must be the inductance of the other coil?

15. Two 25 mH coils and a 910 μH coil are connected in series. Draw the equivalent circuit.

16. The total inductance of 2 series connected coils is 860 mH. If one of the coils has an inductance of 330 mH, what must be the inductance of the other coil?

17. What is the equivalent inductance of a 3 mH coil connected in parallel with a 6 mH coil?

18. The total inductance of 2 parallel connected coils is 56 μH. One of the coils has an inductance of 100 μH, what must be the inductance of the other coil?

19. Two coils connected as in Fig. 12.14*a* have an inductance of 950 μH, and when connected as in Fig. 12.14*b* have an inductance of 510 μH. What is the mutual inductance?

20. If one of the two coils in Problem 19 has an inductance of 410 μH, what must be the inductance of the other coil? Can you work the problem in two ways?

21. Write the equation for the current in Fig. 12.17 if $V_S = 20$ V, $R = 1$ kΩ, and $L = 0.5$ H. Calculate the current for t equal to: (*a*) 100 μsec, (*b*) 200 μsec, (*c*) 500 μsec, (*d*) 2 msec, and plot the results. What is the final value of the current? Plot the voltages across R and L.

22. Repeat Problem 21 if R is changed to 10 kΩ.

23. How long will it take the current in Problem 21 to rise to 15 mA?

24. What is the time constant of the circuit in Problem 21?

25. What will be the value of the current in Problem 21 after one time constant? After 2 time constants? How long will it take to reach its final value?

26. The switch in Fig. 12.21*a* is held in *A* for a long time. Write the equation for the current in Fig. 12.21*b*. Calculate enough points to accurately plot *i*. What are the initial and final values of *i*? Plot the voltages across R and L. $V_S = 25$ V, $R_1 = 100\ \Omega$, $R_2 = 330\ \Omega$, $L = 33$ mH.

27. Repeat Problem 26 if L is changed to 560 mH.
28. How long will it take the current in Problem 26 to decay to 50 mA?
29. What is the time constant of the circuit in Problem 26? What will be the value of i after one time constant? After 2 time constants? How long will it take for i to decay to zero?
30. Make a plot of the inductor current i_L in Fig. 12.25 if the switch is moved from A to B and held for 2.5 msec, and then moved to C.

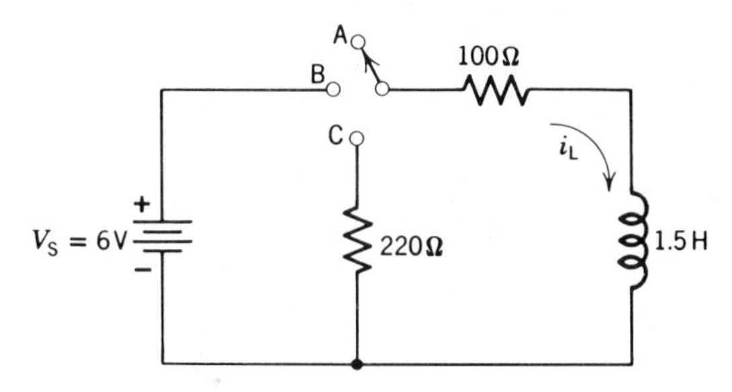

Figure 12.25

31. Repeat Problem 30 if the switch is held on B for 25 msec.
32. If the current in a 700 mH inductor increases from zero to 100 mA, how much energy is stored in the field around the coil?
33. How much energy is transferred from the coil to the resistor in Problem 31 when the switch is in position C? How much power does this represent (Hint: use $t = 5\tau$).

13

CAPACITANCE

No matter how complicated an electric circuit is, it can have at most three basic components, namely, *resistance, inductance,* and *capacitance.* Resistance and inductance have been discussed, and we will devote this chapter to capacitance. In Chapter 5 it was shown that the reciprocal of resistance R is conductance G; thus Ohm's Law can be written in two ways

$$V = IR \qquad I = VG$$

Note carefully that Ohm's Law using G is obtained from $V = IR$ simply by replacing I with V, V with I, and changing R to G. For this reason, conductance G is said to be the *dual* of resistance R. Consider now the v-i relationship developed for inductance in Chapter 12:

$$v = L\frac{di}{dt}$$

If we replace v with i, i with v, and change L to C (the symbol for capacitance), the result is the v-i relationship for capacitance:

$$i = C\frac{dv}{dt}$$

Thus, C is the *dual* of inductance. Since inductance opposes any change in *current*, C must oppose any change in *voltage.* Furthermore, inductance is associated with the *magnetic field* surrounding a current-carrying conductor, and thus C must be associated with an *electric field* between two points of different potential.

13.1 ELECTRIC FIELD

From previous work, we know that it is possible to produce a charged body by creating an *excess* or a *deficit* of electrons as shown in Fig. 13.1. If a *positive test charge** Q_1 is placed near either of the two bodies, there will be a force **F** exerted on the test charge. In Fig. 13.1*a*, the force will be one of *attraction*, since *unlike* charges attract; in Fig. 13.1*b* it will be a force of *repulsion*, since *like* charges repel. These forces can be explained by the presence of an *electric field* surrounding the charged bodies. Thus,

an electric field is said to exist in any space in which a force is exerted on a stationary test charge.

As a result of this definition, there must be an electric field surrounding *any charged* body. The presence of an electric field is shown by drawing *electric field lines* as shown in Fig. 13.1.

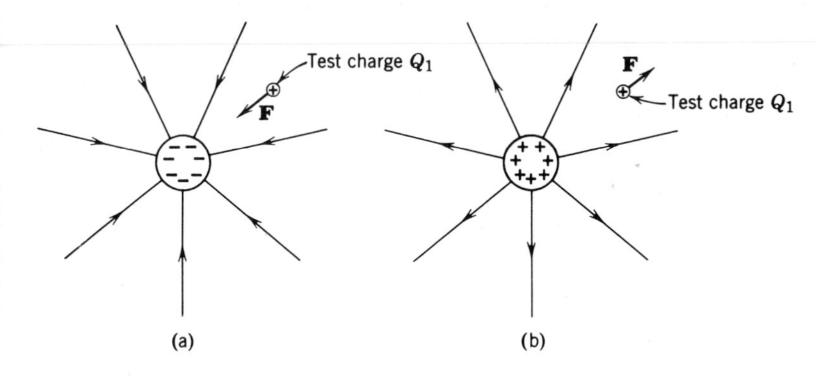

Figure 13.1 Electric field around a charged body (*a*) Negatively charged body (*b*) Positively charged body

Electric field lines can be considered as the paths along which the positive test charge moves. If the test charge in Fig. 13.1 is allowed to move, it will move *directly toward* or *directly away* from the charged body. Thus the field lines around these two charged bodies are *radial* lines, like the spokes of a wheel. The field lines also have an *orientation* shown by the arrowheads which indicate the direction the positive test charge moves.

Consider now the electric field *between* two oppositely charged bodies as shown in Fig. 13.2. The field lines are seen to have an orientation *from* the positive body *toward* the negative body. Thus the field lines seem to *emerge* from the positive body and *enter* the negative body. Furthermore, the field lines always exit and enter changed bodies at *right angles.* We make the

*The test charge is a weightless body having a positive charge.

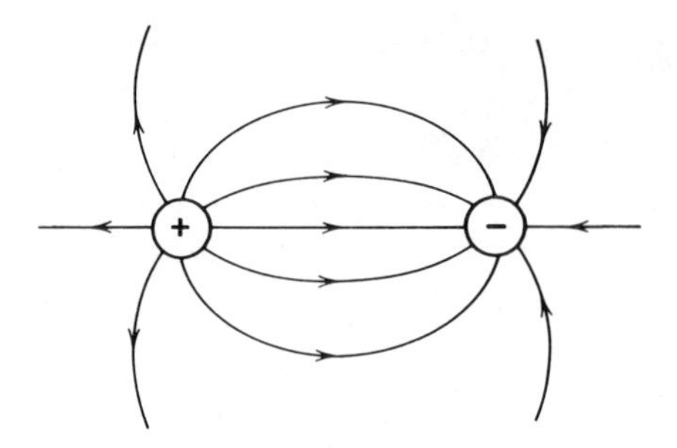

Figure 13.2 Electric field between oppositely charged bodies

following important observations concerning *electric field lines*:

1. The lines always begin and end on a body.*
2. The lines are oriented from the more positive body toward the more negative body.
3. The lines always enter or exit a body at a right angle.
4. The lines never cross.

• • •

Example 13.1 Draw the electric field lines between the outer shield and the inner conductor of a coaxial cable if the shield is more negative than the inner conductor.

Solution The electric field is shown in Fig. 13.3. The lines begin on the inner conductor and end on the shield. Since the lines leave the conductor at right angles (perpendicular to it) and enter the shield at right angles, they resemble the spokes of a wheel.

• • •

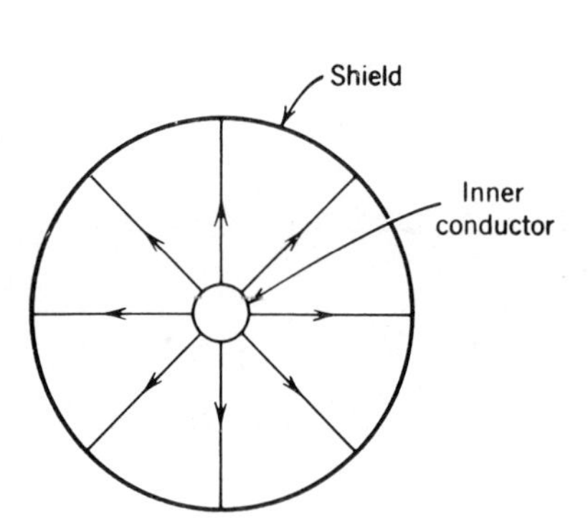

Figure 13.3

The total number of electric field lines is referred to as the *electric flux*, just as with magnetic fields, and the symbol ψ (Greek letter psi) is used. Electric flux ψ is directly proportional to electric charge Q, and thus the MKS unit of electric flux is coulombs, C; hence the magnitude of the flux is

$$\psi\ (\mathrm{C}) \triangleq Q\ (\mathrm{C}) \tag{13.1}$$

*Electric field lines are *not* continuous loops like magnetic field lines.

We define *electric flux density* **D** as the total flux per unit area, and thus

$$\mathbf{D}\ (\text{C/m}^2) = \frac{\psi\ (\text{C})}{A\ (\text{m}^2)} \tag{13.2}$$

The force on the test charge in Fig. 13.1*a* (or Fig. 13.1*b*) is given by Coulomb's Law; in MKS units,

$$\mathbf{F} = k_1 \frac{Q_1 Q_2}{d^2} \tag{13.3}$$

where **F** is the force in newtons
- k_1 is a constant, approximately 9×10^9
- Q_1 is the test body charge in coulombs
- Q_2 is the charge on the large body in coulombs
- d is the distance between the two bodies in meters

Dividing both sides of this equation by the test charge Q_1 yields

$$\frac{\mathbf{F}}{Q_1} = \frac{k_1 Q_2}{d^2}$$

This quantity is defined as the *electric field intensity* **E**. Thus

$$\text{electric field intensity} = \mathbf{E} = \frac{\mathbf{F}}{Q_1} \tag{13.4}$$

A more convenient form for **E** can be obtained by recalling the definition for work W,

$$W = \mathbf{F}d$$

and the definition of the volt,

$$V = \frac{W}{Q}, \text{ or } W = VQ$$

Equating these two relationships,

$$\mathbf{F}d = VQ, \text{ or } \mathbf{F} = \frac{VQ}{d}$$

Substituting this relationship into Eq. 13.4 yields

$$\mathbf{E} = \frac{\mathbf{F}}{Q_1} = \frac{\left(\frac{VQ_1}{d}\right)}{Q_1} = \frac{V}{d}$$

Thus in the MKS system, the electric field intensity can be expressed as

$$\mathbf{E}\ (\text{V/m}) = \frac{V\ (\text{V})}{d\ (\text{m})} \tag{13.5}$$

• • •

Example 13.2 What is the electric field intensity **E** between the two parallel plates shown in Fig. 13.4?

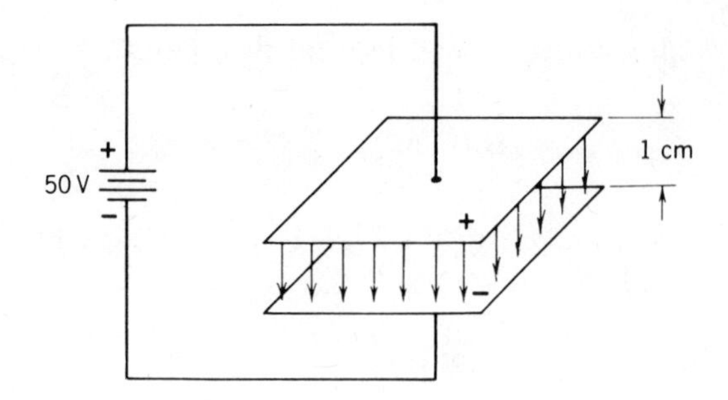

Figure 13.4

Solution Using Eq. 13.5,

$$\mathbf{E} = \frac{V}{d} = \frac{50\ \text{V}}{0.01\ \text{m}} = 5000\ \text{V/m}$$

• • •

Example 13.3 What force is exerted on an electron held between the two plates in Example 13.2?

Solution Equation 13.4 can be rearranged as,

$$\mathbf{F} = Q\mathbf{E}$$

From Example 13.2, $\mathbf{E} = 5000$ V/m, and $Q = 1.6 \times 10^{-19}$ C; thus

$$\mathbf{F} = 1.6 \times 10^{-19}\ \text{C} \times 5 \times 10^{3}\ \text{V/m} = 8 \times 10^{-16}\ \text{newtons}$$

• • •

Equation 13.5 can be rearranged as (E is equal to the magnitude of $\mathbf{E}$)

$$V = E\text{d} \tag{13.6}$$

This equation can be used to find the *potential* with respect to some reference at any point between two charged bodies. Consider for example the two parallel plates shown in Fig. 13.5. The bottom plate is clearly at a potential of 0.0 V, and the *top* plate is at a potential of + 100 V. If we now choose any point between the plates a distance of 5 cm from the bottom plate, Eq. 13.6 yields

$$V\ (\text{V}) = E\ (\text{V/m})\ d\ (\text{m}) = 1000\ \text{V/m} \times 5 \times 10^{-2}\ \text{m} = 50\ \text{V}$$

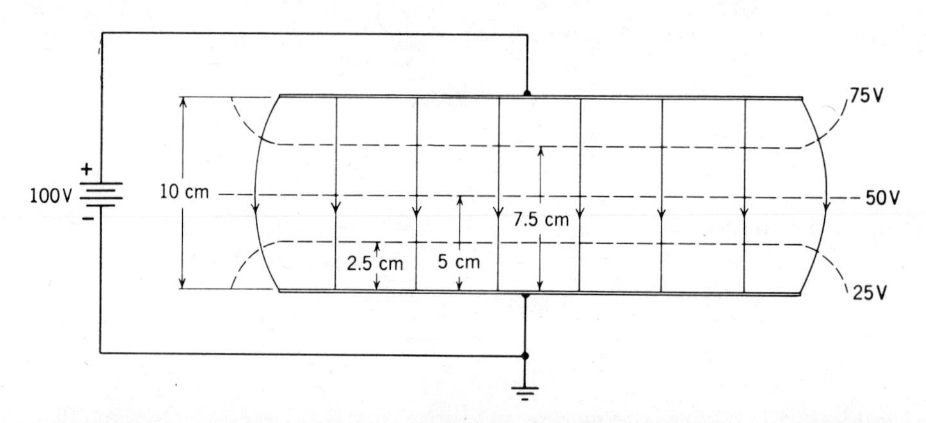

Figure 13.5 Equipotential lines (dashed) and electric field lines (solid)

since $E = V/d = 100\ \text{V}/0.1\ \text{m} = 1000\ \text{V/m}$. Thus the potential at any point along the *dashed line* equally separating the two plates is 50 V. Such a line (dashed) is shown in Fig. 13.5, and it is called an *equipotential* line (constant potential). Another equipotential line 2.5 cm above the bottom plate is found to have a potential of

$$V = E\text{d} = 1000\ \text{V/m} \times 2.5 \times 10^{-2}\ \text{m} = 25\ \text{V}$$

This equipotential line, and another one for 75 V are shown in Fig. 13.5. In fact, there are an infinite number of equipotential lines between the two plates, just as there are an infinite number of electric field lines. We make three important observations concerning *equipotential lines*:

1. They never cross.
2. They are always perpendicular to electric field lines.
3. They form closed loops around charged bodies, but they have no orientation.

• • •

Example 13.4 At what distance from the bottom plate in Fig. 13.5 is the 40 V equipotential line located?

Solution We rearrange Eq. 13.5 as

$$d = \frac{V}{E} = \frac{40\ \text{V}}{1000\ \text{V/m}} = 4\ \text{cm}$$

• • •

Equipotential lines, and thus electric field lines, can be determined experimentally using a special kind of current-conducting paper called Teledeltos® paper as shown in Fig. 13.6. The cross section of each of two bodies is painted on the surface of the paper with a metallic conducting paint, and a voltage is applied between the two painted sections. A voltmeter is used to locate all points on the paper having the same potential (voltage) by touching the voltmeter probe to the paper. All points having the same potential are then connected by drawing a smooth curve between them (1-V line in the figure). The process is then repeated to find other equipotential lines. The electric field lines can then be drawn, since (1) they go from the positive body to the negative body; (2) they enter and leave the two bodies at right angles; and (3) they always cross equipotential lines at right angles.

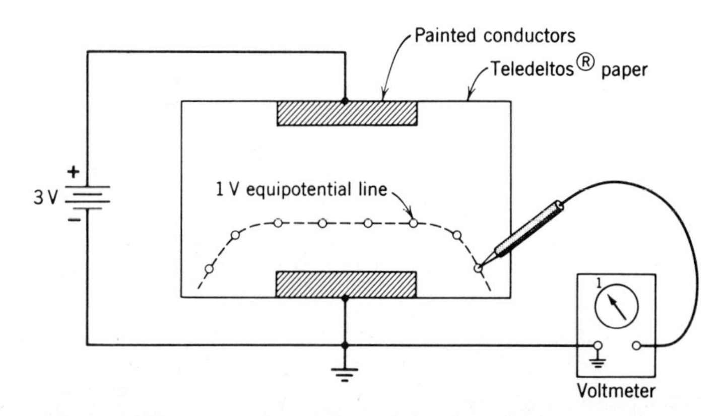

Figure 13.6 Determining equipotential lines

13.2 CAPACITANCE

Capacitance is a measure of the ability of a body to *store* electric charge. Thus a *capacitor* is a circuit element designed specifically to store electric charge. A simple *parallel plate* capacitor is shown in Fig. 13.7. For a given applied voltage V, there will be a certain charge Q stored on the plates. If the applied voltage is doubled to 2 V, the total stored charge will also double to 2 Q. Thus the ratio of stored charge Q to applied voltage V is constant, that is,

$$\frac{Q}{V} = \frac{2Q}{2V} = \text{constant}$$

The ratio of stored charge to applied voltage is defined as capacitance C:

$$C \triangleq \frac{Q}{V} \tag{13.7}$$

When an applied voltage of one volt will store a charge of one C in a capacitor, the capacitance is said to be one *farad F*, named in honor of Michael Faraday. Thus

$$1\ \text{F (farad)} = \frac{1\ \text{C}}{1\ \text{V}}$$

One F is an extremely large unit. More practical units are microfarads μF or picofarads pF.

If we combine Eqs. 13.1 and 13.2, we can obtain

$$Q = \psi = \mathbf{D}A$$

From this relationship, along with Eq. 13.6, we can obtain from the definition of capacitance,

$$C = \frac{Q}{V} = \frac{\mathbf{D}A}{\mathbf{E}d} = \frac{\mathbf{D}}{\mathbf{E}}\frac{A}{d} \tag{13.8}$$

The quantity $\mathbf{D}/\mathbf{E}$ has special significance, and it is defined as the *permittivity* of a material (similar to the permeability $\mu = \mathbf{B}/\mathbf{H}$ in magnetics), and the symbol ε (Greek letter epsilon) is used. Thus

$$\text{permittivity} = \varepsilon \triangleq \frac{\mathbf{D}}{\mathbf{E}} \tag{13.9}$$

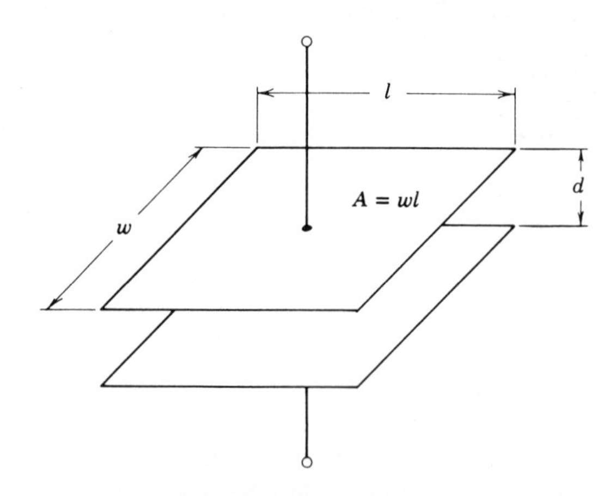

Figure 13.7 A parallel plate capacitor

Permittivity is a constant for any material, and is actually equal to

$$\varepsilon = k\varepsilon_0$$

where ε_0 is the permittivity of free space $= 8.85 \times 10^{-12}$ MKS

k is the dielectric constant given in tables

Some typical dielectric constants for different materials are given in Table 13.1.

Table 13.1.

Dielectric	Dielectric Constant k (typical)
Free space	1.0000
Air	1.0006
Teflon	2.1
Paper	2.5
Rubber	3.0
Transformer oil	4.0
Mica	5.0
Glass	6.0
Porcelain	6.0
Barium-strontium titanate (ceramic)	7500

Equation 13.8 can now be rewritten as

$$C\ (\mathrm{F}) = \varepsilon \frac{A}{d} = \frac{k\varepsilon_0 A}{d} \tag{13.10}$$

We can now specialize this equation for the parallel plate capacitor in Fig. 13.7 as

$$C\ (\mathrm{pF}) = \frac{8.85\ klw}{d} = \frac{8.85\ kA}{d} \tag{13.11}$$

where l, w, and d are in meters, and A is in square meters.

• • •

Example 13.5 Calculate the capacitance of the parallel plate capacitor in Fig. 13.7 if $l = 3$ cm, $w = 10$ cm, $d = 0.1$ cm and the plates are separated by air.

Solution Since the plates are separated by air, $k \approx 1.0$, and we use Eq. 13.11,

$$C\ (\mathrm{pF}) = \frac{8.85 \times 1.0 \times 3 \times 10^{-2}\ \mathrm{m} \times 10^{-1}\ \mathrm{m}}{10^{-3}\ \mathrm{m}} = 26.5\ \mathrm{pF}$$

• • •

Example 13.6 What is the capacitance in Example 13.5 if the two plates are separated by a piece of mica 0.1 cm thick?

Solution The only quantity from the previous solution which changes is the dielectric constant k. It changes from 1.0 to 5.0, and thus the capacitance is increased by this factor. Thus,

$$C\ (\text{pF}) = 26.5\ \text{pF} \times 5.0 = 132.5\ \text{pF}$$

• • •

Equation 13.10 can also be specialized for parallel plate capacitors where the dimensions are measured in *inches*, and A is in square inches. Thus

$$C\ (\text{pF}) = \frac{0.225\ klw}{d} = \frac{0.225\ kA}{d} \tag{13.12}$$

where l, w, and d are measured in inches.

• • •

Example 13.7 A capacitor is constructed by gluing a one-inch square piece of aluminum foil on either side of a piece of paper 3 mils thick. What is the capacitance?

Solution Using Eq. 13.12,

$$C\ (\text{pF}) = \frac{0.225 \times 2.5 \times 1\ \text{in.}^2}{3 \times 10^{-3}\ \text{in.}} = 0.188 \times 10^3\ \text{pF} = 188\ \text{pF}$$

• • •

Example 13.8 A disk capacitor consists of aluminum plating on both sides of a barium-strontium titanate disk 0.5 in. in diameter and 5 mils thick. What is the capacitance?

Solution Using Eq. 13.12,

$$C\ (\text{pF}) = \frac{0.225 \times 7500 \times \dfrac{\pi \times 1/2 \times 1/2}{4}\ \text{in.}^2}{5 \times 10^{-3}\ \text{in.}} = 6.63 \times 10^4\ \text{pF} = 0.0663\ \mu\text{F}$$

• • •

13.3 DIELECTRICS

An examination of Eq. 13.10 shows that the capacitance of a parallel plate capacitor can be increased by placing an insulating material between the plates, since C is directly proportional to the dielectric constant k. Any insulating material used in this way is commonly referred to as a *dielectric*.

To see how a dielectric can increase capacitance, consider the capacitor shown in Fig. 13.8. In Fig. 13.8*a*, there is no applied voltage and no stored charge, and the atoms in the dielectric are unaffected. However, when a voltage is applied across the capacitor and a charge stored on the plates, the *electron orbits* of the atoms in the dielectric are *distorted* slightly, as shown for a single atom in Fig. 13.8*b*. The distortion of the atomic orbits in this fashion is called *polarization*. As a result of polarization, there is *less* opposition to the establishment of an electric field between the capacitor plates, and thus the total capacitance is increased.

When the voltage source is removed from the capacitor in Fig. 13.8*b* and the capacitor terminals are shorted, all of the stored charge is removed from the capacitor plates and the electrons will resume their normal orbits. In certain materials, however, some of the electrons will remain in distorted orbits and thus some stored charge will remain on the capacitor plates. This phenomenon is called *dielectric absorption*, and is analogous to the residual magnetism in a ferromagnetic material.

If the applied voltage V_S in Fig. 13.8*b* is increased, the electron orbits will be

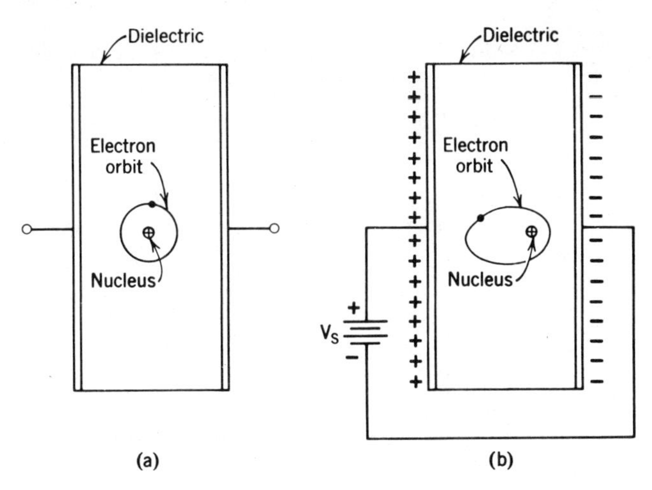

Figure 13.8 Polarization of a dielectric between capacitor plates (*a*) No stored charge; normal electron orbit (*b*) Stored charge on plates; distorted electron orbit

distorted even further, and at some value of applied voltage, the electric field intensity in the dielectric will be great enough to *tear* electrons from their orbits. At this value of applied voltage, the dielectric will act as a conductor and the capacitor will be *shorted.* The maximum electric field intensity which can be applied to any dielectric before shorting would occur is called the *dielectric strength* of the material. Some typical values are given in Table 13.2.

Table 13.2. Typical Dielectric Strengths.

Dielectric	Dielectric strength (V/mil)
Air	75
Barium-strontium titanate	75
Porcelain	200
Bakelite	400
Transformer oil	400
Paper	500
Teflon	1500
Glass	3000
Mica	5000

The maximum voltage which can safely be applied across a capacitor depends upon the *type* of dielectric and the *thickness* of the dielectric.

• • •

Example 13.9 A small disc capacitor with a mica dielectric is shown in Fig. 13.9. What is the maximum voltage which can be applied across the capacitor?

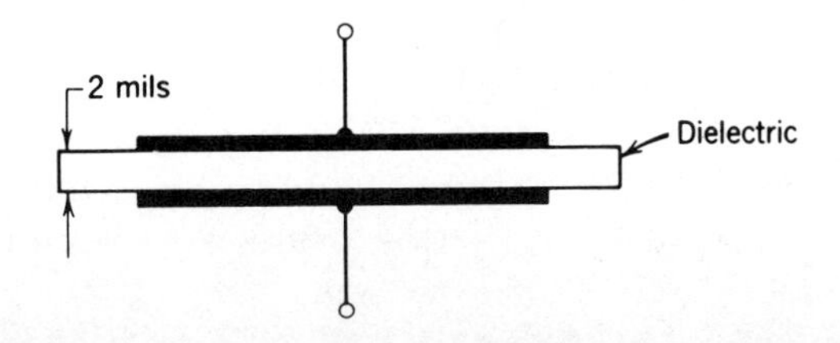

Figure 13.9

Solution From Table 13.2, the dielectric strength of mica is 5000 V/mil. Since the dielectric is 2 mils thick, it will break down at

$$V_{\text{break down}} = 5000 \text{ V/mil} \times 2 \text{ mil} = 10{,}000 \text{ V} = 10 \text{ kV}$$

• • •

Example 13.10 What would be the maximum voltage for the capacitor in Example 13.9 if the dielectric were changed to barium-strontium titanate?

Solution The dielectric strength given for barium-strontium titanate in Table 13.2 is 75 V/mil. Thus

$$V_{\text{break down}} = 75 \text{ V/mil} \times 2 \text{ mil} = 150 \text{ V}$$

• • •

13.4 CAPACITORS

There are many types of capacitors, each constructed differently using different materials. One common type is the *paper* capacitor, made by rolling two long strips of aluminium foil and two long strips of insulating paper into a roll as shown in Fig. 13.10*a*. This roll is then enclosed in a cylindrical waxed cardboard or plastic container as shown in Fig. 13.10*b*. Paper capacitors have typical values from 0.001 μF to 1.0 μF and may have voltage ratings between 100 V to 1500 V. The capacitance of a paper capacitor is sometimes given using color code bands as shown in Fig. 13.10*b*. The color code is exactly the same as the resistor color code. It is important to note that the capacitance value is *always* given in pF when using the color code (Appendix C).

Another common type capacitor is made by stacking alternate strips of metallic foil and mica sheets as shown in Fig. 13.11*a*. The sheets are then enclosed in a bakelite plastic case as shown in Fig. 13.11*b*. Mica capacitors have

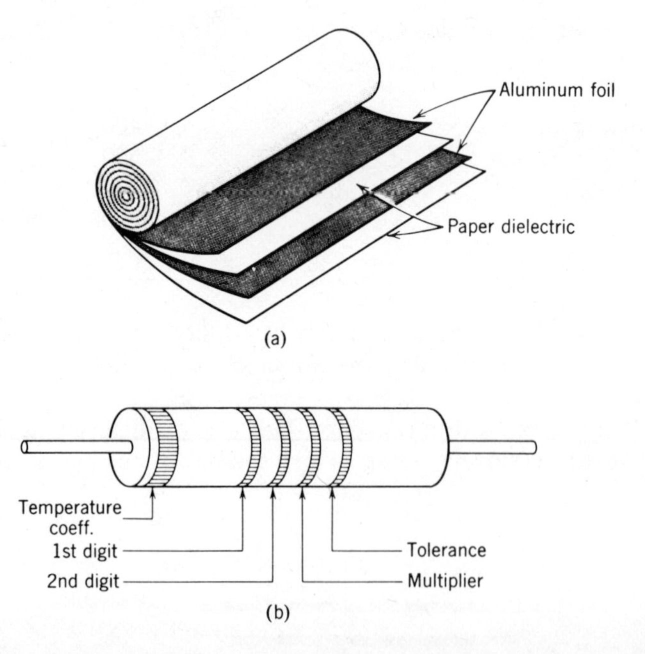

Figure 13.10. Paper capacitors (*a*) Construction (*b*) Typical paper capacitor

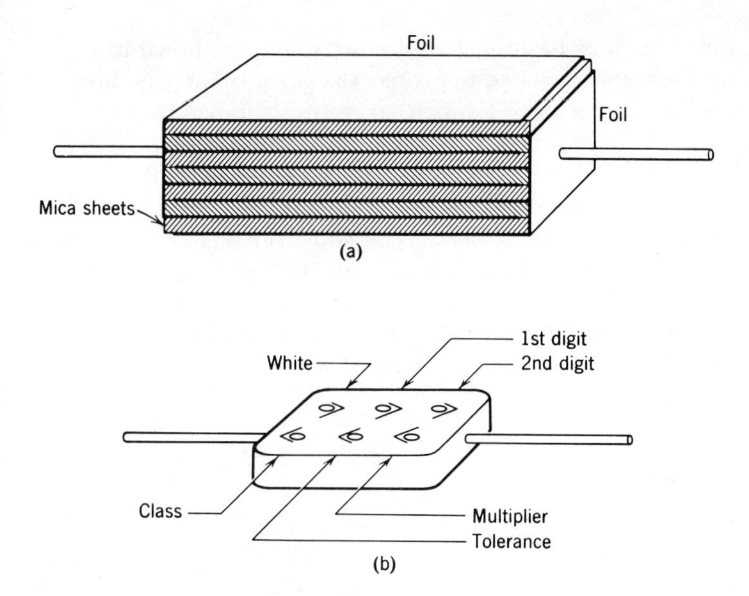

Figure 13.11 Mica capacitor (*a*) Construction (*b*) Typical mica capacitor

typical values between 5 pF and 5000 pF, and have very high voltage ratings—up to 20,000 V. Mica capacitors are usually color coded using the 6-dot system (Fig. 13.11*b*) with the white dot in the upper left corner signifying the EIA coding.* The color code is the same as for resistors, and the coded capacitance value is always given in pF.

Ceramic *disk* and ceramic tubular capacitors are made by silver plating both

*Appendix C.

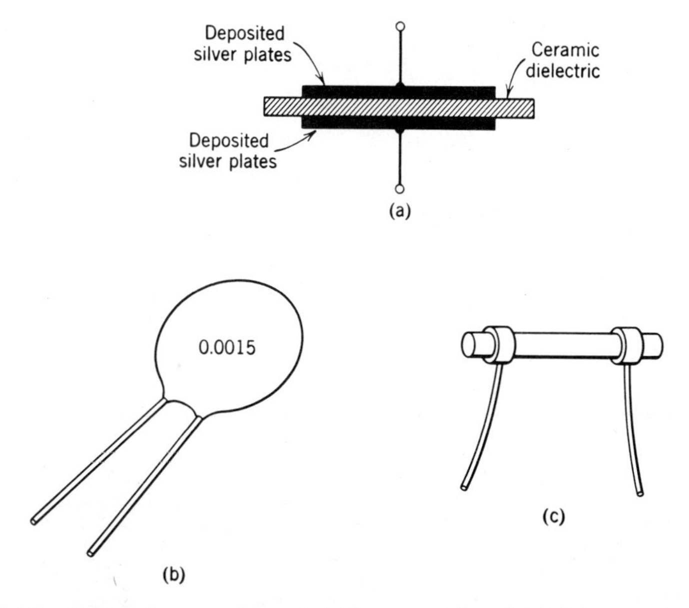

Figure 13.12 Ceramic capacitors (*a*) Construction (*b*) Disk type (*c*) Tubular type

sides of a ceramic form (such as barium-strontium titanate) as shown in Fig. 13.12*a*. The forms are then encapsulated to protect the silver plating as shown in Fig. 13.12*b*. It is possible to achieve a much larger capacitance in a much smaller volume since the dielectric constant of barium-strontium titanate is so large (Table 13.1). Thus ceramic capacitors may have values from 0.5 pF to 0.1 μF and offer very good voltage ratings—up to 20,000 V.

Another common capacitor is the *variable* type shown in Fig. 13.13. The large one is used in radio receivers; the smaller is used as a *trimmer* capacitor, that is, it is placed in parallel with another capacitor so that the total capacitance can be varied slightly. Variable capacitors may have values between 5 pF to 500 pF with breakdown voltages of a few hundred volts.

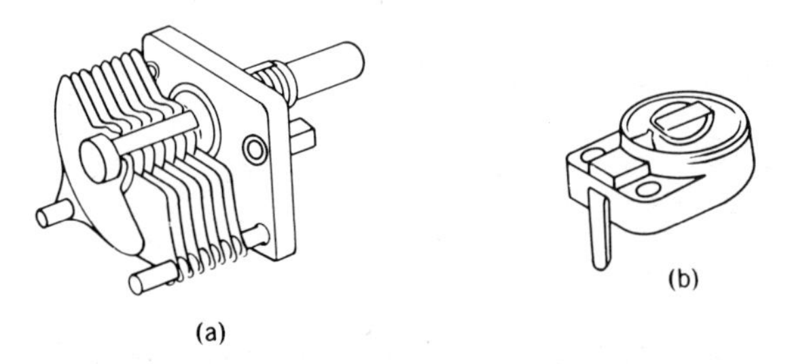

Figure 13.13 Variable capacitors (*a*) Air variable (meshed plate) (*b*) Trimmer

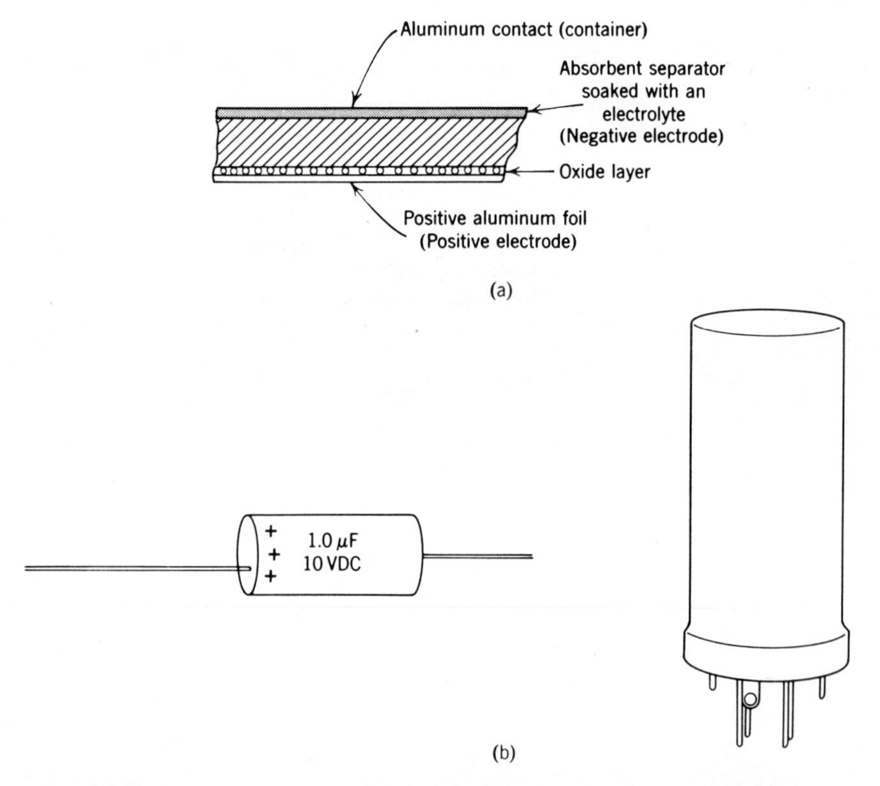

Figure 13.14 Electrolytic capacitors (*a*) Construction (*b*) Typical electrolytic capacitors

An important type of capacitor is the *electrolytic* capacitor, which is formed by separating two aluminum strips with an absorbent material (such as gauze) containing an electrolyte as shown in Fig. 13.14*a*. When a voltage of the proper polarity is applied between the aluminum strips, a thin *aluminum oxide* layer is formed on the positive strip. Aluminum oxide is an excellent insulator, and thus a capacitor is formed. An aluminum strip and the absorbent separator are usually rolled and placed in a cylindrical aluminum can (similar to a rolled paper capacitor), with either a liquid or a paste electrolyte. The foil strip is thus the positive electrode and the aluminum can is the negative electrode. It is most important to *always* use electrolytic capacitors with the correct voltage polarity, since the opposite polarity will decompose the oxide film, and the capacitor will then behave as a conductor. Furthermore, current in an electrolytic capacitor caused by using the wrong polarity will generate gases within the can and the capacitor may burst (explode!). The *capacitance value, polarity,* and voltage ratings are always shown on electrolytic capacitors as seen in Fig. 13.14*b*.

Electrolytic capacitors are important since it is possible to achieve very large capacitance values in fairly small volumes. Electrolytic capacitors may have values ranging from 0.005 μF up to several hundred μF. The voltage ratings may however be low, ranging from 6 V up to a few hundred volts.

The tantalum capacitor is an electrolytic capacitor using tantalum electrodes instead of aluminum. Even greater capacitance values per unit volume are available in tantalum capacitors.

• • •

Example 13.11 What is the value of the mica capacitor shown in Fig. 13.15?

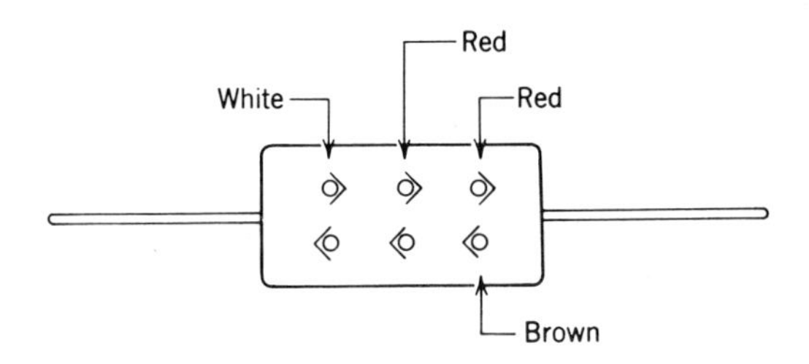

Figure 13.15

Solution The interpretation of the dots are given in Fig. 13.11. Using the standard color code table, the capacitance is seen to be

1st digit	Red $\rightarrow$ 2
2nd digit	Red $\rightarrow$ 2
Multiplier	Brown $\rightarrow \times 10^1$

therefore 220 pF.

• • •

The proper schematic symbols for capacitors are shown in Fig. 13.16. The symbol in Fig. 13.16*a* is used for any capacitor which has no polarity requirements, while the symbol in Fig. 13.16*b* is used for electrolytic capacitors where the proper polarity must be observed. The symbol in Fig. 13.16*c* is used to designate variable capacitors such as meshed plate or trimmer types.

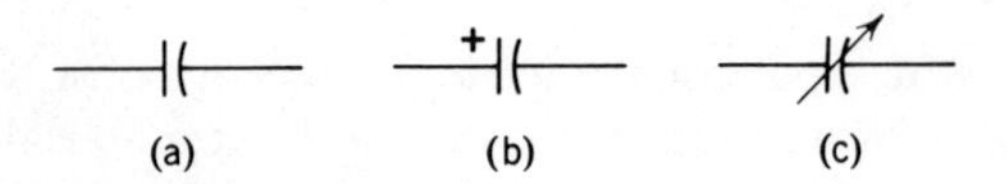

Figure 13.16 Capacitor schematic symbols

Since capacitors must necessarily be used in conjunction with some *insulator* (either the dielectric or the insulation for mounting the plates) and since we know there is no *perfect* insulator, the true equivalent circuit of a capacitor must appear as shown in Fig. 13.17. The resistor R_{leak} represents any imperfect insulator associated with the capacitor. In most cases, R_{leak} is much greater than the other resistances and it can be ignored entirely. We then have the equivalent circuit of a *perfect* capacitor as the symbols in Fig. 13.16 would indicate.

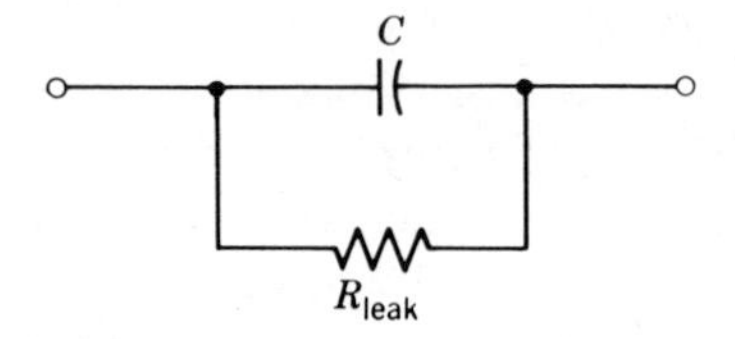

Figure 13.17 Capacitor equivalent circuit

• • •

Example 13.12 Is there anything wrong with the circuit shown in Fig. 13.18?

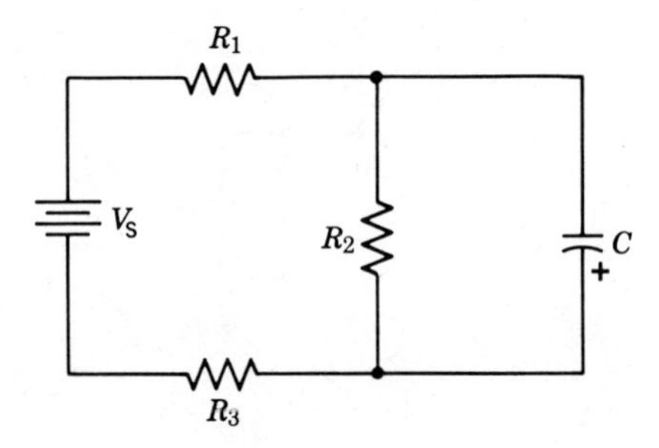

Figure 13.18

Solution The capacitor polarity is incorrect. The capacitor must be reversed or it may be destroyed.

• • •

13.5 CAPACITORS IN SERIES AND PARALLEL

Consider now the connection of a capacitor across a voltage source as shown in Fig. 13.19. When the capacitor is first connected current will flow in the circuit, but after a period of time the charge stored on the capacitor plates will

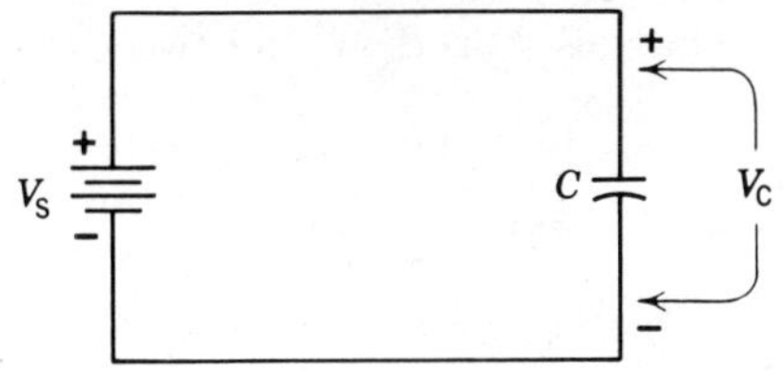

Figure 13.19

increase until the capacitor voltage V_C is equal to the source voltage V_S. At this point, the current in the circuit will be zero. Thus a steady dc current cannot exist through a capacitor. The initial charging current is a transient current whose value changes with time (the topic of the next section) and it is therefore not a steady dc current. Applying KVL to this circuit yields

$$V_S = V_C$$

and thus the charge stored on the capacitor can be found by rearranging Eq. 13.7:

$$Q = CV_C = CV_S \tag{13.13}$$

• • •

Example 13.13 What is the charge stored on the capacitor in Fig. 13.19 if $V_S = 10$ V and $C = 0.1\ \mu$F?

Solution Using the relationship just developed,

$$Q = CV_S = 0.1\ \mu\text{F} \times 10\ \text{V} = 1\ \mu\text{C}$$

• • •

In Fig. 13.20, two capacitors are connected in *parallel* across the voltage source V_S. Clearly the voltage is exactly the same across both capacitors. Thus

$$V_{C_1} = V_{C_2} = V_S$$

Now, the total charge Q_T delivered to the capacitors by the source must be

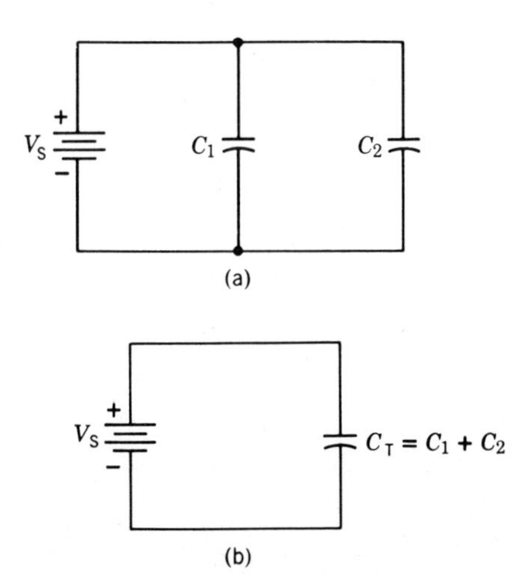

Figure 13.20 Capacitors in parallel (*a*) Original circuit (*b*) Equivalent circuit

equal to the sum of the charges stored on the two capacitors. That is,

$$Q_T = Q_1 + Q_2$$

But, from Eq. 13.13, $Q = CV$, and therefore

$$C_T V_S = C_1 V_{C_1} + C_2 V_{C_2}$$

where C_T is the total equivalent capacitance connected across the source. But we know $V_S = V_{C_1} = V_{C_2}$, and thus

$$C_T = C_1 + C_2$$

Therefore, we see that the total capacitance of two capacitors connected in *parallel* is simply the *sum* of the two capacitances. This is reasonable, since connecting them in parallel is equivalent to increasing the area of the plates, and we know that capacitance is *directly* proportional to the area of the plates. This result can be extended for any number of parallel connected capacitors:

$$\text{(for capacitors in parallel)}, \; C_T = C_1 + C_2 + \cdots + C_N \tag{13.14}$$

• • •

Example 13.14 What is the equivalent circuit of a 1 μF, a 2 μF, and a 0.2 μF capacitor connected in parallel?

Solution The equivalent circuit is simply a capacitor having a value equal to the sum of the three, or

$$C_T = C_1 + C_2 + C_3 = 1\ \mu\text{F} + 2\ \mu\text{F} + 0.2\ \mu\text{F} = 3.2\ \mu\text{F}$$

• • •

Two capacitors C_1 and C_2 are connected in *series* with a voltage source V_S in Fig. 13.21. Applying KVL to this circuit yields

$$V_S = V_{C_1} + V_{C_2}$$

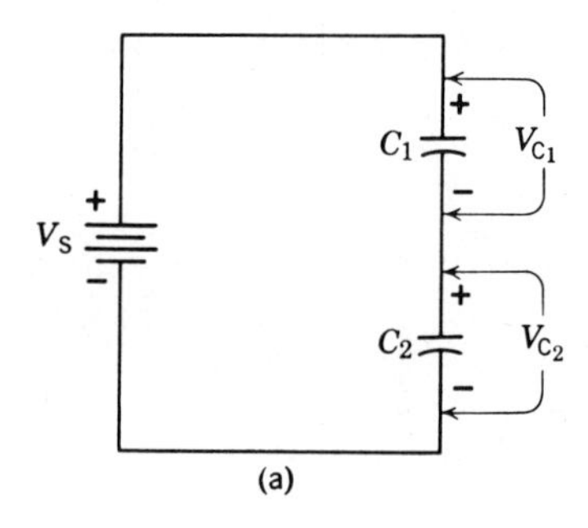

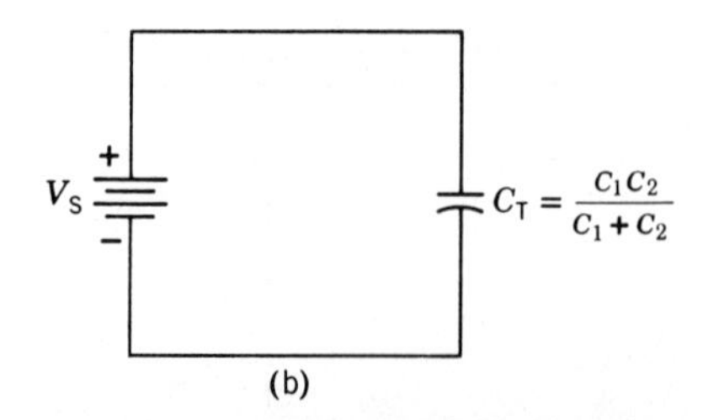

Figure 13.21 Capacitors in series (*a*) Original circuit (*b*) Equivalent circuit

Rearranging Eq. 13.7, we see that $V = Q/C$, and thus

$$V_S = \frac{Q_T}{C_T} \qquad V_{C_1} = \frac{Q_1}{C_1} \qquad V_{C_2} = \frac{Q_2}{C_2}$$

where C_T is the total capacitance connected across the source. Substituting these relationships into the KVL equation,

$$\frac{Q_T}{C_T} = \frac{Q_1}{C_1} + \frac{Q_2}{C_2}$$

Now, this is a *series* circuit and the current must be the same in every element; as a result, the total charge stored on each capacitor must be exactly the same, and this must equal the total charge delivered by the voltage source. That is

$$Q_T = Q_1 = Q_2$$

Therefore, the above equation becomes

$$\frac{Q_T}{C_T} = \frac{Q_T}{C_1} + \frac{Q_T}{C_2}$$

Dividing through by Q_T yields

$$\frac{1}{C_T} = \frac{1}{C_1} + \frac{1}{C_2}, \qquad \text{or} \qquad C_T = \frac{1}{\frac{1}{C_1} + \frac{1}{C_2}} \tag{13.15}$$

which gives the total capacitance of two capacitors in *series*. This equation can be rewritten in a more convenient form as

$$\text{(for two capacitors in series), } C_T = \frac{C_1 C_2}{C_1 + C_2} \tag{13.16}$$

• • •

Example 13.15 What is the value of C_T in Fig. 13.21*b* if $C_1 = 0.1\ \mu F$ and $C_2 = 0.2\ \mu F$?

Solution Using Eq. 13.16,

$$C_T = \frac{C_1 C_2}{C_1 + C_2} = \frac{0.1\ \mu F \times 0.2\ \mu F}{0.1\ \mu F + 0.2\ \mu F} = 0.0667\ \mu F$$

• • •

Equation 13.15 can be extended to cover any member of capacitors connected in series as follows:

$$\text{(capacitors in series), } C_T = \frac{1}{\frac{1}{C_1} + \frac{1}{C_2} + \cdots + \frac{1}{C_N}} \tag{13.17}$$

We make the important observation that capacitors in ***parallel*** combine like resistors in *series*, and capacitors in *series* combine like resistors in ***parallel***.

It is an interesting problem to determine the voltages across the two capacitors in Fig. 13.21*a*. We begin by noting that the charge must be the same on both capacitors:

$$Q_T = Q_1 = Q_2$$

We know from Eq. 13.7 that $Q = CV$, and thus

$$C_T V_S = C_1 V_{C_1} = C_2 V_{C_2}$$

Taking these equations two at a time,

$$V_{C_1} = V_S \frac{C_T}{C_1} \qquad V_{C_2} = V_S \frac{C_T}{C_2} \tag{13.18}$$

Notice the similarity between these equations and the *voltage divider* theorem for resistors.

• • •

Example 13.16 What are the voltages across the capacitors in Example 13.15 if $V_S = 20$ V?

Solution Using Eq. 13.18,

$$V_1 = V_S \frac{C_T}{C_1} = 20\ \text{V} \times \frac{0.0667\ \mu\text{F}}{0.1\ \mu\text{F}} = 13.3\ \text{V}$$

$$V_2 = V_S \frac{C_T}{C_2} = 20\ \text{V} \times \frac{0.0667\ \mu\text{F}}{0.2\ \mu\text{F}} = 6.7\ \text{V}$$

As a check, KVL gives

$$V_S = V_{C_1} + V_{C_2}$$

$$20\ \text{V} = 13.3\ \text{V} + 6.7\ \text{V} = 20\ \text{V}$$

or

$$C_1 V_{C_1} = C_2 V_{C_2}$$

$$0.1\ \mu\text{F} \times 13.34\ \text{V} = 0.2\ \mu\text{F} \times 6.7\ \text{V}$$

• • •

13.6 CHARGING A CAPACITOR

In Fig. 13.22 a switch is used to connect a capacitor C across the terminals of a voltage source V_S. We assume that initially the capacitor is completely *discharged* (it has no charge stored on its places). When the switch is closed, the capacitor voltage v_C will attempt to *rise* to V_S; this will take a short period of time, however, since charge must be moved between the capacitor and the source before a voltage can be developed across the capacitor. As a result, the initial capacitor voltage is zero, and the source voltage equals the voltage drop across both the source resistance and the resistance of the connecting wires. These resistances are very small, and thus the initial *current* surgc to chargc the capacitor will be very large (perhaps thousands of amperes for a short time). This imposes an extremely severe condition on the source, and consequently capacitors are almost always charged through a resistor R, as shown in Fig. 13.23.

If the capacitor C in Fig. 13.23 is initially discharged, a current will flow in the circuit for a period of time *after* the switch is closed. Eventually the

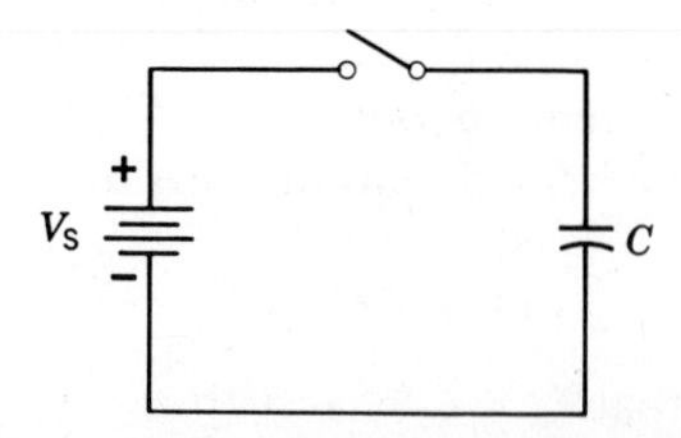

Figure 13.22

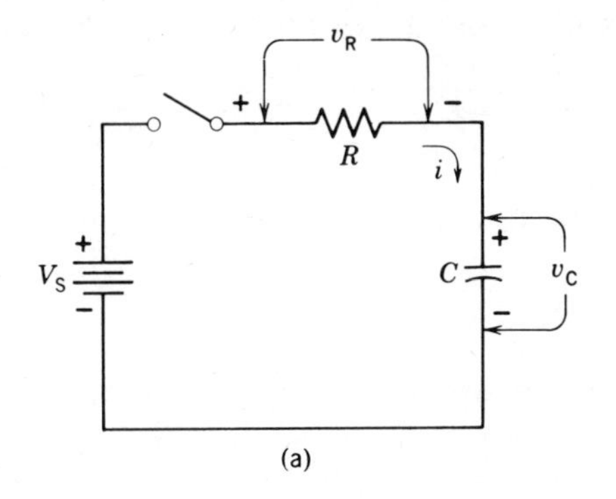

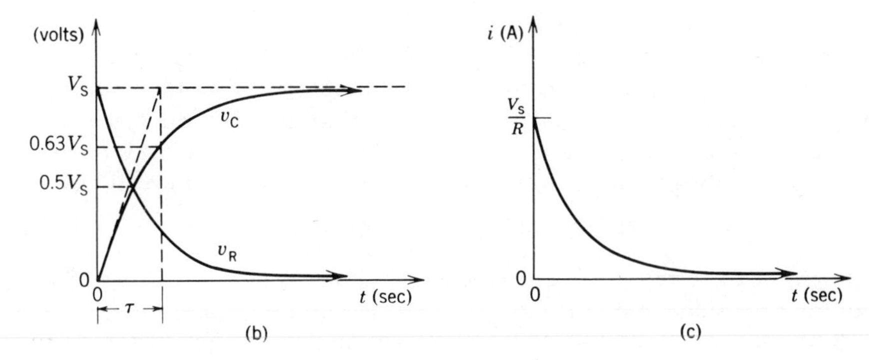

Figure 13.23 (*a*) A series *R-C* circuit (*b*) Element voltages (*c*) Circuit current

capacitor voltage will equal V_S (when the capacitor is completely charged) and the current will then be zero. Thus, this is a transient circuit, and we would like to determine the nature of the circuit current and the element voltages during the transient period.

Applying KVL to this circuit, we obtain

$$V_S = v_R + v_C$$

From Ohm's Law, we know that the voltage across the resistor must be

$$v_R = iR$$

and thus

$$V_S = iR + v_C$$

We now substitute the v-i relationship for a capacitor given at the beginning of this chapter into this equation and obtain

$$V_S = iR + v_C = \left(C\frac{dv_C}{dt}\right) R + v_C = RC\frac{dv_C}{dt} + v_C \tag{13.19}$$

This is again a differential equation and the solution for the voltage across the capacitor v_C is easily found using calculus:

$$v_C = V_S[1 - e^{-(t/RC)}] \tag{13.20}$$

Since KVL for this circuit is

$$V_S = v_R + v_C$$

we can easily find the voltage across the resistor v_R using Eq. 13.20 and KVL. Thus

$$v_R = V_S - v_C = V_S - V_S[1 - e^{-(t/RC)}] = V_S - V_S + V_S e^{-(t/RC)} \qquad (13.21)$$
$$= V_S e^{-(t/RC)}$$

We can now use Ohm's Law to determine the circuit current i as

$$i = \frac{v_R}{R} = \frac{V_S}{R} e^{-(t/RC)} \qquad (13.22)$$

Because of the dual nature between inductance and capacitance, we would expect these equations to be exact duals to the series R-L equations derived in the last chapter. The reader should compare all six equations in detail.

The element voltages v_R and v_C along with the circuit current i for the series R-C circuit in Fig. 13.23*a* are given in Figs. 13.23*b* and *c*. The curves are all exponential in shape just as the series R-L curves plotted in Chapter 12.

If the capacitor voltage were to continue to charge to its final value (as shown by the dashed line in Fig. 13.23*b*), the time required to reach V_S is defined as the R-C time constant τ. The value of the time constant is equal to the reciprocal of the coefficient of t in the exponent of e in Eq. 13.20. Thus

$$RC \text{ time constant} = \tau = RC \text{ (sec)} \qquad (13.23)$$

As might be expected, the value of the capacitor voltage v_C will be 63.2 per cent of the final value in one R-C time constant, and the transient period will have ended in 5τ sec; that is, $v_C \cong V_S$; $v_R \cong 0$; and $i \cong 0$ after 5τ sec.

If we substitute the value $t = 0$ into Eq. 13.20 (i.e., the switch has just been closed), we obtain

$$v_C = V_S[1 - e^{-(0/RC)}] = V_S(1 - 1) = 0 \text{ V}$$

and the capacitor voltage does indeed begin at zero as it should.

Now, we put $t \to \infty$ in Eq. 13.20, and

$$v_C = V_S[1 - e^{-(\infty/RC)}] = V_S(1 - 0) = V_S$$

and the final value of v_C is V_S as it should be.

Similarly, v_R and i will begin at maximum values of V_S and V_S/R for $t = 0$, and end at zero as t becomes large.

• • •

Example 13.17 Calculate and plot the element voltages and circuit current for Fig. 13.23*a* if $V_S = 10$ V, $R = 1$ kΩ, and $C = 2\ \mu$F. What is the time constant? What is the length of the transient period?

Solution We must use Eqs. 13.20, 13.21, and 13.22. The value of the time constant τ is

$$\tau = RC = 10^3\ \Omega \times 2 \times 10^{-6}\ \text{F} = 2 \times 10^{-3} \text{ sec}$$

The proper equations are then

$$v_C = 10[1 - e^{-[t/(2\times 10^{-3})]}] = 10[1 - e^{-5\times 10^2 t}]$$
$$v_R = 10e^{-5\times 10^2 t}$$
$$i = 10^{-2} e^{-5\times 10^2 t}$$

The transient period will end in 5τ sec or $5 \times 2 \times 10^{-3} = 10$ msec. The proper curves are shown in Fig. 13.24.

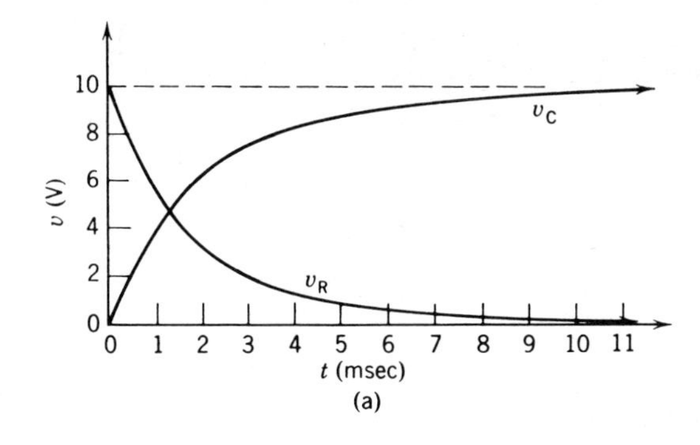

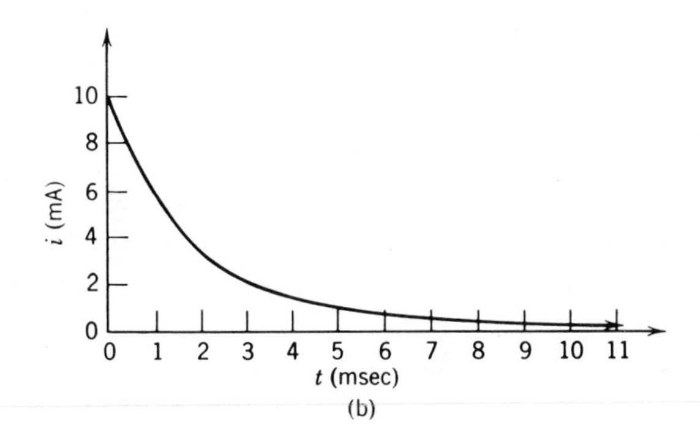

Figure 13.24 (*a*) Element voltages (*b*) Circuit current

• • •

Example 13.18 Write the equation for the voltage across the capacitor after the switch closes in Fig. 13.25*a*. What is the length of the transient period?

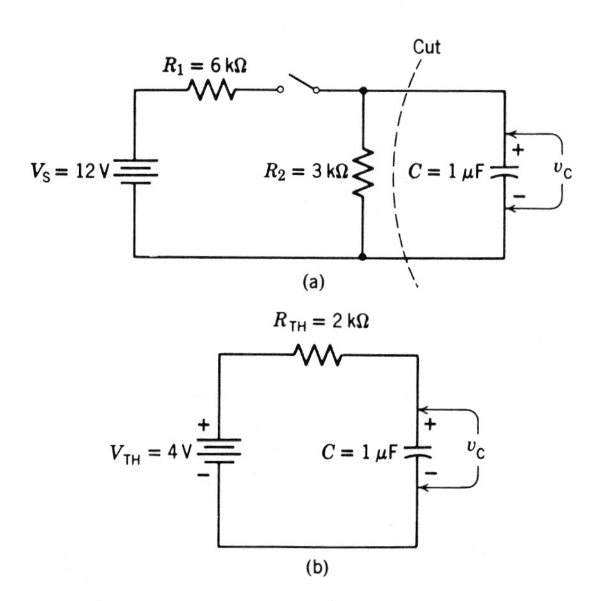

Figure 13.25 (*a*) Original circuit (*b*) Equivalent circuit

Solution This problem is best handled by finding the Thévenin equivalent of the two resistors and the source as shown in Fig. 13.25*b*.

The time constant of the equivalent circuit is

$$\tau = R_{TH}C = 2 \times 10^3\ \Omega \times 10^{-6}\ \text{F} = 2 \times 10^{-3}\ \text{sec}$$

Thus the transient period will end in 5τ sec, or

$$5\tau = 5 \times 2 \times 10^{-3} = 10\ \text{msec}$$

The equation for v_C is (Eq. 13.20),

$$v_C = V_{TH}[1 - e^{-(t/R_{TH}C)}] = 4[1 - e^{-5\times10^2 t}]$$

• • •

13.7 DISCHARGING A CAPACITOR

In Fig. 13.26, the capacitor is initially charged to a voltage of V_0 volts. When the switch is closed, current will flow through R until the capacitor is completely discharged. We can find the equations for the element voltages and the discharge current by setting Eq. 13.19 equal to zero; this corresponds to the KVL equation for this circuit. Thus,

$$0 = RC\frac{dv_C}{dt} + v_C \tag{13.24}$$

Solving this differential equation for v_C yields

$$v_C = V_0 e^{-(t/RC)} \tag{13.25}$$

Since the voltage across the resistor v_R must equal the capacitor voltage in this case, we have

$$v_R = v_C = V_0 e^{-(t/RC)} \tag{13.26}$$

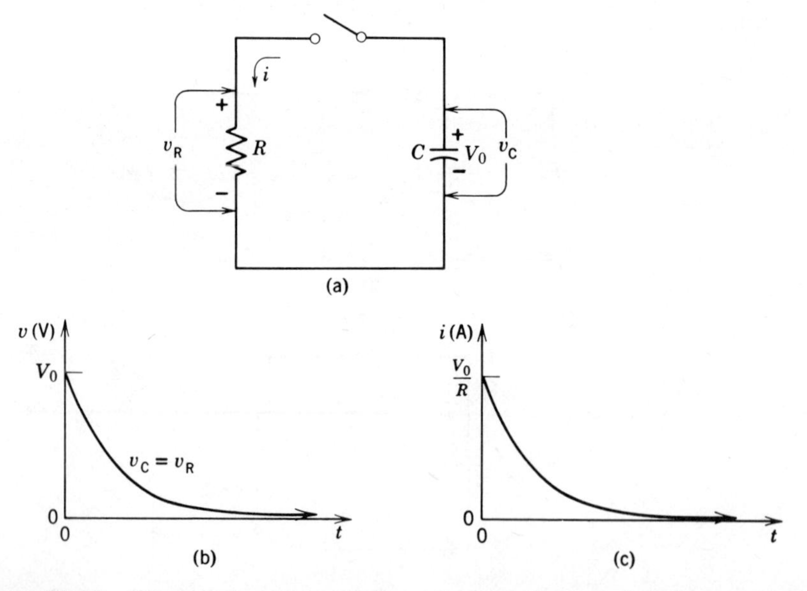

Figure 13.26 Discharging a capacitor (*a*) Discharge circuit (*b*) Element voltages (*c*) Discharge current

The current in found using Ohm's Law as

$$i = \frac{V_0}{R} e^{-(t/RC)} \qquad (13.27)$$

The discharge time constant is again the reciprocal of the coefficient of t in Eq. 13.25 and thus

$$\text{discharge time constant} = \tau = RC \text{ (sec)} \qquad (13.28)$$

The capacitor voltage v_C will discharge to 36.8 per cent of V_0 (down 63.2 per cent) in *one* time constant, and the transient period will essentially be over after *five* time constants. Thus,

$$\text{transient period} \approx 5\tau \text{ (sec)}$$

• • •

Example 13.19 Write the equations for v_R, v_C, and i in Fig. 13.26*a* if $V_0 = 15$ V, $C = 5\ \mu$F, and $R = 100\ \Omega$. What is the length of the transient period?

Solution The discharge time constant τ is

$$\tau = RC = 100\ \Omega \times 5 \times 10^{-6}\ \text{F} = 5 \times 10^{-4}\ \text{sec}$$

The transient period lasts for 5τ sec or $5 \times 5 \times 10^{-4} = 2.5$ msec. From Eqs. 13.25 and 13.26,

$$v_C = v_R = V_0 e^{-(t/RC)} = 15e^{-2\times 10^3 t}$$

From Eq. 13.27

$$i = \frac{V_0}{R} e^{-t/RC} = 0.15e^{-2\times 10^3 t}$$

• • •

Example 13.20 How long will it take the capacitor voltage v_C in Example 13.19 to decay to 5 V?

Solution The equation for v_C is

$$v_C = 15e^{-2\times 10^3 t}$$

Setting $v_C = 5$ V,

$$5 = 15e^{-2\times 10^3 t}$$

Thus,

$$e^{-2\times 10^3 t} = 0.333$$

But $e^{-1.1} = 0.333$, and thus

$$1.1 = 2 \times 10^3 t$$

Therefore,

$$t = \frac{1.1}{2 \times 10^3} = 0.55 \text{ msec}$$

• • •

Example 13.21 The switch in Fig. 13.25*a* has been closed a long time. To what voltage will the capacitor charge? What is the discharge time constant when the switch is opened? What is the length of the transient period? Write the equation for v_C during discharge.

Solution An inspection of the equivalent circuit in Fig. 13.25*b* shows that the capacitor will charge to a voltage of 4 V. When the switch is opened, the circuit is simply R_2 in series with C. Thus the discharge time constant τ is

$$\tau = R_2C = 3 \times 10^3\ \Omega \times 10^{-6}\ F = 3 \times 10^{-3}\ \text{sec}$$

The transient period will then last for 5τ sec, or

$$5\tau = 5 \times 3 \times 10^{-3} = 15\ \text{msec}$$

The equation for V_C is found using Eq. 13.25,

$$v_C = 4e^{-333t}$$

• • •

13.8 ENERGY STORED BY A CAPACITOR

When an ideal capacitor is connected to voltage source, the source must expend energy to charge the capacitor. Since there is no resistance associated with an ideal capacitor, there is no power dissipated and thus energy must be stored in the capacitor. Indeed, just as energy is stored in the *magnetic* field around an inductor, so is energy stored in the *electric* field associated with a capacitor. Again, the derivation of the equation giving the stored energy requires calculus, but the results are

$$W\ (\text{J}) = \tfrac{1}{2}CV_0^2 \tag{13.29}$$

where C is in farads, and V_0 is the total voltage change in volts.

• • •

Example 13.22 How much energy is stored in the capacitor in Fig. 13.25*a* during charging?

Solution Since the capacitor was initially discharged, the total voltage change is 4 V; thus V_0 in Eq. 13.29 is 4 V. The energy stored is then

$$W\ (\text{J}) = \tfrac{1}{2}CV_0^2 = \tfrac{1}{2} \times 10^{-6}\ \text{F} \times (4\ \text{V})^2 = 8 \times 10^{-6}\ \text{J}$$

When the capacitor in Fig. 13.25 is discharged by opening the switch, the energy delivered to C by the source is then dissipated as heat in R_2. From the previous example, the total energy dissipated in R_2 must be 8×10^{-6} J.

• • •

SUMMARY

We have now discussed the three basic properties of electric circuits, resistance, inductance, and capacitance. Just as conductance is the dual of resistance, so capacitance is the dual of inductance. An inductance is always associated with a *magnetic* field, and a capacitance is always associated with an *electric* field. An electric field exists in any space in which a force is exerted on a stationary test charge. An electric field has an orientation, and the field lines begin and end on bodies having a potential difference between them.

Capacitance is a measure of the ability to store electric charge, and any circuit element designed for this purpose is called a capacitor. The capacitance of parallel plate capacitors can be greatly increased by using a dielectric material between the plates. Capacitors in parallel combine like resistors in series; capacitors in series combine like resistors in parallel.

The *voltage* across a capacitor cannot change instantly. Thus opening or closing switches in circuits containing capacitors produces *transient* currents and voltages. These transient currents and voltages are *exponential* in shape, and the time constant of a series R-C circuit is simply RC. For all practical purposes, the transient period is complete after 5τ sec.

GLOSSARY

Capacitance C. *The ability to store electric charge.*
Dielectric. *An insulating material.*
Dielectric absorption. *The retention of charge on a dielectric by the alteration of electron orbits.*
Dielectric constant $k = \varepsilon / \varepsilon_0$. *The ratio of the permittivity of a substance to the permittivity of free space.*
Dielectric strength. *The electric field intensity which will cause breakdown in a dielectric.*
Electric field. *Exists in any space in which a force is exerted on a stationary charge.*
Electric field intensity E. *The strength of an electric field in volts/meter.*
Electric flux ψ. *Electric field lines.*
Electric flux density D. *The electric flux per unit area.*
Equipotential line. *An imaginary line in an electric field connecting all points having the same potential.*
Permittivity $\varepsilon = k\varepsilon_0$. *A measure of the ease of establishing an electric field in a dielectric.*
Polarization. *The distortion of the electron orbits in a dielectric by an electric field.*
Time constant τ **= RC.** *The time required for the transient current and voltages in a series* **R-C** *circuit to reach 63.2% of their final values.*
Transient period. *By agreement, a period of 5τ sec.*

IMPORTANT RELATIONSHIPS

- Capacitor v-i relationship $i = C\dfrac{dv}{dt}$

$$\psi = Q \qquad \mathbf{D} = \frac{\psi}{A}$$

$$\mathbf{F} = k_1 \frac{Q_1 Q_2}{d^2} \qquad \mathbf{E} = \frac{\mathbf{F}}{Q} \qquad E = \frac{V}{d}$$

$$C = \frac{Q}{V} \qquad C = \varepsilon \frac{A}{d} \qquad \varepsilon = k\varepsilon_0 \qquad \varepsilon_0 = 8.85 \times 10^{-12}$$

- Capacitors in parallel

$$C_T = C_1 + C_2 + \cdots + C_N$$

Capacitors in series

$$C_T = \frac{1}{\dfrac{1}{C_1} + \dfrac{1}{C_2} + \cdots + \dfrac{1}{C_N}}$$

- Charging a capacitor

$$v_C = V_S[1 - e^{-(t/RC)}]$$

Discharging a capacitor

$$v_C = V_0 e^{-(t/RC)}$$

$$\text{time constant} = \tau = RC$$

$$W = \tfrac{1}{2}CV_0^2$$

REVIEW QUESTIONS

1. The three basic properties of an electric circuit are ____, ____, and ____.
2. A dual equation can be found by replacing i's with v's, v's with i's, R's with G's, and L's with C's. (TF)
3. Electric field lines form closed loops. (TF)
4. Electric field lines enter and exit bodies at right angles (usually, never, always).
5. The total electric flux leaving a sphere must be equal to the total ____ contained within the sphere.
6. The force exerted between two charged bodies is (directly, inversely) proportional to the square of the distance between the bodies.
7. Equipotential lines are (parallel, perpendicular) to electric field lines.
8. Equipotential lines form closed loops. (TF)
9. The dielectric constant for air is very nearly ____.
10. The capacitance of two parallel plates will be (increased, decreased) by inserting a dielectric between the plate.
11. Polarization means stripping electrons from the atoms in a dielectric. (TF)
12. A residual charge stored on a dielectric is called ____ ____.
13. The breakdown voltage of a capacitor is dependent upon the ____ ____ of the dielectric between the plates.
14. ____ capacitors must always be used with the correct voltage polarity.
15. Mica capacitors can be constructed with very high (voltage ratings, capacitance values, both).
16. When capacitance values are given using color coding, the capacitance is always given in ____.
17. A *real* capacitor always has some leakage ____.
18. Capacitors in parallel are combined like resistors in (series, parallel).
19. Capacitors in series are combined like resistors in (series, parallel).
20. The voltage across a capacitor (can, cannot) change instantly.
21. A current through a capacitor can exist only if the capacitor voltage is ____.
22. The time constant of a series R-C circuit is the product of ____ and ____.
23. The transient period in a series R-C circuit is essentially complete after ____ sec.
24. The energy from a source used to charge a capacitor is stored in the ____ ____ of the capacitor.
25. The voltage across a capacitor in a series R-C circuit will decay to ____ % of its initial value in one time constant.

ROBLEMS

1. Draw the electric field lines between the two bodies in Fig. 13.27*a*.

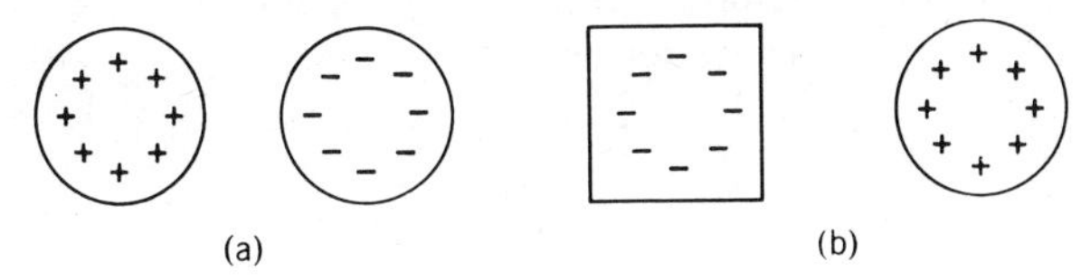

Figure 13.27

2. Draw the electric field lines between the two bodies in Fig. 13.27*b*.
3. What is the electric field intensity between the plates in Fig. 13.4 if the source voltage is 20 V?
4. What is the electric field intensity between two parallel plates separated by a distance of 0.2 cm if the voltage across the plates is 5 V?
5. What is the force exerted on an electron placed half-way between the plates in Problem 4?
6. Draw three of the equipotential lines in Problem 1.
7. Draw five of the equipotential lines in Problem 2.
8. What do the equipotential lines in Fig. 13.3 look like? Draw them.
9. Draw three of the equipotential lines in Fig. 13.4. (Draw as in Fig. 13.5).
10. In Fig. 13.7, $w = 2$ cm, $l = 1$ cm, $d = 0.1$ cm, and the plates are separated by air. Calculate the capacitance.
11. A piece of glass is placed between the plates in Problem 10. Calculate the capacitance.
12. A capacitor is constructed by gluing two pieces of aluminum foil 2 inches square on either side of a piece of mica 2 mils thick. What is the capacitance?
13. What is the capacitance in Problem 12 if paper is used instead of mica?
14. What must be the area of each plate in Problem 12 if a capacitance of 450 pF is desired?
15. One farad is a very large capacitance! What must be the area of each plate in Problem 12 to achieve a capacitance of 1 F?
16. What is the maximum voltage which can be applied to the capacitor in Problem 12?
17. What is the maximum voltage which can be applied to the capacitor in Problem 10?
18. An electrolytic capacitor is to be connected in parallel with R_2 in Fig. 13.28. Show the proper polarity.

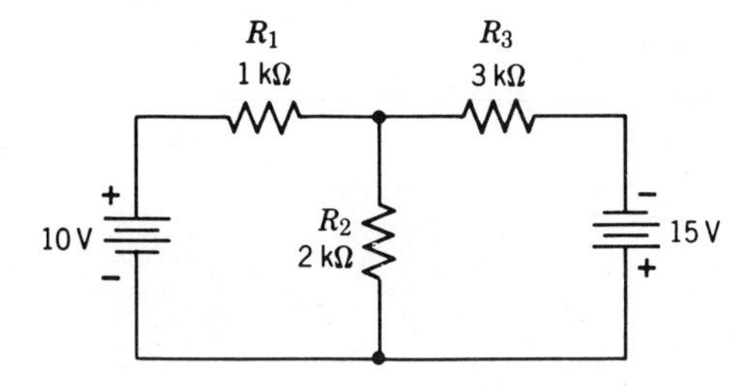

Figure 13.28

19. What will be the total charge stored on the capacitor in Problem 18 if the electrolytic capacitor has a value of 2 μF? What must be the minimum voltage rating of the capacitor?
20. What is the total capacitance of a 0.1-μF capacitor and a 0.5-μF capacitor connected in parallel? In series?
21. What is the total capacitance of the network in Fig. 13.29?

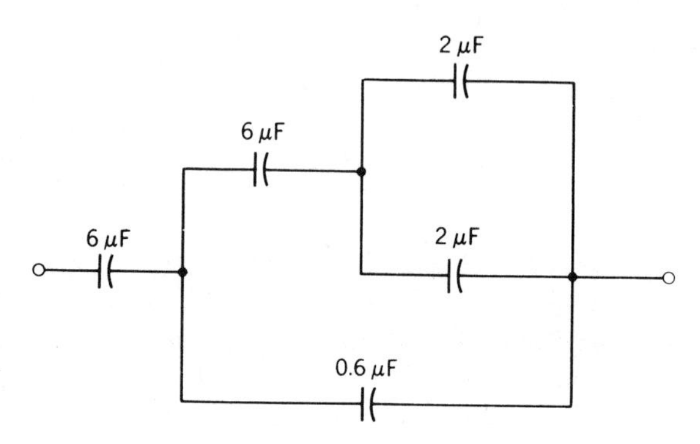

Figure 13.29

22. Find the voltages across the capacitors in Fig. 13.21*a* if $V_S = 28$ V, $C_1 = 0.1\ \mu$F, and $C_2 = 0.3\ \mu$F.
23. In Fig. 13.21*a*, $V_{C_1} = 15$ V, $V_{C_2} = 20$ V, and $C_1 = 400$ pF. What must be the value of C_2?
24. Write the equations for v_C, v_R, and i in Fig. 13.23 if $V_S = 15$ V, $R = 500\ \Omega$, and $C = 4\ \mu$F. Calculate enough points to make a plot of each.
25. What is the time constant in Problem 24? What is the length of the transient period? How long will it take v_C to reach 13 V?
26. What will be the value of i in Fig. 13.23 two time constants after the switch is closed if $V_S = 6$ V, $R = 1$ kΩ, and $C = 0.1\ \mu$F?
27. If the capacitor in Fig. 13.26*a* is initially charged to 45 V, how long will it require to discharge to 16.5 V if $R = 470\ \Omega$ and $C = 100\ \mu$F? How long will it take to discharge completely?
28. Write the equations for v_C, v_R, and i in Problem 27.
29. If v_C must not discharge more than 10% in a period of 1 sec after the switch is opened in Fig. 13.25*a*, what must be the minimum value of C? How could you improve the circuit for this purpose?
30. Find the dual of Eq. 13.20 by replacing i's with v's, v's with i's, R's with G's, and C's with L's, and compare it to the equation for increasing current in an inductance, Eq. 12.17.
31. A 5-μF capacitor is charged from zero to 15 V. What is the energy stored in the capacitor?

14
ac—ALTERNATING CURRENT

In the first half of this book we devoted our attention to circuits in which the voltages and currents did not change with time. Such circuits are called *direct current* or dc circuits. In the previous two chapters, we discussed circuits in which the voltages and currents vary for a *short period* of time, but settle to steady state or dc values after a certain *transient period.* Transient currents are present whenever switches are opened or closed in circuits containing inductance or capacitance, and such circuits are referred to as transient circuits.

In this chapter, we will extend our knowledge to cover circuits in which the currents and voltages *vary continually* with time. Specifically, we will study circuits in which the currents, and thus the voltages, vary *sinusoidally.* The amplitude of a sinusoidal current (or voltage) changes continually and, in addition, the *direction* (or *polarity*) also changes *alternately* from one direction to the opposite direction. Circuits in which the currents and voltages vary sinusoidally are called *alternating current* or ac circuits.

The study of ac circuits will extend our knowledge considerably, since any *periodic* wave form (a wave which *repeats* itself periodically), such as the *square wave,* the *triangular wave,* or the *sawtooth wave* shown in Chapter 1, can be analyzed using sinusoidal waves.* Furthermore, KVL, KCL, and *all* of the circuit theorems we have developed will apply equally well to ac circuits.

*Fourier analysis.

14.1 PERIODIC FUNCTIONS

Any wave which *repeats itself periodically* is said to be a *periodic function.* The triangular wave in Fig. 14.1*a* crosses the horizontal axis with a positive slope at point 0, and reaches a maximum positive value at *a.* It then changes to a negative slope, crosses the axis at *b*, and reaches a maximum negative value at *c*. The slope then changes back to a positive value, and it crosses the axis at *d*. From point *d* to point *h*, the wave *repeats exactly* the same form as between points 0 and *d*. The portion of the wave contained between points 0 and *d* (or *d* and *h*) is defined as one *cycle.* Notice that we could draw this wave with as many cycles as we like by simply drawing one cycle after another (i.e., place the cycles end to end). Thus we see that this wave repeats itself periodically (cycle after cycle) and it is therefore a periodic function.

The distance from point 0 to point *d* on the wave is defined as one *period.* This is exactly the same as the distance from *a* to *e*, or *c* to g, or from any point on one cycle to the *same* point on an adjacent cycle; for example, points *x* and *y* in Fig. 14.1*a*.

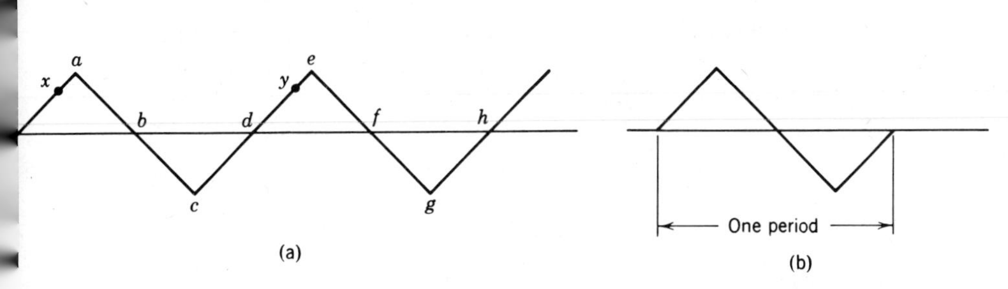

Figure 14.1 A triangular wave (*a*) Wave form (*b*) One cycle

• • •

Example 14.1 Are the waves in Fig. 14.2 periodic?

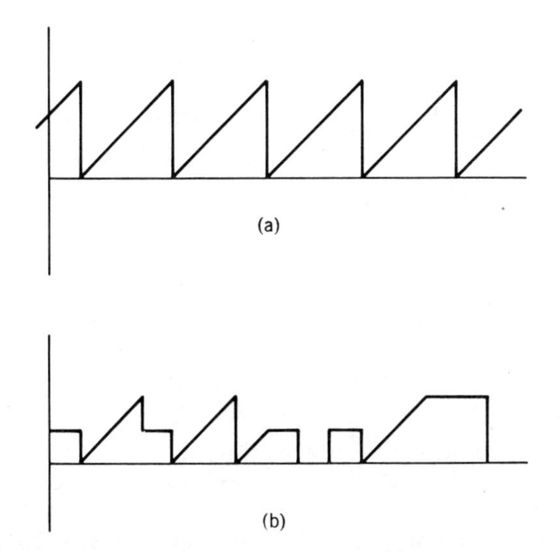

Figure 14.2

Solution Figure 14.2*a*: The wave is periodic. One cycle is a triangle, and one period is the distance between two adjacent points touching the horizontal axis. Figure 14.2*b*; the wave is not periodic, since the wave form does not repeat itself.

• • •

The trigonometric functions of an angle α are defined in terms of a right triangle as shown in Fig. 14.3. The three functions which we will be using are:

$$\sin \alpha = \frac{y}{z} \tag{14.1}$$

$$\cos \alpha = \frac{x}{z} \tag{14.2}$$

$$\tan \alpha = \frac{y}{x} \tag{14.3}$$

These three trigonometric functions are simply the ratios of the length of one side of a right triangle to the length of another side. The lengths of the three sides are of course related by the Pythagorean theorem:

$$z^2 = x^2 + y^2 \tag{14.4}$$

If the lengths x, y, and z in Fig. 14.3 are known, the trigonometric functions in Eqs. 14.1 through 14.3 can be calculated, and the angle α, in degrees, can then be determined using trigonometric tables.*

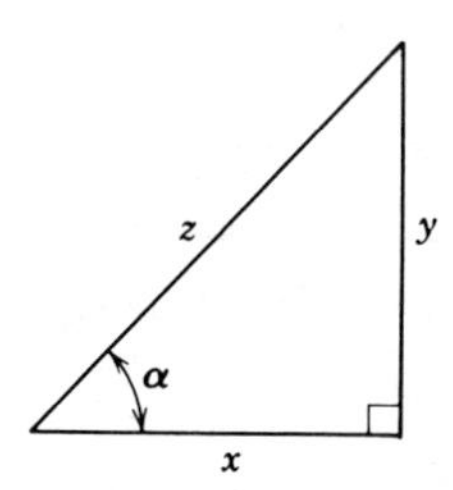

Figure 14.3

• • •

Example 14.2 Find the angle α in Fig. 14.3 when:

(1) $z = 4$ and $y = 2$; (2) $z = 6$ and $x = 3$; (3) $y = 2$ and $x = 2$.

Solution

1. Using Eq. 14.1,

$$\sin \alpha = \frac{y}{z} = \frac{2}{4} = \frac{1}{2} = 0.5 \quad \text{thus, } \alpha = 30°$$

2. Using Eq. 14.2,

$$\cos \alpha = \frac{x}{z} = \frac{3}{6} = 0.5 \quad \text{thus, } \alpha = 60°$$

*Trigonometric tables are given in Appendix B. Slide rules having the S and T scales or electronic calculators can be used in place of tables.

3. Using Eq. 14.3,

$$\tan \alpha = \frac{y}{x} = \frac{2}{2} = 1.0 \quad \text{thus, } \alpha = 45°$$

• • •

On the other hand, if the angle α in Fig. 14.3 is known, along with *one* of the sides, Eqs. 14.1 through 14.4 can be used to find the other sides.

• • •

Example 14.3 Find y in Fig. 14.3 if: (1) $\alpha = 30°$ and $z = 8$.

Solution Equation 14.1 can be rearranged as

$$y = z \sin \alpha$$

Thus,

$$y = z \sin \alpha = 8 \sin 30° = 8 \times 0.5 = 4$$

• • •

Let us now consider the relationship developed by rearranging Eq. 14.1 in the last example, namely,

$$y = z \sin \alpha \tag{14.5}$$

We see that this equation gives us the length of side y in terms of side z for any angle α we may choose. For the moment, assume $z = 1$, so that we get

$$y = \sin \alpha \tag{14.6}$$

If we were to choose a number of different values for α, determine y by looking up $\sin \alpha$, and plot the results, we would obtain the sine wave shown in Fig. 14.4*a*. The shape of this sine wave is *not* circular or elliptical; it has its own exclusive shape, and it is said to be *sinusoidal*. From our previous discussion, it is easy to see that the sine wave is periodic, and one *cycle* and one *period* are shown in Fig. 14.4*b*. Notice that the wave has a *maximum positive* value of +1 and a *maximum negative* value of −1. This maximum value, without regard to

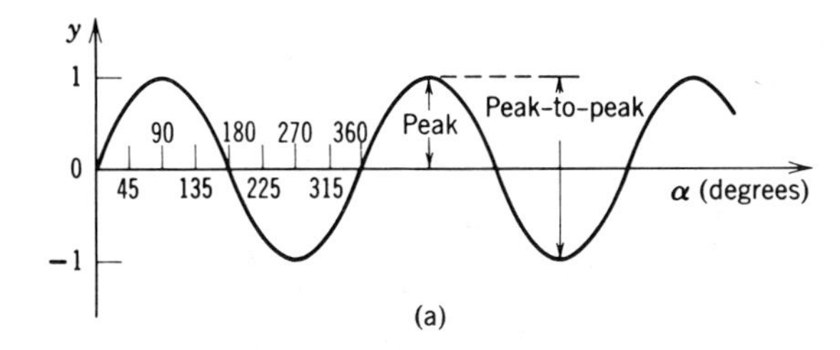

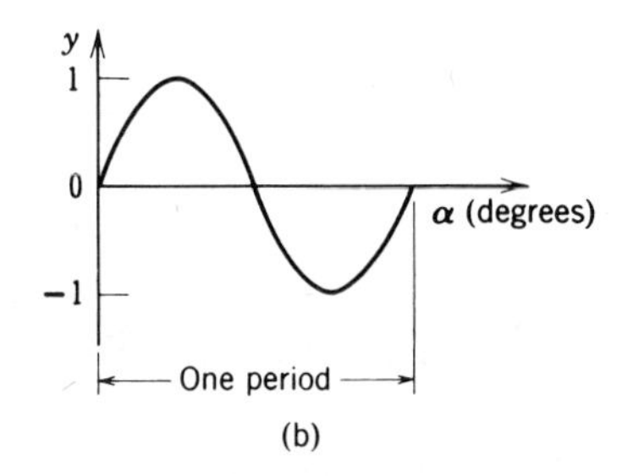

Figure 14.4 (*a*) A sine wave (*b*) One cycle of a sine wave

sign, is called the *amplitude* or *peak* value of the wave. The distance from the positive peak to the negative peak is called the peak-to-peak (P-P) value; obviously, the peak-to-peak value is simply *twice* the peak value.

Referring to Eq. 14.5, we see that the quantity z determines the peak (and thus the P-P) value of a sine wave. For this reason, Eq. 14.5 is often written

$$y = y_P \sin \alpha$$

where y_P is the *maximum* or peak value of the wave.

• • •

Example 14.4 What is the equation of a sine wave having a peak value of 25? What is the P-P value of the wave?

Solution The proper equation is

$$y = y_P \sin \alpha = 25 \sin \alpha$$

The peak value of the wave occurs when $\alpha = 90°$, or $\sin \alpha = 1$, and thus y_P must be 25. The P-P value is *twice* the peak value, and thus $y_{P\text{-}P} = 2 \times 25 = 50$.

• • •

Notice that we could rearrange Eq. 14.2 as

$$x = z \cos \alpha = x_P \cos \alpha \tag{14.7}$$

If we were to plot this equation as a function of α just as we did for the sine wave, we would obtain the curve in Fig. 14.5. Equation 14.7 is the equation of a *cosine wave,* and the curve in Fig. 14.5 is called a cosine wave. This curve has exactly the same shape as a sine wave, and thus it is said to be sinusoidal. The only difference between the two curves is that the sine wave passes through the *origin* of the coordinate system, while the cosine wave does not. We can think of a cosine wave as a sine wave which has been *shifted* 90° to the *left* along the horizontal axis. Indeed, we will show later that a cosine wave simply *leads* a sine wave by 90°.

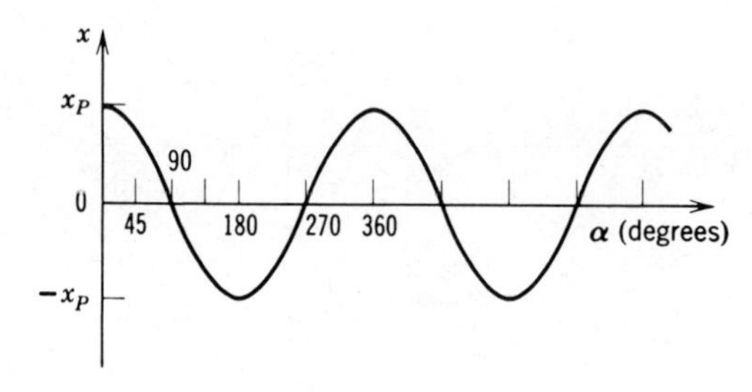

Figure 14.5 A cosine wave

• • •

Example 14.5 Write the equation of a cosine wave having a peak-to-peak value of 250.

Solution The peak value is one-half the P-P value, or 250/2 = 125. Then from Eq. 14.7

$$x = x_P \cos \alpha = 125 \cos \alpha$$

• • •

In solving right triangles, it is most convenient to measure angles in degrees. However, in electrical problems, it is more convenient to measure angles in terms of a different unit of angular measure called the *radian*. An angle of one radian is shown in Fig. 14.6, and

one radian is an angle which subtends an arc, on the circumference of a circle, which is equal to the radius of the circle, when the vertex of the angle is placed at the center of the circle.

Since the circumference of the circle in Fig. 14.6 is equal to 2π times its radius, the circumference must contain 2π arcs of length r. Therefore, there must be exactly 2π radians in the complete circle. Thus

$$2\pi \text{ radians} = 360°$$

Dividing both sides by 2π yields

$$1 \text{ radian} = \frac{360°}{2\pi} = 57.3°$$

On the other hand,

$$1° = \frac{1}{57.3} = 0.0174 \text{ radian}$$

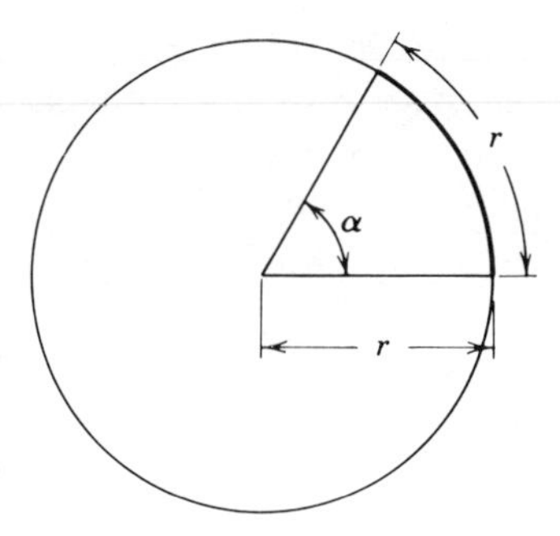

Figure 14.6 An angle α of one radian

• • •

Example 14.6 Change 60° to radians.

Solution We begin with 2π rad = 360°. Dividing by 6,

$$\frac{2\pi}{6} \text{ rad} = \frac{360°}{6} \qquad \text{or} \qquad \frac{\pi}{3} \text{ rad} = 60°$$

• • •

It is sometimes necessary to express α in Eqs. 14.5 and 14.7 in radians rather than degrees. We can still plot both equations just as before; we must however use the trigonometric tables expressing the functions in *radians* rather than degrees.* The sine and cosine curves with the angle α expressed in radians are shown in Fig. 14.7*a* and *b*, respectively. The great advantage in using this method is that the *horizontal* and *vertical* scales *both* have the same units. For example, π on the horizontal axis is 3.14 units on the vertical axis.

*Trigonometric functions for angles expressed in radians are given in Appendix B.

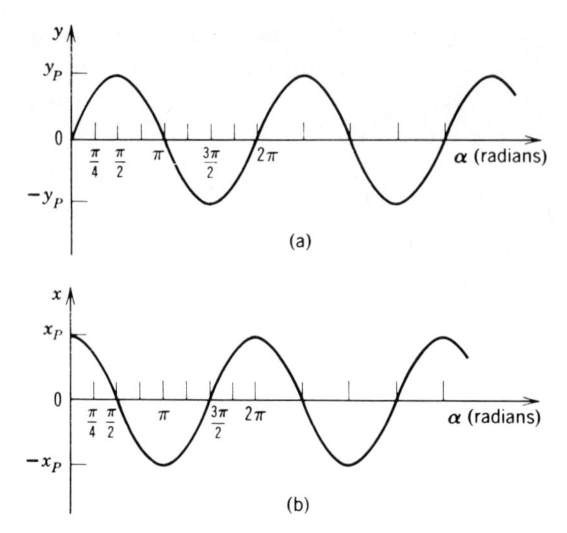

Figure 14.7 (*a*) $y = y_P \sin \alpha$ (*b*) $x = x_P \cos \alpha$

14.2 VECTORS

Suppose we place a vector with *magnitude* z and *angle* α on the x-y plane as shown in Fig. 14.8*a*. The angle α is always measured with respect to the positive x-axis, and counterclockwise is the positive direction of rotation. We will allow this vector to rotate in a counterclockwise direction at a constant rate (the small arc arrow shows this rotation).

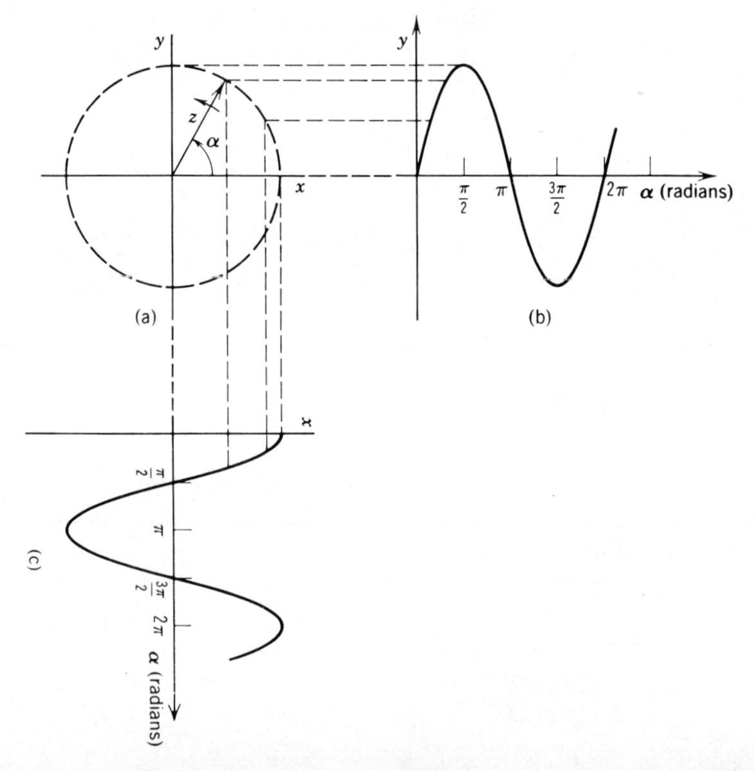

Figure 14.8 (*a*) Rotating vector (*b*) $y = z \sin \alpha$ (*c*) $x = z \cos \alpha$

From the previous section, we see that the *projection* of this vector on the y-axis is $y = z \sin \alpha$, and we can thus plot this projection as a function of α in order to generate a sine wave as shown in Fig. 14.8*b*. Similarly, the projection of this vector on the x-axis is $x = z \cos \alpha$, and we can thus plot this projection to obtain a cosine wave as in Fig. 14.8*c*.

Notice that one complete revolution of the vector will generate exactly one cycle of either a sine wave or a cosine wave. Thus if the vector is rotating at one revolution per second, it will generate sine (or cosine) waves at the rate of one cycle per second (c/s). If it rotates at a rate of two revolutions per second, it will generate waves having two cycles per second, and so on. For a sine or a cosine wave, the *number of cycles per second* is defined as the *frequency* of the wave. It is common practice to measure the frequency of a sinusoidal wave in units of cycles per second (sometimes abbreviated as cps or c/s), and a special unit has been defined for this purpose. Thus a frequency of *one cycle per second* is defined as a frequency of *one Hertz* Hz.

• • •

Example 14.7 What is the frequency of a sine wave if there is one cycle every (*a*) 0.5 sec, (*b*) 1.0 msec?

Solution (*a*) If there is one cycle in 0.5 sec, there must be two cycles every second. Therefore the frequency is 2 Hz. (*b*) If there is one cycle in 1 msec, it will require 1000 cycles to occupy a time of 1 sec. Thus the frequency is 1000 Hz, or 1 kHz.

• • •

Since we specify the frequency of a sinusoidal wave in cycles per unit *time*, it would be more convenient to plot such a wave as a function of time instead of as a function of radians as in Fig. 14.8. What we must do then is express the angle α in terms of time t. We note first of all that the vector in Fig. 14.8 rotates at a constant rate. If it rotates at one revolution per second, it covers an angle of 2π radians per second (or 360° per second). If it rotates at two revolutions per second, it covers $2 \times 2\pi$ radians per second, and so on. The total *angle* covered by the vector in a time of *one second* is the *rate at which the vector rotates*, and this is defined as the *angular velocity*. The angular velocity is a very important quantity and it is given the special symbol ω (Greek letter omega). The angular velocity in radians per second is easily found by simply multiplying the frequency by 2π. Thus

$$\text{angular velocity (rad/sec)} = \omega = 2\pi f\,(\text{Hz}) \tag{14.8}$$

We can now make use of the angular velocity of a vector to determine the angle α in Fig. 14.8. We simply multiply the angular velocity ω by time t. Thus

$$\alpha\,(\text{rad}) = \omega\,(\text{rad/sec})\,t\,(\text{sec})$$

Note carefully the proper units in this relationship. We can now substitute this relationship into the equations for the sine wave and the cosine wave to obtain

$$y = z \sin \alpha = z \sin \omega t \tag{14.9}$$

$$x = z \cos \alpha = z \cos \omega t \tag{14.10}$$

Since the vector has a *constant* angular velocity ($\omega = 2\pi f = \text{constant}$), we can now plot any sine wave or cosine wave as a function of time t by means of Eqs. 14.9 and 14.10; ωt is of course measured in radians (Fig. 14.9).

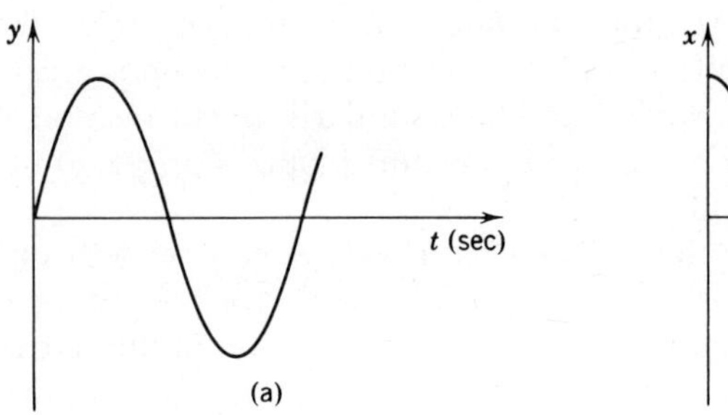

Figure 14.9 (a) $y = z \sin \omega t$ (b) $x = z \cos \omega t$

• • •

Example 14.8 Write the equation for a sine wave having a peak value of 23 and a frequency of 60 Hz.

Solution We use Eqs. 14.8 and 14.9. First,

$$\omega = 2\pi f = 2\pi \times 60 \text{ Hz} \cong 377$$

Then using Eq. 14.9

$$y = z \sin \omega t = 23 \sin 377t$$

• • •

Example 14.9 What is the frequency of the following sine wave?

$$y = 170 \sin 314t$$

Solution Rearranging Eq. 14.8,

$$f = \frac{\omega}{2\pi} = \frac{314}{2\pi} = 50 \text{ Hz}$$

• • •

Example 14.10 Plot the two waves in the previous two examples.

Solution The waves are shown in Fig. 14.10a and b.

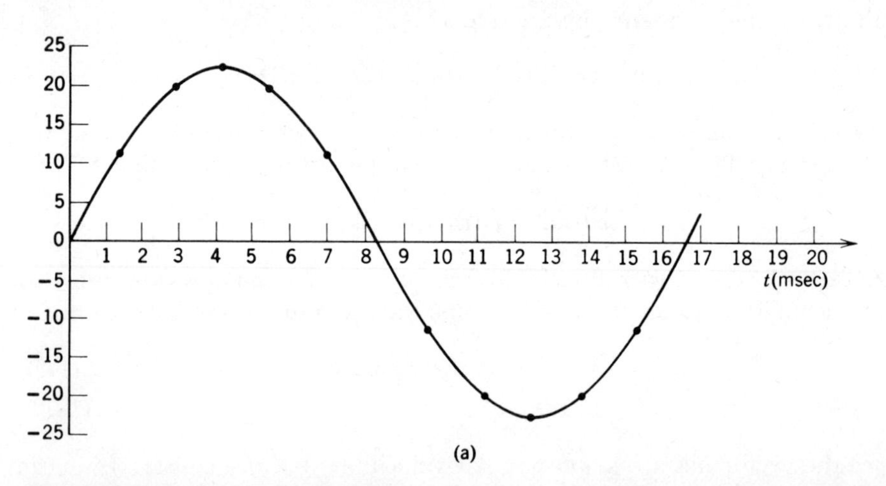

Figure 14.10 (a) $y = 23 \sin 377t$

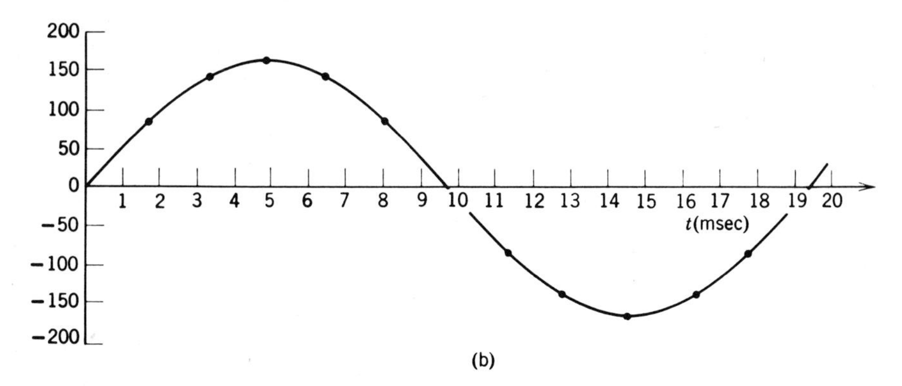

Figure 14.10 (*b*) $x = 170 \sin 314t$

• • •

4.3 ac—CURRENT AND VOLTAGE

We now turn our attention to a rotating coil placed in a magnetic field as shown in Fig. 14.11*a*. We considered this arrangement in Chapter 11, and from Faraday's Law we know that there will be an induced current in the coil as it rotates. Let us consider the current in the upper conductor in the figure, that is, the current which will pass through terminal a. We will draw a vector from the center line (axis) of the coil to the upper conductor as shown in Fig. 14.11*b*. This vector rotates at a constant angular velocity ω and the tip of the arrowhead represents the upper conductor.

At point 0 in Fig. 14.11*b*, the conductor is moving *parallel* to the magnetic field lines, and thus there is no current (remember, the conductor must "cut across" lines in order to induce a current). At point 1, the conductor has begun to "cut across" field lines and there is some induced current i. Using the right-hand-rule, we see that the induced current is coming out of terminal a. We will call this the positive direction, and plot a corresponding point on the curve $i = I_P \sin \omega t$ in the figure. At point 2, the conductor is cutting across field lines at a higher rate resulting in a larger induced current. At point 3, the conductor is cutting directly across the magnetic field lines, and thus the maximum induced current I_P occurs at this time.

As the vector moves from point 3 to point 5, the induced current becomes smaller and is finally zero at point 6. (Why?) Now, as the vector moves from point 6 to point 9, the induced current again builds up to its maximum or peak value, but the current direction is now *into* terminal a. The reader should carefully verify this using the right-hand-rule. Finally, moving from point 9 back to point 0, the induced current decreases to zero.

The current induced in the coil can be shown to be proportional to the sine of the angle of rotation of the coil, and we can therefore simply plot the tip of the vector in Fig. 14.11*b* in order to determine the form of the induced current. From the previous sections we know that this wave form is simply a sine wave, and thus the induced current is

$$i = I_P \sin \omega t \tag{14.11}$$

where I_P is the peak (maximum) value, and ωt is in radians.

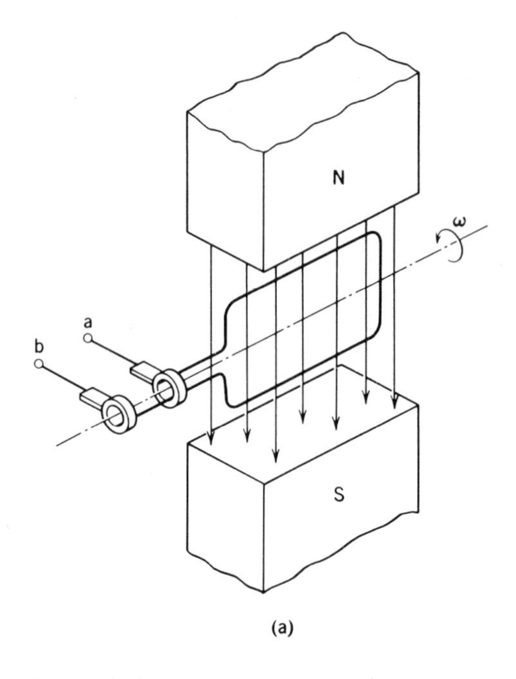

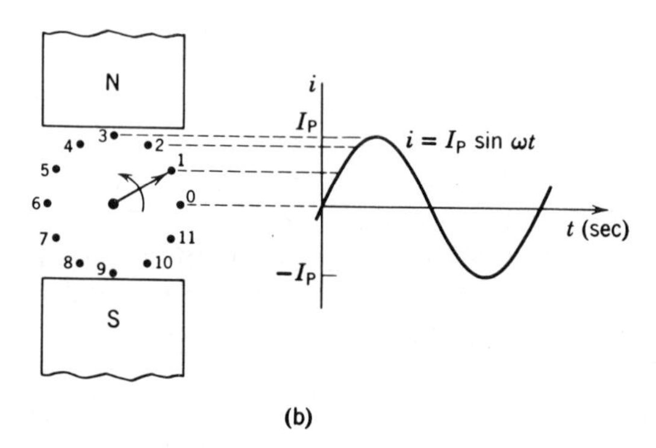

Figure 14.11 ac generator (*a*) Rotating coil in a magnetic field (*b*) Generating an alternating current

Figure 14.11 forms the basis for ac electric generators (alternators), and most of the electric power in the world today is provided in this form. At this point, the difference between dc and ac currents should be quite clear. Notice that we use the lower case letter i in Eq. 14.11 to specify an alternating current, because the amplitude *and* the direction of the current are constantly changing with time. We use the upper case letter I_P for the peak value of the current, since it is constant.

Alternators are not the only sources of alternating currents; there are also electronic circuits (oscillators) capable of producing sinusoidal currents. We will use the symbols shown in Fig. 14.12 to designate any device capable of producing sinusoidal currents and voltages respectively. We will in general call these devices ac *signal generators.*

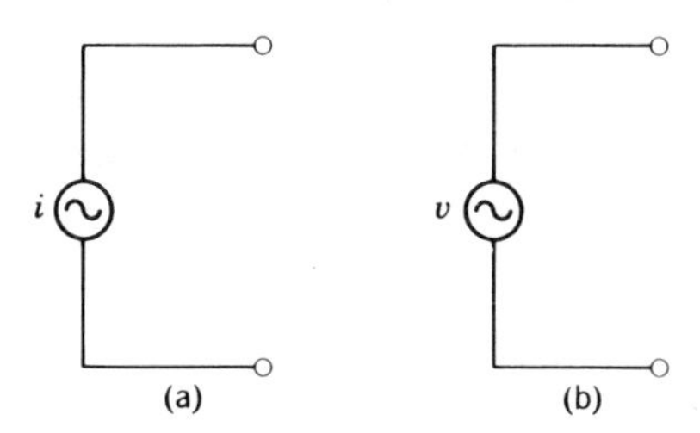

Figure 14.12 ac signal generator symbols (*a*) $i = I_P \sin \omega t$ (*b*) $v = V_P \sin \omega t$

• • •

Example 14.11 What is the peak value of current and the frequency of the ac signal generator in Fig. 14.12*a* if $i = 25 \times 10^{-3} \sin 6.28 \times 10^3 t$?

Solution The peak value is clearly $25 \times 10^{-3} = 25$ mA. We use Eq. 14.8 to find the frequency as

$$f = \frac{\omega}{2\pi} = \frac{6.28 \times 10^3}{2\pi} = 10^3 \text{ Hz} = 1 \text{ kHz}$$

• • •

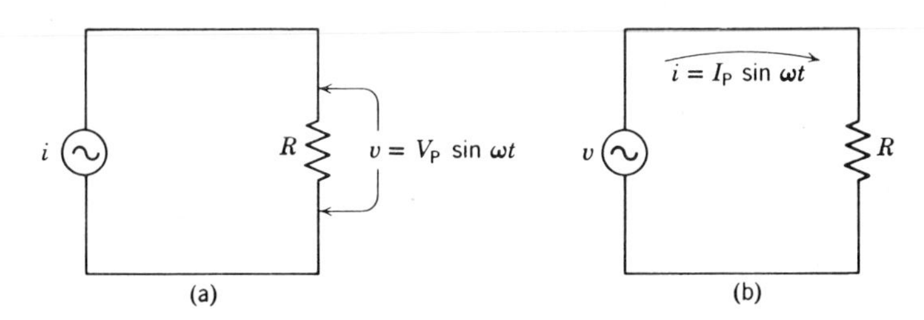

Figure 14.13 (*a*) $i = I_P \sin \omega t$ (*b*) $v = V_P \sin \omega t$

In Fig. 14.13*a*, a resistor R is connected across the terminals of an ac signal generator $i = I_P \sin \omega t$. We can apply Ohm's Law to this simple circuit by simply using *lower case* letters for i and v. Thus

$$v = iR = (I_P \sin \omega t)R = I_P R \sin \omega t$$

Since I_P and R are constant values, their product must be the peak value of the voltage across the resistor. That is,

$$v = V_P \sin \omega t, \qquad \text{where} \qquad V_P = I_P R \tag{14.12}$$

Similarly, we can apply Ohm's Law to determine the current i in the circuit in Fig. 14.13*b*. Thus

$$i = \frac{v}{R} = \frac{V_P \sin \omega t}{R} = \frac{V_P}{R} \sin \omega t$$

Clearly the peak value of the current is

$$I_P = \frac{V_P}{R}$$

and thus

$$i = I_P \sin \omega t, \qquad \text{where} \qquad I_P = \frac{V_P}{R}$$

From the above discussion, we make the very important observation that ***sinusoidal voltages produce sinusoidal currents, and vice versa, in resistors.***

• • •

Example 14.12 What is the voltage v across the resistor R in Fig. 14.13*a* if the signal generator has a peak value of 2 mA, a frequency of 500 Hz, and $R = 10\ \text{k}\Omega$?

Solution The equation for the current generator is,

$$i = I_P \sin \omega t = 2 \sin (2\pi \times 500t)\ \text{mA}$$

Using Ohm's Law, the voltage v is found to be

$$v = iR = 2 \sin (\pi \times 10^3 t)\ \text{mA} \times 10\ \text{k}\Omega$$
$$= 20 \sin (\pi \times 10^3 t)$$

• • •

Example 14.13 What is the peak value of the current in Fig. 14.13*b* if $R = 3\ \text{k}\Omega$ and the equation for the voltage generator is $v = 60 \sin(500t)$?

Solution The peak value of the current is simply the peak voltage divided by the resistance. Thus

$$I_P = \frac{V_P}{R} = \frac{60}{3\ \text{k}\Omega} = 20\ \text{mA}$$

• • •

One of the most useful methods for measuring or observing ac signals is to display them on an oscilloscope. The *vertical* scale of the oscilloscope is usually calibrated in *volts,* and the horizontal scale is calibrated in units of time (seconds, msec, etc.). Thus if we display one cycle of an ac signal on an oscilloscope (Fig. 14.14), we can determine its *peak value,* its *period,* and thus its *frequency.* We can then write an equation for the signal.

The period of a sinusoidal wave T is defined as the *time of one cycle,* as shown in Fig. 14.14, and it is measured in seconds. If the wave in Fig. 14.14

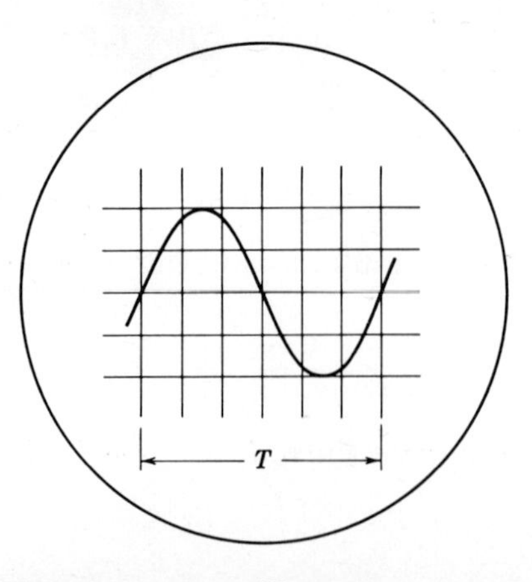

Figure 14.14 An ac signal on an oscilloscope

had a frequency of 1 Hz, there would be 1 cycle each second, and thus its period would be 1 sec. If its frequency were 2 Hz, there would be 2 cycles each second, and thus the time of *one* cycle, or the period, would be 0.5 sec. We can continue this line of reasoning, and after a little thought it will become clear that the period is simply the *reciprocal* of the frequency. Thus

$$\text{period} = T(\text{sec}) = \frac{1}{f(\text{Hz})} \tag{14.13}$$

• • •

Example 14.14 What is the period of the wave in Fig. 14.14 if each horizontal division on the oscilloscope represents 5 msec? What is its frequency?

Solution One cycle of the wave covers 6 horizontal divisions. Thus, the period is

$$T = 6 \text{ divisions} \times \frac{5 \text{ msec}}{\text{division}} = 30 \text{ msec}$$

The frequency is found using Eq. 14.13.

$$f(\text{Hz}) = \frac{1}{T(\text{sec})} = \frac{1}{30 \text{ msec}} = 33.3 \text{ Hz}$$

• • •

Example 14.15 The oscilloscope in Fig. 14.14 is calibrated as follows: vertical = 10 V/division; horizontal = 10 μsec/division. What is the equation for the wave displayed?

Solution The peak value of the wave is 2 divisions. Thus

$$V_P = 2 \text{ divisions} \times \frac{10 \text{ V}}{\text{division}} = 20 \text{ V}$$

The period is

$$T = 6 \text{ divisions} \times \frac{10\ \mu\text{sec}}{\text{division}} = 60\ \mu\text{sec}$$

The frequency is then

$$f(\text{Hz}) = \frac{1}{T(\text{sec})} = \frac{1}{60\ \mu\text{sec}} = 16.7 \text{ kHz}$$

We then find ω as

$$\omega = 2\pi f = 2\pi \times 16.7 \text{ kHz} = 105 \times 10^3 \frac{\text{rad}}{\text{sec}}$$

The voltage equation is then

$$v = V_P \sin \omega t = 20 \sin (105 \times 10^3 t)$$

• • •

PHASE—ϕ

At this point, we have the ability to represent any ac signal (voltage or current) by means of Eqs. 14.11 and 14.12. We assume of course that the signals are sinusoidal, and in fact are sine waves. The fact that these equations are sine waves means that the curves must pass through the origin ($t = 0$ sec) with a positive slope. We will however encounter sinusoidal waves which do not pass through the origin such as those shown in Fig. 14.15. Notice that these waves can be treated as sine waves which have simply been *shifted* to the right or to the left along the t-axis. It is a simple matter to alter Eqs. 14.11 and 14.12 to account for this shifting.

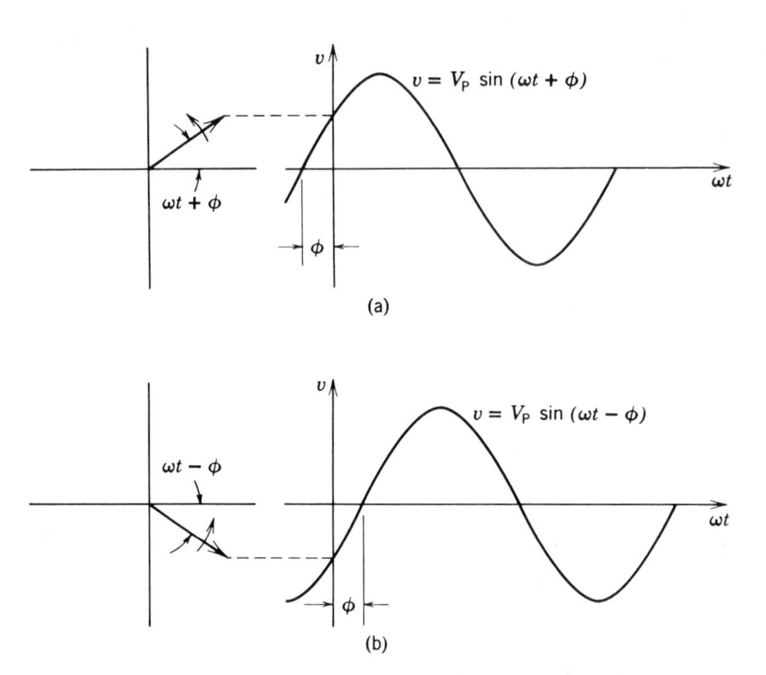

Figure 14.15 Phase angles (a) Leading (b) Lagging

Consider the vector shown in Fig. 14.15a. In order to generate a sine wave, we would ordinarily begin with the vector *on* the horizontal axis at time $t = 0$. Suppose however, we simply move the vector through an angle of ϕ radians before we begin to generate the sine wave. Thus, at time $t = 0$, the wave will have a positive value, and the sine wave has been effectively shifted to the *left*. Notice that the angle of the vector must now be expressed as $(\omega t + \phi)$, where ω is the angular velocity, t is the time, and ϕ is the angle we have shifted the wave. The angle ϕ is defined as the *phase angle*, and in this case it is said to be a *leading* phase angle since the vector is *ahead* of that of a normal sine wave. The complete equation for the wave in Fig. 14.15a is

$$v = V_P \sin (\omega t + \phi) \text{ (leading)} \tag{14.14}$$

In Fig. 14.15b, the phase angle ϕ is negative, and this is seen to shift the wave to the right. Such a negative phase angle is said to be a *lagging* phase angle since the vector is *behind* that of a normal sine wave. Thus the equation for a lagging phase angle as shown in Fig. 14.15b is

$$v = V_P \sin (\omega t - \phi) \text{ (lagging)} \tag{14.15}$$

We must take care to express the phase angle ϕ in *radians* in these equations.

• • •

Example 14.16 Write the equation for the wave in Fig. 14.15a if the peak value of the wave is 150 V, the frequency is 100 Hz, and the phase angle is 45°.

Solution The 45° phase angle must be expressed in radians. Since π radians equals 180°, we have

$$\pi \text{ rad} = 180°, \qquad \text{then} \qquad \frac{\pi}{4}\text{ rad} = \frac{180°}{4} = 45°$$

We can then use Eq. 14.14 to find

$$v = V_P \sin(\omega t + \phi) = 150 \sin\left(200\pi t + \frac{\pi}{4}\right)$$

• • •

Example 14.17 Express a cosine wave using Eq. 14.14.

Solution Examination of a cosine wave reveals that it is simply a sine wave which has been shifted 90° to the left. Thus it is equivalent to Eq. 14.14 with a phase angle of +90° (leading), or $+\pi/2$ rad. Thus

$$v = V_P \cos \omega t = V_P \sin\left(\omega t + \frac{\pi}{2}\right)$$

• • •

The results of this and the previous sections in this chapter can be summarized by noting that we can represent *any* sinusoidal current or voltage with the following two equations:

$$i = I_P \sin(\omega t \pm \phi) \tag{14.16}$$

$$v = V_P \sin(\omega t \pm \phi) \tag{14.17}$$

These two equations are *extremely* important, and the reader should be thoroughly familiar with them.

COMPLEX PLANE

Equations 14.16 and 14.17 can be used to express any sinusoidal current or voltage. However, when writing loop or node equations to analyze complex ac circuits, these equations will become more and more complicated and tedious. There is a much better way to represent ac signals, but we need to define the *complex operator j*, and the *complex plane* first.

We are all familiar with the graphical representation of numbers made by plotting along a linear scale as shown in Fig. 14.16*a*. Positive numbers are plotted along the scale to the right of zero, and negative numbers are plotted along the scale to the left of zero. Using this scale, we can plot any positive or negative number, and this is called the *real* number system.

In certain algebraic operations, we may encounter the problem of finding the square root of a negative number (in using the quadratic formula for example). This will cause difficulties since there is no real number which will give a negative number when multiplied by itself. Thus there is no such thing as the square root of a negative number in the *real* number system.

In order to account for this situation, a new number is defined such that its *square* is *negative*. Such numbers are called *imaginary* numbers, and they are designated by writing the symbol j in front of each number (mathematics books commonly use the symbol i, but in electronics j is chosen in order to avoid any confusion with electric current i). As an example, the number 1 is a

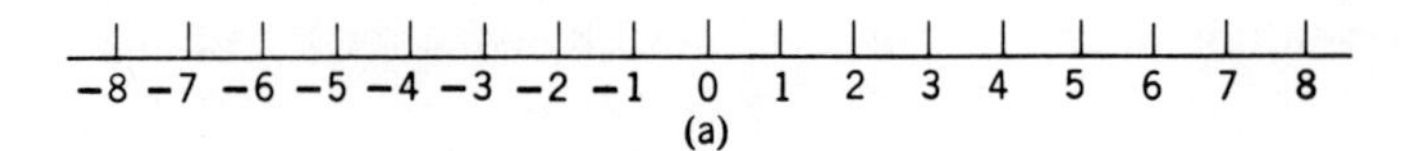

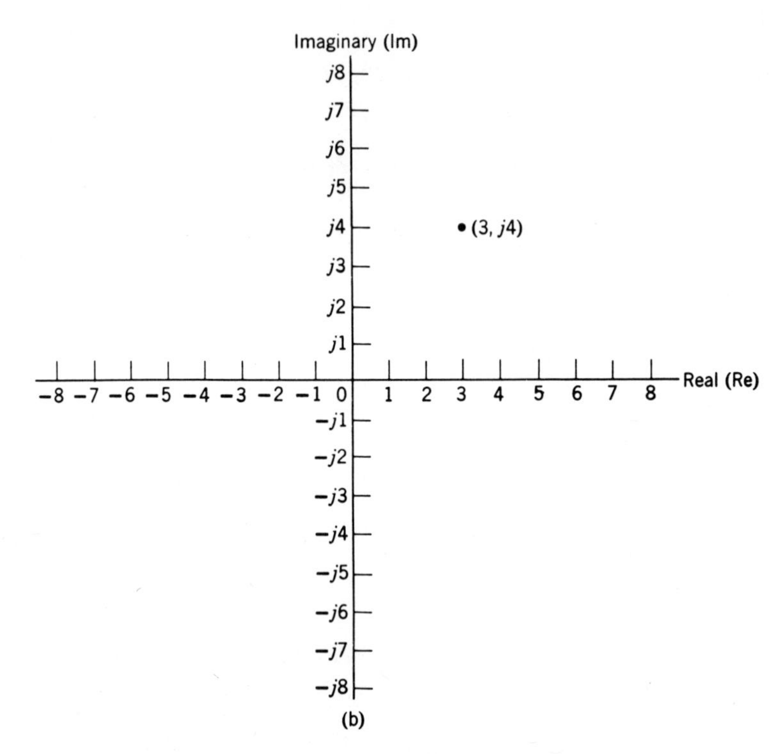

Figure 14.16 (*a*) Real numbers (*b*) Complex numbers. The complex plane

real number, but the number $j1$ is an *imaginary* number. We know that $1^2 = 1$, and thus, $\sqrt{1^2} = \sqrt{1} = 1$. On the other hand, $(j1)^2 = -1$ from the definition of an imaginary number, and thus

$$j = \sqrt{-1}$$

In order to represent complex numbers graphically, another scale is added to the real number scale in Fig. 14.16*a*. The complex number scale is perpendicular to the real number scale as shown in Fig. 14.16*b*. We still plot any real number along the horizontal or *real-axis*, and we plot *imaginary* numbers along the vertical or *imaginary*-axis. It is easy to see that the sum (difference) of any two real numbers is another real number. Similarly, the sum (difference) of any two imaginary numbers is another imaginary number.

The real and imaginary axes in Fig. 14.16*b* remind us immediately of a two-dimensional coordinate system, such as the familiar x-y coordinate system. Figure 14.16*b* can indeed be used as a coordinate system, and this is known as the *complex plane.* We can represent any point in the complex plane by choosing one real number as the horizontal coordinate, and one imaginary number as the vertical coordinate. For example, the point $(3, j4)$ is plotted in the complex plane in Fig. 14.16*b*.

Consider next the result of adding (subtracting) a *real* number and an imaginary number. We could for example add the real number a to the imaginary number jb. The result would simply be (a + jb). This is defined as a *complex* number. We see immediately that any complex number can be plotted as a single point in the complex plane. For example, the point (3, j4) corresponds to the complex number (3 + j4). Thus any complex number has two parts: a real part, and an imaginary part.

$$\text{complex number} = \underset{\substack{\uparrow \\ \text{real part}}}{a} \quad \underset{\substack{\uparrow \\ \text{imaginary part}}}{+jb}$$

Notice that real numbers and imaginary numbers are special cases of complex numbers. That is

$$\text{real number} = a + j0 = a$$

$$\text{imaginary number} = 0 + jb = jb$$

• • •

Example 14.18 Tell whether the following numbers are real, imaginary, or complex, and plot them on the complex plane: (*a*) $2 + j3$, (*b*) $-j4$, (*c*) $+6$, (*d*) $-5 - j2$.

Solution The numbers are all plotted in Fig. 14.17. (*a*) complex, (*b*) imaginary, (*c*) real, (*d*) complex.

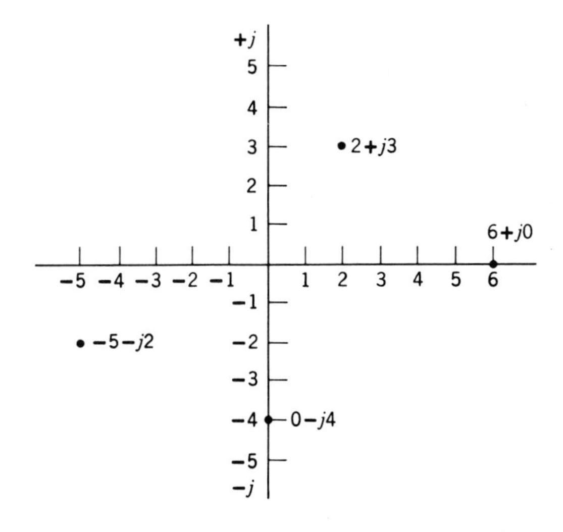

Figure 14.17

• • •

Note from the previous example that any number in the complex plane a + jb can be represented by drawing a *vector* from the origin to the number as shown in Fig. 14.18*a*. Conversely, any vector in the complex plane can be represented by a complex number of the form a + jb, and this is the representation we are interested in developing. When a vector is represented in the complex plane by a complex number a + jb, it is said to be given in *rectangular* form.

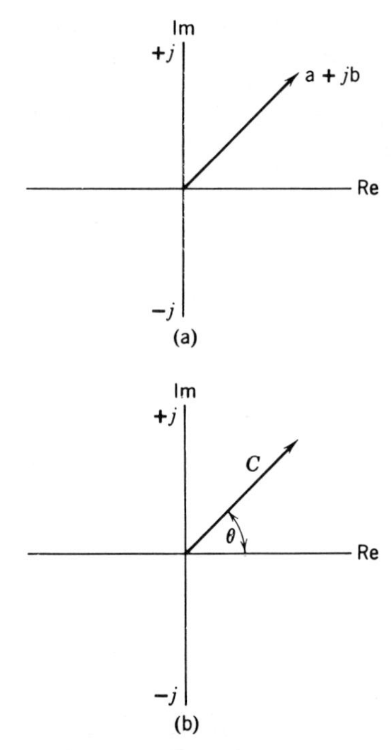

Figure 14.18 A vector in the complex plane (*a*) Rectangular form (*b*) Polar form

Furthermore, any point in the complex plane can be uniquely determined by a vector if we know (1) the length (magnitude) of the vector, and (2) the angle of the vector with respect to the positive real-axis. Thus the vector in Fig. 14.18*b* has a magnitude of C, and an angle θ. This is written as

$$\underset{\text{magnitude}}{\underset{\uparrow}{C}}\underset{\text{angle}}{\underset{\nwarrow}{\angle\theta}}$$

and any vector represented in this way is said to be in *polar* form.

We can develop very simple relationships for changing from polar to rectangular form, or vice versa, by considering the right triangle in Fig. 14.19.

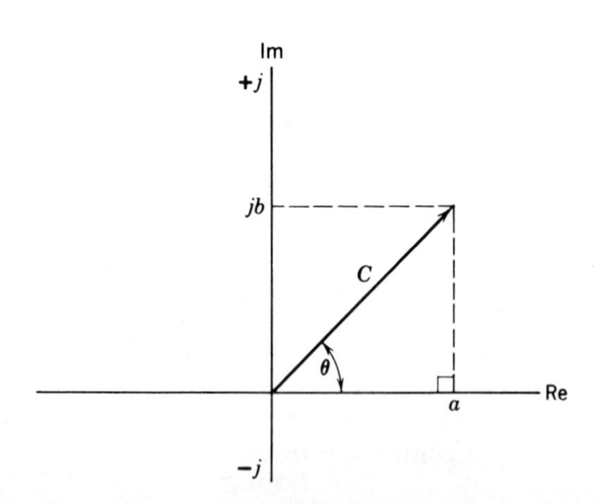

Figure 14.19

The vector in this figure can be represented in two ways:

rectangular—$a + jb$

polar—$C\angle\theta$

Consider first the case when the vector is given in rectangular form and we wish to change it to polar form. The magnitude of the vector C is easily found using the Pythagorean theorem. Thus

$$C = \sqrt{a^2 + b^2} \tag{14.18}$$

The angle is found using the definition of the tangent of an angle (Eq. 14.3). Thus*

$$\theta = \tan^{-1} \frac{b}{a} \tag{14.19}$$

• • •

Example 14.19 Change the following vectors from rectangular to polar form: (*a*) $3 + j4$, (*b*) 5, (*c*) $j6$, (*d*) $4 - j4$.

Solution (*a*) The magnitude is found as

$$C = \sqrt{a^2 + b^2} = \sqrt{3^2 + 4^2} = \sqrt{25} = 5$$

The angle is found to be

$$\theta = \tan^{-1} \frac{b}{a} = \tan^{-1} \frac{4}{3} = \tan^{-1} 1.333 = 53.1°$$

The vector is then

$$3 + j4 = 5\angle 53.1°$$

(*b*) The magnitude is simply 5, and the angle is 0°. Thus

$$5 = 5 + j0 = 5\angle 0°$$

(*c*) The magnitude is simply 6, but the angle is +90°. Thus

$$j6 = 0 + j6 = 6\angle 90°$$

(*d*) The magnitude is

$$C = \sqrt{a^2 + b^2} = \sqrt{4^2 + 4^2} = 5.66$$

The angle is

$$\theta = \tan^{-1} \frac{b}{a} = \tan^{-1} -\frac{4}{4} = \tan^{-1} - 1.0 = -45°$$

Thus,

$$4 - j4 = 5.66\angle -45°$$

• • •

Consider now the case when a vector is given in polar form, and we wish to change it into rectangular form. Since we are given C and θ in Fig. 14.19, it is a simple matter to use the trigonometric formulas, Eqs. 14.1 and 14.2 to find a and b. Thus

$$a = C \cos \theta \tag{14.20}$$

$$b = C \sin \theta \tag{14.21}$$

*The symbol $\tan^{-1}$ means "the angle whose tangent is." The symbol arctan means exactly the same thing.

• • •

Example 14.20 As a check, convert the vectors in the previous example from polar form back into rectangular form.

Solution

$$(a)\ \mathrm{a} = C \cos\theta = 5\cos 53.1° = 5\times 0.6 = 3$$
$$\mathrm{b} = C \sin\theta = 5\sin 53.1° = 5\times 0.8 = 4$$

Thus
$$5\underline{/53.1°} = 3 + j4$$

$$(b)\ \mathrm{a} = C \cos\theta = 5\cos 0° = 5\times 1 = 5$$
$$\mathrm{b} = C \sin\theta = 5\sin 0° = 5\times 0 = 0$$

Thus
$$5\underline{/0°} = 5 + j0 = 5$$

$$(c)\ \mathrm{a} = C \cos\theta = 6\cos 90° = 6\times 0 = 0$$
$$\mathrm{b} = C \sin\theta = 6\sin 90° = 6\times 1 = 6$$

Thus
$$6\underline{/90°} = 0 + j6 = j6$$

$$(d)\ \mathrm{a} = C \cos\theta = 5.66\cos(-45°) = 5.66\times 0.707 = 4$$
$$\mathrm{b} = C \sin\theta = 5.66\sin(-45°) = 5.66\times(-0.707) = -4$$

Thus
$$5.66\underline{/-45°} = 4 - j4$$

• • •

We close this section by noting that any vector in the complex plane can be represented in either rectangular or polar form. It is very easy to change from one form to the other, and it is also very easy to confuse the *signs* in the rectangular form and the *angle* in the polar form. Therefore, it is strongly recommended that a sketch be made whenever *either* of these two conversions are being made. The sketch provides an immediate check on the reasonableness of the solution, and also provides a very easy method for obtaining the correct signs and angles.

14.6 COMPLEX ALGEBRA

Since we now have a new type of number to work with, the complex number, we must learn exactly how to handle these numbers in basic algebraic operations. The extension of the basic algebraic processes to handle complex numbers is straightforward, but there are a few rules which must be observed.

Consider first the addition and subtraction of complex numbers. The sum (or difference) of two complex numbers is another complex number whose real part is simply the algebraic sum (difference) of the two real components, and whose imaginary part is the algebraic sum (difference) of the two imaginary components. Thus to add (subtract) two complex numbers, simply add (subtract) the real parts, and add (subtract) the imaginary parts. For example, the sum of $(\mathrm{a} + j\mathrm{b})$ and $(\mathrm{c} + j\mathrm{d})$ is

$$(\mathrm{a} + j\mathrm{b}) + (\mathrm{c} + j\mathrm{d}) = (\mathrm{a} + \mathrm{c}) + j(\mathrm{b} + \mathrm{d})$$

• • •

Example 14.21 Determine the given sums and differences: (*a*) $(1 + j2) + (3 - j)$; (*b*) $2 - (3 + j4)$; (*c*) $(2 + j6) - j2$.

Solution

$$(a)\ (1+j2)+(3-j)=(1+3)+j(2-1)=4+j1=4+j$$

$$(b)\ 2-(3+j4)=(2-3)+j(0-4)=-1-j4=-(1+j4)$$

$$(c)\ (2+j6)-j2=(2+0)+j(6-2)=2+j4$$

• • •

In considering the multiplication of complex numbers, we note first of all that the product of two *real* numbers is another real number. Similarly, the product of two *imaginary* numbers is a *negative real* number. This must be true, since by definition $j=\sqrt{-1}$, and therefore $j^2=-1$. For example, the product of $j5$ and $j6$ is

$$(j5)\times(j6)=j^2 30=-30$$

since $j^2=-1$.

Note also that the product of a real number and an imaginary number must be an imaginary number. For example

$$(2)\times(j3)=j6$$

Now, the product of two complex numbers is, in general, another complex number. The multiplication proceeds according to the ordinary multiplication rules of algebra, but we use the fact that $j^2=-1$. For example

$$\begin{aligned}(a+jb)\times(c+jd)&=ac+jad+jbc+j^2bd\\&=(ac-bd)+j(ad+bc)\end{aligned}$$

• • •

Example 14.22 Find the products of the following numbers: (*a*) $(2)\times(-3)$; (*b*) $(j3)\times(-4)$; (*c*) $j(j3)$; (*d*) $(1+j2)\times(2+j3)$; (*e*) $(2+j2)\times(3-j2)$.

Solution

$$(a)\ (2)\times(-3)=-6$$

$$(b)\ (j3)\times(-4)=-j12$$

$$(c)\ j(j3)=j^2 3=-3$$

$$\begin{aligned}(d)\ (1+j2)\times(2+j3)&=1\times2+1\times j3+2\times j2+j2\times j3\\&=2+j3+j4+j^2 6\\&=2+j3+j4-6\\&=-4+j7\end{aligned}$$

$$(e)\ (2+j2)\times(3-j2)=(6+4)+j(6-4)=10+j2$$

• • •

The multiplication of complex numbers can be carried out very easily when the numbers are expressed in *polar* form. It is only necessary to *multiply* the *magnitudes* and algebraically *add* the angles. Thus, in general,

$$(A\underline{/\theta_1})\times(B\underline{/\theta_2})=AB\underline{/\theta_1+\theta_2}$$

• • •

Example 14.23 Find the product of the following complex numbers: (*a*) $4\underline{/30°}$, $7\underline{/45°}$; (*b*) $3\underline{/-40°}$, $8\underline{/20°}$; (*c*) $(2+j2)$, $5\underline{/37°}$.

Solution

$$(a)\ 4\underline{/30°}\times7\underline{/45°}=4\times7\underline{/30°+45°}=28\underline{/75°}$$

$$(b)\ 3\underline{/-40°}\times8\underline{/20°}=24\underline{/-20°}$$

(*c*) We first change $2+j2$ into polar form.

Thus $$2+j2=2.83\angle 45^\circ$$

then $$2.83\angle 45^\circ \times 5\angle 37^\circ = 14.1\angle 82^\circ$$

• • •

The division of two complex numbers can also be carried out according to the ordinary rules of algebra, but it is easier if we first alter the problem slightly. Consider the problem of dividing the two complex numbers

$$\frac{a+jb}{c+jd}$$

This fraction will be unchanged if we multiply *both* the numerator and the denominator by the *same* number. Let us choose the number $(c-jd)$. Multiplying yields

$$\frac{a+jb}{c+jd}\times\frac{c-jd}{c-jd}=\frac{(ac+bd)+j(bc-ad)}{c^2+d^2}$$

It may seem that we have complicated the problem, but notice the denominator is now a *real* number. Thus the division process has been greatly simplified since the answer is now obtained by simply dividing the real part and the imaginary part by the denominator.

The mathematical trick used is called *rationalization,* and it provides the means for changing the denominator of a complex fraction into a real number. Thus the product of any complex number and its *conjugate* is a real number. The conjugate of a complex number is another complex number with the *same real* part, and with the *negative* imaginary part. For example, the conjugate of $(c+jd)$ is simply $(c-jd)$.

• • •

Example 14.24 Find the conjugates of the following complex numbers: (*a*) $1+j2$, (*b*) $3-j5$, (*c*) $j7$.

Solution (*a*) $1-j2$, (*b*) $3+j5$; (*c*) since $j7=0+j7$, its conjugate is simply $0-j7$, or $-j7$.

• • •

Example 14.25 Perform the following divisions using conjugates:

$$(a)\ \frac{1+j2}{1+j1} \qquad (b)\ \frac{4-j4}{6-j6}$$

Solution (*a*) Multiply *both* numerator and denominator by the conjugate of the denominator. Thus

$$\frac{1+j2}{1+j1}=\frac{1+j2}{1+j1}\times\frac{1-j1}{1-j1}=\frac{3+j1}{2}=\frac{3}{2}+j\frac{1}{2}$$

$$(b)\ \frac{4-j4}{6-j6}=\frac{4-j4}{6-j6}\times\frac{6+j6}{6+j6}=\frac{48+j0}{72}=\frac{48}{72}=\frac{2}{3}$$

• • •

As might be expected, the division of two complex numbers can be more easily carried out if the numbers are in *polar* form. It is only necessary to *divide* the magnitudes (numerator divided by denominator), and *subtract* the denominator angle from the numerator angle. Thus

$$\frac{A\angle\theta_1}{B\angle\theta_2}=\frac{A}{B}\angle\theta_1-\theta_2$$

• • •

Example 14.26 Perform the following divisions:

$$(a)\ \frac{27\angle 60^\circ}{9\angle 25^\circ}; \qquad (b)\ \frac{15\angle 28^\circ}{7\angle 85^\circ}; \qquad (c)\ \frac{5+j5}{5\angle -32^\circ}$$

Solution

$$(a)\ \frac{27\angle 60^\circ}{9\angle 25^\circ} = \frac{27}{9}\angle 60^\circ - 25^\circ = 3\angle 35^\circ$$

$$(b)\ \frac{15\angle 28^\circ}{7\angle 85^\circ} = \frac{15}{7}\angle 28^\circ - 85^\circ = 2.14\angle -57^\circ$$

(*c*) We first change $5+j5$ into polar form. Thus

$$5+j5 = 7.07\angle 45^\circ$$

Then,

$$\frac{5+j5}{5\angle -32^\circ} = \frac{7.07\angle 45^\circ}{5\angle -32^\circ} = 1.41\angle 77^\circ$$

• • •

SUMMARY

A periodic function is any wave which repeats itself periodically, and any such wave can be represented by a sum of sine and cosine waves. Any sinusoidal wave can be generated by plotting the position of the tip of a rotating vector, and we can write the equation of any sinusoidal wave in terms of an ordinary sine wave plus a phase angle. We must of course know the frequency and the peak value of the wave in order to complete the equation. A phase angle simply shifts an ordinary sine wave to the left (leading) or to the right (lagging). Any sinusoidal current or voltage can also be presented by the sine wave equation along with a phase angle, and we can determine the peak value, period, and thus the frequency of any ac signal by observing it on an oscilloscope. Ohm's Law and all of the previously developed network theorems apply equally well to ac circuits, but the sine wave equations developed will unnecessarily complicate matters when we begin writing loop and node equations for ac circuits. We have therefore developed complex numbers and the complex plane in order to simplify this problem. In the next chapter we will represent ac vectors of voltage and current in the complex plane.

A complex number consists of a real part and an imaginary part. Any point in the complex plane can be uniquely determined by a vector drawn from the origin to the point. This vector can then be expressed in either rectangular or polar form. Complex numbers can be added or subtracted by operating on the real parts and imaginary parts individually. Multiplication of complex numbers in rectangular form can be accomplished by ordinary algebraic methods, but it is usually easier to perform in polar form. Division of complex numbers in rectangular form can be accomplished by using conjugates, but again division is usually easier in polar form.

GLOSSARY

ac. *Alternating current. Used to signify any sinusoidal signal; for example, ac current, ac voltage.**

Angular velocity ω. *The rate of rotation of a phasor, in rad/sec.*

Conjugate. *A complex number formed by changing the sign of the imaginary part of a given complex number.*

Complex number. *A number consisting of a real part added to an imaginary part.* $a + jb$.

Complex plane. *A coordinate system with the real numbers on one axis and the imaginary numbers on an axis perpendicular to the real axis.*

Cycle. *That portion of a periodic wave between two adjacent similar points on the wave.*

Frequency Hz. *The number of cycles of a periodic wave occurring in a time period of one second. Sometimes given in cycles per second, cps or c/s.*

Peak value. *The maximum excursion of a sinusoidal wave above or below the horizontal axis (the axis on which the angle is measured).*

Period T. *The reciprocal of frequency. The time required to complete one cycle of a periodic wave.*

Periodic. *A wave form which repeats itself.*

Phase ϕ. *The angle through which a phasor is shifted from the positive horizontal axis, or, the angle through which a sine wave is shifted (right or left) from that of an ordinary sine wave.*

Radian. *A unit of angle measurement. One radian is approximately 57.3°.*

Sinusoidal. *A wave form having the shape of a sine wave.*

Vector. *A quantity having both magnitude and direction.*

IMPORTANT RELATIONSHIPS

- Trigonometric functions: $\sin \alpha = y/z \qquad \cos \alpha = x/z \qquad \tan \alpha = y/x$

- sinusoidal waves: $y = y_P \sin \alpha \qquad x = x_P \cos \alpha$

  $2\pi \text{ rad} = 360° \qquad 1 \text{ rad} \cong 57.3$

- angular velocity: $\omega = 2\pi f$

- period and frequency: $T = \dfrac{1}{f} \qquad f = \dfrac{1}{T}$

- sinusoidal currents and voltages:

$$i = I_P \sin(\omega t \pm \phi)$$

$$v = V_P \sin(\omega t \pm \phi)$$

- Complex numbers: $j = \sqrt{-1} \qquad j^2 = -1$

rectangular form	polar form
$a + jb$	$C\angle\theta$

**ac current* and *ac voltage* are redundant phrases; nevertheless, they are widely used.

- Converting complex numbers:

rectangular to polar	polar to rectangular
$C = \sqrt{a^2 + b^2}$	$a = C \cos \theta$
$\theta = \tan^{-1} \dfrac{b}{a}$	$b = C \sin \theta$

REVIEW QUESTIONS

1. A wave form which repeats itself after known intervals is said to be ____.
2. All periodic wave forms are sinusoidal. (TF)
3. A sine wave or a cosine wave is said to be (periodic, sinusoidal, both periodic and sinusoidal.)
4. The peak value of a sine wave can be found by multiplying the peak-to-peak value by ____.
5. The two common units for measuring angles are ____ and ____.
6. One radian is approximately equal to ____ degrees.
7. The advantage in plotting a sine wave with the angle expressed in radians is that the horizontal and vertical scales have the same units of length. (TF)
8. The projection of a vector on a vertical axis is proportional to the (sine, cosine) of the angle of rotation, while the projection on a horizontal axis is proportional to the (sin, cos).
9. The angular velocity of a vector can be found by multiplying the frequency of the wave it generates by ____.
10. Use the right-hand-rule to verify the direction of the induced current in Fig. 14.11.
11. Lower case letters i and v are used for ac signals since the currents and voltages vary both in ____ and ____ with time.
12. Two systems capable of producing ac signals are ____ and ____.
13. Using Ohm's Law, we see that sinusoidal currents and voltages produce sinusoidal voltages and currents of the same frequency in resistive circuits. (TF)
14. A positive phase angle ϕ means that the wave form (leads, lags) a pure sine wave, while a *negative* phase angle implies (lead, lag).
15. The phase angle ϕ in Eqs. 14.16 and 14.17 must be given in (degrees, radians).
16. A sine wave can be considered as a cosine wave with 90° of lag. (TF)
17. The complex operator j was defined in order to handle the square root of ____ numbers.
18. A complex number has a ____ part and an ____ part.
19. Real numbers and imaginary numbers are special cases of complex numbers. (TF)
20. The two forms for representing a vector in the complex plane are ____ and ____.
21. The use of a sketch is questionable when converting complex numbers from rectangular to polar form, or vice versa. (TF)
22. Explain how to add (subtract) two complex numbers.
23. In general, it is easier to multiply two complex numbers in rectangular form. (TF)

24. When multiplying two complex numbers in polar form, the angles are _____.
25. The conjugate of the complex number $7-j4$ is _____.
26. The product of any complex number and its conjugate is always a real number. (TF)
27. When dividing two complex numbers in polar form, the angle of the denominator must be (subtracted from, added to) the angle of the numerator.
28. It is generally easier to divide two complex numbers in (rectangular, polar) form.

PROBLEMS

1. State whether or not the waves in Figs. 14.20 are periodic. If they are, draw one complete cycle.
2. Find the angle α in Fig. 14.3 if: (*a*) $y=2$, $z=3$; (*b*) $x=3$, $z=6.5$; (*c*) $x=22$, $y=15$.
3. Find the angle α in Fig. 14.3 if: (*a*) $x=9$, $z=23$; (*b*) $y=7$, $x=3.5$; (*c*) $y=z=4$.
4. In Fig. 14.3, $z=16$ and $\alpha=28°$. Find x and y.
5. Repeat Problem 4 for $\alpha=53°$.
6. On graph paper make a careful plot of the equation $y=10 \sin \alpha$ for one complete cycle. Take α in 15° increments (i.e., 0°, 15°, 30°. . .360°).
7. Convert the following angles into radians: (*a*) 30°, (*b*) 45°, (*c*) 60°, (*d*) 130°.
8. Convert the following angles from radians to degrees: (*a*) $3\pi/4$, (*b*) $5\pi/6$, (*c*) -1.83.
9. Make sketches showing the angles in Problems 7 and 8.
10. What is the frequency of a wave if it has two cycles in a time period of 8 msec?

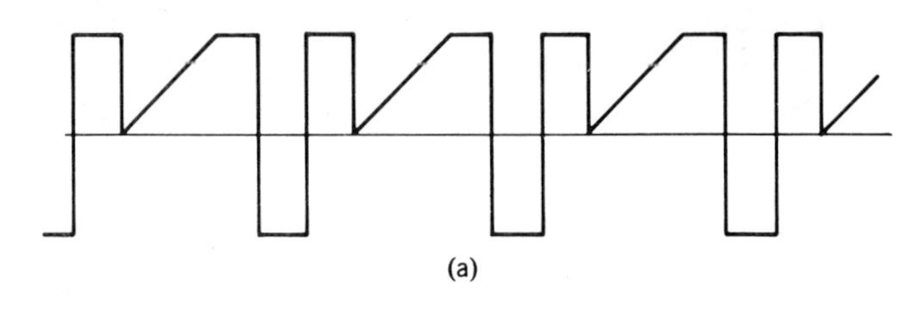

(a)

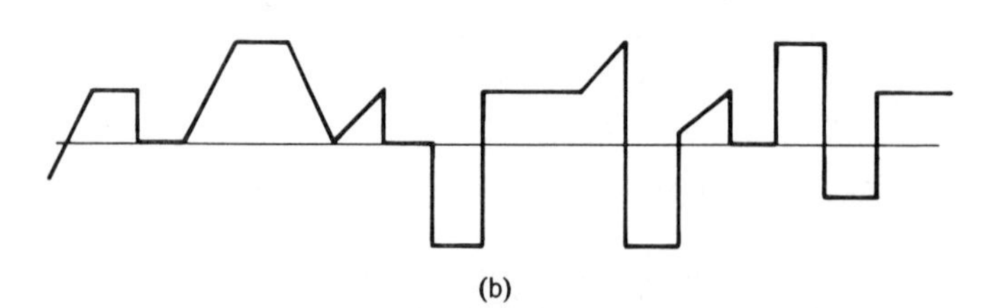

(b)

Figure 14.20.

11. A sinusoidal wave has a frequency of 250 kHz. What is the angular velocity of the vector representing this wave?
12. The angular velocity of a certain vector is 850 rad/sec. What is the frequency of the sinusoidal wave it generates?
13. Write the equation for a sine wave having a peak value of 25 and a frequency of 100 Hz. Use the form $y = y_P \sin \omega t$.
14. Repeat Problem 13 for a peak value of 80 and a frequency of 2 kHz.
15. What is the equation for v in Fig. 14.13*a* if $i = 2 \times 10^{-3} \sin 10^6 t$ A, and $R = 3.3$ kΩ?
16. What is the equation for i in Fig. 14.13*b* if $v = 28 \sin 200t$, and $R = 14\ \Omega$?
17. What is the frequency of the signal in Problem 15?
18. What is the frequency of the signal in Problem 16?
19. The oscilloscope in Fig. 14.14 is calibrated for 5 V/division on the vertical scale, and 1 msec/division on the horizontal scale. Write the equation for the voltage wave. What is the frequency of the wave?
20. Repeat Problem 19 if the oscilloscope is calibrated with 0.2 V/division vertically, and 2 μsec/division horizontally.
21. Write the equation of a voltage wave having a P-P value of 60 V, a period of 2 μsec, and a phase angle of $+30°$. Sketch the wave.
22. Write the equation of a current wave having a peak value of 250 mA, a frequency of 50 Hz, and a lagging phase angle of 45°. Sketch the wave.
23. Make a table showing the values of j, j^2, j^3, j^4, j^5, j^6, j^7, and j^8. Do you notice a pattern? What is it, and how can it be used?
24. Plot the following complex numbers in the complex plane: (*a*) $1 - j7$, (*b*) $4 + j6$, (*c*) $-3 - j8$, (*d*) $-5 + j7$, (*e*) $-j10$, (*f*) $10 + j0$.
25. Change the following complex numbers into polar form: (*a*) $1 + j2$, (*b*) $-2 - j3$, (*c*) $-4 + j4$, (*d*) $3 - j2$. Make careful sketches.
26. Check the results of the Problem 25 by changing back into rectangular form.
27. Change the following complex numbers into rectangular form: (*a*) $7\angle 30°$, (*b*) $20\angle -40°$, (*c*) $10\angle 135°$, (*d*) $5\angle -270°$. Make careful sketches.
28. Check the results of Problem 27 by converting back into polar form.
29. Perform the indicated operations: (*a*) $(1 + j2) + (3 - j)$, (*b*) $(1 + j2) - (3 - j)$, (*c*) $15 - (2 + j6) + (17 - j5)$.
30. Multiply the following complex numbers in rectangular form: (*a*) $(1 + j2)(6.06 + j3.5)$, (*b*) $(-j7)(4 + j4)$.
31. Multiply the numbers in Problem 30 by changing them into polar form first. Change the answers back into rectangular form as a check.
32. Write the conjugates of the following numbers: (*a*) $2 + j2$, (*b*) $j6$, (*c*) $1 - j$, (*d*) 7.
33. Divide the following complex numbers using conjugates: (*a*) $(7 + j7)/(5 - j8.7)$; (*b*) $(-j9)/(2 - j2)$.
34. Perform the divisions in Problem 33 by first changing to polar form. Check your answers.
35. Perform the indicated operations:

$$(a)\ \frac{(3 - j)(2 + j)}{1 - (4 - j)} \qquad (b)\ \frac{1 + 2j}{j(3 - j2)}$$

15

SERIES ac CIRCUITS— IMPEDANCE

We saw in the previous chapter that any sinusoidal current or voltage can be represented by equations of the form $i = I_P \sin(\omega t \pm \phi)$ and $v = V_P \sin(\omega t \pm \phi)$. Analysis of ac circuits will become unnecessarily complicated if we use these equations; therefore we would like to develop a more efficient method. There is indeed a more efficient and consequently easier method for analyzing ac circuits. The technique is to represent sinusoidal waves as rotating vectors (phasors), and to place these phasors in the complex plane. This technique leads to the development of the concept of ac impedance, which is simply the *opposition* which a circuit exhibits to a sinusoidal current.

15.1 IMPEDANCE

In Chapter 14 it was shown that the application of a sinusoidal voltage across the terminals of a resistor would produce a sinusoidal current in that resistor. Furthermore, the current has *exactly* the *same* frequency as the applied voltage, and it is *in phase* with the applied voltage. The current and voltage waveforms could be drawn as shown in Fig. 15.1*a*. To save space, they are more commonly drawn as in Fig. 15.1*b*; in this case, there are really two scales on the vertical axis, one for current and one for voltage.

However, we are more interested in representing these two waves as vectors

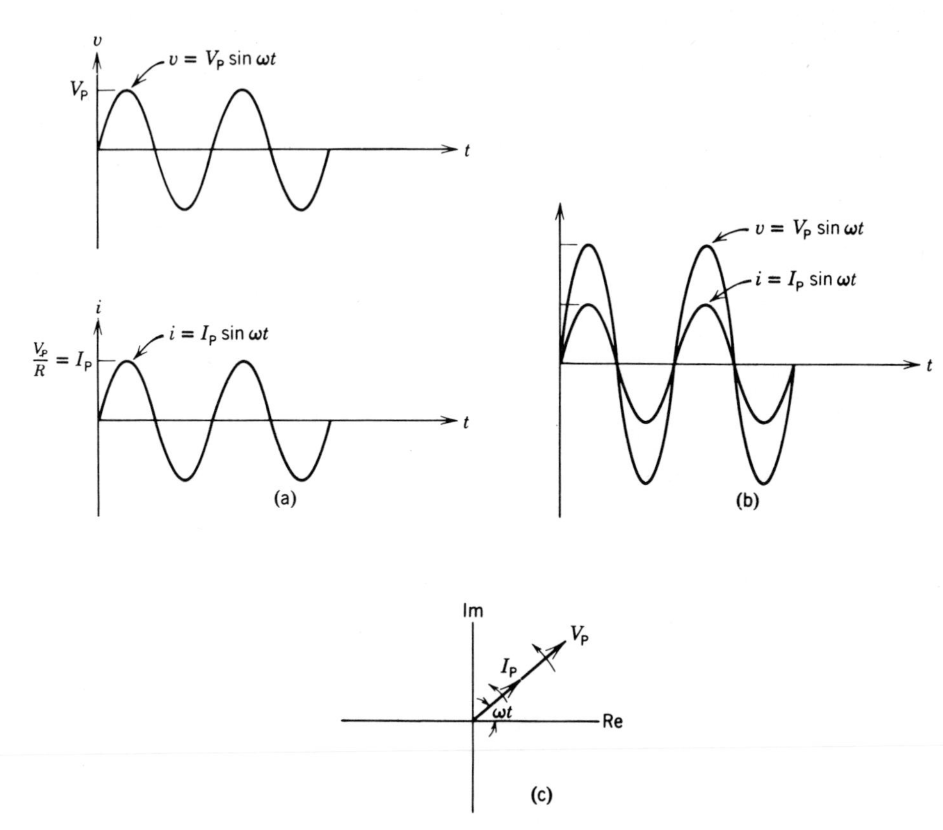

Figure 15.1 Sinusoidal current and voltage for a resistor (*a*) Current and voltage on separate scales (*b*) Current and voltage on the same graph (*c*) Vectors

drawn in the complex plane as shown in Fig. 15.1*c*. Notice that the two waves are *in phase*, and therefore the two vectors are drawn directly on top of each other; that is, there is no phase angle between them. We can represent them in vector form by specifying a magnitude and an angle. Thus these two waves can be written as

$$v = V_P \sin \omega t = V_P \underline{/\omega t}$$

$$i = I_P \sin \omega t = I_P \underline{/\omega t}$$

Since we are dealing only with a resistor this is a special case. In the more general case, a circuit may contain inductors and capacitors as well as resistors, and the current and voltage vectors usually will *not* be in phase with each other, that is, the current and voltage vectors may have a phase angle ϕ between them as shown in Fig. 15.2. This of course means that the current and voltage *wave forms* have a phase angle between them, and they do not cross the horizontal axis (t-axis in Fig. 15.1*b*) at the same point. Thus one of the wave forms is simply shifted to the right or left of the other wave form.

Notice in Fig. 15.2 that the current vector is represented by $I_P \underline{/\omega t}$, and the voltage vector is represented by $V_P \underline{/\omega t + \phi}$. If we choose the current as a

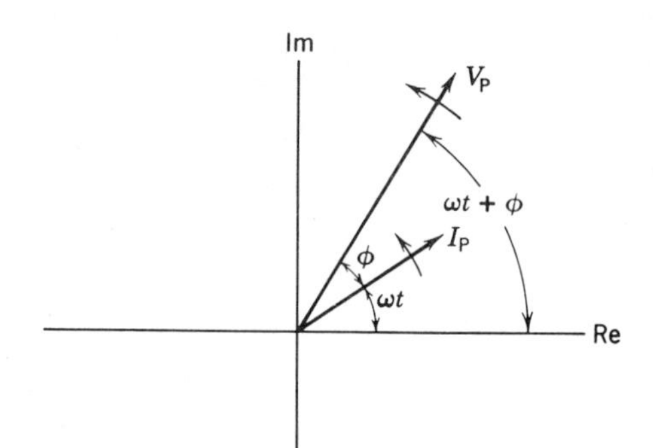

Figure 15.2 General ac vectors

reference, (that is, we assume the current is an ordinary sine wave) we can write the equation for the current wave form as

$$i = I_P \sin \omega t \tag{15.1}$$

The voltage waveform is also a sine wave, but it is simply shifted through a phase angle ϕ; that is, it *leads* the current vector by a phase angle ϕ. Thus the equation for the voltage waveform is simply

$$v = V_P \sin (\omega t + \phi) \tag{15.2}$$

It is important to note that: (1) the current and voltage wave forms have exactly the same frequency $f = \omega/2\pi$; (2) if the current is considered as an ordinary sine wave, the voltage is also a sine wave shifted by an angle ϕ; (3) the two waveforms have different amplitudes, I_P and V_P.

The only information contained in the ωt term is the angular velocity of the vectors, and thus the frequency of the wave forms. However, we know that the currents and voltages will all have the *same* frequencies, and there is therefore no need to carry this term along in our calculations. It is only necessary to note the frequency (or ω) and remember it. Therefore we can *always* choose *any one* of the vectors as a *reference* and simply draw the other vectors by making use of their phase angles. In this way, we can always eliminate the angle ωt. The vectors in Fig. 15.2 could for example be drawn as shown in Fig. 15.3*a* with current as the reference, and we lose no information if we simply remember the value of ω. This representation is called a *phasor diagram,* and these vectors are defined as *phasors.* A voltage or current phasor contains only the magnitude and the phase angle of the vector. Thus the voltage phasor in Fig. 15.2 is written as $\mathbf{V_P} = V_P\underline{/\phi}$, and the current phasor is $\mathbf{I_P} = I_P\underline{/0}$.

Notice that the elimination of the ωt term means that we can *represent* the ac *vectors* as *phasors* in the complex plane in Fig. 15.3*a*: *There is no time-varying term (ωt is eliminated) and thus we can work with vectors and triangles to solve all of our ac circuit problems.* The elimination of the term ωt means that we needn't worry about time in our calculation, and this is indeed a great simplification.

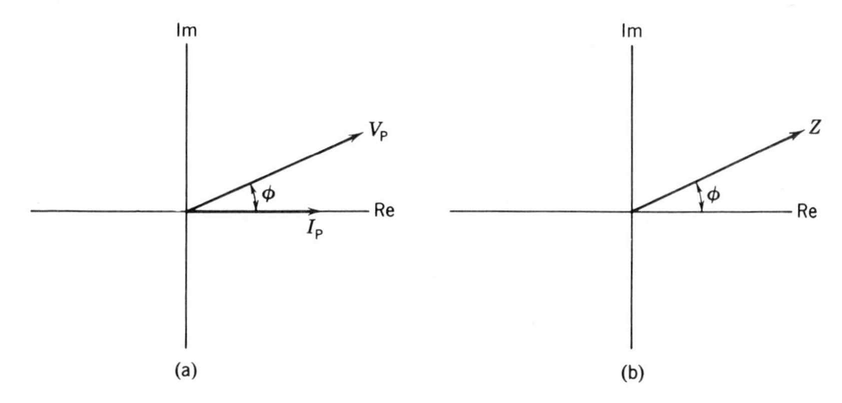

Figure 15.3 (*a*) Phasors in the complex plane (*b*) Impedance vectors in the complex plane

• • •

Example 15.1 For the ac vectors in Fig. 15.2, assume that the voltage is the reference (take the voltage waveform as an ordinary sine wave) and write the current and voltage waveform equations.

Solution Since the current is the reference vector, we assume it has no phase angle. Thus the current waveform equation is

$$i = I_P \sin \omega t$$

Since the voltage vector leads the current vector by a phase angle ϕ, the voltage wave form equation is clearly

$$v = V_P \sin (\omega t + \phi)$$

• • •

If we apply Ohm's Law to the resistive circuit in Fig. 15.1, we obtain

$$\frac{v}{i} = \frac{V_P \sin \omega t}{I_P \sin \omega t} = \frac{V_P \underline{/\omega t}}{I_P \underline{/\omega t}} = \frac{V_P}{I_P} = R$$

In this case, we are dividing a *vector* by a *vector*; but, since the vector angles are equal they cancel (remember, subtract the denominator angle from the numerator angle), and thus the ratio is *not* a vector. It is in fact a scalar, which we recognize immediately as resistance R.

If we now apply Ohm's Law to the more general ac circuit of Fig. 15.2, we obtain

$$\frac{v}{i} = \frac{V_P \sin (\omega t + \phi)}{I_P \sin \omega t} = \frac{V_P \underline{/\omega t + \phi}}{I_P \underline{/\omega t}} = \frac{V_P}{I_P} \underline{/\omega t + \phi - \omega t} = \frac{V_P}{I_P} \underline{/\phi} \tag{15.3}$$

In this case, we divide a vector by a vector; but, since the angles are not equal they do not cancel, and the result is another vector! This is certainly *not* resistance because resistance is a scalar (resistance has magnitude in ohms, but *no direction*). We therefore define this quantity as *impedance*, and we use the symbol Z. Since impedance Z is the ratio of volts divided by amperes, it is measured in the units of ohms. Thus

$$\text{impedance} = \mathbf{Z}(\boldsymbol{\Omega}) = \frac{V_P}{I_P} \underline{/\phi} = Z\underline{/\phi} \tag{15.4}$$

Notice that the magnitude of the impedance Z is the ratio of the applied voltage and the resulting current, or

$$Z = \frac{V_P}{I_P}$$

The angle of the impedance ϕ is simply the *phase angle between* the voltage and current phasors. Notice that ϕ can be measured in either radians or degrees. We will choose the more familiar unit—degrees.

The impedance **Z** associated with the phasors in Fig. 15.3*a* has a magnitude of

$$Z = \frac{V_P}{I_P}$$

and the angle is simply the phase angle ϕ. Therefore we can represent the impedance as a vector in the complex plane as shown in Fig. 15.3*b*.

The results of this section can be summarized by noting that we can use Ohm's Law in ac circuit problems by writing it in the following form:

$$\mathbf{I}_P = \frac{\mathbf{V}_P}{\mathbf{Z}_P} \tag{15.5}$$

where $\mathbf{I}_P$ and $\mathbf{V}_P$ are the current and voltage phasors, and $\mathbf{I}_P$, $\mathbf{V}_P$, and **Z** are all vector quantities having *both magnitude and angle.*

• • •

Example 15.2 In Fig. 15.4*a* an ac signal generator (sine wave) is connected across the terminals of a network containing resistors, inductors, and

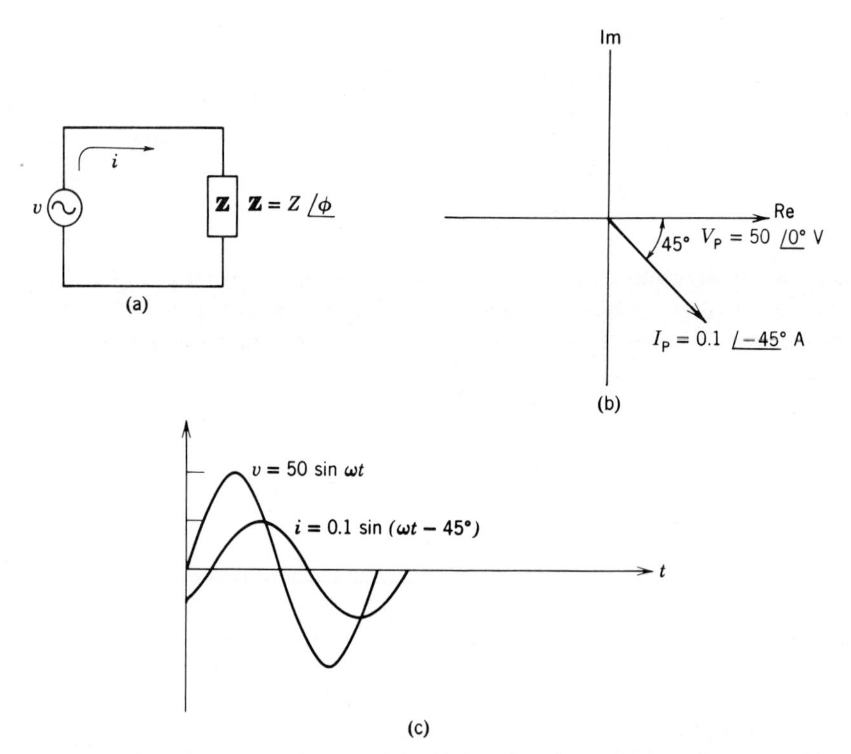

Figure 15.4 (*a*) An ac circuit with impedance (*b*) Phasor diagram (*c*) Current and voltage wave forms

capacitors. The network thus has some impedance $\mathbf{Z} = Z\underline{/\phi}$. Calculate the current in the circuit $\mathbf{I}_P$ if $\mathbf{V}_P = 50\underline{/0^\circ}$ V, and $\mathbf{Z} = 500\underline{/45^\circ}\ \Omega$. Draw the phasor diagram, and sketch the current and voltage wave forms.

Solution We calculate the current using Ohm's Law for ac circuits, Eq. 15.5.

$$\mathbf{I}_P = \frac{\mathbf{V}_P}{\mathbf{Z}} = \frac{50\underline{/0}\ \text{V}}{500\underline{/45^\circ}\ \Omega} = 0.1\underline{/-45^\circ}\ \text{A}$$

The phasor diagram is shown in Fig. 15.4*b*, and the waveforms are shown in Fig. 15.4*c*.

• • •

Example 15.3 Calculate the voltage necessary to produce a current of $15\underline{/30^\circ}$ mA in Fig. 15.4*a*, if $\mathbf{Z} = 4\underline{/-30^\circ}$ kΩ.

Solution Rearranging Eq. 15.5,

$$\mathbf{V}_P = \mathbf{I}_P\mathbf{Z} = 15\underline{/30^\circ}\ \text{mA} \times 4\underline{/-30^\circ}\ \text{k}\Omega = 60\underline{/0^\circ}\ \text{V}$$

• • •

Example 15.4 What must be the impedance of the circuit in Fig. 15.4*a* if a voltage of $\mathbf{V}_P = 25\underline{/40^\circ}$ V produces a current of $\mathbf{I}_P = 5\underline{/15^\circ}$ mA?

Solution Again we rearrange Eq. 15.5 to obtain

$$\mathbf{Z} = \frac{\mathbf{V}_P}{\mathbf{I}_P} = \frac{25\underline{/40^\circ}\ \text{V}}{5\underline{/15}\ \text{A}} = 5\underline{/25^\circ}\ \text{k}\Omega$$

• • •

2 RESISTANCE AND REACTANCE

In the previous section we have shown that the impedance of an ac circuit can be represented as a vector drawn in the complex plane, and as such it has a magnitude Z and an angle ϕ. We recognize this as the *polar* form of a number in the complex plane, and we know that such a number can also be expressed in rectangular form. That is, the impedance vector $\mathbf{Z} = Z\underline{/\phi}$, shown in Fig. 15.5, can be represented in rectangular form, and it will have a *real* component and an *imaginary* component. The real component is drawn on the real-axis, and it is exactly equal to the *resistance* of the circuit. The imaginary component is drawn on the imaginary axis (j-axis), and this component is defined as the reactance X of the circuit. Thus,

$$\mathbf{Z} = Z\underline{/\phi} = R \pm jX \tag{15.6}$$

where $\mathbf{Z}$, R, and X are all measured in ohms. Since R and X are both measured in ohms, the impedance $\mathbf{Z}$ must consist of a resistor R *in series with* a reactance $\pm jX$ as shown in Fig. 15.6. The box $\pm jX$ may contain a number of inductors and capacitors.

Since the reactance X is due to the inductors and capacitors in the circuit, it can be drawn on either the positive or the negative j-axis (depending on the phase angle) as we shall see in the next two sections. The resistance R is of course always positive.

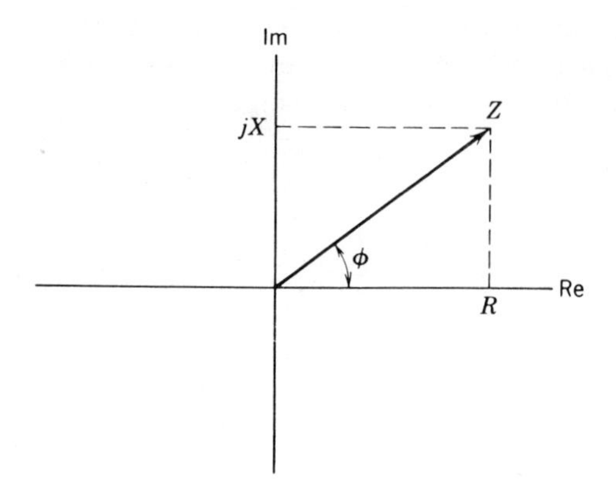

Figure 15.5 Impedance in the complex plane

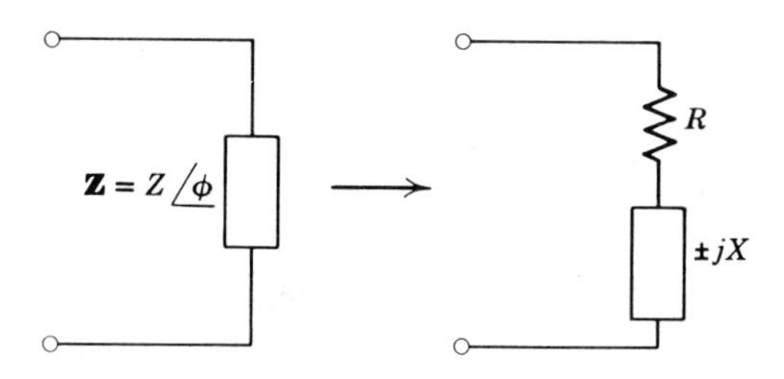

Figure 15.6 The components of impedance

Using the results of the previous chapter, we can find the real and the imaginary components (resistance R and reactance X) of $\mathbf{Z}$ as follows:

$$R = Z \cos \phi \tag{15.7}$$

$$X = Z \sin \phi \tag{15.8}$$

• • •

Example 15.5 The impedance of an ac circuit is known to be $75\underline{/30^\circ}\ \Omega$. What is the resistance and the reactance in the circuit? Draw an equivalent circuit.

Solution We simply use Eqs. 15.7 and 15.8. Thus

$$R = Z \cos \phi = 75 \cos 30^\circ = 75 \times 0.867 = 65\ \Omega$$

$$X = Z \sin \phi = 75 \sin 30^\circ = 75 \times 0.5 = 37.5\ \Omega$$

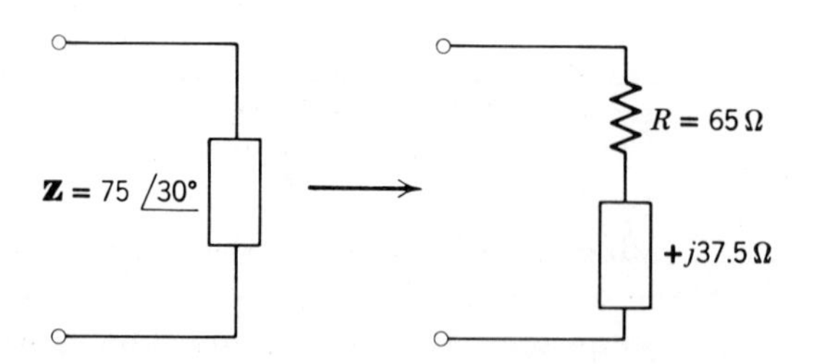

Figure 15.7

The impedance can then be expressed as

$$\mathbf{Z} = R + jX = (65 + j37.5)\ \Omega$$

This impedance is composed of a resistance of 65 Ω in *series* with a reactance of $j37.5\ \Omega$. The equivalent circuit is shown in Fig. 15.7.

• • •

Example 15.6 The impedance of an ac circuit is known to be $\mathbf{Z} = 10\underline{/-45^\circ}\ \text{k}\Omega$. Calculate the resistive and reactive components. Draw the impedance vector and its components; show the equivalent circuit.

Solution Using Eqs. 15.7 and 15.8,

$$R = Z \cos\phi = 10^4 \cos(-45^\circ) = 10^4 \times 0.707 = 7.07\ \text{k}\Omega$$

$$X = Z \sin\phi = 10^4 \sin(-45^\circ) = 10^4(-0.707) = -7.07\ \text{k}\Omega$$

The impedance is then

$$\mathbf{Z} = 10\underline{/-45^\circ}\ \text{k}\Omega = (7.07 - j7.07)\ \text{k}\Omega$$

The impedance vector and the equivalent circuit as shown in Fig. 15.8*a* and *b*.

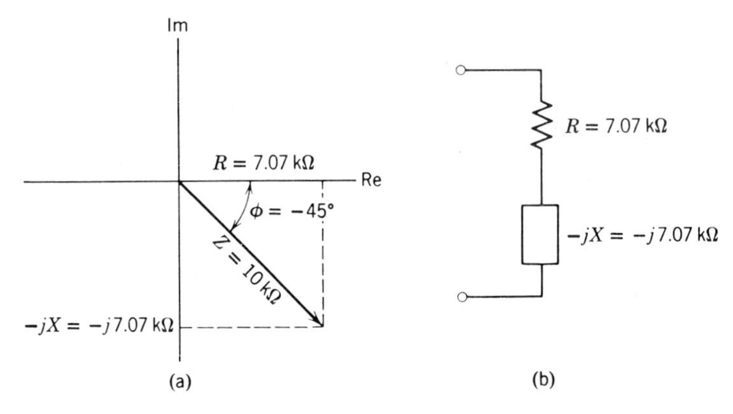

Figure 15.8 (*a*) Impedance diagram (*b*) Equivalent circuit

• • •

We can also use the results of the previous chapter to find the impedance of a resistance R connected in series with a reactance $\pm jX$ as shown in Fig. 15.6. Thus

$$\text{magnitude} = Z = \sqrt{R^2 + X^2} \tag{15.9}$$

$$\text{angle} = \phi = \tan^{-1}\left(\pm\frac{X}{R}\right) \tag{15.10}$$

• • •

Example 15.7 In Fig. 15.6, $R = 1\ \text{k}\Omega$ and $\pm jX = +j700\ \Omega$. Find the magnitude and the angle of the impedance.

Solution We use Eq. 15.9 to obtain the magnitude.

$$Z = \sqrt{R^2 + X^2} = \sqrt{(10^3)^2 + (700)^2} = \sqrt{(100 + 49)10^4} = 1.22\ \text{k}\Omega$$

The angle is obtained from Eq. 15.10 as

$$\phi = \tan^{-1}\frac{X}{R} = \tan^{-1}\frac{700}{1000} = \tan^{-1} 0.7 = 35°$$

The impedance in polar form is then

$$\mathbf{Z} = Z\underline{/\phi} = 1.22\underline{/35°}\ \text{k}\Omega$$

• • •

Example 15.8 Find the series resistance in the circuit in Fig. 15.9.

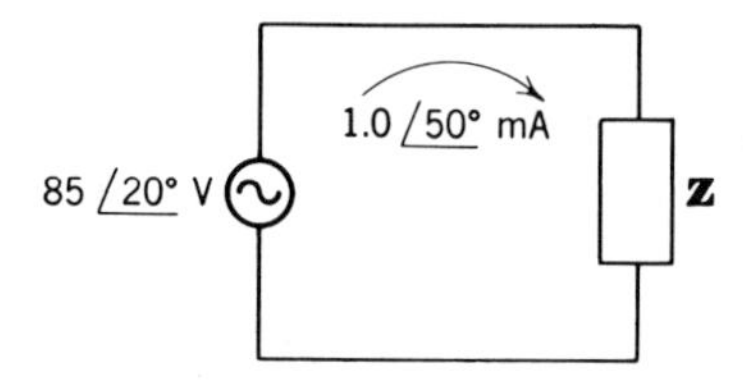

Figure 15.9

Solution We use Ohm's Law for ac circuits to find the impedance **Z**.

$$\mathbf{Z} = \frac{\mathbf{V}_P}{\mathbf{I}_P} = \frac{85\underline{/20°}\ \text{V}}{10^{-3}\underline{/50°}\ \text{A}} = 85\underline{/-30°}\ \text{k}\Omega$$

The resistance R is the real component of the impedance, and from Eq. 15.7 we find,

$$R = Z\cos\phi = 85\ \text{k}\Omega\cos(-30°) = 73.5\ \text{k}\Omega$$

• • •

15.3 INDUCTIVE REACTANCE X_L

In Fig. 15.6, the box representing reactance $\pm jX$ may contain a number of inductors and capacitors. In this section we want to discover the nature of the reactance term $\pm jX$ when *only inductors* are used. To signify that only inductors are used (i.e., no capacitors are included), we use the symbol X_L, and we define this as *inductive reactance.* Inductive reactance is the opposition to a sinusoidal current exhibited by an inductance L.

Let us consider the case of a sinusoidal current produced in an inductance L by a sinusoidal voltage as shown in Fig. 15.10*a*. We will take the current as the reference, and thus we can write

$$i_L = I_{PL}\sin\omega t, \qquad \text{or} \qquad \mathbf{I}_{PL} = I_{PL}\underline{/0°}$$

At this point, recall the v-i relationship for an inductance which was given in Chapter 12 as

$$v_L = L\frac{di_L}{dt} \tag{15.11}$$

where v_L is in volts, L is in henries, and i_L is in amperes. The current i_L is plotted in Fig. 15.10*b*, and we can use this wave form to determine the voltage v_L. Remember that the term di_L/dt is the *slope* of the current at any particular

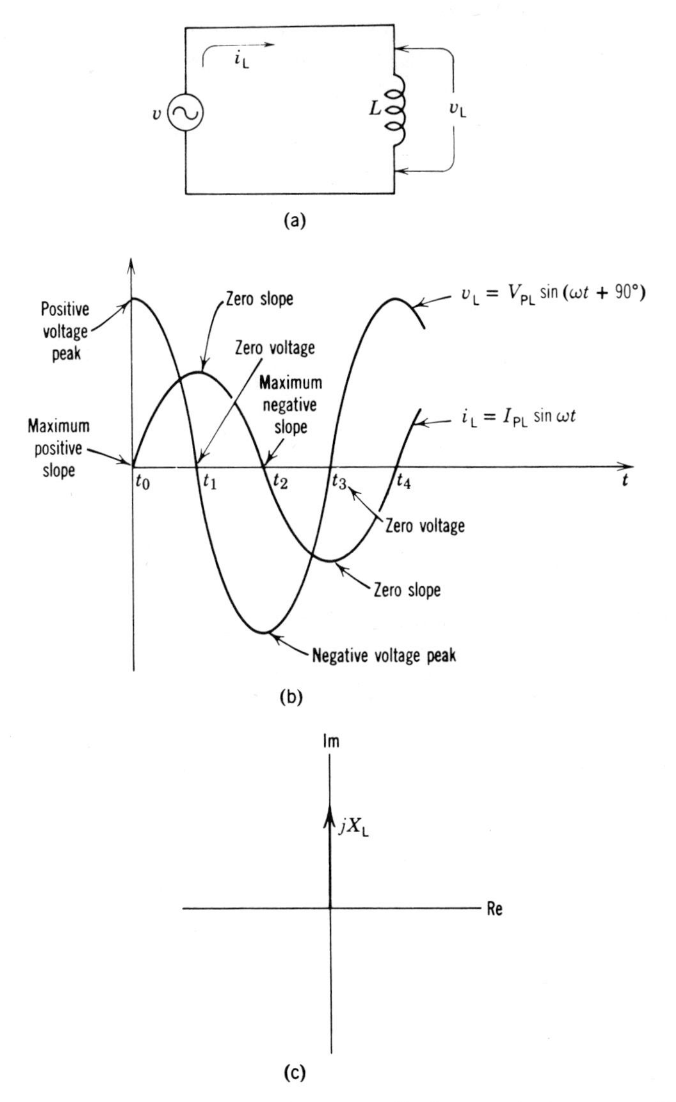

Figure 15.10 (*a*) A sinusoidal voltage applied to an inductance (*b*) Current and voltage wave forms (*c*) Inductive reactance in the complex plane

time t. Now, from Fig. 15.10*b*, we make the following observations:

1. At time t_0: the current wave form is at its *maximum positive* slope. Therefore, from Eq. 15.11, the voltage must be at its *positive peak* value.
2. At time t_1: the slope of the current wave form is *zero*. Therefore from Eq. 15.11 the voltage must be zero.
3. At time t_2: the current wave form is at its *maximum negative* slope. Therefore the voltage must be at its *negative peak* value.
4. At time t_3: The slope of the current wave form is *zero*. Therefore the voltage must be zero.
5. At time t_4 the waves begin to repeat, and we have covered one complete cycle. Since sinusoidal voltages produce sinusoidal currents of the same frequency (and vice versa) in circuits containing R's, L's, and C's, the

voltage wave form v_L must be a sinusoidal wave of the same frequency as the current wave i_L. Having determined four points on v_L, it can simply be drawn in as a sinusoidal wave.

The voltage wave form v_L is then drawn in Fig. 15.10*b* in this manner, and it is seen to be a sine wave which *leads* the current wave form i_L by 90° (it is shifted left by 90°). Thus the equation for the voltage is simply

$$v_L = V_{PL} \sin(\omega t + 90°), \qquad \text{or} \qquad \mathbf{V}_{PL} = V_{PL}\underline{/90°}$$

We make the very important observation

that the voltage across an inductance always leads the current in the inductance by 90°. Alternatively, the current can be said to lag the voltage by 90°.

If we now apply Ohm's Law for ac circuits (Eq. 15.5) to the circuit in Fig. 15.10, we obtain

$$\frac{\mathbf{V}_{PL}}{\mathbf{I}_{PL}} = \frac{V_{PL}\underline{/90°}}{I_{PL}\underline{/0°}} = X_L\underline{/90°}$$

But, $X_L\underline{/90°}$ in the complex plane is a vector on the positive j-axis. Therefore, inductive reactance X_L is *always* drawn on the positive j-axis as $+jX_L$ as shown in Fig. 15.10*c*. The impedance of an ideal inductance is then

$$\mathbf{Z} = R \pm jX = 0 + jX_L = jX_L = X_L\underline{/90°}$$

We can calculate the magnitude of the inductive reactance X_L by taking the ratio of the applied voltage to the element current, but we would like to develop a more convenient expression based on the size of the inductance L in henries. We will do this by using the *average* value of the inductor voltage v_L and the *average rate of change* of inductor current (di_L/dt) in Eq. 15.11. The inductor voltage v_L is a sinusoidal wave and the *average* value of any sinusoidal wave is equal to the peak value multiplied by $2/\pi$.* Thus

$$v_{L_{ave}} = \frac{2V_{PL}}{\pi}$$

Since the inductor current i_L in Fig. 15.10*b* changes from zero to I_P in one-fourth of a cycle, or in a time of $T/4$, its *average* rate of change is

$$\frac{di_L}{dt}\text{ave} = \frac{I_{PL}}{T/4} = \frac{4I_{PL}}{T}$$

But the period T is equal to $1/f$, and thus

$$\frac{di_L}{dt}\text{ave} = \frac{4I_{PL}}{T} = \frac{4I_{PL}}{1/f} = 4fI_{PL}$$

If we now substitute these two average values into Eq. 15.11, we obtain

$$v_{L_{ave}} = L\frac{di_L}{dt}\text{ave}$$

$$\frac{2V_{PL}}{\pi} = L4fI_{PL}$$

*We will demonstrate this in Chapter 17.

But, inductive reactance X_L is V_{PL}/I_{PL}, and thus we solve this relationship as

$$X_L = \frac{V_{PL}}{I_{PL}} = 2\pi f L = \omega L \tag{15.12}$$

where X_L is in ohms, L is in henries, f is in hertz, and ω is in radians/second.

For the reader familiar with calculus, the above results can be obtained very quickly by assuming $i_L = I_{PL} \sin \omega t$, and using Eq. 15.11 to obtain

$$v_L = L\frac{di_L}{dt} = L\frac{d}{dt}(I_{PL} \sin \omega t) = \omega L I_{PL} \cos \omega t = \omega L I_{PL} \sin (\omega t + 90°)$$

The voltage and current phasors in the complex plane are then

$$\mathbf{I}_{PL} = I_{PL}\underline{/0°} \qquad \mathbf{V}_{PL} = V_{PL}\underline{/90°} = \omega L I_{PL}\underline{/90°}$$

and the inductive reactance is

$$\frac{\mathbf{V}_{PL}}{\mathbf{I}_{PL}} = \frac{V_{PL}\underline{/90°}}{I_{PL}\underline{/0°}} = \frac{\omega L I_{PL}\underline{/90°}}{I_{PL}\underline{/0°}} = \omega L\underline{/90°} = jX_L$$

We make the following important observations regarding an inductance in an ac circuit:

1. The opposition which an inductance exhibits to a sinusoidal current is called inductive reactance X_L.
2. Inductive reactance is calculated from the relationship $X_L = \omega L = 2\pi f L$, and it is seen to be directly proportional to the frequency in Hz, and the inductance in henries.
3. Inductive reactance is always plotted on the positive j-axis in the complex plane as $+jX_L$.
4. In an ideal inductor, the voltage always *leads* the current by 90°.

• • •

Example 15.9 What is the inductive reactance of the circuit in Fig. 15.11?

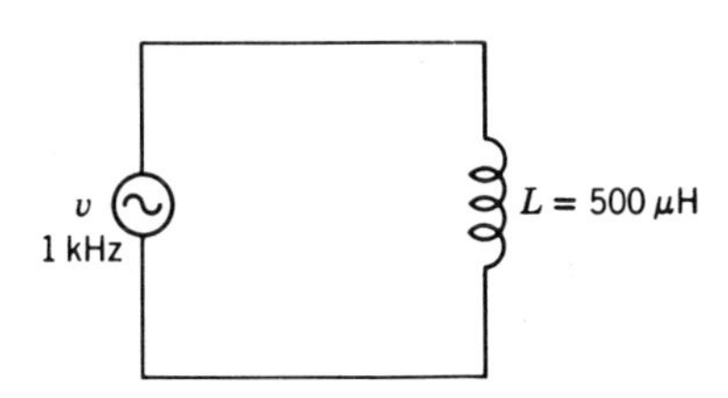

Figure 15.11 Example 15.9

Solution Using Eq. 15.12,

$$X_L = 2\pi f L = 2\pi \times 1 \text{ kHz} \times 500 \ \mu\text{H} = 3.14 \ \Omega$$

• • •

Example 15.10 Calculate the current $\mathbf{I}_{PL}$ in Fig. 15.12*a*. Draw the phasor diagram showing $\mathbf{V}_{PL}$ and $\mathbf{I}_{PL}$, and draw the impedance diagram. What must be the inductance L?

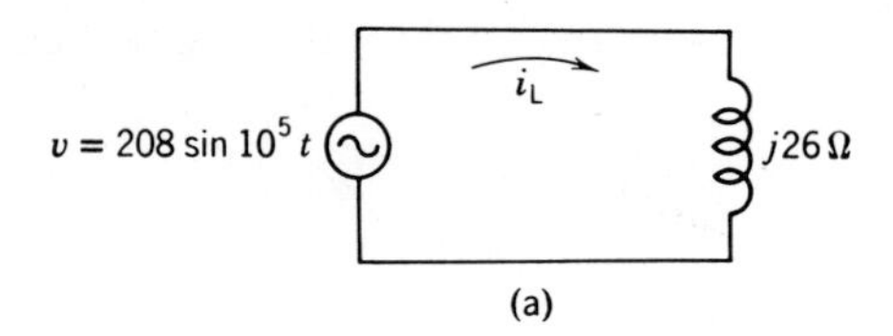

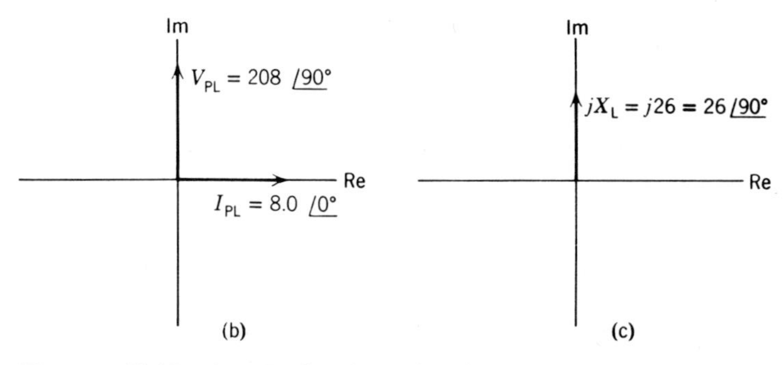

Figure 15.12 (*a*) Inductive circuit (*b*) Phasor diagram (*c*) Impedance diagram

Solution Using Ohm's Law for ac circuits,

$$\mathbf{I}_{PL} = \frac{\mathbf{V}_{PL}}{\mathbf{Z}} = \frac{208\angle 0°\text{ V}}{(0 + j26)\ \Omega} = \frac{208\angle 0°\text{ V}}{26\angle 90°\ \Omega} = 8.0\angle -90°\text{ A}$$

This simply confirms that the current of 8.0 A *lags* the voltage by 90°, or that the voltage *leads* the current by 90°. The phasor diagram is shown in Fig. 15.12*b* with the current taken as the reference. The impedance diagram is shown in Fig. 15.12*c*. The inductive reactance is given by Eq. 15.12 as

$$X_L = \omega L$$

But, X_L is given as 26, from $j26\,\Omega$, and ω is given as 10^5 in the voltage generator equation. That is

$$v = 208 \sin 10^5 t = 208 \sin \omega t$$

Thus, $\omega = 10^5$. Therefore

$$L = \frac{X_L}{\omega} = \frac{26}{10^5} = 260\ \mu\text{H}$$

• • •

15.4 CAPACITIVE REACTANCE X_C

In the previous section it is demonstrated that a reactance composed solely of inductance is plotted as the positive j-axis as jX_L. Because of the dual nature between inductance and capacitance developed earlier, it should not be too surprising to discover that a reactance composed solely of capacitors is always plotted on the *negative* j-axis as $-jX_C$. Where the symbol X_C is defined as capacitive reactance, and signifies the use of capacitors only in the $\pm jX$ box in

Fig. 15.6. Capacitive reactance is the opposition to a sinusoidal current exhibited by a capacitor.

Let us consider the application of a sinusoidal voltage v across the terminals of an ideal capacitor as shown in Fig. 15.13*a*. In this case we will take the voltage as the reference, and thus we can write

$$v_C = V_{PC} \sin \omega t, \qquad \text{or} \qquad \mathbf{V}_{PC} = V_{PC}\underline{/0^\circ}$$

The v-i relationship for a capacitor as given in a previous chapter is

$$i_C = C\frac{dv_C}{dt} \tag{15.13}$$

where i_C is in amperes, C is in farads, and v_C is in volts. We can use this

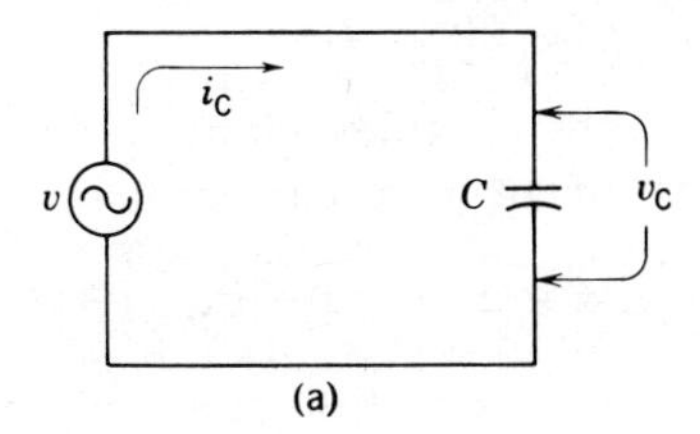

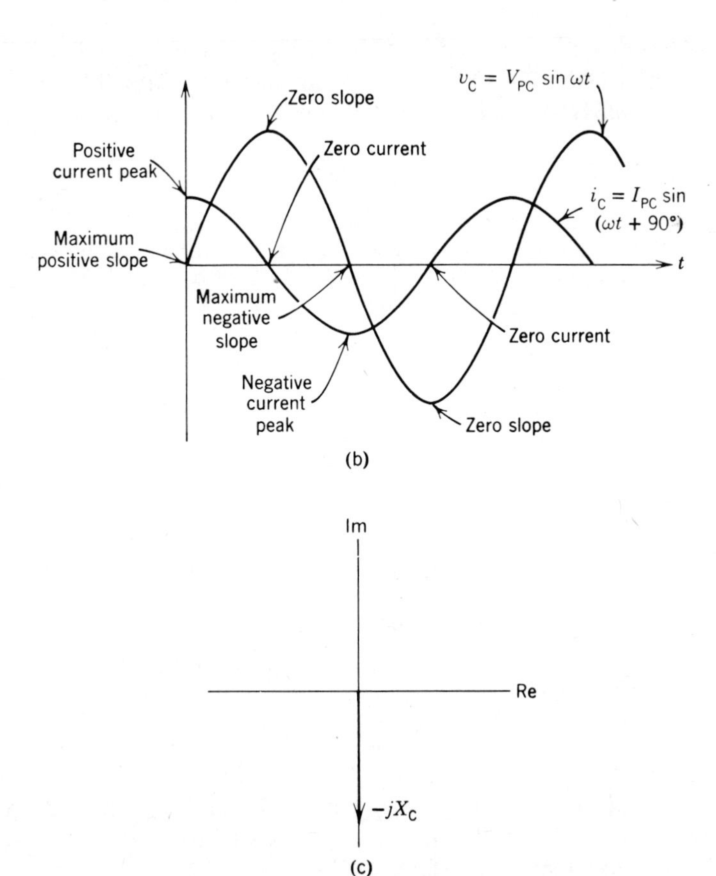

Figure 15.13 (*a*) A sinusoidal voltage applied to a capacitance (*b*) Current and voltage wave forms (*c*) Capacitive reactance in the complex plane

equation and the voltage waveform for v_C plotted in Fig. 15.13*b* to determine the current in the capacitor i_C. Proceeding just as in the previous section, note the current has a *positive* peak when the voltage has a maximum positive slope; it has a *negative* peak when the voltage has a maximum negative slope; and it is zero when the voltage wave has a zero slope. Since we know this current wave i_C must be sinusoidal, we can now draw in the proper wave form as shown in Fig. 15.13*b*. In this capacitive circuit, the current clearly *leads* the voltage by 90°, or we can say the voltage *lags* the current by 90°. Thus the current in the capacitor can be expressed as

$$i_C = I_{PC} \sin(\omega t + 90°), \qquad \text{or} \qquad \mathbf{I}_{PC} = I_{PC}\underline{/90°}$$

Applying Ohm's Law for ac circuits, we obtain

$$\frac{\mathbf{V}_{PC}}{\mathbf{I}_{PC}} = \frac{V_{PC}\underline{/0°}}{I_{PC}\underline{/90°}} = \frac{V_{PC}}{I_{PC}}\underline{/-90°} = X_C\underline{/-90°}$$

But a vector at $-90°$ in the complex plane is plotted on the negative j-axis, and thus the impedance of an ideal capacitor is

$$\mathbf{Z} = R \pm jX = 0 - jX_C = -jX_C$$

Thus capacitive reactance is plotted on the complex plane as shown in Fig. 15.13*c*.

The magnitude of X_C can be found in terms of the frequency and the value of C in farads by using Eq. 15.13 and finding *average* values. The current i_C in Fig. 15.13*b* is clearly sinusoidal, and it therefore has an average value of

$$i_C \text{ ave} = \frac{2I_{PC}}{\pi}$$

Since the voltage v_C goes from zero to V_{PC} in one-fourth of a cycle, $T/4$, its *average* rate of change is given by

$$\frac{dv_C}{dt} \text{ ave} = \frac{V_{PC}}{\frac{T}{4}} = \frac{4V_{PC}}{T} = 4fV_{PC}, \qquad \text{since} \qquad T = \frac{1}{f}$$

Substituting these average values into Eq. 15.13 yields

$$i_C = C\frac{dv_C}{dt} \text{ ave}$$

$$\frac{2I_{PC}}{\pi} = C4fV_{PC}$$

But the capacitive reactance X_C is V_{PC}/I_{PC}, and therefore

$$X_C = \frac{1}{2\pi fC} = \frac{1}{\omega C} \tag{15.14}$$

where X_C is in ohms, C is in farads, f is in Hz, and ω is in rad/sec.

These results can also be obtained very quickly by using calculus. Assuming $v_C = V_{PC} \sin \omega t$, and using Eq. 15.13,

$$i_C = C\frac{dv_C}{dt} = C\frac{d}{dt}(V_{PC} \sin \omega t) = \omega C V_{PC} \cos \omega t = \omega C V_{PC} \sin(\omega t + 90°)$$

The voltage and current phasors in the complex plane are then

$$\mathbf{V}_{PC} = V_{PC}\underline{/0^\circ} \qquad \mathbf{I}_{PC} = I_{PC}\underline{/90^\circ} = \omega C V_{PC}\underline{/90^\circ}$$

and the capacitive reactance is

$$\frac{\mathbf{V}_{PC}}{\mathbf{I}_{PC}} = \frac{V_{PC}\underline{/0^\circ}}{\omega C V_{PC}\underline{/90^\circ}} = \frac{1}{\omega C}\underline{/-90^\circ} = \frac{-j}{\omega C} = -jX_C \tag{15.15}$$

We make the following important observations regarding a capacitance in an ac circuit!

1. The opposition which a capacitance exhibits to a sinusoidal current is called capacitive reactance X_C.
2. Capacitive reactance is calculated from the relationship $X_C = 1/\omega C = 1/2\pi fC$, and it is seen to be *inversely* proportional both to frequency in Hz and to capacitance in farads.
3. Capacitive reactance is always plotted in the negative j axis in the complex plane as $-jX_C$.
4. In an ideal capacitor, the current always *leads* the voltage by 90°.

• • •

Example 15.11 What is the capacitive reactance of the circuit in Fig. 15.14?

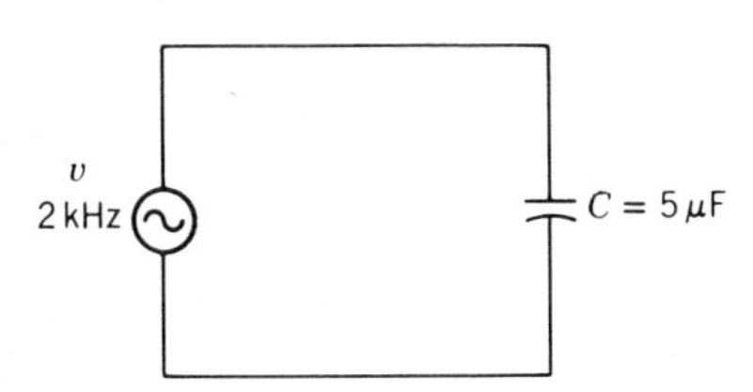

Figure 15.14

Solution Using Eq. 15.14,

$$X_C = \frac{1}{2\pi fC} = \frac{1}{2\pi \times 2\ \text{kHz} \times 5\ \mu\text{F}} = 15.9\ \Omega$$

• • •

Example 15.12 Calculate the current $\mathbf{I}_{PC}$ in Fig. 15.15*a*. Draw the phasor diagram showing $\mathbf{V}_{PC}$ and $\mathbf{I}_{PC}$, and draw the impedance diagram. What must be the capacitance C?

Solution Using Ohm's Law for ac circuits, and choosing $\mathbf{V}_{PC}$ as the reference,

$$\mathbf{I}_{PC} = \frac{\mathbf{V}_{PC}}{\mathbf{Z}} = \frac{5.0\underline{/0^\circ}\ \text{V}}{(0 - j50)\ \Omega} = \frac{5.0\underline{/0^\circ}}{50\underline{/-90^\circ}} = 0.1\underline{/90^\circ}\ \text{A}$$

The phasor diagram is shown in Fig. 15.15*b*, and the impedance diagram is shown in Fig. 15.15*c*. The capacitance C can be found by rearranging Eq. 15.14, and noting that $\omega = 10^6$ from $v_C = 5 \sin 10^6 t$. Thus

$$C = \frac{1}{\omega X_C} = \frac{1}{10^6 \times 50} = 0.02\ \mu\text{F}$$

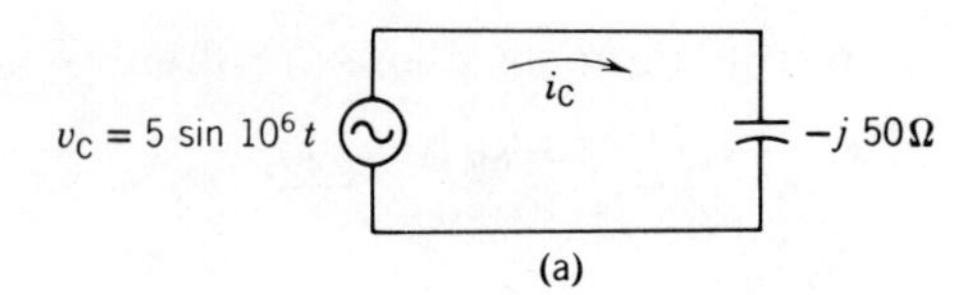

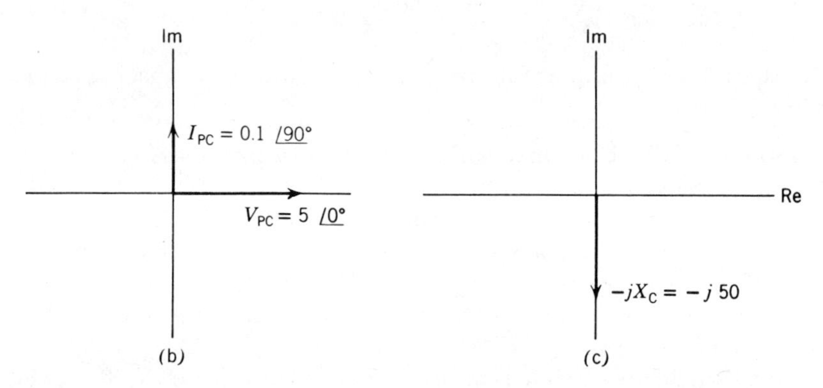

Figure 15.15 (*a*) Capacitive circuit (*b*) Current and voltage phasors (*c*) Impedance diagram

• • •

15.5 SERIES *R*-*L* CIRCUIT

A series ac circuit consisting of a resistance R connected in series with an inductance L is shown in Fig. 15.16*a*. If the current and applied voltage are represented as phasors, we can apply Ohm's Law to the circuit as

$$\mathbf{I}_P = \frac{\mathbf{V}_P}{\mathbf{Z}}$$

Since this is a *series* circuit, we will take the current phasor as a reference, because the current must be the same in all parts of the circuit. Thus, the current can be written as $I_P\underline{/0°}$. From the previous sections, we know that the impedance is the sum of the resistance R and j times the inductive reactance X_L. Thus

$$\mathbf{Z} = R + jX_L \tag{15.16}$$

This impedance can also be written in polar form as

$$\mathbf{Z} = Z\underline{/\phi} = \sqrt{R^2 + X_L^2}\ \underline{/\tan^{-1}\frac{X_L}{R}}$$

The impedance diagram can then be drawn as shown in Fig. 15.16*b*.

The generator voltage can be determined using Ohm's Law as

$$\mathbf{V}_{PS} = \mathbf{I}_P\mathbf{Z} = I_P\underline{/0°}\ Z\underline{/\phi} = I_P Z\underline{/\phi} = V_{PS}\underline{/\phi}$$

and thus the phasor diagram appears as shown in Fig. 15.16*c*. Since this is an *inductive* circuit, the voltage is seen to *lead* the current by some angle ϕ. If the resistance R were zero, the phase angle ϕ would be 90°, but the addition of some resistance R in series with X_L reduces ϕ to some angle between 0° and 90°.

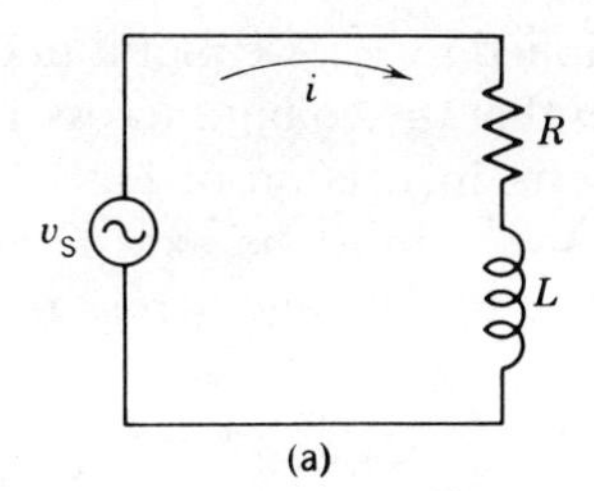

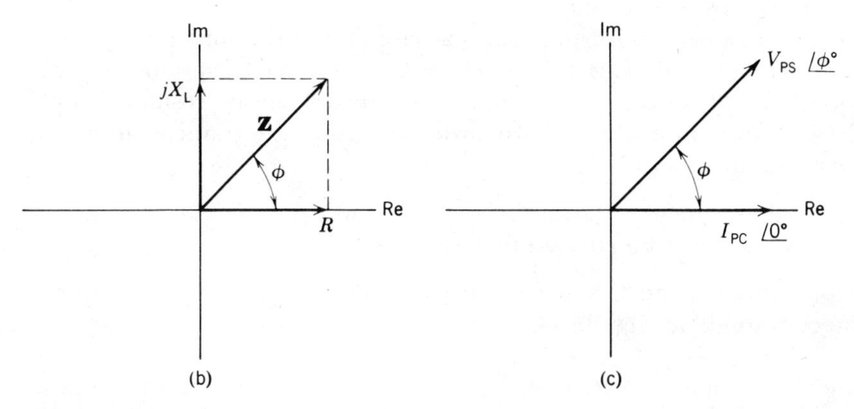

Figure 15.16 A series *R-L* circuit (*a*) Circuit (*b*) Impedance diagram (*c*) Phasor diagram

Let us now calculate the generator voltage again, but this time we will use the impedance **Z** in rectangular form instead of in polar form. The result is

$$\mathbf{V}_{PS} = \mathbf{I}_P\mathbf{Z} = I_P\underline{/0°}(R + jX) = I_P(R + jX_L) = I_PR + jI_PX_L$$

We immediately recognize the form I_PR as the voltage drop across the resistor $\mathbf{V}_{PR}$, and since we know that the current through and the voltage across a resistance are always in phase, we can draw I_PR on the positive real axis as shown in Fig. 15.17. The term jI_PX_L must then be the voltage across the

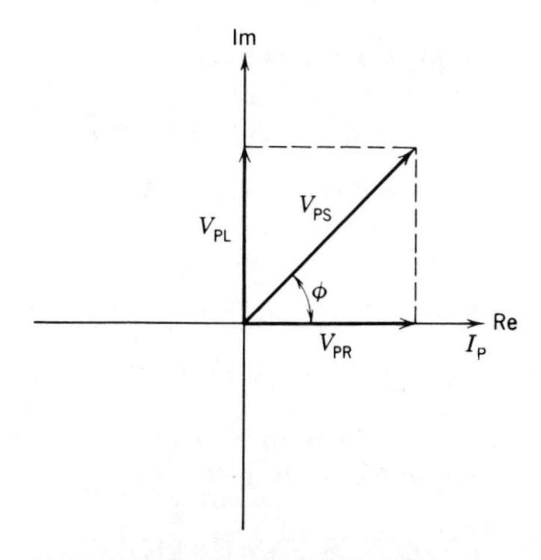

Figure 15.17 The element voltages for Figure 15.16

inductor $\mathbf{V}_{PL}$, and since it is multiplied by $+j$, we plot it in the positive j-axis in Fig. 15.17. This, of course, verifies that the voltage across an inductor *leads* the current through it by 90°. A careful inspection of Fig. 15.17 reveals that the total applied generator voltage $V_{PS}\underline{/\phi}$ must be exactly equal to the voltage across the resistor $V_{PR}\underline{/0^\circ}$ added to the voltage across the inductor $V_{PL}\underline{/90^\circ}$. Thus

$$V_{PS}\underline{/\phi} = V_{PR}\underline{/0^\circ} + V_{PL}\underline{/90^\circ} = V_{PR} + jV_{PL}$$

where V_{PS}, V_{PR}, and V_{PL} are all ***peak*** voltages. This is nothing more than KVL applied to a series R-L ac circuit.

Thus KVL can be applied to a series ac circuit, but we must remember to use *both magnitude and angle* for the element and signal generator voltages. The application of KVL to a series R-L ac circuit will always result in a right triangle as shown in Fig. 15.17. In order to draw this triangle, it is only necessary to remember:

1. $\mathbf{V}_{PR}$ is always in phase with the current, and
2. $\mathbf{V}_{PL}$ always *leads* the current by 90°.

Notice that this voltage triangle is *similar* to (has the same angles as) the impedance triangle in Fig. 15.16*b*.

• • •

Example 15.13 In Fig. 15.16, $R = 1\ \mathrm{k\Omega}$, $L = 1$ mH, and $v_S = 50 \sin 10^6 t$. Calculate the circuit current, and draw a phasor diagram showing V_{PS}, V_{PR}, and V_{PL}. Draw the impedance diagram. What is the generator frequency?

Solution The circuit current is found using Ohm's Law, but we must first determine the impedance $\mathbf{Z}$. R is given as 1 kΩ, and X_L is

$$X_L = \omega L = 10^6 \times 1\ \mathrm{mH} = 10^3\ \Omega, \text{ since } \omega = 10^6 \text{ from } v_S = 50 \sin 10^6\, t$$

Thus

$$\mathbf{Z} = R + jX_L = 10^3 + j10^3 = (1+j)10^3\ \Omega = 1.414\underline{/45^\circ}\ \mathrm{k\Omega}$$

The circuit current is then

$$\mathbf{I}_P = \frac{\mathbf{V}_{PS}}{\mathbf{Z}} = \frac{50\underline{/0^\circ}\ \mathrm{V}}{1.414\underline{/45^\circ}\ \mathrm{k\Omega}} = 35.4\underline{/-45^\circ}\ \mathrm{mA}$$

This shows that the current *lags* the voltage by 45°, *or* the voltage *leads* the current by 45°. Since we want the current as a reference, we will rewrite the generator voltage and the circuit current phasors as

$$\mathbf{V}_{PS} = 50\underline{/45^\circ}\ \mathrm{V} \qquad \mathbf{I}_P = 35.4\underline{/0^\circ}\ \mathrm{mA}$$

The element voltages are now:

$$\mathbf{V}_{PR} = \mathbf{I}_P R = 35.4\underline{/0^\circ}\ \mathrm{mA} \times 1\ \mathrm{k\Omega} = 35.4\underline{/0^\circ}\ \mathrm{V}$$

$$\mathbf{V}_{PL} = \mathbf{I}_P jX_L = 35.4\underline{/0^\circ}\ \mathrm{mA} \times j\mathrm{k\Omega} = 35.4j\ \mathrm{V} = 35.4\underline{/90^\circ}\ \mathrm{V}$$

The phasor diagram and the impedance diagram are shown in Fig. 15.18*a* and *b*. The generator frequency is given by

$$f = \frac{\omega}{2\pi} = \frac{10^6}{2\pi} = 159\ \mathrm{kHz}.$$

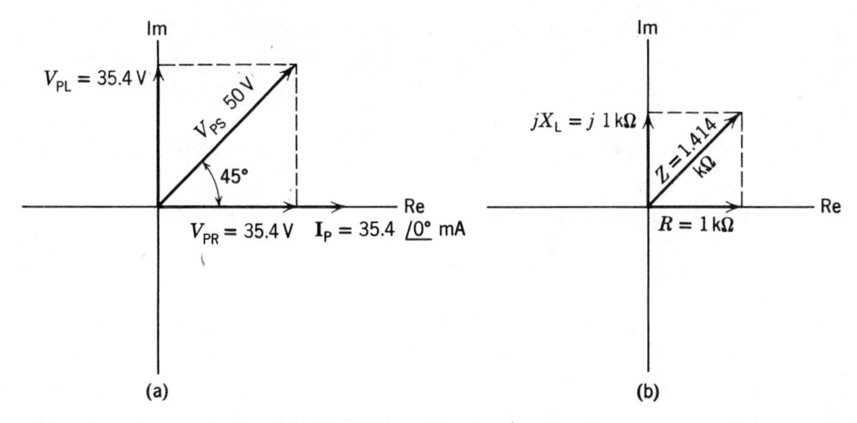

Figure 15.18 (a) Phasor diagram. Current and voltage phasors have different scales (b) Impedance diagram

• • •

.6 SERIES R-C CIRCUIT

An ac circuit consisting of a resistance R connected in *series* with a capacitance C is shown in Fig. 15.19a. Ohm's Law will apply equally well to this circuit if we represent the voltages and the current as phasors. Again, since this is a *series* circuit we will choose the current as the reference phasor. The current is then represented as $I_P/\underline{0°}$. The impedance of this circuit is the sum of the

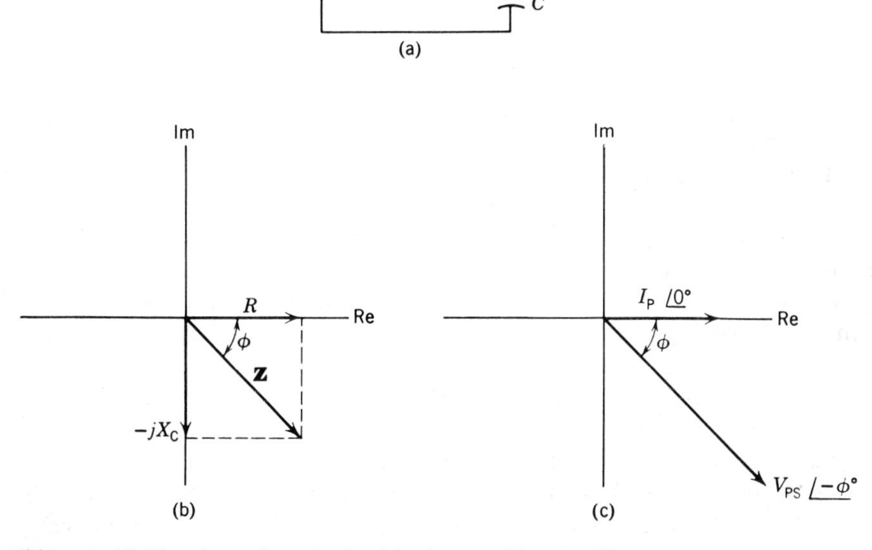

Figure 15.19 A series R-C circuit (a) Circuit (b) Impedance diagram (c) Phasor diagram

resistance R and $-j$ multiplied by the capacitive reactance X_C. Thus

$$\mathbf{Z} = R - jX_C \tag{15.17}$$

In polar form,

$$\mathbf{Z} = Z\underline{/-\phi} = \sqrt{R^2 + X_C^2}\underline{/-\tan^{-1} X_C/R}$$

The impedance diagram can be drawn as shown in Fig. 15.19*b*.

$$\mathbf{V}_{PS} = \mathbf{I}_P\mathbf{Z} = I_P\underline{/0°}\, Z\underline{/-\phi} = I_P Z\underline{/-\phi°} = V_{PS}\underline{/-\phi}$$

The correct phasor diagram then appears as shown in Fig. 15.19*c*, and it is clear that the generator voltage $\mathbf{V}_{PS}$ *lags* the circuit current by some phase angle ϕ. If the resistance R were zero, the voltage would lag by exactly 90°, but the addition of some resistance R will reduce ϕ to some angle between 0° and −90°.

If we now calculate the generator voltage using the impedance in rectangular form, the result is

$$\mathbf{V}_{PS} = \mathbf{I}_P\mathbf{Z} = I_P\underline{/0°}(R - jX_C) = I_P(R - jX_C) = I_P R - jI_P X_C$$

The term $I_P R$ is of course the voltage across the resistor $\mathbf{V}_{PR}$ and it is plotted on the positive real axis in Fig. 15.20 (it is in phase with the current I). The term $-jI_P X_C$ is the voltage across the capacitor $\mathbf{V}_{PC}$ and it must be plotted on the negative j-axis in Fig. 15.20. This of course verifies that the voltage across a capacitor *lags* the current through it by 90°.

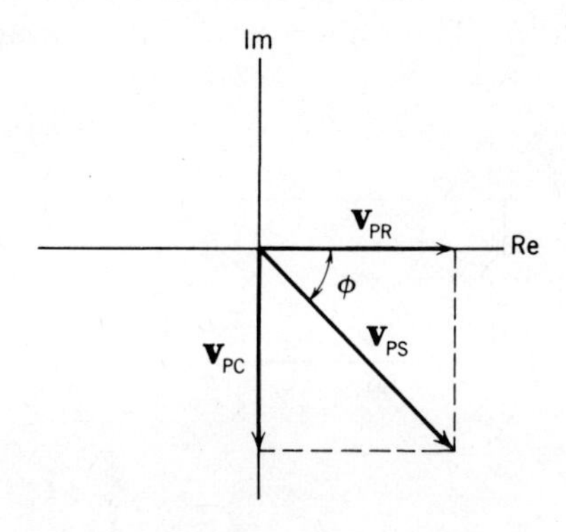

Figure 15.20 The element voltages for Figure 15.19

Just as in the case of a series R-L circuit, the *vector sum* of the element voltages in the series R-C circuit must be equal to the generator voltage. This is again the application of KVL, and in equation form we have

$$V_{PS}\underline{/-\phi} = V_{PR}\underline{/0°} + V_{PC}\underline{/-90°} = V_{PR} - jV_{PC}$$

where V_{PS}, V_{PR}, and V_{PC} are all *peak* voltages.

The application of KVL to the series R-C ac circuit in Fig. 15.19*a* will always lead to a voltage triangle as shown in Fig. 15.20. Again, we must always specify *both magnitude and angle* for the element voltages and the signal generator voltage. The voltage triangle can always be drawn by remembering:

1. $\mathbf{V}_{PR}$ is always in phase with the circuit current, and
2. $\mathbf{V}_{PC}$ always *lags* the circuit current by 90°. Notice that this voltage triangle will always be similar to the impedance triangle in Fig. 15.19*b*.

• • •

Example 15.14 In Fig. 15.19, $R = 1\ \text{k}\Omega$, $C = 1\ \mu\text{F}$, and $v_S = 25 \sin 10^3 t$. Calculate the circuit current, and draw the phasor diagram showing $\mathbf{V}_{PS}$, $\mathbf{V}_{PR}$, and $\mathbf{V}_{PC}$. Draw the impedance diagram. What is the generator frequency?

Solution The total resistance is given as 1 kΩ, and the capacitor reactance is

$$X_C = \frac{1}{\omega C} = \frac{1}{10^3 \times 1\ \mu\text{F}} = 10^3\ \Omega, \text{ since } \omega = 10^3 \text{ from } v_S = 25 \sin 10^3 t$$

The circuit impedance is then

$$\mathbf{Z} = R - jX_C = 10^3 - j10^3 = (1 - j)10^3\ \Omega = 1.414\underline{/-45°}\ \text{k}\Omega$$

The circuit current is then

$$\mathbf{I}_P = \frac{\mathbf{V}_{PS}}{\mathbf{Z}} = \frac{25\underline{/0°}\ \mathbf{V}}{1.414\underline{/-45°}\ \text{k}\Omega} = 17.7\underline{/45°}\ \text{mA}$$

This simply shows that the circuit current *leads* the generator voltage by 45°, or the generator voltage *lags* the circuit current by 45°. Since we want the current as a reference, we rewrite these two phasors as

$$\mathbf{V}_{PS} = 25\underline{/-45°}\ \text{V} \qquad \mathbf{I}_P = 17.7\underline{/0°}\ \text{mA}$$

The element voltages are then:

$$\mathbf{V}_{PR} = \mathbf{I}_P R = 17.7\underline{/0°}\ \text{mA} \times 1\ \text{k}\Omega = 17.7\underline{/0°}\ \text{V}$$

$$\mathbf{V}_{PC} = -\mathbf{I}_P jX_C = -17.7\underline{/0°}\ \text{mA} \times j1\ \text{k}\Omega = -17.7j\text{V} = 17.7\underline{/-90°}\ \text{V}$$

The phasor and impedance diagrams are shown in Fig. 15.21*a* and *b*. The generator frequency must be

$$f = \frac{\omega}{2\pi} = \frac{10^3}{2\pi} = 159\ \text{Hz}$$

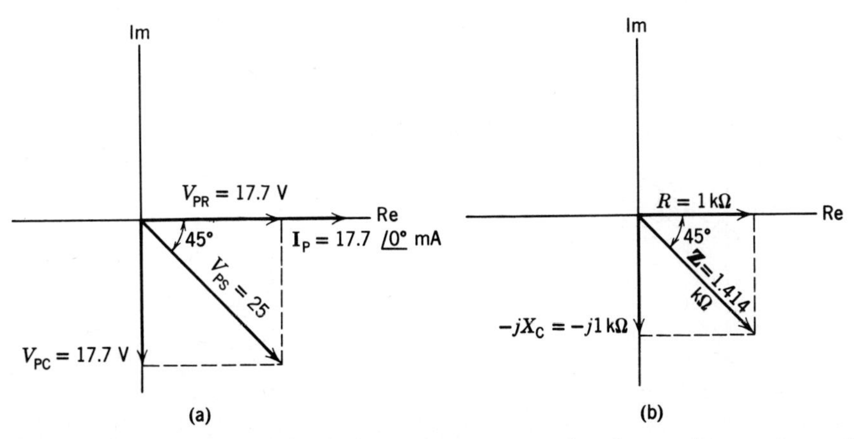

Figure 15.21 (*a*) Phasor diagram. Current and voltage phasors have different scales (*b*) Impedance diagram

• • •

15.7 SERIES *R-L-C* CIRCUIT

Consider the ac circuit in Fig. 15.22*a*. This is a simple ac series circuit since the resistor R, the inductor L, and the capacitor C are all connected in series, and the current must then be the same in all of the elements. We are now interested in determining the impedance of this circuit. In this case, the impedance will consist of a resistive component R, and two reactive components, the inductive reactance jX_L and the capacitive reactance $-jX_C$. Since the elements are all in series, the total impedance is simply the sum of these three components. Thus

$$\mathbf{Z} = R + jX_L - jX_C = R + j(X_L - X_C) \tag{15.18}$$

Equation 15.18 provides the means for determining the impedance $\mathbf{Z}$ of any series R-L-C circuit. The three components of this impedance are shown in Fig. 15.22*b*. It is clear from the figure that the inductive reactance $+jX_L$ and the capacitive reactance $-jX_C$ are two oppositely directed vectors, and their *resultant* (their difference) could be a vector on the $+j$-axis, or it could be a vector on the $-j$-axis. If X_L is greater than X_C, the vector $j(X_L - X_C)$ will be on the positive j-axis as shown in Fig. 15.22*c*. This is said to be an *inductive* circuit because X_L dominates. If X_C is greater than X_L, the vector $j(X_L - X_C)$ will be on the negative j-axis as shown in Fig. 15.22*d*, and this is said to be a *capacitive* circuit.

The impedance diagram in Fig. 15.22*c* is exactly that of a series R-L circuit,

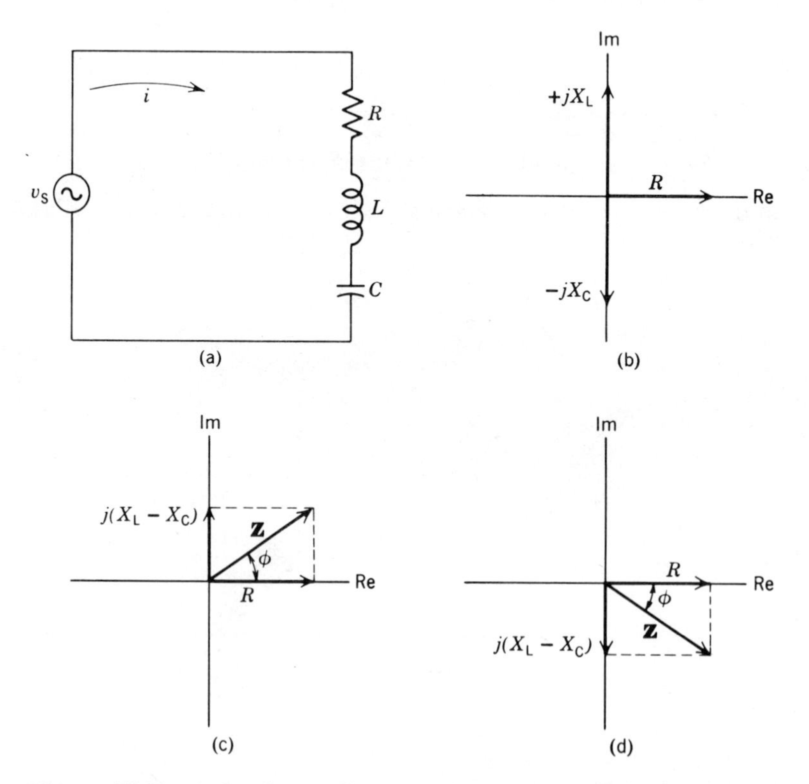

Figure 15.22 Impedance diagrams for a series R-L-C circuit (*a*) A series R-L-C circuit (*b*) The three components of the impedance of a series R-L-C circuit (*c*) Inductive circuit $X_L > X_C$ (*d*) Capacitive circuit $X_C > X_L$

and thus we conclude that a series R-L-C circuit can be reduced to a simple series R-L circuit, as shown in Fig. 15.23*a*, when X_L is greater than X_C. Similarly, the impedance diagram in Fig. 15.22*d* is exactly that of a series R-C circuit, and thus a series R-L-C circuit can be reduced to a simple series R-C circuit, as shown in Fig. 15.23*b*, when X_L is *smaller* than X_C. A very special case occurs when $X_L = X_C$, and the impedance is reduced to

$$\mathbf{Z} = R + j(X_L - X_C) = R + j0 = R$$

In this case, the circuit simply reduces to a resistance R as shown in Fig. 15.23*c*, since the reactive components exactly cancel one another. This condition is called *resonance*, and we will discuss it in detail in Chapter 18.

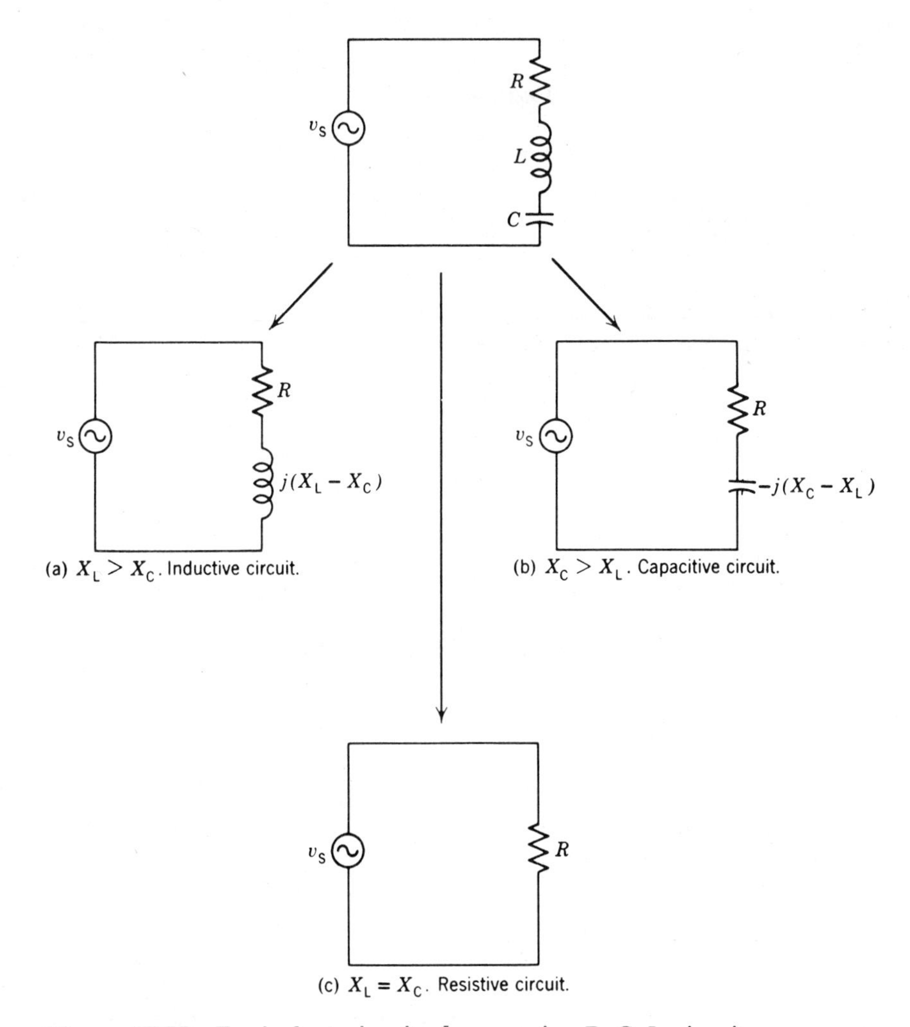

Figure 15.23 Equivalent circuits for a series R-C-L circuit

• • •

Example 15.15 Calculate the impedance of the circuit in Fig. 15.24*a* if $\omega = 2 \times 10^6$ rad/sec. What do the impedance diagram and the equivalent circuit look like?

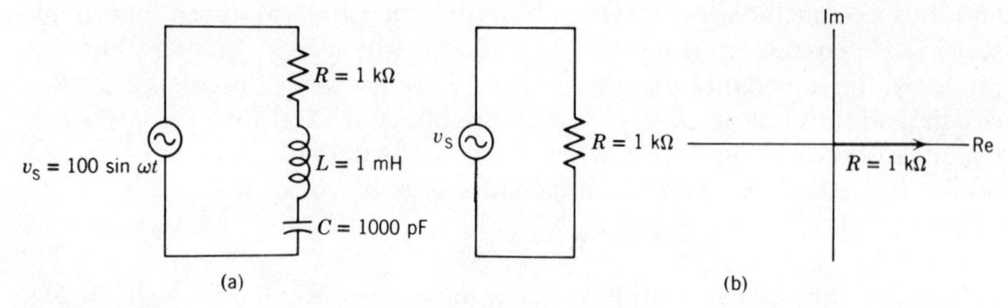

Figure 15.24 A series *R-L-C* circuit (*a*) Circuit values (*b*) Equivalent circuit and impedance diagram for $\omega = 10^6$ rad/sec $X_L = X_C$

Solution The reactances are:

$$X_L = \omega L = 2 \times 10^6 \times 1 \text{ mH} = 2 \text{ k}\Omega$$

$$X_C = \frac{1}{\omega C} = \frac{1}{2 \times 10^6 \times 1000 \text{ pF}} = 0.5 \text{ k}\Omega$$

The impedance is thus

$$\mathbf{Z} = R + j(X_L - X_C) = 1 \text{ k}\Omega + j(2 \text{ k}\Omega - 0.5 \text{ k}\Omega) = (1 + j1.5) \text{ k}\Omega$$

This is an inductive circuit and the impedance diagram looks like Fig. 15.22*c* with $R = 1$ kΩ and $jX = j1.5$ kΩ. The equivalent circuit looks like Fig. 15.23*a* with $R = 1$ kΩ and $j(X_L - X_C) = +j1.5$ kΩ.

• • •

Example 15.16 Repeat Example 15.15 with the frequency reduced by a factor of 4. That is, $\omega = 0.5 \times 10^6$ rad/sec.

Solution The reactances are:

$$X_L = \omega L = 0.5 \times 10^6 \times 1 \text{ mH} = 0.5 \text{ k}\Omega$$

$$X_C = \frac{1}{\omega C} = \frac{1}{0.5 \times 10^6 \times 1000 \text{ pF}} = 2 \text{ k}\Omega$$

The impedance is then

$$\mathbf{Z} = R + j(X_L - X_C) = 1 \text{ k}\Omega + j(0.5 \text{ k}\Omega - 2 \text{ k}\Omega) = (1 - j1.5) \text{ k}\Omega$$

Lowering the frequency has made this circuit appear capacitive, and its impedance diagram and equivalent circuit appear as in Fig. 15.22*d* and Fig. 15.23*b* respectively. Clearly, $R = 1$ kΩ, and $j(X_L - X_C) = -j1.5$ kΩ.

• • •

Example 15.17 Repeat Example 15.15 with the frequency reduced only by a factor of 2. That is, $\omega = 10^6$ rad/sec.

Solution The reactances are:

$$X_L = \omega L = 10^6 \times 1 \text{ mH} = 1 \text{ k}\Omega$$

$$X_C = \frac{1}{\omega C} = \frac{1}{10^6 \times 1000 \text{ pF}} = 1 \text{ k}\Omega$$

The impedance is then

$$\mathbf{Z} = R + j(X - X_C) = 1 \text{ k}\Omega + j(1 - 1) \text{ k}\Omega = 1 \text{ k}\Omega + j0 = 1 \text{ k}\Omega$$

In this case, the reactances exactly cancel one another, and the impedance appears as a pure resistance of $R = 1\ \text{k}\Omega$. The equivalent circuit is simply a 1-kΩ resistor connected across the voltage source v_S, and the impedance diagram is a 1-kΩ resistance drawn on the positive real-axis (Fig. 15.24*b*).

• • •

Since the *R-L-C* circuit in Fig. 15.22*a* is a series circuit, we will use the circuit current as the reference phasor, and thus

$$\mathbf{I}_P = I_P\underline{/0^\circ}$$

We can now use Ohm's Law to find the element voltages as

$$\begin{aligned}\mathbf{V}_{PS} = \mathbf{I}_P\mathbf{Z} &= I_P\underline{/0^\circ}(R + jX_L - jX_C)\\ &= I_PR + jI_PX_L - jI_PX_C\end{aligned}$$

We recognize each of these components as:

$$\text{resistor voltage} = \mathbf{V}_{PR} = I_PR \qquad (15.19a)$$

$$\text{inductor voltage} = \mathbf{V}_{PL} = jI_PX_L \qquad (15.19b)$$

$$\text{capacitor voltage} = \mathbf{V}_{PC} = -jI_PX_C \qquad (15.19c)$$

Thus the KVL equation for the series *R-L-C* circuit is simply

$$\mathbf{V}_{PS} = V_{PR}\underline{/0^\circ} + V_{PL}\underline{/90^\circ} + V_{PC}\underline{/-90^\circ} \qquad (15.20)$$

The element voltages can then be drawn on the phasor diagram for the series *R-L-C* circuit as shown in Fig. 15.25*a*. The capacitor voltage $\mathbf{V}_{PC}$ *lags* the current by 90°, the inductor voltage $\mathbf{V}_{PL}$ *leads* the current by 90°, and the resistor voltage $\mathbf{V}_{PR}$ is in phase with the current.

Now, if X_L is greater than X_C (inductive circuit), $\mathbf{V}_{PL}$ will be greater than

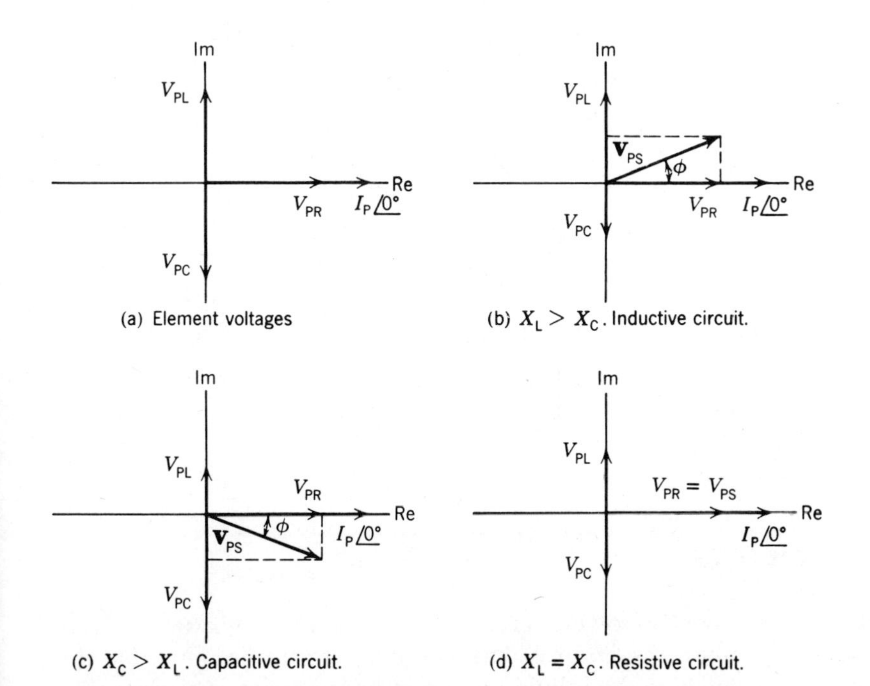

Figure 15.25 Phasor diagrams for a series *R-L-C* circuit

$\mathbf{V}_{PC}$ (why?), and the complete phasor diagram will appear as in Fig. 15.25*b*. On the other hand, if X_C is greater than X_L (capacitive circuit), $\mathbf{V}_{PC}$ will be greater than $\mathbf{V}_{PL}$, and the phasor diagram will appear as in Fig. 15.25*c*. Finally, if $X_L = X_C$ (resistive circuit), then $\mathbf{V}_{PL}$ will equal $\mathbf{V}_{PC}$, and the phasor diagram will appear as in Fig. 15.25*d*. Notice that in this special case $\mathbf{V}_{PR} = \mathbf{V}_{PS}$.

• • •

Example 15.18 Calculate the element voltages in Example 15.15 and draw the complete phasor diagram.

Solution The element voltages are found using Eqs. 15.19. The circuit current is

$$\mathbf{I}_P = \frac{\mathbf{V}_{PS}}{\mathbf{Z}} = \frac{100\underline{/0^\circ}\text{ V}}{(1 + j1.5) \times 1\text{ k}\Omega} = \frac{100\underline{/0^\circ}}{1.8\underline{/56.3^\circ}\text{ k}\Omega} = 55.5\underline{/-56.3^\circ}\text{ mA}$$

Since we want the current as the reference, we rewrite the current and voltage source phasors as ($\mathbf{V}_{PS}$ *leads* $\mathbf{I}_P$ by 56.3°)

$$\mathbf{V}_{PS} = 100\underline{/56.3^\circ}\text{ V} \qquad \mathbf{I}_P = 55.5\underline{/0^\circ}\text{ mA}$$

The element voltages are then:

$$\mathbf{V}_{PR} = I_P R = 55.5\text{ mA} \times 1\text{ k}\Omega = 55.5\text{ V}$$

$$\mathbf{V}_{PL} = jI_P X_L = j55.5 \times \text{mA} \times 2\text{ k}\Omega = j111\text{ V}$$

$$\mathbf{V}_{PC} = -jI_P X_C = -j55.5 \times \text{mA} \times 0.5\text{ k}\Omega = -j27.8\text{ V}$$

The complete phasor diagram is shown in Fig. 15.26.

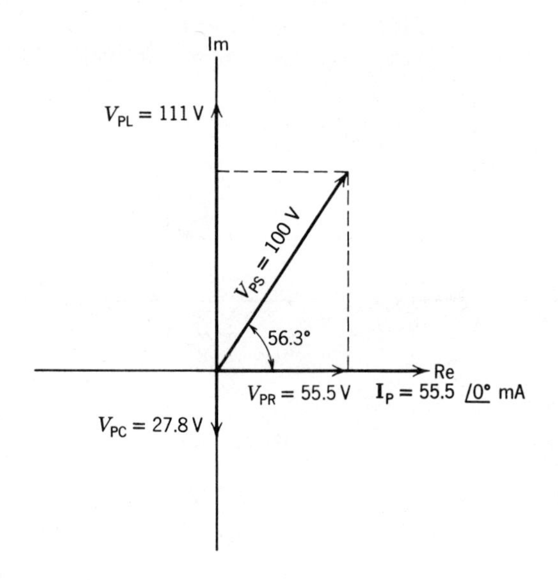

Figure 15.26

• • •

15.8 SERIES ac CIRCUITS

The skill and confidence in solving series ac circuits comes only after studying and working a number of such problems, and it is with this in mind that this section is devoted solely to the solution of a number of series ac circuits. The reader is urged to study each problem carefully.

• • •

Example 15.19 Calculate the circuit current and all of the element voltages for the circuit in Fig. 15.27. In other words, find a *complete solution* for the circuit. Draw the phasor diagram.

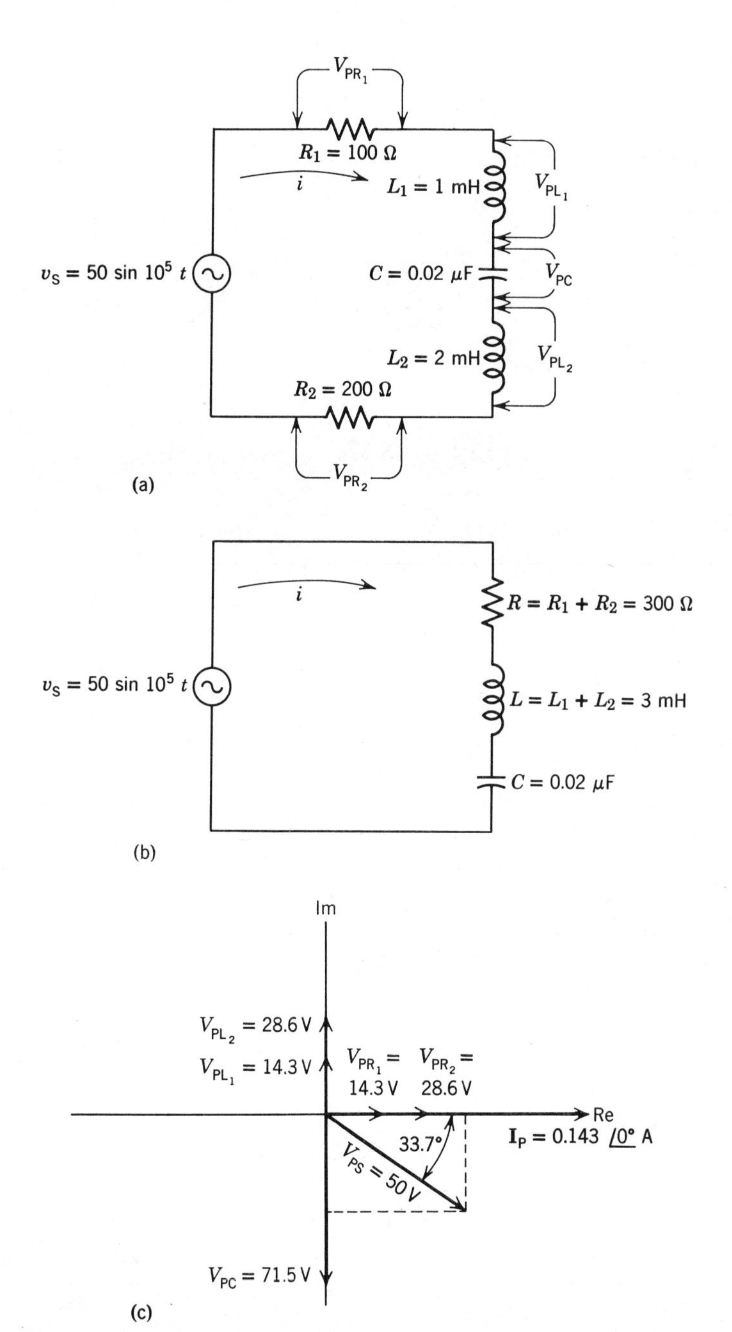

Figure 15.27 (*a*) *R*-*L*-*C* circuit (*b*) Equivalent circuit (*c*) Phasor diagram

Solution Since all of the elements are in *series*, we first simplify to find an equivalent circuit as shown in Fig. 15.27*b*. Notice that the order of the equivalent elements makes no difference here, since we are simply trying to find the total impedance **Z** and thus the circuit current $\mathbf{I}_P$. The reactances in the equivalent circuit are:

$$X_L = \omega L = 10^5 \times 3 \text{ mH} = 300\ \Omega$$

$$X_C = \frac{1}{\omega C} = \frac{1}{10^5 \times 0.02\ \mu\text{F}} = 500\ \Omega$$

The total impedance is then

$$\mathbf{Z} = R + j(X_L - X_C) = 300 + j(300 - 500) = 300 - j200\ \Omega$$

The circuit current is thus

$$\mathbf{I}_P = \frac{\mathbf{V}_{PS}}{\mathbf{Z}} = \frac{50/\underline{0^\circ}\text{ V}}{300 - j200\ \Omega} = \frac{50/\underline{0^\circ}}{351/\underline{-33.7^\circ}} = 0.143/\underline{33.7^\circ}\text{ A}$$

Using the current phasor as a reference, we write

$$\mathbf{V}_{PS} = 50/\underline{-33.7^\circ} \qquad \mathbf{I}_P = 0.143/\underline{0^\circ}$$

We now return to the original circuit in Fig. 15.27*a* to calculate the element voltages.

$$\mathbf{V}_{PR_1} = I_P R_1 = 0.143\text{ A} \times 100\ \Omega = 14.3\text{ V}$$

$$\mathbf{V}_{PL_1} = jI_P X_{L_1} = j0.143\text{ A} \times 10^5 \times 1\text{ mH} = j14.3\text{ V}$$

$$\mathbf{V}_{PC} = -jI_P X_C = -j0.143\text{ A} \times 500\ \Omega = -j71.5\text{ V}$$

$$\mathbf{V}_{PL_2} = jI_P X_{L_2} = j0.143\text{ A} \times 10^5 \times 2\text{ mH} = j28.6\text{ V}$$

$$\mathbf{V}_{PR_2} = I_P R_2 = 0.143\text{ A} \times 200\ \Omega = 28.6\text{ V}$$

The complete phasor diagram is shown in Fig. 15.27*c*.

• • •

Example 15.20 The peak values of the voltages in Fig. 15.28*a* are measured with an oscilloscope and they are

$$V_{PR} = 60\text{ V} \qquad V_{PL} = 50\text{ V} \qquad V_{PC} = 70\text{ V}$$

Calculate the peak value of the circuit current, and the peak value of the signal generator voltage. What must be the frequency of the signal generator?

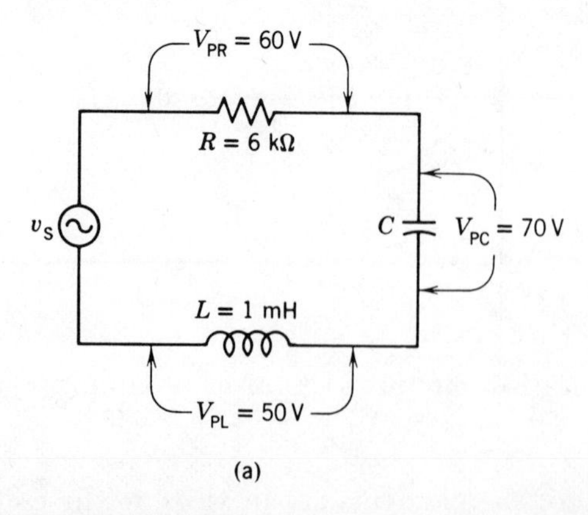

Figure 15.28 (*a*) *R-L-C* circuit

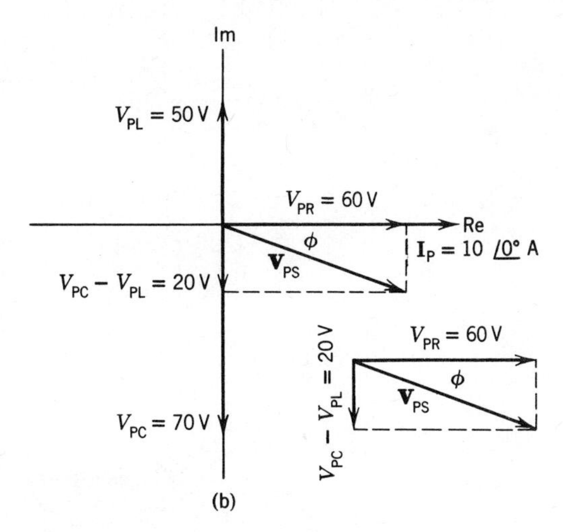

Figure 15.28 (*b*) Phasor diagram

Solution The circuit current can be found using R and V_{PR}. Thus

$$\mathbf{I}_P = \frac{\mathbf{V}_{PR}}{R} = \frac{60\ V}{6\ k\Omega} = 10\underline{/0^\circ}\ \text{mA}$$

We are using $\mathbf{I}_P$ as the reference. The generator voltage $\mathbf{V}_{PS}$ can be found from the phasor diagram in Fig. 15.28*b*. We must solve the right triangle shown for V_{PS} and ϕ. Thus

$$V_{PS} = \sqrt{V_{PR}^2 + (V_{PC} - V_{PL})^2} = \sqrt{60^2 + 20^2} = \sqrt{4000} = 63.3\ \text{V}$$

The phase angle ϕ is

$$\phi = \tan^{-1}\frac{(V_{PC} - V_{PL})}{V_{PR}} = \tan^{-1}\frac{20}{60} = \tan^{-1}\frac{1}{3} = 18.4^\circ$$

Thus

$$\mathbf{V}_{PS} = V_{PS}\underline{/-\phi} = 63.3\underline{/-18.4^\circ}\ \text{V}$$

We can find the frequency using the voltage across the inductor, since

$$V_{PL} = I_P X_L = I_P \omega L = I_P 2\pi f L$$

Thus

$$f = \frac{V_{PL}}{2\pi I_P L} = \frac{50\ \text{V}}{2\pi \times 10\ \text{mA} \times 1\ \text{mH}} = 794\ \text{kHz}$$

• • •

Example 15.21 The peak values of the voltages in Fig. 15.29 are measured with an oscilloscope, and they are

$$V_{PC} = 26\ \text{V} \qquad V_{PL} = 63\ \text{V} \qquad V_{PR} = 10\ \text{V}$$

What are the values of the capacitor C and the inductor L?

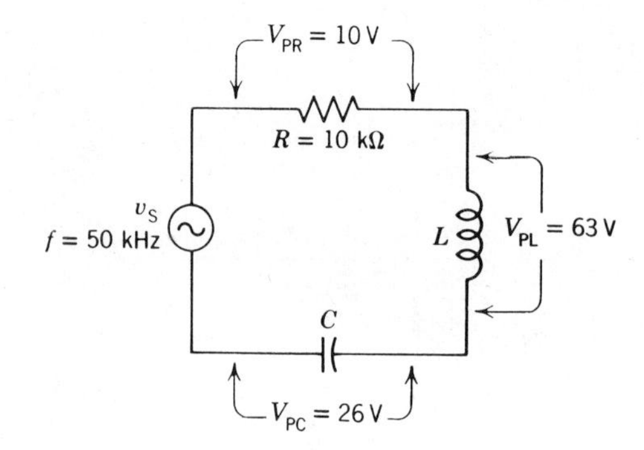

Figure 15.29

Solution The circuit current is

$$\mathbf{I}_P = \frac{\mathbf{V}_{PR}}{R} = \frac{10\text{ V}}{10\text{ k}\Omega} = 1.0\underline{/0^\circ}\text{ mA}$$

The value of L is found from

$$V_{PL} = I_P X_L = 2\pi f I_P L$$

Thus

$$L = \frac{V_{PL}}{2\pi f I_P} = \frac{63\text{ V}}{2\pi \times 50\text{ kHz} \times 1\text{ mA}} = 0.2\text{ H}$$

The value of C is found from

$$V_{PC} = I_P X_C = \frac{I_P}{\omega C} = \frac{I_P}{2\pi f C}$$

Thus

$$C = \frac{I_P}{2\pi f V_{PC}} = \frac{1\text{ mA}}{2\pi \times 50\text{ kHz} \times 26\text{ V}} = 122\text{ pF}$$

• • •

SUMMARY

The opposition to a sinusoidal current exhibited by a resistor is called *resistance*, and since the resistor voltage and current are *in phase*, resistance can be represented by a vector drawn along the positive real-axis in the complex plane. An inductor also exhibits an opposition to a sinusoidal current and this is defined as *inductive reactance* X_L. The voltage across an ideal inductor *leads* the inductor current by 90°, and this 90° phase angle is accounted for by plotting X_L on the *positive* j-axis in the complex plane. The opposition to a sinusoidal current exhibited by a capacitor is defined as *capacitive reactance* X_C, and since the capacitor voltage *lags* the capacitor current by 90°, X_C is plotted on the negative j-axis. The vector sum of R, X_L, and X_C in the complex plane is defined as *impedance* $\mathbf{Z}$, and this is the total opposition to a sinusoidal current exhibited by a circuit consisting of resistance, inductance, and capacitance connected in *series*. Impedance is a vector in the complex plane having *both magnitude* Z *and angle* ϕ. The *phase* angle ϕ shows the phase between the applied sinusoidal voltage and the sinusoidal current in a series R-L-C circuit. ϕ may have any value between $+90°$ and $-90°$ in a series R-L-C circuit.

Since the current is the same in all parts of a series *R-L-C* circuit, the current phasor is usually chosen as a reference. The element voltages can then be calculated and drawn in the complex plane along with the current phasor; this is called a *phasor diagram.* Ohm's Law and KVL can be applied to a series ac circuit, provided that all quantities are treated as *vectors.* Since sinusoidal voltages produce sinusoidal currents of the *same* frequencies in *R-L-C* circuits, and vice versa, the ωt term can be omitted and the voltages and currents can be represented as phasors in the complex plane. This means that most ac circuit problems reduce to finding solutions for right triangles, and this is of course a great simplification.

We must always remember that the voltage and current phasors in the complex plane really represent voltage and current vectors which have an angular velocity ω, and thus they represent *sinusoidal* voltage and current *wave forms* having a frequency $f = \omega/2\pi$ which are shifted along the time scale with respect to one another.

GLOSSARY

Capacitive reactance X_C. *The opposition to a sinusoidal current exhibited by a capacitance.*

Impedance Z. *The vector sum of resistance, inductive reactance, and capacitive reactance.*

Inductive reactance X_L. *The opposition to a sinusoidal current exhibited by an inductance.*

Lagging phase angle. *An angle in the clockwise direction from a reference phasor in the complex plane.*

Leading phase angle. *An angle in the counterclockwise direction from a reference phasor in the complex plane.*

Phase angle ϕ. *The angle between two phasors on the complex plane.*

Phasor. *A vector in the complex plane which represents a sinusoidal wave.*

Phasor diagram. *The phasors representing the voltages and currents in an ac circuit.*

Reactance X. *The imaginary part of impedance due to capacitance and inductance.*

Reference phasor. *Any phasor arbitrarily given a phase angle of 0°; it is drawn on the positive real-axis in the complex plane on the phasor diagram.*

Resonance. *An R-L-C series circuit in which $X_L = X_C$.*

IMPORTANT RELATIONSHIPS

- Ohm's Law for ac circuits: $\mathbf{I}_P = \dfrac{\mathbf{V}_P}{\mathbf{Z}}$
- Impedance: $\mathbf{Z} = Z\underline{/\phi} = R \pm jX$

$$R = Z \cos \phi \qquad Z = \sqrt{R^2 + X^2}$$

$$X = Z \sin \phi \qquad \phi = \tan^{-1}\frac{X}{R}$$

- Reactance:

$$X_L = \omega L = 2\pi f L \qquad X_C = \frac{1}{\omega C} = \frac{1}{2\pi f C}$$

- Series R-L circuit:

$$\mathbf{Z} = R + jX_L \qquad \mathbf{V}_{PS} = \mathbf{I}_P\mathbf{Z}$$
$$\mathbf{V}_{PR} = \mathbf{I}_P R \qquad \mathbf{V}_{PL} = j\mathbf{I}_P X_L$$

- Series R-C circuit:

$$\mathbf{Z} = R - jX_C \qquad \mathbf{V}_{PS} = \mathbf{I}_P\mathbf{Z}$$
$$\mathbf{V}_{PR} = \mathbf{I}_P R \qquad \mathbf{V}_{PC} = j\mathbf{I}_P X_C$$

- Series R-L-C circuit:

$$\mathbf{Z} = R + j(X_L - X_C) \qquad \mathbf{V}_{PS} = \mathbf{I}_P\mathbf{Z}$$
$$\mathbf{V}_{PR} = \mathbf{I}_P R \qquad \mathbf{V}_{PL} = j\mathbf{I}_P X_L \quad \mathbf{V}_{PC} = -j\mathbf{I}_P X_C$$

REVIEW QUESTIONS

1. A sinusoidal voltage applied to an R-L-C circuit will always produce currents and voltages having the same frequency as the applied voltage. (TF)
2. The phase angle between the voltage across and the current through a resistor is _____ degrees.
3. A phasor which is shifted clockwise from a reference phasor is said to (lead, lag) the reference phasor.
4. Since the currents and voltages in an R-L-C circuit all have the same frequency, the term _____ can be eliminated in the phasor diagram.
5. The *real* component of impedance is _________.
6. The *imaginary* component of impedance is _____.
7. An impedance with a reactance of $-jX$ will have a (positive, negative) phase angle.
8. The voltage across an inductance always (leads, lags) the current in the inductance.
9. The opposition to a sinusoidal current exhibited by an inductor is called _____ _____.
10. Inductive reactance X_L is plotted on the _____ j-axis in the complex plane.
11. Inductive reactance X_L will increase with increasing frequency. (TF)
12. The opposition to a sinusoidal current exhibited by a capacitor is called _____ _____.
13. Capacitive reactance X_C is plotted on the _____ j-axis in the complex plane.
14. The voltage across a capacitor always (leads, lags) the capacitor current.
15. Capacitive reactance X_C will increase with increasing frequency. (TF)
16. In a series R-L circuit, the applied voltage always leads the circuit current by some angle between _____ degrees and _____ degrees.
17. In a series R-C circuit, the phase angle between the applied voltage and the circuit current must be between _____ degrees and _____ degrees.
18. In a series R-L-C circuit, the phase angle between the applied voltage and the circuit current must be between _____ degrees and _____ degrees.
19. A series R-L-C circuit can always be reduced to a series R-L circuit, a series R-C circuit, or a simple resistive circuit. (TF)

20. When $X_L = X_C$ in a series R-L-C circuit, this condition is called ______.
21. KVL can be used with series R-L-C circuits if the voltages are all represented as ______.

PROBLEMS

1. Draw the phasors and the sinusoidal wave forms for $\mathbf{I}_P = 10\underline{/0^\circ}$ A and $\mathbf{V}_P = 50\underline{/30^\circ}$ V.
2. Draw the phasors and the sinusoidal wave forms for $\mathbf{I}_P = 6\underline{/10^\circ}$ mA, and $\mathbf{V}_P = 25\underline{/-40^\circ}$ V.
3. Calculate the impedance $\mathbf{Z}$ and draw the impedance diagram for Problem 1.
4. Calculate the impedance $\mathbf{Z}$ and draw the impedance diagram for Problem 2.
5. In Fig. 15.4*a*, $\mathbf{Z} = 200\underline{/45^\circ}$ and $\mathbf{V}_P = 10\underline{/0^\circ}$. Calculate the circuit current. Draw a phasor diagram and an impedance diagram.
6. What must be the impedance $\mathbf{Z}$ in Fig. 15.4*a* if $\mathbf{I}_P = 0.2\underline{/0^\circ}$ mA and $\mathbf{V}_P = 18\underline{/-30^\circ}$ V? Draw the phasor and impedance diagrams.
7. Calculate the voltage necessary to produce a current of $15\underline{/0^\circ}$ mA in Fig. 15.4*a* if $\mathbf{Z} = 6\underline{/25^\circ}$ kΩ. Draw the phasor and impedance diagrams.
8. Calculate the resistance and reactance in Problem 5. Draw the impedance diagram and an equivalent circuit.
9. Calculate the resistance and reactance in Problem 6. Draw the impedance diagram and an equivalent circuit.
10. Calculate the resistance and reactance in Problem 7. Draw the impedance diagram and an equivalent circuit.
11. Change $\mathbf{Z} = 90 + j60$ into a polar form, and draw an impedance diagram.
12. Change $\mathbf{Z} = (3 - j4)$ kΩ into polar form, and draw an impedance diagram.
13. What is the inductive reactance of a 4 mH inductor at 25 kHz?
14. At what frequency will a 100 μH inductor have an inductive reactance of 753 Ω?
15. What must be the inductance of an inductor having an inductive reactance of 1 kΩ at 50 kHz?
16. What is the capacitive reactance of a 0.01 μF capacitor at 1 kHz? At 1 MHz?
17. At what frequency will a 2000 pF capacitor have a capacitive reactance of 1 kΩ?
18. What value of capacitance must be used to provide a capacitive reactance of 800 Ω at 20 kHz?
19. Calculate the circuit current and the element voltages in Fig. 15.16*a* if $v_S = 40 \sin 10^3 t$, $R = 500\ \Omega$, and $L = 200$ mH. Draw the phasor diagram.
20. What must be the value of X_L in Fig. 15.16*a* if $v_S = 25 \sin 10^5 t$, $\mathbf{I}_P = 10\underline{/-30^\circ}$ mA, and $R = 2.16$ kΩ? What is L? Draw the phasor diagram.
21. The peak voltage across R and L in Fig. 15.16*a* are measured with an oscilloscope as $V_{PR} = 28$ V and $V_{PL} = 15$ V. Calculate the source voltage $\mathbf{V}_{PS}$, the circuit current, the inductive reactance, and the impedance if $R = 3$ kΩ. (Hint: phasor and impedance diagrams will be a great aid.)
22. Calculate the circuit current and the element voltages in Fig. 15.19*a* if $v_S = 40 \sin 10^5 t$, $R = 500\ \Omega$ and $C = 0.05\ \mu$F. Draw the phasor diagram.

23. What must be the value of X_C and C in Fig. 15.19*a* if $v_S = 25 \sin 10^5 t$, $\mathbf{I}_P = 10\underline{/30^\circ}$ mA? What is the generator frequency? Draw the phasor diagram.

24. The peak voltage across R and C in Fig. 15.19*a* are measured with an oscilloscope as $V_{PR} = 31$ V and $V_{PC} = 31$ V. Calculate the source voltage $\mathbf{V}_{PS}$, the circuit current, the capacitive reactance, and the impedance if $R = 10$ kΩ.

25. Draw the simplest equivalent circuit for the R-L-C circuit in Fig. 15.22*a* if $R = 2$ kΩ, $L = 10$ mH, $C = 125$ pF, and $v_S = 65 \sin 10^6 t$. Draw the impedance diagram.

26. Calculate the circuit current and the element voltages for Problem 25. Draw a complete phasor diagram.

27. Repeat Problem 25 if $v_S = 65 \sin (8 \times 10^5)t$.

28. Calculate the circuit current and the element voltages for Problem 27. Draw a complete phasor diagram.

16

SERIES-PARALLEL ac CIRCUITS

Ohm's Law can be applied to ac circuits provided we use phasors (vectors in the complex plane) to represent voltages and currents, and a vector quantity called impedance to represent the opposition to a sinusoidal current. KVL can also be applied to ac circuits, and this involves finding the vector sum of the element voltages; this is most easily accomplished by drawing a phasor diagram. In this chapter we will extend our knowledge of ac circuits by discovering how to combine impedances in series and parallel. This will allow us to solve more complex ac circuits by *simplification*; recall that this was the first technique used to solve dc circuits containing *resistances.* We will then define the reciprocal of impedance, admittance. Admittance can be used to simplify the analysis of *parallel* circuits, and it is used in the solution of ac circuits by node equations. We can apply KCL, the voltage divider theorem, the current divider theorem, and mesh and node equations to ac circuits, provided vector quantities are used, and we will demonstrate the use of these techniques in solving ac circuits.

16.1 IMPEDANCES IN SERIES

The impedance $\mathbf{Z}$ of the series R-L-C circuit in Fig. 16.1a is known to be

$$\mathbf{Z} = R + jX_{\mathrm{L}} - jX_{\mathrm{C}} \tag{16.1}$$

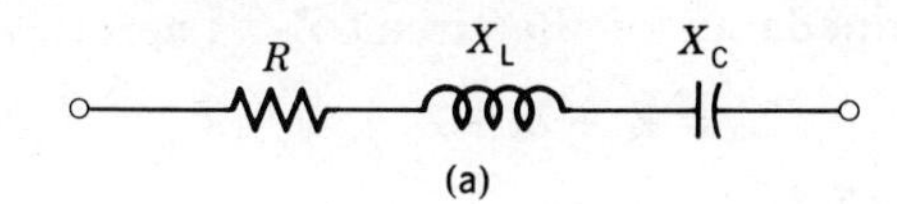

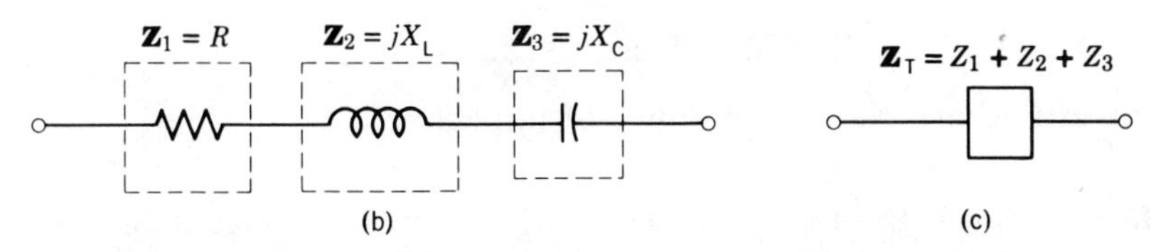

Figure 16.1 (*a*) A series *R*-*L*-*C* circuit (*b*) Three impedances in series (*c*) Equivalent circuit of (*a*) and (*b*)

Suppose we treat each of the elements in this figure as a *separate* impedance; there are then three impedances as shown in Fig. 16.1*b*, and they are:

$$\mathbf{Z}_1 = R + j0 = R;\ \mathbf{Z}_2 = 0 + jX_L = jX_L;\ \mathbf{Z}_3 = 0 - jX_C = -jX_C$$

$\mathbf{Z}_1$ is the special case of an impedance consisting of a resistance only, $\mathbf{Z}_2$ is an impedance consisting of inductive reactance only, and $\mathbf{Z}_3$ is an impedance composed solely of capacitive reactance. But, from Eq. 16.1, it is clear that the *total* series impedance is simply the sum of these terms, that is,

$$\mathbf{Z}_T = \mathbf{Z}_1 + \mathbf{Z}_2 + \mathbf{Z}_3$$

for a series ac circuit, and we can thus use the equivalent circuit shown in Fig. 16.1*c*.

This idea can be extended to any number of impedances connected in *series* by applying KVL to the circuit in Fig. 16.2*a*. Notice first of all that the

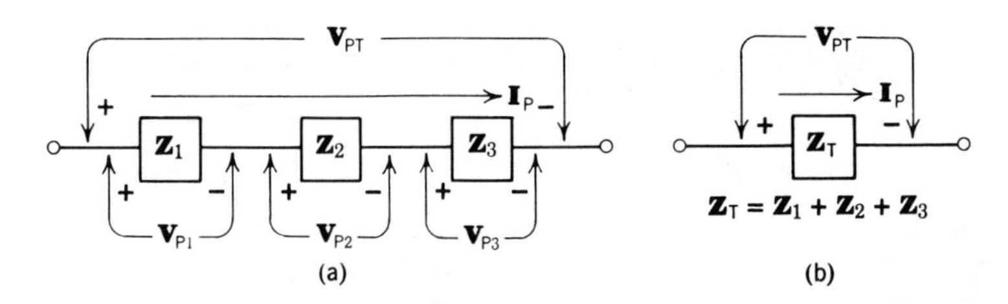

Figure 16.2 (*a*) Impedances connected in series (*b*) Equivalent circuit of (*a*)

impedances are connected in series, and therefore the current must be the *same* in every impedance. The voltage across each impedance is

$$\mathbf{V}_{P1} = \mathbf{I}_P\mathbf{Z}_1 \qquad \mathbf{V}_{P2} = \mathbf{I}_P\mathbf{Z}_2 \qquad \mathbf{V}_{P3} = \mathbf{I}_P\mathbf{Z}_3$$

and the application of KVL yields

$$\begin{aligned}\mathbf{V}_{PT} &= \mathbf{V}_{P1} + \mathbf{V}_{P2} + \mathbf{V}_{P3}\\ &= \mathbf{I}_P\mathbf{Z}_1 + \mathbf{I}_P\mathbf{Z}_2 + \mathbf{I}_P\mathbf{Z}_3\\ &= \mathbf{I}_P(\mathbf{Z}_1 + \mathbf{Z}_2 + \mathbf{Z}_3)\end{aligned}$$

Thus

$$\frac{\mathbf{V}_{PT}}{\mathbf{I}_P} = \mathbf{Z}_1 + \mathbf{Z}_2 + \mathbf{Z}_3$$

But, $\mathbf{V}_{PT}/\mathbf{I}_P$ is the total impedance of the circuit $\mathbf{Z}_T$. Therefore,

$$\mathbf{Z}_T = \mathbf{Z}_1 + \mathbf{Z}_2 + \mathbf{Z}_3$$

This idea can be immediately extended to a series circuit having any number of impedances, and thus the total impedance $\mathbf{Z}_T$ is

$$\mathbf{Z}_T = \mathbf{Z}_1 + \mathbf{Z}_2 + \cdots + \mathbf{Z}_N \tag{16.2}$$

for *series circuits*, where N is the number of impedances.

• • •

Example 16.1 What is the total impedance of the circuit in Fig. 16.3?

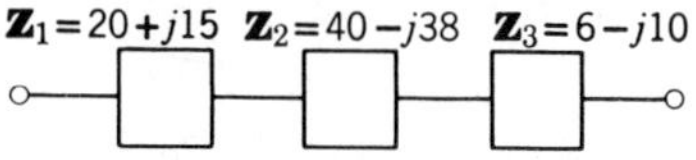

$\mathbf{Z}_1 = 20 + j15$ $\mathbf{Z}_2 = 40 - j38$ $\mathbf{Z}_3 = 6 - j10$

Figure 16.3 Example 16.1

Solution Using Eq. 16.2,

$$\mathbf{Z}_T = \mathbf{Z}_1 + \mathbf{Z}_2 + \mathbf{Z}_3 = (20 + j\,15) + (40 - j\,38) + (6 - j\,10) = (66 - j\,33)$$
$$= 33(2 - j)\ \Omega$$

• • •

Example 16.2 What is the total impedance of the circuit in Fig. 16.4*a*?

Solution (*a*) The circuit can be simplified by finding the impedance of each

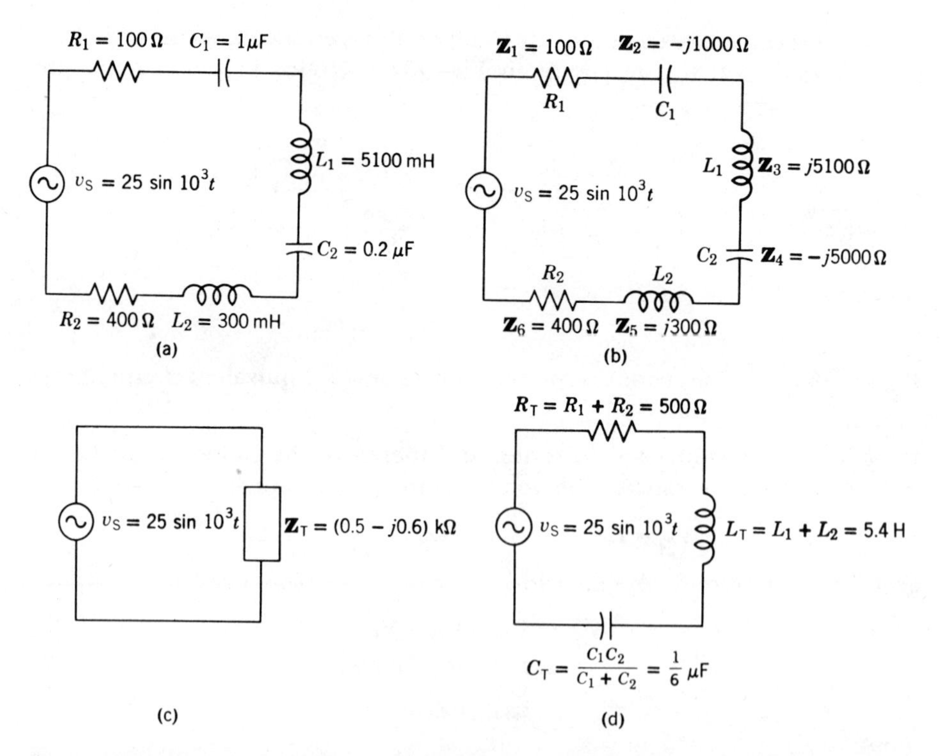

Figure 16.4 Example 16.2 (*a*) A series ac circuit (*b*) Finding impedances (*c*) Equivalent circuit (*d*) Combining elements

element as shown in Fig. 16.4*b*. These impedances are:

$$\mathbf{Z}_1 = R_1 = 100\ \Omega \qquad \mathbf{Z}_2 = -jX_{C_1} = -j\,1000\ \Omega \qquad \mathbf{Z}_3 = jX_{L_1} = j5100\ \Omega$$
$$\mathbf{Z}_4 = -jX_{C_2} = -j5000\ \Omega \qquad \mathbf{Z}_5 = jX_{L_2} = j\,300\ \Omega \qquad \mathbf{Z}_6 = R_2 = 400\ \Omega$$

Then, using Eq. 16.2

$$\begin{aligned}\mathbf{Z}_T &= \mathbf{Z}_1 + \mathbf{Z}_2 + \mathbf{Z}_3 + \mathbf{Z}_4 + \mathbf{Z}_5 + \mathbf{Z}_6 \\ &= 100 - j\,1000 + j5100 - j5000 + j300 + 400 = 500 - j600 \\ &= (0.5 - j0.6)\ \text{k}\Omega\end{aligned}$$

The equivalent circuit is shown in Fig. 16.4*c*.
(*b*) The total impedance can also be found by first combining ***like*** elements in series as shown in Fig. 16.4*d*. The total resistance is

$$R_T = R_1 + R_2 = 100 + 400 = 500\ \Omega$$

The total inductance is

$$L_T = L_1 + L_2 = 5100\ \text{mH} + 300\ \text{mH} = 5400\ \text{mH} = 5.4\ \text{H}$$

The total capacitance is

$$C_T = \frac{C_1 C_2}{C_1 + C_2} = \frac{1\ \mu\text{F} \times 0.2\ \mu\text{F}}{1\ \mu\text{F} + 0.2\ \mu\text{F}} = \frac{1}{6}\ \mu\text{F}$$

The total impedance of this series *R-L-C* circuit is then,

$$\begin{aligned}\mathbf{Z}_T &= R_T + j(X_{L_T} - X_{C_T}) = 500 + j\left(10^3 \times 5.4 - \frac{1}{10^3 \times 1/6 \times 10^6}\right) \\ &= 500 + j\,(5.4 - 6.0)10^3 = (0.5 - j\,0.6)\ \text{k}\Omega\end{aligned}$$

• • •

.2 IMPEDANCES IN PARALLEL

The circuit in Fig. 16.5*a* shows two impedances connected in parallel. Each impedance, $\mathbf{Z}_1$ or $\mathbf{Z}_2$, may consist of a resistance, an inductance, or a capacitance *alone*, or of a *combination* of these elements. Since the impedances

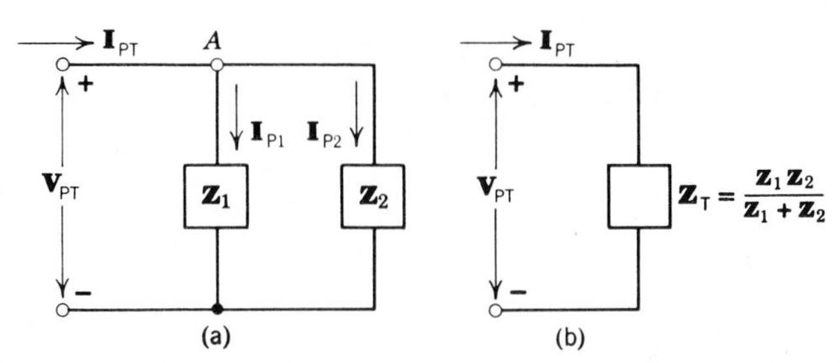

Figure 16.5 (*a*) Two impedances connected in parallel (*b*) Equivalent circuit of (*a*)

are connected in parallel, it is clear that the voltage is the same across each of them. The current in each impedance can then be found as:

$$\mathbf{I}_{P1} = \frac{\mathbf{V}_{PT}}{\mathbf{Z}_1} \qquad \mathbf{I}_{P2} = \frac{\mathbf{V}_{PT}}{\mathbf{Z}_2}$$

The application of KCL at node A yields,

$$\mathbf{I}_{PT} = \mathbf{I}_{P1} + \mathbf{I}_{P2}$$

Now, by substituting the above relationships for $\mathbf{I}_{P1}$ and $\mathbf{I}_{P2}$ into the KCL equation, we find

$$\mathbf{I}_{PT} = \mathbf{I}_{P1} + \mathbf{I}_{P2} = \frac{\mathbf{V}_{PT}}{\mathbf{Z}_1} + \frac{\mathbf{V}_{PT}}{\mathbf{Z}_2} = \mathbf{V}_{PT}\left(\frac{1}{\mathbf{Z}_1} + \frac{1}{\mathbf{Z}_2}\right)$$

Dividing both sides by $\mathbf{V}_{PT}$,

$$\frac{\mathbf{I}_{PT}}{\mathbf{V}_{PT}} = \frac{1}{\mathbf{Z}_1} + \frac{1}{\mathbf{Z}_2}$$

But $\mathbf{V}_{PT}/\mathbf{I}_{PT}$ is the total impedance of the circuit $\mathbf{Z}_T$. Thus we take the reciprocal of this equation to obtain the total impedance $\mathbf{Z}_T$.

$$\mathbf{Z}_T = \frac{1}{\dfrac{1}{\mathbf{Z}_1} + \dfrac{1}{\mathbf{Z}_2}} = \frac{\mathbf{Z}_1\mathbf{Z}_2}{\mathbf{Z}_1 + \mathbf{Z}_2} \tag{16.3}$$

where $\mathbf{Z}_1$ and $\mathbf{Z}_2$ are in *parallel*. The equivalent circuit is shown in Fig. 16.5*b*.

We notice that *impedances* in *parallel* combine exactly like *resistances* in *parallel*, and we can therefore generalize this result. Thus, the total impedance $\mathbf{Z}_T$ of any number (N) of impedances connected in parallel is

$$\mathbf{Z}_T = \frac{1}{\dfrac{1}{\mathbf{Z}_1} + \dfrac{1}{\mathbf{Z}_2} + \cdots + \dfrac{1}{\mathbf{Z}_N}} \tag{16.4}$$

Just as in the case of parallel resistances, it will usually be easier to combine parallel impedances by taking them two at a time and using Eq. 16.3.

• • •

Example 16.3 What is the total impedance of the circuit in Fig. 16.5*a* if $\mathbf{Z}_1 = 3 + j4$ and $\mathbf{Z}_2 = 2 - j3$?

Solution We use Eq. 16.3,

$$\mathbf{Z}_T = \frac{\mathbf{Z}_1\mathbf{Z}_2}{\mathbf{Z}_1 + \mathbf{Z}_2} = \frac{(3 + j4)(2 - j3)}{(3 + j4) + (2 - j3)} = \frac{18 - j}{5 + j} = \frac{18 - j}{5 + j} \times \frac{5 - j}{5 - j}$$

$$= \frac{89 - j23}{26} = 3.42 - j0.885$$

• • •

Example 16.4 Find the total impedance of the circuit in Fig. 16.6*a*.

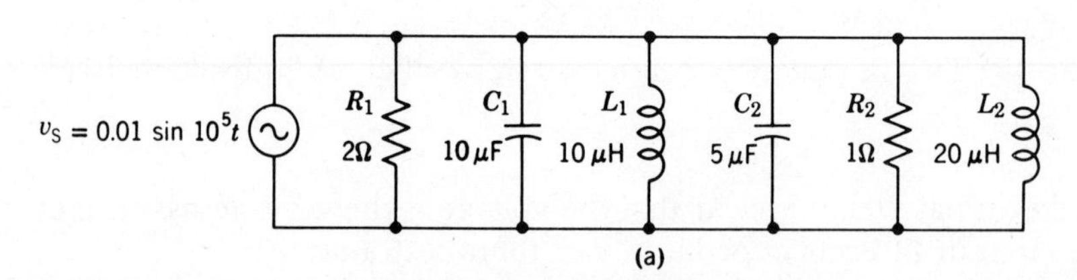

Figure 16.6 Example 16.4 (*a*) A parallel circuit

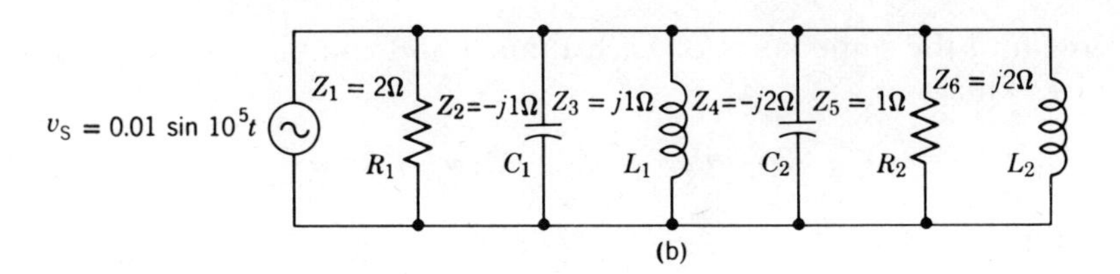

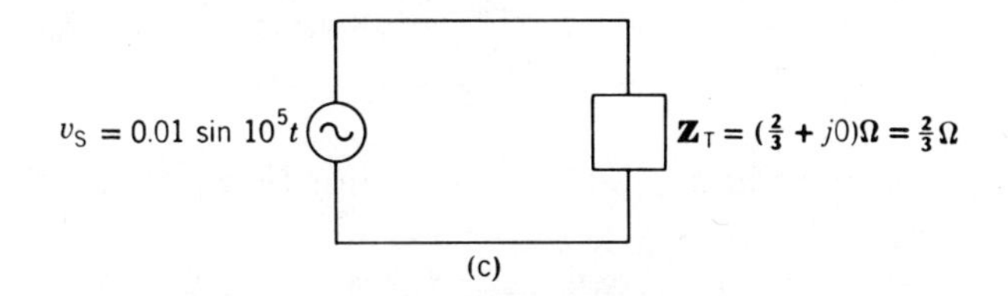

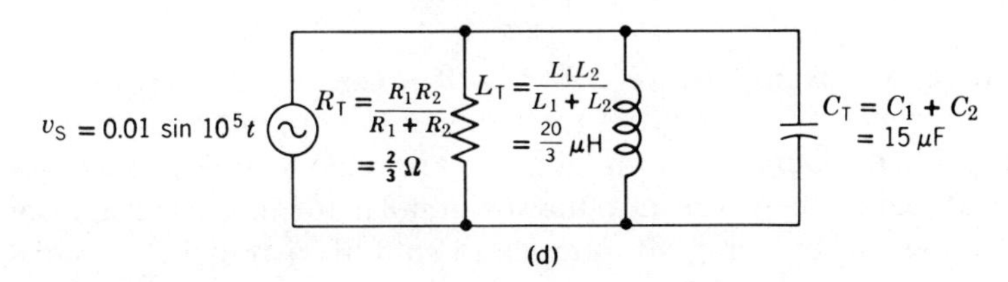

Figure 16.6 Example 16.4 (*b*) Finding impedances (*c*) Equivalent circuit (*d*) Combining elements

Solution (*a*) The impedance of each element can be found as shown in Fig. 16.6*b*. These impedances are:

$$\mathbf{Z}_1 = R_1 = 2\ \Omega \qquad \mathbf{Z}_2 = -jX_{C_1} = -j1\ \Omega \qquad \mathbf{Z}_3 = jX_{L_1} = j1\ \Omega$$
$$\mathbf{Z}_4 = -jX_{C_2} = -j2\ \Omega \qquad \mathbf{Z}_5 = R_2 = 1\ \Omega \qquad \mathbf{Z}_6 = jX_{L_2} = j2\ \Omega$$

The total impedance $\mathbf{Z}_T$ is then found using Eq. 16.4,

$$\mathbf{Z}_T = \frac{1}{\dfrac{1}{\mathbf{Z}_1} + \dfrac{1}{\mathbf{Z}_2} + \dfrac{1}{\mathbf{Z}_3} + \dfrac{1}{\mathbf{Z}_4} + \dfrac{1}{\mathbf{Z}_5} + \dfrac{1}{\mathbf{Z}_6}}$$
$$= \frac{1}{\dfrac{1}{2} + \dfrac{1}{-j1} + \dfrac{1}{j1} + \dfrac{1}{-j2} + \dfrac{1}{1} + \dfrac{1}{j2}}$$

Multiplying numerator *and* denominator by $j2$,

$$\mathbf{Z}_T = \frac{j2}{j - 2 + 2 - 1 + j2 + 1} = \frac{j2}{j3} = \left(\frac{2}{3} + j0\right)\Omega = \frac{2}{3}\Omega$$

The equivalent circuit is shown in Fig. 16.6*c*.
(*b*) The total impedance can also be found by first combining *like* elements in *parallel* as shown in Fig. 16.6*d*. The combinations are:

$$R_T = \frac{R_1R_2}{R_1 + R_2} = \frac{2 \times 1}{2 + 1} = \frac{2}{3}\Omega$$
$$L_T = \frac{L_1L_2}{L_1 + L_2} = \frac{10\ \mu\text{H} \times 20\ \mu\text{H}}{10\ \mu\text{H} + 20\ \mu\text{H}} = \frac{20}{3}\mu\text{H}$$
$$C_T = C_1 + C_2 = 10\ \mu\text{F} + 5\ \mu\text{F} = 15\ \mu\text{F}$$

We must now find the impedance of each branch and combine them using Equation 16.4. Thus,

$$\mathbf{Z}_1 = R_T = \tfrac{2}{3}\,\Omega \qquad \mathbf{Z}_2 = jX_{L_T} = j\tfrac{2}{3} \qquad \mathbf{Z}_3 = -jX_{C_T} = -j\tfrac{2}{3}$$

Then using Eq. 16.4,

$$\mathbf{V}_T = \frac{1}{\dfrac{1}{\mathbf{Z}_1} + \dfrac{1}{\mathbf{Z}_2} + \dfrac{1}{\mathbf{Z}_3}} = \frac{1}{\dfrac{1}{\frac{2}{3}} + \dfrac{1}{j\frac{2}{3}} + \dfrac{1}{-j\frac{2}{3}}}$$

multiplying both numerator and denominator by $j\frac{2}{3}$,

$$\mathbf{Z}_T = \frac{j\frac{2}{3}}{j + 1 - 1} = (\tfrac{2}{3} + j0)\,\Omega = \tfrac{2}{3}\,\Omega$$

• • •

From the above examples, we note that it is considerably more tedious to combine *impedances* in *parallel* than it is to combine parallel resistances. This of course is due to the fact that impedances are vectors, and thus we must use complex numbers either in rectangular or in polar form. We will alleviate this problem somewhat by using admittance to simplify parallel ac circuits.

16.3 ADMITTANCE

When working with dc circuits containing resistance only, the resistance is given by Ohm's Law as

$$R = \frac{V}{I}$$

The reciprocal of resistance is defined as conductance G. In a similar fashion, Ohm's Law for ac circuits is written as

$$\mathbf{Z} = \frac{\mathbf{V}_P}{\mathbf{I}_P}$$

where $\mathbf{Z}$ has been defined as the impedance. We now define *admittance* $\mathbf{Y}$ as the reciprocal of impedance. Thus

$$\text{admittance} = \mathbf{Y} = \frac{1}{\mathbf{Z}} = \frac{\mathbf{I}_P}{\mathbf{V}_P} \tag{16.5}$$

Just as with conductance, the units for admittance are mhos ℧. Impedance $\mathbf{Z}$ is a vector quantity, $\mathbf{Z} = Z\underline{/\phi}$, and therefore admittance must also be a vector quantity. That is,

$$\mathbf{Y} = Y\underline{/\phi} \tag{16.6}$$

Equation 16.6 expresses admittance in *polar* form, and it must also be expressible in *rectangular* form. In order to discover the nature of the *rectangular components* of admittance, let us consider the *R-L-C* parallel circuit in Fig. 16.7*a*. Since the elements are in parallel, the voltage must be the same across each element. The element currents are then

$$\mathbf{I}_{PR} = \frac{\mathbf{V}_{PT}}{R} \qquad \mathbf{I}_{PL} = \frac{\mathbf{V}_{PT}}{jX_L} \qquad \mathbf{I}_{PC} = \frac{\mathbf{V}_{PT}}{-jX_C}$$

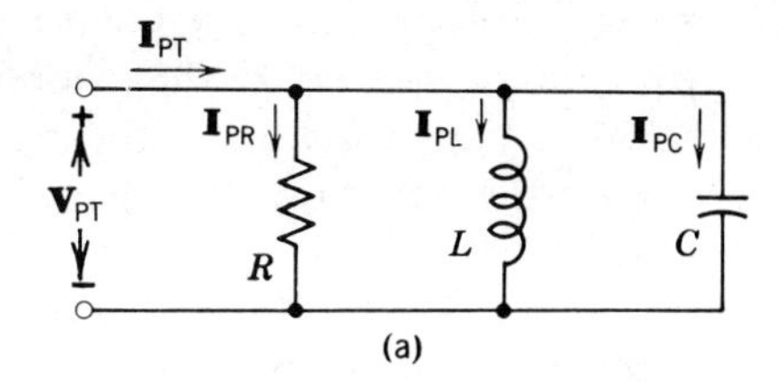

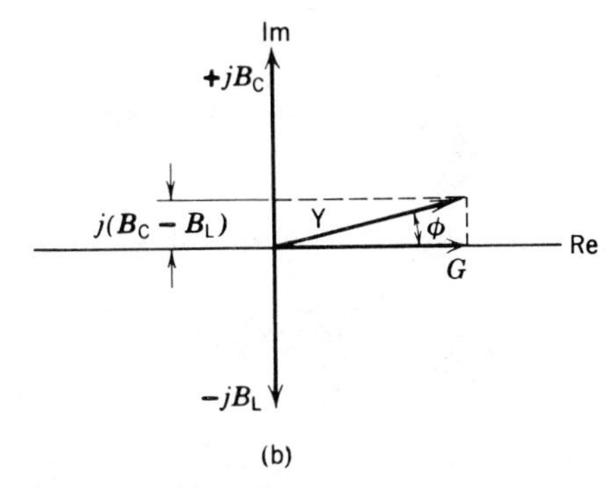

Figure 16.7 (*a*) A parallel *R-L-C* circuit (*b*) Admittance diagram in the complex plane

Using KCL, and substituting values we have

$$\mathbf{I}_{PT} = \mathbf{I}_{PR} + \mathbf{I}_{PL} + \mathbf{I}_{PC} = \frac{\mathbf{V}_{PT}}{R} + \frac{\mathbf{V}_{PT}}{jX_L} + \frac{\mathbf{V}_{PT}}{-jX_C} = \mathbf{V}_{PT}\left(\frac{1}{R} + \frac{1}{jX_L} + \frac{1}{-jX_C}\right)$$

Thus

$$\frac{\mathbf{I}_{PT}}{\mathbf{V}_{PT}} = \left(\frac{1}{R} + \frac{1}{jX_L} + \frac{1}{-jX_C}\right)$$

But the quantity $\mathbf{I}_{PT}/\mathbf{V}_{PT}$ has been defined as admittance in Eq. 16.5. Thus we can write

$$\mathbf{Y}_T = \left(\frac{1}{R} + \frac{1}{jX_L} + \frac{1}{-jX_C}\right)$$

We multiply the numerator and denominator of the two j-terms by j to obtain

$$\mathbf{Y} = \frac{1}{R} + j\frac{1}{X_C} - j\frac{1}{X_L} = G + jB_C - jB_L = G + j(B_C - B_L) \qquad (16.7)$$

Equation 16.7 expresses the admittance of a circuit in rectangular form, and the individual components are:

$$\text{conductance} = G = \frac{1}{R} \qquad (16.8)$$

$$\text{capacitive susceptance} = B_C = \frac{1}{X_C} = \omega C \qquad (16.9)$$

$$\text{inductive susceptance} = B_L = \frac{1}{X_L} = \frac{1}{\omega L} \qquad (16.10)$$

Conductance, capacitive susceptance, and *inductive susceptance* are all the reciprocals of quantities measured in ohms, and they must therefore all have units of mhos.

The admittance of a circuit expressed in either the polar form of Eq. 16.6 or the rectangular form of Eq. 16.7 can be plotted on the complex plane to form an admittance triangle as shown in Fig. 16.7*b*. From the relationships of the right triangle formed, we develop the equations for changing from polar to rectangular form or vice versa. Thus:

rectangular to polar

$$Y = \sqrt{G^2 + (B_C - B_L)^2} \tag{16.11}$$

$$\phi = \tan^{-1} \frac{(B_C - B_L)}{G} \tag{16.12}$$

where ϕ is again the phase angle between voltage and current.

polar to rectangular

$$B = Y \sin \phi \tag{16.13}$$

$$G = Y \cos \phi \tag{16.14}$$

The term $j(B_C - B_L)$ is the total *susceptance* of the circuit; it is plotted on the positive j-axis if the capacitor dominates, and on the negative j-axis if the inductor dominates. At this point, we note that admittance and impedance are duals of each other. This makes the task of utilizing admittance an easier one, since we can always relate it to the more familiar quantity, impedance.

• • •

Example 16.5 Calculate the admittance of the circuit in Fig. 16.7*a* for an ω of 10^4 rad/sec if $R = 1\ \text{k}\Omega$, $L = 0.2$ H, and $C = 0.1\ \mu$F. Find both the polar and rectangular forms, and draw an admittance diagram.

Solution Using Eqs. 16.8, 16.9, and 16.10,

$$G = \frac{1}{R} = \frac{1}{1\ \text{k}\Omega} = 1\ \text{m℧}$$

$$B_C = \omega C = 10^4 \times 0.1\ \mu\text{F} = 1\ \text{m℧}$$

$$B_L = \frac{1}{\omega L} = \frac{1}{10^4 \times 0.2\ \mu\text{H}} = 0.5\ \text{m℧}$$

Then using Eq. 16.7,

$$\mathbf{Y} = G + j(B_C - B_L) = 1\ \text{m℧} + j(1 - 0.5)\ \text{m℧} = (1 + j0.5)\ \text{m℧}$$

This admittance is changed into polar form using Eqs. 16.11 and 16.12.

$$Y = \sqrt{G^2 + B^2} = \sqrt{(10^{-3})^2 + (0.5 \times 10^{-3})^2} = \sqrt{1.25} \times 10^{-3} = 1.12 \times 10^{-3}$$

$$\phi = \tan^{-1} \frac{(B_C - B_L)}{G} = \tan^{-1} \frac{0.5\ \text{m℧}}{1\ \text{m℧}} = \tan^{-1} 0.5 = 26.6°$$

The admittance diagram is shown in Fig. 16.8.

• • •

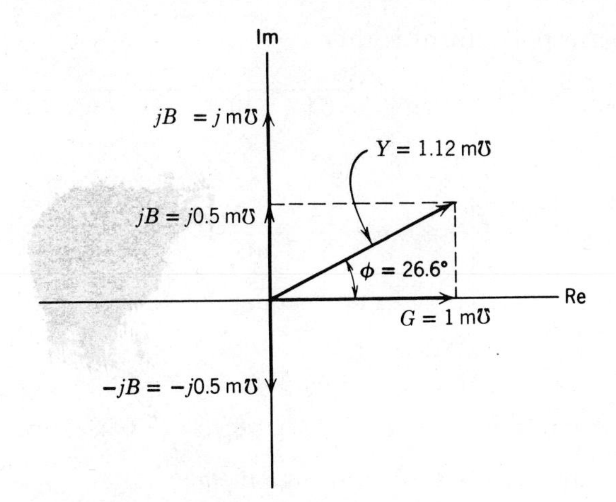

Figure 16.8 Admittance diagram for Example 16.5

• • •

Example 16.6 Calculate the total admittance of the circuit in Fig. 16.9 if $\omega = 10^3$. Draw the admittance diagram.

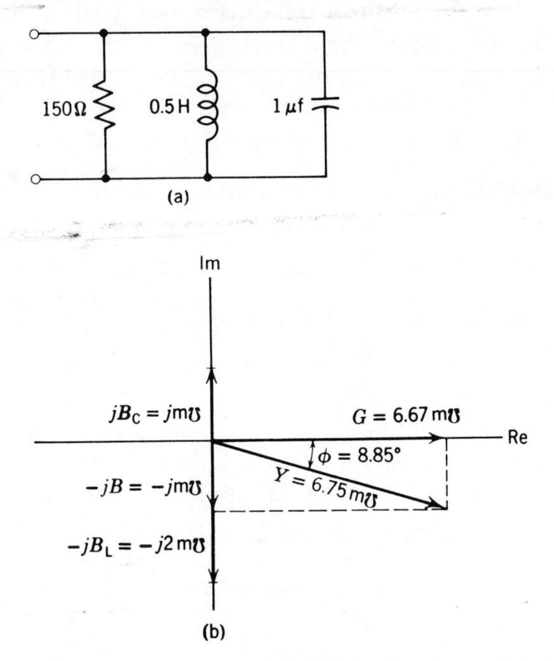

Figure 16.9 Example 16.6 (*a*) A parallel *R-L-C* circuit (*b*) Admittance diagram for (*a*)

Solution The conductance and susceptance are found using Eqs. 16.8, 16.9, and 16.10 as

$$G = \frac{1}{R} = \frac{1}{150} = 6.67 \text{ m℧}$$

$$B_C = \omega C = 10^3 \times 1\ \mu\text{F} = 1 \text{ m℧}$$

$$B_L = \frac{1}{\omega L} = \frac{1}{10^3 \times 0.5\text{H}} = 2 \text{ m℧}$$

The admittance in polar form is then

$$Y = \sqrt{G^2 + (B_C - B_L)^2} = \sqrt{(6.67 \times 10^{-3})^2 + (10^{-3} - 2 \times 10^{-3})^2}$$

$$= \sqrt{(44.5 + 1)10^{-6}} = 6.75 \text{ m℧}$$

$$\phi = \tan^{-1}\frac{(B_C - B_L)}{G} = \tan^{-1}\frac{(1-2)\text{ m℧}}{6.67\text{ m℧}} = \tan^{-1}\frac{-1}{6.67} = -8.55°$$

The admittance can be changed into rectangular form using Eqs. 16.13 and 16.14.

$$B = Y \sin\phi = 6.75 \text{ m℧} \times \sin(-8.55°) = 6.75 \times (-0.148) \text{ m℧} = -1 \text{ m℧}$$

$$G = Y \cos\phi = 6.75 \text{ m℧} \times \cos(-8.55°) = 6.75 \times (+0.989) \text{ m℧} = 6.67 \text{ m℧}$$

The admittance diagram is shown in Fig. 16.9*b*.

• • •

16.4 ADMITTANCES IN PARALLEL AND SERIES

The circuit in Fig. 16.10 shows three admittances connected in parallel. Each admittance could be a single resistor, inductor, or capacitor, or a combination of these three elements. Since the admittances are all connected in parallel, the voltage must be the same across each of them, and thus the individual currents are:

$$\mathbf{I}_{P1} = \mathbf{V}_{PT}\mathbf{Y}_1 \qquad \mathbf{I}_{P2} = \mathbf{V}_{PT}\mathbf{Y}_2 \qquad \mathbf{I}_{P3} = \mathbf{V}_{PT}\mathbf{Y}_3$$

By applying KCL at node A, and substituting the above current relationships, we obtain

$$\mathbf{I}_{PT} = \mathbf{I}_{P1} + \mathbf{I}_{P2} + \mathbf{I}_{P3} = \mathbf{V}_{PT}\mathbf{Y}_1 + \mathbf{V}_{PT}\mathbf{Y}_2 + \mathbf{V}_{PT}\mathbf{Y}_3 = \mathbf{V}_{PT}(\mathbf{Y}_1 + \mathbf{Y}_2 + \mathbf{Y}_3)$$

Dividing both sides by $\mathbf{V}_{PT}$,

$$\frac{\mathbf{I}_{PT}}{\mathbf{V}_{PT}} = \mathbf{Y}_1 + \mathbf{Y}_2 + \mathbf{Y}_3$$

But $\mathbf{I}_{PT}/\mathbf{V}_{PT}$ is the total admittance of the circuit as shown in Fig. 16.10*b*. We can immediately extend this result to a parallel circuit consisting of any number of admittances. Thus, the total admittance of a parallel circuit is

$$\mathbf{Y}_T = \mathbf{Y}_1 + \mathbf{Y}_2 + \cdots + \mathbf{Y}_N \qquad (16.15)$$

where N is the total number of admittances.

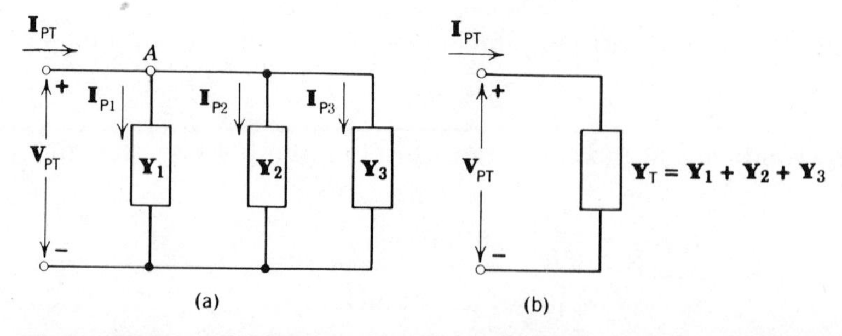

Figure 16.10 Combining parallel admittances (*a*) Admittances in parallel (*b*) Equivalent circuit of (*a*)

We see from Eq. 16.15 that admittances can be combined in parallel by simply adding them, just as conductances in parallel combine by adding. The relationship for combining admittances in *series* can be found in a similar manner (using KVL), and the result is

$$\mathbf{Y}_T = \frac{1}{\dfrac{1}{\mathbf{Y}_1} + \dfrac{1}{\mathbf{Y}_2} + \cdots + \dfrac{1}{\mathbf{Y}_N}} \tag{16.16}$$

This is similar to the equation for combining *impedances* in *parallel*, and will generally be quite tedious to use.

• • •

Example 16.7 What is the total admittance of the circuit in Fig. 16.10*a* if

$$\mathbf{Y}_1 = 0.2 + j0.33,\ \mathbf{Y}_2 = 0.04 - j0.67,\ \text{and } \mathbf{Y}_3 = 0.13 + j0.02?$$

What is the equivalent impedance?

Solution The total admittance is found using Eq. 16.15.

$$\begin{aligned}\mathbf{Y}_T &= \mathbf{Y}_1 + \mathbf{Y}_2 + \mathbf{Y}_3 = (0.2 + j0.33) + (0.04 - j0.67) + (0.13 + j0.02)\\ &= (0.37 - j0.32)\mho\end{aligned}$$

The equivalent impedance $\mathbf{Z}_T$ is

$$\mathbf{Z}_T = \frac{1}{\mathbf{Y}_T} = \frac{1}{0.37 - j0.32} = \frac{1}{0.49\underline{/-40.9^\circ}} = \underline{2.04\underline{/40.9^\circ}\ \Omega}$$

• • •

Example 16.8 Find the total impedance of the circuit in Fig. 16.11 if $\omega = 10^6$.

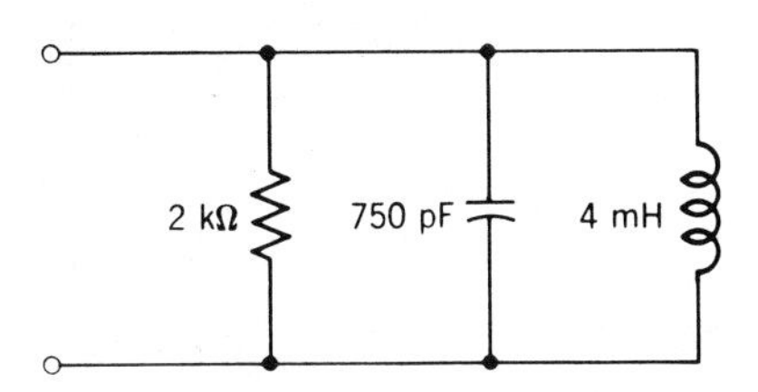

Figure 16.11 Example 16.8

Solution We will find the admittance of each branch, add them to find $\mathbf{Y}_T$, and then take the reciprocal to find $\mathbf{Z}_T$.

$$G = \frac{1}{R} = \frac{1}{2\ \text{k}\Omega} = 0.5\ \text{m}\mho$$

$$B_C = \omega C = 10^6 \times 750\ \text{pF} = 0.75\ \text{m}\mho$$

$$B_L = \frac{1}{\omega L} = \frac{1}{10^6 \times 4\ \text{mH}} = 0.25\ \text{m}\mho$$

Thus

$$\begin{aligned}\mathbf{Y}_T &= G + j(B_C - B_L) = [0.5 + j(0.75 - 0.25)]\ \text{m}\mho = 0.5 + j0.5\ \text{m}\mho\\ &= 0.707 \times 10^{-3}\underline{/45^\circ}\ \mho\end{aligned}$$

The impedance is then

$$\mathbf{Z}_T = \frac{1}{\mathbf{Y}_T} = \frac{1}{0.707 \times 10^{-3}/\underline{45^\circ}} = 1414/\underline{-45^\circ}\ \Omega$$

• • •

16.5 ac EQUIVALENT CIRCUITS

The solution of ac circuits is necessarily more complicated than the solution of dc circuits, since we must work with vector quantities. In general, the problem of finding solutions for ac circuits will be simplified if the circuit is first reduced to its simplest equivalent circuit. This can be done by combining impedances, or combining admittances as shown in the previous sections. Alternatively, we can simply combine element values (R's, L's, and C's) following the methods developed earlier.

Any combination of *impedances* can always be reduced to the simple equivalent circuit shown in Fig. 16.12, and the impedance for this circuit is

$$\mathbf{Z} = R \pm jX \tag{16.17}$$

where X will be $-X_C$ if the circuit is capacitive, and it will be $+X_L$ for an inductive circuit. Since both of the terms in Eq. 16.17 are measured in units of ohms, they *must always* represent two elements connected in *series* as shown in Fig. 16.12. Notice that this equivalent impedance can also be expressed in polar form as

$$\mathbf{Z} = Z/\underline{\pm\,\phi}$$

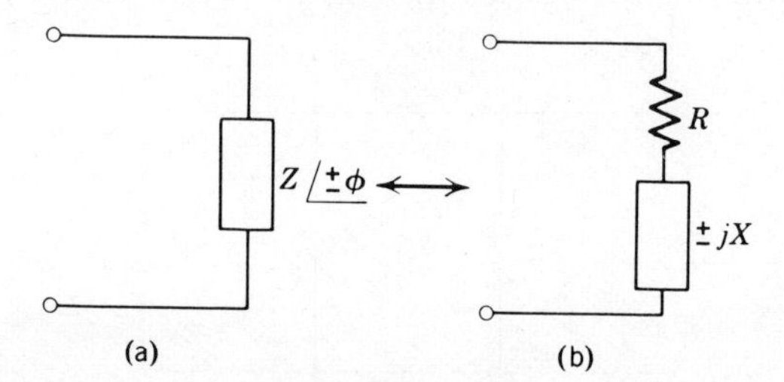

Figure 16.12 Simplest series equivalent circuit (*a*) Polar form (*b*) Rectangular form

Similarly, any combination of *admittances* can be reduced to the simple equivalent circuit shown in Fig. 16.13, and the admittance for this circuit is

$$\mathbf{Y} = G \pm jB \tag{16.18}$$

where B will be $+B_C$ if the capacitive susceptance dominates, and it will be $-B_L$ if the inductive susceptance dominates. Since both of the terms in Eq. 16.18 are measured in mhos, they *must always* represent two elements connected in *parallel* as shown in Fig. 16.13. The equivalent admittance can of course always be expressed in polar form as

$$\mathbf{Y} = Y/\underline{\pm\,\phi}$$

We will find it most convenient to be able to change a *series* equivalent

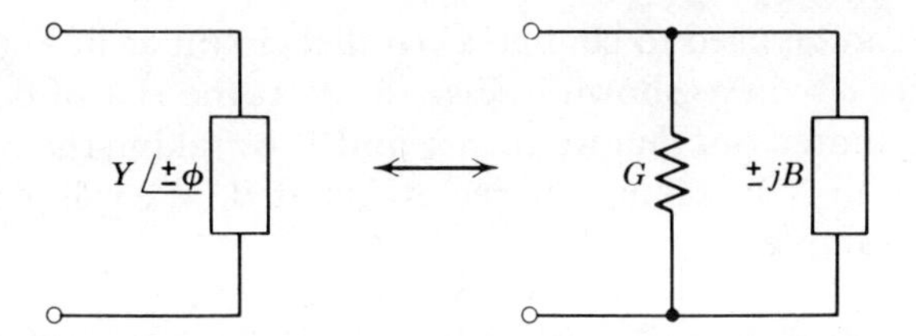

Figure 16.13 Simplest parallel equivalent circuit

circuit into its *parallel* equivalent, and vice versa. Since admittance has been defined as the reciprocal of impedance, we can easily find the parallel equivalent of a series circuit by

$$\mathbf{Y} = \frac{1}{\mathbf{Z}} \tag{16.19}$$

Notice that when using Eq. 16.19, we are changing a circuit from that shown in Fig. 16.12 (series) into that shown in Fig. 16.13 (parallel). It may not be obvious at first, but it is certainly true that $G \neq 1/R$ and $B \neq 1/X$ when making this transformation. An example will best illustrate this point.

• • •

Example 16.9 Find the parallel equivalent circuit of the series circuit in Fig. 16.14*a*.

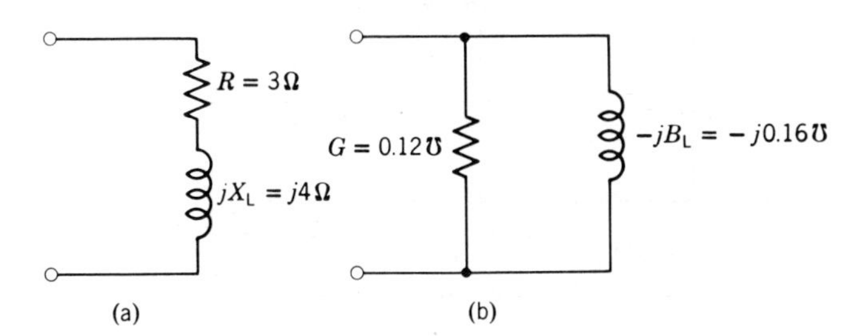

Figure 16.14 Example 16.9 (*a*) A series impedance (*b*) Parallel equivalent of (*a*)

Solution Using Eq. 16.19,

$$\mathbf{Y} = \frac{1}{\mathbf{Z}} = \frac{1}{3 + j4} = \frac{1}{5/53.1^\circ} = 0.2/-53.1^\circ = 0.12 - j0.16$$

The proper parallel equivalent circuit is shown in Fig. 16.14*b*. Notice the *incorrect* result of trying to find G and B as

$$G = \frac{1}{R} = \frac{1}{3} = 0.333 \qquad B = \frac{1}{jX} = \frac{1}{j4} = -0.25j$$

• • •

The series equivalent circuit of a parallel circuit can be found by simply taking the reciprocal of Eq. 16.19. Thus

$$\mathbf{Z} = \frac{1}{\mathbf{Y}} \tag{16.20}$$

This equation can be used to change a parallel circuit as in Fig. 16.13 into an equivalent series circuit as shown in Fig. 16.12. At the risk of being tedious, it must again be pointed out that we *cannot* find R by taking the reciprocal of G, and we *cannot* find X by taking the reciprocal of B! We will demonstrate this with another example.

• • •

Example 16.10 Find the series equivalent circuit of the parallel circuit in Fig. 16.14*b* (this provides a check on Example 16.9).

Solution We use Eq. 16.20 to obtain,

$$\mathbf{Z} = \frac{1}{\mathbf{Y}} = \frac{1}{0.12 - j0.16} = \frac{1}{0.2\underline{/-53.1^\circ}} = 5\underline{/53.1^\circ} = 3 + j4$$

Notice the *incorrect* result of trying to find R and X as

$$R = \frac{1}{G} = \frac{1}{0.12} = 8.34 \qquad X = \frac{1}{-jB_L} = \frac{1}{-j0.16} = j6.25$$

• • •

The methods of changing series circuits to parallel equivalents and vice versa will be very useful in reducing more complex ac circuits. Consider, for example the inductor connected in parallel with the capacitor in Fig. 16.15*a*.

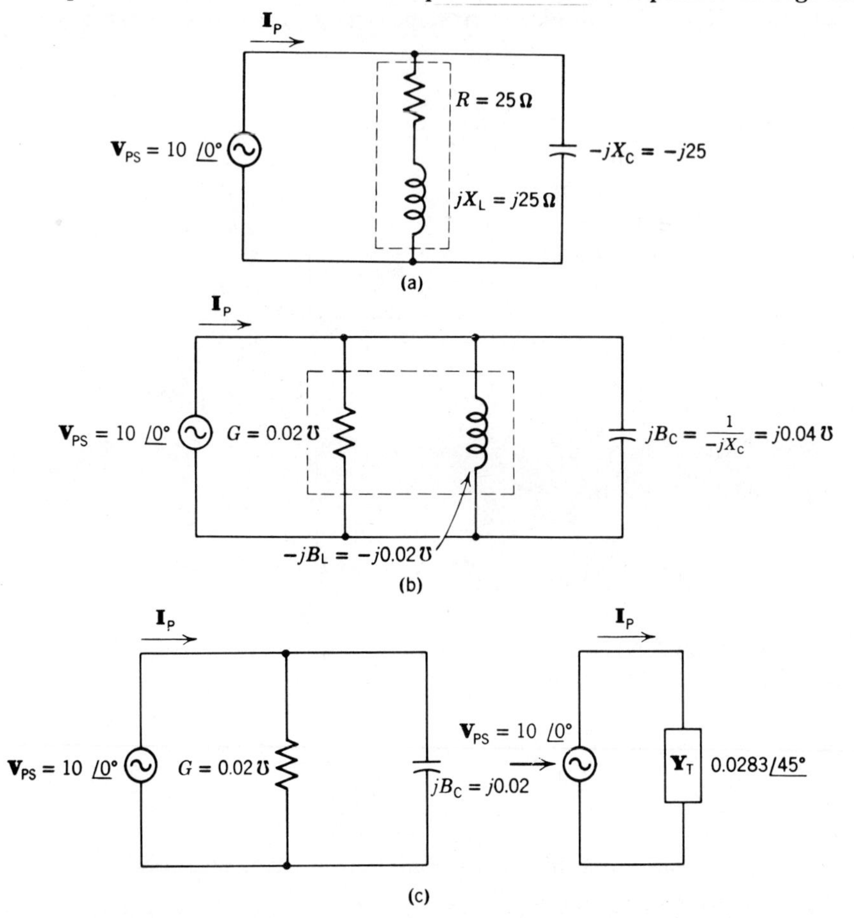

Figure 16.15 (*a*) A real inductor in parallel with a capacitor (*b*) Parallel equivalent circuit of (*a*) (*c*) Simplest parallel equivalent circuit

An *ideal* inductor has no resistance, but this inductor has a resistance of 25 Ω (the resistance of the coil wire). Suppose we would like to find the total current delivered by the generator. This is easily done by first replacing the inductor with its equivalent parallel representation (admittance), adding the admittances to find a total admittance, and then using Ohm's Law. The parallel equivalent circuit for the inductor is (Fig. 16.15*b*)

$$\mathbf{Y} = \frac{1}{\mathbf{Z}} = \frac{1}{25 + j25} = \frac{1}{35.4\angle 45^\circ} = 0.0283\angle -45^\circ$$
$$= 0.020 - j0.020$$

the total admittance of the circuit is then (Fig. 16.15*c*)

$$\mathbf{Y}_T = \mathbf{Y} + jB_C = 0.020 - j0.020 + j0.040 = 0.020 + j0.020$$
$$= 0.0283\angle 45^\circ \ \mho$$

the generator current is then

$$\mathbf{I}_P = \mathbf{V}_{PS}\mathbf{Y}_T = 10\angle 0^\circ \times 0.0283\angle 45^\circ \text{ A} = 0.283\angle 45^\circ \text{ A}$$

16.6 PARALLEL ac CIRCUITS

In the previous chapter, we solved a number of series ac circuits for the circuit current and the element voltages. In these problems the current phasor is used as the reference, since it is common to all the impedances. In the parallel circuit shown in Fig. 16.16, the voltage $\mathbf{V}_{PS}$ is the same across all the elements, and we will therefore use this voltage as the reference phasor. In the parallel ac circuit, we will be interested in finding the current in each branch and the signal generator current $\mathbf{I}_{PS}$.

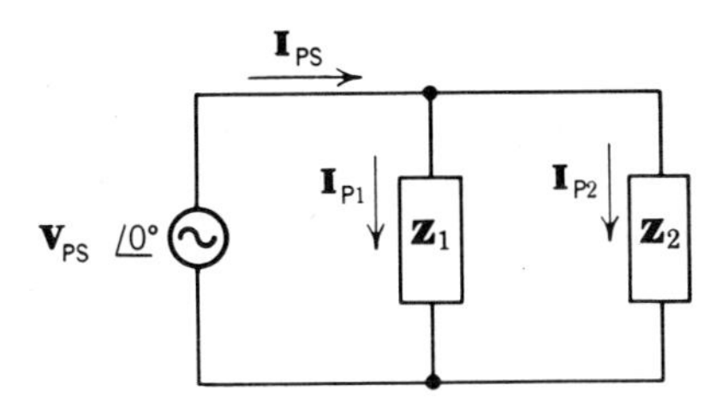

Figure 16.16 A parallel ac circuit

In Fig. 16.16, the branch currents are easily found using Ohm's Law. Thus,

$$\mathbf{I}_{P1} = \frac{\mathbf{V}_{PS}}{\mathbf{Z}_1} \qquad \mathbf{I}_{P2} = \frac{\mathbf{V}_{PS}}{\mathbf{Z}_2}$$

The generator current is then found using KCL. That is,

$$\mathbf{I}_{PS} = \mathbf{I}_{P1} + \mathbf{I}_{P2}$$

• • •

Example 16.11 Find all of the currents in Fig. 16.16 if $\mathbf{V}_{PS} = 100\angle 0^\circ$, $\mathbf{Z}_1 = 6000\angle 45^\circ$ Ω, and $\mathbf{Z}_2 = 10{,}000\angle -30^\circ$ Ω. Draw the phasor diagram.

Solution Applying Ohm's Law,

$$\mathbf{I}_{P1} = \frac{\mathbf{V}_{PS}}{\mathbf{Z}_1} = \frac{100\underline{/0^\circ}\text{ V}}{6000\underline{/45^\circ}\ \Omega} = 16.7\underline{/-45^\circ}\text{ mA}$$

$$\mathbf{I}_{P2} = \frac{\mathbf{V}_{PS}}{\mathbf{Z}_2} = \frac{100\underline{/0^\circ}\text{ V}}{10{,}000\underline{/-30^\circ}\ \Omega} = 10\underline{/30^\circ}\text{ mA}$$

The generator current is then found using KCL. Thus

$$\mathbf{I}_{PS} = \mathbf{I}_{P1} + \mathbf{I}_{P2} = 16.7\underline{/-45^\circ}\text{ mA} + 10\underline{/30^\circ}\text{ mA}$$
$$= [(11.8 - j11.8) + (8.66 + j5)]\text{ mA} = (20.5 - j6.8)\text{ mA} = 21.6\underline{/-18.4^\circ}\text{ mA}$$

The phasor diagram is shown in Fig. 16.17.

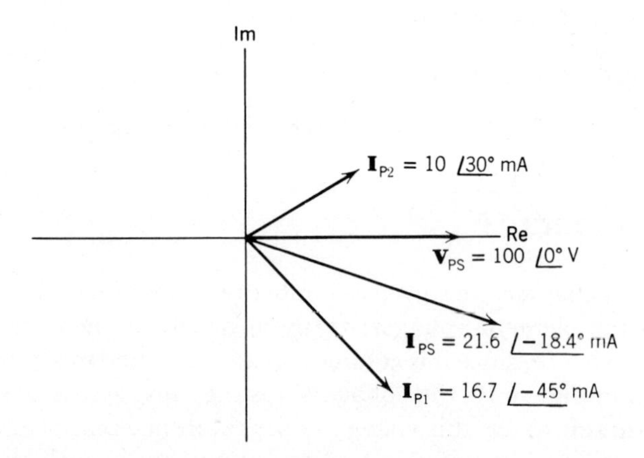

Figure 16.17 Phasor diagram for Example 16.11

• • •

Any parallel ac circuit can be solved in terms of the impedances as was done in the previous example. However, we know that any parallel circuit can be simplified to a conductance G in parallel with a susceptance B (either B_C or B_L). Therefore let us consider the parallel R-L-C circuit shown in Fig. 16.18. If we apply Ohm's Law to each element in order to obtain the element currents, the results are

$$\mathbf{I}_{PR} = \frac{\mathbf{V}_{PS}}{R} = \frac{V_{PS}\underline{/0^\circ}}{R} = \frac{V_{PS}}{R}\underline{/0^\circ} \tag{16.21}$$

$$\mathbf{I}_{PL} = \frac{\mathbf{V}_{PS}}{jX_L} = \frac{V_{PS}\underline{/0^\circ}}{X_L\underline{/90^\circ}} = \frac{V_{PS}}{X_L}\underline{/-90^\circ} \tag{16.22}$$

$$\mathbf{I}_{PC} = \frac{\mathbf{V}_{PS}}{-jX_C} = \frac{V_{PS}\underline{/0^\circ}}{X_C\underline{/-90^\circ}} = \frac{V_{PS}}{X_C}\underline{/90^\circ} \tag{16.23}$$

From these relationships, we note the following:

1. The current in the resistance is *in phase* with the voltage.
2. The current in the inductance *lags* the voltage by 90°.
3. The current in the capacitance *leads* the voltage by 90°.

These results are really the same as those obtained previously, but here we

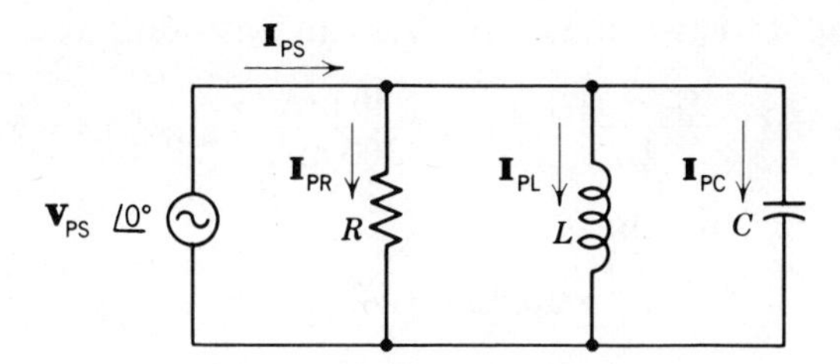

Figure 16.18 A parallel R-L-C circuit

wish to express the currents with reference to the applied voltage. This will aid in drawing the phasor diagrams, and we will again have right triangles to work with.

• • •

Example 16.12 Solve for the currents in Fig. 16.18 if $\mathbf{V}_{PS} = 20\underline{/0°}$, $R = 200\ \Omega$, $L = 500\ \mu\text{H}$, $C = 4000$ pF, and $\omega = 10^6$ rad/sec. Draw the phasor diagram.

Solution The element currents are found using Eqs. 16.21, 16.22, and 16.23.

$$\mathbf{I}_{PR} = \frac{V_{PS}}{R}\underline{/0°} = \frac{20}{200}\underline{/0°} = 0.1\underline{/0°}\ \text{A}$$

$$\mathbf{I}_{PL} = \frac{V_{PS}}{X_L}\underline{/-90°}\frac{20}{10^6 \times 500\ \mu\text{H}}\underline{/-90°} = 0.04\underline{/-90°}\ \text{A}$$

$$\mathbf{I}_{PC} = \frac{V_{PS}}{X_C}\underline{/90°} = \frac{20}{\dfrac{1}{10^6 \times 4000\ \text{PF}}}\underline{/90°} = 0.08\ \underline{/90°}\ \text{A}$$

These three element currents, along with the generator voltage as a reference, are drawn on the phasor diagram in Fig. 16.19. The generator current can now be found by applying the Pythagorean theorem to the right triangle on the phasor diagram. Thus

$$I_{PS} = \sqrt{(I_{PC} - I_{PL})^2 + I_{PR}^2} = \sqrt{(40 \times 10^{-3})^2 + (100 \times 10^{-3})^2}$$
$$= \sqrt{(16 + 100) \times 10^{-4}} = 108\ \text{mA}$$

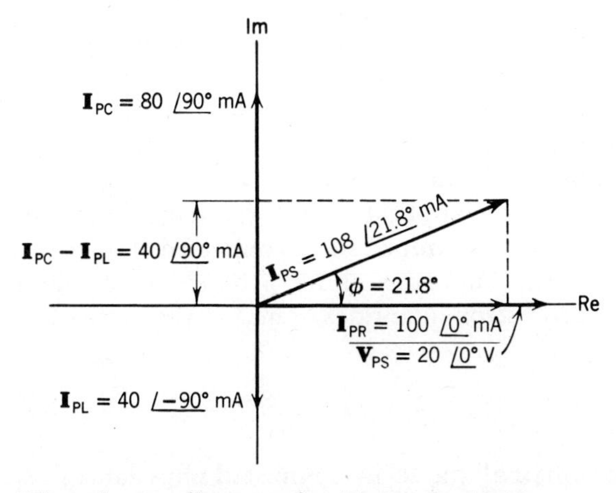

Figure 16.19 The phasor diagram for Example 16.12

The phase angle ϕ between $\mathbf{I}_{PS}$ and $\mathbf{V}_{PS}$ can be found as

$$\phi = \tan^{-1}\frac{(I_{PC} - I_{PL})}{I_{PR}} = \tan^{-1}\frac{40 \text{ mA}}{100 \text{ mA}} = \tan^{-1} 0.4 = 21.8°$$

Thus the generator current is

$$\mathbf{I}_{PS} = I_{PS}\underline{/\phi} = 108\underline{/21.8°} \text{ mA}$$

• • •

Example 16.13 Check the solution for $\mathbf{I}_{PS}$ in the previous example by finding the total admittance of the circuit.

Solution

$$G = \frac{1}{R} = \frac{1}{200} = 5 \text{ m℧}$$

$$B_C = \omega C = 10^6 \times 4000 \ \mu\text{F} = 4 \text{ m℧}$$

$$B_L = \frac{1}{\omega L} = \frac{1}{10^6 \times 500 \ \mu\text{H}} = 2 \text{ m℧}$$

The total admittance is then

$$\mathbf{Y} = G + j(B_C - B_L) = [5 + j(4 - 2)] \text{ m℧} = [5 + j2] \text{ m℧} = 5.39\underline{/21.8°} \text{ m℧}$$

The generator current $\mathbf{I}_{PS}$ is then found using Ohm's Law. Thus

$$\mathbf{I}_{PS} = \mathbf{V}_{PS}\mathbf{Y} = 20\underline{/0°} \text{ V} \times 5.39\underline{/21.8°} \text{ m℧} = 108\underline{/21.8°} \text{ mA}$$

and the results check.

• • •

16.7 SERIES-PARALLEL CIRCUITS

Since Ohm's Law, KVL, and KCL, all can be applied to ac circuits provided vector quantities are used, we can use the voltage divider theorem, the current divider theorem, and mesh and node equations: It is only necessary to continue using *phasors* for the currents and voltages.

We will devote this entire section to the solution of various ac circuits, and in the process we will demonstrate these theorems and equations. Before proceeding, it should be pointed out that it is usually easier to solve ac problems in terms of the $\mathbf{I}_P$'s, $\mathbf{V}_P$'s, and $\mathbf{Z}$'s (or $\mathbf{Y}$'s) first, and then enter the numerical data as the *last step* in the solution. If the numerical values are entered into the equations at the start, the equations may become too unwieldy to handle, and thus the chance of error is greatly increased.

Let us first consider the application of the voltage divider theorem to an ac circuit having a number of impedances connected in series. The theorem is applied in exactly the same way as for a dc circuit containing resistances, except we must use vector quantities. The voltage across any element $\mathbf{Z}_A$ is then given by

$$\mathbf{V}_{PA} = \frac{\mathbf{V}_{PS}\mathbf{Z}_A}{\mathbf{Z}_T} \qquad (16.24)$$

where $\mathbf{Z}_T$ is the sum of all the series connected impedances, $\mathbf{V}_{PS}$ is the applied voltage, and $\mathbf{Z}_A$ is the impedance of the element in question.

• • •

Example 16.14 Use the voltage divider theorem to find the voltage across $\mathbf{Z}_2$ in Fig. 16.20.

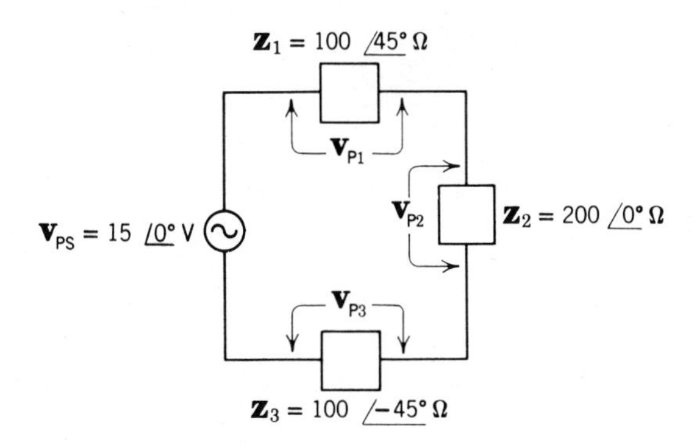

Figure 16.20 Example 16.14

Solution From Eq. 16.24,

$$\mathbf{V}_{P2} = \frac{\mathbf{V}_{PS}\mathbf{Z}_2}{\mathbf{Z}_1 + \mathbf{Z}_2 + \mathbf{Z}_3} = \frac{15\underline{/0^\circ}\text{ V} \times 200\underline{/0^\circ}\ \Omega}{100\underline{/45^\circ}\ \Omega + 200\underline{/0^\circ}\ \Omega + 100\underline{/-45^\circ}\ \Omega}$$

$$= \frac{15\underline{/0^\circ}\text{ V} \times 200\underline{/0^\circ}}{(70.7 + j70.7) + (200) + (70.7 - j70.7)} = \frac{3000\underline{/0^\circ}}{341\underline{/0^\circ}} = 8.8\underline{/0^\circ}\text{ V}$$

• • •

The current divider theorem was applied to dc circuits containing two resistors connected in parallel. It can be applied equally well to ac circuits consisting of two impedances connected in parallel. The general form of the equation as applied to the circuit in Fig. 16.21 is

$$\mathbf{I}_{P1} = \frac{\mathbf{I}_{PS}\mathbf{Z}_2}{\mathbf{Z}_1 + \mathbf{Z}_2} \quad \text{and} \quad \mathbf{I}_{P2} = \frac{\mathbf{I}_{PS}\mathbf{Z}_1}{\mathbf{Z}_1 + \mathbf{Z}_2} \tag{16.25}$$

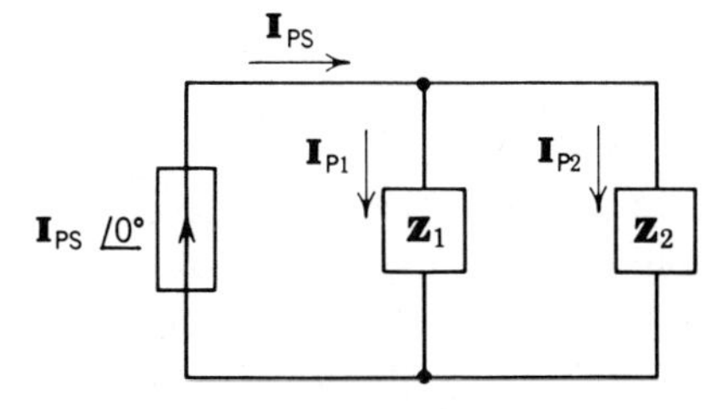

Figure 16.21 A parallel ac circuit driven by a current generator

• • •

Example 16.15 Use the current divider theorem to find $\mathbf{I}_{P1}$ in Fig. 16.21 if: $\mathbf{I}_{PS} = 21\underline{/0^\circ}$ mA, $\mathbf{Z}_1 = 10\underline{/30^\circ}\ \Omega$, and $\mathbf{Z}_2 = 20\underline{/60^\circ}\ \Omega$. Notice that the signal generator here is a *current generator* and not a voltage generator.

Solution Using Eq. 16.25,

$$\mathbf{I}_{P1} = \frac{\mathbf{I}_{PS}\mathbf{Z}_2}{\mathbf{Z}_1 + \mathbf{Z}_2} = \frac{21\underline{/0^\circ}\text{ mA} \times 20\underline{/60^\circ}\ \Omega}{10\underline{/30^\circ}\ \Omega + 20\underline{/60^\circ}\ \Omega} = \frac{21\underline{/0^\circ}\text{ mA} \times 20\underline{/60^\circ}}{(8.67 + j5) + (10 + j17.7)}$$

$$= \frac{420\underline{/60^\circ}\text{ mA}}{18.67 + j22.7} = \frac{420\underline{/60^\circ}\text{ mA}}{29.4\underline{/50.6^\circ}} = 14.3\underline{/9.4^\circ}\text{ mA}$$

• • •

The more general, and therefore more powerful, techniques of mesh and node equations can be applied to ac circuits. We will be content here to work one problem using both methods in order to illustrate the procedure. In the event an ac circuit contains *more than one* signal generator, we can use the superposition theorem, or mesh equations, or node equations. We must however make the assumption that both generators have *exactly the same frequency*! This is necessary if we are to continue to use phasors, since we have arrived at the impedance diagrams and phasor diagrams by eliminating the ωt term. In the event that the generators have different frequencies we can find a solution for each *generator separately*, but we *cannot* find the complete solution since we have not developed the techniques for adding sinusoidal waves of different frequencies.

The circuit in Fig. 16.22 has two signal generators, but they have *exactly the same frequency*. There may, however, be a phase angle between the two generator voltages, and we thus designate one of them as a reference (zero phase angle) and the other with some phase angle θ.

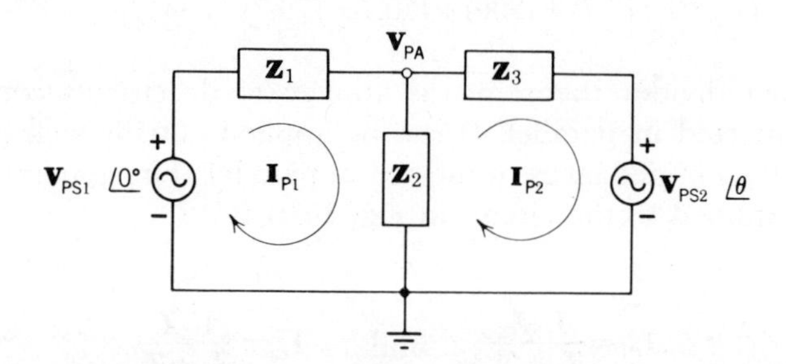

Figure 16.22

Since the circuit is planar and has two meshes (windows), we can find a complete solution for the circuit by assigning two mesh currents $\mathbf{I}_{P1}$ and $\mathbf{I}_{P2}$, as shown in the figure, and solving the two mesh equations. The mesh equations are written just as they were for a dc circuit, except we use vector quantities. Notice that we have assigned polarities to the voltage generators. This simply means that we are examining the circuit at one particular instant in time. In other words, we have momentarily *stopped* the rotation of the phasors at a point where $\mathbf{V}_{PS_1}$ has the polarity shown. If the other voltage generator has a polarity the opposite of that shown, or if either of the currents is actually going in the opposite direction to that assigned, the solutions will simply have negative signs on them. The two mesh equations are then

$$\text{mesh 1} \qquad \mathbf{V}_{PS_1} = (\mathbf{Z}_1 + \mathbf{Z}_2)\mathbf{I}_{P1} - \mathbf{Z}_2\mathbf{I}_{P2}$$

$$\text{mesh 2} \qquad -\mathbf{V}_{PS_2} = -\mathbf{Z}_2\mathbf{I}_{P1} + (\mathbf{Z}_2 + \mathbf{Z}_3)\mathbf{I}_{P2}$$

• • •

Example 16.16 Solve for the mesh currents in Fig. 16.22 if: $\mathbf{V}_{PS_1} = 10\underline{/0°}$ V, $\mathbf{V}_{PS_2} = 5\underline{/45°}$ V, $\mathbf{Z}_1 = (1 + j2)\ \Omega$, $\mathbf{Z}_2 = (2 + j2)\ \Omega$, and $\mathbf{Z}_3 = (2 + j1)\ \Omega$.

Solution We first solve the mesh equations for $\mathbf{I}_{P1}$ and $\mathbf{I}_{P2}$. Using determinants:

$$\mathbf{I}_{P1} = \frac{\begin{vmatrix} \mathbf{V}_{PS_1} & -\mathbf{Z}_2 \\ -\mathbf{V}_{PS_2} & (\mathbf{Z}_2 + \mathbf{Z}_3) \end{vmatrix}}{\begin{vmatrix} (\mathbf{Z}_1 + \mathbf{Z}_2) & -\mathbf{Z}_2 \\ -\mathbf{Z}_2 & (\mathbf{Z}_2 + \mathbf{Z}_3) \end{vmatrix}} = \frac{\mathbf{V}_{PS_1}(\mathbf{Z}_2 + \mathbf{Z}_3) - \mathbf{V}_{PS_2}\mathbf{Z}_2}{\mathbf{Z}_1\mathbf{Z}_2 + \mathbf{Z}_1\mathbf{Z}_3 + \mathbf{Z}_2\mathbf{Z}_3} = \frac{\mathbf{V}_{PS_1}(\mathbf{Z}_2 + \mathbf{Z}_3) - \mathbf{V}_{PS_2}\mathbf{Z}_2}{\mathbf{Z}_1(\mathbf{Z}_2 + \mathbf{Z}_3) + \mathbf{Z}_2\mathbf{Z}_3}$$

$$\mathbf{I}_{P2} = \frac{\begin{vmatrix} (\mathbf{Z}_1 + \mathbf{Z}_2) & \mathbf{V}_{PS_1} \\ -\mathbf{Z}_2 & -\mathbf{V}_{PS_2} \end{vmatrix}}{\begin{vmatrix} (\mathbf{Z}_1 + \mathbf{Z}_2) & -\mathbf{Z}_2 \\ -\mathbf{Z}_2 & (\mathbf{Z}_2 + \mathbf{Z}_3) \end{vmatrix}} = \frac{\mathbf{V}_{PS_1}\mathbf{Z}_2 - \mathbf{V}_{PS_2}(\mathbf{Z}_1 + \mathbf{Z}_2)}{\mathbf{Z}_1(\mathbf{Z}_2 + \mathbf{Z}_3) + \mathbf{Z}_2\mathbf{Z}_3}$$

Notice that the denominator of both expressions is the same. We thus evaluate it first,

$$\begin{aligned} \mathbf{Z}_1(\mathbf{Z}_2 + \mathbf{Z}_3) + \mathbf{Z}_2\mathbf{Z}_3 &= (1 + j2)[(2 + j2) + (2 + j1)] + (2 + j2)(2 + j1) \\ &= (1 + j2)(4 + j3) + (2 + j6) = (-2 + j11) + (2 + j6) = j17 \end{aligned}$$

Then,

$$\mathbf{I}_{P1} = \frac{10\underline{/0°}[(2 + j2) + (2 + j1)] - 5\underline{/45°}(2 + j2)}{j17} = \frac{40 + 15.9j}{j17} = 0.933 - j2.35$$

$$\mathbf{I}_{P2} = \frac{10\underline{/0°}(2 + j2) - 5\underline{/45°}[(1 + j2) + (2 + j2)]}{j17} = \frac{23.54 - 4.7j}{j17}$$

$$= -(0.276 + j1.38)$$

Notice that $\mathbf{I}_{P2}$ is going in a direction opposite to that assigned.

• • •

The solution for the circuit in Fig. 16.22 would be somewhat easier using node equations, since there is only one unknown node voltage $\mathbf{V}_{PA}$, and thus it only requires one equation. The node equation at node A is then

$$0 = \left(\frac{1}{\mathbf{Z}_1} + \frac{1}{\mathbf{Z}_2} + \frac{1}{\mathbf{Z}_3}\right)\mathbf{V}_{PA} - \frac{\mathbf{V}_{PS_1}}{\mathbf{Z}_1} - \frac{\mathbf{V}_{PS_2}}{\mathbf{Z}_3}$$

Solving for $\mathbf{V}_{PA}$ yields,

$$\mathbf{V}_{PA} = \frac{\dfrac{\mathbf{V}_{PS_1}}{\mathbf{Z}_1} + \dfrac{\mathbf{V}_{PS_2}}{\mathbf{Z}_3}}{\dfrac{1}{\mathbf{Z}_1} + \dfrac{1}{\mathbf{Z}_2} + \dfrac{1}{\mathbf{Z}_3}} = \frac{\mathbf{V}_{PS_1}\mathbf{Z}_2\mathbf{Z}_3 + \mathbf{V}_{PS_2}\mathbf{Z}_1\mathbf{Z}_2}{\mathbf{Z}_2\mathbf{Z}_3 + \mathbf{Z}_1(\mathbf{Z}_2 + \mathbf{Z}_3)}$$

• • •

Example 16.17 Solve for the node voltage $\mathbf{V}_{PA}$ in Fig. 16.22 using the same circuit as in Example 16.16.

Solution Notice that the denominator in the node voltage equation is exactly the same as in the mesh current equations (Example 16.16). The node voltage is then

$$\mathbf{V}_{\mathrm{PA}} = \frac{10\underline{/0^\circ}(2+j2)(2+j1) + 5\underline{/45^\circ}(1+j2)(2+j2)}{j17}$$

$$= \frac{-8.28 + j74.1}{j17} = (4.36 + j0.486)\ \mathrm{V}$$

• • •

For the circuit in Fig. 16.22, we must bear in mind that the impedances $\mathbf{Z}_1$, $\mathbf{Z}_2$, and $\mathbf{Z}_3$ can be resistors, inductors, capacitors, or any combination of the three. Furthermore, the two voltage sources must have the same frequency.

Let us now consider the problem of finding a complete solution for the circuit shown in Fig. 16.23. After comparison, we recognize that this circuit

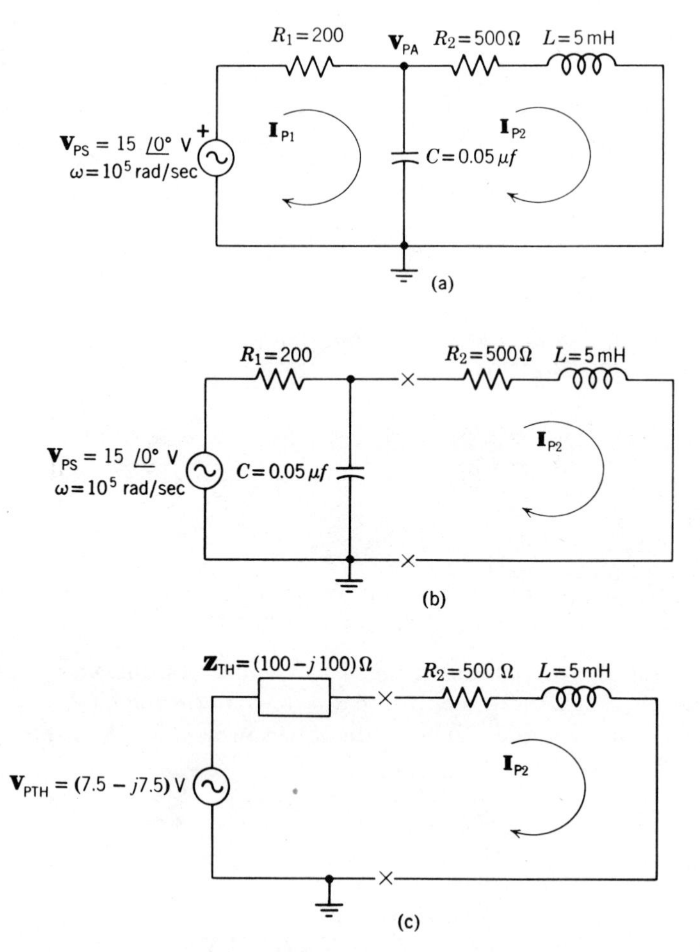

Figure 16.23 (*a*) A two-mesh circuit (*b*) Cutting the circuit in (a) (*c*) Thévenin equivalent of (a)

has exactly the same form as that of Fig. 16.22, if we make

$$\mathbf{V}_{\mathrm{PS}_1} = \mathbf{V}_{\mathrm{PS}} = 15\underline{/0^\circ}\ \mathrm{V} \qquad \mathbf{V}_{\mathrm{PS}_2} = 0$$

$$\mathbf{Z}_1 = R_1 = 200 = 200\underline{/0^\circ}\ \Omega$$

$$\mathbf{Z}_2 = -jX_C = \frac{-j}{10^5 \times 0.05\ \mu\text{F}} = -j200 = 200\underline{/-90^\circ}\ \Omega$$

$$\mathbf{V}_3 = R_2 + jX_L = 500 + j(10^5 \times 5\ \text{mH}) = 500 + j500$$

$$= 500\sqrt{2}\underline{/45^\circ}\ \Omega$$

If we substitute these values into the two mesh equations for this circuit

$$15 = (200 - j200)\mathbf{I}_{P1} - (-j200)\mathbf{I}_{P2}$$
$$0 = -(-j200)\mathbf{I}_{P1} + (-j200 + 500 + j500)\mathbf{I}_{P2}$$

Simplifying,

$$15 = (200 - j200)\mathbf{I}_{P1} + j200\mathbf{I}_{P2}$$
$$0 = j200\mathbf{I}_{P1} + (500 + j300)\mathbf{I}_{P2}$$

We use determinants to solve for the loop currents $\mathbf{I}_{P1}$ and $\mathbf{I}_{P2}$.

$$\mathbf{I}_{P1} = \frac{\begin{vmatrix} 15 & j200 \\ 0 & (500+j300) \end{vmatrix}}{\begin{vmatrix} (200-j200) & j200 \\ j200 & (500+j300) \end{vmatrix}} = \frac{15(500+j300)}{(200-j200)(500+j300) - j200 \times j200}$$

$$= \frac{1.5 \times 10^3(5+j3)}{4 \times 10^4(5-j)} = 0.0375\frac{5.84\underline{/31^\circ}}{5.10\underline{/-11.3^\circ}}$$

$$= 42.9\underline{/42.3^\circ}\ \text{mA}$$

$$\mathbf{I}_{P2} = \frac{\begin{vmatrix} (200-j200) & 15 \\ j200 & 0 \end{vmatrix}}{4 \times 10^4(5-j)} = \frac{-15 \times j200}{4 \times 10^4(5-j)}$$

$$= \frac{3 \times 10^3\underline{/-90^\circ}}{4 \times 10^4 \times 5.10\underline{/-11.3^\circ}} = 14.7\underline{/-78.7^\circ}\ \text{mA}$$

We now solve the same circuit by using the node voltage equation. Thus the voltage $\mathbf{V}_{PA}$ is

$$\mathbf{V}_{PA} = \frac{\mathbf{V}_{PS_1}\mathbf{Y}_1}{\mathbf{Y}_1 + \mathbf{Y}_2 + \mathbf{Y}_3}$$

The admittances are:

$$\mathbf{Y}_1 = \frac{1}{\mathbf{Z}_1} = \frac{1}{200\ \Omega} = 5\ \text{m℧}$$

$$\mathbf{Y}_2 = \frac{1}{\mathbf{Z}_2} = \frac{1}{-j200\ \Omega} = j5\ \text{m℧}$$

$$\mathbf{Y}_3 = \frac{1}{\mathbf{Z}_3} = \frac{1}{500\sqrt{2}\underline{/45^\circ}} = 1.414\underline{/-45^\circ}\ \text{m℧} = (1-j)\ \text{m℧}$$

The node voltage is then

$$\mathbf{V}_{PA} = \frac{15 \times 5\ \text{m℧}}{5\ \text{m℧} + j5\ \text{m℧} + (1-j)\ \text{m℧}} = \frac{75}{6+j4}$$

$$= \frac{75}{7.2\underline{/33.7^\circ}} = 10.4\underline{/-33.7^\circ}\ \text{V}$$

The technique for finding a general solution for ac networks is clearly demonstrated by the solution of the circuit in Fig. 16.23. There are times however, when a general solution is neither desired nor necessary. For example, consider the problem of finding the current through R_2 in Fig. 16.23. This of course can be accomplished with either mesh or node equations, but we could also use Thévenin's theorem to obtain a solution.

Suppose we cut the circuit as shown in Fig. 16.23*b*, and find the Thévenin equivalent of the circuit to the *left* of the cut. We use exactly the same steps to find a Thévenin equivalent of an ac circuit as for a dc circuit. The only difference is that we must use *phasor voltages* and *impedances.*

If the voltage source in Fig. 16.23*b* is shorted, the Thévenin impedance is simply R_1, in parallel with C. Thus

$$\mathbf{Z}_{\mathrm{TH}} = \frac{R_1(-jX_{\mathrm{C}})}{R_1 - jX_{\mathrm{C}}} = \frac{200(-j200)}{200 - j200} = \frac{200 \times 200\underline{/-90^\circ}}{200\sqrt{2}\underline{/-45^\circ}} = \frac{200}{\sqrt{2}}\underline{/-45^\circ}\ \Omega = (100 - j100)\ \Omega$$

where

$$X_{\mathrm{C}} = \frac{1}{\omega C} = \frac{1}{10^5 \times 0.05\ \mu\mathrm{F}} = 200\ \Omega$$

The Thévenin voltage is found as the voltage across C using the voltage divider theorem. Thus

$$\mathbf{V}_{\mathrm{PTH}} = \mathbf{V}_{\mathrm{PS}}\frac{(-jX_{\mathrm{C}})}{R_1 - jX_{\mathrm{C}}} = \frac{15\cdot 0^\circ(-j200)}{200 - j200} = \frac{15\underline{/0^\circ} \times 200\underline{/-90^\circ}}{200\sqrt{2}\underline{/-45^\circ}}$$

$$= \frac{15}{\sqrt{2}}\underline{/-45^\circ} = (7.5 - j7.5)\ \mathrm{V}$$

The Thévenin equivalent circuit is shown in Fig. 16.23*c*. From this equivalent circuit, the current $\mathbf{I}_{\mathrm{P2}}$ is easily calculated as

$$\mathbf{I}_{\mathrm{P2}} = \frac{\mathbf{V}_{\mathrm{PTH}}}{\mathbf{Z}_{\mathrm{T}}} = \frac{(7.5 - j7.5)\ \mathrm{V}}{\mathbf{Z}_{\mathrm{TH}} + R_2 + jX_{\mathrm{L}}} = \frac{7.5 - j7.5}{100 - j100 + 500 + j500}$$

$$= \frac{7.5 - j7.5}{600 + j400} = \frac{7.5\sqrt{2}\underline{/-45^\circ}}{721\underline{/33.7^\circ}} = 14.7\underline{/-78.7^\circ}\ \mathrm{mA}$$

This result agrees with the previous result. Norton's theorem will apply equally well to ac circuits and it is most useful when there are ac current generators present (in the equivalent circuits of transistors, for example).

16.8 ac BRIDGE CIRCUITS

A general ac bridge has the same form as a resistance bridge except that the resistances are replaced by impedances and the voltage source is an ac source as shown in Fig. 16.24*a*. The meter for detecting when the bridge is balanced must also be capable of measuring an ac signal. When the bridge is balanced, the voltage $\mathbf{V}_{\mathrm{PA}}$ must be equal to the voltage $\mathbf{V}_{\mathrm{PB}}$. This of course means that there is no current through the meter, and when balanced the meter can be removed as shown in Fig. 16.24*b*. Applying the voltage divider theorem to Fig.

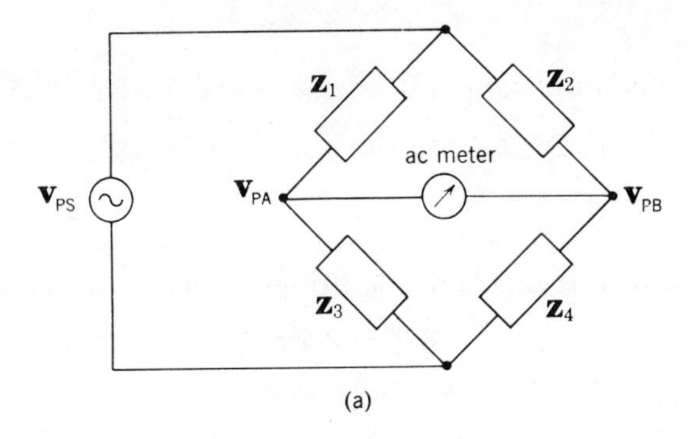

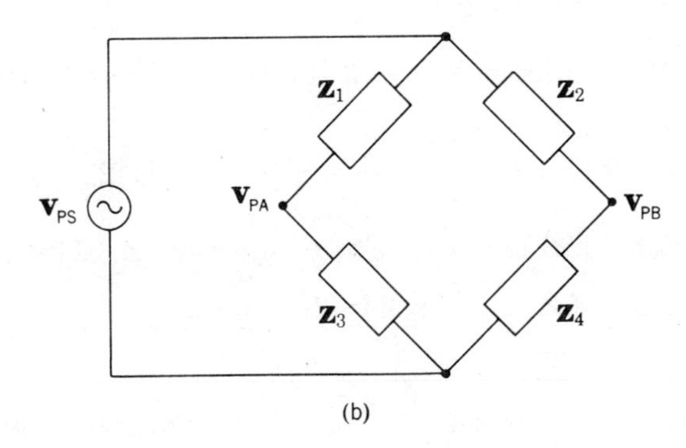

Figure 16.24 (*a*) A general ac bridge circuit (*b*) The bridge circuit in (*a*) under a balanced condition

16.24*b*, we have

$$\mathbf{V}_{PA} = \mathbf{V}_{PS}\frac{\mathbf{Z}_3}{\mathbf{Z}_1+\mathbf{Z}_3} \qquad \mathbf{V}_{PB} = \mathbf{V}_{PS}\frac{\mathbf{Z}_4}{\mathbf{Z}_2+\mathbf{Z}_4}$$

Since $\mathbf{V}_{PA} = \mathbf{V}_{PB}$, we equate these two relationships to obtain

$$\mathbf{V}_{PS}\frac{\mathbf{Z}_3}{\mathbf{Z}_1+\mathbf{Z}_3} = \mathbf{V}_{PS}\frac{\mathbf{Z}_4}{\mathbf{Z}_2+\mathbf{Z}_4}$$

Simplifying,

$$\frac{\mathbf{Z}_3}{\mathbf{Z}_1+\mathbf{Z}_3} = \frac{\mathbf{Z}_4}{\mathbf{Z}_2+\mathbf{Z}_4}$$

Thus

$$\mathbf{Z}_3(\mathbf{Z}_2+\mathbf{Z}_4) = \mathbf{Z}_4(\mathbf{Z}_1+\mathbf{Z}_3)$$

$$\mathbf{Z}_3\mathbf{Z}_2 + \cancel{\mathbf{Z}_3\mathbf{Z}_4} = \mathbf{Z}_4\mathbf{Z}_1 + \cancel{\mathbf{Z}_4\mathbf{Z}_3} \tag{16.26}$$

$$\mathbf{Z}_2\mathbf{Z}_3 = \mathbf{Z}_1\mathbf{Z}_4$$

Equation 16.26 is the *balance equation* for the ac bridge circuit, and it is seen to be similar to the balance equation for a basic resistance bridge. If Eq. 16.26 is satisfied, the bridge is *balanced,* otherwise it is not balanced. Notice that the balancing of the bridge *does not* depend on either the *magnitude* or *frequency* of the voltage source.

• • •

Example 16.18 The impedances in the ac bridge in Fig. 16.24a are:

$$\mathbf{Z}_1 = 1 + j2 \qquad \mathbf{Z}_2 = 2 + j4 \qquad \mathbf{Z}_3 = -j1 \qquad \mathbf{Z}_4 = -j2$$

Is the bridge balanced?

Solution If Eq. 16.26 is satisfied, the bridge is balanced. Thus,

$$\mathbf{Z}_2\mathbf{Z}_3 = \mathbf{Z}_1\mathbf{Z}_4$$
$$(2 + j4)(-j1) = (1 + j2)(-j2)$$
$$4 - j2 = 4 - j2$$

and the bridge is indeed balanced.

• • •

Example 16.19 What must be the value of $\mathbf{Z}_1$ in order to balance the bridge in Fig. 16.24a if:

$$\mathbf{Z}_2 = 1 + j1 \qquad \mathbf{Z}_3 = 4 - 2j \qquad \mathbf{Z}_4 = j2$$

Solution Equation 16.26 is rearranged to solve for $\mathbf{Z}_1$. Thus,

$$\mathbf{Z}_1 = \frac{\mathbf{Z}_2\mathbf{Z}_3}{\mathbf{Z}_4} = \frac{(1 + j1)(4 - 2j)}{j2} = \frac{6 + 2j}{j2} = 1 - j3$$

• • •

The ac bridge in Fig. 16.24a can be used to measure the value of an unknown capacitance C_X by constructing the bridge as shown in Fig. 16.25.

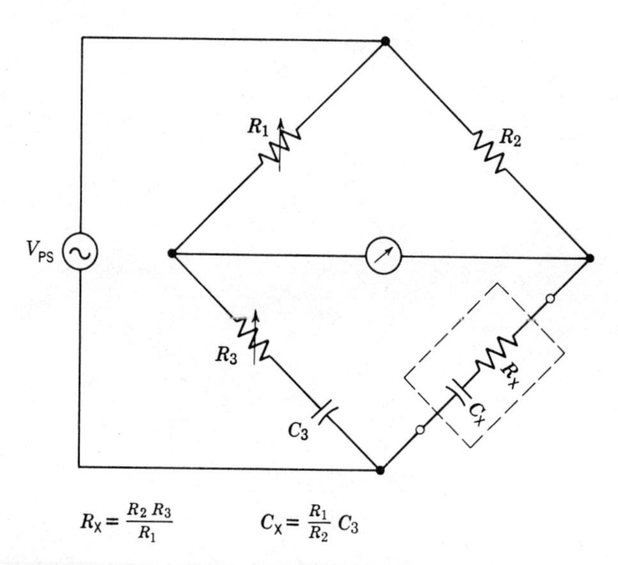

Figure 16.25 Series capacitance-comparison bridge

This is known as a *series capacitance-comparison* bridge. The balance equation still holds here (Eq. 16.26) and in this case we have

$$\mathbf{Z}_1 = R_1 \qquad \mathbf{Z}_2 = R_2 \qquad \mathbf{Z}_3 = R_3 + jX_{C_3} \qquad \mathbf{Z}_4 = R_X + jX_{C_X}$$

where R_X and C_X are the unknown quantities. Solving Eq. 16.26 for $\mathbf{Z}_4$, and substituting values, we have

$$\mathbf{Z}_4 = \frac{\mathbf{Z}_2\mathbf{Z}_3}{\mathbf{Z}_1} = \frac{R_2(R_3 + jX_{C_3})}{R_1} = \frac{R_2R_3}{R_1} + j\frac{R_2}{R_1}X_{C_3}$$

In order for two complex numbers to be equal, the real parts must be equal and the imaginary parts must be equal. Thus from the above equation we have,

$$R_X = \frac{R_2R_3}{R_1}$$

$$X_{C_X} = \frac{R_2}{R_1}X_{C_3} \tag{16.27}$$

We know that $X_{C_X} = 1/\omega C_X$ and $X_{C_3} = 1/\omega C_3$. Thus by substitution,

$$\frac{1}{\omega C_X} = \frac{R_2}{R_1} \times \frac{1}{\omega C_3}$$

Solving for C_X, we have

$$C_X = \frac{R_1}{R_2}C_3 \tag{16.28}$$

Equations 16.27 and 16.28 can be used to determine the value of an unknown capacitance C_X when the bridge in Fig. 16.25 is balanced. The resistors R_1 and R_3 are variable so that the bridge can be balanced by adjusting them.

• • •

Example 16.20 When the bridge in Fig. 16.25 is balanced,

$$R_1 = 10\ \text{k}\Omega,\ R_2 = 5\ \text{k}\Omega,\ R_3 = 1\ \Omega,\ \text{and}\ C_3 = 0.01\ \mu\text{F}.$$

What are the values of the unknown R_X and C_X?

Solution Using Eqs. 16.27 and 16.28,

$$R_X = \frac{R_2R_3}{R_1} = \frac{5\ \text{k}\Omega \times 1\ \Omega}{10\ \text{k}\Omega} = 0.5\ \Omega$$

$$C_X = \frac{R_1}{R_2}C_3 = \frac{10\ \text{k}\Omega}{5\ \text{k}\Omega} \times 0.01\ \mu\text{F} = 0.02\ \mu\text{F}$$

• • •

A slightly different form of an ac bridge can be used to measure an inductance in series with a resistance. Such a bridge is shown in Fig. 16.26 and it is known as a Maxwell bridge. We can again solve for the unknown inductance L_X and resistance R_X by beginning with the balance equation for the general ac bridge. That is,

$$\mathbf{Z}_4 = \frac{\mathbf{Z}_2\mathbf{Z}_3}{\mathbf{Z}_1} = \frac{R_2R_3}{\dfrac{-R_1jX_{C_1}}{R_1 - jX_{C_1}}} = \frac{R_2R_3(R_1 - jX_{C_1})}{-R_1jX_{C_1}} = \frac{R_2R_3}{R_1X_{C_1}}(X_{C_1} + jR_1)$$

$$= \frac{R_2R_3}{R_1} + j\frac{R_2R_3}{X_{C_1}}$$

But, $\mathbf{Z}_4 = R_X + jX_{L_X}$ in this case, and equating the real and imaginary parts yields

$$R_X = \frac{R_2R_3}{R_1} \tag{16.29}$$

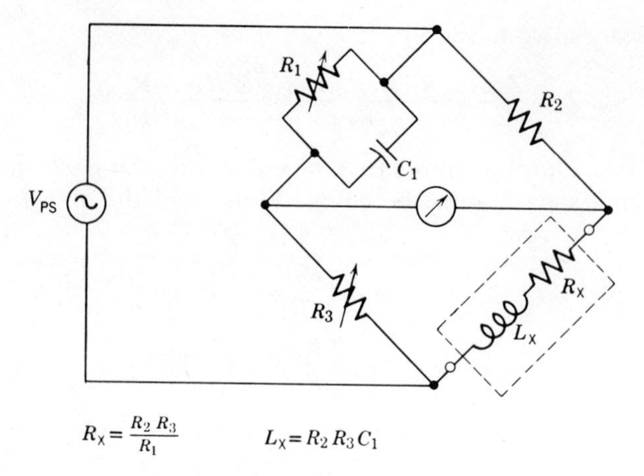

Figure 16.26 Maxwell bridge

and

$$X_{L_X} = \frac{R_2 R_3}{X_{C_1}}$$

We now substitute for X_L and X_{C_1} to obtain

$$\omega L_X = \frac{R_2 R_3}{\dfrac{1}{\omega C_1}} = \omega R_2 R_3 C_1$$

Thus

$$L_X = R_2 R_3 C_1 \tag{16.30}$$

Equations 16.29 and 16.30 can be used in conjunction with the Maxwell bridge in Fig. 16.26 to determine the value of an unknown inductance; R_1 and R_3 are variable, so the bridge can be balanced.

• • •

Example 16.21 The Maxwell bridge in Fig. 16.26 is balanced with:

$R_1 = 250\ \text{k}\Omega \qquad R_2 = 1.0\ \text{k}\Omega \qquad R_3 = 5.0\ \text{k}\Omega \qquad C_1 = 2000\ \text{pF}$

What are the values of the unknown resistance R_X and inductance L_X?

Solution Using Eqs. 16.29 and 16.30,

$$R_X = \frac{R_2 R_3}{R_1} = \frac{1\ \text{k}\Omega \times 5\ \text{k}\Omega}{250\ \text{k}\Omega} = 20\ \Omega$$

$$L_X = R_2 R_3 C_1 = 1\ \text{k}\Omega \times 5\ \text{k}\Omega \times 2000\ \text{pF} = 10\ \text{mH}$$

• • •

We end this section by noting that there are many other forms of ac bridge circuits, but these topics are more appropriately covered in a basic instrumentation course.

SUMMARY

One of the most straightforward ways of analyzing ac circuits is to simplify the circuit by combining impedances. In this chapter we have demonstrated that impedances can be combined in exactly the same manner used to combine resistances, that is, impedances in *series* simply add, and impedances in parallel may be combined by taking the *product over the sum* (two impedances at a time). Combining impedances in parallel can become tedious because they are vector quantities, but the use of admittance offers help. Admittance is defined as the reciprocal of impedance, and admittances can be combined in exactly the same manner as conductances. Thus, admittances in parallel are simply added to determine the equivalent admittance. Circuits can also be simplified by first combining *like* elements.

Any R-L-C circuit can be reduced to an impedance whose equivalent circuit is a resistance in *series* with a reactance. Similarly, any R-L-C ac circuit can be reduced to an admittance whose equivalent circuit is a conductance in *parallel* with a susceptance. Furthermore, the parallel equivalent circuit of a series impedance can be determined by simply using the fact that admittance is the reciprocal of impedance. The reverse process of finding the series equivalent circuit of an admittance can also be accomplished.

Parallel ac circuits can be solved for branch currents by simply applying Ohm's Law. The total circuit current can then be determined by using KCL. Phasor diagrams for parallel circuits are used to solve for branch currents, or the total circuit current in terms of right triangle relationships. In parallel circuits, the voltage across the elements is used as the reference phasor.

All laws and theorems previously used for dc circuits composed of resistances are equally valid for use with ac R-L-C circuits provided we use phasors throughout.

A complete solution for an ac circuit can be obtained by using the general methods of mesh or node equations. Thévenin's and Norton's theorems can be used to simplify ac circuits in exactly the same way they are used in dc circuits; we must of course use impedances and phasors for the voltages and currents.

The balance equation for an ac bridge circuit is similar to that for the dc resistance bridge. Bridges for measuring capacitance and inductance can be constructed by selecting the proper elements (R, L, or C) for different legs of the bridge.

Multiple-source ac circuits can be solved using mesh or node equations, or the Principle of Superposition can be used.

GLOSSARY

Admittance Y. *The reciprocal of impedance.*

Capacitive circuit. *A circuit in which the capacitance term is greater than the inductance term.*

Capacitive susceptance B_C. *The reciprocal of capacitive reactance X_C.*

Conductance G. *The reciprocal of resistance.*

Inductive circuit. *A circuit in which the inductance term is greater than the capacitance term.*

Inductive susceptance B_L. *The reciprocal of inductive reactance X_L.*

Maxwell bridge. *An ac bridge circuit for measuring inductance.*

Parallel equivalent circuit. *The simplest equivalent circuit of an R-L-C combination consisting of a conductance G in parallel with a susceptance B.*

Series capacitance comparison bridge. *An ac bridge for measuring capacitance.*

Series equivalent circuit. *The simplest equivalent circuit of an R-L-C combination consisting of a resistance R in series with a reactance X.*

IMPORTANT RELATIONSHIPS

- Series impedances: $\mathbf{Z}_T = \mathbf{Z}_1 + \mathbf{Z}_2 + \cdots + \mathbf{Z}_N$.
- Parallel impedances:

$$\mathbf{Z}_T = \frac{\mathbf{Z}_1\mathbf{Z}_2}{\mathbf{Z}_1 + \mathbf{Z}_2} \qquad \mathbf{Z}_T = \frac{1}{\dfrac{1}{\mathbf{Z}_1} + \dfrac{1}{\mathbf{Z}_2} + \cdots + \dfrac{1}{\mathbf{Z}_N}}$$

- Admittance: $\mathbf{Y} = G + j(B_C - B_L)$ $\qquad \mathbf{Y} = Y\underline{/\phi}$

$$G = \frac{1}{R} \qquad B_C = \omega C \qquad B_L = \frac{1}{\omega L}$$

- Rectangular to polar / Polar to rectangular

$$Y = \sqrt{G^2 + (B_C - B_L)^2} \qquad G = Y\cos\phi$$

$$\phi = \tan^{-1}\frac{(B_C - B_L)}{G} \qquad B = Y\sin\phi$$

- parallel admittances: $\mathbf{Y}_T = \mathbf{Y}_1 + \mathbf{Y}_2 + \cdots + \mathbf{Y}_N$
- admittance/impedance: $\mathbf{Y} = \dfrac{1}{\mathbf{Z}}$ $\qquad \mathbf{Z} = \dfrac{1}{\mathbf{Y}}$
- voltage divider theorem:

$$\mathbf{V}_{PA} = \frac{\mathbf{V}_{PS}\mathbf{Z}_A}{\mathbf{Z}_T}$$

- current divider theorem:

$$\mathbf{I}_{P1} = \frac{\mathbf{I}_{PS}\mathbf{Z}_2}{\mathbf{Z}_1 + \mathbf{Z}_2}$$

- ac bridge balance equation $\qquad \mathbf{Z}_2\mathbf{Z}_3 = \mathbf{Z}_1\mathbf{Z}_4$
- capacitance bridge equations $\qquad R_X = \dfrac{R_2R_3}{R_1} \qquad C_X = \dfrac{R_1}{R_2}C_3$
- Maxwell bridge equations $\qquad R_X = \dfrac{R_2R_3}{R_1} \qquad L_X = R_2R_3C_1$

REVIEW QUESTIONS

1. The total impedance of a number of impedances connected in series is equal to the sum of the individual impedances. (TF)
2. Impedances are (scalar, vector) quantities.
3. In a series *R-L-C* circuit, the (voltage, current) is used as the reference phasor.
4. Impedances can be combined according to the same rules used for combining ______.

5. The reference phasor for impedances connected in parallel is (voltage, current).
6. Combining impedances in parallel may become quite tedious since impedances are vector quantities. (TF)
7. The units of conductance and susceptance are ____.
8. In an admittance triangle, capacitive susceptance is drawn on the ____ j-axis.
9. The negative j-axis quantity in an admittance triangle is ____ susceptance.
10. The simplest series equivalent circuit consists of a ____ in series with a ____.
11. A conductance connected in parallel with a susceptance is the form of a parallel equivalent circuit, and this is an ____.
12. In any phasor diagram, the current through a resistor is *always* in phase with the voltage across the resistor. (TF)
13. The capacitor current phasor in a parallel *R-L-C* circuit (leads, lags) the applied voltage phasor.
14. The inductor current phasor in a parallel *R-L-C* circuit (leads, lags) the applied voltage phasor.
15. KVL and KCL can be applied to ac circuits provided the voltages and currents are represented as ____.
16. The voltage divider theorem is useful in (series, parallel) circuits.
17. The current divider theorem is useful in (series, parallel, both) circuits.
18. We can solve ac circuits having more than one signal generator provided the wave forms have *exactly* the same ____.
19. Explain the reasoning behind assigning polarities to ac voltage generators.
20. An inductive circuit cannot contain any capacitance. (TF)
21. The Thévenin equivalent of an ac circuit consists of an ac voltage generator in series with an ____.
22. The meter in an ac bridge must be capable of reading ac currents (or voltages). (TF)
23. An ac bridge is balanced when the meter current is ____.
24. The balance equations for an ac bridge (do, do not) depend on the amplitude of the voltage source.
25. The equation for the unknown inductance in a Maxwell bridge depends on the frequency of the bridge voltage source. (TF)
26. Two complex numbers are equal if and only if the two real parts are equal and the two imaginary parts are equal. (TF)
27. One ac bridge for measuring capacitance is called a ____ ____ ____ bridge.
28. A Maxwell bridge is used to measure ____.
29. The Superposition theorem is useful in solving ac circuits containing multiple ____.

PROBLEMS

1. Calculate the total impedance of Fig. 16.1*a* if $R = 20\ \text{k}\Omega$, $X_L = 50\ \text{k}\Omega$, and $X_C = 78\ \text{k}\Omega$. Draw the impedance diagram.
2. Find the total impedance of Fig. 16.2*a* if $\mathbf{Z}_1 = 10 + j6$, $\mathbf{Z}_2 = 4 - j30$, and $\mathbf{Z}_3 = 12 - j4$. Draw the impedance diagram.

3. What is the total impedance in Problem 2 in polar form?
4. What is the total capacitance needed to connect in series with an impedance of $400 + j600\ \Omega$ in order to obtain a total impedance of $400 - j400$ at $\omega = 10^5$ rad/sec?
5. Calculate the total impedance of the circuit in Fig. 16.5*a* if $\mathbf{Z}_1 = 10 + j10\ \Omega$ and $\mathbf{Z}_2 = 5 + j5\ \Omega$.
6. What is the total impedance of a 2000 pF capacitor connected in parallel with a 40 μH inductor at $\omega = 10^8$ rad/sec?
7. Find the total admittance of the circuit in Fig. 16.7*a* if: $R = 10\ \Omega$, $L = 40\ \mu$H, $C = 3.3\ \mu$F, and $\omega = 10^5$ rad/sec. Draw the admittance diagram.
8. Repeat Problem 7 for $\omega = 10^6$ rad/sec.
9. What is the total admittance in Fig. 16.10*a* if $\mathbf{Y}_1 = 0.02\underline{/30^\circ}$, $\mathbf{Y}_2 = 0.1\underline{/45^\circ}$, and $\mathbf{Y}_3 = 0.01\underline{/-45^\circ}$?
10. What is the total admittance in Fig. 16.10*a* if $\mathbf{Y}_1 = 0.2 - j0.4$, $\mathbf{Y}_2 = 0.09 + j0.04$, and $\mathbf{Y}_3 = 0.1\underline{/-45^\circ}$?
11. The admittance of a circuit is known to be $0.04\underline{/30^\circ}$ ℧. Change this admittance to rectangular form, draw the admittance diagram, and draw an equivalent circuit.
12. Repeat Problem 11 for an admittance of $0.2\underline{/90^\circ}$ ℧.
13. Find the impedance, and draw the series equivalent for the admittance in Problem 11.
14. Find the admittance, and draw the parallel equivalent circuit for the total impedance in Problem 2.
15. A coil having an inductance of 10 mH and a resistance of 40 Ω is connected in parallel with a 10 μF capacitor. What is the equivalent admittance of the circuit at $\omega = 4 \times 10^3$ rad/sec?
16. Find the currents in the circuit in Fig. 16.16 if $\mathbf{V}_{PS} = 120\underline{/0^\circ}$, $\mathbf{Z}_1 = 3\underline{/30^\circ}$ kΩ, and $\mathbf{Z}_2 = 6\underline{/-45^\circ}$ kΩ. Draw the phasor diagram.
17. Solve for the currents in Fig. 16.18 if $R = 10\ \Omega$, $L = 100\ \mu$H, $C = 0.5\ \mu$F, and $V_{PS} = 25 \sin 2 \times 10^5 t$ mV. Draw the phasor diagram.
18. In Fig. 16.18, $\mathbf{V}_{PS} = 28\underline{/0^\circ}$V and $\mathbf{I}_{PS} = 7\underline{/30^\circ}$ mA. What are the individual element currents if $X_L = 7000\ \Omega$? Draw the phasor diagram.
19. Find the generator current $\mathbf{I}_{PS}$ in Fig. 16.27 by simplification.

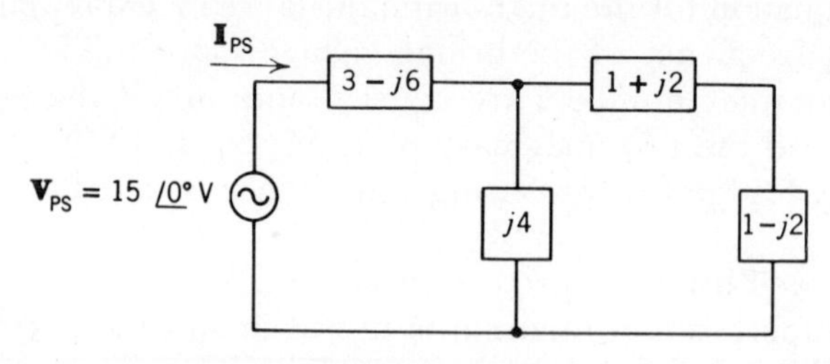

Figure 16.27 Impedance given in ohms

20. Find the generator current $\mathbf{I}_{PS}$ in Fig. 16.28. Draw the phasor diagram.
21. Find $\mathbf{V}_{P1}$ and $\mathbf{V}_{P3}$ in Example 16.14 (use the voltage divider theorem). Draw the phasor diagram and verify the solution using KVL.
22. Find $\mathbf{I}_{P2}$ in Example 16.15 using the current divider theorem. Check your solution by any method.

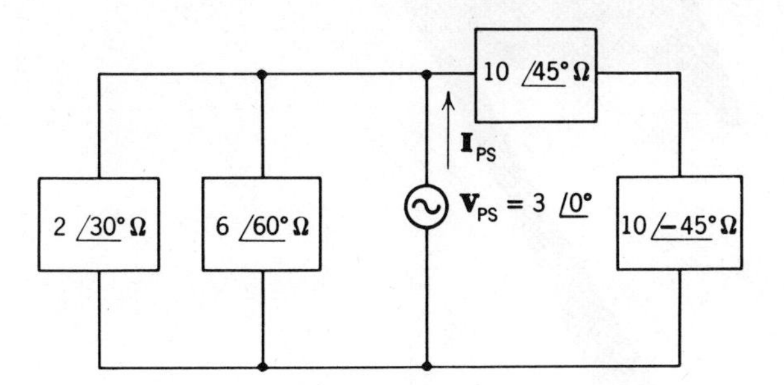

Figure 16.28

23. Use mesh equations to find the current in the $2+j2$ impedance in Example 16.16 (Fig. 16.22) if $\mathbf{V}_{PS_2}$ is changed to $10\underline{/0°}$ V.

24. Solve Problem 23 using node equations.

25. Solve Problem 23 using superposition (and simplification).

26. Verify the node voltage $\mathbf{V}_{PA} = 10.4\underline{/-33.7°}$ V calculated for the circuit in Fig. 16.23*a*.

27. Solve for $\mathbf{I}_{P1}$ and $\mathbf{I}_{P2}$ in Fig. 16.22 if: $\mathbf{V}_{PS_1} = 20\underline{/0°}$ $\mathbf{V}_{PS_2} = 15\underline{/0°}$, $\mathbf{Z}_1 = 2\ \Omega$, $\mathbf{Z}_2 = (3+j3)\ \Omega$, and $\mathbf{Z}_3 = -j2\ \Omega$.

28. Solve Problem 27 using node equations.

29. Repeat Problem 27 if $V_{PS_1} = 10\underline{/0°}$.

30. Repeat Problem 28 if $V_{PS_1} = 10\underline{/0°}$.

31. Use Thévenin's theorem to solve for $\mathbf{I}_{P_1}$ in Problem 27.

32. Use Thévenin's theorem to solve for $\mathbf{I}_{P_2}$ in Problem 27.

33. Use Thévenin's theorem to solve for $\mathbf{I}_{P1}$ in Problem 29.

34. What must be the value of $\mathbf{Z}_1$ in Fig. 16.24*a* in order to balance the bridge if $\mathbf{Z}_2 = 25\ \Omega$, $\mathbf{Z}_3 = 100\ \Omega$, and $\mathbf{Z}_4 = (15+j10)\Omega$?

35. What components could be used for $\mathbf{Z}_1$ in Problem 34, and how would you connect them?

36. Is the bridge in Fig. 16.24*a* balanced if $\mathbf{Z}_1 = j5\ \Omega$, $\mathbf{Z}_2 = (3+j4)\Omega$, $\mathbf{Z}_3 = (3-j4)\Omega$, and $\mathbf{Z}_4 = -j5\ \Omega$?

37. The capacitance bridge in Fig. 16.25 is balanced when $R_1 = 10\ \text{k}\Omega$, $R_2 = 100\ \text{k}\Omega$, $R_3 = 0.0\ \Omega$, and $C_3 = 4000$ pF. What is the value of the unknown capacitance?

38. What is the value R_X and C_X in the capacitance bridge of Fig. 16.25 if the bridge balances with: $R_1 = 100\ \Omega$, $R_2 = 200\ \Omega$, $R_3 = 500\ \Omega$, and $C_3 = 0.1\ \mu$F?

39. What is the value of R_X and L_X in Fig. 16.26 if the bridge balances when: $R_1 = 200\ \Omega$, $R_2 = 300\ \Omega$, $R_3 = 400\ \Omega$, and $C_1 = 100$ pF?

40. What is the value of the unknown inductance in Fig. 16.26 if the bridge balances with: $R_1 = \infty$, $R_2 = 10\ \Omega$, $R_3 = 7\ \Omega$, and $C_1 = 150$ pF?

41. Solve Problem 27 using Superposition.

42. Solve Problem 29 using Superposition.

17

POWER IN ac CIRCUITS

The power in a resistance in a dc circuit can be found very simply by taking the product of the current through and the voltage across the resistance. Since the current and voltage *do not* vary with time, the power is constant. The power in a resistance in an ac circuit can also be found by taking the product of the current and voltage, but since the current and voltage both vary sinusoidally, the power is *not* constant. The power in a resistance found by taking the product of a sinusoidal current and voltage is called *instantaneous* power p (the power at any instant of time), and it varies periodically between zero and some maximum value. It is more convenient to calculate the *average* power in an ac circuit, and we will do this by determining the *effective* or RMS values of sinusoidal currents and voltages.

17.1 POWER IN A RESISTANCE

In Fig. 17.1*a*, a resistance R is connected across the terminals of a voltage source $v = V_p \sin \omega t$. The current in the circuit is $i = I_p \sin \omega t$, where $I_p = V_p/R$, and the current and voltage wave forms are shown in Fig. 17.1*b*. The *instantaneous* power p in the resistance R is found by taking the product of the voltage and current,

$$p = vi = V_p \sin \omega t \, I_p \sin \omega t = V_p I_p \sin^2 \omega t = P_p \sin^2 \omega t \qquad (17.1)$$

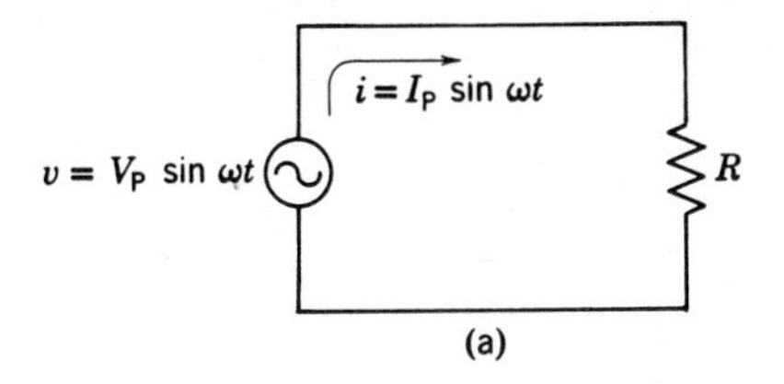

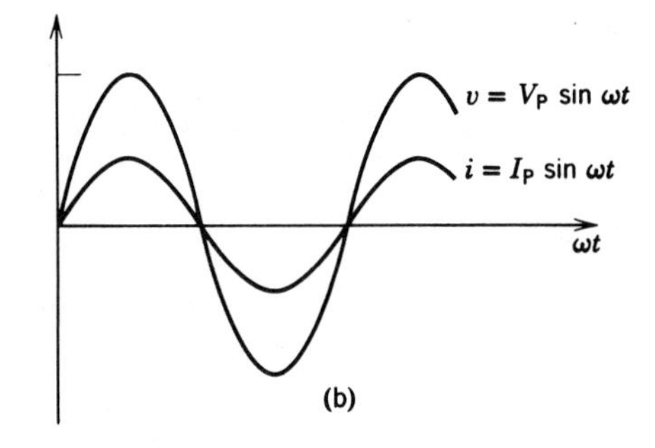

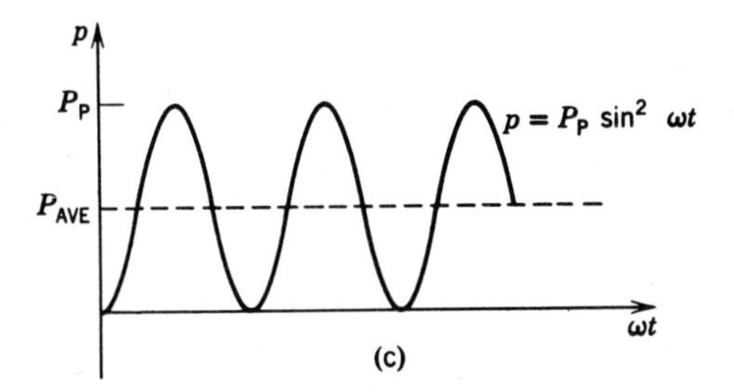

Figure 17.1 (*a*) Resistive ac circuit (*b*) Current and voltage in the circuit (*c*) Instantaneous power in R

This wave form is plotted in Fig. 17.1*c*, and since it is a $\sin^2 \omega t$ wave, it is *always* positive. This simply means that the source is always delivering power to the resistor. Notice, however, that the wave has a minimum value of zero and a maximum value of $P_P = V_P I_P$. Thus the power varies between zero and P_P and it is therefore not constant. It is zero whenever the current and voltage are zero, and maximum when the current and voltage are at their peak values.

It is clear from Fig. 17.1*c* that the average value of the wave, or the *average power*, is *one-half* the peak value; that is,

$$P_{AVE} = \tfrac{1}{2} P_P \tag{17.2}$$

This must be so since the portion of the wave *above* P_{AVE} (the dotted line) is equal to the portion of the wave *below* P_{AVE}. This result can also be shown mathematically by using the trigonometric identity

$$P_P \sin^2 \omega t = \tfrac{1}{2} P_P (1 - \cos 2\omega t)$$

The average value of $\cos 2\omega t$ (or any sinusoidal wave) is zero, and thus

$$P_{AVE} = \tfrac{1}{2} P_P (1 - 0) = \tfrac{1}{2} P_P$$

Since we will generally be interested in average power rather than peak power, we will drop the subscript AVE and use the symbol P for average power. Thus the symbol

$$P \triangleq P_{\text{AVE}}$$

• • •

Example 17.1 Find the peak power P_p and the average power P in Fig. 17.1 if $v = 80 \sin 377t$ and $R = 1\ \text{k}\Omega$.

Solution The current in the circuit is

$$i = \frac{v}{R} = \frac{80 \sin 377t \text{ V}}{1 \text{ k}\Omega} = 80 \sin 377t \text{ mA}$$

The instantaneous power is then

$$p = vi = 80 \sin 377t \times 80 \times 10^{-3} \sin 377t = 6.4 \sin^2 377t \text{ W}$$

The peak power is then

$$P_\text{p} = 6.4 \text{ W}$$

The average power is found from Eq. 17.2 as

$$P = \tfrac{1}{2} P_\text{p} = \tfrac{1}{2} \times 6.4 \text{ W} = 3.2 \text{ W}$$

• • •

From Fig. 17.1 it is clear that the peak power P_p in a resistance can be found as the product of the peak voltage V_p and the peak current I_p. Combining this with Eq. 17.2, we have the result

$$P = \tfrac{1}{2} P_\text{p} = \tfrac{1}{2} V_\text{p} I_\text{p}$$

But we know that $V_\text{p} = I_\text{p} R$ and thus

$$P = \tfrac{1}{2} V_\text{p} I_\text{p} = \tfrac{1}{2}(I_\text{p} R) I_\text{p} = \tfrac{1}{2} I_\text{p}^2 R \tag{17.3}$$

Equation 17.3 can always be used to calculate the ac power in a resistance in terms of the *peak* value of the current. However, many ac meters provide ac voltage and current measurements in terms of *effective* or RMS values rather than peak values. Rearranging Eq. 17.3 into the form

$$P = \frac{1}{2} I_\text{p}^2 R = \left(\frac{I_\text{p}}{\sqrt{2}}\right)^2 R = I^2 R \tag{17.4}$$

The current I in Eq. 17.4 is known as the *effective* or RMS value of the current, and it is seen that

$$I = I_{\text{eff}} = I_{\text{RMS}} = \frac{I_\text{p}}{\sqrt{2}} = 0.707 I_\text{p} \tag{17.5}$$

This result is valid for *any sinusoidal* wave whether it be voltage or current. Equation 17.4 can be used to calculate the power in a resistance in terms of the peak current I_p or the RMS value of the current I.

• • •

Example 17.2 Calculate the RMS value of the current wave form shown in Fig. 17.2.

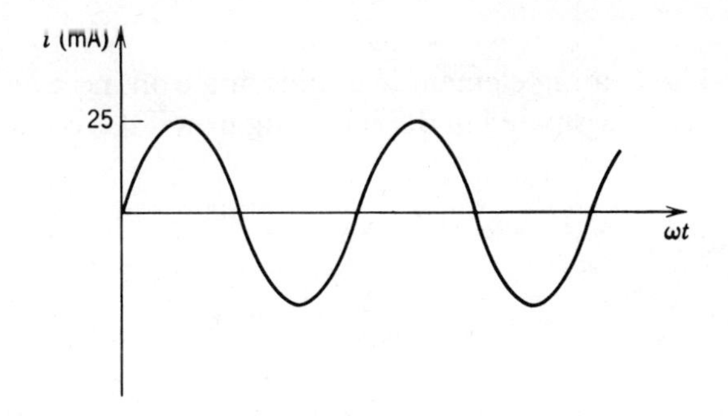

Figure 17.2

Solution The peak value is $I_p = 25$ mA. Using Eq. 17.5, the RMS value is

$$I = 0.707 I_p = 0.707 \times 25 \text{ mA} = 17.7 \text{ mA}$$

• • •

Example 17.3 Calculate the peak power and the average power in a 600 Ω resistor if the current is as shown in Fig. 17.2.

Solution The peak power is found as

$$P_p = I_p^2 R = (25 \times 10^{-3})^2 \times 600 = 0.375 \text{ W}$$

Since the average power is one-half the peak power, it is clearly,

$$P = \frac{1}{2} P_p = \frac{0.375 \text{ W}}{2} = 0.188 \text{ W}$$

• • •

If we make the substitution $I_p = V_p/R$ in Eq. 17.3, we obtain the result

$$P = \frac{1}{2} V_p I_p = \frac{1}{2} V_p \left(\frac{V_p}{R}\right) = \frac{1}{2} \frac{V_p^2}{R}$$

This equation can also be rearranged into the form

$$P = \frac{1}{2} \frac{V_p^2}{R} = \frac{\left(\frac{V_p}{\sqrt{2}}\right)^2}{R} = \frac{V^2}{R} \tag{17.6}$$

The voltage V in Eq. 17.6 is the *effective* or RMS value of the voltage and it is seen that

$$V = V_{\text{eff}} = V_{\text{RMS}} = \frac{V_p}{\sqrt{2}} = 0.707 V_p \tag{17.7}$$

Equation 17.6 can be used to calculate the average power in a resistor in terms of either the peak value of the voltage V_p or the RMS voltage V.

• • •

Example 17.4 The commercial ac power supplied to a home is specified as 117 V ac. This is actually the RMS voltage. What is the peak value of the voltage?

Solution Rearranging Eq. 17.7,

$$V_P = \sqrt{2}\, V = 1.414\, V = 1.414 \times 117 = 165 \text{ V}$$

• • •

• • •

Example 17.5 The heating element in a soldering iron has a *hot* resistance of 391 Ω. What power is dissipated in the soldering iron if it is connected to a 117-V-ac source?

Solution Using Eq. 17.6,

$$P = \frac{V^2}{R} = \frac{(117\text{ V})^2}{391\ \Omega} = 35\text{ W}$$

• • •

Example 17.6 What is the ***peak*** value of the current in the soldering iron in Example 17.5?

Solution The RMS current is

$$I = \frac{V}{R} = \frac{117\text{ V}}{391\ \Omega} = 0.299\text{ A}$$

The peak value of the current is found by rearranging Eq. 17.5 as

$$I_p = \sqrt{2}I = 1.414I = 1.414 \times 0.299\text{ A} = 0.423\text{ A}$$

• • •

From Eqs. 17.5 and 17.7, it is clear that

$$I_p = \sqrt{2}I \qquad V_p = \sqrt{2}V$$

If these two relationships are substituted into Eq. 17.2, we have

$$P = \frac{1}{2}P_p = \frac{1}{2}(V_pI_p) = \frac{1}{2}(\sqrt{2}V)(\sqrt{2}I) = \frac{\sqrt{2}\sqrt{2}}{2}VI = VI \tag{17.8}$$

Thus the average power is found as the product of the RMS voltage and current. We can now summarize the results of this section by noting that the ac power in a resistance can be found as follows:

1. Using peak values,

$$P(\text{W}) = \frac{1}{2}P_p = \frac{V_pI_p}{2} = \frac{I_p^2R}{2} = \frac{V_p^2}{2R}$$

2. Using RMS values,

$$P(\text{W}) = \frac{1}{2}P_p = VI = I^2R = \frac{V^2}{R}$$

where $I = 0.707I_p$ and $V = 0.707V_p$.*

From this point on, whenever an ac voltage or current is specified it will be taken as an RMS value unless explicitly stated otherwise.

• • •

Example 17.7 Calculate the peak voltage, peak current, RMS current, and power in the resistance in Fig. 17.3.

*Two very useful numbers to memorize are: (1) $\sqrt{2} = 1.414$, and (2) $1/\sqrt{2} = 0.707$.

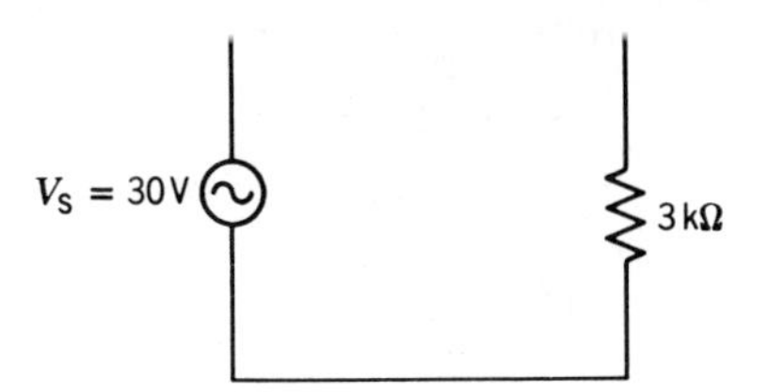

Figure 17.3

Solution The peak voltage is

$$V_p = 1.414\,V_S = 1.414 \times 30\,V = 42.4\,V$$

The peak current is

$$I_p = \frac{V_P}{R} = \frac{42.4\text{ V}}{3\text{ k}\Omega} = 14.1\text{ mA}$$

The RMS current is

$$I = 0.707 I_P = 10\text{ mA}$$

The power is

$$P = V_S I = 30\text{ V} \times 10\text{ mA} = 300\text{ mW}$$

• • •

2 POWER IN AN INDUCTANCE

In Fig. 17.4*a*, an ideal inductance L having an inductive reactance $X_L = \omega L$ is connected across a voltage source. If we take the current as the reference, $i = I_P \sin \omega t$, then the voltage must *lead* by 90° or $\pi/2$ rad. Thus, $v = V_p \sin(\omega t + \pi/2)$. These two wave forms are shown in Fig. 17.4*b*.

We can now calculate the instantaneous power p in the inductance by taking the product of the instantaneous voltage v and current i. Thus

$$p = vi = V_p \sin\left(\omega t + \frac{\pi}{2}\right) I_p \sin \omega t = V_p I_p \sin\left(\omega t + \frac{\pi}{2}\right) \sin \omega t$$

This can be simplified since we know that $\sin(\omega t + \pi/2) = \cos \omega t$. Thus

$$p = V_p I_p \cos \omega t \sin \omega t$$

We now make use of the trigonometric identity

$$\sin \alpha \cos \alpha = \tfrac{1}{2} \sin 2\alpha$$

Thus

$$p = V_p I_p \cos \omega t \sin \omega t = \tfrac{1}{2} V_p I_p \sin 2\,\omega t = P_p \sin 2\,\omega t \tag{17.9}$$

Equation (17.9) is plotted in Fig. 17.4*c* and it is clear that the power in the inductance varies sinusoidally at a frequency twice that of the applied voltage, and it has a peak value of $\frac{1}{2} V_p I_p$.

In Fig. 17.4*c* the instantaneous power is seen to be alternately ***positive*** and ***negative.*** This means that the inductor ***alternately absorbs*** power (stores energy in its magnetic field) from the voltage source and then ***returns*** power to the voltage source. When the inductor current and voltage in Fig. 17.4*b* are ***both positive,*** or ***both negative,*** energy is being stored in the inductance. On the other hand, when the inductor current and voltage have ***unlike*** signs, the inductance is returning energy to the source.

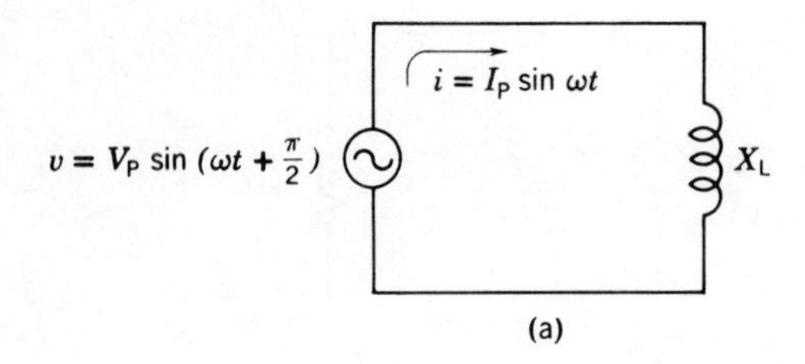

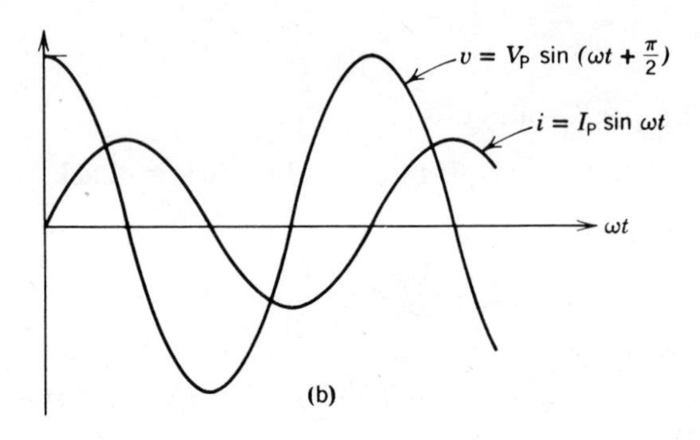

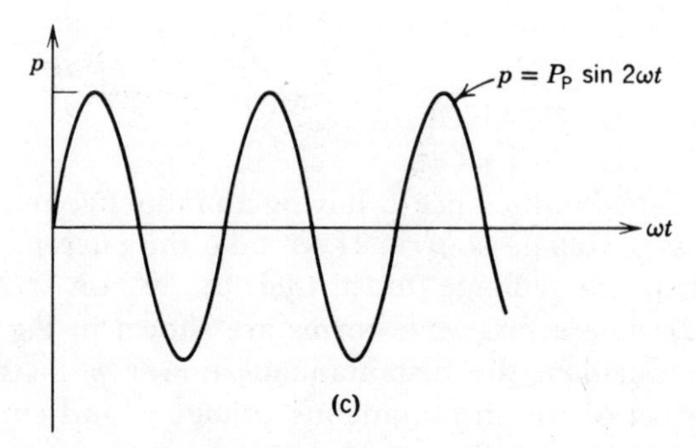

Figure 17.4 (*a*) A circuit containing an ideal inductance (*b*) Current and voltage in the circuit (*c*) Instantaneous power in L

Since the instantaneous power wave form in Fig. 17.4*c* is sinusoidal, its *average* value is zero. In other words, the portion of the wave above the horizontal-axis is equal to the portion of the wave below the axis. This means that the energy *stored* by the inductance is exactly equal to the energy *returned* by the inductance, and thus there is no power dissipated in the inductance. Thus this power is not a *true* power in Watts, since it does not do any work; that is, it simply circulates between the inductance and the source. For this reason it is designated as *reactive power* P_Q. In order to avoid any confusion between the true power in watts dissipated in a resistance and the reactive power in an inductance, the latter is measured in units of *volt amperes reactive,* or var.

The reactive power P_Q in an inductance can be found by using RMS values of current and voltage just as before, and thus

$$\text{reactive power} = P_Q(\text{var}) = V_L I_L = I_L^2 X_L = \frac{V_L^2}{X_L} \qquad (17.10)$$

where V_L and I_L are the RMS values of the inductor voltage and current.

• • •

Example 17.8 Calculate the instantaneous power and the reactive power P_Q in Fig. 17.4*a* if $v = 40 \sin(377t + \pi/2)$ and $L = 10$ H.

Solution We must first calculate the inductive reactance X_L.

$$X_L = \omega L = 377 \times 10 \text{ H} = 3.77 \text{ k}\Omega$$

The circuit current is then

$$\mathbf{I}_P = \frac{\mathbf{V}_P}{jX_L} = \frac{V_P/90^\circ}{j3.77 \text{ k}\Omega} = \frac{40/90^\circ}{3.77 \text{ k}\Omega/90^\circ} = 10.6/0^\circ \text{ mA}$$

Thus,

$$i = 10.6 \sin 377t \text{ mA}$$

The instantaneous power p is then (using Eq. 17.9),

$$p = vi = 40 \sin\left(377t + \frac{\pi}{2}\right) \times 10.6 \times 10^{-3} \sin 377t$$

$$= \frac{40 \times 10.6 \times 10^{-3}}{2} \sin 2 \times 377t = 0.212 \sin 754t \text{ var}$$

The RMS voltage and current are

$$V_L = 0.707\ V_p = 0.707 \times 40 = 28.3 \text{ V}$$
$$I_L = 0.707\ I_p = 0.707 \times 10.6 \text{ mA} = 7.5 \text{ mA}$$

The reactive power is then found using Eq. 17.10,

$$P_Q = V_L I_L = 28.3 \text{ V} \times 7.5 \text{ mA} = 0.212 \text{ var}$$

As a check, we calculate P_Q using the other two forms of Eq. 17.10.

$$P_Q = I_L^2 X_L = (7.5 \times 10^{-3})^2 \times 3.77 \text{ k}\Omega = 0.212 \text{ var}$$

$$P_Q = \frac{V_L^2}{X_L} = \frac{28.3^2}{3.77 \text{ k}\Omega} = 0.212 \text{ var}$$

• • •

17.3 POWER IN A CAPACITANCE

In Fig. 17.5*a* an ideal capacitor is connected across the terminals of a voltage source. The circuit current is taken as a reference, and since the capacitor voltage must *lag* the current by 90° or $\pi/2$ rad, the current and voltage are given as

$$i = I_p \sin \omega t$$

$$v = V_p \sin\left(\omega t - \frac{\pi}{2}\right)$$

These two wave forms are shown in Fig. 17.5*b*.

We again calculate the power in the capacitor by taking the product of the capacitor voltage and current. Thus, the instantaneous power p is

$$p = vi = V_p \sin\left(\omega t - \frac{\pi}{2}\right) I_p \sin \omega t = V_p I_p \sin\left(\omega t - \frac{\pi}{2}\right) \sin \omega t$$

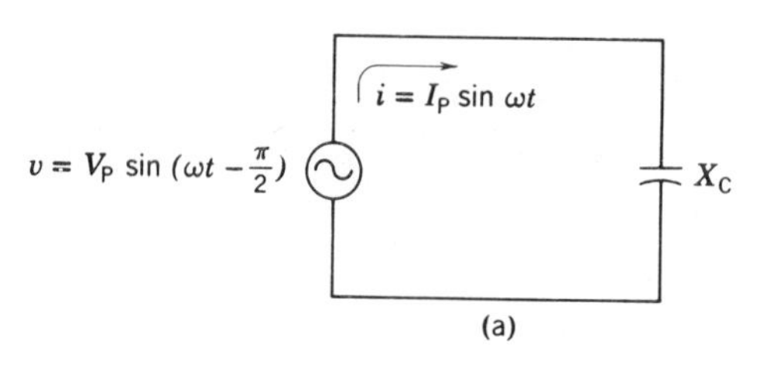

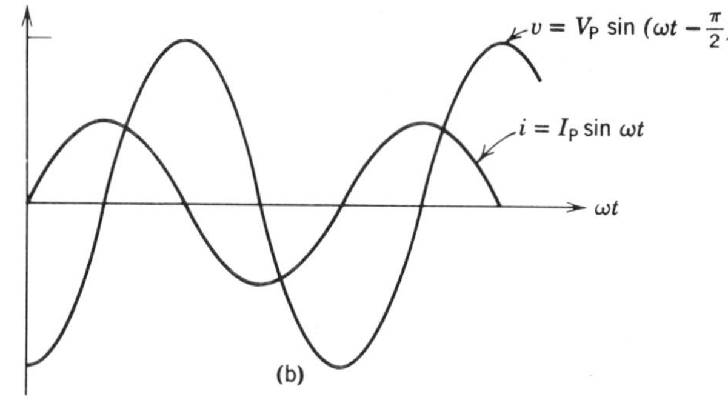

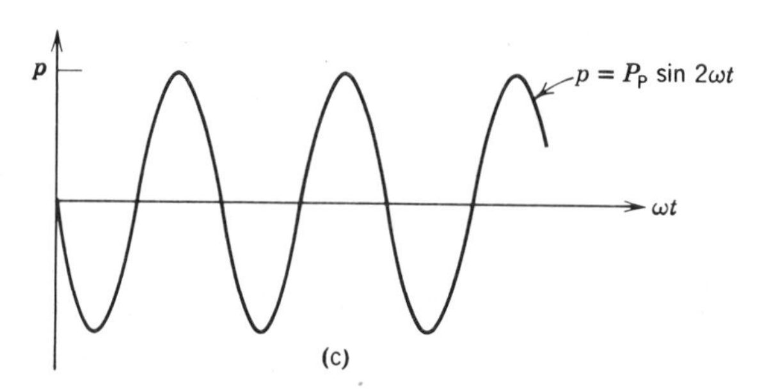

Figure 17.5 (*a*) A circuit containing an ideal capacitor (*b*) Current and voltage in the circuit (*c*) Instantaneous power in C

But, we know that $\sin(\omega t - \pi/2)$ is equal to $-\cos \omega t$, and thus

$$p = -V_p I_p \cos \omega t \sin \omega t$$

We again use the trigonometric identity, $\cos \alpha \sin \alpha = \frac{1}{2} \sin 2\alpha$ to obtain

$$p = -V_p I_p \cos \omega t \sin \omega t = -\tfrac{1}{2} V_p I_p \sin 2\omega t = -P_p \sin 2\omega t \qquad (17.11)$$

Equation 17.11 is plotted in Fig. 17.5*c*, and it is clear that the power in the capacitor varies sinusoidally at a frequency twice that of the applied voltage and it has a peak value of $\frac{1}{2} V_p I_p$ (just as with an inductance). The instantaneous power is alternately *positive* and *negative*, and the capacitor is thus seen to alternately absorb power from the source and then return it to the source. When the capacitor current and voltage are *both positive* or *both negative* in Fig. 17.5*b*, energy is being stored in the capacitor. When the capacitor voltage and current have opposite signs, the energy stored in the capacitor is returned to the voltage source. This is exactly analogous to the circulating power in an ideal inductance.

Since the instantaneous power wave form in Fig. 17.5*c* is sinusoidal, the average value of the wave must be zero. Thus *all* of the energy stored by the capacitor is returned to the source periodically, and there is no power dissipated in the capacitor. For this reason, this power is again referred to as reactive power P_Q, and it is also measured in vars. The reactive power in a capacitance can be calculated using RMS values of voltage and current, and thus

$$\text{reactive power} = P_Q(\text{var}) = V_C I_C = I_C^2 X_C = \frac{V_C^2}{X_C} \quad (17.12)$$

where V_C and I_C are the RMS values of capacitor voltage and current.

• • •

Example 17.9 Calculate the instantaneous power p and the reactive power P_Q in Fig. 17.5*a* if $v = 50 \sin (10^4 t - \pi/2)$ and $C = 0.1\ \mu\text{F}$.

Solution We first calculate the capacitive reactance X_C.

$$X_C = \frac{1}{\omega C} = \frac{1}{10^4 \times 0.1\ \mu\text{F}} = 10^3\ \Omega$$

The circuit current is then

$$\mathbf{I}_P = \frac{\mathbf{V}_P}{-jX_C} = \frac{V_P\underline{/-90^\circ}}{-j10^3\ \Omega} = \frac{50\underline{/-90^\circ}}{10^3\underline{/-90^\circ}} = 50\underline{/0^\circ}\ \text{mA}$$

$$i_p = 50 \sin 10^4 t\ \text{mA}$$

Using Eq. 17.11, the instantaneous power is

$$p = vi = -50 \sin \left(10^4 t - \frac{\pi}{2}\right) \times 50 \times 10^{-3} \sin 10^4 t$$

$$= -1.25 \sin 2 \times 10^4 t\ \text{var}$$

The negative sign simply shows that the power wave form is a *negative* sine wave with respect to the current wave form (the current wave form serves as the reference). The RMS voltage and current are

$$V_C = 0.707\, V_p = 0.707 \times 50\ \text{V} = 35.4\ \text{V}$$

$$I_C = 0.707 I_p = 0.707 \times 50\ \text{mA} = 35.4\ \text{mA}$$

The reactive power is then found using Eq. 17.12,

$$P_Q = V_C I_C = 35.4\ \text{V} \times 35.4\ \text{mA} = 1.25\ \text{var}$$

As a check, we calculate P_Q using the other two forms of Eq. 17.12,

$$P_Q = I_C^2 X_C = (35.4 \times 10^{-3})^2 \times 1\ \text{k}\Omega = 1.25\ \text{var}$$

$$P_Q = \frac{V_C^2}{X_C} = \frac{(35.4)^2}{1\ \text{k}\Omega} = 1.25\ \text{var}$$

• • •

Example 17.10 What value of capacitance has a reactive power of 200 var when connected to a 240 V, 60 Hz source?

Solution Rearranging Eq. 17.12,

$$X_C = \frac{V_C^2}{P_Q} = \frac{(240\ \text{V})^2}{200\ \text{var}} = 288\ \Omega$$

Thus

$$C = \frac{1}{\omega X_C} = \frac{1}{2\pi \times 60 \times 288} = \frac{1}{377 \times 278} = 9.2\ \mu\text{F}$$

• • •

17.4 POWER IN AN IMPEDANCE

In Fig. 17.6*a* an impedance $Z\underline{/\phi}$ is connected across the terminals of a voltage source $\mathbf{V}_S$. The applied voltage and the circuit current are related by Ohm's Law as

$$\mathbf{V}_S = \mathbf{IZ}$$

and if we take the current as the reference, we have the phasor diagram as shown in Fig. 17.6*b*. The voltage phasor is given as

$$\mathbf{V}_S = \mathbf{IZ} = I\underline{/0°}Z\underline{/\phi} = IZ\underline{/\phi} = V_S\underline{/\phi}$$

The voltage phasor $\mathbf{V}_S$ can then be broken down into its two components to form the triangle shown in Fig. 17.6*c*. Remember that $\mathbf{V}_S$ and $\mathbf{I}_S$ are RMS values.

If we multiply each side of the triangle in Fig. 17.6*c* by the magnitude of the circuit current I, the result is a similar triangle as shown in Fig. 17.7. The side of this triangle along the real-axis is $V_R I$, and since this is the product of the voltage across a resistance V_R and the current through it I, this must be the *real power P* in the circuit. The vertical leg of the triangle is the product of the voltage across a reactance V_X and the current through it I, and this must therefore be the reactive power P_Q. The hypotenuse of the triangle is the

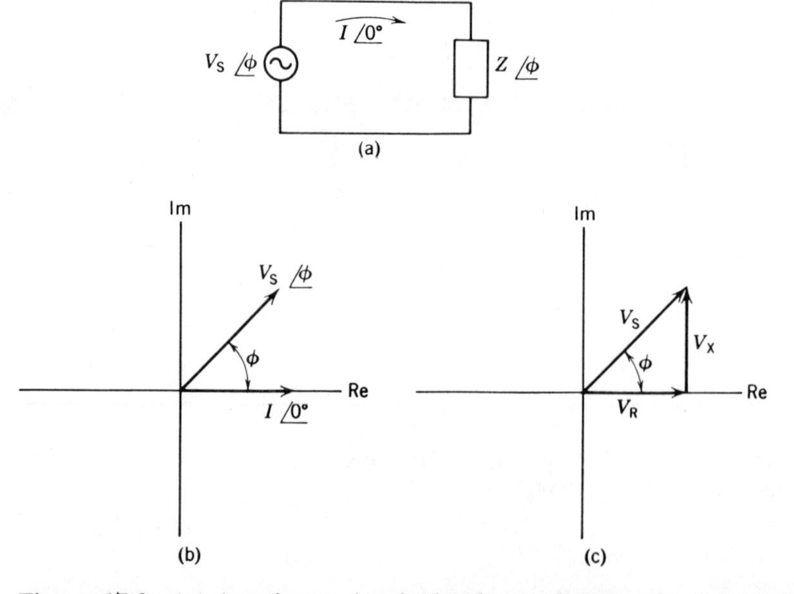

Figure 17.6 (*a*) A series ac circuit (*b*) Phasor diagram for (*a*) (*c*) The voltage phasor diagram for (*a*)

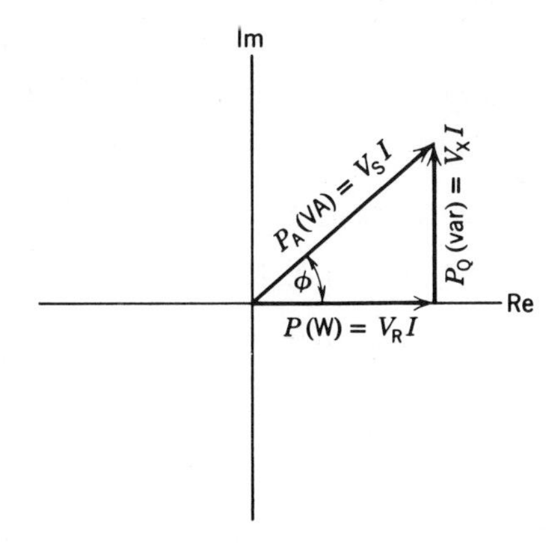

Figure 17.7 A power triangle

product of the applied voltage V_S and the circuit current I, and this is defined as the *apparent power* P_A. The unit of apparent power is volt-amperes VA. Since the three legs of this triangle represent *real power, reactive power,* and *apparent power,* this is called a *power triangle.* Notice that the angle ϕ of the power triangle is exactly equal to the phase angle ϕ of the impedance $\mathbf{Z} = Z\underline{/\phi}$.

Since the power triangle is drawn in the complex plane, we can write the following relationship.

$$\mathbf{P}_A(\text{VA}) = P_A\underline{/\phi} = V_S I\underline{/\phi} = P(W) \pm jP_Q(\text{var}) \tag{17.13}$$

In the event the impedance is inductive, we use $+j$ in Eq. 17.13 and the reactive power P_Q is drawn on the positive vertical-axis. On the other hand, if the impedance is capacitive, we use $-j$ in Eq. 17.13 and P_Q is drawn on the negative vertical-axis.

We can also apply the simple trigonometric relationships of a right triangle to the power triangle in Fig. 17.7 to obtain alternate expressions for the real and reactive power. Thus

$$P(\text{W}) = V_S I \cos\phi \tag{17.14}$$

$$P_Q(\text{var}) = V_S I \sin\phi \tag{17.15}$$

Using these two expressions, the apparent power can be written as

$$\mathbf{P}_A = P(\text{W}) \pm jP_Q(\text{var}) = V_S I \cos\phi \pm jV_S I \sin\phi \tag{17.16}$$

• • •

Example 17.11 Calculate the apparent power, real power, and reactive power for the circuit in Fig. 17.6 if $\mathbf{V}_S = 20\underline{/30^\circ}$ V and $\mathbf{Z} = 400\underline{/30^\circ}\ \Omega$. Draw the power triangle.

Solution The circuit current $\mathbf{I}$ is

$$\mathbf{I} = \frac{\mathbf{V}_S}{\mathbf{Z}} = \frac{20\underline{/30^\circ}\text{ V}}{400\underline{/30^\circ}\ \Omega} = 0.05\underline{/0^\circ}\text{ A}$$

The apparent power is found using Eq. 17.13,

$$\mathbf{P}_A(\text{VA}) = V_S I\underline{/\phi} = 20\text{ V} \times 0.05\text{ A}\underline{/30^\circ} = 1.0\underline{/30^\circ}\text{ VA}$$

The real power is found from Eq. 17.14.

$$P(\text{W}) = V_S I \cos\phi = 20\text{ V} \times 0.05\text{ A} \cos 30^\circ = 1.0 \times 0.867 = 0.867\text{ W}$$

The reactive power is found using Eq. 17.15.

$$P_Q(\text{var}) = V_S I \sin\phi = 20\text{ V} \times 0.05\text{ A} \sin 30^\circ = 1.0 \times 0.5 = 0.5\text{ var}$$

The power triangle is shown in Fig. 17.8.

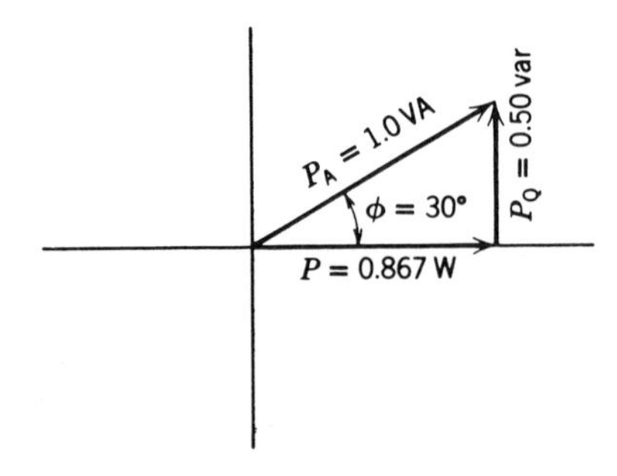

Figure 17.8 Power triangle for Example 17.11

• • •

Example 17.12 Calculate the apparent power, real power, and reactive power for the circuit in Fig. 17.9*a*. Draw the power triangle.

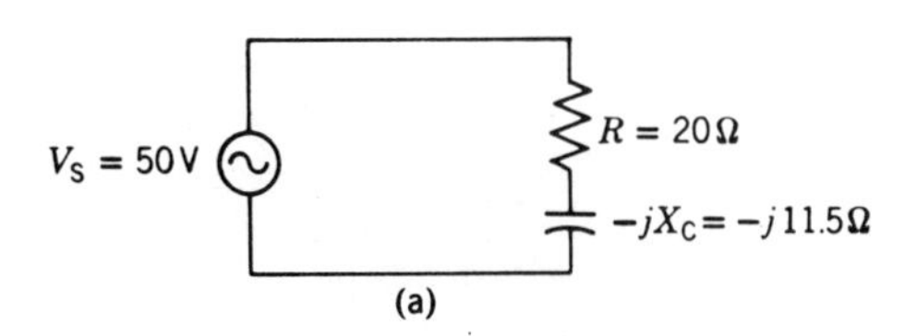

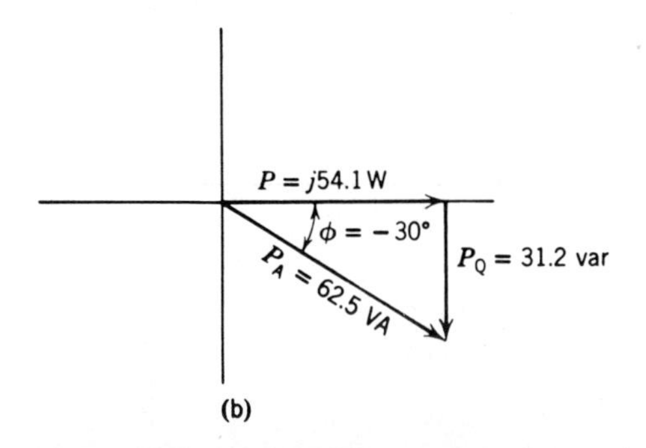

Figure 17.9 (*a*) An *R*-*C* circuit (*b*) Power triangle for (*a*)

Solution The circuit impedance is

$$\mathbf{Z} = R - jX_C = 20 - j11.5 = 40\underline{/-30°}\ \Omega$$

The circuit current is then

$$\mathbf{I} = \frac{\mathbf{V}_S}{\mathbf{Z}} = \frac{50\underline{/0°}\ \text{V}}{40\underline{/-30°}\ \Omega} = 1.25\underline{/30°}\ \text{A}$$

If we take the circuit current as a reference, we can write

$$\mathbf{I} = 1.25\underline{/0°}\ \text{A} \qquad \mathbf{V}_S = 50\underline{/-30°}\ \text{V}$$

The apparent power is found from Eq. 17.13

$$\mathbf{P}_A = V_S I\underline{/\phi} = 50\ \text{V} \times 1.25\ \text{A}\underline{/-30°} = 62.5\underline{/-30°}\ \text{VA}$$

The real power is found from Eq. 17.14

$$P(\text{W}) = V_S I \cos\phi = 50\ \text{V} \times 1.25\ \text{A} \cos(-30°) = 50 \times 1.25 \times 0.867 = 54.1\ \text{W}$$

The reactive power is found from Eq. 17.15

$$P_Q(\text{var}) = V_S I \sin\phi = 50\ \text{V} \times 1.25\ \text{A} \sin(-30°) = 50 \times 1.25(-0.5) = -31.2\ \text{var}$$

The negative sign on P_Q shows that the circuit is capacitive, and the power triangle is shown in Fig. 17.9*b*.

• • •

Since the three components of power, apparent, real and reactive, are added in the complex plane in Fig. 17.7, all of the components of power in any ac circuit can be added as vectors in this plane. Thus in any circuit, the real power in watts from all of the branches can be added to find the total real power delivered by the source; similarly, the reactive power from each branch can be added (+ for inductive and − for capacitive) to find the total reactive power delivered by the source. An example will best illustrate the procedure.

• • •

Example 17.13 Determine the apparent power, real power, and reactive power delivered by the source in Fig. 17.10*a*. Draw the power triangle.

Solution The impedance of the *R*-*L* branch is

$$\mathbf{Z}_1 = R + jX_L = 200 + j200 = 283\underline{/45°}\ \Omega$$

Thus the phase angle ϕ_1 is seen to be $+45°$. The current in this branch is then (the voltage $\mathbf{V}_S$ is taken as the reference)

$$\mathbf{I}_1 = \frac{\mathbf{V}_S}{\mathbf{Z}_1} = \frac{50\underline{/0°}\ \text{V}}{283\underline{/45°}\ \Omega} = 0.177\underline{/-45°}\ \text{A}$$

The true power in this branch is then

$$P(\text{W}) = V_S I_1 \cos\phi_1 = 50\ \text{V} \times 0.177\ \text{A} \cos 45° = 50 \times 0.177 \times 0.707 = 6.25\ \text{W}$$

The reactive power in this branch is

$$P_{Q_1}(\text{var}) = V_S I_1 \sin\phi_1 = 50\ \text{V} \times 0.177\ \text{A} \sin 45° = 50 \times 0.177 \times 0.707 = 6.25\ \text{var}$$

The reactive power in the capacitor is

$$P_{Q_2}(\text{var}) = V_S I_2 \sin\phi_2$$

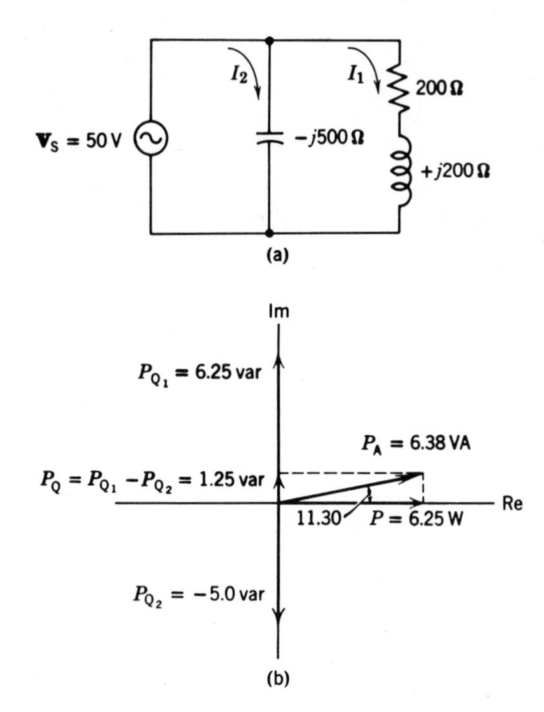

Figure 17.10 Example 17.13 (*a*) A parallel circuit (*b*) Power triangle for (*a*)

Where the phase angle ϕ_2 is $-90°$ since the voltage across the capacitor *lags* the current by 90°. The current in the capacitor is

$$\mathbf{I}_2 = \frac{\mathbf{V}_S}{-jX_C} = \frac{50\underline{/0°}\ \text{V}}{500\underline{/-90°}\ \Omega} = 0.1\underline{/90°}\ \text{A}$$

Thus

$$P_{Q_2}(\text{var}) = V_S I_2 \sin\phi_2 = 50\ \text{V} \times 0.1\ \text{A} \sin(-90°) = 50 \times 0.1(-1.0) = -5.0\ \text{var}$$

Notice that P_{Q_2} could also be calculated as

$$P_{Q_2}(\text{var}) = \frac{-V_S^2}{X_C} = \frac{-(50\ \text{V})^2}{500} = \frac{-2500}{500} = -5.0\ \text{var}$$

The total real power is then 6.25 W, and the total reactive power is

$$P_Q = P_{Q_1} + P_{Q_2} = 6.25 - 5.0 = 1.25\ \text{var}$$

The apparent power is found using Eq. 17.13,

$$\mathbf{P}_A = P + jP_Q = 6.25 + j1.25 = 6.38\underline{/11.3°}$$

The power triangle is shown in Fig. 17.10*b*.

• • •

17.5 POWER FACTOR

The angle ϕ in Eq. 17.13 comes from an impedance and is the angle by which the voltage *leads* the current in that impedance. It is also the specified angle in the power triangle. The cosine of this angle is a very useful quantity which is defined as the *power factor* PF. Thus

$$\text{power factor} = \text{PF} \stackrel{\Delta}{=} \cos\phi \tag{17.17}$$

Since the angle of an impedance can have any value from $+90°$ to $-90°$, the value of the power factor PF must always be between 0.0 and 1.0 [remember that the $\cos(-\phi) = \cos\phi$]. For an inductive circuit, the angle ϕ is between $0°$ and $+90°$, and this is traditionally referred to as a *lagging* power factor. For a capacitive circuit, the angle ϕ is between $0°$ and $-90°$, and this is traditionally referred to as a *leading* power factor.

• • •

Example 17.14 What is the power factor for the total load connected to the voltage source in Fig. 17.10 (Example 17.13)?

Solution The angle in the power triangle is $+11.3°$. Thus the power factor is

$$\text{PF} = \cos\phi = \cos 11.3° = 0.98 \text{ lagging}$$

• • •

Since the reactive power P_Q in a circuit is simply a component of power which circulates between the load and the source and serves no useful purpose, it is desirable to minimize this component. An examination of Eq. 17.15 shows that making $\phi = 0°$ would reduce P_Q to zero, since $\sin 0° = 0$. Furthermore, $\cos 0° = 1$, and thus a phase angle of $0°$ would maximize the true power as given in Eq. 17.14. Therefore, the ideal case is where the power factor is 1.0; that is $\phi = 0°$, and

$$\text{PF} = \cos\phi = \cos 0° = 1.0$$

A power factor of 1.0, or a phase angle of $0°$, simply means that the load appears as a pure resistance, and thus *all* of the power delivered by the source is dissipated in the load.

Since reactive power P_Q circulates between a load and the voltage source, it causes undesirable losses in the power lines connecting the load to the source. For this reason it is desirable to *correct* the power factor of a reactive load such that it is near unity (1.0).

As an example of power factor correction, let us consider the ac load shown in Fig. 17.11*a*. This is an inductive load and the total impedance is

$$\mathbf{Z} = R + jX_L = 50 + j86.7 = 100\underline{/60°}\ \Omega$$

The power factor is then

$$\text{PF} = \cos\phi = \cos 60° = 0.5 \text{ lagging}$$

The circuit current is

$$\mathbf{I} = \frac{\mathbf{V}_S}{\mathbf{Z}} = \frac{117\underline{/0°}\text{ V}}{100\underline{/60°}\ \Omega} = 1.17\underline{/-60°}\text{ A}$$

Taking the current as the reference, we have

$$\mathbf{I} = 1.17\underline{/0°}\text{ A} \qquad \mathbf{V}_S = 117\underline{/60°}\text{ V}$$

The true power is

$$P\text{ (W)} = V_S I\cos\phi = 117\text{ V} \times 1.17\text{ A}\cos 60° = 117 \times 1.17 \times 0.5 = 68.4\text{ W}$$

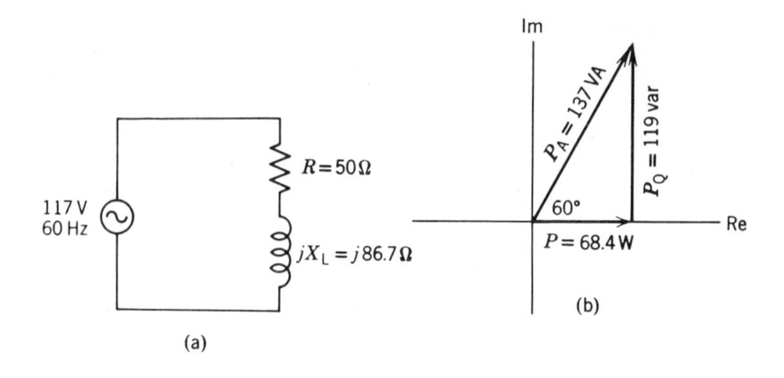

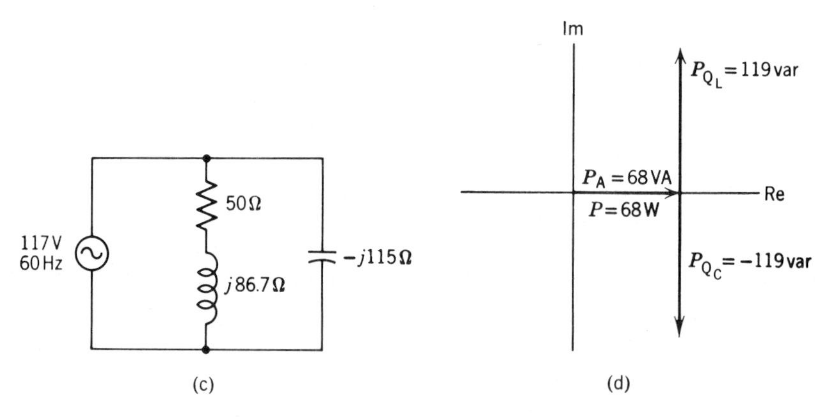

Figure 17.11 (*a*) An inductive circuit (*b*) Power triangle for (*a*) (*c*) Correcting for unity power factor (*d*) Power triangle for (*c*)

The reactive power is

$$P_Q\,(\text{var}) = V_S I \sin\phi = 117\text{ V} \times 1.17\text{ A} \sin 60° = 117 \times 1.17 \times 0.867 = 119\text{ var}$$

The apparent power is then

$$\mathbf{P}_A\,(\text{VA}) = V_S I \underline{/\phi} = 117\text{ V} \times 1.17\text{ A}\underline{/60°} = 137\underline{/60°}\text{ VA}$$

and the power triangle is shown in Fig. 17.11*b*. From the power triangle, it is clear that there is a relatively large component of reactive power P_Q.

We can obtain a unity power factor for the circuit in Fig. 17.11*a* by simply reducing the power factor angle ϕ to 0°. This can be accomplished by adding a capacitor to the circuit which will require −119 var of reactive power. The capacitive power and the inductive power will then cancel each other in the power triangle resulting in a unity power factor. We must, however, add the capacitor in *parallel* with the original load as shown in Fig. 17.11*c*. The new power triangle is shown in Fig. 17.11*d*, and the total load now appears as a resistive load on the source. The proper value of capacitive reactance X_C is found from

$$X_C = \frac{V_S^2}{P_Q} = \frac{(117\text{ V})^2}{119\text{ var}} = 115\ \Omega$$

and

$$C = \frac{1}{\omega X_C} = \frac{1}{377 \times 115} = 23.1\ \mu\text{F}$$

The reactive component of power needed by the inductance is now supplied by the capacitance, and all of the power delivered by the source is real power.

We might now consider the effect of connecting the 23.1-μF capacitor in *series* with the inductive load in Fig. 17.11*a* in order to correct the power factor. In this case, the total circuit impedance would be

$$\mathbf{Z}_T = R + jX_L - jX_C = 50 + j86.7 - j115 = 50 - j28.3 = 57.5\underline{/-30°}\ \Omega$$

The circuit current is then

$$\mathbf{I} = \frac{\mathbf{V}_S}{\mathbf{Z}_T} = \frac{117\underline{/0°}\ \text{V}}{57.5\underline{/-30°}\ \Omega} = 2.04\underline{/30°}\ \text{A}$$

This is obviously not an acceptable method since the circuit current through the original *R-L* load has almost doubled from 1.17 A to 2.04 A. Furthermore, the voltage across the original *R-L* load has increased to

$$\mathbf{V} = \mathbf{IZ} = 2.04\underline{/30°}\ \text{A} \times 100\underline{/60°}\ \Omega = 204\underline{/90°}\ \text{V}$$

Thus the voltage across the original load has almost doubled, and this connection is entirely unsatisfactory. Consider the consequences if the original load represents a 117-V-ac motor! Thus we conclude that the power factor correction of an inductive load must be accomplished by connecting a capacitance in *parallel* with the initial load. This insures that the voltage across the load will remain unchanged.

6 MAXIMUM POWER TRANSFER

An ideal voltage source has zero internal impedance, but a real voltage source will in general have some internal impedance $\mathbf{Z}_S$ as shown in Fig. 17.12. Since the real power in watts delivered to the load by the source is given by

$$P\ (\text{W}) = I^2R$$

it is clear that maximizing I will guarantee that the maximum power is delivered to the load. Since the load impedance $\mathbf{Z}_L$ and the source impedance $\mathbf{Z}_S$ are connected in series, the total impedance of the circuit is

$$\mathbf{Z}_T = \mathbf{Z}_S + \mathbf{Z}_L = R_S \pm jX_S + R_L \pm jX_L = R_S + R_L \pm j(X_S + X_L)$$

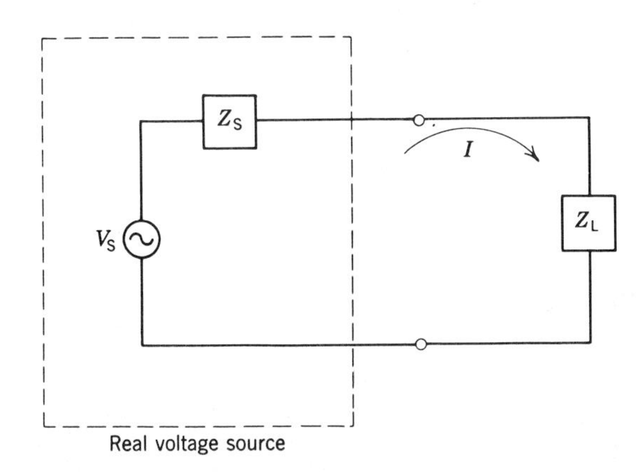

Figure 17.12

We can insure that the circuit current is maximum if the total impedance is minimum, and thus we would like to make X_L have an *equal* magnitude and an *opposite* sign of X_S. This would result in the cancellation of these two reactive terms. Under this condition, we have $X_S = -X_L$ and thus

$$\mathbf{Z}_T = R_S + R_L \pm j(X_S - X_S) = R_S + R_L + j0 = R_S + R_L$$

The circuit now appears as a resistive circuit, and from our previous work we know that maximum power transfer occurs when $R_S = R_L$. These two conditions for maximum power transfer can be summarized by stating that the load impedance must be the *complex conjugate* of the source impedance in order to obtain maximum power transfer. That is, if $\mathbf{Z}_S = R_S + jX_S$, then the load must be $\mathbf{Z}_L = R_S - jX_S$.

• • •

Example 17.15 What must be the value of $\mathbf{Z}_L$ in Fig. 17.13 in order to obtain maximum power transfer to the load? What is the power delivered to the load?

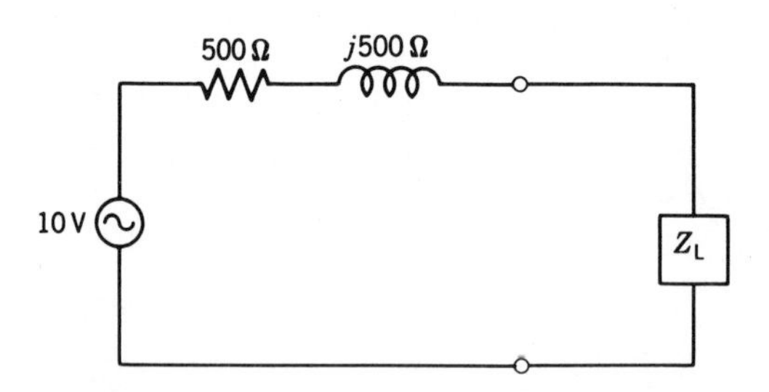

Figure 17.13 Example 17.15

Solution The load $\mathbf{Z}_L$ must be the complex conjugate of the source impedance $\mathbf{Z}_S$. Since $\mathbf{Z}_S = 500 + j500\ \Omega$, the load must be

$$\mathbf{Z}_L = 500 - j500\ \Omega$$

Under this condition, the total circuit impedance is

$$\mathbf{Z}_T = \mathbf{Z}_S + \mathbf{Z}_L = 500 + j500 + 500 - j500 = 1000\ \Omega$$

The circuit current is then

$$\mathbf{I} = \frac{\mathbf{V}_S}{\mathbf{Z}_T} = \frac{10\underline{/0°}\ \text{V}}{1000\underline{/0°}\ \Omega} = 10\underline{/0°}\ \text{mA}$$

The power delivered to the load is then

$$P\ (\text{W}) = I^2 R_L = (10 \times 10^{-3})^2\ \text{A} \times 500\ \Omega = 50\ \text{mW}$$

• • •

17.7 MULTIPLE SOURCE ac CIRCUITS

A *multiple source* ac circuit is one which contains two or more sources. The circuit in Fig. 17.14*a* contains two voltage sources, $\mathbf{V}_{S_1}$ and $\mathbf{V}_{S_2}$, and it is therefore a multiple source circuit. Such circuits can be solved by the Principle of Superposition provided we use phasors for the voltages and currents. This

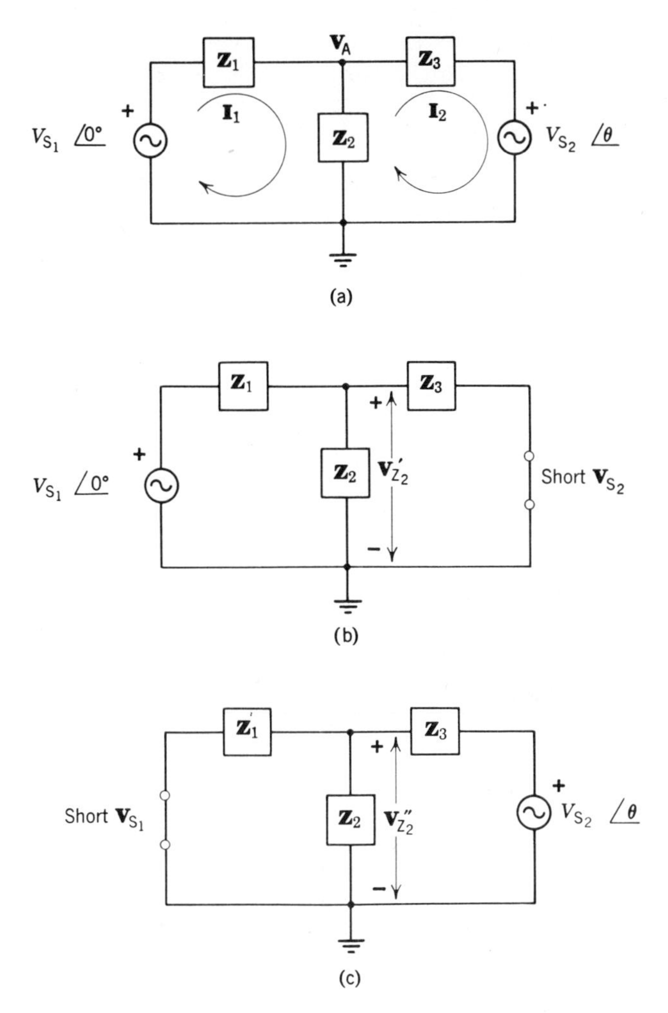

Figure 17.14 Solving a circuit by Superposition Theorem (*a*) A general two-mesh ac network (*b*) Voltage across $\mathbf{Z}_2$ due to $\mathbf{V}_{S_1}$ alone (*c*) Voltage across $\mathbf{Z}_2$ due to $\mathbf{V}_{S_2}$ alone

of course requires that the voltage sources have exactly the same frequencies. Again, note the use of RMS values.

The procedure is exactly the same as that used to solve dc circuits. Suppose that we desire to solve for the voltage across $\mathbf{Z}_2$. We begin by replacing $\mathbf{V}_{S_2}$ with its internal impedance (zero for an ideal source) as shown in Fig. 17.14*b*. The voltage across $\mathbf{Z}_2$ due to $\mathbf{V}_{S_1}$ is found by using the voltage divider theorem.

$$\mathbf{V}'_{Z_2} = \mathbf{V}_{S_1} \frac{\mathbf{Z}_2 \| \mathbf{Z}_3}{\mathbf{Z}_1 + \mathbf{Z}_2 \| \mathbf{Z}_3} \tag{17.18}$$

where the *prime* shows that this is the voltage due to $\mathbf{V}_{S_1}$ alone. We then replace $\mathbf{V}_{S_1}$ with its internal impedance as shown in Fig. 17.14*c*, and solve for the voltage across $\mathbf{Z}_2$ due to $\mathbf{V}_{S_2}$. Using the voltage divider theorem again, we have

$$\mathbf{V}''_{Z_2} = \mathbf{V}_{S_2} \frac{\mathbf{Z}_1 \| \mathbf{Z}_2}{\mathbf{Z}_3 + \mathbf{Z}_1 \| \mathbf{Z}_2} \tag{17.19}$$

where the *double prime* shows that this is the voltage due to $\mathbf{V}_{S_2}$ alone. The voltage across $\mathbf{Z}_2$, $\mathbf{V}_A$, is then the sum of these two components. Thus

$$\mathbf{V}_A = \mathbf{V}'_{Z_2} + \mathbf{V}''_{Z_2} \tag{17.20}$$

• • •

Example 17.16 Use Superposition to find $\mathbf{V}_A$ in Fig. 17.14 if: $\mathbf{V}_{S_1} = 10\underline{/0°}$ V, $\mathbf{V}_{S_2} = 10\underline{/45°}$ V, $\mathbf{Z}_1 = j2\ \Omega$, $\mathbf{Z}_2 = 2\ \Omega$, and $\mathbf{Z}_3 = -j1\ \Omega$.

Solution The voltage due to $\mathbf{V}_{S_1}$ is found from Eq. 17.18.

$$\mathbf{V}'_{Z_2} = 10\underline{/0°}\frac{\dfrac{2(-j)}{2-j}}{j2+\dfrac{2(-j)}{2-j}} = \frac{10(0.4-j0.8)}{j2+0.4-j0.8} = \frac{4-j8}{0.4+j1.2} = (-5-j5)\text{ V}$$

The voltage due to $\mathbf{V}_{S_2}$ is found from Eq. 17.19.

$$\mathbf{V}''_{Z_2} = 10\underline{/45°}\frac{\dfrac{2(j2)}{2+j2}}{-j1+\dfrac{2(j2)}{2+j2}} = \frac{(7.07+j7.07)(1+j)}{-j1+1+j1} = j14.14\text{ V}$$

The voltage $\mathbf{V}_A$ is then

$$\mathbf{V}_A = \mathbf{V}'_{Z_2} + \mathbf{V}''_{Z_2} = -5-j5+j14.14 = (-5+j9.14)\text{ V}$$

• • •

A widely used two-source ac system is shown in Fig. 17.15. The source voltage $\mathbf{V}_{S_1}$ is equal in magnitude, but 180° out of phase, to the source voltage $\mathbf{V}_{S_2}$. This can be accomplished by winding two coils with an opposite sense on the same rotor of an ac generator. The induced currents and thus the induced voltages will be 180° out of phase with one another. The two voltage sources can then be represented as

$$\mathbf{V}_{S_1} = V_S \sin \omega t \qquad \mathbf{V}_{S_2} = -V_S \sin \omega t$$

If each source voltage has an RMS value of 117 V ac, this represents the widely used three-wire system for distributing power to residences. Notice that the voltage taken *across* lines A and B is equal to the *difference* between the two

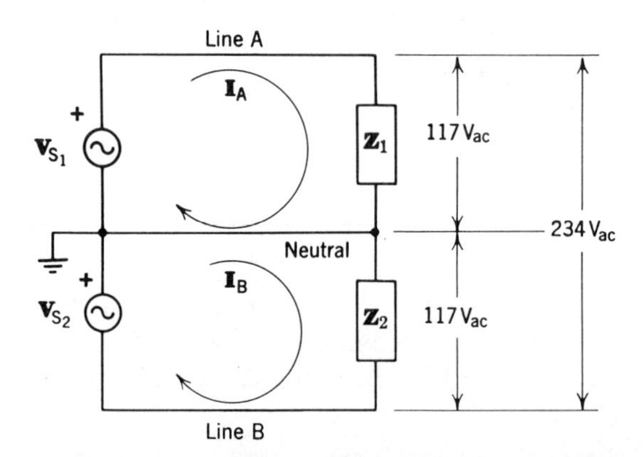

Figure 17.15 The three-wire single phase Edison power distribution system

source voltages and is thus

$$\mathbf{V}_{AB} = (117 \sin \omega t) - (-117 \sin \omega t) = 234 \sin \omega t$$

Thus lighting and appliance outlets are connected betwen A and *neutral* or B and neutral, and electric dryers, ovens and ranges, and air conditioners (234 V) are connected between lines A and B. This is known as the *three-wire single phase* Edison system. Notice that if the two loads $\mathbf{Z}_1$ and $\mathbf{Z}_2$ are *equal,* the neutral current will be zero.

• • •

Example 17.17 Calculate the neutral line current in Fig. 17.15 if $\mathbf{Z}_1 = 20\ \Omega$ and $\mathbf{Z}_2 = 20\ \Omega$.

Solution From the circuit it is clear that the line currents are:

$$\mathbf{I}_A = \frac{\mathbf{V}_{S_1}}{\mathbf{Z}_1} = \frac{117\underline{/0^\circ}\text{ V}}{20\underline{/0^\circ}\ \Omega} = 5.85\underline{/0^\circ}\text{ A}$$

$$\mathbf{I}_B = \frac{\mathbf{V}_{S_2}}{\mathbf{Z}_2} = \frac{117\underline{/0^\circ}\text{ V}}{20\underline{/0^\circ}\ \Omega} = 5.85\underline{/0^\circ}\text{ A}$$

The neutral current $\mathbf{I}_N$ is then

$$\mathbf{I}_N = \mathbf{I}_A - \mathbf{I}_B = 5.85\underline{/0^\circ} - 5.85\underline{/0^\circ} = 0.0\text{ A}$$

• • •

A system that is universally used for ac power distribution is the *three-phase* system shown in Fig. 17.16. The three source voltages are obtained from three coils wound on the rotor of an ac generator. The coils are placed physically with 120° angles between them and the three induced voltages are then 120° apart in phase.* The voltages can be represented as

$$\mathbf{V}_A = 117\underline{/0^\circ}\text{ V} \qquad \mathbf{V}_B = 117\underline{/-120^\circ}\text{ V} \qquad \mathbf{V}_C = 117\underline{/+120^\circ}\text{ V}$$

The three-load impedances are connected in a *four-wire Wye* connection in Fig. 17.16. If the three impedances are equal (both in magnitude and phase), the system is said to be balanced, and the neutral wire current will be zero. The three source voltages of this three-phase system can be used to produce a

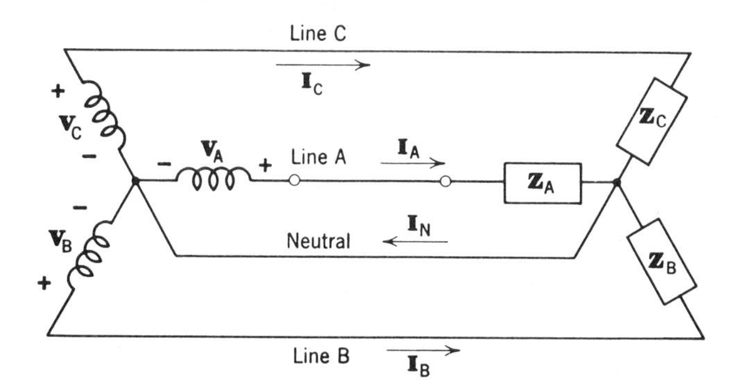

Figure 17.16 A three-phase Wye-connected ac power system

*Three-phase power systems are discussed in Chapter 20.

magnetic field that *rotates* around a number of coils, and this is precisely the action required for an ac motor. Furthermore, the total power delivered by a three-phase system is *constant*, and does not pulsate sinusoidally as in a single phase ac system. This provides an advantage for large electric motors.

• • •

Example 17.18 Calculate the neutral current in Fig. 17.16 if the three impedances all have a value of

$$\mathbf{Z}_A = \mathbf{Z}_B = \mathbf{Z}_C = 10\underline{/25°}\ \Omega$$

Draw the phasor diagram for the system.

Solution The line currents are:

$$\mathbf{I}_A = \frac{\mathbf{V}_A}{\mathbf{Z}_A} = \frac{117\underline{/0°}\ \text{V}}{10\underline{/25°}\ \Omega} = 11.7\underline{/-25°}\ \text{A}$$

$$\mathbf{I}_B = \frac{\mathbf{V}_B}{\mathbf{Z}_B} = \frac{117\underline{/-120°}\ \text{V}}{10\underline{/25°}\ \Omega} = 11.7\underline{/-145°}\ \text{A}$$

$$\mathbf{I}_C = \frac{\mathbf{V}_C}{\mathbf{Z}_C} = \frac{117\underline{/120°}\ \text{V}}{10\underline{/25°}\ \Omega} = 11.7\underline{/95°}\ \text{A}$$

The neutral current is the sum of the three line currents. Thus

$$\begin{aligned}\mathbf{I}_N &= \mathbf{I}_A + \mathbf{I}_B + \mathbf{I}_C = 11.7\underline{/-25°} + 11.7\underline{/-145°} + 11.7\underline{/95°} \\ &= 10.6 - j4.95 - 9.58 - j6.71 - 1.02 + j11.66 = 0.0\ \text{A}\end{aligned}$$

The phasor diagram is shown in Fig. 17.17.

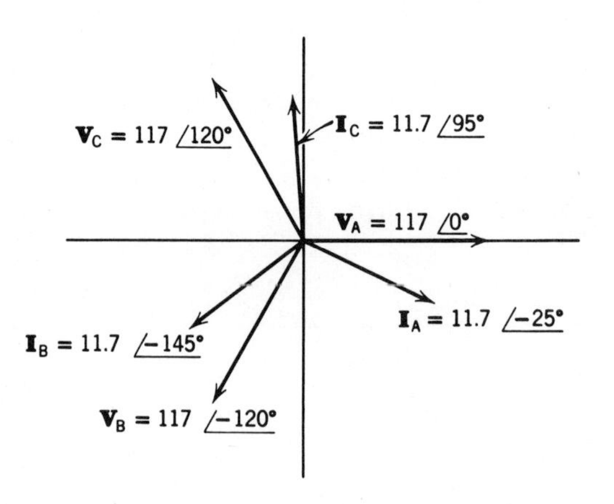

Figure 17.17 Phasor diagram for Example 17.18

• • •

SUMMARY

Instantaneous ac power is found as the product of the ac voltage and current. The instantaneous power in a resistor varies sinusoidally at a frequency twice that of the applied voltage, and has a peak value equal to the product of the peak voltage and the peak current. The average value of the power in a resistance is equal to one-half the peak power, and average power is easily calculated by using RMS values of voltage and current. For sinusoidal waves, the RMS value is equal to 0.707 times the peak value of the wave. The average power dissipated in a resistance is known as real power (or true power) and it is measured in watts.

The instantaneous power in an inductor or a capacitor also varies sinusoidally at twice the frequency of the applied voltage, but the average power is zero. As a result, no power is dissipated in an ideal inductor or capacitor, and the power which circulates between the reactance and the source is known as reactive power. Reactive power is measured in vars, and it is equal to the product of the RMS voltage across and the RMS current through a reactance.

The product of the RMS voltage across and current through an impedance is known as the apparent power measured in VA. Apparent power can be drawn as a vector in the complex plane, and it is composed of two components; the real power drawn on the positive horizontal-axis, and the reactive power drawn on the vertical-axis. Inductive power is drawn on the positive vertical-axis, and capacitive power is drawn on the negative vertical-axis.

The triangle formed by the apparent power, the real power, and the reactive power drawn in a complex plane is known as a power triangle, and the cosine of the angle between the apparent power and the positive horizontal-axis is defined as the power factor. An ideal power factor is unity, 1.0, and this implies making the load appear as a pure resistance to the source. A unity power factor is desirable since this reduces the reactive power delivered by the source to zero.

Maximum power will be delivered to a load by a source when the load impedance is equal to the complex conjugate of the source impedance. Systems for ac power distribution usually are multiple source ac circuits such as the Edison three-wire single-phase system or the three-phase Wye system. The power delivered by the Edison system varies sinusoidally, but in the four-wire three-phase Wye system it is constant. In either of these two power systems, the neutral line current is zero if the loads are balanced.

GLOSSARY

Apparent power P_A. *The product of the RMS voltage across and the RMS current through an impedance.*

Edison system. *A three-wire single-phase power distribution system.*

Four-wire Wye system. *A four-wire three-phase power distribution system.*

Instantaneous power p. *The power at any instant in time; equal to the product of instantaneous voltage and current.*

Maximum power transfer. *Maximum power is delivered to a load by a source when the load and source impedances are the complex conjugates of one another.*

Power factor PF. *Defined as the cosine of the impedance angle ϕ. Also defined as the ratio of apparent power to real power.*

Power triangle. *The triangle formed by drawing apparent power, real power, and reactive power in the complex plane.*

Reactive power P_Q. *The circulating power in a reactance given in vars.*

Real power P. *The true power dissipated in a resistance; measured in watts.*

RMS Root **M**ean **S**quare. *The effective value of a sinusoidal wave measured by most ac meters. Equal to 0.707 times the peak value of the wave.*
VA Volt **A**mpere. *The unit used to measure apparent power.*
Var Volt **A**mpere **R**eactive. *The unit used to measure reactive power.*

IMPORTANT RELATIONSHIPS

- Power in a resistance using peak values:

$$P\ (\text{W}) = \tfrac{1}{2}P_\text{P} = \frac{V_\text{P}I_\text{P}}{2} = \frac{I_\text{P}^2R}{2} = \frac{V_\text{P}^2}{2R}$$

- RMS (effective) values:

$$I = I_\text{RMS} = 0.707I_\text{p} \qquad V = V_\text{RMS} = 0.707V_\text{p}$$

- Power using RMS voltages and currents:

Resistance $P\ (\text{W}) = VI = I^2R = \dfrac{V^2}{R}$

Inductance $P_\text{Q}\ (\text{var}) = V_\text{L}I_\text{L} = I_\text{L}^2X_\text{L} = \dfrac{V_\text{L}^2}{X_\text{L}}$

Capacitance $P_\text{Q}\ (\text{var}) = V_\text{C}I_\text{C} = I_\text{C}^2X_\text{C} = \dfrac{V_\text{C}^2}{X_\text{C}}$

Impedance $P_\text{A}\ (\text{VA}) = P\ (\text{W}) \pm jP_\text{Q}\ (\text{var}) = P_\text{A}\angle\phi$

$P\ (\text{W}) = VI\cos\phi \qquad P_\text{Q}\ (\text{var}) = VI\sin\phi$

- Power factor: $\text{PF} = \cos\phi$

REVIEW QUESTIONS

1. The instantaneous ac power in a resistance varies sinusoidally. (TF)
2. The average ac power in a resistance is equal to ____ the peak power.
3. The average or real power in a resistance is measured in ____.
4. The RMS value of a sinusoidal wave is found by multiplying the peak value by ____.
5. The peak value of a 117 V ac wave is ____ V.
6. The power in an inductance varies sinusoidally at a frequency ____ that of the applied voltage.
7. The average power in an inductance is zero. (TF)
8. The reactive power in an inductance circulates between the inductance and the load, and it (does, does not) perform useful work.
9. The reactive power in a capacitance varies sinusoidally at a frequency equal to that of the applied voltage. (TF)
10. The average power in a capacitance is ____.
11. The power which circulates between a capacitance and a source is called ____ ____ power.
12. Reactive power is measured in units of ____.
13. The product of the RMS voltage across and the RMS current through an impedance is ____ power.
14. Apparent power is measured in ____ ____.

15. In a power triangle, the reactive power in a capacitor is taken as ____ while the reactive power in an inductor is taken as ____.
16. The cosine of the angle between the positive horizontal-axis and the apparent power in a power triangle is defined as the ____ ____.
17. The power factor is equal to the cosine of the angle of an impedance. (TF)
18. A capacitive circuit is said to have a (leading, lagging) power factor.
19. An inductive circuit is said to have a (leading, lagging) power factor.
20. The most desirable power factor is (0.0, 1.0).
21. Correcting the PF of a load essentially means making the total impedance angle of the load as seen by the source equal to ____ degrees.
22. A resistive load has a PF of ____.
23. The power factor of an inductive circuit can be corrected by connecting the proper value of capacitance in (series, parallel) with the original load.
24. Explain why the capacitor in (23) above should or should not be connected in series with the original load.
25. A voltage source will deliver maximum power to a load when the impedance of the load is equal to the ____ ____ of the source impedance.
26. The Edison system is a ____ phase power distribution system.
27. The neutral line current in an Edison power system is ____ if the loads are equal.
28. The voltages in a three-phase Wye power system have a phase angle of ____ degrees between them.
29. The neutral line current in a three-phase Wye power system is zero if the three-load impedances are ____.

ROBLEMS

1. Find the peak power and the average power dissipated in a 200 Ω resistor connected across the terminals of a voltage source of $v = 10 \sin^5 t$ V.
2. The average power dissipated in a 1 kΩ resistor is 1.5 W. What is the peak power, and what must be the peak value of the applied voltage?
3. What is the RMS value of a sinusoidal wave having a peak value of 28 V?
4. An RMS reading voltmeter is used to measure the voltage across a resistor, and it reads 6.3 V. What is the peak value of the voltage?
5. What is the equation for the voltage in Problem 4 if the frequency is 60 Hz?
6. The heating element in an electric heater has a hot resistance of 10 Ω. What is the power dissipated in the heater when connected to a source of 117 V ac?
7. The RMS voltage across a resistor is 78 V, and the RMS current is 18 mA. What is the power dissipated in the resistor?
8. The RMS voltage across a 6.8 kΩ resistor is 63 V. What is the power dissipated in the resistor?
9. What is the power dissipated in the resistor in Problem 8 if the voltage is 63 V peak?
10. What is the maximum RMS voltage that can safely be applied to a 2.2 kΩ, $\frac{1}{2}$ W resistor?

11. Calculate the instantaneous and reactive power, in a 4 mH inductor connected across the terminals of a voltage source $v = 50 \sin 10^6 t$.
12. What must be the voltage across an inductive reactance of $j300\ \Omega$ if the reactive power is 2.8 var?
13. What must be the inductor current in Problem 12?
14. Calculate the instantaneous and reactive power in a 1 μF capacitor connected across the terminals of a voltage source $v = 50 \sin 10^6 t$.
15. What must be the voltage across a capacitive reactance of $-j500$ if the reactive power is 0.81 var?
16. What must be the capacitor current in Problem 15?
17. Calculate the apparent power, real power, and reactive power if an impedance of $90\underline{/45^\circ}\ \Omega$ is connected across the terminals of a voltage source having an RMS voltage of 80 V. Draw the power triangle.
18. Repeat Problem 17 for an impedance of $45\underline{/-45^\circ}\ \Omega$.
19. An impedance of $(300 + j400)\ \Omega$ in parallel with a capacitive reactance of $-j500\ \Omega$ are connected across the terminals of a 220 V ac source. Calculate the total power delivered by the source and draw the power triangle.
20. Repeat Problem 19 if the capacitive reactance is changed to $-j830\ \Omega$.
21. Show that the power factor is equal to the ratio of the real power P to the apparent power P_A.
22. What is the PF in Problem 17? Problem 18?
23. An electric motor draws a current of 2.4 A when connected across the terminals of a 117 V ac source. If the real power delivered by the source is 250 W, what is the PF?
24. A source delivers 0.138 VA of apparent power to a load with a PF of 0.8 leading. Calculate the real and reactive power and draw the power triangle.
25. If the voltage applied to the load in Problem 24 is 23 V, what is the load impedance?
26. What must be the value of X_C in Problem 19 in order to correct the PF to 1.0?
27. To verify that the capacitor in Fig. 17.11*c* has indeed corrected the PF to unity, calculate the total impedance seen by the generator.
28. A fluorescent lamp and its ballast have an overall impedance of $(100 + j200)\ \Omega$. What capacitance must be connected in parallel with the lamp in order to achieve a unity PF? It is a 117 V ac, 60 Hz lamp.
29. A certain voltage source has an internal impedance of $(600 - j600)\ \Omega$. What must be the impedance of the load in order to insure maximum power transfer from the source to the load?
30. What is the power delivered to the load in Problem 29 if the source voltage is 15 V?
31. In the power system shown in Fig. 17.15, $\mathbf{Z}_1 = 80\underline{/30^\circ}\ \Omega$ and $\mathbf{Z}_2 = 80\underline{/30^\circ}\ \Omega$. Calculate the neutral line current.
32. Repeat Problem 31 if $\mathbf{Z}_2$ is changed to $80\underline{/-30^\circ}\ \Omega$.
33. Calculate the neutral line current in the power system in Fig. 17.16 if: $\mathbf{V}_A = 117\underline{/0^\circ}$ V, $\mathbf{V}_B = 117\underline{/-120^\circ}$ V, $\mathbf{V}_C = 117\underline{/+120^\circ}$ V, and $\mathbf{Z}_1 = \mathbf{Z}_2 = \mathbf{Z}_3 = 50\ \Omega$.
34. Repeat Problem 33 if $\mathbf{Z}_3$ is removed.

RESONANCE

The impedance of a series R-L-C circuit is equal to the sum of the resistance and the operator j multiplied by the circuit reactance. The inductive and capacitive reactances are both frequency-dependent, and at one particular frequency they cancel each other. This is called the *resonant* frequency, and under this condition the impedance is entirely resistance. Such an R-L-C combination is called a *resonant circuit* and it has some very special properties and applications.

Similarly, a parallel R-L-C circuit becomes resonant at some particular frequency, and the parallel resonance circuit also has some interesting properties and applications.

In this chapter we will investigate the properties of both series and parallel resonant circuits, and we will study the use of such circuits as electric filters.

18.1 SERIES RESONANCE

A series R-L-C circuit is connected to the terminals of an ac voltage generator in Fig. 18.1. The total impedance of the circuit is given by

$$\mathbf{Z} = R + j\left(\omega L - \frac{1}{\omega C}\right) \tag{18.1}$$

Clearly, as the frequency is varied (i.e., ω is varied) the inductive and

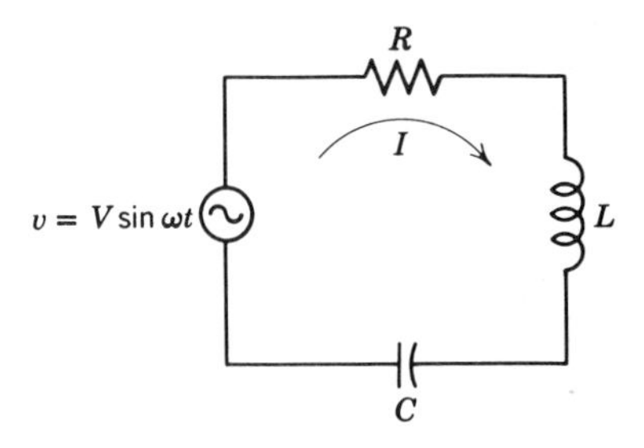

Figure 18.1 A series R-L-C circuit

capacitive reactances will vary. A plot of these reactances is shown in Fig. 18.2*a*. The inductive reactance is zero at a frequency of zero (dc), and it increases linearly with increasing frequency. It is plotted as a *positive* quantity since it is positive in Eq. 18.1. The capacitive reactance is plotted as a *negative* quantity, and it approaches $-\infty$ at a frequency of zero (dc). It then becomes smaller in magnitude as the frequency increases and it approaches zero at very high frequencies.

There is one particular frequency where the inductive reactance is equal to but opposite in sign from the capacitive reactance. This is defined as the *resonant frequency* f_r. At the frequency f_r, the circuit is said to be *in resonance*

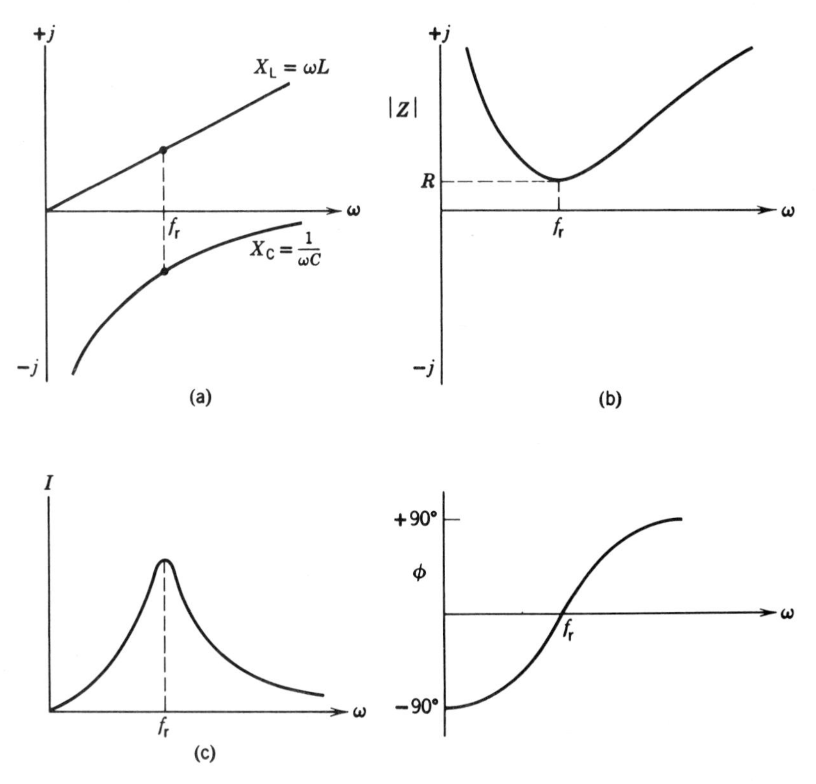

Figure 18.2 Varying the frequency in an R-L-C series circuit (a) Reactances (b) Magnitude $|Z|$ and angle ϕ of the impedance (c) The circuit current

and the impedance of the circuit is seen to be purely resistive. That is, at resonance,

$$\mathbf{Z} = R + j(X_L - X_C) = R$$

A plot of the magnitude and angle of the impedance $\mathbf{Z} = Z\underline{/\phi}$ as a function of frequency is shown in Fig. 18.2*b*. At very low frequencies, the magnitude of $\mathbf{Z}$ is large and the phase angle ϕ is negative since a *capacitor* has a very large reactance at low frequencies. At high frequencies, the magnitude of $\mathbf{Z}$ is again large, but the phase angle ϕ is positive since the inductive reactance is dominant. At the resonant frequency f_r, the two reactances cancel and as a result the phase angle ϕ is zero and the magnitude of the impedance is simply equal to the resistance of the circuit R.

If the magnitude of the voltage source is held *constant* as ω is varied, the circuit current can be calculated using Ohm's Law. Thus

$$\mathbf{I} = \frac{\mathbf{V}}{\mathbf{Z}}$$

and it is clear that the current will be zero at dc (because of the capacitor), will increase until f_r is reached, and will then decrease as the frequency goes above f_r. A plot of the circuit current is shown in Fig. 18.2*c*. It is clear that the *maximum* circuit current occurs at the resonant frequency f_r.

• • •

Example 18.1 Calculate the impedance and the circuit current for the R-L-C circuit in Fig. 18.1 if $\mathbf{V} = 10\underline{/0°}$ V, $R = 1$ kΩ, $L = 2$H, and $C = 2$ μF.

Solution We will calculate $\mathbf{Z}$ and $\mathbf{I}$ for values of ω from 100 rad/sec to 900 rad/sec in 200 rad/sec increments.

$\omega = 100$ rad/sec

$$\mathbf{Z} = R + j\left(\omega L - \frac{1}{\omega C}\right) = 1\text{ k}\Omega + j\left(100 \times 2 - \frac{1}{100 \times 2\ \mu\text{F}}\right) = 1\text{ k}\Omega - j4.8\text{ k}\Omega$$
$$= 4.9\underline{/-78.2°}\text{ k}\Omega$$

$$\mathbf{I} = \frac{\mathbf{V}}{\mathbf{Z}} = \frac{10\underline{/0°}\text{ V}}{4.9\underline{/-78.2°}\text{ k}\Omega} = 2.02\underline{/78.2°}\text{ mA}$$

$\omega = 300$ rad/sec

$$\mathbf{Z} = 1\text{ k}\Omega + j\left(300 \times 2 - \frac{1}{300 \times 2\ \mu\text{F}}\right) = 1\text{ k}\Omega - j1.07\text{ k}\Omega = 1.46\underline{/-46.9°}\text{ k}\Omega$$

$$\mathbf{I} = \frac{10\underline{/0°}\text{ V}}{1.46\underline{/-46.9°}\text{ k}\Omega} = 6.84\underline{/46.9°}\text{ mA}$$

$\omega = 500$ rad/sec

$$\mathbf{Z} = 1\text{ k}\Omega + j\left(500 \times 2 - \frac{1}{500 \times 2\ \mu\text{F}}\right) = 1\text{ k}\Omega + j(1000 - 1000) = 1\underline{/0°}\text{ k}\Omega$$

$$\mathbf{I} = \frac{10\underline{/0°}\text{ V}}{1\underline{/0°}\text{ k}\Omega} = 10\underline{/0°}\text{ mA}$$

$\omega = 700$ rad/sec

$$\mathbf{Z} = 1\ \text{k}\Omega + j\left(700 \times 2 - \frac{1}{700 \times 2\ \mu\text{F}}\right) = 1\ \text{k}\Omega + j685 = 1.21\underline{/34.4^\circ}\ \text{k}\Omega$$

$$\mathbf{I} = \frac{10\underline{/0^\circ}\ \text{V}}{1.21\underline{/34.4^\circ}\ \text{k}\Omega} = 8.26\underline{/-34.4^\circ}\ \text{mA}$$

$\omega = 900$ rad/sec

$$\mathbf{Z} = 1\ \text{k}\Omega + j\left(900 \times 2 - \frac{1}{900 \times 2\ \mu\text{F}}\right) = 1\ \text{k}\Omega + j1.25\ \text{k}\Omega = 1.60\underline{/51.3^\circ}\ \text{k}\Omega$$

$$\mathbf{I} = \frac{10\underline{/0^\circ}\ \text{V}}{1.60\underline{/51.3^\circ}\ \text{k}\Omega} = 6.25\underline{/-51.3^\circ}\ \text{mA}$$

The impedance and the magnitude of the current are plotted in Fig. 18.3.

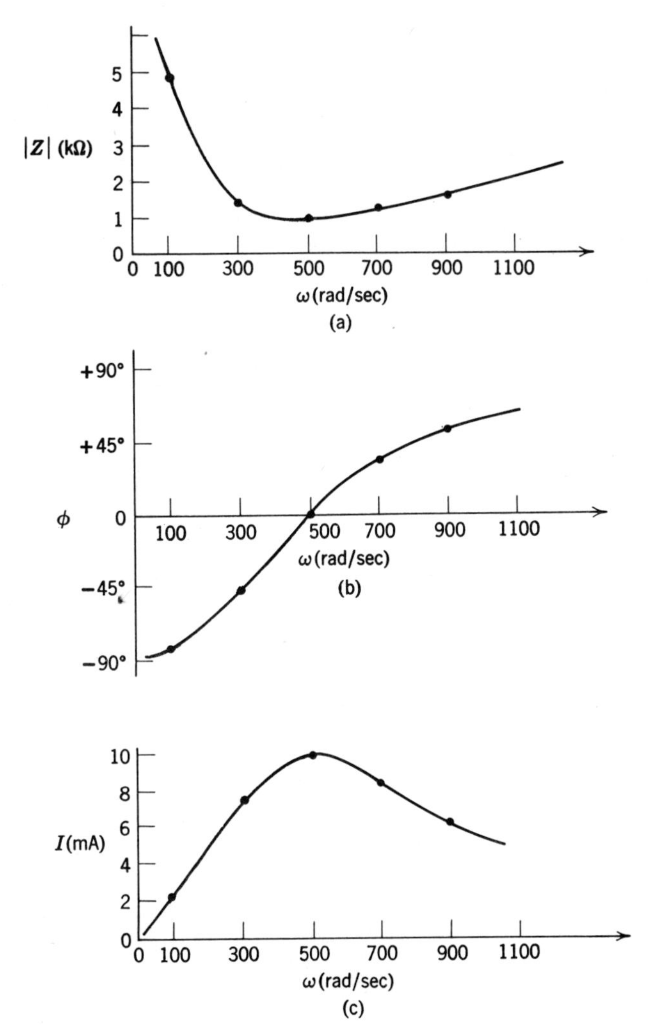

Figure 18.3 Example 18.1 (*a*) The magnitude of the impedance (*b*) The phase angle of the impedance (*c*) The magnitude of the current

• • •

In Example 18.1, the resonant frequency ω_r is obviously 500 rad/sec, or $f_r = \omega/2\pi = 500/2\pi = 79.5$ Hz. It would be nice if we could calculate the resonant frequency in terms of the circuit elements. This can be accomplished quite easily by noting that at f_r, the inductive and capacitive reactances are equal. That is

$$X_L = X_C$$

or,

$$\omega_r L = \frac{1}{\omega_r C}$$

We can readily solve this relationship for ω_r, and we obtain (the reader should work it out step by step)

$$\omega_r = \frac{1}{\sqrt{LC}}$$

This is the angular velocity and we are more interested in the frequency. But, we know that $\omega_r = 2\pi f_r$, and thus

$$\omega_r = 2\pi f_r = \frac{1}{\sqrt{LC}}$$

Solving for the resonant frequency f_r, we have

$$f_r = \frac{1}{2\pi\sqrt{LC}} \tag{18.2}$$

where L is in henries, C is in farads, and f_r is in Hz. Equation (18.2) can be used to solve for the resonant frequency of any series R-L-C circuit.

• • •

Example 18.2 Use Eq. 18.2 to calculate the resonant frequency of the circuit in Example 18.1.

Solution

$$f_r = \frac{1}{2\pi\sqrt{LC}} = \frac{1}{2\pi\sqrt{2 \times 2 \times 10^{-6}}} = \frac{1}{2\pi \times 2 \times 10^{-3}} = 79.5 \text{ Hz}$$

• • •

Example 18.3 The circuit in Fig. 18.1 is required to be resonant at 10 kHz. What value of C must be used if $L = 2.5$ mH?

Solution We solve Eq. 18.2 for C. Thus, multiplying both sides by 2π, and then squaring both sides yields

$$4\pi^2 f_r^2 = \frac{1}{LC}$$

The value of C is then found to be

$$C = \frac{1}{4\pi^2 f_r^2 L} = \frac{1}{4\pi^2 \times (10^4)^2 \times 2.5 \times 10^{-3}} = 0.102\ \mu\text{F}$$

• • •

18.2 Q OF A SERIES CIRCUIT

From the preceding section we see that the impedance of an R-L-C series circuit at f_r is simply equal to the resistance R in the circuit. Thus if we make the resistance of the circuit very small, the circuit current will be *very large*! In an *ideal* circuit, we could make the total resistance equal to zero, in which case the current would become infinite. In real circuits we cannot however realize zero resistance, but we can realize very small resistances.

As an example, consider the R-L-C circuit in Example 18.3 driven by a 1.0 V source as shown in Fig. 18.4. If the circuit resistance is 100 Ω, the current at resonance is clearly

$$\mathbf{I} = \frac{\mathbf{V}_S}{R} = \frac{1.0\underline{/0^\circ}\text{ V}}{100\ \Omega} = 10\underline{/0^\circ}\text{ mA}$$

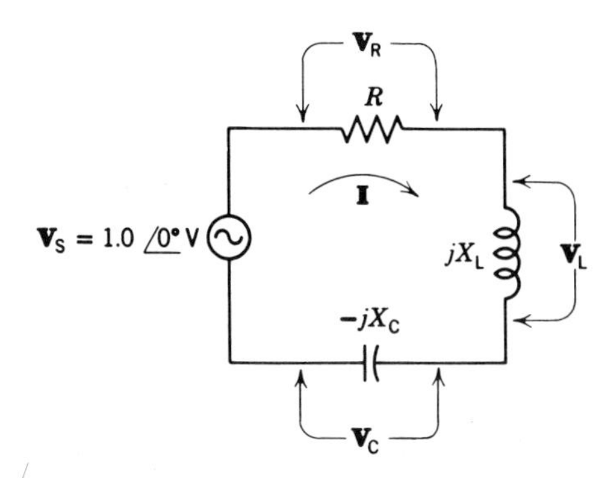

Figure 18.4 A tuned R-L-C circuit. $X_L = X_C$

On the other hand, if the total resistance is only 10 Ω, the circuit current is

$$\mathbf{I} = \frac{1.0\underline{/0^\circ}\text{ V}}{10\ \Omega} = 0.1\underline{/0^\circ}\text{ A}$$

Let us examine the case where $R = 10\ \Omega$. The inductive and capacitive reactances are equal in magnitude and have a value of

$$X = \omega_r L = 2\pi \times 10\text{ kHz} \times 2.5\text{ mH} = 157\ \Omega$$

Since this is a series circuit, the voltages across each of the elements are found to be

$$\mathbf{V}_R = \mathbf{I}R = 0.1\underline{/0^\circ}\text{ A} \times 10\ \Omega = 1.0\underline{/0^\circ}\text{ V}$$

$$\mathbf{V}_L = \mathbf{I}X_L\underline{/90^\circ} = 0.1\underline{/0^\circ}\text{ A} \times 157\underline{/90^\circ}\ \Omega = 15.7\underline{/90^\circ}\text{ V}$$

$$\mathbf{V}_C = \mathbf{I}X_C\underline{/-90^\circ} = 0.1\underline{/0^\circ}\text{ A} \times 157\underline{/-90^\circ}\ \Omega = 15.7\underline{/-90^\circ}\text{ V}$$

Notice that the voltages across the inductance and the capacitance are both 15.7 times the source voltage! This phenomenon is known as *resonant voltage rise,* and is one of the valuable characteristics of a *tuned* (resonant) R-L-C circuit. It is exactly this property which is used in the *tuning* of a radio or television circuit. KVL is still satisfied however, since $\mathbf{V}_C$ and $\mathbf{V}_L$ cancel one another as shown in the phasor diagram in Fig. 18.5.

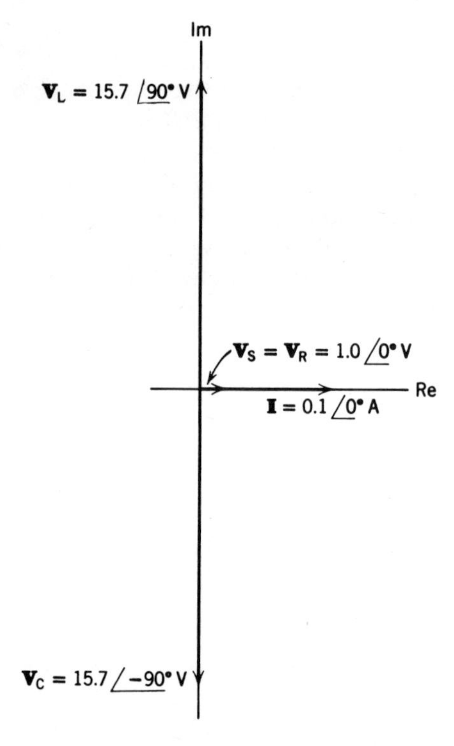

Figure 18.5 Phasor diagram for the circuit in Figure 18.4

Notice that the *smaller* the value of R, the larger the circuit current, and therefore the *greater* the resonant voltage rise. For this reason, we always like to have series R-L-C resonant circuits with the smallest possible resistance. The lowest value of resistance obtainable is usually limited by the resistance of the wire used to form the inductor in the circuit. The smaller the circuit resistance, the better the *quality* of the circuit. A measure of the circuit quality, or Q,* is defined as the ratio of the reactive power P_Q to the real power P. Thus

$$\text{circuit quality} = Q \triangleq \frac{P_Q}{P}$$

However, these two powers can be calculated as

$$P_Q = I^2 X_L \qquad P = I^2 R$$

Substituting these expressions into the above definition, we obtain

$$Q = \frac{P_Q}{P} = \frac{I^2 X_L}{I^2 R} = \frac{X_L}{R} = \frac{\omega_r L}{R} = \frac{2\pi f_r L}{R} \tag{18.3}$$

From this expression, it is clear that the *greater* the value of Q, the *better* the quality of the circuit.

• • •

Example 18.4 Calculate the Q of the circuit in Example 18.3 if $R = 10\ \Omega$.

*The symbol Q for quality must not be confused with Q for charge (coulombs).

Solution Using Eq. 18.3, we have

$$Q = \frac{\omega_r L}{R} = \frac{2\pi \times 10\ \text{kHz} \times 2.5\ \text{mH}}{10} = 15.7$$

• • •

Suppose we take Eq. 18.3 and divide both the numerator and the denominator of the right hand side by I. The result is

$$Q = \frac{\dfrac{I^2 X_L}{I}}{\dfrac{I^2 R}{I}} = \frac{IX_L}{IR} = \frac{V_L}{V_R} = \frac{V_C}{V_R}$$

since $IX_L = IX_C = V_L = V_C$, and $IR = V_R$. From this equation, we can calculate the resonant voltage rise in terms of the Q of the circuit. That is,

$$V_C = V_L = QV_R = QV_S \tag{18.4}$$

since $V_R = V_S$ at resonance. It must be noted that this is the resonant voltage rise, and thus gives the capacitor and inductor voltages at the resonant frequency f_r only.

• • •

Example 18.5 Calculate the resonant frequency, the Q, and the resonant voltage rise across L (and C) for the circuit in Fig. 18.6.

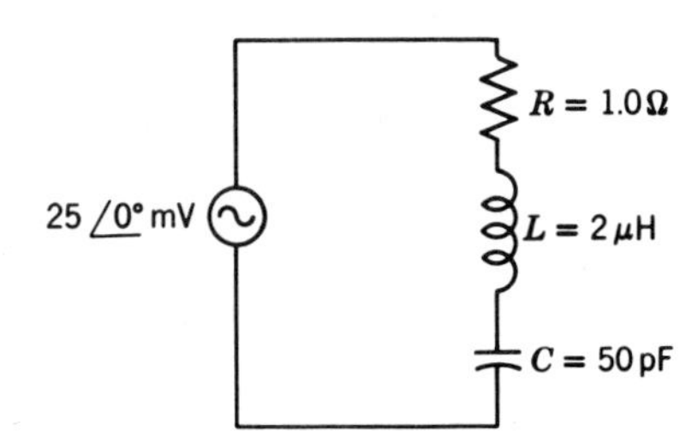

Figure 18.6

Solution The resonant frequency is found to be

$$f_r = \frac{1}{2\pi\sqrt{LC}} = \frac{1}{2\pi\sqrt{2 \times 10^{-6} \times 50 \times 10^{-12}}} = \frac{1}{2\pi\sqrt{10^{-16}}} = 15.9\ \text{MHz}$$

The circuit Q is

$$Q = \frac{\omega_r L}{R} = \frac{2\pi \times 15.9\ \text{MHz} \times 2\ \mu\text{H}}{1.0} = 200$$

The resonant voltage rise is then

$$V_C = V_L = QV_S = 200 \times 25\ \text{mV} = 5.0\ \text{V}$$

• • •

Example 18.6 What would be the resonant voltage rise in Example 18.5 if the circuit resistance were increased to $R = 20\ \Omega$?

Solution The circuit Q would be reduced to

$$Q = \frac{\omega_r L}{R} = \frac{200}{20} = 10$$

The resonant voltage rise would then be

$$V_C = V_L = QV_S = 10 \times 25 \text{ mV} = 250 \text{ mV}$$

• • •

Since the circuit current at resonance is dependent on the total resistance R, and since the Q is also dependent on R, we can say that the circuit current peak at resonance is dependent on the Q of the circuit. The current in Fig. 18.1 might appear as shown in Fig. 18.7 for a number of different values of Q. The higher the Q, the higher the current peak at resonance, and the *steeper* the sides of the curve.

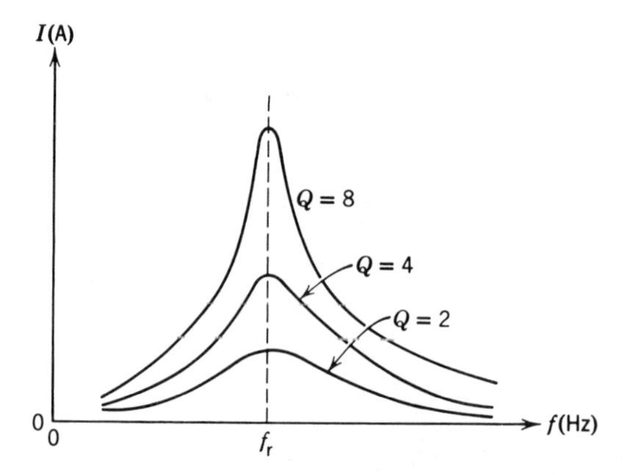

Figure 18.7 The current in an R-L-C series circuit versus frequency for different values of Q

Consider now the curve of current versus frequency for a series R-L-C circuit as shown in Fig. 18.8. At the resonant frequency f_r the current has a value of I_p. At some frequency f_1, below f_r, the current has fallen to a value of $0.707 I_p$. Similarly, at a frequency f_2, above f_r, the current has also fallen to

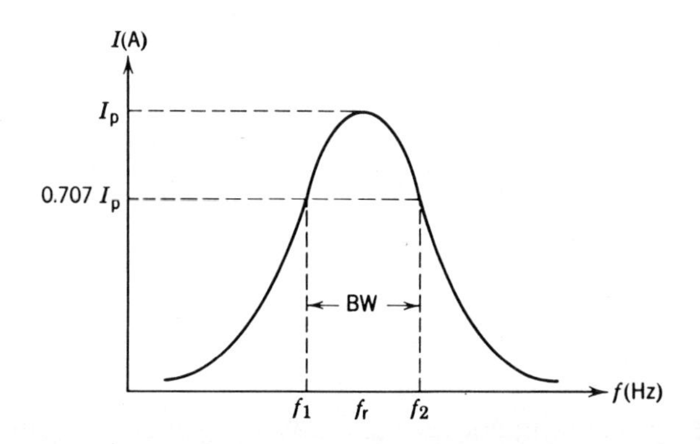

Figure 18.8 The bandwidth of an R-L-C series circuit

$0.707I_p$. Since the current has been reduced by a factor of 0.707 at these two points, the power has been reduced to

$$P = (0.707I_p)^2R = \left(\frac{I_p}{\sqrt{2}}\right)^2 R = \frac{I_p^2R}{2} = 0.5I_p^2R$$

This is clearly *one-half* the power at f_r; that is, the power at f_r is $P = I_p^2R$. Therefore these two points are called the *half-power points* (also called the 3db points). Note that f_r is approximately halfway between f_1 and f_2.*

The *bandwidth* BW of the *R-L-C* circuit is defined as the difference between the two half-power-point frequencies. Thus

$$\text{BW} = f_2 - f_1 \tag{18.5}$$

It can also be shown that the bandwidth can be calculated as

$$\text{BW} = \frac{f_r}{Q} = \frac{R}{2\pi L} \tag{18.6}$$

The bandwidth is usually calculated from Eq. 18.6, and the upper and lower half-power frequencies are then found from Eq. 18.5.

• • •

Example 18.7 Calculate the bandwidth and the upper and lower half-power frequencies for the resonant circuit in Example 18.5.

Solution From Eq. 18.6,

$$\text{BW} = \frac{f_r}{Q} = \frac{15.9 \text{ MHz}}{200} = 79.5 \text{ kHz}$$

The upper and lower half-power frequencies are then

$$f_1 = 15{,}900 \text{ kHz} - \frac{79.5}{2} \text{ kHz} = 15{,}900 - 39.8 = 15.86 \text{ MHz}$$

$$f_2 = 15{,}900 \text{ kHz} + \frac{79.5}{2} \text{ kHz} = 15{,}900 + 39.8 = 15.94 \text{ MHz}$$

As a check on the bandwidth, we use

$$\text{BW} = \frac{R}{2\pi L} = \frac{1.0}{2\pi \times 2 \ \mu\text{H}} = 79.5 \text{ kHz}$$

• • •

IDEAL PARALLEL RESONANCE

A parallel *R-L-C* circuit driven by a voltage source is shown in Fig. 18.9. The total admittance of the circuit is given by

$$\mathbf{Y} = G + j(B_C - B_L) \tag{18.7}$$

The circuit is said to be resonant when the capacitive susceptance B_C is equal to the inductive susceptance B_L. This is the same thing as having the capacitive reactance equal to the inductive reactance, however, since

$$B_C = \frac{1}{X_C} \qquad B_L = \frac{1}{X_L}$$

* The error in assuming f_r midway between f_1 and f_2 is less than 5% for a circuit having a Q greater than 1.58.

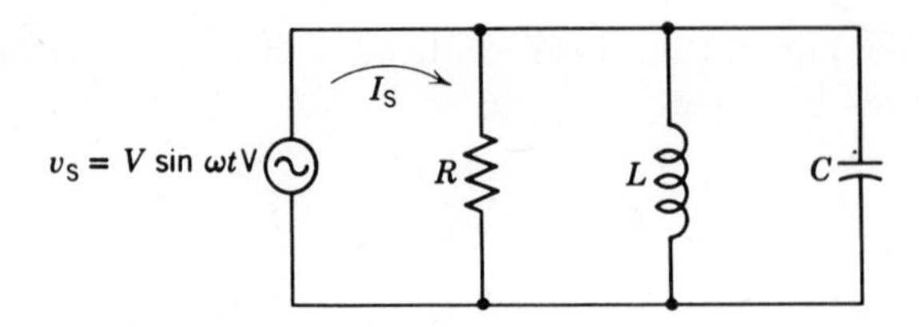

Figure 18.9 A parallel R-L-C circuit

Thus the condition for resonance in a *parallel* R-L-C circuit is the *same* as for a *series* R-L-C circuit.

The resonant frequency of the parallel R-L-C circuit is sometimes called the *antiresonant* frequency f_{ar}, simply to show that this is a parallel circuit. Since the condition for resonance is the same as for the series circuit ($X_L = X_C$), the antiresonant frequency is given by

$$f_{ar} = \frac{1}{2\pi\sqrt{LC}} \tag{18.8}$$

• • •

Example 18.8 Calculate the resonant frequency of the parallel R-L-C circuit in Fig. 18.9 if: $L = 0.3$ mH and $C = 0.333\ \mu$F.

Solution Using Eq. 18.8,

$$f_{ar} = \frac{1}{2\pi\sqrt{LC}} = \frac{1}{2\pi\sqrt{0.3 \times 10^{-3} \times 0.333 \times 10^{-6}}}$$

$$= \frac{1}{2\pi\sqrt{10^{-10}}} = 15.9 \text{ kHz}$$

• • •

From Eq. 18.7, the admittance of the parallel R-L-C circuit at resonance is

$$\mathbf{Y} = G + j(B - B) = G$$

This is the *minimum* value of admittance, or the *maximum* value of impedance since $\mathbf{Z} = 1/\mathbf{Y}$. If we were to calculate the impedance of the parallel circuit as a function of frequency, it would appear as shown in Fig. 18.10. The maximum valuc of the impedance occurs at resonance, and it is clearly equal to the

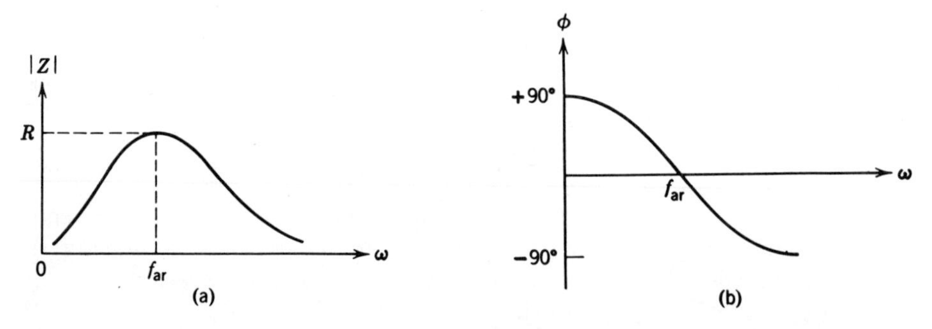

Figure 18.10 The impedance of the parallel circuit in Figure 18.9 as a function of frequency (a) The magnitude of the impedance (b) The phase angle of the impedance

resistance in the circuit, R. For the parallel resonant circuit, we like to have this resistance as *high* as possible in order to obtain the *sharpest* resonant curve as shown in Fig. 18.10*a*. By comparison, we see that the impedance of a *series* R-L-C circuit is *very small* at resonance, while the impedance of a *parallel* R-L-C circuit is *very large* at resonance.

A measure of the *quality* of a parallel resonant circuit is again given as the ratio of the reactive power to the real power, but since this is a *parallel* circuit, the voltage is common to each element. The quality factor Q is thus expressed as

$$Q = \frac{P_Q}{P} = \frac{V_S^2/X_L}{V_S^2/R} = \frac{R}{X_L} = \frac{R}{\omega_{ar} L} = \frac{R}{2\pi f_{ar} L} \tag{18.9}$$

• • •

Example 18.9 Calculate the Q of the circuit in Example 18.8 if $R = 10\ \text{k}\Omega$.

Solution From Example 18.8, the angular frequency at resonance ω_{ar} is clearly

$$\omega_{ar} = 10^5\ \text{rad/sec}$$

Thus, from Eq. 18.9, we have

$$Q = \frac{R}{\omega_{ar} L} = \frac{10\ \text{k}\Omega}{10^5 \times 0.3\ \text{mH}} = 333$$

• • •

Since the impedance of a parallel R-L-C circuit has a magnitude as shown in Fig. 18.10*a*, the total circuit current delivered by the source must appear as shown in Fig. 18.11. It is clear that the minimum current is delivered by the source at the antiresonant frequency f_{ar}. Let us consider the resonant circuit shown in Fig. 18.12*a* as a specific example. Since the voltage is the same across every element, the element currents can be calculated as

$$\mathbf{I}_R = \frac{\mathbf{V}_S}{R} = \frac{1.0\underline{/0^\circ}\ \text{V}}{1\ \text{k}\Omega} = 1.0\ \text{mA}\underline{/0^\circ}$$

$$\mathbf{I}_L = \frac{\mathbf{V}_S}{jX_L} = \frac{1.0\underline{/0^\circ}\ \text{V}}{10^5 \times 0.3\ \text{mH}\underline{/90^\circ}\ \Omega} = \frac{1.0\underline{/0^\circ}\ \text{V}}{30\underline{/90^\circ}\ \Omega} = 33.3\underline{/-90^\circ}\ \text{mA}$$

$$\mathbf{I}_C = \frac{\mathbf{V}_S}{-jX_C} = \frac{1.0\underline{/0^\circ}\ \text{V}}{\dfrac{1}{10^5 \times 0.333\ \mu\text{F}}\underline{/-90^\circ}\ \Omega} = \frac{1.0\underline{/0^\circ}\ \text{V}}{30\underline{/-90^\circ}\ \Omega} = 33.3\underline{/90^\circ}\ \text{mA}$$

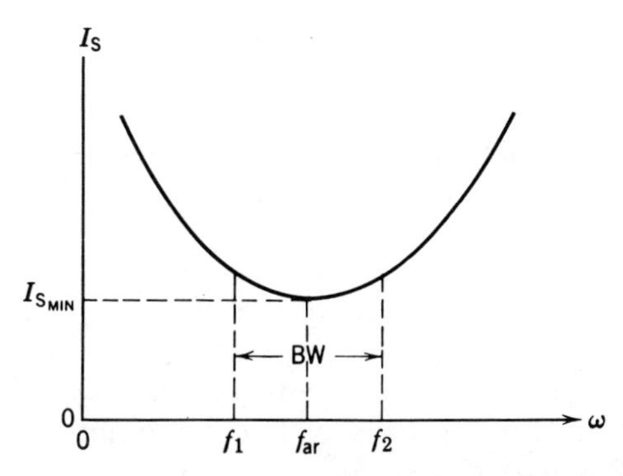

Figure 18.11 The source current delivered to a parallel R-L-C circuit

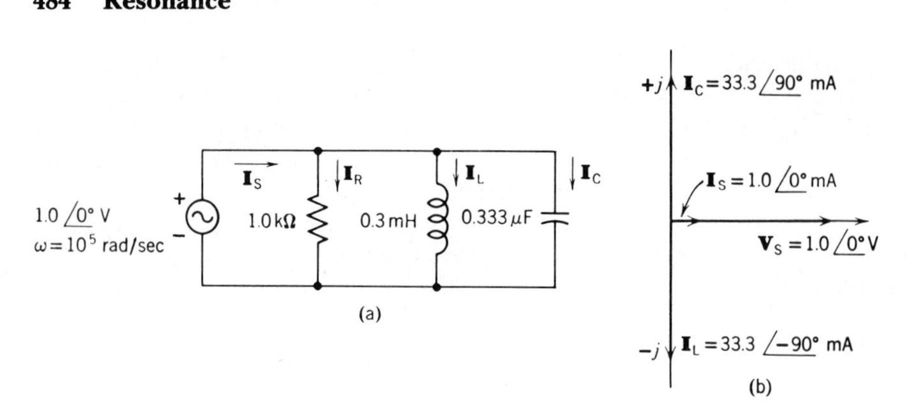

Figure 18.12 A resonant parallel *R-L-C* circuit (*a*) Parallel *R-L-C* circuit (*b*) The phasor diagram for (*a*)

The total source current is then the vector sum of these three element currents. Thus,

$$\mathbf{I}_S = \mathbf{I}_R + \mathbf{I}_L + \mathbf{I}_C = 1.0\angle 0° + 33.3\angle -90° + 33.3\angle 90°$$
$$= 1.0 - j33.3 + j33.3 = 1.0 \text{ mA}$$

The phasor diagram for the circuit is shown in Fig. 18.12*b*. It is clear from the phasor diagram that KCL is satisfied, but notice that the inductor and capacitor currents are both 33.3 times as great at the source current! This is known as the *resonant current rise* in a parallel circuit. Since these two currents circulate back and forth between the inductor and the capacitor, the parallel combination of L and C is often called the *tank circuit*, and the circulating current in the tank circuit is called the *tank current*.

Notice that the Q of the circuit in Fig. 18.12*a* is

$$Q = \frac{R}{\omega_{ar}L} = \frac{1 \text{ k}\Omega}{10^5 \times 0.3 \text{ mH}} = 33.3$$

But the inductor (or capacitor) current is exactly 33.3 times greater than the source current. As might be suspected, we can state that the tank current in a parallel *R-L-C* circuit is equal to the Q multiplied by the source current. That is,

$$I_C = I_L = QI_S \tag{18.10}$$

An examination of the source current delivered to a parallel *R-L-C* circuit shown in Fig. 18.11 reveals that this circuit also has a bandwidth. The bandwidth for the parallel circuit is defined just as it is for the series circuit, and it is the difference between the upper and lower half-power frequencies. It is also equal to the antiresonant frequency divided by the circuit Q. Thus,

$$\text{BW} = f_2 - f_1 = \frac{f_{ar}}{Q} \tag{18.11}$$

• • •

Example 18.10 Calculate the antiresonant frequency, the Q, the bandwidth, and the tank current for the parallel *R-L-C* circuit in Fig. 18.9 if:

$\mathbf{V}_S = 20\angle 0°$ mV, $R = 50$ kΩ, $L = 2.5$ mH, and $C = 400$ pF.

Solution The antiresonant frequency is found to be

$$f_{\text{ar}} = \frac{1}{2\pi\sqrt{LC}} = \frac{1}{2\pi\sqrt{2.5\times10^{-3}\times4\times10^{-10}}} = \frac{1}{2\pi\sqrt{10^{-12}}} = 159\text{ kHz}$$

The circuit Q is

$$Q = \frac{R}{\omega_{\text{ar}}L} = \frac{50\text{ k}\Omega}{10^6\times2.5\text{ mH}} = 20$$

The bandwidth is then

$$\text{BW} = \frac{f_{\text{ar}}}{Q} = \frac{159\text{ kHz}}{20} = 7.95\text{ kHz}$$

The current delivered by the source is simply the resistor current. Thus

$$\mathbf{I}_{\text{S}} = \mathbf{I}_{\text{R}} = \frac{\mathbf{V}_{\text{S}}}{R} = \frac{20\,/\underline{0^\circ}\text{ mV}}{50\text{ k}\Omega} = 0.4\,/\underline{0^\circ}\ \mu\text{A}$$

The tank current is then

$$I_{\text{L}} = I_{\text{C}} = QI_{\text{S}} = 20\times0.4\ \mu\text{A} = 8\ \mu\text{A}$$

• • •

8.4 PRACTICAL PARALLEL RESONANT CIRCUITS

When constructing a parallel R-L-C circuit, it is desired to have the resistance R as high as possible in order to provide a high Q. For this reason, it is common to construct a parallel resonant circuit by simply connecting a capacitor in parallel with an inductor. Since an inductor always has some resistance R_{L}, the equivalent circuit of a practical parallel resonant circuit appears as shown in Fig. 18.13*a*. This circuit is a series-parallel combination, but we can change it into a simple parallel R-L-C circuit by finding the parallel equivalent of the inductor resistance and inductance. Since the impedance of the inductor alone is

$$\mathbf{Z}_{\text{L}} = R_{\text{L}} + jX_{\text{L}}$$

its admittance can be found simply as

$$\mathbf{Y}_{\text{L}} = \frac{1}{\mathbf{Z}_{\text{L}}} = \frac{1}{R_{\text{L}}+jX_{\text{L}}} = \frac{1}{R_{\text{L}}+jX_{\text{L}}}\times\frac{R_{\text{L}}-jX_{\text{L}}}{R_{\text{L}}-jX_{\text{L}}} = \frac{R_{\text{L}}-jX_{\text{L}}}{R_{\text{L}}^2+X_{\text{L}}^2}$$

$$= \frac{R_{\text{L}}}{R_{\text{L}}^2+X_{\text{L}}^2} - j\frac{X_{\text{L}}}{R_{\text{L}}^2+X_{\text{L}}^2} = G_{\text{L}} - jB_{\text{L}}$$

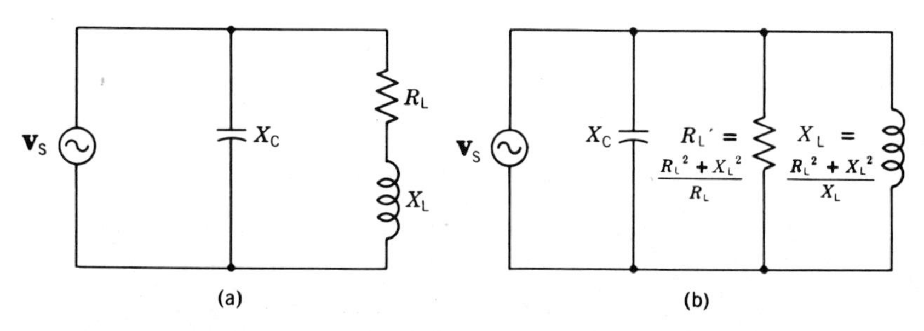

Figure 18.13 A practical parallel resonant circuit (*a*) A capacitor in parallel with an inductor (*b*) Parallel equivalent of the circuit in (*a*)

Therefore, the parallel equivalent circuit of the inductor is a resistance R'_L connected in parallel with an inductance X'_L as shown in Fig. 18.13*b*. The equivalent circuit values are

$$R'_L = \frac{1}{G_L} = \frac{R_L^2 + X_L^2}{R_L}$$

$$X'_L = \frac{1}{B_L} = \frac{R_L^2 + X_L^2}{X_L}$$

Since the circuit in Fig. 18.13*b* is a simple parallel *R-L-C* circuit, we can apply all of the relationships developed in the previous section. The anti-resonant frequency is determined by equating X_C and X'_L in Fig. 18.13*b*. Thus

$$X_C = X'_L$$

$$\frac{1}{\omega_{ar} C} = \frac{R_L^2 + (\omega_{ar} L)^2}{\omega_{ar} L}$$

Simplifying,

$$\frac{L}{C} - R_L^2 = (\omega_{ar} L)^2$$

The resonant frequency is then

$$f_{ar} = \frac{\sqrt{\frac{L}{C} - R_L^2}}{2\pi L}$$

We can arrange this equation into a slightly more convenient form, and after some algebra we obtain

$$f_{ar} = \frac{1}{2\pi\sqrt{LC}} \sqrt{1 - \frac{R_L^2 C}{L}} \tag{18.12}$$

From this equation, we can make a number of observations about the resonant frequency of a practical *L-G* circuit. First of all, we notice that the resonant frequency depends on the resistance of the coil R_L. Secondly, the resonant frequency is exactly the same as an ideal parallel circuit (or a series circuit) modified by the term $\sqrt{1 - R_L^2 C/L}$. Finally, we note that if the term $R_L^2 C/L$ is greater than 1.0, there is no resonant frequency for the circuit.

• • •

Example 18.11 Calculate the resonant frequency for the circuit in Fig. 18.13*a* if $C = 1\ \mu\text{F}$, $L = 0.1$ mH, and $R_L = 7\ \Omega$.

Solution Using Eq. 18.12,

$$f_{ar} = \frac{1}{2\pi\sqrt{10^{-4} \times 10^{-6}}} \sqrt{1 - \frac{7^2 \times 10^{-6}}{10^{-4}}} = \frac{10^5}{2\pi} \sqrt{1 - 0.49} = 11.4\ \text{kHz}$$

• • •

Example 18.12 Repeat example 18.11 if $R_L = 1.0\ \Omega$.

Solution The only quantity which will change is $\sqrt{1 - R_L^2 C/L}$, and this term becomes

$$\sqrt{1 - \frac{R_L^2 C}{L}} = \sqrt{1 - \frac{1^2 \times 10^{-6}}{10^{-4}}} = \sqrt{1 - 0.01} = \sqrt{0.99} = 0.995$$

Thus,

$$f_{ar} = \frac{10^5 \times 0.995}{2\pi} = 15.8 \text{ kHz}$$

• • •

From the above two examples, we note that if the term $R_L^2 C/L$ is small, the resonant frequency is nearly that of an ideal circuit. Equation 18.12 can be rewritten in a slightly different form as

$$f_{ar} = \frac{1}{2\pi\sqrt{LC}} \sqrt{\frac{1}{1 + 1/Q^2}} \tag{18.13}$$

From this equation it is clear that the modifying term $\sqrt{1/(1 + 1/Q^2)}$ will be negligibly small for large values of Q. Thus we can say that if the Q is greater than 10 or so, the resonant frequency is approximately

$$f_{ar} \cong \frac{1}{2\pi\sqrt{LC}} \tag{18.14}$$

The Q of the practical parallel circuit in Fig. 18.13 is found quite simply as

$$Q = \frac{R_L'}{X_L'} = \frac{\dfrac{R_L^2 + X_L^2}{R_L}}{\dfrac{R_L^2 + X_L^2}{X_L}} = \frac{X_L}{R_L} = \frac{\omega_{ar} L}{R_L} = \frac{2\pi f_{ar} L}{R_L} \tag{18.15}$$

From this equation it is seen that the Q of a practical parallel circuit is simply equal to the Q of the coil itself.

• • •

Example 18.13 Use Eqs. 18.13 and 18.14 to calculate the resonant frequency of the circuit in Example 18.12. What is the percent error if Eq. 18.14 is used?

Solution The Q of the circuit is found from Eq. 18.15 as

$$Q = \frac{\omega_{ar} L}{R} = \frac{0.995 \times 10^5 \times 10^{-4}}{1.0} = 9.95$$

The resonant frequency is then

$$f_{ar} = \frac{1}{2\pi\sqrt{LC}} \sqrt{\frac{1}{1 + 1/Q^2}} = \left|\frac{1}{2\pi\sqrt{10^{-4} \times 10^{-6}}}\right| \sqrt{\frac{1}{1 + 1/9.95^2}} = \frac{10^5}{2\pi} \sqrt{\frac{1}{1 - 0.01}}$$
$$= 15.8 \text{ kHz}$$

Using Eq. 18.14, the resonant frequency is

$$f_{ar} \cong \frac{1}{2\pi\sqrt{LC}} = \frac{1}{2\pi\sqrt{10^{-4} \times 10^{-6}}} = 15.9 \text{ kHz}$$

The percent error is then

$$100 \times \frac{15.9 - 15.8}{15.8} = 100 \times \frac{0.1}{15.8} = 0.633\%$$

• • •

The bandwidth of the practical parallel circuit is the same as for the parallel circuit, and it is

$$\text{BW} = \frac{f_{ar}}{Q} \tag{18.16}$$

• • •

Example 18.14 Calculate the bandwidth of the circuit in Example 18.12.

Solution Using Eq. 18.16,

$$\mathrm{BW} = \frac{f_{\mathrm{ar}}}{Q} = \frac{15.8\ \mathrm{kHz}}{9.95} = 1.59\ \mathrm{kHz}$$

• • •

18.5 BAND PASS FILTER

Very often R-L-C circuits are used in electronic systems to separate a signal of one particular frequency from a number of available signals. An example of this is the tuning of an R-L-C circuit in the input of a radio or television receiver. Such a circuit allows us to select only one-desired signal out of a number of signals being broadcast. Resistors, inductors, and capacitors can be connected in a great many different ways in order to achieve many different desired results. In any case, such a circuit is generally called an *electric filter*.

The filter circuit shown in Fig. 18.14*a* consists of an R-L-C series connection. The input voltage to the filter is $\mathbf{V}_\mathrm{S}$, and the output is $\mathbf{V}_\mathrm{L}$. Using the voltage divider theorem, we see that the output voltage is

$$\mathbf{V}_\mathrm{L} = \mathbf{V}_\mathrm{S} \frac{R}{R + j(X_\mathrm{L} - X_\mathrm{C})} \tag{18.17}$$

At the resonant frequency of the circuit, $X_\mathrm{L} = X_\mathrm{C}$, and the output voltage is equal to the input voltage. That is, at f_r,

$$\mathbf{V}_\mathrm{L} = \mathbf{V}_\mathrm{S} \frac{R}{R + jX_\mathrm{L} - jX_\mathrm{C}} = \mathbf{V}_\mathrm{S} \frac{R}{R} = \mathbf{V}_\mathrm{S}$$

Thus, the filter does not *attenuate* (reduce) the signal at a frequency of f_r. However, at a much lower or a much higher frequency the reactance of the circuit dominates and the output voltage is greatly attenuated. A plot of output voltage versus frequency might appear as shown in Fig. 18.14*b*. Such a plot is often referred to as the frequency response for the filter.

Since this filter readily passes any signal having a frequency equal to the

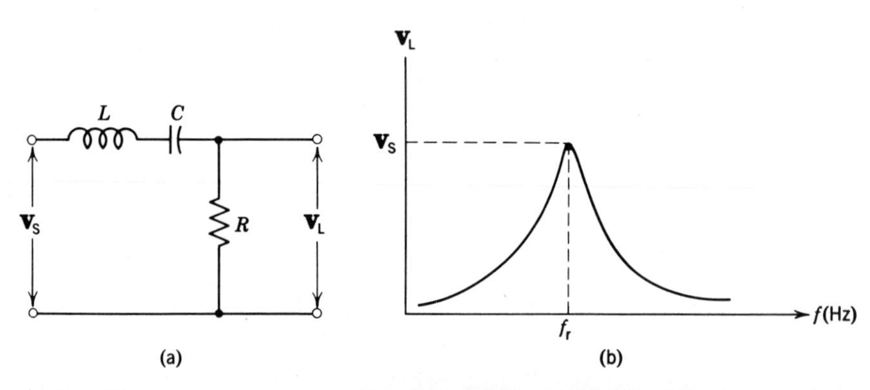

Figure 18.14 A Band pass filter (*a*) Filter circuit (*b*) Frequency response

resonant frequency, but greatly attenuates signals having frequencies above or below f_r, it is called a *band pass filter.*

• • •

Example 18.15 Calculate the resonant frequency and the bandwidth of the filter in Fig. 18.14*a* if $R = 500\ \Omega$, $L = 100\ \mu H$, and $C = 4.0$ pF. What would be the value of $\mathbf{V}_S$ at a frequency of 3 MHz if $\mathbf{V}_S = 1.0$ V?

Solution Since it is a series R-L-C circuit, the resonant frequency is

$$f_r = \frac{1}{2\pi\sqrt{LC}} = \frac{1}{2\pi\sqrt{10^{-4}\times 4\times 10^{-12}}} = \frac{10^8}{4\pi} = 7.95 \text{ MHz}$$

The circuit Q is

$$Q = \frac{\omega_r L}{R} = \frac{10^8 \times 100\ \mu\text{H}}{500\ \Omega} = 20$$

The bandwidth is then

$$\text{BW} = \frac{f_r}{Q} = \frac{7.95 \text{ MHz}}{20} = 398 \text{ kHz}$$

The output voltage at 3 MHz is found using Eq. 18.17,

$$\omega = 2\pi f = 2\pi \times 3 \times 10^6 = 1.89 \times 10^7$$

$$\mathbf{V}_S = 1.0\underline{/0^\circ} \text{ V} \frac{500}{500 + j\left(1.89\times 10^7 \times 100\ \mu\text{H} - \dfrac{1}{1.89\times 10^7 \times 4 \text{ pF}}\right)}$$

$$= \frac{500\underline{/0^\circ}}{500 - j11.3\times 10^3} = \frac{500\underline{/0^\circ}}{11.3\times 10^3\underline{/-80.5^\circ}} = 0.0442\underline{/87.5^\circ} \text{ V}$$

• • •

BAND STOP FILTER

Another type of filter is shown in Fig. 18.15*a*. Since we know that the impedance of a parallel L-C circuit is very large at its antiresonant frequency, we would expect this circuit to have a very low output voltage at f_{ar}. At frequencies below f_{ar}, the inductor has a very small reactance and thus all signals below f_{ar} will pass through the filter with very little attenuation. Similarly, at frequencies above f_{ar},

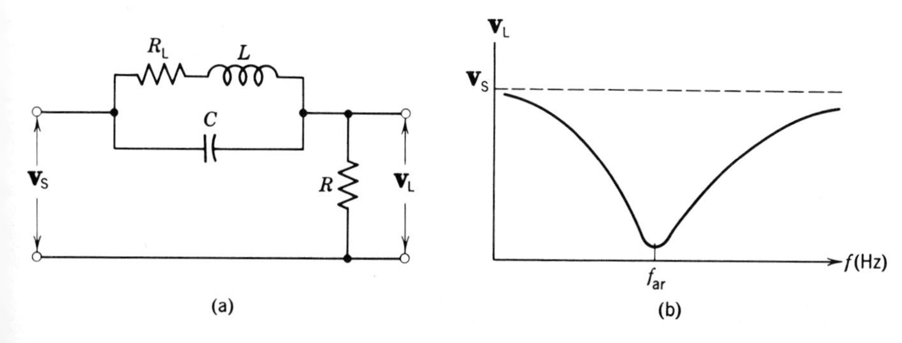

Figure 18.15 A band stop filter (*a*) Filter circuit (*b*) Frequency response for the filter

the capacitor has a very small reactance, and signals above f_{ar} will also pass through the filter with little attenuation. As a result, this filter will pass signals of all frequencies except those near the antiresonant frequency, and it is therefore called a *band stop filter.* The frequency response curve for the band stop filter is shown in Fig. 18.15*b*.

• • •

Example 18.16 Calculate the resonant frequency for the circuit in Fig. 18.15*a* if: $R = 1\ \text{k}\Omega$, $C = 1.0\ \mu\text{F}$, $L = 1\ \text{mH}$, and the coil resistance is $R_L = 10\ \Omega$.

Solution We are here concerned with a practical parallel resonant circuit. The antiresonant frequency is

$$f_{ar} = \frac{1}{2\pi\sqrt{LC}}\sqrt{1-\frac{R_L^2 C}{L}} = \frac{1}{2\pi\sqrt{10^{-3}\times 10^{-6}}}\sqrt{1-\frac{10^2\times 10^{-6}}{10^{-3}}}$$

$$= \frac{10^5}{2\pi\sqrt{10}}\sqrt{1-0.1} = 4.77\ \text{kHz}$$

• • •

18.7 DOUBLE TUNED FILTER

It is sometimes convenient to construct a circuit capable of both *passing* one frequency, and at the same time capable of *rejecting* a different frequency. Such a circuit is shown in Fig. 18.16*a*, and it is said to be *double-tuned.* The capacitor C_S and its series inductor L_S are tuned to resonate at the frequency which it is desired to pass through the circuit. The series combination of C_S and L_S will provide a very low impedance at the resonant frequency, and the desired signal will pass through the circuit with very little attenuation.

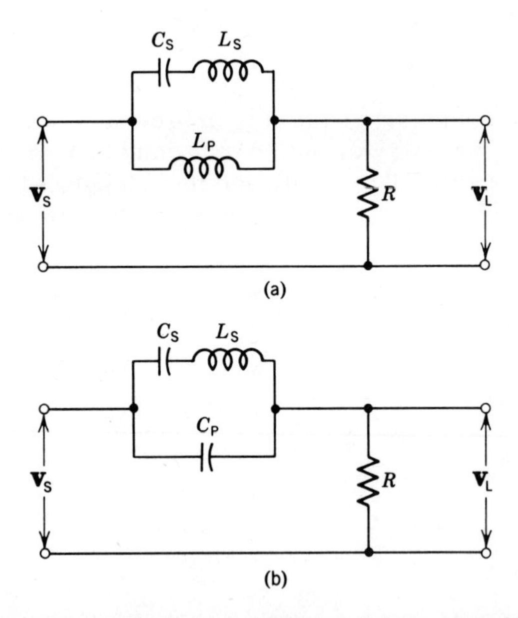

Figure 18.16 Double tuned filters (*a*) Rejected frequency below the desired frequency (*b*) Rejected frequency above the desired frequency

If the frequency to be rejected is *lower* than the desired frequency, the combination of C_S and L_S will appear *capacitive*. We will therefore place its *inductance* L_P in *parallel* with C_S and L_S (Fig. 18.16*a*) and make this combination resonant at the frequency to be rejected. Since the impedance of a parallel resonant circuit is high, the undesired frequency will be attenuated by the filter.

If the frequency to be rejected is *higher* than the desired frequency, the combination of C_S and L_S will appear *inductive*. We will therefore place a *capacitance* C_P in parallel with C_S and L_S (Fig. 18.16*b*), and make this combination resonant at the frequency to be rejected.

• • •

Example 18.17 The capacitor C_S in Fig. 18.16*a* has a value of 100 pF. Calculate the values of L_S and L_P needed to pass a frequency of 1 MHz, and reject a frequency of 500 kHz.

Solution C_S and L_S must be resonant at 1 MHz. Thus

$$1\text{ MHz} = \frac{1}{2\pi\sqrt{L_S C_S}}$$

and

$$L_S = \frac{1}{4\pi^2 f^2 C} = \frac{1}{4\pi^2 \times 10^{12} \times 10^{-10}} = 252\ \mu\text{H}$$

At 500 kHz, the series combination of C_S and L_S appears as

$$\begin{aligned} X_S &= j(X_{L_S} - X_{C_S}) \\ &= j\left(2\pi \times 500\text{ kHz} \times 252\ \mu\text{H} - \frac{1}{2\pi \times 500\text{ kHz} \times 100\text{ pF}}\right) \\ &= j(791 - 3180) = -j2390\ \Omega \end{aligned}$$

Thus the value of $X_P = \omega L_P$ must be equal to 2390 Ω at 500 kHz. Therefore

$$L_P = \frac{X_P}{\omega} = \frac{2390}{2\pi \times 500 \times 10^3} = 0.760\text{ mH}$$

• • •

SUMMARY

A series R-L-C circuit is resonant at a particular frequency, called the resonant frequency. At resonance the inductive and capacitive reactances cancel each other; as a result, the impedance of a series R-L-C circuit at resonance appears as a pure resistance; since this resistance is usually made as small as possible, the impedance is very nearly zero Ω. Therefore the current is maximum at resonance, and this produces a voltage rise across the inductor and the capacitor which may be many times greater than the applied source voltage. This phenomenon is known as resonant voltage rise. A measure of the quality of a series R-L-C circuit is the Q, and it can be used to determine the resonant voltage rise as well as the bandwidth of the circuit.

A parallel R-L-C circuit will also be resonant at a particular frequency, called the antiresonant frequency, to emphasize that it is a parallel circuit. The resonant frequency of an ideal circuit is calculated with the identical formula used for the series circuit. A practical parallel *R-L-C* circuit takes into consideration the resistance of the inductor (and any external resistance added to it), and the resonant frequency of this circuit is slightly lower than that of the ideal circuit. The *Q* of the practical parallel circuit is exactly equal to the *Q* of the coil used.

Both series and parallel *R-L-C* circuits can be used to construct electric filters, and there are many different circuit arrangements. The band pass filter is used to select a particular signal frequency while the band stop filter is used to reject a particular frequency. A double-tuned filter is really a combination of a band pass and a band stop filter constructed as one single filter. It will pass a particular frequency and at the same time will reject a particular different frequency.

GLOSSARY

Antiresonant frequency f_{ar}. *The resonant frequency of a parallel circuit.*

Band pass filter. *A filter that will accept one frequency and attenuate all others.*

Band stop filter. *A filter that will reject one frequency and pass all others.*

Bandwidth BW. *The difference between the upper and lower half-power frequencies.*

Circuit quality *Q*. *A measure of the quality of a tuned circuit. The ratio of reactive power to real power.*

Double-tuned filter. *A filter that will simultaneously accept one frequency and reject another frequency.*

Half-power frequencies. *The frequency where the filter output signal is down to 0.707 of its maximum output.*

Resonant circuit. *A circuit in which the inductive and capacitive reactances have equal magnitudes.*

Resonant frequency. *The frequency at which the inductive and capacitive reactances in a circuit have equal magnitudes.*

Resonant voltage rise. *The increase in the capacitor or inductor voltage that occurs in a series R-L-C circuit at resonance.*

Tank circuit. *A capacitor connected in parallel with an inductor.*

Tank current. *The circulating current in a resonant tank circuit.*

IMPORTANT RELATIONSHIPS

- *R-L-C* series circuit

$$f_r = \frac{1}{2\pi\sqrt{LC}} \qquad Q = \frac{P_Q}{P} = \frac{2\pi f_r L}{R}$$

$$\text{resonant voltage rise} = QV_S \qquad \text{BW} = \frac{R}{2\pi L} = \frac{f_r}{Q}$$

- Ideal *R-L-C* parallel circuit

$$f_{ar} = \frac{1}{2\pi\sqrt{LC}} \qquad Q = \frac{P_Q}{P} = \frac{R}{2\pi f_{ar} L}$$

$$\text{resonant current rise} = QI_S \qquad \text{BW} = \frac{f_{ar}}{Q}$$

- Real R-L-C parallel circuit

$$f_{ar} = \frac{1}{2\pi\sqrt{LC}}\sqrt{1 - \frac{R_L^2 C}{L}} = \frac{1}{2\pi\sqrt{LC}}\sqrt{\frac{1}{1 + 1/Q^2}}$$

$$Q = \frac{2\pi f_{ar} L}{R} \qquad \text{BW} = \frac{f_{ar}}{Q}$$

VIEW QUESTIONS

1. The frequency at which the inductive and capacitive reactances in a series R-L-C circuit have equal magnitudes is called the ____ frequency.
2. The impedance of a series R-L-C circuit at resonance is equal to ____.
3. The impedance of a series R-L-C circuit is capacitive (above, below) f_r, and it is inductive (above, below) f_r.
4. The impedance of a series R-L-C circuit is at its (maximum, minimum) at f_r.
5. It is usually desired to have the smallest possible value of R in a series R-L-C circuit. (TF)
6. The impedance of an ideal L-C series circuit would be ____ at the resonant frequency.
7. The ratio of reactive power to real power in a series R-L-C circuit is defined as ____.
8. The Q of a series R-L-C circuit is directly proportional to R. (TF)
9. The voltage across the inductor or the capacitor in a series R-L-C circuit at resonance is equal to the source voltage multiplied by the circuit ____.
10. The frequency at which the voltage across the capacitor in a series R-L-C circuit has fallen to 0.707 of its value at resonance is called the ____ ____ frequency.
11. The difference between the two half-power frequencies of a series R-L-C circuit is called the ____.
12. The resonant frequency of a series R-L-C circuit is exactly half-way between the two half-power frequencies. (TF)
13. The frequency at which an ideal parallel R-L-C circuit is resonant is called the ____ frequency.
14. The impedance of an ideal parallel R-L-C circuit at its antiresonant frequency is very (large, small).
15. It is usually desired to have the largest possible value of R in an ideal parallel R-L-C circuit. (TF)
16. The impedance of an ideal parallel L-C circuit at its antiresonant frequency would be ____.
17. The Q of an ideal parallel R-L-C circuit is directly proportional to R. (TF)
18. The tank current in an ideal parallel R-L-C circuit at its antiresonant frequency is equal to the source current multiplied by the circuit ____.
19. The quantity that distinguishes a practical parallel L-C circuit from an ideal parallel L-C circuit is the ____ of the inductor.
20. The antiresonant frequency of a practical parallel L-C circuit is slightly (smaller, greater) than the antiresonant frequency of an ideal parallel L-C circuit.

21. For values of Q greater than 10, the f_{ar} of a practical parallel L-C circuit is very close to that of an ideal parallel L-C circuit. (TF)
22. The Q of a practical parallel L-C circuit is simply equal to the Q of the coil. (TF)
23. A band _____ filter can be used to select a signal of one particular frequency. (TF)
24. A band stop filter will attenuate signals of any frequency except the frequency to which it is tuned. (TF)
25. A filter that will simultaneously pass a signal of one particular frequency and reject a signal of one other frequency is called a _____ _____ filter.

PROBLEMS

1. Calculate the resonant frequency of a series R-L-C circuit if: $R = 10\ \Omega$, $L = 1$ mH, and $C = 0.02\ \mu$F.
2. Repeat Problem 1 if C is changed to 300 pF.
3. What value of inductance must be connected in series with a 500-pF capacitor if the desired resonant frequency is 100 kHz?
4. What must be the value of L in Problem 3 for a resonant frequency of 2 MHz?
5. What value capacitance must be connected in series with a 200 μH inductance in order to obtain a resonant frequency of 1.5 MHz?
6. What must be the value of C in Problem 5 for an f_r of 250 kHz?
7. Calculate the Q of the series circuit in Problem 1.
8. A 400-μH inductor having a resistance of 20 Ω is connected in series with a 1000-pF capacitor. Calculate the circuit Q.
9. What must be the maximum resistance of the coil in Problem 8 in order to obtain a circuit Q of at least 50?
10. A series R-L-C circuit consists of a 5000-pF capacitor connected in series with a 200 μH inductor having a resistance of 25 Ω. Calculate the resonant frequency, the circuit Q, and the voltage across the capacitor at resonance if the circuit is connected to a 50 mV signal generator.
11. Repeat Problem 10 using a capacitor having a value of 500 pF.
12. A capacitor and an inductor are connected in series across the terminals of a signal generator. The frequency of the signal generator is adjusted to obtain a maximum voltage of 3.75 V across the capacitor. If the signal generator has a voltage of 64 mV under this condition. What is the Q of the circuit?
13. What is the bandwidth of the circuit in Problem 1?
14. What is the bandwidth of the circuit in Problem 10? What are the half-power frequencies?
15. What must be the Q of a series R-L-C circuit in order to obtain a bandwidth of 15 kHz if the circuit is resonant at 2.3 MHz?
16. Calculate the antiresonant frequency of an ideal parallel R-L-C circuit if: $R = 2$ kΩ, $L = 1$ mH, and $C = 0.02\ \mu$F.
17. Repeat Problem 16 if C is changed to 300 pF.
18. What value of capacitance must be connected in parallel with an *ideal* inductance of 5.06 mH if the desired antiresonant frequency is 0.1 MHz?

19. Calculate the Q of the circuit in Problem 16.
20. What would be the tank current of the circuit in Problem 16 if the circuit is connected to a 75-V signal generator at the antiresonant frequency?
21. Calculate the antiresonant frequency, the circuit Q, and the bandwidth of a 200-kΩ resistor, a 100-μH inductor (ideal), and a 400 pF capacitor connected in parallel.
22. A 0.03-μF capacitor is connected in parallel with a 700-μH inductor having a resistance of 100 Ω. Calculate the antiresonant frequency, the circuit Q, and the bandwidth. Is this a very good tuned circuit?
23. Repeat Problem 22 if the inductor resistance is reduced to 10 Ω.
24. If the bandwidth in Problem 22 must be less than 1.0 kHz, what must be the maximum value of the inductor resistance?
25. Calculate the value of C in Fig. 18.14*a* to pass a frequency of 455 kHz if $L = 560$ μH (assume an ideal inductor).
26. Repeat Problem 25 for a frequency of 1 MHz.
27. Calculate the value of C in Fig. 18.15*a* to reject a frequency of 455 kHz if $L = 560$ μH (assume an ideal inductor).
28. Repeat Problem 27 for a frequency of 2 MHz.
29. Calculate the values of L_S and L_P in Fig. 18.16*a* to pass a frequency of 1 MHz and reject a frequency of 500 kHz if $C_S = 100$ pF.
30. Calculate the values of L_S and C_P in Fig. 18.16*b* to pass a frequency of 1 MHz and reject a frequency of 2 MHz if $C_S = 100$ pF.

TRANSFORMERS

A *transformer* is a device used to transfer electric energy from one circuit to another, where the circuits are electrically connected by a common magnetic field. In general, transformers can be grouped into two main categories according to their construction—*air core* or *iron core.* Transformers constructed by forming windings on a core of magnetic material (iron) provide close coupling between the circuits, and this is the technique used to build power transformers, audio frequency transformers, and other transformers in which close coupling between windings is required. Windings can also be formed on a nonmagnetic core (wood or bakelite) to provide loose coupling between the circuits, and this is the technique used to construct the high frequency transformers so widely used in radio and television receivers.

19.1 TRANSFORMER FUNDAMENTALS

The standard symbols for air core and iron core transformers are shown in Figs. 19.1*a* and *b* respectively. The distinguishing feature between the two symbols is the set of parallel lines drawn between the two windings to show the presence of a magnetic core material. Traditionally, the winding on the left is referred to as the *primary* and is generally considered as the input, while the winding on the right is referred to as the *secondary* and serves as the output of the transformer.

Figure 19.1 Transformer symbols (*a*) Air core (*b*) Iron core

It is occasionally necessary to know the phase of the output voltage with respect to the input voltage. The relative phase is, of course, determined according to the manner in which the two windings are constructed. In Fig. 19.2*a*, the two windings are constructed in such a way that the secondary appears to be a continuation of the primary; they are said to be wound in the same sense, since a flux in one would aid the flux in the other. In this case, a current into the top terminal of the primary will set up a flux linking both coils. This flux will induce a voltage in the secondary such that the top terminal of the secondary is positive with respect to the bottom terminal (transformer action). The positive terminals of the primary and the secondary are thus identified with dots in Fig. 19.2*a*.

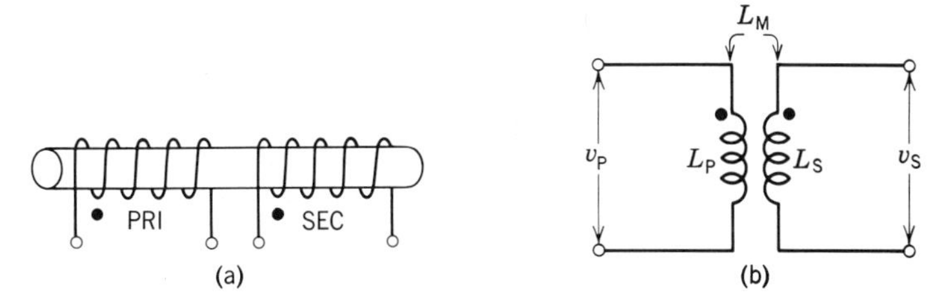

Figure 19.2 Transformer "dot" convention (*a*) Windings with same sense (*b*) Equivalent circuit

It is not always practical to examine the winding details of a transformer, and thus the dot notation shown in Fig. 19.2*b* is used. It is only necessary to remember that the positive terminal is identified by the dot.

The windings in Fig. 19.3*a* are wound in the opposite sense, and the dots have been changed accordingly. The equivalent circuit is shown in Fig. 19.3*b*.

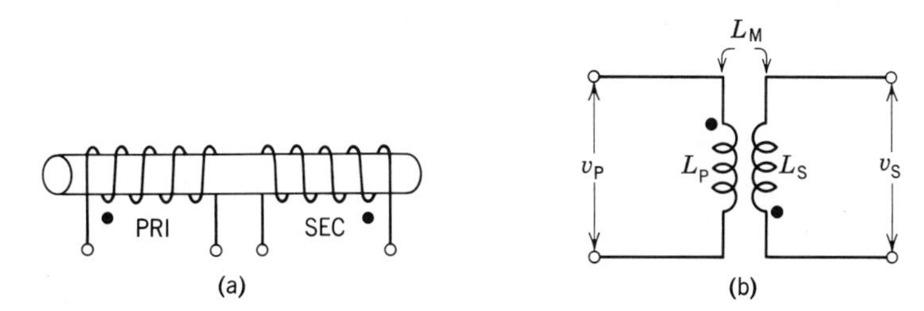

Figure 19.3 Transformer "dot" convention (*a*) Windings with opposite sense (*b*) Equivalent circuit

The windings appear to have changed, but the dotted end of the secondary still has exactly the same voltage polarity as the dotted end of the primary.

• • •

Example 19.1 If $v_P = 10 \sin \omega t$ in Figs. 19.2*b* and 19.3*b*, determine the phase of v_S with respect to v_P in each figure.

Solution In each case, the voltage at the dot end of the secondary is *in phase* with the voltage at the dot and of the primary.

• • •

In Chapter 12, the *coefficient of coupling, k,* was defined as the ratio of the flux linking two coils to the total flux in one of the coils. In Fig. 19.4, the flux ϕ_{12} links all of the turns in both coils, while ϕ_1 links only the primary turns, and ϕ_2 links only the secondary turns. The coefficient of coupling is thus

$$k = \frac{\phi_{12}}{\phi_1 + \phi_{12}} \tag{19.1}$$

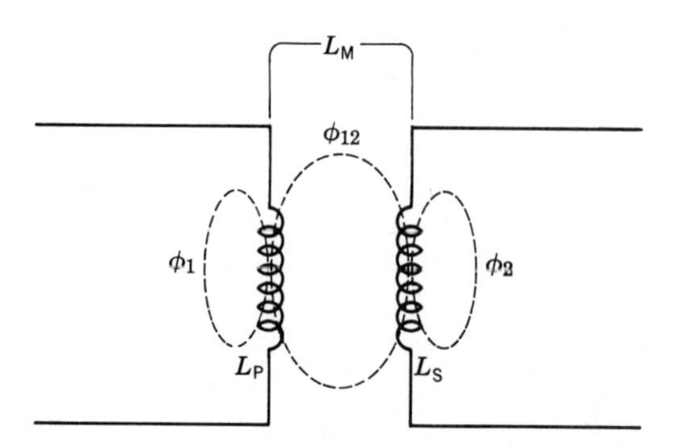

Figure 19.4 Transformer windings showing fluxes

If the two windings in Fig. 19.4 are wound on a magnetic core, both ϕ_1 and ϕ_2 are negligibly small compared to ϕ_{12}, and the coefficient of coupling k is very nearly equal to 1.0. Such coils are said to be *tightly coupled,* and this is the usual case for all iron core transformers such as power transformers, audio transformers, and so forth.

If the two windings in Fig. 19.4 are placed on a nonmagnetic core, ϕ_1 and ϕ_2 are not negligible compared to ϕ_{12}. The coefficient of coupling k may then be considerably less than 1.0. This is the technique used to construct radio frequency transformers, and such windings are said to be *loosely coupled.*

Whenever two coils are arranged such that there is coupling between them (as in Figs. 19.2, 19.3, and 19.4), a current in one coil is capable of inducing a voltage in the other coil. This is transformer action—that is, energy is being coupled from one coil to another. This coupling can be accounted for by the presence of *mutual inductance,* L_M, in the equivalent circuits. Recall from Chapter 12, *mutual inductance* is defined as $L_M = k\sqrt{L_1 L_2}$. For transformers, we can write $L_M = k\sqrt{L_P L_S}$.

• • •

Example 19.2 In Fig. 19.4, $\phi_1 = 30\ \mu$Wb and $\phi_{12} = 3$ mWb. What is the coefficient of coupling k? If $L_P = 100\ \mu$H and $L_S = 200\ \mu$H, what is L_M?

Solution Using Fig. 19.1,

$$k = \frac{\phi_{12}}{\phi_1 + \phi_2} = \frac{3\ \text{mWb}}{30\ \mu\text{Wb} + 3\ \text{mWb}} = \frac{3}{3.03} = 0.99 \cong 1$$

$$L_M = k\sqrt{L_P L_S} = 0.99\sqrt{100\ \mu\text{H} \times 200\ \mu\text{H}} = 140\ \mu\text{H}$$

Clearly the windings are tightly coupled, and the core must be a magnetic material.

• • •

All *real* transformers have losses as shown in the equivalent circuits in Fig. 19.5. The resistors R_1 and R_2 represent the *winding resistances* of the primary and secondary. R_C is an equivalent resistance which represents the power lost in *eddy currents* and *hysteresis* effects in the core. These losses can be appreciable in iron core transformers, but are essentially zero in an air core transformer.

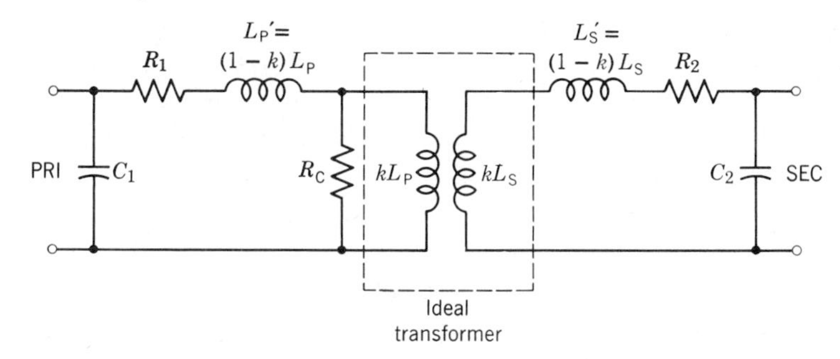

R_1 = primary winding resistance
R_2 = secondary winding resistance
R_C = core losses
L_P' = primary leakage inductance
L_S' = secondary leakage inductance
kL_P = coupled primary inductance
kL_S = coupled secondary inductance
L_P = primary incremental inductance
L_S = secondary incremental inductance
C_1 = equivalent primary capacitance
C_2 = equivalent secondary capacitance

Figure 19.5 Complete transformer equivalent circuit

When the coefficient of coupling is less than 1.0, there is some *leakage flux* (ϕ_1 or ϕ_2 in Fig. 19.4), and this represents an inductance which does not take part in the energy coupling between the primary and the secondary. The *leakage inductance* of the primary L_P' and the *secondary leakage inductance* L_S' account for these effects. The actual *incremental inductance* of the primary and the secondary are represented by L_P and L_S respectively. kL_P and kL_S represent the *coupled inductance* which provides for an energy transfer from primary to secondary.

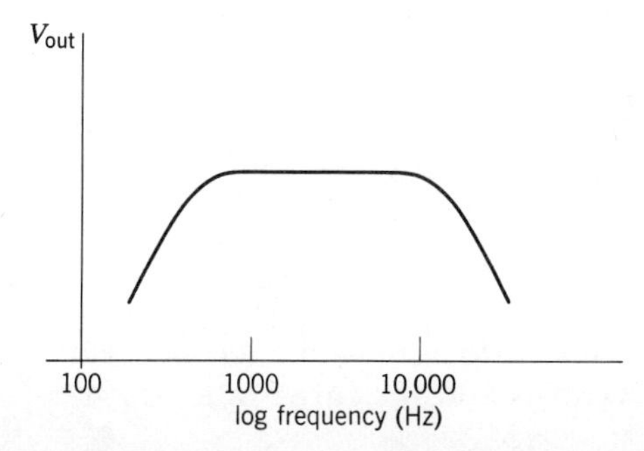

Figure 19.6 Frequency response for a typical audio transformer

There must be some capacitance associated with the transformer windings, and C_1 and C_2 represent an *equivalent capacitance* associated with the primary and the secondary as shown. These capacitances have relatively no effect on power transformers, or in applications at frequencies lower than a few hundred kHz. They do, however, limit the upper frequency response of a transformer. For example, an audio transformer being used in high fidelity equipment might be capable of coupling signals between the input and output over a range of frequencies as shown in Fig. 19.6.

Notice in Fig. 19.5 the *ideal transformer* is one in which all the losses are negligible, $k = 1$, and we need only consider the primary (L_P) and the secondary (L_S) as shown inside the dashed area.

19.2 IDEAL IRON CORE TRANSFORMERS

As seen in Fig. 19.5, a *real* transformer has a number of losses associated with it; among them are the resistances of the windings, core losses, leakage reactances, and a certain amount of capacitance. In our present discussion, we will consider these losses negligibly small and we will represent the iron core transformer as an *ideal* element as shown in Fig. 19.7.

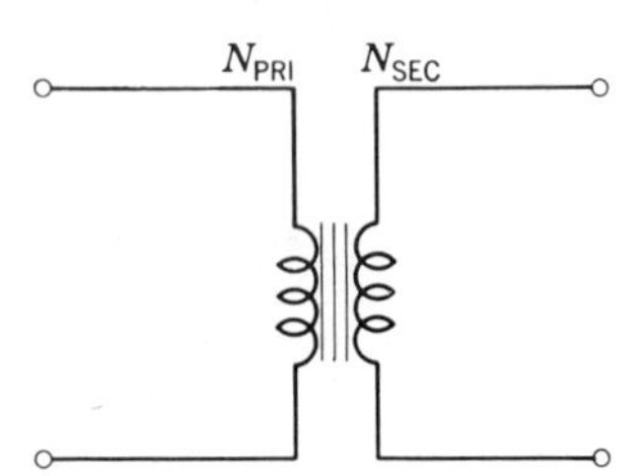

Figure 19.7 An ideal transformer

N_{PRI} represents the total number of turns on the *primary* winding, and N_{SEC} represents the total number of turns on the *secondary* winding. The *turns ratio* n of the transformer is defined as the ratio of turns in the secondary to turns in the primary. Thus

$$\text{turns ratio} = n \triangleq \frac{N_{SEC}}{N_{PRI}} \tag{19.2}$$

• • •

Example 19.3 If the transformer in Fig. 19.7 has 700 turns on the primary and 2100 turns on the secondary, what is the turns ratio?

Solution Using Eq. 19.2,

$$n = \frac{N_{SEC}}{N_{PRI}} = \frac{2100}{700} = 3$$

This ratio is sometimes specified as 3:1.

• • •

Example 19.4 How many turns must the secondary of a transformer have if the primary has 170 turns and the turns ratio is 7:1?

Solution Equation 19.2 can be solved for N_{SEC} as

$$N_{SEC} = nN_{PRI} = 7 \times 170 = 1190 \text{ turns}$$

• • •

At this point, we recall some of the material on magnetics covered in Chapter 11, and some of the material on inductance from Chapter 12. Specifically, we know that a voltage induced in an inductance is proportional to the number of turns, and the rate of change of flux. That is

$$v = N\frac{d\phi}{dt} \tag{19.3}$$

If we apply this equation to the primary winding of the ideal iron-core transformer shown in Fig. 19.8, we obtain

$$v_{PRI} = N_{PRI}\frac{d\phi_{PRI}}{dt} \tag{19.4}$$

Similarly, at the secondary we have

$$v_{SEC} = N_{SEC}\frac{d\phi_{SEC}}{dt} \tag{19.5}$$

But, since this is an ideal transformer, $\phi_{PRI} = \phi_{SEC}$ and thus

$$\frac{d\phi_{PRI}}{dt} = \frac{d\phi_{SEC}}{dt}$$

Thus we can solve Eqs. 19.4 and 19.5 for these quantities and equate the results to obtain

$$\frac{d\phi_{PRI}}{dt} = \frac{v_{PRI}}{N_{PRI}} = \frac{d\phi_{SEC}}{dt} = \frac{v_{SEC}}{N_{SEC}}$$

v_{PRI} and v_{SEC} represent sinusoidal voltages here, and thus we will use phasors $\mathbf{V}_{PRI}$ and $\mathbf{V}_{SEC}$ for them. As a result we have

$$\frac{\mathbf{V}_{SEC}}{\mathbf{V}_{PRI}} = \frac{N_{SEC}}{N_{PRI}} = n \tag{19.6}$$

Equation 19.6 relates the primary and secondary voltages of the transformer in terms of the turns ratio.

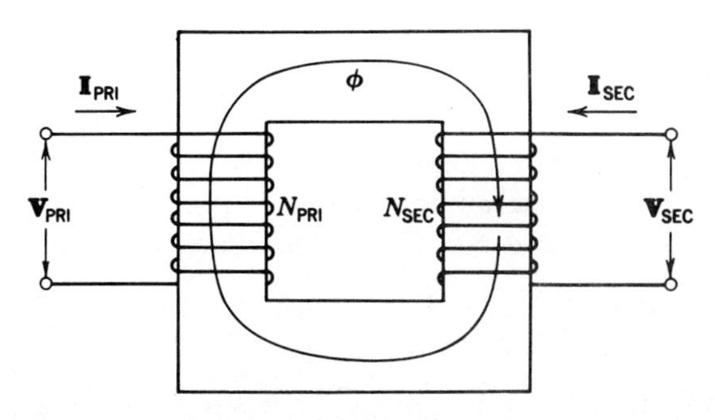

Figure 19.8 An ideal iron core transformer

• • •

Example 19.5 What is the secondary voltage of a transformer having a turns ratio of 20 if the primary voltage is 25 V?

Solution Using Eq. 19.6, we have

$$\mathbf{V}_{SEC} = n\mathbf{V}_{PRI} = 25 \times 20 = 500 \text{ V}$$

• • •

Example 19.6 What must be the turns ratio of a transformer whose output voltage is 6.3 V if the input voltage is 117 V?

Solution From Eq. 19.6,

$$n = \frac{\mathbf{V}_{SEC}}{\mathbf{V}_{PRI}} = \frac{6.3 \text{ V}}{117 \text{ V}} = \frac{1}{18.6}$$

• • •

The transformer in Example 19.5 is called a *step-up* transformer since the output voltage is *greater* than the input voltage; on the other hand, the transformer in Example 19.6 is called a *step-down* transformer since the output voltage is less than the input voltage.

We now recall Ohm's Law for magnetic circuits, and we have

$$\phi = \frac{\mathcal{F}}{\mathcal{R}} = \frac{NI}{\dfrac{l}{\mu A}} = \frac{I\mu NA}{l}$$

In order to have a changing flux $\Delta\phi$, there must be a changing current ΔI, and thus

$$\Delta\phi = \Delta I \frac{\mu NA}{l}$$

Now, from the ideal iron core transformer in Fig. 19.8, we see that

$$\Delta\phi_{PRI} = \Delta I_{PRI} \frac{\mu N_{PRI} A}{l} \qquad \text{and} \qquad \Delta\phi_{SEC} = \Delta I_{SEC} \frac{\mu N_{SEC} A}{l}$$

Since this is an ideal transformer, $\phi_{PRI} = \phi_{SEC}$ and thus

$$\Delta\phi_{PRI} = \Delta\phi_{SEC}$$

We can therefore equate the two expressions above to obtain

$$\Delta I_{PRI} \frac{\mu N_{PRI} A}{l} = \Delta I_{SEC} \frac{\mu N_{SEC} A}{l}$$

Cancelling common terms,

$$\Delta I_{PRI} N_{PRI} = \Delta I_{SEC} N_{SEC}$$

ΔI_{PRI} and ΔI_{SEC} represent a change in currents, but since we are concerned with sinusoidal currents here (they change continuously), we simply substitute the phasor values $\mathbf{I}_{PRI}$ and $\mathbf{I}_{SEC}$ for the currents. Thus we have

$$\frac{\mathbf{I}_{PRI}}{\mathbf{I}_{SEC}} = \frac{N_{SEC}}{N_{PRI}} = n \tag{19.7}$$

This equation gives the relationship between the primary and secondary currents in a transformer in terms of the turns ratio.

• • •

Example 19.7 A certain transformer has a secondary current of 50 mA when the primary current is 0.8 A. What is the turns ratio?

Solution From Eq. 19.7,

$$n = \frac{\mathbf{I}_{PRI}}{\mathbf{I}_{SEC}} = \frac{800 \text{ mA}}{50 \text{ mA}} = 16, \text{ or } 16:1$$

• • •

Example 19.8 Will the secondary current in the 6.3 V transformer in Example 19.6 be larger or smaller than the primary current?

Solution The turns ratio was found to be 1/18.6. Using this in Eq. 19.7 we have

$$\mathbf{I}_{SEC} = \frac{\mathbf{I}_{PRI}}{n} = \frac{\mathbf{I}_{PRI}}{\frac{1}{18.6}} = 18.6\, \mathbf{I}_{PRI}$$

The secondary current is clearly 18.6 times greater than the primary current.

• • •

From Example 19.8, we make the important observation that the current is larger in the secondary than in the primary of a step-down transformer. We can also easily show that the secondary current in a step-up transformer is smaller than the primary current.

19.3 IDEAL TRANSFORMER CIRCUITS

In Fig. 19.9 a voltage source is connected to the primary of a transformer and a load R_L is connected to the secondary. Since this is an ideal transformer, there are no losses, and *all* the power delivered by the source must be dissipated in the load R_L. We can demonstrate this by equating Eqs. 19.6 and 19.7. That is

$$n = \frac{\mathbf{V}_{SEC}}{\mathbf{V}_{PRI}} = \frac{\mathbf{I}_{PRI}}{\mathbf{I}_{SEC}}$$

Cross-multiplying, we obtain

$$\mathbf{V}_{SEC}\mathbf{I}_{SEC} = \mathbf{V}_{PRI}\mathbf{I}_{PRI}$$

But, $\mathbf{V}_{SEC}\mathbf{I}_{SEC}$ is the power in the secondary $\mathbf{P}_{SEC}$, and $\mathbf{V}_{PRI}\mathbf{I}_{PRI}$ is the power in the primary $\mathbf{P}_{PRI}$. Thus

$$\mathbf{P}_{SEC} = \mathbf{P}_{PRI}$$

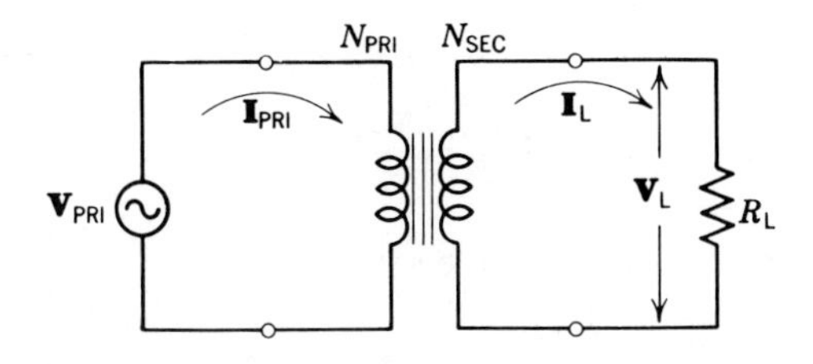

Figure 19.9

• • •

Example 19.9 The transformer in Fig. 19.9 has a turns ratio of $n = N_{SEC}/N_{PRI} = 10$. Verify that the power delivered to the load is equal to the power delivered from the source if the source voltage is $\mathbf{V}_{PRI} = 15$ V and $R_L = 300\ \Omega$.

Solution The secondary voltage is

$$\mathbf{V}_L = n\mathbf{V}_{PRI} = 10 \times 15\text{ V} = 150\text{ V}$$

The secondary current is then

$$\mathbf{I}_L = \frac{\mathbf{V}_L}{R_L} = \frac{150\text{ V}}{300\ \Omega} = 0.5\text{ A}$$

The power delivered to the load is seen to be

$$\mathbf{P}_L = \mathbf{V}_L\mathbf{I}_L = 150\text{ V} \times 0.5\text{ A} = 75\text{ W}$$

The primary current is then found to be

$$\mathbf{I}_{PRI} = n\mathbf{I}_{SEC} = 10 \times 0.5\text{ A} = 5.0\text{ A}$$

The power delivered by the source is then

$$\mathbf{P} = \mathbf{V}_{PRI}\mathbf{I}_{PRI} = 15\text{ V} \times 5.0\text{ A} = 75\text{ W}$$

• • •

Quite often it is possible to simplify ac circuits containing ideal transformers by replacing the transformer with its equivalent circuit. The transformer and load $\mathbf{Z}_L$ in Fig. 19.10*a* can be replaced with an equivalent circuit as shown in Fig. 19.10*b*. The process involves finding the impedance *looking into* the primary of the transformer. The load impedance $\mathbf{Z}_L$ will be *reflected* into the primary according to the turns ratio *n*.

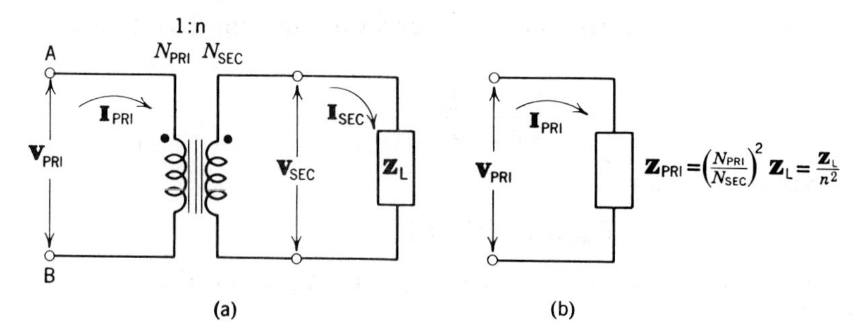

Figure 19.10 (*a*) An ideal transformer with a load $\mathbf{Z}_L$ (*b*) The equivalent circuit of the transformer and load in (*a*)

The input impedance to the primary in Fig. 19.10*a* is clearly

$$\mathbf{Z}_{PRI} = \frac{\mathbf{V}_{PRI}}{\mathbf{I}_{PRI}}$$

But from Eqs. 19.6 and 19.7, we have

$$\mathbf{V}_{PRI} = \frac{\mathbf{V}_{SEC}}{n} \qquad \mathbf{I}_{PRI} = n\mathbf{I}_{SEC}$$

Substituting these two relationships into the above,

$$\mathbf{Z}_{\text{PRI}} = \frac{\mathbf{V}_{\text{PRI}}}{\mathbf{I}_{\text{PRI}}} = \frac{\frac{\mathbf{V}_{\text{SEC}}}{n}}{n\mathbf{I}_{\text{SEC}}} = \frac{1}{n^2}\frac{\mathbf{V}_{\text{SEC}}}{\mathbf{I}_{\text{SEC}}} = \left(\frac{N_{\text{PRI}}}{N_{\text{SEC}}}\right)^2 \frac{\mathbf{V}_{\text{SEC}}}{\mathbf{I}_{\text{SEC}}}$$

But $\mathbf{V}_{\text{SEC}}/\mathbf{I}_{\text{SEC}} = \mathbf{Z}_{\text{SEC}} = \mathbf{Z}_{\text{L}}$, and thus the input impedance to the transformer looks like,

$$\mathbf{Z}_{\text{PRI}} = \left(\frac{N_{\text{PRI}}}{N_{\text{SEC}}}\right)^2 \mathbf{Z}_{\text{L}} = \frac{\mathbf{Z}_{\text{L}}}{n^2} \tag{19.9}$$

Equation 19.9 shows that a load impedance connected to the secondary of a transformer appears at the input terminals of the transformer divided by the square of the turns ratio.

• • •

Example 19.10 Calculate the primary current in Example 19.9 by finding the transformer equivalent circuit.

Solution The equivalent circuit is shown in Fig. 19.11. The primary load impedance is found using Eq. 19.9. Thus

$$\mathbf{Z}_{\text{PRI}} = \frac{\mathbf{Z}_{\text{L}}}{n^2} = \frac{300\ \Omega}{10^2} = 3\ \Omega$$

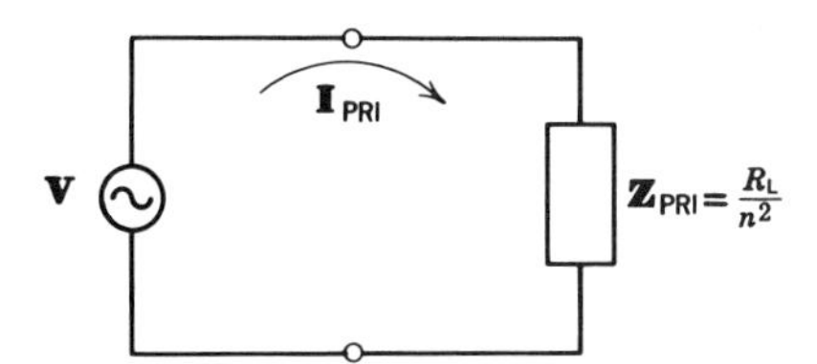

Figure 19.11 Example 19.9. Equivalent circuit for Figure 19.9

The primary current is then

$$\mathbf{I}_{\text{PRI}} = \frac{\mathbf{V}}{\mathbf{Z}_{\text{PRI}}} = \frac{15\ \text{V}}{3\ \Omega} = 5.0\ \text{A}$$

and this agrees with the previous result.

• • •

Transformers are often used in ac circuits as *impedance-matching* devices. Consider for example the problem of selecting R_L to match R_S in Fig. 19.12 to obtain maximum power transfer from the source to the load. The source ($\mathbf{V}_S$

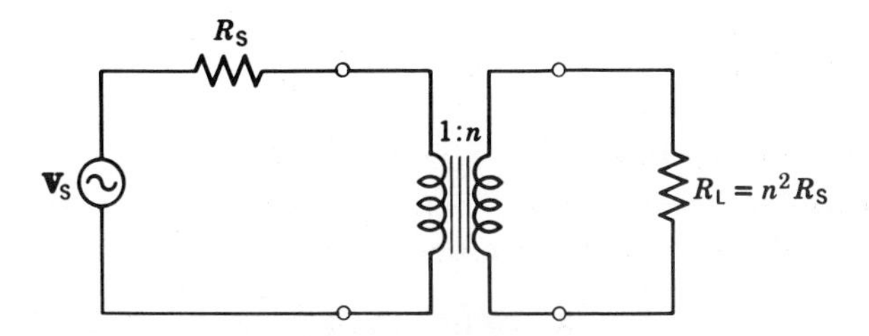

Figure 19.12 Matching R_L and R_S for maximum power tansfer

and R_S) might represent the output Thévenin equivalent of an output amplifier in a stereo system. The transformer will be used to match R_L to R_S in order to obtain the maximum output power.

Recall that for maximum power transfer, the load resistance must equal the source resistance. Thus, the resistance looking into the input terminals of the transformer must be equal to R_S. But from Eq. 19.9, we know this resistance is given in terms of R_L as

$$\mathbf{Z}_{PRI} = \frac{R_L}{n^2}$$

Equating this to R_S, we have

$$R_S = \frac{R_L}{n^2}$$

or

$$R_L = n^2 R_S \tag{19.10}$$

• • •

Example 19.11 Find the proper value of R_L in Fig. 19.12 for maximum power transfer if $R_S = 100\ \Omega$ and $n = 0.2$.

Solution Using Eq. 19.10, we have

$$R_L = n^2 R_S = (0.2)^2 \times 100\ \Omega = 0.04 \times 100\ \Omega = 4\ \Omega$$

• • •

Frequently *multiple tap* transformers such as shown in Fig. 19.13 are used for impedance matching. This transformer is designed for an 8 Ω load

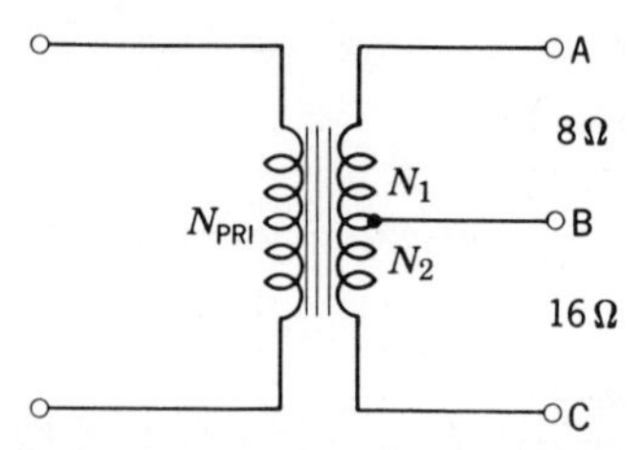

Figure 19.13

connected between terminals A and B, or a 16 Ω load connected between terminals B and C. One might be tempted to conclude that it would be proper to connect a load of (8 + 16) = 24 Ω between terminals A and C. This is *not* the case, however; and we can find the proper load by considering each output terminal pair separately. Using Eq. 19.9 we have

$$\mathbf{Z}_{PRI} = \left(\frac{N_{PRI}}{N_1}\right)^2 8\ \Omega = \left(\frac{N_{PRI}}{N_2}\right)^2 16\ \Omega = \left(\frac{N_{PRI}}{N_1 + N_2}\right)^2 R_{AC}$$

where R_{AC} is the proper load between terminals A and C. Notice that the total number of turns in the secondary between terminals A and C is $N_1 + N_2$. Simplifying the above equation, we obtain

$$\frac{\sqrt{8}}{N_1} = \frac{\sqrt{16}}{N_2} = \frac{\sqrt{R_{AC}}}{N_1 + N_2}$$

Taking the first two,

$$\frac{2\sqrt{2}}{N_1} = \frac{4}{N_2}$$

and

$$N_1 = \frac{N_2}{\sqrt{2}}$$

Substituting this into the second two, we obtain

$$\frac{4}{N_2} = \frac{\sqrt{R_{AC}}}{\frac{N_2}{\sqrt{2}} + N_2}$$

or

$$4 = \frac{\sqrt{R_{AC}}}{0.707 + 1} = \frac{\sqrt{R_{AC}}}{1.707}$$

Thus, R_{AC} is found to be

$$R_{AC} = (6.83)^2 = 46.5\ \Omega$$

From this result it is clear that the load impedance of a multiple tap transformer *cannot* simply be added or subtracted. They must always be calculated on the basis of the squares of the turns ratios.

At this point, it must be emphasized that the preceding discussion of transformers is based on an *ideal model* which neglects all transformer losses as well as the inductances associated with the transformer. An examination of Figs. 19.8 and 19.10 shows that there must be some primary inductance L_{PRI} and some secondary inductance L_{SEC} associated with any transformer. Furthermore, the primary and secondary are coupled by means of the iron core, and thus there must be some mutual inductance L_M. Accounting for these three inductances, the mesh equations for Fig. 19.10*a* are

$$\mathbf{V}_{PRI} = j\omega L_{PRI}\mathbf{I}_{PRI} - j\omega L_M\mathbf{I}_{SEC} \qquad (19.11a)$$

$$0 = -j\omega L_M\mathbf{I}_{PRI} + (\mathbf{Z}_L + j\omega L_{SEC})\mathbf{I}_{SEC} \qquad (19.11b)$$

In this case the mutual inductance terms ($j\omega L_M$) have negative values because one of the mesh currents goes into the dot(PRI) while the other mesh current comes out of the dot(SEC). If the mesh currents are assigned such that both currents enter (or leave) the dots, the sign of the $j\omega L_M$ term is positive in each mesh equation.

Solving for the input impedance at the primary of the transformer, we obtain

$$\mathbf{Z}_{in} = \frac{\mathbf{V}_{PRI}}{\mathbf{I}_{PRI}} = j\omega L_{PRI} + \frac{\omega^2 L_M^2}{\mathbf{Z}_L + j\omega L_{SEC}}$$

$$= \frac{j\omega L_{PRI}\mathbf{Z}_L - \omega^2 L_{PRI}L_{SEC} + \omega^2 L_M^2}{Z_L + j\omega L_{SEC}} \qquad (19.12)$$

Now, assuming the iron core ensures that all the flux links both the primary and the secondary, the relationship $L_M^2 = L_{PRI}L_{SEC}$ holds, and thus Eq. 19.12

becomes

$$\mathbf{Z}_{\text{in}} = \frac{j\omega L_{\text{PRI}}\mathbf{Z}_{\text{L}} - \omega^2 L_{\text{M}}^2 + \omega^2 L_{\text{M}}^2}{\mathbf{Z}_{\text{L}} + j\omega L_{\text{SEC}}} = \frac{j\omega L_{\text{PRI}}\mathbf{Z}_{\text{L}}}{\mathbf{Z}_{\text{L}} + j\omega L_{\text{SEC}}}$$

However, the turns ratio fixes the ratio of L_{PRI} and L_{SEC} according to $L_{\text{SEC}} = n^2 L_{\text{PRI}}$. Substituting this into the above expression, we have

$$\mathbf{Z}_{\text{in}} = \frac{j\omega L_{\text{PRI}}\mathbf{Z}_{\text{L}}}{\mathbf{Z}_{\text{L}} + j\omega n^2 L_{\text{PRI}}}$$

Dividing numerator and denominator of this expression by $j\omega L_{\text{PRI}}$ yields

$$\mathbf{Z}_{\text{in}} = \frac{\mathbf{Z}_{\text{L}}}{\dfrac{\mathbf{Z}_{\text{L}}}{j\omega L_{\text{PRI}}} + n^2} \tag{19.13}$$

Now, consider the case when $j\omega L_{\text{PRI}}$ is very large compared with $\mathbf{Z}_{\text{L}}$. Under this condition, the term $\mathbf{Z}_{\text{L}}/j\omega L_{\text{PRI}}$ is small compared with n^2, and Eq. 19.13 reduces to

$$\mathbf{Z}_{\text{in}} \cong \frac{\mathbf{Z}_{\text{L}}}{n^2} \tag{19.14}$$

which is exactly the relationship developed for the ideal model of the transformer. Thus the ideal model is very useful in practical applications where the transformer reactance $j\omega L_{\text{PRI}}$ is considerably larger than the load impedance $\mathbf{Z}_{\text{L}}$. As a rule of thumb, if $j\omega L_{\text{PRI}}$ is at least 10 times greater than $\mathbf{Z}_{\text{L}}$ the ideal transformer model is sufficient in most practical situations.

19.4 LOOSELY COUPLED TRANSFORMERS

An air core transformer similar to those used in radio frequency circuits is shown in Fig. 19.14*a*, and the appropriate equivalent circuit with a load $\mathbf{Z}_{\text{L}}$ is shown in Fig. 19.14*b*. R_{P} and R_{S} represent any resistance present in the primary or the secondary respectively, while L_{P} and L_{S} are the total primary and secondary inductances. L_{M} is the mutual inductance term and $\mathbf{Z}_{\text{L}}$ represents the load connected to the secondary. The two mesh equations for

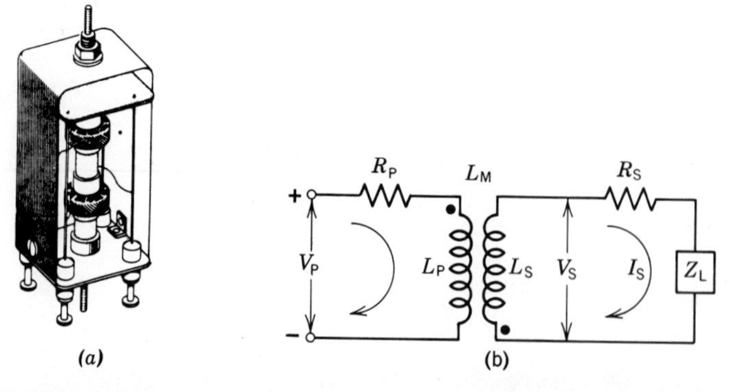

Figure 19.14 (*a*) Loosely coupled transformer (*b*) Equivalent circuit

the equivalent circuit in Fig. 19.14*b* are

$$\mathbf{V}_P = (R_P + j\omega L_P)\mathbf{I}_P + j\omega L_M\mathbf{I}_S = \mathbf{Z}_P\mathbf{I}_P + j\omega L_M\mathbf{I}_S$$
$$0 = +j\omega L_M\mathbf{I}_P + (R_S + j\omega L_S + \mathbf{Z}_L)\mathbf{I}_S = +j\omega L_M\mathbf{I}_P + \mathbf{Z}_S\mathbf{I}_S$$

where

$$\mathbf{Z}_P = R_P + j\omega L_P \quad \text{and} \quad \mathbf{Z}_S = R_S + j\omega L_S + \mathbf{Z}_L$$

We can easily solve these two equations for the transformer input impedance $\mathbf{Z}_{in} = \mathbf{V}_P/\mathbf{I}_P$ as follows:

$$\mathbf{Z}_{in} = \frac{\mathbf{V}_P}{\mathbf{I}_P} = \mathbf{Z}_P + \frac{\omega^2 L_M^2}{\mathbf{Z}_S} \tag{19.15}$$

From Eq. 19.15, we see that the total input impedance to the transformer is the primary impedance, $\mathbf{Z}_P$, in series with an additional term, $\omega^2 L_M^2/\mathbf{Z}_S$. This term represents the impedance reflected from the secondary into the primary, $\mathbf{Z}_{REF}$. Thus

$$\mathbf{Z}_{REF} = \frac{\omega^2 L_M^2}{\mathbf{Z}_S} \tag{19.16}$$

• • •

Example 19.12 For the transformer in Fig. 19.14*b*, $\omega = 10^4$, $R_S = R_P = 0$, $L_P = L_S = 2$ mH, $k = 0.5$, and $\mathbf{Z}_L = 100\ \Omega$. Find the impedance reflected into the primary $\mathbf{Z}_{REF}$, and the total primary input impedance $\mathbf{Z}_{in}$.

Solution

$$L_M = k\sqrt{L_P L_S} = 0.5\sqrt{2\text{ mH} \times 2\text{ mH}} = 1\text{ mH}$$
$$\omega^2 L_M^2 = (10^4)^2(1\text{ mH})^2 = 100$$
$$\omega L_P = \omega L_S = 10^4 \times 2\text{ mH} = 20$$
$$\mathbf{Z}_S = \mathbf{Z}_L + j\omega L_S = 100 + j20\ \Omega$$
$$\mathbf{Z}_P = j\omega L_P = j20\ \Omega$$

then using Eq. 19.16

$$\mathbf{Z}_{REF} = \frac{100}{100 + j20} = 0.96 - j0.19\ \Omega$$

The total input impedance from Eq. 19.15 is then

$$\mathbf{Z}_{in} = j20 + 0.96 - j0.19 = 0.96 + j19.8\ \Omega$$

• • •

Notice that $\mathbf{Z}_{in}$ for a loosely coupled transformer is considerably different than for an ideal iron core transformer. Compare Eqs. 19.14 and 19.15.

The analysis of an air core transformer, or any loosely coupled transformer, is somewhat more complicated than the ideal iron core transformer, but we can establish a simple procedure to aid in computation.

First of all, the effect of the load $\mathbf{Z}_L$ connected to the secondary is simply to reflect an impedance $\mathbf{Z}_{REF} = \omega^2 L_M^2/\mathbf{Z}_S$ into the primary in series with $\mathbf{Z}_P$. Therefore, the primary current $\mathbf{I}_P$ due to an applied voltage $\mathbf{V}_P$ is found to be

$$\mathbf{I}_P = \frac{\mathbf{V}_P}{\mathbf{Z}_{in}}$$

Secondly, the mesh equation for the secondary is

$$0 = +j\omega L_M \mathbf{I}_P + \mathbf{Z}_S \mathbf{I}_S$$

As a result, it is clear that a primary current $\mathbf{I}_p$ will induce a voltage in the secondary

$$\mathbf{V}_S = -j\omega L_M \mathbf{I}_P$$

Finally, this induced secondary voltage will result in a secondary current.

$$\mathbf{I}_S = \frac{\mathbf{V}_S}{\mathbf{Z}_S}$$

To summarize, we can compute the behavior of a coupled circuit by

1. Determine the input impedance $\mathbf{Z}_{in}$.

$$\mathbf{Z}_{in} = \mathbf{Z}_P + \frac{\omega^2 L_M^2}{\mathbf{Z}_S} \tag{19.17a}$$

2. Determine the primary current $\mathbf{I}_P$

$$\mathbf{I}_P = \frac{\mathbf{V}_P}{\mathbf{Z}_{in}} \tag{19.17b}$$

3. Determine the secondary induced voltage $\mathbf{V}_S$

$$\mathbf{V}_S = j\omega L_M \mathbf{I}_P \tag{19.17c}$$

4. Determine the secondary current $\mathbf{I}_S$

$$\mathbf{I}_S = \frac{\mathbf{V}_S}{\mathbf{Z}_S} = \frac{j\omega L_M \mathbf{I}_P}{\mathbf{Z}_S} \tag{19.17d}$$

• • •

Example 19.13 Determine the secondary current $\mathbf{I}_S$ for the transformer in Fig. 19.14*b* if: $\omega = 10^3$, $L_P = 1$ mH, $L_S = 2$ mH, $L_M = 0.5$ mH, $\mathbf{Z}_L = 2 + j4$, $R_S = 1\ \Omega$, $R_P = 0$, and $\mathbf{V}_P = 10\underline{/0°}$ V.

Solution We use the four steps above.

1. $\mathbf{Z}_{in} = \mathbf{Z}_P + \dfrac{\omega^2 L_M^2}{\mathbf{Z}_S} = j\omega L_P + \dfrac{\omega^2 L_M^2}{R_S + j\omega L_S + \mathbf{Z}_L}$

$$= j(10^3)(1\text{ mH}) + \frac{(10^3)^2(0.5\text{ mH})^2}{1 + j(10^3)(2\text{ mH}) + (2 + j4)}$$

$$= j + \frac{0.25}{3 + j6} = \frac{1 + j58}{60} = 0.967\underline{/89°}\ \Omega$$

2. $\mathbf{I}_P = \dfrac{\mathbf{V}_P}{\mathbf{Z}_{in}} = \dfrac{10\underline{/0°}}{0.967\underline{/89°}} = 10.34\underline{/-89°}$ A

3. $\mathbf{V}_S = -j\omega L_M \mathbf{I}_P = (10^3)(0.5\text{ mH})\ \underline{/-90°} \times 10.34\underline{/-89°} = 5.17\underline{/-179°}$ V

4. $\mathbf{I}_S = \dfrac{\mathbf{V}_S}{\mathbf{Z}_S} = \dfrac{5.17\underline{/-179°}}{3 + j6} = 0.77\underline{/-242.4°}$ A

.5 TUNED TRANSFORMER CIRCUITS

Quite often the effects of a tuned circuit are used in conjunction with a transformer. For example, connecting a capacitor C across the secondary of a transformer as shown in Fig. 19.15 results in series R-L-C circuit which has a

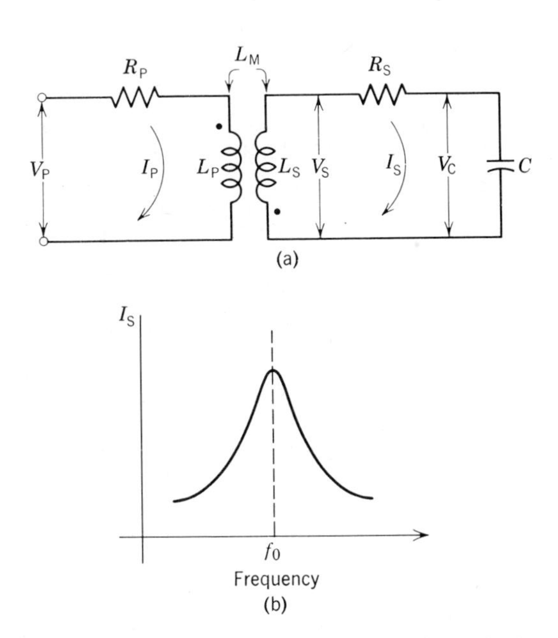

Figure 19.15 (*a*) Transformer with a tuned secondary (*b*) Secondary current versus frequency

resonant frequency f_0 given by

$$f_0 = \frac{1}{2\pi\sqrt{L_S C}} \tag{19.18}$$

From our knowledge of series resonant circuits, we know that the total impedance of the secondary at resonance must be $\mathbf{Z}_S = R_S$—that is, it is simply the resistance present in the secondary circuit. Therefore, at resonance the secondary current $\mathbf{I}_S$ is

$$\mathbf{I}_S = \frac{-j\omega L_M \mathbf{I}_P}{\mathbf{Z}_S} = \frac{-j\omega L_M \mathbf{I}_P}{R_S}$$

Above and below resonance, $|\mathbf{Z}_S| > R_S$, and thus $\mathbf{I}_P$ is maximum at f_0 as shown in Fig. 19.15*b*. Now, at resonance the output voltage across the capacitor $\mathbf{V}_C$ is

$$\mathbf{V}_C = -j\mathbf{I}_S X_C = \frac{-j\mathbf{I}_S}{\omega_0 C}$$

But, $1/\omega_0 C = \omega_0 L_S$ at f_0, and therefore

$$\mathbf{V}_C = -j\omega_0 L_S \mathbf{I}_S = -(\omega_0 L_M \mathbf{I}_P)\left(\frac{\omega_0 L_S}{R_S}\right)$$

Now, since $\mathbf{V}_S = \omega L_M \mathbf{I}_P$, and the Q of the secondary is $Q_S = \omega_0 L_S / R_S$, we see that the output voltage across the capacitor is Q_S times the induced secondary voltage at resonance. That is, at f_0,

$$\mathbf{V}_C = -Q_S \mathbf{V}_S \tag{19.19}$$

• • •

Example 19.15 For the circuit in Fig. 19.15*a*, $R_P = 0$, $L_P = L_S = 0.1$ mH, $k = 0.1$, $R_S = 10\ \Omega$, and $C = 0.01\ \mu$F. Calculate the output voltage across C at resonance if $\mathbf{V}_P = 1\underline{/0^\circ}$ V.

Solution

$$f_0 = \frac{1}{2\pi\sqrt{L_S C}} = \frac{1}{2\pi\sqrt{0.1 \text{ mH} \times 0.1\ \mu\text{F}}} = 159 \text{ kHz}$$

$$\omega_0 = 2\pi f_0 = 10^6 \text{ rad/sec}$$

$$Q_S = \omega_0 L_S / R_S = 10^6 \times 0.1 \text{ mH}/10 = 10$$

$$L_M = k\sqrt{L_P L_S} = 0.1\sqrt{0.1 \text{ mH} \times 0.1 \text{ mH}} = 0.01 \text{ mH}$$

Now, using the four-step procedure (Eqs. 19.17*a, b, c, d*)

1. $\mathbf{Z}_{in} = \mathbf{Z}_P + \dfrac{\omega^2 L_M^2}{\mathbf{Z}_S} = j10^6 \times 0.1 \text{ mH} + \dfrac{(10^6)^2(0.01 \text{ mH})^2}{10}$

   $= j100 + 10 = 100.5\underline{/5.7^\circ}\ \Omega$

2. $\mathbf{I}_P = \mathbf{V}_P/\mathbf{Z}_{in} = 10\underline{/0^\circ}/100.5\underline{/5.7} = 0.0995\underline{/-5.7^\circ}$ A

3. $\mathbf{V}_S = -j\omega L_M \mathbf{I}_P = 10^6 \times 0.01 \text{ mH}\underline{/-90^\circ} \times 0.0995\underline{/-5.7^\circ} = 0.995\underline{/-95.7^\circ}$ V

Then using Eq. 19.19

4. $\mathbf{V}_C = -Q_S \mathbf{V}_S = -10 \times 0.995\underline{/-95.7^\circ} = -9.95\underline{/-95.7^\circ}$ V

• • •

Consider now a transformer having a tuned secondary and a tuned primary as shown in Fig. 19.16*a*. If the secondary and the primary are both separately tuned to the *same* resonant frequency f_0, then at resonance $\mathbf{Z}_S = R_S$, $\mathbf{Z}_P = R_P$ and

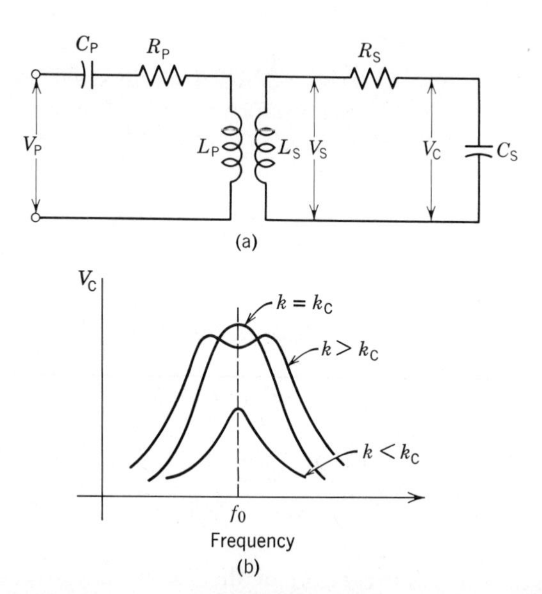

Figure 19.16 (*a*) Transformer with tuned primary and tuned secondary (*b*) Response curve

$\omega_0 = 1/\sqrt{L_P C_P} = 1/\sqrt{L_S C_S}$. Then from Eq. 19.17

$$\mathbf{Z}_{in} = R_P + \frac{\omega_0^2 L_M^2}{R_S} = \frac{R_P R_S + \omega_0^2 L_M^2}{R_S}$$

$$\mathbf{I}_P = \frac{\mathbf{V}_P}{\mathbf{Z}_{in}} = \frac{R_S \mathbf{V}_P}{R_P R_S + \omega_0^2 L_M^2}$$

$$\mathbf{I}_S = \frac{-j\omega_0 L_M \mathbf{I}_P}{\mathbf{Z}_S} = \frac{-j\omega_0 L_M \mathbf{V}_P}{R_P R_S + \omega_0^2 L_M^2}$$

Then the output voltage across the capacitor is $\mathbf{I}_S(1/j\omega_0 C_S) = -j\omega_0 L_S \mathbf{I}_S$ at resonance. Thus, we can write*

$$\frac{\mathbf{V}_C}{\mathbf{V}_P} = \sqrt{\frac{L_S}{L_P}} \times \frac{k}{k^2 + \left(\dfrac{1}{Q_P Q_S}\right)} \tag{19.20}$$

where $Q_P = \omega_0 L_P/R_P$, $Q_S = \omega_0 L_S/R_S$.

Since the secondary is still a series tuned circuit, a plot of $\mathbf{I}_S$ versus frequency, and therefore $\mathbf{V}_C$ versus frequency, reaches a maximum at f_0 as shown in Fig. 19.16*b*. However, $\mathbf{V}_C$ depends strongly on the value of k since both the primary and the secondary are tuned. For very small values of k (perhaps 0.01 or less) a plot of $\mathbf{V}_C$ versus frequency will appear as shown in Fig. 19.16*b* for the curve labeled $k < k_C$.

As the coupling is increased, k becomes larger, the value of $\mathbf{V}_C$ at f_0 increases until it has its maximum value when critical coupling is reached; that is, $k = k_C$ in Fig. 19.16*b*.

If the coupling is increased still further, k is made closer to 1.0, the curve for $\mathbf{V}_C$ in Fig. 19.16*b* develops a double hump ($k > k_C$). Furthermore, the value of $\mathbf{V}_C$ at f_0, is now reduced below the maximum value achieved with critical coupling (at $k = k_C$).

The value for k for critical coupling is found from

$$k_C = \frac{1}{\sqrt{Q_P Q_S}} \tag{19.21}$$

When the coefficient of coupling is set at the critical value, and if $Q_P = Q_S$, the curve for $\mathbf{V}_C$ has maximum flatness around f_0, and has a 3 *db* (0.707) bandwidth that is $\sqrt{2}$ times greater than a single tuned circuit. A plot of $\mathbf{V}_C$ versus frequency together with a plot of a single tuned circuit is shown for comparison in Fig. 19.17.

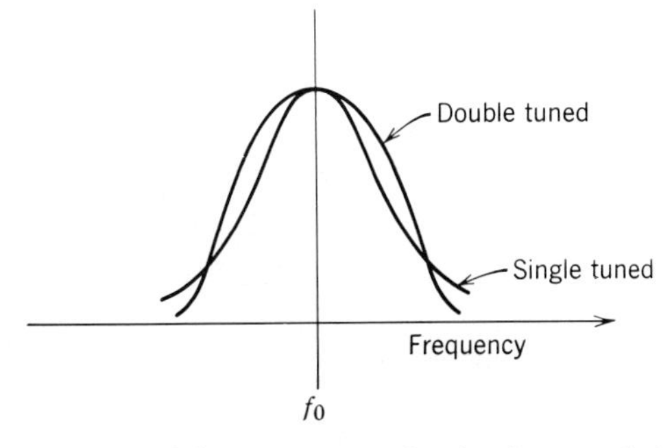

Figure 19.17 Comparison of the response of a single tuned circuit with a double tuned circuit

*See Problem 35.

• • •

Example 19.16 The circuit in Fig. 19.16*a* is designed such that $Q_P = Q_S = 20$. What is the value of critical coupling k_C?

Solution Using Eq. 19.21

$$k_C = \frac{1}{\sqrt{Q_P Q_S}} = \frac{1}{\sqrt{20 \times 20}} = 0.05$$

• • •

Example 19.17 Calculate $\mathbf{V}_C/\mathbf{V}_P$ for $k = 0.01$, 0.05 and 0.1 for the circuit in Example 19.16. $L_S = L_P = 1$ mH.

Solution Using Eq. 19.20.

1. for $k = 0.01$ $$\frac{\mathbf{V}_C}{\mathbf{V}_P} = \sqrt{\frac{1\text{ mH}}{1\text{ mH}}} \times \frac{0.01}{(0.01)^2 + \dfrac{1}{(20 \times 20)}} = 3.85$$

2. for $k = k_C = 0.05$ $$\frac{\mathbf{V}_C}{\mathbf{V}_P} = 1 \times \frac{0.05}{(0.05)^2 + \dfrac{1}{400}} = 10$$

3. for $k = 0.1$ $$\frac{\mathbf{V}_C}{\mathbf{V}_P} = 1 \times \frac{0.1}{(0.1)^2 + \dfrac{1}{400}} = 8$$

• • •

In certain applications, the primary is fed by a voltage source in shunt with the primary tuned circuit as shown in Fig. 19.18*a*. In this case, we can replace R, C_P and $\mathbf{V}_P$ by its Thévenin equivalent as shown in Fig. 19.18*b*. The equivalent

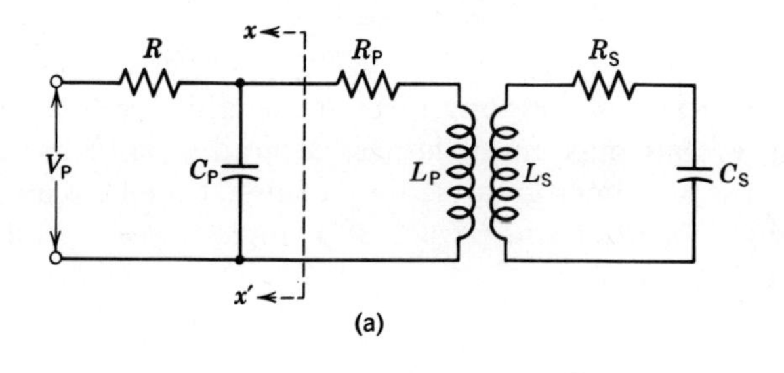

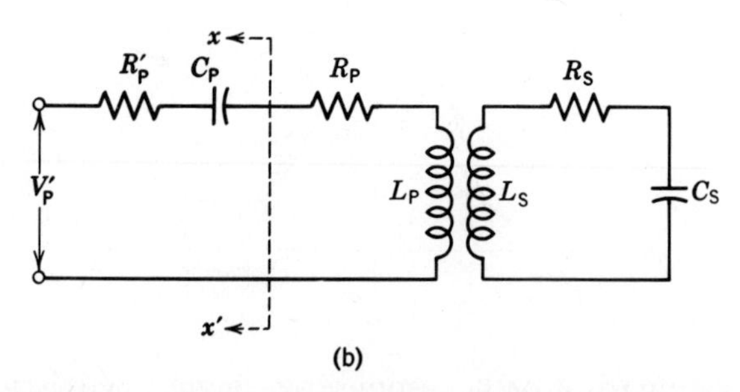

Figure 19.18 Tuned primary, tuned secondary circuit (*a*) Shunt excited primary (*b*) Thévenin equivalent

voltage is found using the voltage divider theorem,

$$\mathbf{V}_P' = \frac{\dfrac{-\mathbf{V}_P j}{\omega C_P}}{R - j/\omega C_P} \cong \frac{-j\mathbf{V}_P}{\omega R C_P} \quad \text{if} \quad R \gg \frac{1}{\omega C_P} \tag{19.22}$$

The equivalent Thévenin impedance is R in parallel with C_P, and if $R \gg 1/\omega C_P$

$$\mathbf{Z}_{th} = R_P' - jX_C = \frac{\dfrac{R}{j\omega C_P}}{R - j/\omega C_P} = \frac{1}{\omega^2 C_P^2 R} - \frac{j}{\omega C_P}$$

where $R_P' = 1/\omega^2 C_P^2 R$.

From the Thévenin equivalent in Fig. 19.18*b*, it is clear that the only differences between this circuit and the series excited primary in Fig. 19.16 are:

1. There is an additional primary resistance R_P' which reduces the primary Q.
2. The voltage applied at the primary terminals in the equivalent circuit is the Thévenin voltage $\mathbf{V}_P'$.

Otherwise the two circuits in Fig. 19.16 and 19.18 have essentially the same characteristics provided $R \gg 1/\omega C_P$ in Fig. 19.18.

• • •

Example 19.18 For the circuit in Fig. 19.18 $R = 10\ \text{k}\Omega$, $C_p = 0.1\ \mu\text{F}$ and $\omega = 10^5$. Find the values of $\mathbf{V}_P'$ and R_P'. Find $\mathbf{V}_C/\mathbf{V}_P$ if $R_P = 1\ \Omega$, $L_P = 1\ \text{mH}$, $k = 0.0447$ and $Q_S = 10$.

Solution

$$R = 10\ \text{k}\Omega \qquad \frac{1}{\omega C_P} = \frac{1}{(10^5 \times 10^{-7})} = 100$$

thus

$$R \gg \frac{1}{\omega C_P}$$

From Eq. 19.22

$$\mathbf{V}_P' \cong \frac{-j\mathbf{V}_P}{10^5 \times 10\ \text{k}\Omega \times 0.1\ \mu\text{F}} = 10^{-2}\ \mathbf{V}_P \underline{/-90^\circ}$$

$$R_P' = \frac{1}{(10^5)^2 (0.1\ \mu\text{F})^2 \times 10\ \text{k}\Omega} = 1\ \Omega$$

$$Q_P = \frac{\omega L_P}{R_P + R_P'} = \frac{10^5 \times 1\ \text{mH}}{1+1} = 50 \qquad k_C = \frac{1}{\sqrt{Q_P Q_S}} = \frac{1}{\sqrt{50 \times 10}} = 0.0447$$

Therefore, we have $k = k_C$. Using Eq. 19.20, $\dfrac{\mathbf{V}_C}{\mathbf{V}_P'} = 11.18$. But, $\mathbf{V}_P' = 10^{-2} \underline{/90^\circ}\ \mathbf{V}_P$. Thus,

$$\frac{\mathbf{V}_C}{\mathbf{V}_P} = 11.18 \times 10^{-2} \underline{/-90^\circ} = 0.112 \underline{/-90^\circ}$$

• • •

SUMMARY

The two general categories of transformers are air core and iron core. Iron core transformers have coupling coefficients very nearly equal to 1.0, and their electrical characteristics can be predicted by means of an ideal model having no losses. The ideal model can be used to predict voltage and current ratios as well as impedance transformations based solely on the turns ratio n.

Loosely coupled transformers have characteristics vastly different from the ideal device, and it is necessary to consider the coefficient of coupling k when dealing with them. Loosely coupled transformers are frequently used in conjunction with tuned circuits in order to take advantage of their combined characteristics. Transformers having a tuned secondary, as well as transformers having both a tuned primary and tuned secondary, are widely used in communications circuits.

GLOSSARY

Coefficient of coupling k. *The ratio of flux linking two coils to flux in one of the coils.*

Critical coupling k_C. *That value of coupling which provides maximum output voltage in the secondary of a double tuned transformer.*

Dot convention. *A technique for showing the voltage polarities across the primary and the secondary of a transformer.*

Ideal transformer. *A transformer having no losses and a coefficient of coupling equal to 1.0.*

Leakage inductance. *An inductance in the equivalent circuit of a transformer which accounts for uncoupled flux.*

Loose coupling. *Values of k much less than 1.0.*

Mutual inductance. *A measure of the inductive coupling between two coils.*

Primary. *The input winding of a transformer.*

Secondary. *The output winding of a transformer.*

Step down transformer. *A transformer having fewer turns on the secondary than on the primary.*

Step up transformer. *A transformer having more turns on the secondary than on the primary.*

Tight coupling. *Values of k very close to 1.0.*

Transformer. *A device capable of transferring energy from one coil to another by means of magnetic coupling.*

Turns ratio n. *The ratio of the number of turns on the secondary of a transformer to the number of turns on the primary. (Note: sometimes defined in the opposite way.)*

IMPORTANT RELATIONSHIPS

$$k = \frac{\phi_{12}}{\phi_1 + \phi_{12}} \qquad L_M = k\sqrt{L_P L_S} \qquad f_0 = \frac{1}{2\pi\sqrt{LC}}$$

- ideal transformer

$$\text{turns ratio} = n = \frac{N_{SEC}}{N_{PRI}} \qquad n = \frac{\mathbf{V}_{SEC}}{\mathbf{V}_{PRI}} \qquad n = \frac{\mathbf{I}_{PRI}}{\mathbf{I}_{SEC}} \qquad \mathbf{Z}_{PRI} = \frac{\mathbf{Z}_L}{n^2}$$

- loosely coupled transformers

tuned output | double tuned

$$\mathbf{V}_C = -Q_S\mathbf{V}_S \qquad k_C = \frac{1}{\sqrt{Q_SQ_P}} \qquad \frac{\mathbf{V}_C}{\mathbf{V}_P} = \sqrt{\frac{L_S}{L_P}} \times \frac{k}{\frac{1}{k^2} + \frac{1}{Q_SQ_P}}$$

REVIEW QUESTIONS

1. What is the significance of the parallel lines between the windings of a transformer symbol? (_____).
2. The dots on a transformer schematic deal with the voltage polarity. (TF)
3. The coefficient of coupling k has a maximum numerical value of _____.
4. Is it possible for L_M to be greater than $\sqrt{L_pL_s}$? (_____).
5. In Fig. 19.5, the (leakage, incremental, coupled) inductance corresponds to the flux ϕ_1 in Fig. 19.4.
6. Hysterysis and eddy current losses are represented by a _____ in Fig. 19.5.
7. Hysterysis and eddy current losses (are, are not) negligible in an air core transformer.
8. In an ideal transformer, k = _____.
9. A transformer with no losses is called an _____ transformer.
10. In a step-up transformer, the secondary voltage is (larger, smaller) than the primary voltage.
11. In a step-down transformer, the secondary voltage is (larger, smaller) than the primary voltage.
12. The secondary current in a step-up transformer is greater than the primary current. (TF)
13. The primary current in a step-down transformer is greater than the secondary current. (TF)
14. The power in the primary of an ideal transformer is (equal to, greater than, less than) the power in the secondary.
15. The load impedance connected to the secondary of a transformer is reflected into the primary (multiplied, divided) by the square of the turns ratio.
16. In order to match a 1-kΩ source to a 10-Ω load, one would use a (step-up, step-down) transformer.
17. The primary reason for matching impedances with a transformer is to obtain maximum power transfer. (TF)
18. A multiple tap transformer has a 4-Ω winding and an 8-Ω winding on its secondary. It would be proper to connect a $(4+8) = 12$-Ω load across the two output windings. (TF)
19. For an *ideal* iron core transformer, a capacitive load is reflected into the primary as (capacitive, inductive).
20. For a loosely coupled transformer, a capacitive load is reflected into primary as (capacitive, inductive).
21. At resonance, the voltage $\mathbf{V}_C$ in Fig. 19.15 is _____ times greater than $\mathbf{V}_s$.
22. For the double tuned transformer in Fig. 19.16, $\mathbf{V}_C/\mathbf{V}_p$ is equal to Q_s. (TF)

23. The magnitude of $\mathbf{V}_C$ at resonance is maximum for a double tuned transformer when $k =$ ____.
24. For the transformer in Fig. 19.16, the bandwidth of $\mathbf{V}_C$ versus frequency is (greater, less) than for a single tuned circuit.
25. For a double tuned transformer, double peaks occur in the output response when k is (greater, less) than k_c.
26. Equation 19.20 is valid only at resonance, f_0. (TF)

PROBLEMS

1. Draw the schematic symbol for an iron core transformer. An air core transformer.
2. Make a sketch showing V_P and V_S for Example 19.1. Assume V_S has a peak value of 20 V.
3. State the range of values that k may have (upper and lower limits).
4. Calculate the coefficient of coupling for the transformer in Fig. 19.4 if $\phi_{12} = 10$ mWb and $\phi_1 = 2$ mWb.
5. What is the value of L_M in Fig. 19.4 if $k = 0.2$, $L_P = 50\ \mu$H and $L_S = 100\ \mu$H?
6. For a certain transformer, the following are measured: $L_M = 1$ mH, $L_P = 3$ mH, $L_S = 10$ mH. What is the coefficient at coupling?
7. For an ideal iron core transformer having identical windings, $L_P = L_S$. Find the value of L_P if L_M is measured to be 2.7 mH.
8. What is the turns ratio of a transformer which has 120 turns on the primary and 3600 turns on the secondary?
9. A transformer commonly used in TV sets is shown in Fig. 19.19. Calculate the turns ratio for each winding.

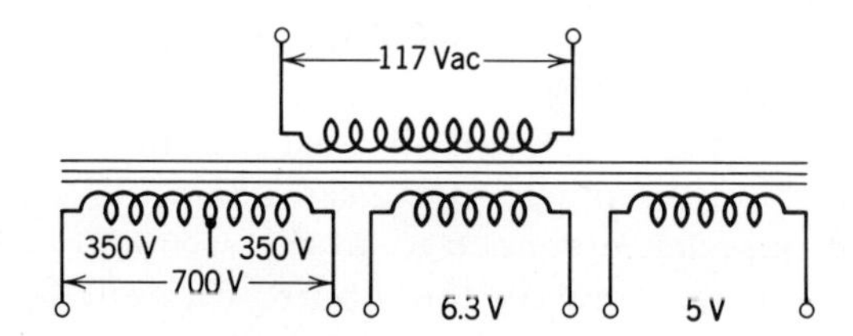

Figure 19.19 Typical TV power transformer

10. What is the turns ratio of a step-down transformer whose output voltage is 12.6 V with a primary voltage of 117 V?
11. If the transformer in Problem 10 has a primary current of 100 mA, what must the secondary current be?
12. Calculate the primary and secondary power for the transformer in Problems 10 and 11.
13. The transformer in Fig. 19.12 has a turns ratio of $n = 20$. Calculate the load voltage $\mathbf{V}_L$, the load current $\mathbf{I}_L$, the primary current $\mathbf{I}_{PRI}$, and the power delivered to the load if $\mathbf{V}_S = 8.5$ V and $R_L = 2000\ \Omega$.
14. Calculate the reflected load and draw the equivalent circuit for the transformer in Problem 13. Check your circuit by calculating the primary current and comparing it with that obtained in Problem 13.

15. Draw the equivalent circuit for the transformer in Fig. 19.12 if R_L is replaced with an impedance of $(6-j7)\Omega$ and turns ratio is 1/10.
16. What source impedance is required to match the load in Problem 15?
17. If the source resistance R_S in Fig. 19.15 is 750 Ω, what turns ratio must the transformer have in order to match a 30-Ω load?
18. What impedance can be matched between terminals A and C in Fig. 19.16 if *both* the output windings are 8-Ω windings?
19. What impedance can be matched between terminals A and C in Fig. 19.16 if both the output windings are 16-Ω windings?
20. For the transformer in Fig. 19.14, $\omega = 10^5$, $R_S = R_P = 0$, $L_P = 1$ mH, $L_S = 3$ mH, $k = 0.2$, and $\mathbf{Z}_L = 10\ \Omega$. Calculate $\mathbf{Z}_{REF}$ and $\mathbf{Z}_{in}$.
21. Calculate $\mathbf{Z}_{REF}$ and $\mathbf{Z}_{in}$ for the transformer in Example 19.12 if k is changed to 0.1.
22. Determine $\mathbf{I}_S$ for the transformer in Fig. 19.14*b* if $\omega = 10^4$, $L_P = 0.7$ mH, $L_S = 1$ mH, $L_M = 0.5$ mH, $\mathbf{Z}_L = 20\ \Omega$, $R_S = 3\ \Omega$, $R_P = 0$, and $\mathbf{V}_P = 20\underline{/0^\circ}$ V.
23. Repeat Example 19.13 if $R_P = 1\ \Omega$.
24. Repeat Problem 22 if both R_S and $R_P = 0$.
25. Repeat Example 19.15 if $R_S = 5\ \Omega$, and then 20 Ω.
26. In Example 19.15, the capacitor is changed to 0.1 μF. Find f_0 and $\mathbf{V}_C$.
27. What value of C is needed for an f_0 of 10 kHz in Example 19.15?
28. In Fig. 19.15*a*, $R_P = 2\ \Omega$, $L_P = L_S = 2$ mH, $k = 0.2$, $R_S = 8\ \Omega$, and $C = 15$ *n*F. Find $\mathbf{V}_C$ in terms of $\mathbf{V}_P$.
29. The circuit in Fig. 19.16 has $Q_P = 20$ and $Q_S = 10$. Calculate k for critical coupling.
30. Repeat Example 19.17 if $L_P = 1$ mH and $L_S = 4$ mH.
31. Calculate the value for $\mathbf{V}_C/\mathbf{V}_P$ in Fig. 19.16 if $k = 0.1$, $Q_P = 50$ and $Q_S = 5$. Assume $L_S = L_P$.
32. Repeat Problem 31 if k is adjusted to its critical value.
33. Show whether or not Example 19.18 is still valid if R is changed to 5 kΩ.
34. Repeat Example 19.18 with $R = 5$ kΩ.
35. Derive Eq. 19.20. Use $\mathbf{V}_C = \mathbf{I}_S(1/j\omega_0 C_S) = \mathbf{I}_S(j\omega_0 L_S)$.
36. In Example 19.18, Q_S is changed from 10 to 50, and at the same time k is varied to maintain critical coupling. What is the new value for k_C, and what is $\mathbf{V}_C/\mathbf{V}_P$?

20

THREE-PHASE CIRCUITS

Our study of ac circuits up to this point has been largely confined to circuits containing a single ac voltage source. Such circuits or systems are referred to as *single-phase* ac. Circuits or systems containing more than one ac voltage source of the same frequency in which there is a phase difference between the source voltages are called *polyphase* ac. Such systems are especially important because virtually all power is generated and distributed as polyphase power at a frequency of 50 Hz or 60 Hz.

One of the most common polyphase systems, the *three-phase* system, was discussed briefly in Chapter 17. The three-phase system can be considered as a combination of three ac generators, each of which produces a sinusoidal voltage. The three source voltages are equal in magnitude, but the generators are arranged in such a way that there is a phase angle of 120° between any two of the voltages.

A distinct advantage of such a system is the fact that the power delivered to a *balanced load* (for example, the three identical windings of a three-phase motor) is constant. You will recall that the power delivered to a load by a single-phase source "pulsates" with time. Since the power delivered is constant, the torque developed in rotating machines is more nearly constant. As a result, there is less vibration associated with three-phase machines. There are also important economical advantages associated with a three-phase power distribution system. For example, the total quantity of wire necessary for a three-phase power system is almost always less than that required for an equivalent single-phase system.

In nearly all practical situations, the generators in a three-phase power system can be approximated as *ideal* voltage sources. In this chapter, we will study three-phase power systems by considering the three interconnected voltage sources and the common methods of connecting loads to them.

THREE-PHASE GENERATORS

The generation of three-phase voltages can be accomplished in exactly the same manner as that for single-phase voltages as discussed in Chapter 14. To generate three voltages, it is only necessary to use three windings on the rotor as shown in Fig. 20.1*a*. If the rotor revolves at a constant speed, there will be a sinusoidal voltage induced across each of the three sets of windings aa′, bb′, and cc′. Since the coils are mechanically fixed with respect to one another, the

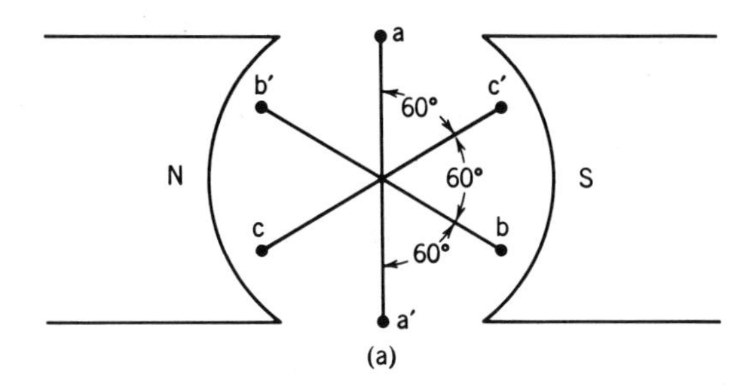

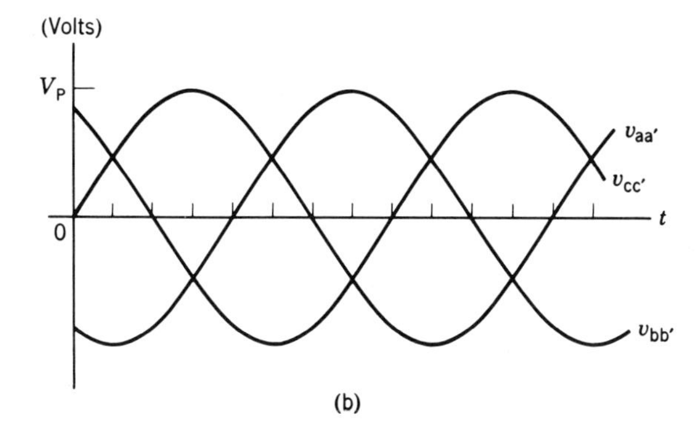

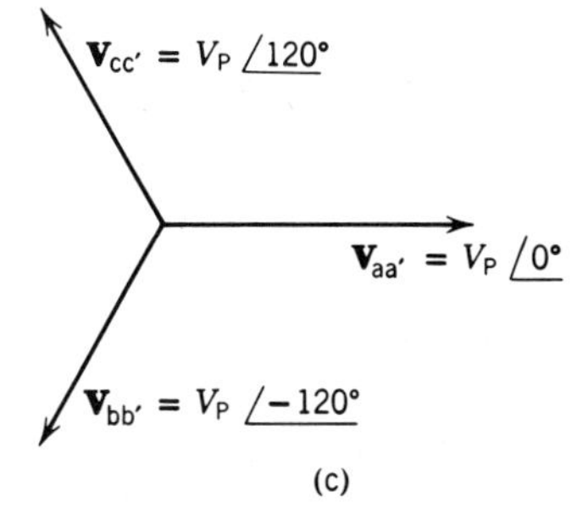

Figure 20.1 Three-phase voltage generation (*a*) Three-phase generation (*b*) Instantaneous voltages (*c*) Phasor voltages

induced voltages will be of exactly the same frequency. Furthermore, if the coils are constructed identically, the induced voltages will also have the same magnitudes. Thus the only difference between the three voltages will be their *phases*; that is, each induced voltage will reach its peak value at a different instant in time.

Notice that each coil has two terminals, a and a′, b and b′, and c and c′. The *unprimed* terminals are called the "starts," and the *primed* terminals are called the "finishes." Three-phase voltages and currents are more easily described using a system called *double-subscript notation.* This is simply a method for precisely designating a voltage or current. Thus, by definition, a voltage V_{ab} is the voltage at point a *with respect to* point b. In Fig. 20.1*a*, the voltage across coil aa′ can be written as $V_{aa'}$. This means the voltage measured at point a with respect to point a′. Obviously, $V_{aa'} = -V_{a'a}$. The current through coil aa′ can be written as $I_{aa'}$, and this means the current is in a direction from terminal a to terminal a′. Again it is clear that $I_{aa'} = -I_{a'a}$.

Now, using double subscript notation, let us investigate the three voltages produced by the generator in Fig. 20.1*a*. Assume the rotor is rotating in a *counterclockwise* direction at a constant speed. Clearly there will be a sinusoidal voltage induced across each of the three coil terminals. If we take the rotor in the position shown in Fig. 20.1*a* at time $t = 0$, the voltage $v_{aa'}$ will be zero at $t = 0$. The voltage $v_{bb'}$ will be approaching its negative peak after just passing zero; the voltage $v_{cc'}$ will be approaching zero after just passing its positive peak. The three voltage wave forms will then appear as shown in Fig. 20.1*b*. From the rotor coil positions, and from the three voltage wave forms we can see that $v_{bb'}$ lags $v_{aa'}$ by 120°, and $v_{cc'}$ lags $v_{bb'}$ by 120°. The equations for the three voltages can then be written as

$$v_{aa'} = V_P \sin \omega t \tag{20.1}$$

$$v_{bb'} = V_P \sin (\omega t - 120°) \tag{20.2}$$

$$v_{cc'} = V_P \sin (\omega t - 240°) \tag{20.3}$$

where V_P is the peak value of each voltage.

Since the frequency is the same for each voltage ($\omega = 2\pi f$), we can treat the three voltages as phasors, and the proper equations are

$$\mathbf{V}_{aa'} = \frac{V_P}{\sqrt{2}}\underline{/0°} = V\underline{/0°} \tag{20.4}$$

$$\mathbf{V}_{bb'} = \frac{V_P}{\sqrt{2}}\underline{/-120°} = V\underline{/-120°} \tag{20.5}$$

$$\mathbf{V}_{cc'} = \frac{V_P}{\sqrt{2}}\underline{/-240°} = V\underline{/120°} \tag{20.6}$$

where V is the RMS value of V_P.

Notice that a phase angle of $-240°$ is exactly the same as $+120°$. The phasor diagram for these three voltages will thus appear as shown in Fig. 20.1*c*. It is interesting to note that the sum of the three coil voltages at any instant in time is zero, since

$$\begin{aligned}\mathbf{V}_{aa'} + \mathbf{V}_{bb'} + \mathbf{V}_{cc'} &= V\underline{/0°} + V\underline{/-120°} + V\underline{/120°}\\ &= V(1.0 - 0.50 - j0.87 - 0.50 + j0.87)\\ &= V(0 + j0) = 0\end{aligned}$$

When the rotor in Fig. 20.1*a* revolves in a counterclockwise direction, the three voltages reach their positive peaks in the order $\mathbf{V}_{aa'}$, $\mathbf{V}_{bb'}$, $\mathbf{V}_{cc'}$. The generator is then said to have a *phase sequence* of abc. If the rotor were turned in the opposite direction, the coil voltages would reach their positive peaks in the order $\mathbf{V}_{aa'}$, $\mathbf{V}_{cc'}$, $\mathbf{V}_{bb'}$. Under this condition, the phase sequence is said to be acb, or cba. If the three coil voltages are used independently of one another, the phase sequence is of little importance. However, if the coils are interconnected to form a three-phase generator supplying a three-phase load, the phase sequence becomes of great importance. For example, phase sequence determines the direction of rotation of a motor.

Each of the three coils in the generator in Fig. 20.1*a* can be used as a single-phase ac source. But in practice, the three coils are usually connected together to form a three-phase ac source. One method of forming a three-phase system is to simply connect the three *finishes* of the coils together. That is, connect a′, b′, and c′ together to form a common node designated as n. The equivalent circuit of such a three-phase generator is shown in Fig. 20.2*a* and it is known as the Y-connection (sometimes called the star-connection). Since the three coil voltages are all measured with respect to their *primed* terminals, the node n becomes the common termination and it is called the *neutral.* The three voltages can now be referred to as *phase* voltages and they are designated as $\mathbf{V}_{an}$, $\mathbf{V}_{bn}$, and $\mathbf{V}_{cn}$.

There are three other important voltages produced by the generator in Fig. 20.2*a*. They are measured between the three terminals a, b, and c and they are called the *line-to-line* or simply the *line* voltages, $\mathbf{V}_{ab}$, $\mathbf{V}_{bc}$, and $\mathbf{V}_{ca}$. The line voltages can be found in terms of the phase voltages by making use of the phasor diagram in Fig. 20.2*b*. The line voltage $\mathbf{V}_{ab}$ is clearly the difference between the two phase voltages $\mathbf{V}_{an}$ and $\mathbf{V}_{bn}$. That is

$$\mathbf{V}_{ab} = \mathbf{V}_{an} - \mathbf{V}_{bn}$$

If we take phase voltage $\mathbf{V}_{an}$ as the reference, then

$$\mathbf{V}_{an} = V\underline{/0^\circ} \qquad \mathbf{V}_{bn} = V\underline{/-120^\circ}$$

$$= V(1 + j0) \qquad = V(-0.50 - j0.87)$$

Thus,

$$\mathbf{V}_{ab} = \mathbf{V}_{an} - \mathbf{V}_{bn} = V(1 + j0) - V(-0.50 - j0.87)$$

$$= V(1.5 + j0.87) = 1.73\,V\underline{/30^\circ}.$$

Similarly, it can be shown that

$$\mathbf{V}_{bc} = \mathbf{V}_{bn} - \mathbf{V}_{cn} = 1.73\,V\underline{/-90^\circ}$$

$$\mathbf{V}_{ca} = \mathbf{V}_{cn} - \mathbf{V}_{an} = 1.73\,V\underline{/-210^\circ}$$

These three line voltages are shown along with the three-phase voltages in Fig. 20.2*b*. Notice that $1.73 = \sqrt{3}$, and thus the relationship for the magnitude of the line voltage can be given in terms of the phase voltage as

$$V_{\text{line}} = \sqrt{3}\ V_{\text{phase}} \tag{20.7}$$

Thus the line voltages have phase angles of 120° between them and they are greater than the phase voltages by a factor of $\sqrt{3}$.

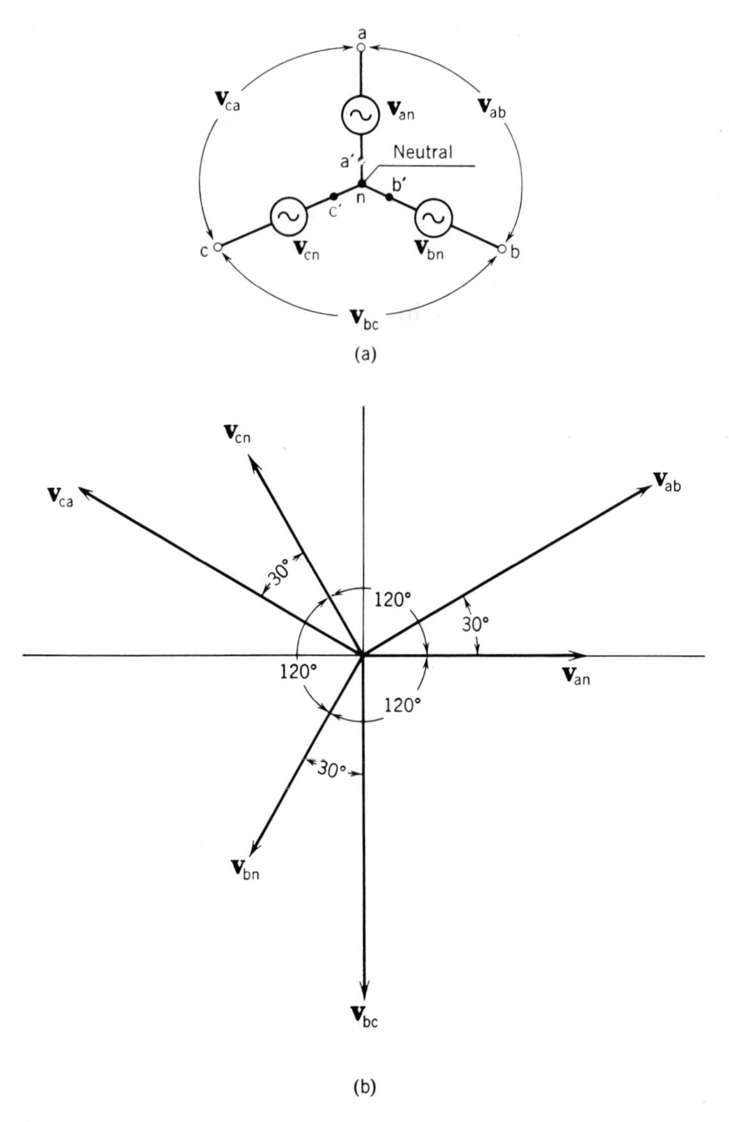

Figure 20.2 Three-phase Y-connected generator (*a*) Equivalent circuit (*b*) Phase and line voltages

• • •

Example 20.1 A three-phase generator is connected as a Y as shown in Fig. 20.2*a*. The RMS voltage of each phase is 120 V. Write the equations describing the three phase voltages. What is the magnitude of the line voltages?

Solution The three phase voltages are given by Eqs. 20.4 through 20.6. Thus

$$\mathbf{V}_{an} = 120\underline{/0^\circ} \qquad \mathbf{V}_{bn} = 120\underline{/-120^\circ} \qquad \mathbf{V}_{cn} = 120\underline{/120^\circ}$$

The line voltage is found from Eq. 20.7 as

$$V_{\text{line}} = \sqrt{3}\, V_{\text{phase}} = \sqrt{3} \times 120 = 208 \text{ V RMS}$$

• • •

A different three-phase system can be constructed by connecting the coils of the generator in Fig. 20.1*a* as shown in Fig. 20.3. In this case, the *start* of one winding is simply connected to the *finish* of another winding. This is called a Δ-connection (mesh-connection) and it will operate satisfactorily so long as the phase voltages have exactly equal magnitudes. Under this condition, the sum of the voltage sources is zero as was previously shown and there will be no circulating current around the Δ-mesh formed by the coils. This system, however, has the disadvantage of a large circulating current around the coils in the event of even a slight unbalance of the phase voltages. At the very least, this will reduce power output capacity and lower the system efficiency. A second disadvantage is the lack of a connection from the system to ground. Such a connection is desirable for safety and system protection (the neutral point in the Y-connection is ordinarily used as ground). Because of these disadvantages the generator itself is seldom connected in a Δ. It is important to note that in a Δ-connected source the phase and line voltages are exactly the same. An inspection of Fig. 20.3 makes this quite clear.

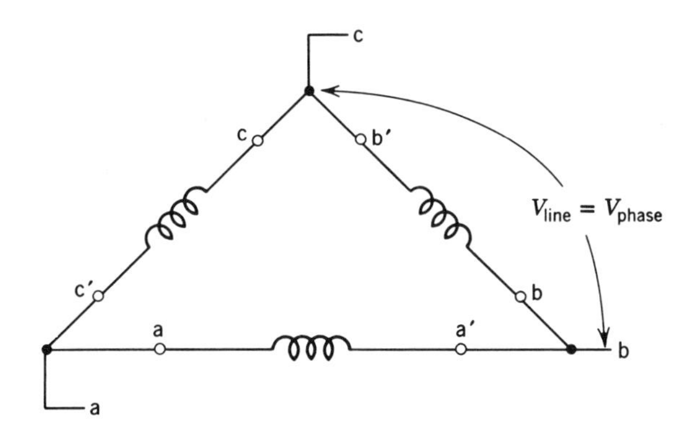

Figure 20.3 Three-phase Δ-connected generator

2 THREE-PHASE POWER DISTRIBUTION

Nearly all three-phase power systems use transformers in their transmission and distribution systems. The generator voltages of perhaps 10 kV or 20 kV are *stepped up* by means of transformers to perhaps 110 kV or 220 kV for transmission. At the receiving end of the transmission line, transformers are used to *step down* the high voltages to a few thousand volts for local distribution. Finally, transformers are again used to *step down* these voltages in order to supply customers with the usual 240 V and 120 V.

The transformers used can be a *bank* of three single-phase transformers as shown in Fig. 20.4*a*. Three-phase transformers in which the windings have a common core are also used to advantage. The primary and secondary windings can then be connected to form either a Y or a Δ system. Typically, a Y-connected generator is connected to a step-up transformer with its primary in a Δ-connection and its secondary in a Y-connection as shown in Fig. 20.4*b*. The Y-connected secondary is then used to feed a high voltage transmission line system. At the receiving end of the transmission line, the high voltage is stepped down to a few thousand volts for local distribution. Thus the final

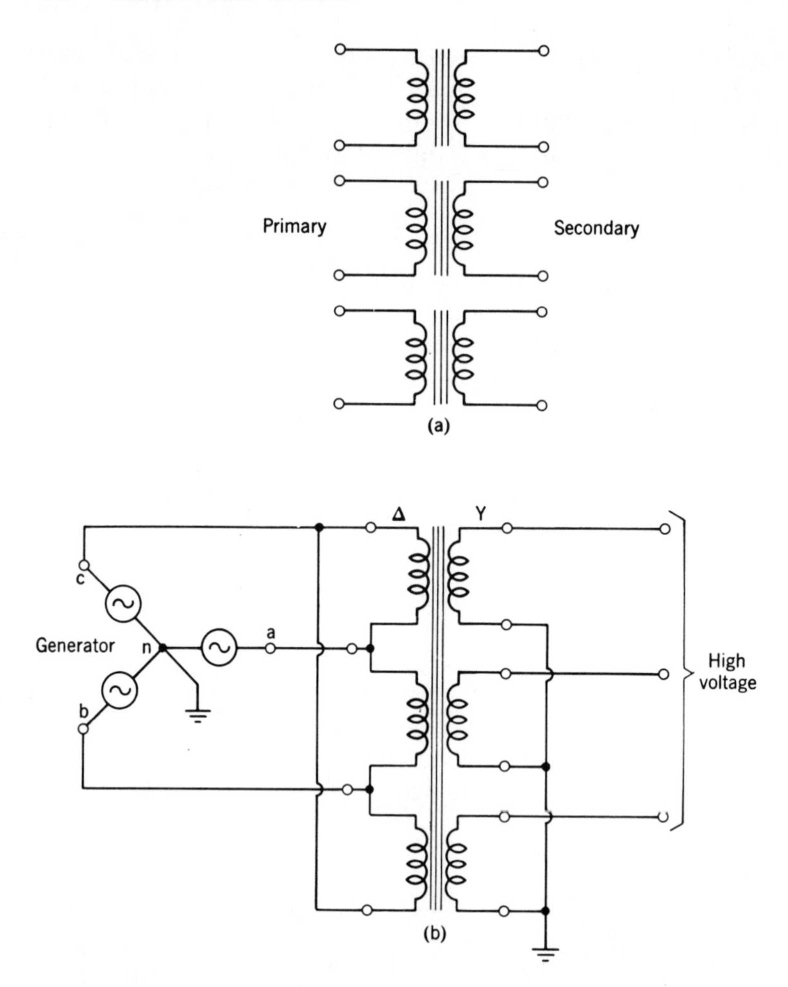

Figure 20.4 A three-phase step up transformer for high voltage transmission (*a*) Three-phase transformer (*b*) Transformer connections

transformers in the system may have a few thousand volts on their primary windings, and they will be used to step down the voltage to 240 V and 120 V.

Three different methods for connecting the final distribution transformers are shown in Fig. 20.5. In Fig. 20.5*a*, the transformer primary is Y-connected and the secondary is Δ-connected. The voltages available to the consumer are then 240 V line-to-line single-phase, as well as 240 V line-to-line three-phase. The secondary windings of the transformer may also be *center-tapped* to provide 120 V between the center tap and either line.

A second method is shown in Fig. 20.5*b*, where the primary is Δ-connected and the secondary is Y-connected. In this system, the line-to-line voltages are 208 V and the phase voltages are 120 V. This system also makes available 208 V single-phase and 208 V three-phase voltages. The letters a, b, and c show the phase sequence.

The two connections described above are widely used in industrial applications requiring three-phase power, but in the average residence the three-phases are not usually needed. Thus the three transformer secondary windings can be used individually to supply 240 V across the outside lines and

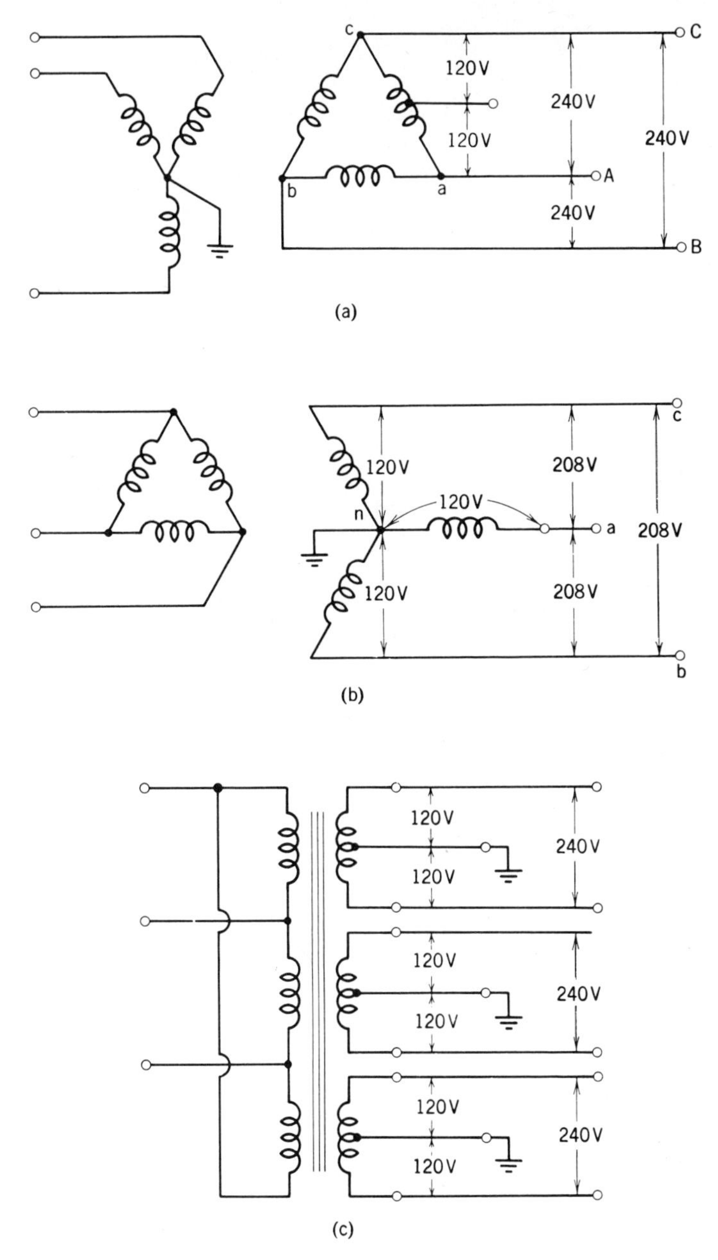

Figure 20.5 Local power distribution systems (*a*) Y–Δ transformer connection (*b*) Δ-Y transformer connection (*c*) Three single-phase circuits

120 V between the center tap and either outside line as shown in Fig. 20.5*c*. The center tap is usually considered neutral, and acts as a ground. The two 120 V sources serve the lighting and small appliance requirements, while the 240 V source is applied to ovens, ranges, clothes dryers, and so forth.

• • •

Example 20.2 A 208 V three-phase motor and a 120 V lamp are shown in Fig. 20.6. Describe the proper connections of these loads to the Δ-Y transformer in Fig. 20.5*b*.

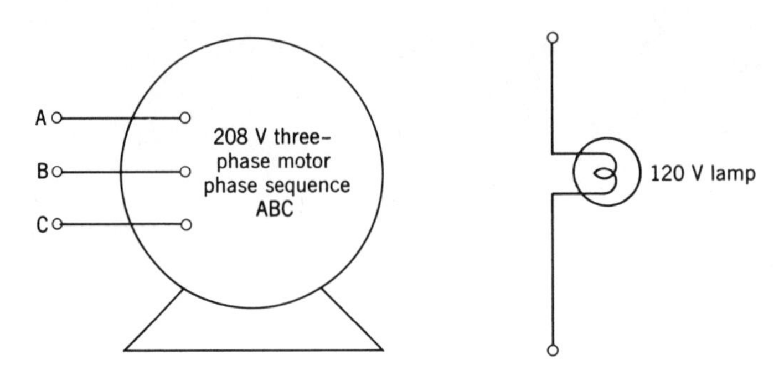

Figure 20.6

Solution The motor must be connected to the three terminals of the transformer labeled a, b, and c. For proper phase sequence, connect a-A, b-B, and c-C. The lamp can be connected between the neutral (ground) and any one of the three phase voltages (a, b, or c).

• • •

20.3 BALANCED THREE-PHASE Y-LOADS

As far as the consumer is concerned, the three transformer configurations shown in Fig. 20.5 (or other configurations) serve as the available sources of ac power. The transformers themselves of course have some internal impedances, but they are usually small compared to the required loads. We will therefore assume the transformers are *ideal* and we will simply treat them as ideal voltage sources.

Consider now the connection of a *balanced* Y-connected three-phase load to a Y-connected source as shown in Fig. 20.7*a*. The load is said to be balanced since the three phase impedances $\mathbf{Z}_P$ are equal. If phase an is taken as the reference, the phase voltages are

$$\mathbf{V}_{an} = V\underline{/0^\circ} \qquad \mathbf{V}_{bn} = V\underline{/-120^\circ} \qquad \mathbf{V}_{cn} = V\underline{/120^\circ}$$

Each of the loads is connected between a line and neutral; thus this is exactly the same as three single-phase circuits. The line currents are then easily found as

$$\mathbf{I}_{aA} = \frac{\mathbf{V}_{an}}{\mathbf{Z}_P} \qquad \mathbf{I}_{bB} = \frac{\mathbf{V}_{bn}}{\mathbf{Z}_P} = \mathbf{I}_{aA}\underline{/-120^\circ} \qquad \mathbf{I}_{cC} = \frac{\mathbf{V}_{cn}}{\mathbf{Z}_P} = \mathbf{I}_{aA}\underline{/120^\circ}$$

since $V_{an} = V_{bn} = V_{cn}$.

The three line currents therefore have equal magnitudes but are separated by a 120° phase angle just like the phase voltages. As a result, the vector sum of the three currents is zero as shown in Fig. 20.7*b*. From KCL, the neutral current $\mathbf{I}_n$ is equal to the sum of the three line currents. That is

$$\mathbf{I}_n = \mathbf{I}_{aA} + \mathbf{I}_{bB} + \mathbf{I}_{cC} = 0$$

Therefore, in a *balanced* Y-load, the neutral wire can be eliminated. Notice in this case that the *line* currents are exactly the same as the *phase* currents.

The power delivered to the Y-load in Fig. 20.7*a* is simply the sum of the powers delivered to each phase. Since the phase voltages and currents all have

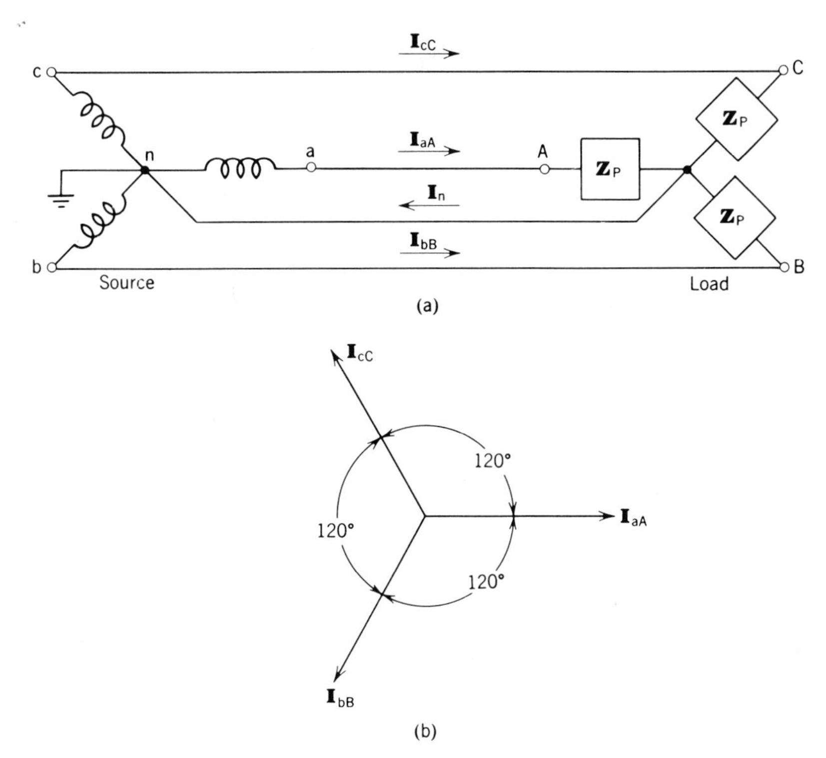

Figure 20.7 A Y-connected generator and a balanced Y-connected load (*a*) Source and load connections (*b*) Line (phase) currents

the same magnitudes, the power in each phase is given as

$$P_{\text{phase}} = V_{\text{phase}} I_{\text{phase}} \cos \theta \tag{20.8}$$

where θ is the power factor angle, or the angle between V_{phase} and I_{phase} (it is also the angle of $\mathbf{Z}_P$). The total power delivered to the Y-load is therefore

$$P = 3P_{\text{phase}} = 3V_{\text{phase}} I_{\text{phase}} \cos \theta \tag{20.9}$$

• • •

Example 20.3 Calculate the line currents, the power delivered to each phase, the total power delivered to the load, and draw the phasor diagram for the load in Fig. 20.7*a* if $\mathbf{Z}_P = 15\underline{/45°}\ \Omega$ and $V_{\text{line}} = 208$ V.

Solution It is easier to solve the problem for each phase. The phase voltages are (taking $\mathbf{V}_{an}$ as the reference)

$$V_{\text{phase}} = \frac{V_{\text{line}}}{\sqrt{3}} = \frac{208}{\sqrt{3}} = 120 \text{ V}$$

$$\mathbf{V}_{an} = 120\underline{/0°} \qquad \mathbf{V}_{bn} = 120\underline{/-120°} \qquad \mathbf{V}_{cn} = 120\underline{/120°}$$

The three line (or phase) currents are then

$$\mathbf{I}_{aA} = \frac{\mathbf{V}_{an}}{\mathbf{Z}_P} = \frac{120\underline{/0°}}{15\underline{/45°}} = 8.0\underline{/-45°} \text{ A}$$

$$\mathbf{I}_{bB} = \frac{\mathbf{V}_{bn}}{\mathbf{Z}_P} = \frac{120\underline{/-120^\circ}}{15\underline{/45^\circ}} = 8.0\underline{/-165^\circ}\text{ A}$$

$$\mathbf{I}_{cC} = \frac{\mathbf{V}_{cn}}{\mathbf{Z}_P} = \frac{120\underline{/120^\circ}}{15\underline{/45^\circ}} = 8.0\underline{/75^\circ}\text{ A}$$

The phasor diagram is shown in Fig. 20.8. The power delivered to each phase is

$$P_{\text{phase}} = V_{\text{phase}} I_{\text{phase}} \cos\theta = 120 \times 8.0 \times \cos 45^\circ$$
$$= 120 \times 8.0 \times 0.707 = 680\text{ W}$$

The total power is then

$$P = 3P_{\text{phase}} = 3 \times 680 = 2040\text{ W} = 2.04\text{ kW}$$

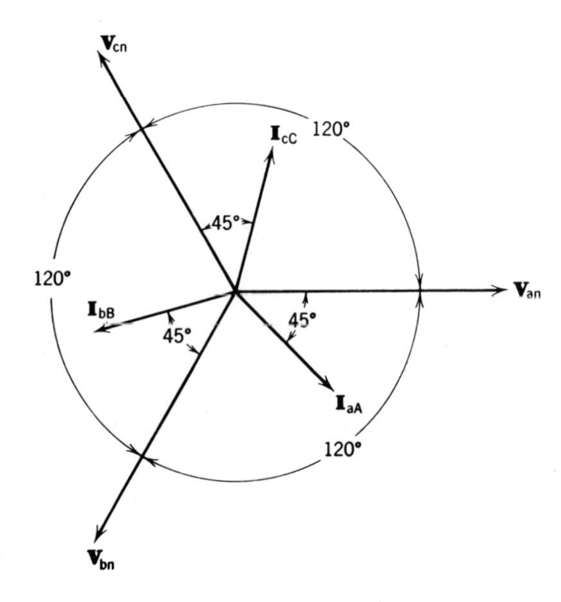

Figure 20.8

• • •

A balanced Y-load can also be connected to a Δ-connected generator as shown in Fig. 20.9. The calculation of line currents and power will be exactly

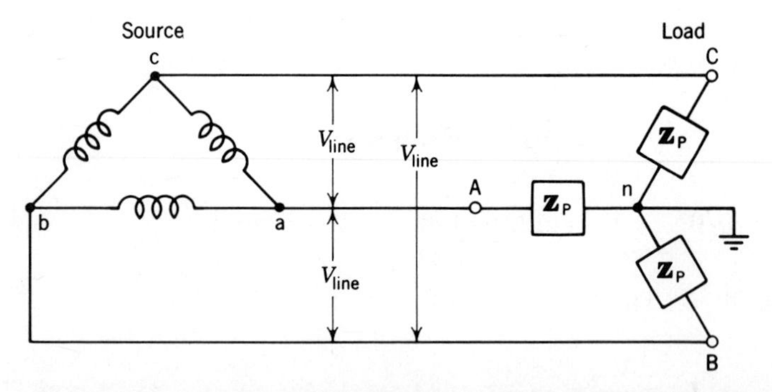

Figure 20.9 A Δ-connected generator and a balanced Y-load

the same as for the Y-connected load. The only additional step is to determine the phase voltage from the line voltage using Eq. 20.7.

• • •

Example 20.4 Calculate the line currents and the power delivered to the load in Fig. 20.9 if $\mathbf{Z}_P = 21\underline{/-15^\circ}\ \Omega$, and $V_{line} = 240$ V.

Solution The phase voltages for the load are found using Eq. 20.7.

$$V_{phase} = \frac{V_{line}}{\sqrt{3}} = \frac{240}{\sqrt{3}} = 138 \text{ V}$$

Taking phase an as the reference, the line currents are

$$\mathbf{I}_{aA} = \frac{138\underline{/0^\circ}}{21\underline{/-15^\circ}} = 6.6\underline{/15^\circ} \text{ A}$$

$$\mathbf{I}_{bB} = \frac{138\underline{/-120^\circ}}{21\underline{/-15^\circ}} = 6.6\underline{/-105^\circ} \text{ A}$$

$$\mathbf{I}_{cC} = \frac{138\underline{/120^\circ}}{21\underline{/-15^\circ}} = 6.6\underline{/135^\circ} \text{ A}$$

The power delivered to each phase is

$$P_{phase} = V_{phase} I_{phase} \cos\theta = 138 \times 6.6 \times \cos 15^\circ$$
$$= 138 \times 6.6 \times 0.965 = 880 \text{ W}$$

The total power delivered to the load is then

$$P = 3P_{phase} = 3 \times 880 = 2640 \text{ W} = 2.64 \text{ kW}$$

• • •

4 BALANCED THREE-PHASE Δ-LOADS

Three-phase loads can also be connected in a Δ-configuration, and this is sometimes an advantage since it is relatively easy to alter the loads in any one phase of the Δ-load. In Fig. 20.10*a*, a balanced Δ-load is connected to a Δ-source. The voltage across each phase of the load is simply the line voltage, and thus the phase currents are found as ($\mathbf{V}_{ab}$ is taken as the reference)

$$\mathbf{I}_{AB} = \frac{\mathbf{V}_{ab}}{\mathbf{Z}_P} = \frac{V}{\mathbf{Z}_P}\underline{/0^\circ}$$

$$\mathbf{I}_{BC} = \frac{\mathbf{V}_{bc}}{\mathbf{Z}_P} = \frac{V}{\mathbf{Z}_P}\underline{/-120^\circ} \qquad \mathbf{I}_{CA} = \frac{\mathbf{V}_{ca}}{\mathbf{Z}_P} = \frac{V}{\mathbf{Z}_P}\underline{/120^\circ}$$

The three-phase currents therefore have equal magnitudes and are separated by 120° phase angles as shown in Fig. 20.10*b*. However, in the case of the Δ-load the line currents are *not* equal to the phase currents. Applying KCL to node A in Fig. 20.10*a*, we obtain

$$\mathbf{I}_{aA} = \mathbf{I}_{AB} - \mathbf{I}_{CA}$$

Let us assume for the moment that $\mathbf{Z}_P$ is resistive (has no phase angle); the line

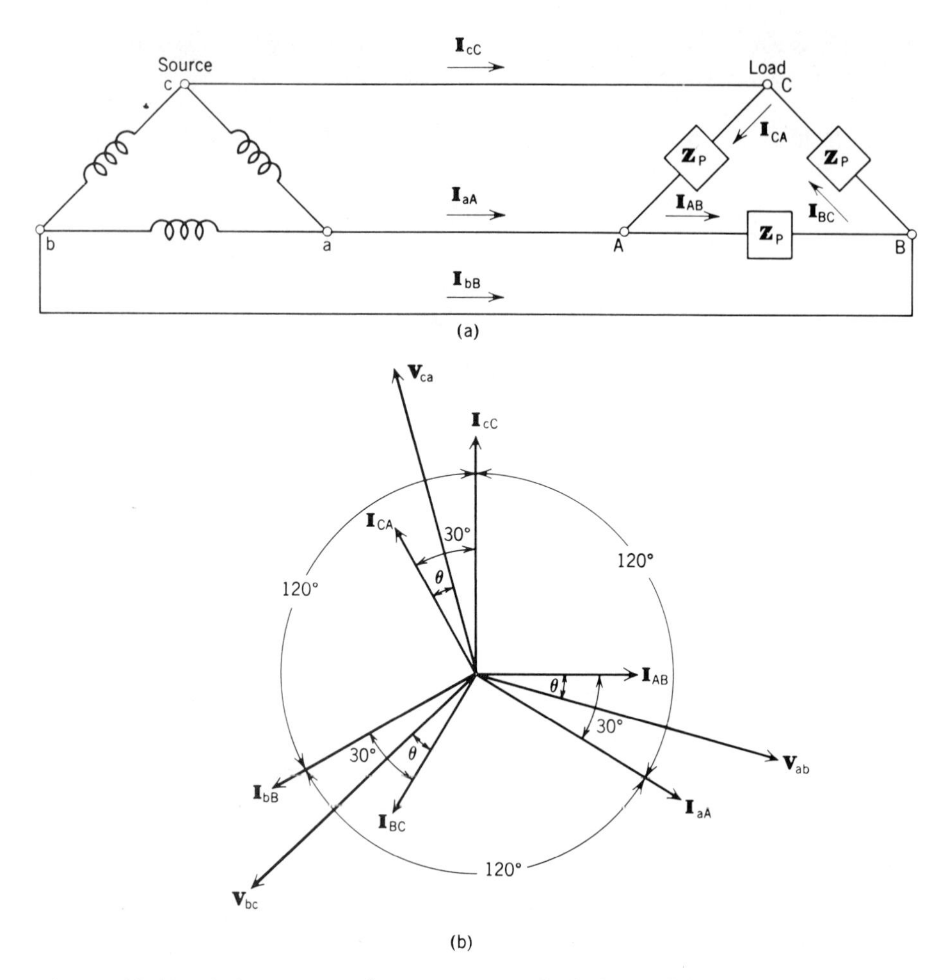

Figure 20.10 A Δ-connected generator and a balanced Δ-load (*a*) Source and load connections (*b*) Δ-load phasor diagram

current $\mathbf{I}_{aA}$ can then be determined as

$$\begin{aligned}\mathbf{I}_{aA} = \mathbf{I}_{AB} - \mathbf{I}_{CA} &= I_{\text{phase}}\underline{/0^\circ} - I_{\text{phase}}\underline{/120^\circ} \\ &= I_{\text{phase}}(1 + j0) - I_{\text{phase}}(-0.5 + j0.87) \\ &= I_{\text{phase}}(1.5 - j0.87) = 1.73 I_{\text{phase}}\underline{/-30^\circ} \\ &= \sqrt{3}\, I_{\text{phase}}\underline{/-30^\circ}\end{aligned}$$

Similarly, the other line currents can be found as

$$\mathbf{I}_{bB} = \sqrt{3}\, I_{\text{phase}}\underline{/-150^\circ} \qquad \mathbf{I}_{cC} = \sqrt{3}\, I_{\text{phase}}\underline{/90^\circ}$$

Thus the three line currents are seen to lag the phase currents by a 30° phase angle. Furthermore, the line currents all have the same magnitudes and they are $\sqrt{3}$ times larger than the phase currents. That is,

$$I_{\text{line}} = \sqrt{3}\, I_{\text{phase}} \tag{20.10}$$

The phasor diagram for the Δ-load is given in Fig. 20.10*b*. If a load $\mathbf{Z}_P$ having some phase angle θ is introduced in each phase, the current phasors will simply be shifted through an angle θ with respect to the voltage phasors; however, the 120° phase angles between the line currents (and the phase currents) will remain fixed.

The power delivered to the Δ-load in Fig. 20.10 is found in the same manner as for a Y-load. Notice that the line voltage is exactly the same as the phase voltage, and thus the power in each phase is simply

$$P_{\text{phase}} = V_{\text{phase}} I_{\text{phase}} \cos\theta \tag{20.11}$$

which is exactly the same as Eq. 20.9. The total power delivered to the load is then

$$P = 3\,P_{\text{phase}} = 3\,V_{\text{phase}} I_{\text{phase}} \cos\theta \tag{20.12}$$

• • •

Example 20.5 Calculate the phase currents, line currents, power, and draw the phasor diagram for the Δ-load in Fig. 20.10*a* if $\mathbf{Z}_P = 17\underline{/60^\circ}\ \Omega$ and $V_{\text{phase}} = 240$ V.

Solution The phase currents are found to be

$$\mathbf{I}_{AB} = \frac{\mathbf{V}_{ab}}{\mathbf{Z}_P} = \frac{240\underline{/0^\circ}}{17\underline{/60^\circ}} = 14.1\underline{/-60^\circ}\text{ A}$$

$$\mathbf{I}_{BC} = \frac{\mathbf{V}_{bc}}{\mathbf{Z}_P} = \frac{240\underline{/-120^\circ}}{17\underline{/60^\circ}} = 14.1\underline{/-180^\circ}\text{ A}$$

$$\mathbf{I}_{CA} = \frac{\mathbf{V}_{ca}}{\mathbf{Z}_P} = \frac{240\underline{/120^\circ}}{17\underline{/60^\circ}} = 14.1\underline{/60^\circ}\text{ A}$$

The line currents all have equal magnitudes, and from Eq. 20.10 they are

$$I_{\text{line}} = \sqrt{3}\,I_{\text{phase}} = 1.73 \times 14.1 = 24.5\text{ A}$$

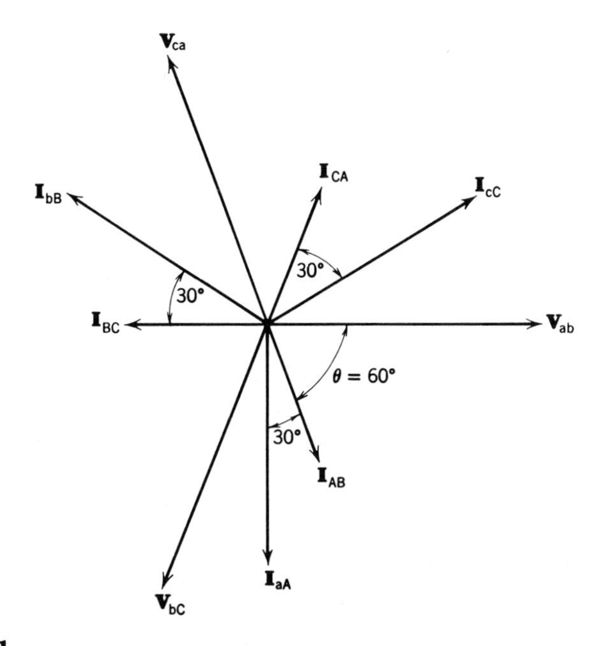

Figure 20.11

The line currents lag the phase currents by 30° and they are

$$\mathbf{I}_{aA} = 24.5\underline{/-90°}\text{ A};\ \mathbf{I}_{bB} = 24.5\underline{/-210°}\text{ A};\ \mathbf{I}_{cC} = 24.5\underline{/30°}\text{ A}$$

The phasor diagram is given in Fig. 20.11. The power in each phase is

$$\begin{aligned} P_{\text{phase}} &= V_{\text{phase}} I_{\text{phase}} \cos\theta = 240 \times 14.1 \cos 60° \\ &= 240 \times 14.1 \times 0.5 = 1690 \text{ W} = 1.69 \text{ kW} \end{aligned}$$

The total power is then

$$P = 3\,P_{\text{phase}} = 3 \times 1.69 \text{ kW} = 5.08 \text{ kW}$$

• • •

A balanced Δ-load can also be connected to a Y-source as shown in Fig. 20.5*b*. The calculations for line currents and power will be exactly the same as in the previous example. It is of course necessary to use the line source voltage as the phase voltage for the Δ-load.

We have seen that the power delivered to a Y-load or a Δ-load is found from exactly the same relationship, Eq. 20.9 and Eq. 20.12. These two expressions are in terms of phase voltages and currents. It is usually easier to measure line voltages and currents, and thus we would like an expression for power in terms of line quantities. For the Y-load, $V_{\text{phase}} = V_{\text{line}}/\sqrt{3}$, $I_{\text{line}} = I_{\text{phase}}$, and thus

$$\begin{aligned} P &= 3V_{\text{phase}} I_{\text{phase}} \cos\theta = 3\frac{V_{\text{line}}}{\sqrt{3}} I_{\text{line}} \cos\theta \\ &= \sqrt{3}\, V_{\text{line}} I_{\text{line}} \cos\theta \end{aligned}$$

Now, for the Δ-load, $V_{\text{phase}} = V_{\text{line}}$, $I_{\text{phase}} = I_{\text{line}}/\sqrt{3}$, and thus

$$\begin{aligned} P &= 3V_{\text{phase}} I_{\text{phase}} \cos\theta = 3V_{\text{line}} \frac{I_{\text{line}}}{\sqrt{3}} \cos\theta \\ &= \sqrt{3}\, V_{\text{line}} I_{\text{line}} \cos\theta \end{aligned}$$

Therefore, for either a Y or Δ load, the total power is given as

$$P = \sqrt{3}\, V_{\text{line}} I_{\text{line}} \cos\theta \qquad (20.13)$$

• • •

Example 20.6 A three-phase, 5-hp motor with a power factor of 0.9 lagging is connected to a three-phase source having line voltages of 240 V. Calculate the line currents required.

Solution Since 1 hp = 746 W, the 5 hp motor requires a total power of

$$P = 5 \times 746 = 3730 \text{ W}$$

The line currents are then found from Eq. 20.13 as

$$I_{\text{line}} = \frac{P}{\sqrt{3}\, V_{\text{line}} \cos\theta} = \frac{3730}{1.73 \times 240 \times 0.9} = 9.97 \text{ A}$$

• • •

At this point a summary of the line and phase voltages and currents for both *Y* and Δ loads might be helpful. Table 20.1 gives such a summary. In the power equations, note carefully that θ is the angle between phase voltage and phase current, or it is the power factor of the load; *it is not* the angle between line voltage and line current as can be seen in Fig. 20.11.

Table 20.1.

Y	Δ
$V_{\text{line}} = \sqrt{3}\, V_{\text{phase}}$	$V_{\text{line}} = V_{\text{phase}}$
$I_{\text{line}} = I_{\text{phase}}$	$I_{\text{line}} = \sqrt{3}\, I_{\text{phase}}$

$$P = 3\, V_{\text{phase}} I_{\text{phase}} \cos\theta = \sqrt{3}\, V_{\text{line}} I_{\text{line}} \cos\theta$$

20.5 UNBALANCED THREE-PHASE LOADS

Let us now consider the case of a three-phase Y-load in which the impedances of the three phases are not equal. Such a load is said to be *unbalanced.* An unbalanced load can be considered on a *phase-at-a-time* basis provided the neutral wire is present, and it can be assumed that the impedance of the neutral wire is zero. If the neutral wire is not present, the problem is considerably more complicated. We will assume the presence of a neutral wire in the following discussion.

An unbalanced three-phase Y-load is shown in Fig. 20.12. The individual phase currents are found by simply dividing each phase impedance into the phase voltage. The phase currents are of course equal to the line currents. The neutral current is then found as the sum of the individual phase currents. The power in each phase is found according to Eq. 20.8 and the total power is the sum of the power delivered to each phase. Note especially that the total power is *not* simply three times the power in each phase.

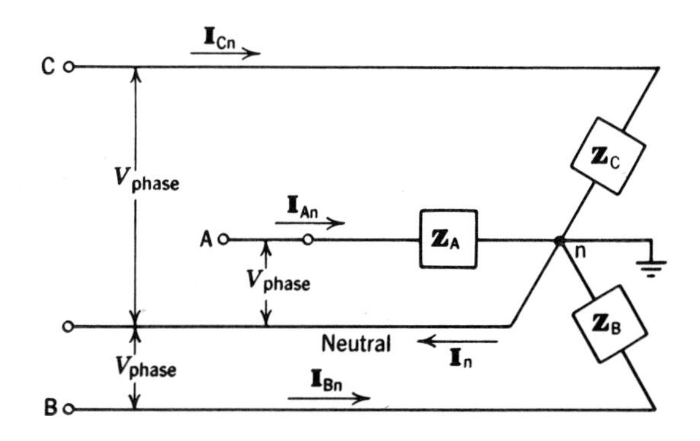

Figure 20.12 An unbalanced three-phase Y-load

• • •

Example 20.7 Calculate the line currents, neutral current, power in each phase, and the total power for the Y-load in Fig. 20.12. $V_{\text{phase}} = 120$ V, $\mathbf{Z}_A = 10\ \Omega$, $\mathbf{Z}_B = (3 + j4)\ \Omega$, and $\mathbf{Z}_C = (10 - j10)\ \Omega$.

Solution The line currents are

$$\mathbf{I}_{An} = \frac{V_{\text{phase}}}{10\underline{/0^\circ}} = \frac{120}{10}\underline{/0^\circ} = 12\underline{/0^\circ}\text{ A}$$

$$\mathbf{I}_{Bn} = \frac{120\underline{/-120^\circ}}{3+j4} = \frac{120\underline{/-120}}{5\underline{/53.1^\circ}} = 24\underline{/-173.1^\circ} \text{ A}$$

$$\mathbf{I}_{Cn} = \frac{120\underline{/120^\circ}}{10-j10} = \frac{120\underline{/120^\circ}}{14.1\underline{/-45^\circ}} = 8.5\underline{/165^\circ} \text{ A}$$

The neutral current is then

$$\begin{aligned}\mathbf{I}_n &= \mathbf{I}_{An} + \mathbf{I}_{Bn} + \mathbf{I}_{Cn} = 12\underline{/0^\circ} + 24\underline{/-173.1^\circ} + 8.5\underline{/165^\circ} \\ &= 12 + j0 - 23.8 - j2.88 - 7.70 + j3.59 \\ &= -19.5 + j0.71 = 19.5\underline{/178^\circ} \text{ A}\end{aligned}$$

The power in each phase is found to be

Phase A $\quad P = V_{\text{phase}} I_{\text{phase}} \cos\theta = 120 \times 12 \times \cos 0^\circ$

$\quad = 120 \times 12 \times 1.0 = 1.44 \text{ kW}$

Phase B $\quad P = 120 \times 24 \times \cos 53.1^\circ = 120 \times 24 \times 0.6 = \mathbf{1.73\ kW}$

Phase C $\quad P = 120 \times 8.5 \times \cos 45^\circ = 120 \times 8.5 \times 0.707 = \mathbf{0.72\ kW}$

The total power is then

$$P = 1.44 + 1.73 + 0.72 = 3.89 \text{ kW}$$

• • •

The three unequal loads in Fig. 20.12 can also be connected in a Δ to form an unbalanced Δ-load as shown in Fig. 20.13. Since the line voltage appears across each of the three loads, the phase currents and the power in each phase can be found easily. The line currents are then found by KCL as in Section 20.4. The total power is of course the sum of the power delivered to each phase. An example will best illustrate the procedure.

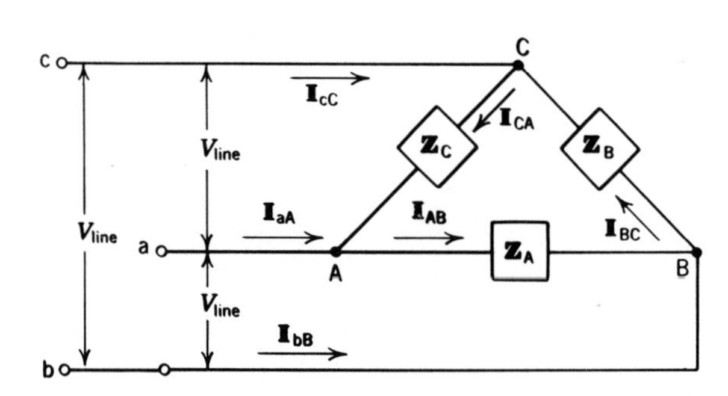

Figure 20.13 An unbalanced three-phase Δ-load

• • •

Example 20.8 Calculate the three line currents for the unbalanced Δ-load in Fig. 20.13. The line voltage is 240 V. $\mathbf{Z}_A$ represents a 1.5 kW lighting load, $\mathbf{Z}_B$ and $\mathbf{Z}_C$ are equal and they represent 2.0 kW loads with lagging power factors of 0.8.

Solution The lighting load has a PF of 1.0, and the current $\mathbf{I}_{AB}$ is thus found as

$$I_{AB} = \frac{P}{V_{\text{line}} \cos\theta} = \frac{1500}{240 \times 1.0} = 6.25 \text{ A}$$

If the line voltage V_{AB} is taken as the reference, we can write

$$\mathbf{I}_{AB} = 6.25\underline{/0^\circ} \text{ A}$$

The currents $\mathbf{I}_{BC}$ and $\mathbf{I}_{CA}$ have equal magnitudes which are found to be

$$I_{BC} = I_{CA} = \frac{P}{V_{\text{line}} \cos \theta} = \frac{2000}{240 \times 0.8} = 10.4 \text{ A}$$

The current $\mathbf{I}_{BC}$ lags the voltage $\mathbf{V}_{BC}$ by a phase angle of

$$\theta = \cos^{-1} 0.8 = 36.9^\circ$$

Since $\mathbf{V}_{BC} = 240\underline{/-120^\circ}$,

$$\mathbf{I}_{BC} = 10.4\underline{/-156.9^\circ} \text{ A}$$

Similarly, $\mathbf{I}_{CA}$ lags $\mathbf{V}_{CA}$ by 36.9°, and since $\mathbf{V}_{CA} = 240\underline{/120^\circ}$,

$$\mathbf{I}_{CA} = 10.4\underline{/83.1^\circ} \text{ A}$$

The three line currents are then found to be

$$\begin{aligned}
\mathbf{I}_{aA} &= \mathbf{I}_{AB} - \mathbf{I}_{CA} = 6.25\underline{/0^\circ} - 10.4\underline{/83.1^\circ} \\
&= 6.25 - 1.20 - j10.3 = 5.05 - j10.3 \\
&= 11.5\underline{/-63.9^\circ} \text{ A} \\
\mathbf{I}_{bB} &= \mathbf{I}_{BC} - \mathbf{I}_{AB} = -9.57 - j4.08 - 6.25 \\
&= -15.8 - j4.08 \\
&= 16.4\underline{/-165.5^\circ} \text{ A} \\
\mathbf{I}_{cC} &= \mathbf{I}_{CA} - \mathbf{I}_{BC} = 1.20 + j10.3 + 9.57 + j4.08 \\
&= 10.8 + j14.4 \\
&= 18.0\underline{/53^\circ} \text{ A}
\end{aligned}$$

The phasor diagram for this example should be drawn (Problem 16).

• • •

THREE-PHASE POWER MEASUREMENT

An instrument used for measuring power is the *wattmeter*, and it is very similar to the basic D'Arsonval meter used as a voltmeter. The basic difference is that the permanent magnet of the D'Arsonval meter is replaced with a large current-carrying coil in the wattmeter as shown in Fig. 20.14*a*. As a result, the wattmeter is seen to have two separate coils; the large coil used to measure the load current is called the *current coil*, and the coil with the series resistance is called the *potential* or *voltage coil*. The resistance may be the actual resistance of the fine wire used for this coil, or it may be an external resistor. The purpose of the resistance is to make the current in the potential coil proportional to the voltage across the coil terminals.

The torque applied to the pointer is proportional to the product of the instantaneous currents in the two coils. However, the mechanical inertia of the system prevents the rapid movement of the pointer, and thus the meter actually responds to the *average* value of the applied torque. We will use the symbol shown in Fig. 20.14*b* for the wattmeter.

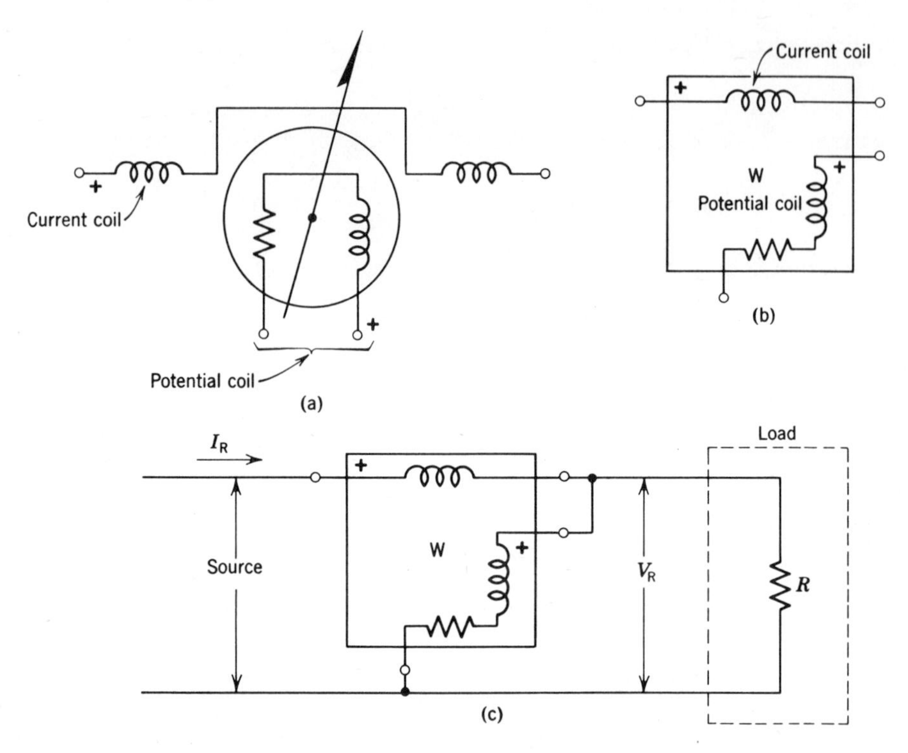

Figure 20.14 The wattmeter

Now, if the wattmeter is connected to a load (Fig. 20.14*c*) such that the load current passes through the current coil and the load voltage appears across the potential coil, the wattmeter will read the *average power* delivered to the load. Notice that the wattmeter is connected so that the load current I_R *enters* the + terminal of the current coil, and the + terminal of the potential coil is at a higher voltage than the other end of the potential coil. This connection will provide an *upscale* deflection on the wattmeter. Reversing either one, but not both, of the coils will result in a *downscale* reading.

The wattmeter described above is known as an *electrodynamometer* instrument, and it can be used to measure both dc and ac power. In fact, it can be used to measure the power delivered to a load even if the wave forms are not sinusoidal.

The most straightforward method for measuring the total power delivered to a three-phase load is to connect three wattmeters as shown in Fig. 20.15*a* or 20.15*b*. Clearly, each wattmeter measures the voltage across and the current through a load, and thus the total power is simply the sum of the readings given by the meters. Unfortunately, these connections are not practical since only the points A, B, and C are available for connection in a typical installation. We must therefore seek a more practical method.

In Fig. 20.16, the three wattmeters are connected to read the total power delivered to the load. Each current coil measures a line current which is the same as the phase current in each load. The + terminals of the potential coils are connected to the lines, and the other potential coil terminals are connected together at some arbitrary point O. As a result, the potential coils

INDUCTIVE REACTANCE $(X_L) = 2\pi f L$

FOR SERIES INDUCTORS $L_T = L_1 + L_2 + L_3$

FOR PARALLEL INDUCTORS $L_T = \frac{1}{1/L_1 + 1/L_2 + 1/L_3 \text{ etc}}$

CAPACITIVE REACTANCE $(X_C) = 1/2\pi f C$

FOR SERIES CAPACITORS $C_T = \frac{1}{1/C_1 + 1/C_2 + 1/C_3 \text{ etc}}$

FOR PARALLEL CAPACITORS $C_T = C_1 + C_2 + C_3$

IN AN RL CIRCUIT IMPEDANCE $(Z) = \sqrt{R^2 + X_L^2}$

IN AN RC CIRCUIT IMPEDANCE $(Z) = \sqrt{R^2 + X_C^2}$

L_1 250 mH

L_2 45 mH

L_3 5,000 μH

A

FIG- A

S

L_1 60mH

L_2 400mH

L_3 2H

A

FIG B

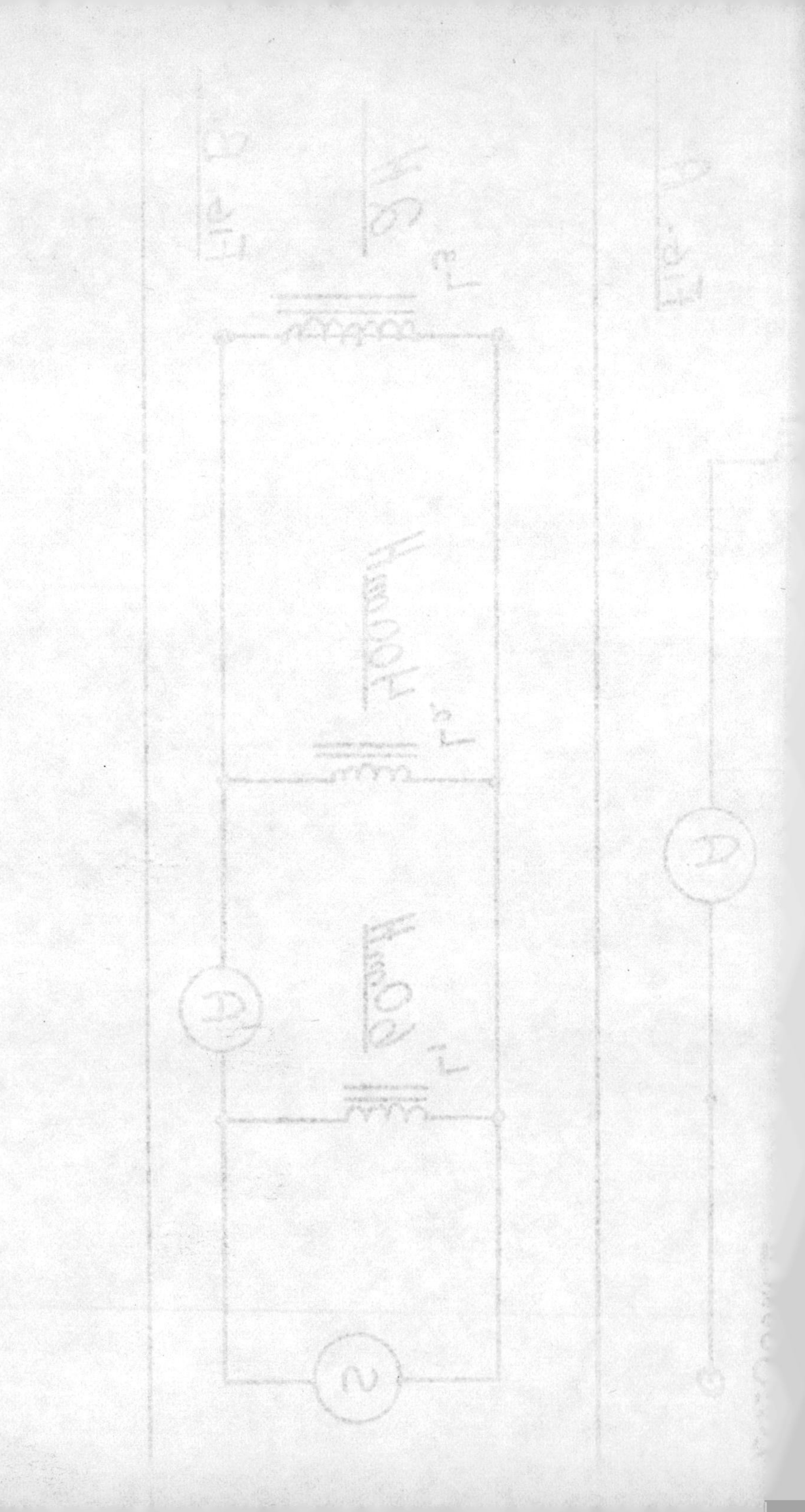

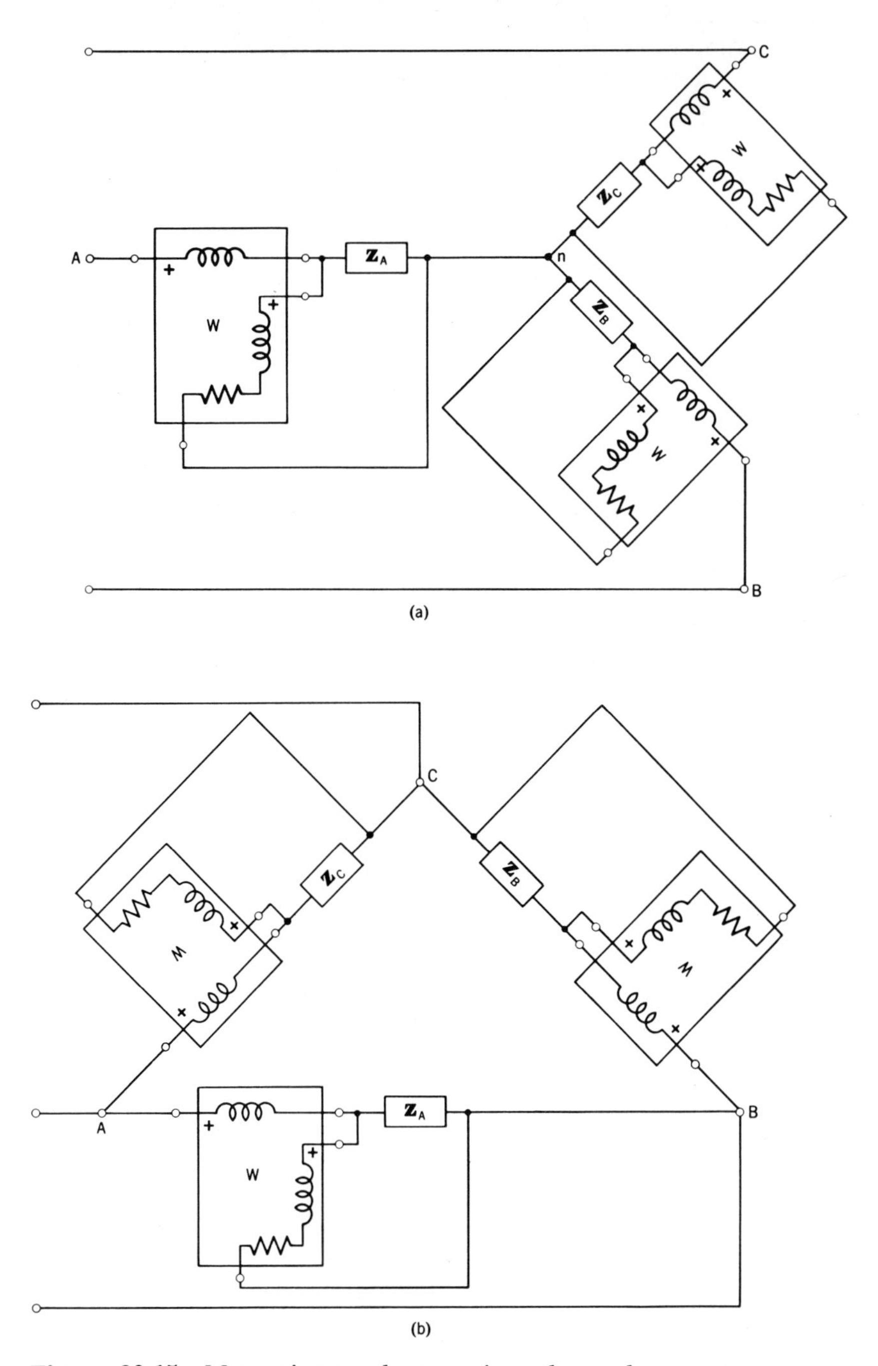

Figure 20.15 Measuring total power in a three-phase system

are connected as a Y and each coil therefore reads phase voltage. When connected this way, the currents through the wattmeters will appear as phase currents, and the total power delivered to the load is then algebraic sum of the three wattmeter readings. This method will work whether the load is Y or Δ-connected and it is called the *three-wattmeter method.*

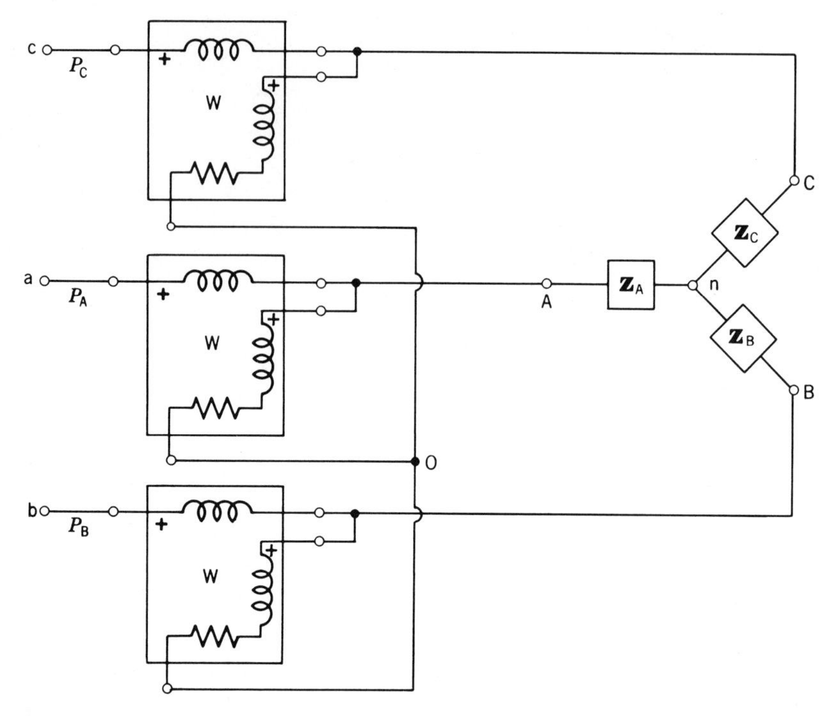

Figure 20.16 Three-wattmeter method

• • •

Example 20.9 What is the total power delivered to the load in Fig. 20.16 if the wattmeter readings are 840 W, 915 W, and −175 W?

Solution Notice that one of the wattmeters has a negative (downscale) reading. The power is the algebraic sum of the three readings, and thus

$$P = 840 + 915 - 175 = 1580 \text{ W} = 1.58 \text{ kW}$$

• • •

Let us now show that the point O in Fig. 20.16 is truly arbitrary, since this will lead to a second method for measuring the total power delivered to a three-phase load. Each of the three wattmeters reads the average value of the instantaneous power. The instantaneous power can be found as the product of instantaneous current i and voltage v, and we shall designate the average value of this product as $(vi)_{\text{AVE}}$. Now, from Fig. 20.16 the power in each wattmeter can be written as

$$P_A = (v_{AO}i_{aA})_{\text{AVE}}; \; P_B = (v_{BO}i_{bB})_{\text{AVE}}; \; P_C = (v_{CO}i_{cC})_{\text{AVE}}$$

The total power is then

$$P = P_A + P_B + P_C = (v_{AO}i_{aA})_{\text{AVE}} + (v_{BO}i_{bB})_{\text{AVE}} + (v_{CO}i_{cC})_{\text{AVE}} \qquad (20.14)$$

Now, notice that

$$v_{AO} = v_{An} + v_{nO}; \; v_{BO} = v_{Bn} + v_{nO}; \; v_{CO} = v_{Cn} + v_{nO}$$

If we substitute these relationships in Eq. 20.14, the result is

$$\begin{aligned} P &= P_A + P_B + P_C = \\ &= [(v_{An} + v_{nO})i_{aA}]_{AVE} + [(v_{Bn} + v_{nO})i_{bB}]_{AVE} + [(v_{Cn} + v_{nO})i_{cC}]_{AVE} \\ &= (v_{An}i_{aA})_{AVE} + (v_{Bn}i_{bB})_{AVE} + (v_{Cn}i_{cC})_{AVE} + v_{nO}(i_{aA} + i_{bB} + i_{cC})_{AVE} \end{aligned}$$

But, from KCL, $(i_{aA} + i_{bB} + i_{cC}) = 0$. Therefore, the total power is given as

$$P = (v_{An}i_{aA})_{AVE} + (v_{Bn}i_{bB})_{AVE} + (v_{Cn}i_{cC})_{AVE} \qquad (20.15)$$

which is exactly the same as Eq. 20.13. Therefore, the point O is truly arbitrary; that is, it can be connected to any desired point.

Since point O can be connected to any desired point, let us connect it to point A in Fig. 20.16. The wattmeter reading P_A will then read zero since the voltage across its potential coil is zero. We can therefore remove this wattmeter and the total power is then simply the sum of P_B and P_C. This connection is known as the *two-wattmeter method* and it is shown in Fig. 20.17. The load in Fig. 20.17 is Δ-connected, but this is of no concern since the two-wattmeter method is valid for either Y or Δ loads.

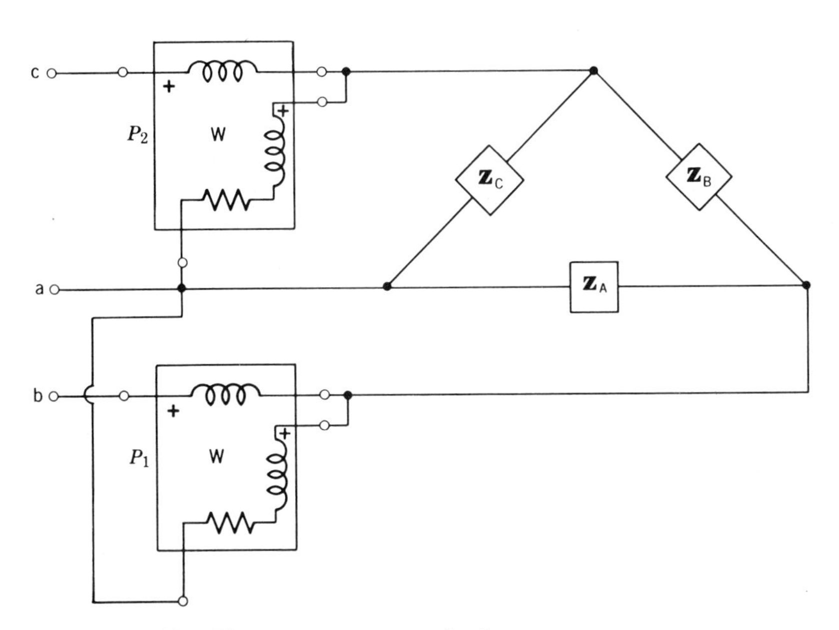

Figure 20.17 Two-wattmeter method

• • •

Example 20.10 Calculate the total power delivered to the load in Fig. 20.17 if P_1 reads 1340 W and P_2 reads −750 W.

Solution Taking the algebraic sum, the total power is

$$P = 1340 - 750 = 590 \text{ W}$$

• • •

SUMMARY

Circuits or systems containing more than one ac voltage source of the same frequency where there is a phase difference between the source voltages are called polyphase ac. One of the most important polyphase ac systems is the three-phase system which is the technique used in virtually all power distribution systems. A three-phase power system can be considered as three ideal voltage generators supplying equal-amplitude voltages of the same frequency with a 120° phase angle between them. The sources can be connected in a Y or a Δ configuration, but Y is the more common method. Power distribution systems almost always make use of transformers, and these transformers can be considered as ideal voltage sources connected in either a Y or Δ configuration. Three-phase loads can also be connected in either a Y or Δ configuration. In a balanced Y-load, the line and phase currents are the same, while the line voltage is $\sqrt{3}$ times greater than the phase voltage. In a balanced Δ-load, the line and phase voltages are equal, but the line current is $\sqrt{3}$ times greater than the phase current. The power delivered to a three-phase load is simply a sum of the power in each phase. The wattmeter is an instrument for measuring power, and it can be used in either the three-wattmeter or the two-wattmeter connections. Either of these two connections can be used to measure the total power delivered to Y or Δ loads.

GLOSSARY

Current coil. *The coil used to measure the current in a wattmeter.*

Δ-connection. *The connection of three sources or loads such that one terminal of each element is connected to only one terminal of another element in order to form a mesh.*

Electrodynamometer. *A basic meter movement having two current coils in which the pointer deflection is proportional to the product of the instantaneous currents in the two coils.*

Neutral. *The common or ground wire in a three-phase system. The common node in a Y-connection.*

Phase sequence. *The order in which the voltages in a three-phase source reach their positive peaks.*

Potential coil. *The coil having a series resistance in a wattmeter; used to measure a current which is proportional to voltage.*

Single-phase ac. *A circuit containing a single ac voltage source.*

Three-phase ac. *A circuit containing three equal amplitude voltage sources of the same frequency in which there is a phase angle of 120° between each pair of the three voltages.*

Three-wattmeter method. *A method for connecting three wattmeters in order to measure the total power delivered to a three-phase load.*

Two-wattmeter method. *A method for connecting two wattmeters in order to measure the total power delivered to a three-phase load.*

Wattmeter. *An electrodynamometer used to measure average power.*

Y-connection. *The connection of three sources or loads such that one terminal from each element is connected to a common node.*

IMPORTANT RELATIONSHIPS

Y	Δ
$V_{\text{line}} = \sqrt{3}\, V_{\text{phase}}$	$V_{\text{line}} = V_{\text{phase}}$
$I_{\text{line}} = I_{\text{phase}}$	$I_{\text{line}} = \sqrt{3}\, I_{\text{phase}}$

Y and Δ

$$P = P_A + P_B + P_C$$
$$P = 3 V_{\text{phase}} I_{\text{phase}} \cos\theta$$
$$P = \sqrt{3}\, V_{\text{line}} I_{\text{line}} \cos\theta$$

:VIEW QUESTIONS

1. Explain why the voltages from the generator in Fig. 20.1*a* all have the same frequency and are separated by 120° phase angles.
2. Double subscript notation is a convenient means of identifying voltage polarities and current directions. (TF)
3. The direction of rotation of a three-phase motor can be determined from ______ ______.
4. A Y-connection is sometimes referred to as a ______-connection.
5. In a Y-connection, the line voltage is equal to the (sum, difference) of two adjacent phase voltages.
6. In a Y-connection, the line and phase currents are the same. (TF)
7. Explain why a three-phase generator is seldom connected in a Δ-configuration.
8. The line and phase voltages in a Δ-connected generator are equal. (TF)
9. A Δ-connection is also called a ______.
10. Transformers are used in power distribution systems to ______ ______ and ______ ______ the voltages.
11. The secondary windings of a three-phase transformer can be connected in a ______, or ______, or they can be used as three individual ______ ______ sources.
12. The neutral current in a balanced Y-load is zero. (TF)
13. The line and phase currents in a Δ-load (are, are not) equal.
14. The power delivered to a balanced Y-load is equal to three times the power delivered to each phase. (TF)
15. The total power delivered to a balanced Δ-load is found in exactly the same way as for a balanced Y-load. (TF)
16. The phase currents in an unbalanced three-phase Y-load can be found on a phase-at-a-time basis if the ______ is present.
17. A wattmeter has two coils, and they are called the ______ coil and the ______ coil.
18. The torque produced in a wattmeter is proportional to the product of the average currents in the two coils. (TF)
19. The wattmeter reads ______ power.
20. A wattmeter can be used to measure either ac or dc power. (TF)

21. The coil polarities on a wattmeter are important since the coils can be connected to give either an ____ or a ____ reading.
22. The three-wattmeter and the two-wattmeter methods can both be used to measure the total power delivered to a Y or a Δ load. (TF)
23. The total power delivered to a three-phase load is equal to the ____ sum of the wattmeter readings.

PROBLEMS

1. Write the equations for the three generator voltages in Fig. 20.1*a* if the frequency is 60 Hz and the voltages have a peak value of 294 V.
2. Write the phasor expressions for the voltages in Problem 1.
3. A Y-connected three-phase generator has line voltages of 208 V. What are the magnitudes of the phase voltages?
4. Draw the diagram for the three-phase generator in Problem 3. Draw the complete phasor diagram showing all line and phase voltages.
5. Draw the diagram of a three-phase Δ-connected generator. Show the coil "starts" and "finishes" clearly. Draw the phasor diagram for the voltages if $V_{line} = 240$ V.
6. Show the connections for the three-phase motor in Example 20.2 if the direction of rotation is to be reversed.
7. Draw the complete diagram for connecting three 1.5-kW lighting loads (120 V) and two three-phase motors (208 V) to the Δ-Y transformer in Fig. 20.5*b*. Keep the load symmetrical.
8. The transformer in Fig. 20.5*c* is used to supply three residences. Each home has an oven, a range, and a dryer (all 240 V), and four 120 V appliance circuits. Show the complete wiring diagram.
9. Calculate the line currents, power delivered to each phase, total power, and draw the phasor diagram for the load in Fig. 20.7*a* if each impedance represents a 120 V lighting load of 3 kW.
10. A three-phase motor can be considered as a balanced Y-load. Calculate the line currents required for a 15 hp three-phase motor having a PF of 0.9 lagging if it is connected to the transformer in Fig. 20.5*b*.
11. Calculate the line currents if the motor in Problem 10 is connected to the transformer in Fig. 20.5*a*.
12. Calculate the line currents for the loads in Problem 7 if both motors are 5 hp with lagging PF of 0.8.
13. Calculate the phase currents, line currents, phase power, and the total power delivered to a Δ-load composed of three 4 kW, 240 V lighting circuits connected as in Fig. 20.10*a*. Draw the complete phasor diagram.
14. Calculate the magnitudes of the phase and line currents if the load in Fig. 20.10*a* represents a 7.5 hp motor with a lagging PF of 0.85. The line voltage is 440 V.
15. Calculate the line currents, the power in each phase, and the total power in Fig. 20.12 if: $\mathbf{Z}_A$ is a 4 kW lighting load, $\mathbf{Z}_B$ is a heating load of $(7 + j5)\Omega$, and $\mathbf{Z}_C$ is an appliance load of $(5 + j5)\Omega$. V_{phase} is 120 V.
16. Draw the phasor diagram for Example 20.8.
17. Calculate the phase currents and the line currents if the loads in Problem 15 are connected in a Δ as in Fig. 20.13. $V_{line} = 240$ V, and the phase sequence is ABC.

18. A three-phase, 10 hp motor with a PF of 0.9 lagging is connected to the transformer in Fig. 20.5*a*. A 2 kW lighting load is also connected across terminals A–C. Calculate the three line currents.

19. Calculate the load power if the wattmeter readings in Fig. 20.16 are 1230 W, −720 W, and 1300 W.

20. Calculate the load power if the wattmeter readings in Fig. 20.17 are 760 W and 340 W.

21. The three-wattmeter method is used to measure the total power delivered to the motor in Problem 10. Calculate the reading shown on each wattmeter.

21

ELECTRONIC CIRCUITS

In the previous chapters we have established a firm background which will prove to be invaluable in the analysis and design of electric circuits. In applying the laws and techniques presented, we have confined our attention to *passive* circuits containing only *linear* elements (R's, L's, and C's) and *ideal* sources. There is almost an infinite variety of circuits containing *electronic* elements such as diodes, vacuum tubes, and transistors, and we are now in a position to apply the principles and techniques we have learned to such circuits. Circuits containing electronic elements are commonly called *electronic circuits,* and we will devote this chapter to the discussion of a number of simple electronic circuits.

21.1 THE SEMICONDUCTOR DIODE

When discussing resistance in Chapter 5, it was determined that a resistor is a *linear, bilateral* element. It is bilateral since current will pass through a resistor in either direction with equal ease. It is said to be linear since a plot of V versus I results in a straight line as shown in Fig. 21.1. This V-I plot is called a *characteristic curve* and such curves are widely used in describing the behavior of electronic devices. Notice that the slope of the resistor characteristic curve is equal to the resistance R. That is,

$$\text{slope} = \frac{\Delta V}{\Delta I} = R$$

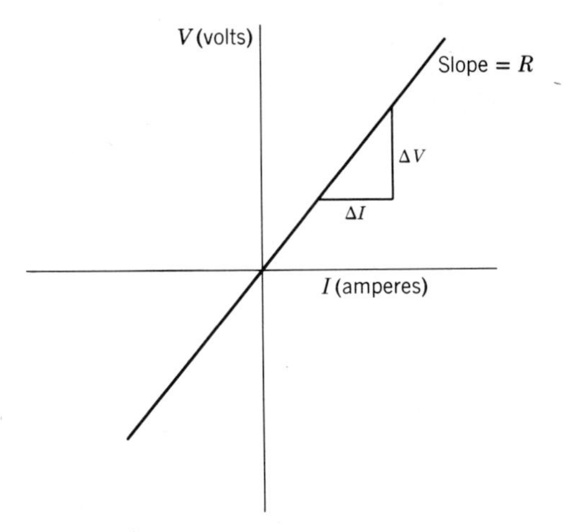

Figure 21.1 The V-I plot or characteristic curve of a resistor

A semiconductor diode will conduct current easily in one direction, but it offers a very high resistance to current in the opposite direction. It is therefore *not* a bilateral element. The symbol for a semiconductor diode is shown in Fig. 21.2.

Figure 21.2 Semiconductor diode symbol

When a voltage source is connected across the diode as shown in Fig. 21.3*a*, the diode offers very little opposition to current flow, and it is said to be *forward biased.* When it is forward biased, the diode may have a resistance of only a few Ω, and thus a resistor R is usually connected in series with it to limit the current to a safe value. If the source voltage V_S in Fig. 21.3*a* is increased from zero V to some positive voltage, the diode current will begin at zero and increase very slowly up to a point. The diode current will then increase very rapidly with only a very small change in diode voltage V_d, as shown in Fig. 21.3*b*. The point at which the diode current begins to increase rapidly is called the ***knee*** and the diode voltage at this point is called the ***knee voltage.*** The plot of diode current I_d versus diode voltage V_d in Fig. 21.3*b* is called the *forward characteristic* curve of the diode.

If the polarity of the voltage source V_S is now reversed as shown in Fig. 21.3*c*, the diode will strongly oppose any current flow, and the diode is said to be *reverse biased.* As V_S is increased from zero, the diode current will be very small (picoamperes to microamperes) even for fairly large values of V_S. At some point however, the applied voltage will exceed the electrical properties of the diode and it will *break down.* The diode will then appear essentially as a *short circuit,* and the current will increase rapidly as shown in Fig. 21.3*d.* The point at which the diode breaks down in the reverse direction is called the

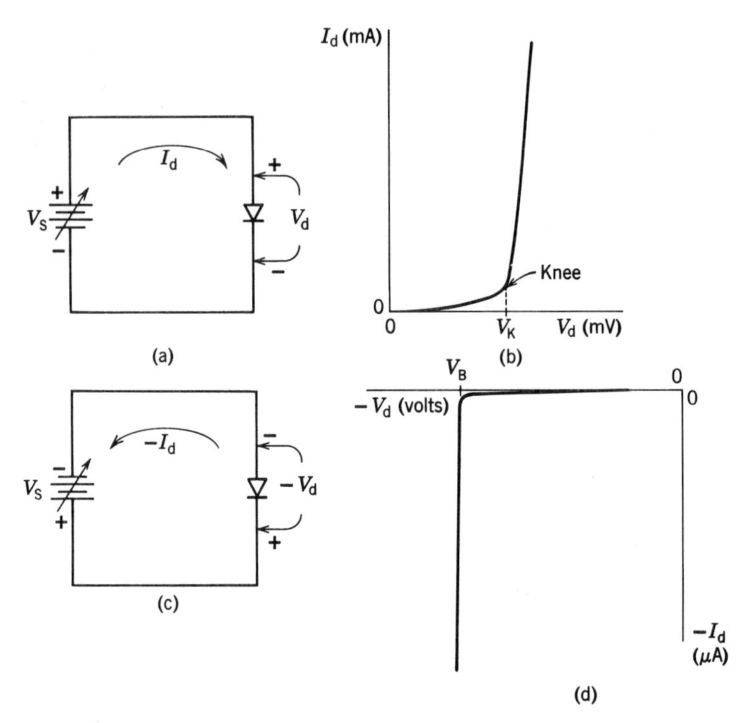

Figure 21.3 (*a*) Forward biased diode (*b*) Characteristic curve for a forward biased diode (*c*) Reverse biased diode (*d*) Characteristic curve for a reverse biased diode

breakdown voltage V_B. The plot of diode current I_d versus diode voltage V_d in Fig. 21.3*d* is called the *reverse characteristic* curve of the diode. Notice carefully that the application of a voltage great enough to break the diode down in the reverse direction may very well result in destruction of the device.

The complete characteristic curve of a diode can be obtained by joining the forward characteristic and the reverse characteristic as shown in Fig. 21.4. From this curve, it is easy to see that current will readily flow in the forward

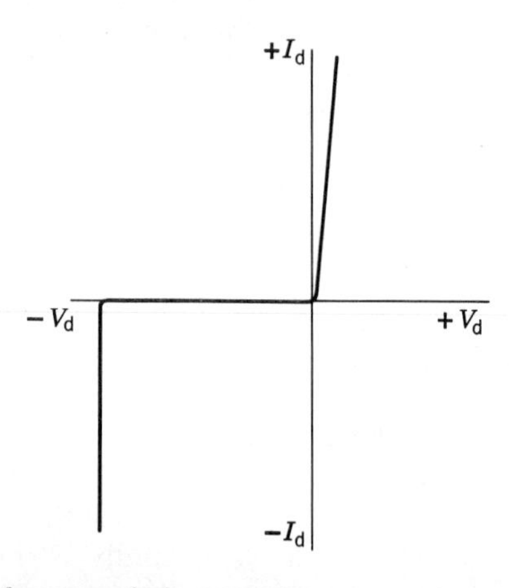

Figure 21.4 Diode characteristic curve

direction, but it will not flow in the reverse direction. An easy way to remember the forward and reverse directions is to notice that current passes easily through the diode in the direction that the triangle points on the diode symbol, but it does not pass through the diode in the opposite direction.

An equation which describes the behavior of the diode, called the *diode equation*, is

$$I_d = I_S(e^{40V_d} - 1) \tag{21.1}$$

where I_d is the diode current
I_S is the reverse leakage current of the diode
V_d is the diode voltage
e is the natural logarithm base

This equation can be used to calculate the diode current if the diode voltage is known (I_S is usually specified in manufacturers data sheets).

• • •

Example 21.1 Use the diode equation to calculate I_d in Fig. 21.3*a* if: (*a*) $V_d = 10$ mV, (*b*) $V_d = 0.1$ V, (*c*) $V_d = 0.4$ V. Assume $I_S = 1\ \mu$A.

Solution

(*a*) $I_d = I_S(e^{40V_d} - 1) = 10^{-6}(e^{40\times 0.01} - 1) = 10^{-6}(e^{0.4} - 1)$
$= 10^{-6}(1.49 - 1) = 0.49\ \mu\text{A}$

(*b*) $I_d = 10^{-6}(e^{40\times 0.1} - 1) = 53.6\ \mu\text{A}$

(*c*) $I_d = 10^{-6}(e^{40\times 0.4} - 1) = 8.93\ \text{A}$

This example shows clearly that the diode current increases very rapidly with increasing diode voltage.

• • •

Example 21.2 Calculate the reverse diode current in Fig. 21.3*c* if $I_S = 10^{-6}$ A and $V_S = 1.0$ V.

Solution Notice first of all that the diode is reverse biased. Thus, V_d in the diode equation will have a negative sign.

$$I_d = 10^{-6}(e^{40(-1.0)} - 1) = 10^{-6}(e^{-40} - 1)$$

The term e^{-40} is equal to $1/e^{40}$, and this is a very small number. As a result,

$$I_d \cong 10^{-6}(0 - 1) = -1\ \mu\text{A}$$

• • •

From this example, it is clear that the reverse current through the diode will be very nearly equal to I_S for any value of reverse voltage smaller than the reverse breakdown voltage.

Since the diode conducts current readily in the forward direction, but not in the reverse direction, we can to a first approximation treat it as a switch. We can therefore make use of the equivalent circuit of the ideal diode as shown in Fig. 21.5*a*. When the diode is forward biased, the switch is closed and the diode simply appears as a small resistance R_d. When the diode is reverse biased, the switch is open and the diode conducts no current. This equivalent circuit leads to the characteristic curve shown in Fig. 21.5*b*, and this curve does indeed approximate the true diode characteristic curve.

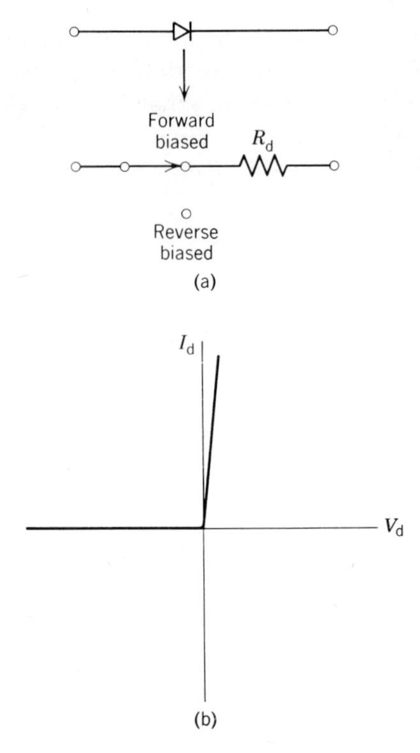

Figure 21.5 (*a*) Ideal diode equivalent circuit (*b*) Ideal diode characteristic curve

• • •

Example 21.3 Calculate the voltage across R_L in Fig. 21.6*a* if the diode forward resistance R_d is 10Ω.

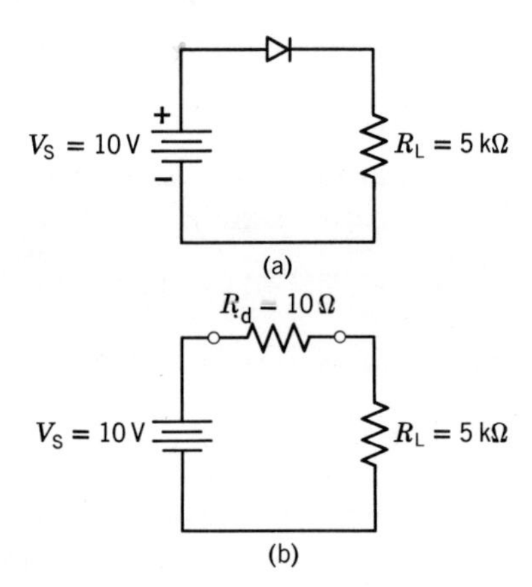

Figure 21.6 (*a*) A simple diode circuit (*b*) Equivalent circuit of *a*

Solution Since the diode is forward biased, we can replace it with its equivalent circuit of $R_d = 10\Omega$ as shown in Fig. 21.6*b*. The voltage across R_L is then found by a simple application of the voltage divider theorem. Thus

$$V_{R_L} = V_S \frac{R_L}{R_L + R_d} = 10\text{V}\frac{5\text{k}\Omega}{5\text{k}\Omega + 10\Omega} \cong 5\text{V}$$

• • •

• • •

Example 21.4 Calculate the voltage across R_L in Fig. 21.7*a*.

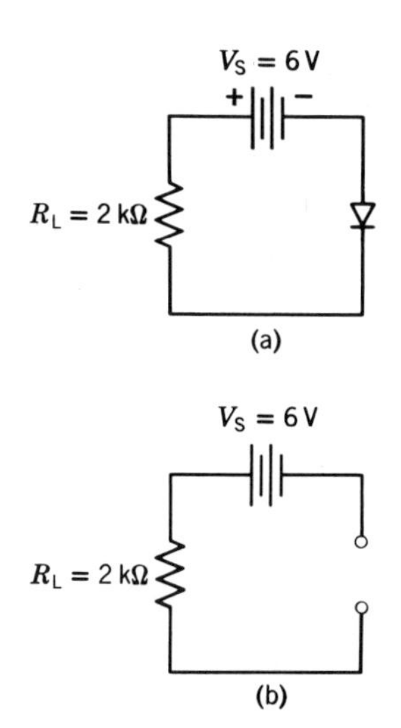

Figure 21.7 Example 21.4(*a*) A simple diode circuit (*b*) Equivalent circuit of *a*

Solution In this circuit, the diode is reverse biased. Its equivalent circuit is then an *open* switch as shown in Fig. 21.7*b*. Since there is no current in the circuit, the voltage across R_L is clearly zero volts.

• • •

.2 THE DIODE AS A RECTIFIER

In Fig. 21.8*a*, a diode in series with a load resistor R_L is connected across the terminals of a sine wave voltage generator. When the signal generator voltage is positive, the diode is forward biased, and there will be a circuit current. However, when the generator voltage is negative, the diode is reverse biased and there is no circuit current. As a result, the circuit current appears as a series of positive pulses as shown in Fig. 21.8*b*. These current pulses are very nearly sinusoidal in form, and since the voltage across R_L is simply R_L multiplied by the circuit current, the voltage v_{R_L} appears as shown in Fig. 21.8*b*. In fact, if R_L is considerably larger than R_d, the voltage across R_L is very nearly equal to the *positive half cycles* of the generator voltage. This circuit is then called a *half wave rectifier.*

Notice that the voltage v_{R_L} is no longer a sine wave, nor is it a dc voltage. It is however *always* positive. The diode in this circuit has changed an ac voltage v_S, which has *both* positive and negative values, into a voltage v_{R_L}, which has *only* positive values. In this case, the diode is said to *rectify* the source voltage v_S, and v_{R_L} is called the *rectified* output voltage. Thus, when a diode is used in this fashion it is often called a *rectifier.*

The half wave rectifier can be used to construct a simple dc *power supply* by connecting a capacitor C across the load resistor R_L as shown in Fig. 21.9*a*. The capacitor will charge up to the peak value of the applied voltage when the

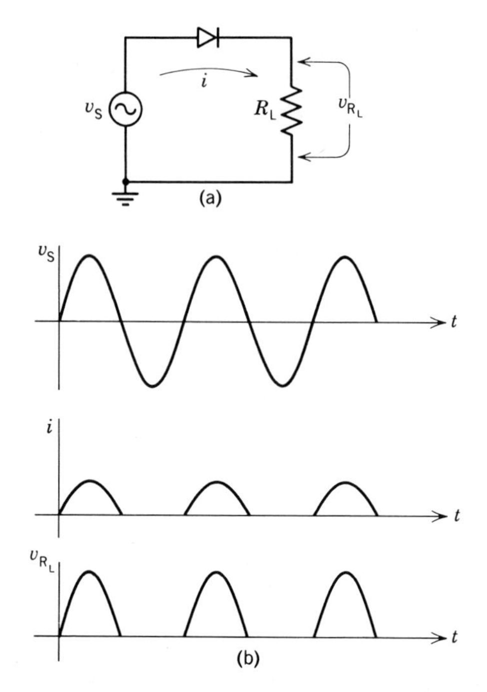

Figure 21.8 Half wave rectifier (*a*) Basic circuit (*b*) Circuit current and voltage

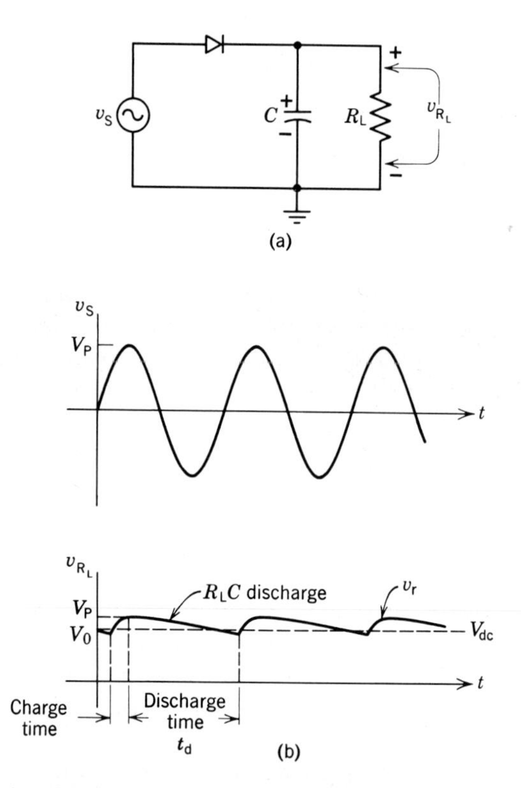

Figure 21.9 A simple dc power supply (*a*) Half wave rectifier circuit (*b*) Circuit voltage wave forms

diode is forward biased. Then when v_S begins to decrease, the diode becomes reverse biased and acts as an open switch. During this time, the capacitor C will discharge through R_L as a simple series R-C circuit (recall our discussion of R-C transient circuits in Chapter 13). The voltage across R_L will then appear as shown in Fig. 21.9*b*.

The voltage across the load v_{R_L} can be considered as the *sum* of a dc voltage V_{dc} and a ***ripple voltage*** v_r as shown in Fig. 21.9*b*. If the time constant of the circuit R_LC is large enough, the capacitor will not discharge appreciably during the discharge time, and v_r will be very small. Under this condition, the output voltage can simply be considered as a dc voltage whose value is only slightly less than the peak source voltage V_p.

Recall from Chapter 13 that the voltage across a discharging capacitor is given by

$$v = V_p e^{-t/RC}$$

Applying this equation to the circuit in Fig. 21.9*a* during the time the diode is reverse biased (during the discharge time), we have

$$V_0 = V_p e^{-t_d/R_LC} \tag{21.2}$$

where V_p is the peak value of the voltage, and V_0 is the value to which v_{R_L} discharges. From Fig. 21.9*b*, it is clear that the discharge time is fixed, and it is approximately equal to the ***period*** of the source voltage wave. That is, t_d in Eq. 21.2 is approximately

$$\text{discharge time} = t_d \cong T = \frac{1}{f} \tag{21.3}$$

It is also clear from Fig. 21.9*b* that the ***peak-to-peak*** value of the ripple voltage $V_{p\text{-}p}$ is given by

$$V_{p\text{-}p} = V_p - V_0$$

Substituting Eq. 21.2 into this relationship, we have

$$\text{peak-to-peak ripple voltage} = V_{p\text{-}p} = V_p - V_0 = V_p - V_p e^{-t_d/R_LC}$$
$$= V_p(1 - e^{-t_d/R_LC}) \tag{21.4}$$

• • •

Example 21.5 A simple dc power supply using a half wave rectifier is shown in Fig. 21.10*a*. Calculate the output ripple voltage, and give an approximate value for the dc output voltage.

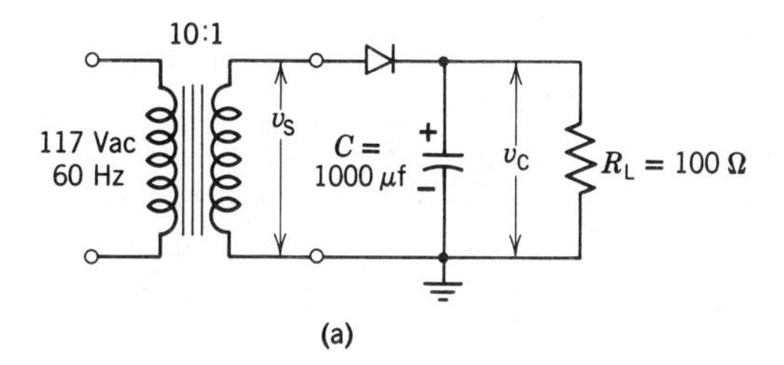

Figure 21.10 A simple dc power supply (*a*) Half wave rectifier circuit

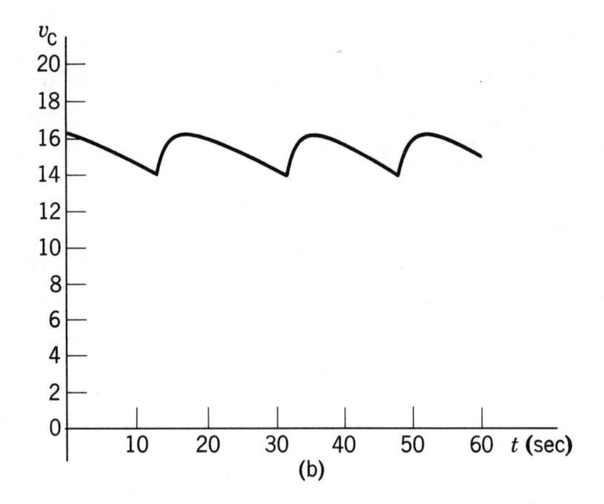

Figure 21.10 (*continued*) (*b*) Output voltage

Solution A transformer is used to isolate the power supply from the ac power lines; one side of the circuit can then safely be grounded as a reference as shown in the figure. The voltage supplied to the rectifier circuit is the secondary voltage v_S of the transformer. This is clearly

$$V_S = \frac{V_{pri}}{N} = \frac{117\text{V ac}}{10} = 11.7\text{ V}$$

This is an RMS voltage, and the peak voltage V_p is then

$$V_p = \sqrt{2}\,11.7 = 1.414 \times 11.7 = 16.5\text{ V}$$

In order to determine the peak-to-peak ripple voltage, we will use Eq. 21.4. We already have determined V_p as 16.5 V. The time constant is

$$R_L C = 100 \times 1000 \times 10^{-6} = 0.1$$

Since the frequency of the source voltage is 60 Hz, the discharge time is approximately

$$t_d \cong \frac{1}{f} = \frac{1}{60\text{ Hz}} = 16.7 \times 10^{-3}$$

Placing these values into Eq. 21.4, we now have

$$V_{p\text{-}p} = V_p(1 - e^{-t_d/R_L C}) = 16.5(1 - e^{-(16.7\times 10^{-3})/0.1}) = 2.54\text{ V}$$

The approximate output voltage wave form is shown in Fig. 20.10*b*, and the dc output voltage might be approximated as +15 V dc. The output ripple voltage can of course be reduced by increasing either R_L or C, or both.

• • •

An interesting and useful variation of the basic half wave rectifier is the *voltage doubler* circuit in Fig. 21.11. During the positive half cycles of v_S diode D_1 conducts and capacitor C_1 charges to the peak value of V_p, with the polarity shown on C_1. During the negative half cycles of v_S, diode D_2 conducts and capacitor C_2 charges to a value of V_p with the polarity shown in the figure. The net output voltage V_{R_L} is then the sum of the two capacitor voltages. Thus, for the voltage doubler,

$$V_{R_L} \cong 2\,V_p \tag{21.5}$$

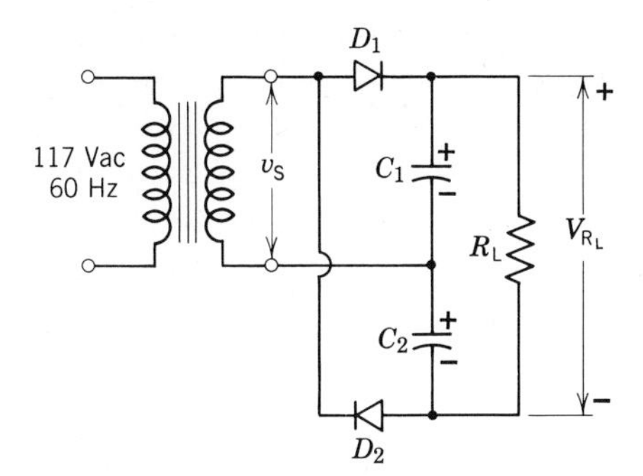

Figure 21.11 A basic voltage doubler circuit

Equation 21.5 is a good approximation if the ripple voltage is small enough to be ignored.

From Fig. 21.11, it is clear that the voltage doubler is constructed by connecting two half wave rectifiers in *series*, or in *cascade*. Voltage triplers and quadruplers are also possible by properly interconnecting basic half wave rectifiers.

• • •

Example 21.6 Calculate the approximate dc output voltage V_{R_L} for the voltage doubler in Fig. 21.11 if $V_{PRI} = 117$ V ac, 60 Hz, and the transformer is a 1:2 step-up transformer.

Solution Since we are using a 1:2 step-up transformer, the transformer secondary voltage V_{SEC} is

$$V_{SEC} = 2 \times 117 \text{ V} = 234 \text{ V RMS}$$

The peak value of the secondary voltage is then

$$V_p = 1.414 \times 234 = 331 \text{ V}$$

If we assume that the ripple voltage is negligible, the dc load voltage will be approximately twice the peak voltage. Thus,

$$V_{R_L} 2 V_p = 2 \times 331 = 662 \text{ V}$$

The actual value of V_{R_L} will of course be slightly less than 662 V due to the ripple voltage.

• • •

.3 ZENER DIODES

The reverse breakdown voltage of a diode can be altered by varying the physical properties of the diode during the manufacturing process. As a result it is possible to construct a diode with any desired reverse breakdown voltage, within certain physical limits. When using a diode as a rectifier, we do not wish it to break down in the reverse direction, and we will therefore choose a diode having a reverse breakdown voltage greater than the maximum reverse voltage it will be subjected to.

On the other hand, there are a number of applications in which diodes are intentionally reverse biased beyond their reverse breakdown voltages. Diodes which are specifically designed for this purpose are called *zener diodes,* and the proper symbol is shown in Fig. 21.12*a*. The characteristic curve for a zener diode is shown in Fig. 21.12*b*, and the reverse breakdown voltage is referred to as the *zener voltage* V_Z. Practical values of zener voltages range from around 3.0 V up to perhaps 100 V.

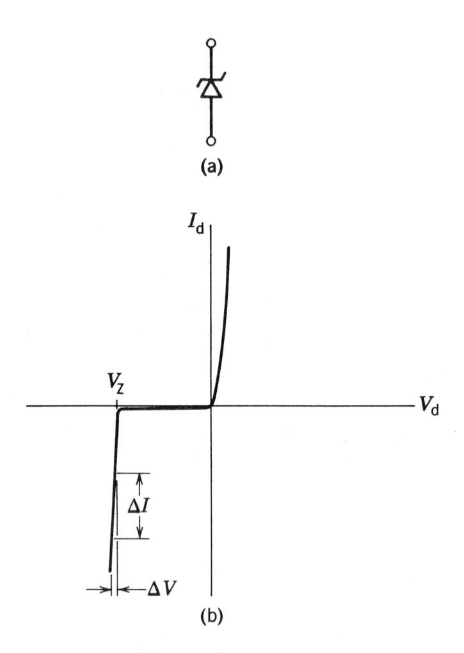

Figure 21.12 Zener diode (*a*) Symbol (*b*) Characteristic curve

A careful examination of the characteristic curve in Fig. 21.12*b* shows that the curve is not quite vertical in the *breakdown region.* It does in fact have some slope, and the value of the slope is given by

$$\text{slope} = \frac{\Delta I}{\Delta V}$$

The units in this relationship are A/V, which we recognize as the definition of admittance in ℧ (from Ohm's Law). If we were to use the *reciprocal* of the slope, we would have units of V/A which of course is the same as resistance in Ω. Thus in the reverse direction the diode exhibits some resistance R_Z given by

$$R_Z = \frac{1}{\text{slope}} = \frac{\Delta V}{\Delta I} \tag{21.6}$$

The value of R_Z will normally be very small since a fairly large change in I_d on the characteristic curve ΔI, will result in only a very small value of ΔV. For an *ideal* zener diode, ΔV will always be zero and the diode will therefore have zero resistance.

• • •

Example 21.7 Calculate the reverse resistance in the breakdown region for the zener diode in Fig. 21.12 if $\Delta I = 10$ mA and $\Delta V = 100$ mV.

Solution The resistance is found from Eq. 21.6 as

$$R_Z = \frac{\Delta V}{\Delta I} = \frac{100\ \text{mV}}{10\ \text{mA}} = 10\Omega$$

• • •

The behavior of a zener diode can be approximated by using the equivalent circuit shown in Fig. 21.13 (we assume that the diode is always reverse biased and that there is some current through it). We see immediately that this is exactly the same circuit used to represent a *real voltage source.* The diode resistance R_Z corresponds to the source resistance, and the zener voltage corresponds to the ideal source voltage. Furthermore, since a value of $R_Z = 0\ \Omega$ corresponds to an *ideal* zener, an ideal zener is seen to correspond exactly to an ideal voltage source. For these reasons, zener diodes are frequently used to construct voltage sources.

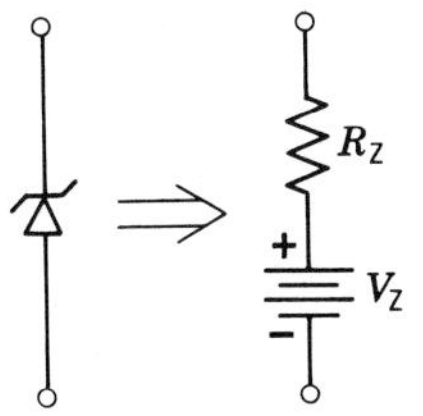

Figure 21.13 Equivalent circuit of a zener diode operating in the breakdown region

Consider for example the circuit shown in Fig. 21.14*a*. We begin with a voltage source of V_S, and we wish to construct a voltage source having a lower voltage. If we replace the zener with its equivalent circuit as shown in Fig. 21.14*b*, we see that the total load voltage V_L is

$$V_L = V_S - V_Z$$

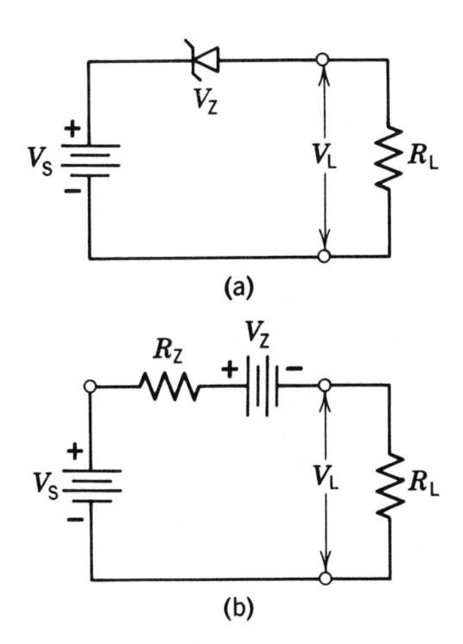

Figure 21.14 Using a zener diode to construct a voltage source (*a*) Circuit diagram (*b*) Replacing the zener with its equivalent circuit

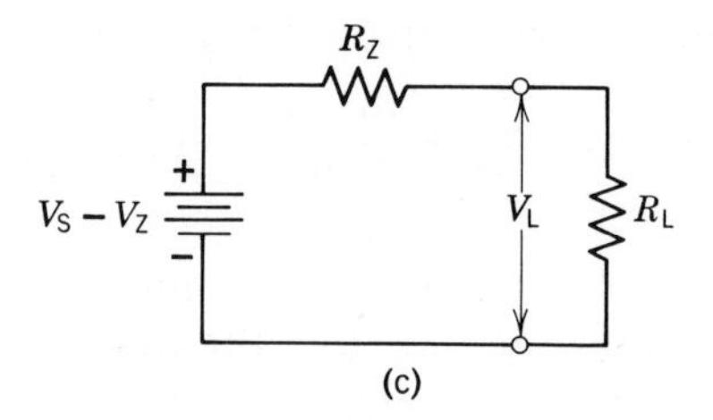

Figure 21.14 (*continued*) (*c*) Final equivalent circuit

Thus we can draw the final equivalent circuit as shown in Fig. 21.14*c*. This is a very efficient voltage source since power is dissipated in the zener only when there is a load current. The power dissipated in the zener P_Z is approximately

$$P_Z = I_Z V_Z \tag{21.7}$$

Furthermore, if the zener resistance R_Z is very small (an ideal zener) the load voltage V_L is very nearly equal to $(V_S - V_Z)$ for any value of load resistance R_L.

• • •

Example 21.8 Show a method for constructing a 6 V dc power supply capable of supplying 100 mA using a 20 V dc power supply and a zener diode.

Solution The circuit will appear exactly as shown in Fig. 21.14*a*. The proper zener voltage is

$$V_Z = V_S - V_L = 20 \text{ V} - 6 \text{ V} = 14 \text{ V}$$

Since the supply must be capable of delivering 100 mA to the load, the power dissipated in the zener at full load is found from Eq. 21.7. Thus,

$$P_Z = I_Z V_Z = 100 \text{ mA} \times 14 \text{ V} = 1.4 \text{ W}$$

We must therefore use a zener having a zener voltage of 14 V, and it must have a power rating greater than 1.4 W.

• • •

Example 21.9 If the zener diode selected in Example 21.8 is found to have a resistance of $R_Z = 1.0\ \Omega$, what will be the actual output voltage with a load of $R_L = 100\ \Omega$?

Solution The value of V_L can be found by applying the voltage divider theorem to the equivalent circuit in Fig. 21.14*c*. Thus,

$$V_L = (V_S - V_Z)\frac{R_L}{R_L + R_Z}$$

Substituting the known values into this equation yields

$$V_L = 6 \text{ V}\frac{100\ \Omega}{100\ \Omega + 1.0\ \Omega} = \frac{6 \times 100}{101} = 5.94 \text{ V}$$

• • •

21.4 A ZENER REGULATOR

A zener diode can also be used to construct a dc voltage source as shown in Fig. 21.15*a*. The source voltage consists of the two voltages v_S and V_S. V_S represents the dc component of voltage, and v_S represents a ripple voltage. The total source voltage is then the sum of V_S and v_S and is representative of the output voltage of a filtered half wave rectifier as in Fig. 21.9. The total load voltage is the sum of a dc component V_L and an ac component v_L.

In order to determine the total load voltage, we replace the zener with its equivalent circuit as shown in Fig. 21.15*b*. Since there are two types of voltage sources in the equivalent circuit (ac and dc), we can apply the Principle of Superposition to determine the load voltages.

In order to calculate V_L, we replace v_S with its internal impedance (a short) and form the dc equivalent circuit shown in Fig. 21.15*c*. From this equivalent circuit, the value of V_L can be found by simply applying Millman's theorem. Thus

$$V_L = \frac{\dfrac{V_S}{R_S} + \dfrac{V_Z}{R_Z}}{\dfrac{1}{R_S} + \dfrac{1}{R_Z} + \dfrac{1}{R_L}} \tag{21.8}$$

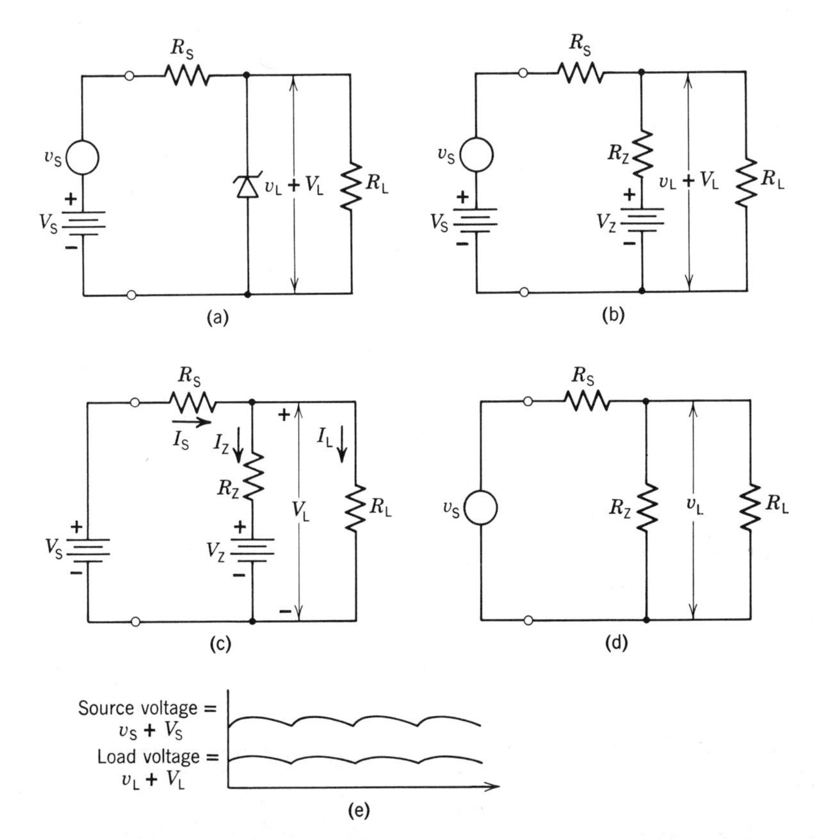

Figure 21.15 (*a*) A zener diode voltage regulator (*b*) Complete equivalent circuit (*c*) dc equivalent circuit (*d*) ac equivalent circuit (*e*) The source and load voltages

The ripple voltage across the load v_L can be found by replacing the dc voltage sources with their internal resistances (shorts) to create the ac equivalent circuit as shown in Fig. 21.15*d*. A simple application of the voltage divider theorem yields

$$v_L = v_S \frac{R_Z \| R_L}{R_Z \| R_L + R_S}$$

It is almost always true that R_Z is much smaller than the load R_L, and therefore the parallel combination of R_Z and R_L is very nearly equal to R_Z. That is,

$$R_Z \| R_L \cong R_Z$$

As a result, the value of v_L is very nearly

$$v_L \cong v_S \frac{R_Z}{R_Z + R_S} \tag{21.9}$$

The total load voltage is therefore the sum of the dc component V_L, given by Eq. 21.8, and the ac component v_L, given by Eq. 21.9. The load voltage is thus at a lower dc level than the source voltage, and the ripple voltage is greatly reduced as shown in Fig. 21.15*e*.

• • •

Example 21.10 Calculate the total load voltage for the circuit in Fig. 21.15*a* if: $V_S = 10\text{ V}$, $v_S = 0.5\text{ V}_{\text{P-P}}$, $V_Z = 6\text{ V}$, $R_S = 100\ \Omega$, $R_Z = 1\ \Omega$, and $R_L = 1\text{ k}\Omega$.

Solution We use Eqs. 21.8 and 21.9 to calculate the dc and ac components respectively. Thus,

$$V_L = \frac{\frac{10}{100} + \frac{6}{1}}{\frac{1}{100} + \frac{1}{1} + \frac{1}{1000}} = 6.02\text{ V dc}$$

$$v_L \cong 0.5 \frac{1}{1 + 100} = 4.95\text{ mV}_{\text{P-P}}$$

• • •

In this example, notice that the output voltage is a 6.02 V dc level with a 4.95 mV peak to peak ripple voltage added to it. The ripple voltage is so small compared to the 6.02 V dc level that it would barely be discernable; for all practical purposes it can be neglected. Notice however that the ripple voltage in the source v_S is quite discernable. The fact that the output ripple voltage has been reduced shows that the zener acts as a *voltage regulator.* That is, it tends to *prevent* any change in the source voltage from appearing at the output. The factor by which changes in the source voltage are attenuated before reaching the output is called the *attenuation factor* μ, and from Eq. 21.9 it is clearly

$$\mu = \frac{v_L}{v_S} \cong \frac{R_Z}{R_Z + R_S} \tag{21.10}$$

• • •

Example 21.11 Calculate the attenuation factor for the zener regulator in Example 21.10.

Solution Using Eq. 21.10,

$$\mu \cong \frac{R_Z}{R_Z + R_S} = \frac{1}{1 + 100} = 0.99 \times 10^{-2}$$

Notice that we can use this attenuation factor to calculate the output ripple voltage v_L. That is,

$$v_L = v_S\mu = 0.5\ V_{p\text{-}p} \times 0.99 \times 10^{-2} = 4.95\ mV_{p\text{-}p}$$

This result of course checks with the solution to Example 21.10 as it should.

• • •

When using the zener regulator in Fig. 21.15, we must always be certain that there is some zener current. That is, in Fig. 21.15*c*, we must never allow the zener current I_Z to become zero. If it does, the zener moves out of the breakdown region and simply appears as an open circuit. In this case, we no longer have a zener regulator, but simply a resistive voltage divider. We can insure that the zener is always operated properly by simply limiting the load current I_L to a maximum value. From Fig. 21.15*c*, it is clear that the total source current I_S is very nearly

$$I_S \cong \frac{V_S - V_Z}{R_S}$$

If the load is reduced to the point where it takes all of the source current, there will be none left for the zener. We must therefore never allow this to occur, and the maximum load current is then

$$I_{L_{MAX}} \cong \frac{V_S - V_Z}{R_S} \tag{21.11}$$

Since the load voltage is approximately equal to V_Z, the minimum value of load resistance is then

$$R_{L_{MIN}} \cong \frac{V_L}{I_{L_{MAX}}} \cong \frac{V_Z}{\dfrac{V_S - V_Z}{R_S}} = \frac{V_Z R_S}{V_S - V_Z} \tag{21.12}$$

• • •

Example 21.12 Calculate the maximum load current and the minimum load resistance for the regulator in Example 21.10.

Solution The maximum load current is found from Eq. 21.11 as

$$I_{L_{MAX}} \cong \frac{V_S - V_Z}{R_S} = \frac{10 - 6}{100} = \frac{4}{100} = 40\ mA$$

The minimum value of R_L is found from, Eq. 21.12. Thus

$$R_{L_{MIN}} \cong \frac{V_Z}{I_{L_{MAX}}} = \frac{6\ V}{40\ mA} = 150\ \Omega$$

• • •

21.5 ac VOLTMETERS

The basic D'Arsonval meter movement and its applications in constructing dc ammeters, dc voltmeters, and ohmmeters were discussed in detail in Chapter 9. We can use this basic meter movement in conjunction with a diode to construct two different types of voltmeters capable of measuring ac voltages. The first type is called an *average reading* voltmeter since it responds to the *average* value of ac voltage. The second type is called a *peak reading* voltmeter since it responds to the *peak* value of the voltage.

The circuit shown in Fig. 21.16*a* forms the basis for constructing an average reading ac voltmeter. We recognize the circuit as a basic half wave rectifier with a series resistance R to limit the current and a D'Arsonval meter movement. If an ac voltage source is connected across the terminals of the circuit, the current will appear as shown in Fig. 21.16*b*.

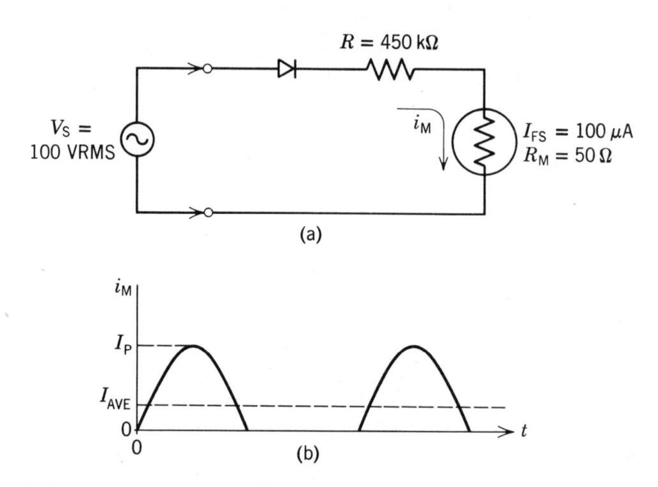

Figure 21.16 (*a*) An average reading ac voltmeter (*b*) The meter current in *a*

Now, if the frequency of the ac voltage source is low enough, the meter needle will have to move faster and faster in order to respond to the current move from zero up to a maximum value and back down to zero each time a current pulse occurs. In between pulses, the needle will remain at zero.

On the other hand, as the source voltage frequency is increased, the meter needle will have to move faster and faster in order to respond to the current pulses. The mass of the moving portion of the meter will prevent the needle from moving too fast. As a result, the meter will not have time to move up to I_p during one quarter of a cycle, nor will it have time to move back down to zero between pulses. Eventually a frequency will be reached where the needle will simply settle at a steady value between I_p and zero. At this frequency and all frequencies above this the meter will actually respond to the *average* value of the current. It can be shown that the average value of the half wave rectified current is equal to the peak value divided by π. Thus

$$I_{AVE} = \frac{I_P}{\pi} = 0.318 I_p \tag{21.13}$$

Alternatively, for the half wave rectifier,

$$I_p = \pi I_{AVE} \tag{21.14}$$

The meaning of I_p and I_{AVE} are shown in Fig. 21.16*b*.

The meter in Fig. 21.16*a* has a fullscale current I_{FS} of 100 μA. This means that the *maximum* ac voltage which we can measure with this meter will produce an average current I_{AVE} of 100 μA. But, from Eq. 21.14 we see that this corresponds to a current having a peak value of

$$I_{P_{MAX}} = \pi I_{AVE_{MAX}} = \pi I_{FS} = \pi \times 100\ \mu A = 314\ \mu A$$

Such a current will be produced if we apply to the meter terminals an ac voltage having a peak value of approximately

$$V_{P_{MAX}} \cong I_{P_{MAX}}R = 314\ \mu\text{A} \times R = 314R \times 10^{-6}\ \text{V}$$

We are assuming that the diode resistance and the meter resistance R_M are very small compared to R and that the diode voltage is small compared to the applied voltage. Thus $V_{P_{MAX}}$ corresponds to the full scale peak voltage of the meter. Most meters of this type are calibrated in terms of RMS voltages, and since $V_{RMS} = 0.707\ V_p$, it is clear that the full-scale voltage of the meter is

$$V_{FS} = 0.707\ V_{P_{MAX}} = 0.707 \times 314R \times 10^{-6} = 222 \times 10^{-6}R\ \text{V(RMS)}$$

If we now make the value of R equal to 450 kΩ, we have a voltmeter with a full-scale voltage of

$$V_{FS} = 222 \times 10^{-6} \times 450 \times 10^{3} = 100\ V_{RMS}$$

From this example, we can write an equation for the full-scale RMS voltage of the meter in Fig. 21.16*a* as

$$V_{FS}(\text{RMS}) = 0.707 \times \pi \times I_{FS} \times R = 2.22 I_{FS} R \qquad (21.15)$$

• • •

Example 21.13 What must be the value of R in the voltmeter in Fig. 21.16*a* in order to change the full-scale voltage to 150 V_{RMS}?

Solution From Eq. 21.15,

$$R = \frac{V_{FS}(\text{RMS})}{2.22 I_{FS}} = \frac{150\ \text{V}}{2.22 \times 10^{-4}\ \text{A}} = 675\ \text{k}\Omega$$

It must be emphasized that this average reading voltmeter is based on a *sinusoidal* voltage, and if it is used to measure other than sinusoidal voltages, the readings will obviously be incorrect.

• • •

The construction of a peak reading ac voltmeter is quite simple as shown in Fig. 21.17*a*. When an ac voltage v_S is applied to the terminals of the meter, the capacitor will charge to the peak value of the applied voltage the first time it goes positive as shown in Fig. 21.17*b*. If the resistance of the dc voltmeter is very high, the capacitor voltage v_C will not discharge appreciably between positive peaks of the source voltage v_S. As a result, the capacitor voltage will remain essentially constant at a value equal to the peak voltage of the applied signal V_p; that is

$$v_C \cong V_p$$

As a result, the voltage measured by the dc voltmeter is very nearly equal to the peak value of the source voltage.

It should be pointed out that the two voltmeters discussed in this section are theoretically correct. However, most commercial instruments are somewhat more complicated since they contain additional refinements to account for such things as diode leakage current, the voltage drop across the forward biased diode, and so forth.

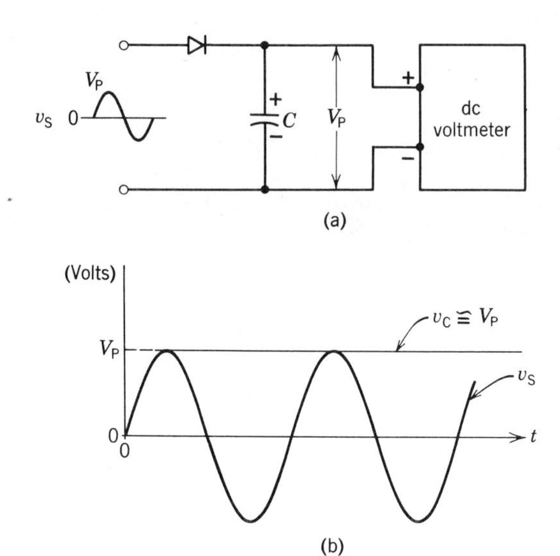

Figure 21.17 (*a*) A peak reading ac voltmeter (*b*) The applied voltage v_S and the capacitor voltage v_C

• • •

Example 21.14 Could the peak reading ac voltmeter in Fig. 21.17*a* be used to measure RMS voltages?

Solution The meter can be used to measure RMS voltages by simply multiplying the dc voltmeter scales by 0.707, since $V_{RMS} = 0.707\ V_p$.

• • •

21.6 AMPLIFIERS

An electronic circuit which has been designed specifically for the purpose of increasing the value of a current or voltage can be called an *amplifier*. An amplifier can be represented in block diagram form as a box with a pair of input terminals and a pair of output terminals as shown in Fig. 21.18. The circuit inside the box may contain a number of *passive* elements (R's, L's, and C's) as well as *active* elements such as vacuum tubes or transistors.

The *gain* A of an amplifier is a measure of how much the output signal has been *increased* over the input signal. Thus, if we are interested in amplifying a voltage, the *voltage gain* A_V of the amplifier is

$$A_V = \frac{v_{out}}{v_{in}} \tag{21.16}$$

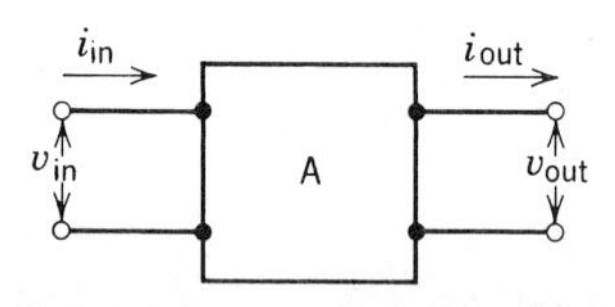

Figure 21.18 The block diagram of an amplifier

This quantity is called the *open circuit* voltage gain since it is measured with the output terminals open circuited (that is, no load resistance is connected to the output). An *ideal amplifier* will provide at its output a faithful reproduction of the signal at its input. Thus if the input signal is a sine wave, the output will be an amplified sine wave; if the input is triangular, the output will be triangular, and so on.

• • •

Example 21.15 What is the open circuit voltage gain of the amplifier in Fig. 21.18 if a 5 mV signal at the input produces a 1.0 V output signal?

Solution The open circuit voltage gain is the ratio of output voltage to input voltage as given by Eq. 21.16. Thus

$$A_V = \frac{v_{out}}{v_{in}} = \frac{1.0\text{ V}}{5\text{ mV}} = 200$$

• • •

A generalized equivalent circuit of a *voltage amplifier* can be represented as shown in Fig. 21.19*a*. An amplifier will in general have some input resistance, and this is accounted for by the resistor R_{in} in the figure. Since the purpose here is to amplify voltages, we can simply place a voltage source in series with a resistance to represent the amplifier output. The resistor R_{out} represents the output resistance of the amplifier. The voltage source $A_V v_{in}$ is called a *dependent source,* since its value *depends* on the open circuit gain A_V of the amplifier.

In Fig. 21.19*b* a signal voltage v_S to be amplified is connected to a voltage amplifier with a load resistance of R_L connected to its output. The overall voltage gain of the amplifier is now given by

$$A_{V_L} = \frac{v_L}{v_S} \tag{21.17}$$

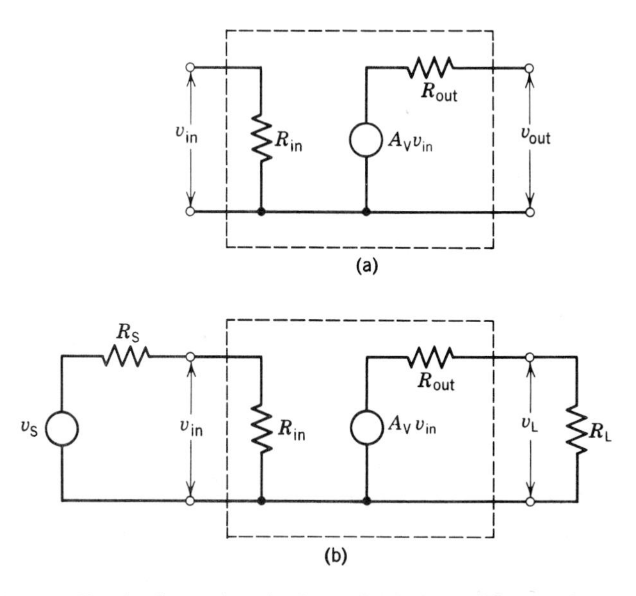

Figure 21.19 (*a*) Equivalent circuit (*b*) Voltage amplifier with a signal voltage v_S and a load R_L

and it will be somewhat less than the open circuit voltage gain A_V because of the presence of R_S and R_{out}.

In order to determine A_{V_L}, we begin by finding v_{in}. An application of the voltage divider theorem to the input of the amplifier yields

$$v_{in} = v_S \frac{R_{in}}{R_{in} + R_S} \tag{21.18}$$

The load voltage v_L is found by applying the voltage divider theorem at the output, and

$$v_L = A_V v_{in} \frac{R_L}{R_L + R_{out}} \tag{21.19}$$

Substituting Eq. 21.18 into this relationship yields

$$v_L = A_V \left(v_S \frac{R_{in}}{R_{in} + R_S} \right) \times \frac{R_L}{R_L + R_{out}}$$

We then obtain the amplifier voltage gain by dividing both sides of this equation by v_S. Thus

$$A_{V_L} = \frac{v_L}{v_S} = A_V \frac{R_{in}}{R_{in} + R_S} \times \frac{R_L}{R_L + R_{out}} \tag{21.20}$$

From Eq. 21.20 it is clear that the actual amplifier voltage gain A_{V_L} is reduced from the open circuit value A_V because of the presence of R_S and R_{out}. If we could make $R_S = 0$ or make R_{in} very large compared to R_S, the term $R_{in}/(R_{in} + R_S)$ would become 1.0. Similarly, if we could make $R_{out} = 0$, the term $R_L/(R_L + R_{out})$ would also become 1.0. Under these conditions, the amplifier voltage gain would always be equal to the open circuit value. That is, $A_{V_L} = A_V$. These are the conditions necessary for an *ideal voltage amplifier*, and voltage amplifiers are therefore designed to have a ***high input resistance*** and a ***low output resistance.***

• • •

Example 21.16 Calculate the voltage gain of the amplifier in Fig. 21.19*b* if: $R_S = 1\ \text{k}\Omega$, $R_{in} = 4\ \text{k}\Omega$, $R_{out} = 500\ \Omega$, $R_L = 5\ \text{k}\Omega$, and $A_V = 120$.

Solution We use Eq. 21.20 to obtain

$$A_{V_L} - A_V \times \frac{R_{in}}{R_{in} + R_S} \times \frac{R_L}{R_L + R_{out}} = 120 \times \frac{4\ \text{k}\Omega}{4\ \text{k}\Omega + 1\ \text{k}\Omega} \times \frac{5\ \text{k}\Omega}{5\ \text{k}\Omega + 0.5\ \text{k}\Omega}$$

$$= 120 \times 0.8 \times 0.91 = 87.4$$

• • •

Example 21.17 Calculate the voltage gain in Example 21.16 with $R_S = 0\ \Omega$, then with $R_{out} = 0\ \Omega$, and finally with both R_S and $R_{out} = 0\ \Omega$.

Solution

with $R_S = 0\ \Omega$,

$$A_{V_L} = 120 \times \frac{4\ \text{k}\Omega}{4\ \text{k}\Omega} \times \frac{5\ \text{k}\Omega}{5\ \text{k}\Omega + 0.5\ \text{k}\Omega} = 120 \times 1.0 \times 0.91 = 109$$

with $R_{out} = 0\ \Omega$,

$$A_{V_L} = 120 \times \frac{4\ \text{k}\Omega}{4\ \text{k}\Omega + 1\ \text{k}\Omega} \times \frac{5\ \text{k}\Omega}{5\ \text{k}\Omega} = 120 \times 0.8 \times 1.0 = 96$$

with $R_S = R_{out} = 0\ \Omega$,

$$A_{V_L} = 120 \times \frac{4\ \text{k}\Omega}{4\ \text{k}\Omega} \times \frac{5\ \text{k}\Omega}{5\ \text{k}\Omega} = 120 \times 1.0 \times 1.0 = 120$$

• • •

SUMMARY

In this chapter we have demonstrated the application of the basic principles of circuit analysis to electronic circuits. Electronic circuits may contain passive elements such as R's, L's, and C's as well as electronic elements such as diodes, vacuum tubes, and transistors. The diode is a nonlinear device, and we can predict its behavior in a circuit by using the diode equivalent circuits developed from a characteristic curve. The zener diode has a specially designed reverse breakdown voltage, the zener voltage, and it is normally operated in the reverse breakdown region. Zener diodes are useful in constructing dc power sources, and in providing voltage regulation. A single half wave rectifier circuit can be used in conjunction with a D'Arsonval meter to construct a simple average reading ac voltmeter. A variation of the half wave rectifier (sometimes called a *peak detector*) can be used with a dc voltmeter to form a peak reading ac voltmeter. Amplifiers are used to amplify signals, and they usually contain active devices such as vacuum tubes or transistors. The equivalent circuit of a basic voltage amplifier takes into account the input resistance and the open circuit voltage gain. An ideal voltage amplifier has a very high input resistance and a very low output resistance. Under these conditions, the amplifier voltage gain will remain constant regardless of the load resistance or the source resistance.

GLOSSARY

Amplifier. *A circuit used to increase a current or voltage.*

Attenuation factor. *The factor by which a voltage regulator reduces voltage variations.*

Average reading meter. *A meter that responds to the average value of a signal.*

Bilateral element. *An element through which current will pass with equal ease in either direction.*

Breakdown voltage. *The reverse voltage beyond which a diode appears as a very low resistance.*

Characteristic curve. *A plot of the current through versus the voltage across an element.*

Dependent source. *In a voltage amplifier, a source whose value depends on the open circuit voltage gain.*

Knee voltage. *The forward voltage across a diode beyond which the current increases rapidly.*

Linear element. *An element whose characteristic curve is a straight line.*

Peak reading meter. *A meter that responds to the peak value of a signal.*

Ripple voltage. *In a dc power supply, the periodic voltage superimposed on the dc output voltage.*

Voltage regulator. *A circuit designed to maintain a constant value of dc voltage.*

Zener diode. *A diode normally operated in its reverse breakdown region.*

Zener voltage. *The reverse breakdown voltage of a zener diode.*

IMPORTANT RELATIONSHIPS

- Diode equation:

$$I = I_S(e^{40V_d} - 1)$$

- Half wave rectifier:

$$\text{ripple voltage} = V_{p\text{-}p} = V_p(1 - e^{-t_d/R_L C})$$

where

$$t_d \cong \frac{1}{f}$$

- Zener diode:

$$R_Z = \frac{\Delta V}{\Delta I}$$

$$P_Z = I_Z V_Z$$

- Zener regulator:

$$\mu = \frac{v_L}{v_S} \cong \frac{R_Z}{R_Z + R_S}$$

$$I_{L_{MAX}} \cong \frac{V_S - V_Z}{R_S} \qquad R_{L_{MIN}} \cong \frac{V_Z R_S}{V_S - V_Z}$$

- Average reading ac voltmeter using a half-wave rectifier:

$$V_{FS}(\text{RMS}) = 2.22 I_{FS} R$$

- Amplifier:

$$A_V = \frac{v_{out}}{v_{in}} \qquad A_{V_L} = \frac{v_L}{v_S} = A_V \frac{R_{in}}{R_{in} + R_S} \times \frac{R_L}{R_L + R_{out}}$$

REVIEW QUESTIONS

1. A semiconductor diode is a bilateral element. (TF)
2. A semiconductor diode (is, is not) a linear circuit element.
3. The approximate equivalent circuit of a (forward, reverse) biased diode is simply a small resistance.
4. The behavior of a reverse biased diode can be approximated by an (open, closed) switch.
5. A half-wave rectifier can be used to change an ac voltage to a dc voltage. (TF)
6. The ripple voltage at the output of the dc power supply in Fig. 21.9*a* is due to the discharge of the capacitor through the diodes. (TF)
7. The discharge time for the dc power supply in Fig. 21.9*a* is approximately equal to the ____ of the applied voltage wave form.
8. The ripple voltage in Fig. 21.9 can be reduced by increasing either ____ or ____.
9. The dc output voltage of the voltage doubler in Fig. 21.11 is approximately equal to twice the input (RMS, peak) voltage.
10. The reverse breakdown voltage of a zener diode is called the ____ voltage.
11. A zener diode is normally operated in the ____ region.
12. Zener voltages from around 3.0 V up to 100 V are easily obtained. (TF)

13. The approximate equivalent circuit of a zener diode in the breakdown region is the same as that of a voltage source. (TF)
14. To be a good voltage regulator, a zener diode should have a (large, small) zener resistance.
15. The attenuation factor of a zener voltage regulator should be as (large, small) as possible.
16. Will the zener voltage source in Fig. 21.14 reduce any ripple voltage present in the source voltage?
17. Why must there always be zener current I_Z in Fig. 21.15c?
18. The ac voltmeter in Fig. 21.16a is called an average reading meter since it responds to the average value of the current through the meter movement. (TF)
19. The average value of the meter current in Fig. 21.16 is equal to the peak value divided by ____.
20. The ac voltmeter in Fig. 21.16 will give proper RMS voltage values only if the input signal is ____.
21. The peak reading voltmeter in Fig. 21.17 will give correct readings of peak voltage if the input signal is triangular. (TF)
22. An ideal voltage amplifier should have a (high, low) input resistance and a (high, low) output resistance.
23. The voltage source in the output of the general equivalent circuit of a voltage amplifier is called a ____ source.
24. The actual voltage gain A_{V_L} of a voltage amplifier is usually somewhat (less than, greater than) the open circuit voltage gain A_V.

ROBLEMS

1. Calculate I_d in Fig. 21.3a if $I_S = 0.02\ \mu A$ and $V_d = 50$ mV and 500 mV.
2. Repeat Problem 1 if $I_S = 4$ nA.
3. What must be the value of the leakage current I_S in a diode if a forward diode voltage of 0.3 V causes a diode current of 2 mA?
4. Draw the equivalent circuit for a forward biased diode.
5. Draw the equivalent circuit for a reverse biased diode.
6. Calculate the voltage across R_L in Fig. 21.6 if R_L is changed to 100 Ω.
7. If the reverse biased diode in Fig. 21.7 were simulated by a 1 MΩ resistor instead of an open switch (to account for leakage current I_S), what would be the voltage across R_L?
8. Calculate the ripple voltage and the approximate dc voltage for the half-wave rectifier in Fig. 21.9 if $v_S = 25 \sin 377t$, $C = 500\ \mu F$, and $R_L = 200\ \Omega$.
9. Repeat Problem 8 if C is changed to 5000 μF.
10. What would be the ripple voltage for the power supply in Fig. 21.10 if the transformer were changed to a 1:1 turns ratio?
11. What is the maximum reverse voltage applied across the diode in Fig. 21.10?
12. What must be the turns ratio of the transformer in Fig. 21.11 in order to produce an output voltage of approximately 65 V dc?
13. Draw the equivalent circuit of a zener diode being operated in the breakdown region.
14. What must be the zener resistance of the diode in Fig. 21.12 if $\Delta I = 10$ mA and $\Delta V = 23$ mV?

15. Draw the complete circuit (similar to Fig. 21.14*a*) for constructing a 10 V dc power supply capable of supplying 250 mA from a 50 V dc source and a zener diode.
16. Calculate the actual voltage delivered to a load of 120 Ω by the source in Problem 15 if the zener has a resistance of 14 Ω.
17. Repeat Problem 15 if the source voltage is changed to 25 V dc.
18. Calculate the total load voltage for the regulator in Fig. 21.15 if: $V_S = 28$ V, $v_S = 0.3$ $V_{P\text{-}P}$, $V_Z = 10$ V, $R_S = 50$ Ω, $R_Z = 5$ Ω, and $R_L = 700$ Ω.
19. Calculate the attenuation factor for the regulator in Problem 18.
20. Repeat Problems 18 and 19 if V_Z is changed to 5 V and $R_Z = 2$ Ω.
21. What is the minimum value of load resistance which can be connected to the regulator in Problem 18? What is the maximum load current?
22. Calculate the value of R in Fig. 21.16 to construct a 50 V and a 200 V full-scale meter.
23. If the value of R in Fig. 21.16 is given as 675 kΩ, and the full-scale RMS voltage with that value of resistance is 75 V, what must be the full-scale meter current?
24. Draw the generalized equivalent circuit of a voltage amplifier.
25. What is the voltage gain of the amplifier in Fig. 21.18 if a 25 mV input produces a 1.95 V output?
26. The open circuit voltage gain of the amplifier in Fig. 21.19 is 75. What will be its overall voltage gain if a load of 1 kΩ is connected to its output, and: $R_S = R_{in} = 500$ Ω, $R_{out} = 1$ kΩ?
27. Find the overall voltage gain in Problem 26 if R_{in} is increased to 500 kΩ and R_{out} is reduced to 100 Ω.
28. An input signal of $v_S = 0.03 \sin \omega t$ is applied to the input of the amplifier in Problem 27. Write the equation for the output voltage.

DETERMINANTS AND MATRICES

In mathematics, the problem of finding a simultaneous solution for a number of equations is covered quite thoroughly in the study of *matrix algebra.** In the study of electric circuits, this problem is encountered when attempting to solve a circuit by mesh equations or node equations. Simultaneous algebraic equations can of course be solved by the traditional methods of *elimination* or *substitution*, but these methods may become unwieldy as the number of equations increases or if the coefficients are other than simple integers. The methods of *determinants* developed in matrix algebra provides a very simple and systematic procedure for solving such equations, and it will be well worthwhile to master this technique.

In solving mesh or node equations, we will always have the *same* number of *equations* as *unknowns*, and we will therefore concentrate our attention on solving a system of n equations for n unknowns. Let us first consider the problem of solving two mesh equations for the two unknown mesh currents. The equations can be written in general form as

$$V_{11} = R_{11}I_1 - R_{12}I_2 \tag{A-1}$$

$$V_{22} = -R_{21}I_1 + R_{22}I_2 \tag{A-2}$$

In these equations, the voltages V_{11} and V_{22} are the voltage sources around meshes 1 and 2 respectively, and they are considered as *constants*. The mesh currents I_1 and I_2 are the unknowns which we are trying to solve for, and we call them the *variables*. The resistances R_{11}, R_{12}, R_{21}, and R_{22} are the *coefficients* of the variables.

The constants and the coefficients in Eqs. A-1 and A-2 will simply be numbers, and

*For example, see *Matrix Algebra*, Hohn, F. E., Macmillan, New York, 1967.

these are the quantities which must be manipulated to find solutions for the mesh currents. These two equations can be written in *matrix form* (a mathematical shorthand) without any loss of information as

$$\underset{\text{constants' matrix}}{\begin{bmatrix} V_{11} \\ V_{22} \end{bmatrix}} = \underset{\text{coefficient matrix}}{\underbrace{\overbrace{\begin{bmatrix} \overset{\text{Col. 1}}{\downarrow} & \overset{\text{Col. 2}}{\downarrow} \\ R_{11} & -R_{12} \\ -R_{21} & R_{22} \end{bmatrix}}^{\text{Columns}}}} \underset{\text{variable matrix}}{\begin{bmatrix} I_1 \\ I_2 \end{bmatrix}}. \quad \left.\begin{matrix} \xleftarrow{\text{Row 1}} \\ \xleftarrow{\text{Row 2}} \end{matrix}\right\} \text{Rows} \qquad \text{(A-3)}$$

Each of the quantities within a pair of brackets [ ] is called a *matrix*, and Eq. A-3 is seen to consist of three different matrices as labeled. A matrix is then simply a rectangular array of quantities having a number of rows (horizontal) and a number of columns (vertical). The coefficient matrix in Eq. A-3 clearly has two rows and two columns, while the constants matrix has two rows and one column. Notice that all of the information necessary to determine I_1 and I_2 is contained within the coefficient matrix and the constants matrix.

In order to evaluate I_1 and I_2, we must determine the numerical values of three matrices. The numerical value of a square matrix is its *determinant.* The coefficient matrix in Eq. A-3 has a determinant since it is square (it has two rows and two columns); it therefore has a numerical value which we will designate as Δ.* A determinant is distinguished from a matrix by using brackets of the form | | instead of [ ], and thus

$$\Delta = \begin{vmatrix} R_{11} & -R_{12} \\ -R_{21} & R_{22} \end{vmatrix}$$

We can form a second determinant by replacing the 1st column of the coefficient matrix with the constants' matrix. We call this determinant Δ_1, and thus

$$\Delta_1 = \begin{vmatrix} V_{11} & -R_{12} \\ V_{22} & R_{22} \end{vmatrix}$$

The third determinant Δ_2 is formed in a similar fashion by replacing the 2nd column of the coefficient matrix with the constants' matrix, and thus

$$\Delta_2 = \begin{vmatrix} R_{11} & V_{11} \\ -R_{21} & V_{22} \end{vmatrix}$$

The values of I_1 and I_2 are easily calculated from the values of these three determinants, and by Cramer's rule they are

$$I_1 = \frac{\Delta_1}{\Delta} \qquad \text{(A-4)}$$

$$I_2 = \frac{\Delta_2}{\Delta} \qquad \text{(A-5)}$$

Notice that Δ_1 is formed by replacing ***the coefficients of*** I_1 (col. 1) with the constants, and Δ_2 is formed by replacing ***the coefficients of*** I_2 (col. 2) with the constants.

We must now state the rules for finding the numerical value of a 2 by 2 (2 rows and 2 columns) determinant. Very simply, we take the product of the *forward diagonal* and *subtract* from it the product of the *reverse diagonal.* For example, the value of the

*A matrix has a determinant (unique numerical value) only if it is square.

determinant below is

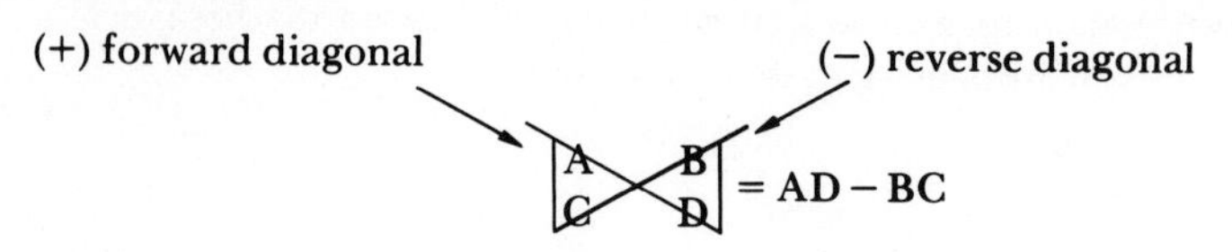

As an example, the numerical value of the determinant below is

$$\begin{vmatrix} 1 & 3 \\ 2 & 7 \end{vmatrix} = (1 \times 7) - (2 \times 3) = 7 - 6 = 1$$

Let us now apply this technique to solve the following two mesh equations for I_1 and I_2.

$$12 = 10I_1 - 5I_2$$
$$6 = -5I_1 + 15I_2$$

In matrix form these equations are

$$\begin{bmatrix} 12 \\ 6 \end{bmatrix} = \begin{bmatrix} 10 & -5 \\ -5 & 15 \end{bmatrix} \begin{bmatrix} I_1 \\ I_2 \end{bmatrix}$$

$$\Delta = \begin{vmatrix} 10 & -5 \\ -5 & 15 \end{vmatrix} = (10 \times 15) - (-5)(-5) = 150 - 25 = 125$$

$$\Delta_1 = \begin{vmatrix} 12 & -5 \\ 6 & 15 \end{vmatrix} = (12 \times 15) - (-5)(6) = 180 + 30 = 210$$

$$\Delta_2 = \begin{vmatrix} 10 & 12 \\ -5 & 6 \end{vmatrix} = (10 \times 6) - (12)(-5) = 60 + 60 = 120$$

The two mesh currents are then

$$I_1 = \frac{\Delta_1}{\Delta} = \frac{210}{125} = 1.68$$

$$I_2 = \frac{\Delta_2}{\Delta} = \frac{120}{125} = 0.96$$

A system of three simultaneous equations can also be solved very easily by using determinants. Three mesh equations can be written in general form as

$$V_{11} = R_{11}I_1 - R_{12}I_2 - R_{13}I_3$$
$$V_{22} = -R_{21}I_1 + R_{22}I_2 - R_{23}I_3$$
$$V_{33} = -R_{31}I_1 - R_{32}I_2 + R_{33}I_3$$

These three equations in matrix form are,

$$\begin{bmatrix} V_{11} \\ V_{22} \\ V_{33} \end{bmatrix} = \begin{bmatrix} R_{11} & -R_{12} & -R_{13} \\ -R_{21} & R_{22} & -R_{23} \\ -R_{31} & -R_{32} & R_{33} \end{bmatrix} \begin{bmatrix} I_1 \\ I_2 \\ I_3 \end{bmatrix}$$

For this system we must evaluate four 3 by 3 (3 rows and 3 columns) determinants. The first determinant is simply the coefficient determinant and we still use the symbol Δ. Thus

$$\Delta = \begin{vmatrix} R_{11} & -R_{12} & -R_{13} \\ -R_{21} & R_{22} & -R_{23} \\ -R_{31} & -R_{32} & R_{33} \end{vmatrix}$$

The other three determinants are formed quite easily by replacing each of the columns, one at a time, with the constants. Thus

$$\Delta_1 = \begin{vmatrix} V_{11} & -R_{12} & -R_{13} \\ V_{22} & R_{22} & -R_{23} \\ V_{33} & -R_{32} & R_{33} \end{vmatrix}$$

$$\Delta_2 = \begin{vmatrix} R_{11} & V_{11} & -R_{13} \\ -R_{21} & V_{22} & -R_{23} \\ -R_{31} & V_{33} & R_{33} \end{vmatrix}$$

$$\Delta_3 = \begin{vmatrix} R_{11} & -R_{12} & V_{11} \\ -R_{21} & R_{22} & V_{22} \\ -R_{31} & -R_{32} & V_{33} \end{vmatrix}$$

We then use Cramer's rule to calculate the mesh currents, and they are

$$I_1 = \frac{\Delta_1}{\Delta} \tag{A-6}$$

$$I_2 = \frac{\Delta_2}{\Delta} \tag{A-7}$$

$$I_3 = \frac{\Delta_3}{\Delta} \tag{A-8}$$

We must now turn our attention to the technique for calculating the value of a 3 by 3 determinant. There are a number of methods,* but the one presented here is straightforward and easy to apply. The technique is best described graphically. We take the product of the three *forward diagonals* as shown in Eq. A-9, and *subtract* the product of the three *reverse diagonals* as shown in Eq. A-10.

(+) (+) (+)

$$\begin{vmatrix} A & B & C \\ D & E & F \\ G & H & J \end{vmatrix} \quad +AEJ + BFG + CHD \tag{A-9}$$

$$\begin{vmatrix} A & B & C \\ D & E & F \\ G & H & J \end{vmatrix} \quad -CEG - BDJ - AHF \tag{A-10}$$

(−) (−) (−)

Thus, the value of the determinant shown is

$$\begin{vmatrix} A & B & C \\ D & E & F \\ G & H & J \end{vmatrix} = AEJ + BFG + CHD - CEG - BDJ - AHF$$

As an example, consider the solution of the three mesh equations below by using determinants.

$$\begin{aligned} 6 &= 3I_1 - 2I_2 - 0I_3 \\ 0 &= -2I_1 + 6I_2 - 1I_3 \\ -3 &= 0I_1 - 1I_2 + 3I_3 \end{aligned}$$

The four required determinants are,

$$\Delta = \begin{vmatrix} 3 & -2 & 0 \\ -2 & 6 & -1 \\ 0 & -1 & 3 \end{vmatrix} = (3)(6)(3) + (-2)(-1)(0) + (0)(-2)(-1)$$
$$- (0)(6)(0) - (-2)(-2)(3) - (3)(-1)(-1)$$
$$= 54 + 0 + 0 - 0 - 12 - 3 = 39$$

*Hohn, F. E., *Matrix Algebra.*

$$\Delta_1 = \begin{vmatrix} 6 & -2 & 0 \\ 0 & 6 & -1 \\ -3 & -1 & 3 \end{vmatrix} = (6)(6)(3) + (-2)(-1)(-3) + (0)(0)(-1)$$
$$- (0)(6)(-3) - (-2)(0)(3) - (6)(-1)(-1)$$
$$= 108 - 6 + 0 - 0 - 0 - 6 = 96$$

$$\Delta_2 = \begin{vmatrix} 3 & 6 & 0 \\ -2 & 0 & -1 \\ 0 & -3 & 3 \end{vmatrix} = (3)(0)(3) + (6)(-1)(0) + (0)(-2)(-3)$$
$$- (0)(0)(0) - (6)(-2)(3) - (3)(-1)(-3)$$
$$= 0 + 0 + 0 - 0 + 36 - 9 = 27$$

$$\Delta_3 = \begin{vmatrix} 3 & -2 & 6 \\ -2 & 6 & 0 \\ 0 & -1 & -3 \end{vmatrix} = (3)(6)(-3) + (-2)(0)(0) + (6)(-2)(-1)$$
$$- (6)(6)(0) - (-2)(-2)(-3) - (3)(-1)(0)$$
$$= -54 + 0 + 12 - 0 + 12 + 0 = -30$$

Then from Eqs. A-6, A-7, and A-8 we have

$$I_1 = \frac{\Delta_1}{\Delta} = \frac{96}{39} = 2.46$$

$$I_2 = \frac{\Delta_2}{\Delta} = \frac{27}{39} = 0.693$$

$$I_3 = \frac{\Delta_3}{\Delta} = \frac{-30}{39} = -0.769$$

The method of determinants is widely used to solve simultaneous equations having any number of unknowns. However, the physical labor required increases rapidly as the number of equations increases. For example, it requires the computation of 720 terms to evaluate a 6 by 6 determinant by this method! For this reason, digital computers are widely used for solving such problems. The digital computer can evaluate determinants with great speed and accuracy. In this book, we will not concern ourselves with determinants larger than 3 by 3, and the interested reader is referred to other texts devoted to this subject.

B

MATHEMATICAL TABLES AND CONVERSION FACTORS

Common Logarithms of Numbers 1–100

N	0	1	2	3	4	5	6	7	8	9
0	....	0000	3010	4771	6021	6990	7782	8451	9031	9542
1	0000	0414	0792	1139	1461	1761	2041	2304	2553	2788
2	3010	3222	3424	3617	3802	3979	4150	4314	4472	4624
3	4771	4914	5051	5185	5315	5441	5563	5682	5798	5911
4	6021	6128	6232	6335	6435	6532	6628	6721	6812	6902
5	6990	7076	7160	7243	7324	7404	7482	7559	7634	7709
6	7782	7853	7924	7993	8062	8129	8195	8261	8325	8388
7	8451	8513	8573	8633	8692	8751	8808	8865	8921	8976
8	9031	9085	9138	9191	9243	9294	9345	9395	9445	9494
9	9542	9590	9638	9685	9731	9777	9823	9868	9912	9956
10	0000	0043	0086	0128	0170	0212	0253	0294	0334	0374
11	0414	0453	0492	0531	0569	0607	0645	0682	0719	0755
12	0792	0828	0864	0899	0934	0969	1004	1038	1072	1106
13	1139	1173	1206	1239	1271	1303	1335	1367	1399	1430
14	1461	1492	1523	1553	1584	1614	1644	1673	1703	1732
15	1761	1790	1818	1847	1875	1903	1931	1959	1987	2014
16	2041	2068	2095	2122	2148	2175	2201	2227	2253	2279
17	2304	2330	2355	2380	2405	2430	2455	2480	2504	2529
18	2553	2577	2601	2625	2648	2672	2695	2718	2742	2765
19	2788	2810	2833	2856	2878	2900	2923	2945	2967	2989
20	3010	3032	3054	3075	3096	3118	3139	3160	3181	3201
21	3222	3243	3263	3284	3304	3324	3345	3365	3385	3404
22	3424	3444	3464	3483	3502	3522	3541	3560	3579	3598
23	3617	3636	3655	3674	3692	3711	3729	3747	3766	3784
24	3802	3820	3838	3856	3874	3892	3909	3927	3945	3962
25	3979	3997	4014	4031	4048	4065	4082	4099	4116	4133
26	4150	4166	4183	4200	4216	4232	4249	4265	4281	4298
27	4314	4330	4346	4362	4378	4393	4409	4425	4440	4456
28	4472	4487	4502	4518	4533	4548	4564	4579	4594	4609
29	4624	4639	4654	4669	4683	4698	4713	4728	4742	4757
30	4771	4786	4800	4814	4829	4843	4857	4871	4886	4900
31	4914	4928	4942	4955	4969	4983	4997	5011	5024	5038
32	5051	5065	5079	5092	5105	5119	5132	5145	5159	5172
33	5185	5198	5211	5224	5237	5250	5263	5276	5289	5302
34	5315	5328	5340	5353	5366	5378	5391	5403	5416	5428
35	5441	5453	5465	5478	5490	5502	5514	5527	5539	5551
36	5563	5575	5587	5599	5611	5623	5635	5647	5658	5670
37	5682	5694	5705	5717	5729	5740	5752	5763	5775	5786
38	5798	5809	5821	5832	5843	5855	5866	5877	5888	5899
39	5911	5922	5933	5944	5955	5966	5977	5988	5999	6010
40	6021	6031	6042	6053	6064	6075	6085	6096	6107	6117
41	6128	6138	6149	6160	6170	6180	6191	6201	6212	6222
42	6232	6243	6253	6263	6274	6284	6294	6304	6314	6325
43	6335	6345	6355	6365	6375	6385	6395	6405	6415	6425
44	6435	6444	6454	6464	6474	6484	6493	6503	6513	6522
45	6532	6542	6551	6561	6571	6580	6590	6599	6609	6618
46	6628	6637	6646	6656	6665	6675	6684	6693	6702	6712
47	6721	6730	6739	6749	6758	6767	6776	6785	6794	6803
48	6812	6821	6830	6839	6848	6857	6866	6875	6884	6893
49	6902	6911	6920	6928	6937	6946	6955	6964	6972	6981
50	6990	6998	7007	7016	7024	7033	7042	7050	7059	7067
N	0	1	2	3	4	5	6	7	8	9

N	0	1	2	3	4	5	6	7	8	9
50	6990	6998	7007	7016	7024	7033	7042	7050	7059	7067
51	7076	7084	7093	7101	7110	7118	7126	7135	7143	7152
52	7160	7168	7177	7185	7193	7202	7210	7218	7226	7235
53	7243	7251	7259	7267	7275	7284	7292	7300	7308	7316
54	7324	7332	7340	7348	7356	7364	7372	7380	7388	7396
55	7404	7412	7419	7427	7435	7443	7451	7459	7466	7474
56	7482	7490	7497	7505	7513	7520	7528	7536	7543	7551
57	7559	7566	7574	7582	7589	7597	7604	7612	7619	7627
58	7634	7642	7649	7657	7664	7672	7679	7686	7694	7701
59	7709	7716	7723	7731	7738	7745	7752	7760	7767	7774
60	7782	7789	7796	7803	7810	7818	7825	7832	7839	7846
61	7853	7860	7868	7875	7882	7889	7896	7903	7910	7917
62	7924	7931	7938	7945	7952	7959	7966	7973	7980	7987
63	7993	8000	8007	8014	8021	8028	8035	8041	8048	8055
64	8062	8069	8075	8082	8089	8096	8102	8109	8116	8122
65	8129	8136	8142	8149	8156	8162	8169	8176	8182	8189
66	8195	8202	8209	8215	8222	8228	8235	8241	8248	8254
67	8261	8267	8274	8280	8287	8293	8299	8306	8312	8319
68	8325	8331	8338	8344	8351	8357	8363	8370	8376	8382
69	8388	8395	8401	8407	8414	8420	8426	8432	8439	8445
70	8451	8457	8463	8470	8476	8482	8488	8494	8500	8506
71	8513	8519	8525	8531	8537	8543	8549	8555	8561	8567
72	8573	8579	8585	8591	8597	8603	8609	8615	8621	8627
73	8633	8639	8645	8651	8657	8663	8669	8675	8681	8686
74	8692	8698	8704	8710	8716	8722	8727	8733	8739	8745
75	8751	8756	8762	8768	8774	8779	8785	8791	8797	8802
76	8808	8814	8820	8825	8831	8837	8842	8848	8854	8859
77	8865	8871	8876	8882	8887	8893	8899	8904	8910	8915
78	8921	8927	8932	8938	8943	8949	8954	8960	8965	8971
79	8976	8982	8987	8993	8998	9004	9009	9015	9020	9025
80	9031	9036	9042	9047	9053	9058	9063	9069	9074	9079
81	9085	9090	9096	9101	9106	9112	9117	9122	9128	9133
82	9138	9143	9149	9154	9159	9165	9170	9175	9180	9186
83	9191	9196	9201	9206	9212	9217	9222	9227	9232	9238
84	9243	9248	9253	9258	9263	9269	9274	9279	9284	9289
85	9294	9299	9304	9309	9315	9320	9325	9330	9335	9340
86	9345	9350	9355	9360	9365	9370	9375	9380	9385	9390
87	9395	9400	9405	9410	9415	9420	9425	9430	9435	9440
88	9445	9450	9455	9460	9465	9469	9474	9479	9484	9489
89	9494	9499	9504	9509	9513	9518	9523	9528	9533	9538
90	9542	9547	9552	9557	9562	9566	9571	9576	9581	9586
91	9590	9595	9600	9605	9609	9614	9619	9624	9628	9633
92	9638	9643	9647	9652	9657	9661	9666	9671	9675	9680
93	9685	9689	9694	9699	9703	9708	9713	9717	9722	9727
94	9731	9736	9741	9745	9750	9754	9759	9763	9768	9773
95	9777	9782	9786	9791	9795	9800	9805	9809	9814	9818
96	9823	9827	9832	9836	9841	9845	9850	9854	9859	9863
97	9868	9872	9877	9881	9886	9890	9894	9899	9903	9908
98	9912	9917	9921	9926	9930	9934	9939	9943	9948	9952
99	9956	9961	9965	9969	9974	9978	9983	9987	9991	9996
100	0000	0004	0009	0013	0017	0022	0026	0030	0035	0039
N	0	1	2	3	4	5	6	7	8	9

Squares, Cubes, Square and Cube Roots of Numbers 1–100

N	N^2	N^3	$\sqrt{N}$	$\sqrt[3]{N}$	N	N^2	N^3	$\sqrt{N}$	$\sqrt[3]{N}$
1	1	1	1.0000	1.0000	51	2601	132651	7.1414	3.7084
2	4	8	1.4142	1.2599	52	2704	140608	7.2111	3.7325
3	9	27	1.7321	1.4422	53	2809	148877	7.2801	3.7563
4	16	64	2.0000	1.5874	54	2916	157464	7.3485	3.7798
5	25	125	2.2361	1.7100	55	3025	166375	7.4162	3.8030
6	36	216	2.4495	1.8171	56	3136	175616	7.4833	3.8259
7	49	343	2.6458	1.9129	57	3249	185193	7.5498	3.8485
8	64	512	2.8284	2.0000	58	3364	195112	7.6158	3.8709
9	81	729	3.0000	2.0801	59	3481	205379	7.6811	3.8930
10	100	1000	3.1623	2.1544	**60**	3600	216000	7.7460	3.9149
11	121	1331	3.3166	2.2240	61	3721	226981	7.8102	3.9365
12	144	1728	3.4641	2.2894	62	3844	238328	7.8740	3.9579
13	169	2197	3.6056	2.3513	63	3969	250047	7.9373	3.9791
14	196	2744	3.7417	2.4101	64	4096	262144	8.0000	4.0000
15	225	3375	3.8730	2.4662	65	4225	274625	8.0623	4.0207
16	256	4096	4.0000	2.5198	66	4356	287496	8.1240	4.0412
17	289	4913	4.1231	2.5713	67	4489	300763	8.1854	4.0615
18	324	5832	4.2426	2.6207	68	4624	314432	8.2462	4.0817
19	361	6859	4.3589	2.6684	69	4761	328509	8.3066	4.1016
20	400	8000	4.4721	2.7144	**70**	4900	343000	8.3666	4.1213
21	441	9261	4.5826	2.7589	71	5041	357911	8 4261	4.1408
22	484	10648	4.6904	2.8020	72	5184	373248	8.4853	4.1602
23	529	12167	4.7958	2.8439	73	5329	389017	8.5440	4.1793
24	576	13824	4.8990	2.8845	74	5476	405224	8.6023	4 1983
25	625	15625	5.0000	2.9240	75	5625	421875	8.6603	4.2172
26	676	17576	5.0990	2.9625	76	5776	438976	8.7178	4.2358
27	729	19683	5.1962	3.0000	77	5929	456533	8.7750	4 2543
28	784	21952	5.2915	3.0366	78	6084	474552	8.8318	4.2727
29	841	24389	5.3852	3.0723	79	6241	493039	8.8882	4.2908
30	900	27000	5.4772	3.1072	**80**	6400	512000	8.9443	4.3089
31	961	29791	5.5678	3.1414	81	6561	531441	9.0000	4.3267
32	1024	32768	5.6569	3.1748	82	6724	551368	9.0554	4.3445
33	1089	35937	5.7446	3.2075	83	6889	571787	9 1104	4.3621
34	1156	39304	5.8310	3.2396	84	7056	592704	9.1652	4.3795
35	1225	42875	5.9161	3.2711	85	7225	614125	9.2195	4.3968
36	1296	46656	6.0000	3.3019	86	7396	636056	9.2736	4.4140
37	1369	50653	6.0828	3.3322	87	7569	658503	9.3274	4.4310
38	1444	54872	6.1644	3.3620	88	7744	681472	9.3808	4.4480
39	1521	59319	6.2450	3.3912	89	7921	704969	9.4340	4.4647
40	1600	64000	6.3246	3.4200	**90**	8100	729000	9.4868	4.4814
41	1681	68921	6.4031	3.4482	91	8281	753571	9.5394	4.4979
42	1764	74088	6.4807	3.4760	92	8464	778688	9.5917	4.5144
43	1849	79507	6.5574	3.5034	93	8649	804357	9.6437	4.5307
44	1936	85184	6.6332	3.5303	94	8836	830584	9.6954	4.5468
45	2025	91125	6.7082	3.5569	95	9025	857375	9.7468	4.5629
46	2116	97336	6.7823	3.5830	96	9216	884736	9.7980	4.5789
47	2209	103823	6.8557	3.6088	97	9409	912673	9.8489	4.5947
48	2304	110592	6.9282	3.6342	98	9604	941192	9.8995	4.6104
49	2401	117649	7.0000	3.6593	99	9801	970299	9.9499	4.6261
50	2500	125000	7.0711	3.6840	**100**	10000	1000000	10.0000	4.6416

Values of e^x and e^{-x}

x	Function	0.00	0.01	0.02	0.03	0.04	0.05	0.06	0.07	0.08	0.09
0.0	ϵ^x	1.0000	1.0101	1.0202	1.0305	1.0408	1.0513	1.0618	1.0725	1.0833	1.0942
	ϵ^{-x}	1.0000	0.9900	0.9802	0.9704	0.9608	0.9512	0.9418	0.9324	0.9231	0.9139
0.1	ϵ^x	1.1052	1.1163	1.1275	1.1388	1.1503	1.1618	1.1735	1.1853	1.1972	1.2093
	ϵ^{-x}	0.9048	0.8958	0.8869	0.8781	0.8694	0.8607	0.8521	0.8437	0.8353	0.8270
0.2	ϵ^x	1.2214	1.2337	1.2461	1.2586	1.2712	1.2840	1.2969	1.3100	1.3231	1.3364
	ϵ^{-x}	0.8187	0.8106	0.8025	0.7945	0.7866	0.7788	0.7711	0.7634	0.7558	0.7483
0.3	ϵ^x	1.3499	1.3634	1.3771	1.3910	1.4049	1.4191	1.4333	1.4477	1.4623	1.4770
	ϵ^{-x}	0.7408	0.7334	0.7261	0.7189	0.7118	0.7047	0.6977	0.6907	0.6839	0.6771
0.4	ϵ^x	1.4918	1.5068	1.5220	1.5373	1.5527	1.5683	1.5841	1.6000	1.6161	1.6323
	ϵ^{-x}	0.6703	0.6637	0.6570	0.6505	0.6440	0.6376	0.6313	0.6250	0.6188	0.6126
0.5	ϵ^x	1.6487	1.6653	1.6820	1.6989	1.7160	1.7333	1.7507	1.7683	1.7860	1.8040
	ϵ^{-x}	0.6065	0.6005	0.5945	0.5886	0.5827	0.5769	0.5712	0.5655	0.5599	0.5543
0.6	ϵ^x	1.8221	1.8404	1.8589	1.8776	1.8965	1.9155	1.9348	1.9542	1.9739	1.9939
	ϵ^{-x}	0.5488	0.5434	0.5379	0.5326	0.5273	0.5220	0.5169	0.5117	0.5066	0.5017
0.7	ϵ^x	2.0138	2.0340	2.0544	2.0751	2.0959	2.1170	2.1383	2.1598	2.1815	2.2034
	ϵ^{-x}	0.4966	0.4916	0.4868	0.4819	0.4771	0.4724	0.4677	0.4630	0.4584	0.4538
0.8	ϵ^x	2.2255	2.2479	2.2705	2.2933	2.3164	2.3396	2.3632	2.3869	2.4109	2.4351
	ϵ^{-x}	0.4493	0.4449	0.4404	0.4360	0.4317	0.4274	0.4232	0.4190	0.4148	0.4107
0.9	ϵ^x	2.4596	2.4843	2.5093	2.5345	2.5600	2.5857	2.6117	2.6379	2.6645	2.6912
	ϵ^{-x}	0.4066	0.4025	0.3985	0.3946	0.3906	0.3867	0.3829	0.3791	0.3753	0.3716
1.0	ϵ^x	2.7183	2.7456	2.7732	2.8011	2.8292	2.8577	2.8864	2.9154	2.9447	2.9743
	ϵ^{-x}	0.3679	0.3642	0.3606	0.3570	0.3535	0.3499	0.3465	0.3430	0.3396	0.3362
1.1	ϵ^x	3.0042	3.0344	3.0649	3.0957	3.1268	3.1582	3.1899	3.2220	3.2544	3.2871
	ϵ^{-x}	0.3329	0.3296	0.3263	0.3230	0.3198	0.3166	0.3135	0.3104	0.3073	0.3042
1.2	ϵ^x	3.3201	3.3535	3.3872	3.4212	3.4556	3.4903	3.5254	3.5609	3.5966	3.6328
	ϵ^{-x}	0.3012	0.2982	0.2952	0.2923	0.2894	0.2865	0.2837	0.2808	0.2780	0.2753
1.3	ϵ^x	3.6693	3.7062	3.7434	3.7810	3.8190	3.8574	3.8962	3.9354	3.9749	4.0149
	ϵ^{-x}	0.2725	0.2698	0.2671	0.2645	0.2618	0.2592	0.2567	0.2541	0.2516	0.2491
1.4	ϵ^x	4.0552	4.0960	4.1371	4.1787	4.2207	4.2631	4.3060	4.3492	4.3929	4.4371
	ϵ^{-x}	0.2466	0.2441	0.2417	0.2393	0.2369	0.2346	0.2322	0.2299	0.2276	0.2254
1.5	ϵ^x	4.4817	4.5267	4.5722	4.6182	4.6646	4.7115	4.7588	4.8066	4.8550	4.9037
	ϵ^{-x}	0.2231	0.2209	0.2187	0.2165	0.2144	0.2122	0.2101	0.2080	0.2060	0.2039
1.6	ϵ^x	4.9530	5.0028	5.0531	5.1039	5.1552	5.2070	5.2593	5.3122	5.3656	5.4195
	ϵ^{-x}	0.2019	0.1999	0.1979	0.1959	0.1940	0.1920	0.1901	0.1882	0.1864	0.1845
1.7	ϵ^x	5.4739	5.5290	5.5845	5.6407	5.6973	5.7546	5.8124	5.8709	5.9299	5.9895
	ϵ^{-x}	0.1827	0.1809	0.1791	0.1773	0.1755	0.1738	0.1720	0.1703	0.1686	0.1670
1.8	ϵ^x	6.0496	6.1104	6.1719	6.2339	6.2965	6.3598	6.4237	6.4883	6.5535	6.6194
	ϵ^{-x}	0.1653	0.1637	0.1620	0.1604	0.1588	0.1572	0.1557	0.1541	0.1526	0.1511
1.9	ϵ^x	6.6859	6.7531	6.8210	6.8895	6.9588	7.0287	7.0993	7.1707	7.2427	7.3155
	ϵ^{-x}	0.1496	0.1481	0.1466	0.1451	0.1437	0.1423	0.1409	0.1395	0.1381	0.1367

Values of e^x and e^{-x}

x	Function	0.00	0.01	0.02	0.03	0.04	0.05	0.06	0.07	0.08	0.09
2.0	ϵ^x	7.3891	7.4633	7.5383	7.6141	7.6906	7.7679	7.8460	7.9248	8.0045	8.0849
	ϵ^{-x}	0.1353	0.1340	0.1327	0.1313	0.1300	0.1287	0.1275	0.1262	0.1249	0.1237
2.1	ϵ^x	8.1662	8.2482	8.3311	8.4149	8.4994	8.5849	8.6711	8.7583	8.8463	8.9352
	ϵ^{-x}	0.1225	0.1212	0.1200	0.1188	0.1177	0.1165	0.1153	0.1142	0.1130	0.1119
2.2	ϵ^x	9.0250	9.1157	9.2073	9.2999	9.3933	9.4877	9.5831	9.6794	9.7767	9.8749
	ϵ^{-x}	0.1108	0.1097	0.1086	0.1075	0.1065	0.1054	0.1044	0.1033	0.1023	0.1013
2.3	ϵ^x	9.9742	10.074	10.176	10.278	10.381	10.486	10.591	10.697	10.805	10.913
	ϵ^{-x}	0.1003	0.0993	0.0983	0.0973	0.0963	0.0954	0.0944	0.0935	0.0926	0.0916
2.4	ϵ^x	11.023	11.134	11.246	11.359	11.473	11.588	11.705	11.822	11.941	12.061
	ϵ^{-x}	0.0907	0.0898	0.0889	0.0880	0.0872	0.0863	0.0854	0.0846	0.0837	0.0829
2.5	ϵ^x	12.182	12.305	12.429	12.554	12.680	12.807	12.936	13.066	13.197	13.330
	ϵ^{-x}	0.0821	0.0813	0.0805	0.0797	0.0789	0.0781	0.0773	0.0765	0.0758	0.0750
2.6	ϵ^x	13.464	13.599	13.736	13.874	14.013	14.154	14.296	14.440	14.585	14.732
	ϵ^{-x}	0.0743	0.0735	0.0728	0.0721	0.0714	0.0707	0.0699	0.0693	0.0686	0.0679
2.7	ϵ^x	14.880	15.029	15.180	15.333	15.487	15.643	15.800	15.959	16.119	16.281
	ϵ^{-x}	0.0672	0.0665	0.0659	0.0652	0.0646	0.0639	0.0633	0.0627	0.0620	0.0614
2.8	ϵ^x	16.445	16.610	16.777	16.945	17.116	17.288	17.462	17.637	17.814	17.993
	ϵ^{-x}	0.0608	0.0602	0.0596	0.0590	0.0584	0.0578	0.0573	0.0567	0.0561	0.0556
2.9	ϵ^x	18.174	18.357	18.541	18.728	18.916	19.106	19.298	19.492	19.688	19.886
	ϵ^{-x}	0.0550	0.0545	0.0539	0.0534	0.0529	0.0523	0.0518	0.0513	0.0508	0.0503
3.0	ϵ^x	20.086	20.287	20.491	20.697	20.905	21.115	21.328	21.542	21.758	21.977
	ϵ^{-x}	0.0498	0.0493	0.0488	0.0483	0.0478	0.0474	0.0469	0.0464	0.0460	0.0455
3.1	ϵ^x	22.198	22.421	22.646	22.874	23.104	23.336	23.571	23.807	24.047	24.288
	ϵ^{-x}	0.0450	0.0446	0.0442	0.0437	0.0433	0.0429	0.0424	0.0420	0.0416	0.0412
3.2	ϵ^x	24.533	24.779	25.028	25.280	25.534	25.790	26.050	26.311	26.576	26.843
	ϵ^{-x}	0.0408	0.0404	0.0400	0.0396	0.0392	0.0388	0.0384	0.0380	0.0376	0.0373
3.3	ϵ^x	27.113	27.385	27.660	27.938	28.219	28.503	28.789	29.079	29.371	29.666
	ϵ^{-x}	0.0369	0.0365	0.0362	0.0358	0.0354	0.0351	0.0347	0.0344	0.0340	0.0337
3.4	ϵ^x	29.964	30.265	30.569	30.877	31.187	31.500	31.817	32.137	32.460	32.786
	ϵ^{-x}	0.0334	0.0330	0.0327	0.0324	0.0321	0.0317	0.0314	0.0311	0.0308	0.0305
3.5	ϵ^x	33.115	33.448	33.784	34.124	34.467	34.813	35.163	35.517	35.874	36.234
	ϵ^{-x}	0.0302	0.0299	0.0296	0.0293	0.0290	0.0287	0.0284	0.0282	0.0279	0.0276
3.6	ϵ^x	36.598	36.966	37.338	37.713	38.092	38.475	38.861	39.252	39.646	40.045
	ϵ^{-x}	0.0273	0.0271	0.0268	0.0265	0.0263	0.0260	0.0257	0.0255	0.0252	0.0250
3.7	ϵ^x	40.447	40.854	41.264	41.679	42.098	42.521	42.948	43.380	43.816	44.256
	ϵ^{-x}	0.0247	0.0245	0.0242	0.0240	0.0238	0.0235	0.0233	0.0231	0.0228	0.0226
3.8	ϵ^x	44.701	45.150	45.604	46.063	46.525	46.993	47.465	47.942	48.424	48.911
	ϵ^{-x}	0.0224	0.0221	0.0219	0.0217	0.0215	0.0213	0.0211	0.0209	0.0207	0.0204
3.9	ϵ^x	49.402	49.899	50.400	50.907	51.419	51.935	52.457	52.985	53.517	54.055
	ϵ^{-x}	0.0202	0.0200	0.0198	0.0196	0.0195	0.0193	0.0191	0.0189	0.0187	0.0185

x	Function	0.00	0.01	0.02	0.03	0.04	0.05	0.06	0.07	0.08	0.09
4.0	ϵ^x	54.598	55.147	55.701	56.261	56.826	57.397	57.974	58.557	59.145	59.740
	ϵ^{-x}	0.0183	0.0181	0.0180	0.0178	0.0176	0.0174	0.0172	0.0171	0.0169	0.0167
4.1	ϵ^x	60.340	60.947	61.559	62.178	62.803	63.434	64.072	64.715	65.366	66.023
	ϵ^{-x}	0.0166	0.0164	0.0162	0.0161	0.0159	0.0158	0.0156	0.0155	0.0153	0.0151
4.2	ϵ^x	66.686	67.357	68.033	68.717	69.408	70.105	70.810	71.522	72.240	72.966
	ϵ^{-x}	0.0150	0.0148	0.0147	0.0146	0.0144	0.0143	0.0141	0.0140	0.0138	0.0137
4.3	ϵ^x	73.700	74.440	75.189	75.944	76.708	77.478	78.257	79.044	79.838	80.640
	ϵ^{-x}	0.0136	0.0134	0.0133	0.0132	0.0130	0.0129	0.0128	0.0127	0.0125	0.0124
4.4	ϵ^x	81.451	82.269	83.096	83.931	84.775	85.627	86.488	87.357	88.235	89.121
	ϵ^{-x}	0.0123	0.0122	0.0120	0.0119	0.0118	0.0117	0.0116	0.0114	0.0113	0.0112
4.5	ϵ^x	90.017	90.922	91.836	92.759	93.691	94.632	95.583	96.544	97.514	98.494
	ϵ^{-x}	0.0111	0.0110	0.0109	0.0108	0.0107	0.0106	0.0105	0.0104	0.0103	0.0102
4.6	ϵ^x	99.484	100.48	101.49	102.51	103.54	104.58	105.64	106.70	107.77	108.85
	ϵ^{-x}	0.0101	0.0100	0.0099	0.0098	0.0097	0.0096	0.0095	0.0094	0.0093	0.0092
4.7	ϵ^x	109.95	111.05	112.17	113.30	114.43	115.58	116.75	117.92	119.10	120.30
	ϵ^{-x}	0.0091	0.0090	0.0089	0.0088	0.0087	0.0087	0.0086	0.0085	0.0084	0.0083
4.8	ϵ^x	121.51	122.73	123.97	125.21	126.47	127.74	129.02	130.32	131.63	132.95
	ϵ^{-x}	0.0082	0.0081	0.0081	0.0080	0.0079	0.0078	0.0078	0.0077	0.0076	0.0075
4.9	ϵ^x	134.29	135.64	137.00	138.38	139.77	141.17	142.59	144.03	145.47	146.94
	ϵ^{-x}	0.0074	0.0074	0.0073	0.0072	0.0072	0.0071	0.0070	0.0069	0.0069	0.0068
5.0	ϵ^x	148.41	149.90	151.41	152.93	154.47	156.02	157.59	159.17	160.77	162.39
	ϵ^{-x}	0.0067	0.0067	0.0066	0.0065	0.0065	0.0064	0.0063	0.0063	0.0062	0.0062
5.1	ϵ	164.02	165.67	167.34	169.02	170.72	172.43	174.16	175.91	177.68	179.47
	ϵ^{-x}	0.0061	0.0060	0.0060	0.0059	0.0059	0.0058	0.0057	0.0057	0.0056	0.0056
5.2	ϵ^x	181.27	183.09	184.93	186.79	188.67	190.57	192.48	194.42	196.37	198.34
	ϵ^{-x}	0.0055	0.0055	0.0054	0.0054	0.0053	0.0052	0.0052	0.0051	0.0051	0.0050
5.3	ϵ^x	200.34	202.35	204.38	206.44	208.51	210.61	212.72	214.86	217.02	219.20
	ϵ^{-x}	0.0050	0.0049	0.0049	0.0048	0.0048	0.0047	0.0047	0.0047	0.0046	0.0046
5.4	ϵ^x	221.41	223.63	225.88	228.15	230.44	232.76	235.10	237.46	239.85	242.26
	ϵ^{-x}	0.0045	0.0045	0.0044	0.0044	0.0043	0.0043	0.0043	0.0042	0.0042	0.0041
5.5	ϵ^x	244.69	247.15	249.64	252.14	254.68	257.24	259.82	262.43	265.07	267.74
	ϵ^{-x}	0.0041	0.0040	0.0040	0.0040	0.0039	0.0039	0.0038	0.0038	0.0038	0.0037
5.6	ϵ^x	270.43	273.14	275.89	278.66	281.46	284.29	287.15	290.03	292.95	295.89
	ϵ^{-x}	0.0037	0.0037	0.0036	0.0036	0.0036	0.0035	0.0035	0.0034	0.0034	0.0034
5.7	ϵ^x	298.87	301.87	304.90	307.97	311.06	314.19	317.35	320.54	323.76	327.01
	ϵ^{-x}	0.0033	0.0033	0.0033	0.0032	0.0032	0.0032	0.0032	0.0031	0.0031	0.0031
5.8	ϵ^x	330.30	333.62	336.97	340.36	343.78	347.23	350.72	354.25	357.81	361.41
	ϵ^{-x}	0.0030	0.0030	0.0030	0.0029	0.0029	0.0029	0.0029	0.0028	0.0028	0.0028
5.9	ϵ^x	365.04	368.71	372.41	376.15	379.93	383.75	387.61	391.51	395.44	399.41
	ϵ^{-x}	0.0027	0.0027	0.0027	0.0027	0.0026	0.0026	0.0026	0.0026	0.0025	0.0025

Trigonometric Functions in Radian Measure

Rad	Sin	Tan	Ctn	Cos
.00	.0000	.0000	....	1.0000
.01	.0100	.0100	99.997	1.0000
.02	.0200	.0200	49.993	.9998
.03	.0300	.0300	33.323	.9996
.04	.0400	.0400	24.987	.9992
.05	.0500	.0500	19.983	.9988
.06	.0600	.0601	16.647	.9982
.07	.0699	.0701	14.262	.9976
.08	.0799	.0802	12.473	.9968
.09	.0899	.0902	11.081	.9960
.10	.0998	.1003	9.967	.9950
.11	.1098	.1104	9.054	.9940
.12	.1197	.1206	8.293	.9928
.13	.1296	.1307	7.649	.9916
.14	.1395	.1409	7.096	.9902
.15	.1494	.1511	6.617	.9888
.16	.1593	.1614	6.197	.9872
.17	.1692	.1717	5.826	.9856
.18	.1790	.1820	5.495	.9838
.19	.1889	.1923	5.200	.9820
.20	.1987	.2027	4.933	.9801
.21	.2085	.2131	4.692	.9780
.22	.2182	.2236	4.472	.9759
.23	.2280	.2341	4.271	.9737
.24	.2377	.2447	4.086	.9713
.25	.2474	.2553	3.916	.9689
.26	.2571	.2660	3.759	.9664
.27	.2667	.2768	3.613	.9638
.28	.2764	.2876	3.478	.9611
.29	.2860	.2984	3.351	.9582
.30	.2955	.3093	3.233	.9553
.31	.3051	.3203	3.122	.9523
.32	.3146	.3314	3.018	.9492
.33	.3240	.3425	2.920	.9460
.34	.3335	.3537	2.827	.9428
.35	.3429	.3650	2.740	.9394
.36	.3523	.3764	2.657	.9359
.37	.3616	.3879	2.578	.9323
.38	.3709	.3994	2.504	.9287
.39	.3802	.4111	2.433	.9249
.40	.3894	.4228	2.365	.9211
.41	.3986	.4346	2.301	.9171
.42	.4078	.4466	2.239	.9131
.43	.4169	.4586	2.180	.9090
.44	.4259	.4708	2.124	.9048
.45	.4350	.4831	2.070	.9004
.46	.4439	.4954	2.018	.8961
.47	.4529	.5080	1.969	.8916
.48	.4618	.5206	1.921	.8870
.49	.4706	.5334	1.875	.8823
.50	.4794	.5463	1.830	.8776

Rad	Sin	Tan	Ctn	Cos
.50	.4794	.5463	1.830	.8776
.51	.4882	.5594	1.788	.8727
.52	.4969	.5726	1.747	.8678
.53	.5055	.5859	1.707	.8628
.54	.5141	.5994	1.668	.8577
.55	.5227	.6131	1.631	.8525
.56	.5312	.6269	1.595	.8473
.57	.5396	.6410	1.560	.8419
.58	.5480	.6552	1.526	.8365
.59	.5564	.6696	1.494	.8309
.60	.5646	.6841	1.462	.8253
.61	.5729	.6989	1.431	.8196
.62	.5810	.7139	1.401	.8139
.63	.5891	.7291	1.372	.8080
.64	.5972	.7445	1.343	.8021
.65	.6052	.7602	1.315	.7961
.66	.6131	.7761	1.288	.7900
.67	.6210	.7923	1.262	.7838
.68	.6288	.8087	1.237	.7776
.69	.6365	.8253	1.212	.7712
.70	.6442	.8423	1.187	.7648
.71	.6518	.8595	1.163	.7584
.72	.6594	.8771	1.140	.7518
.73	.6669	.8949	1.117	.7452
.74	.6743	.9131	1.095	.7385
.75	.6816	.9316	1.073	.7317
.76	.6889	.9505	1.052	.7248
.77	.6961	.9697	1.031	.7179
.78	.7033	.9893	1.011	.7109
.79	.7104	1.009	.9908	.7038
.80	.7174	1.030	.9712	.6967
.81	.7243	1.050	.9520	.6895
.82	.7311	1.072	.9331	.6822
.83	.7379	1.093	.9146	.6749
.84	.7446	1.116	.8964	.6675
.85	.7513	1.138	.8785	.6600
.86	.7578	1.162	.8609	.6524
.87	.7643	1.185	.3437	.6448
.88	.7707	1.210	.8267	.6372
.89	.7771	1.235	.8100	.6294
.90	.7833	1.260	.7936	.6216
.91	.7895	1.286	.7774	.6137
.92	.7956	1.313	.7615	.6058
.93	.8016	1.341	.7458	.5978
.94	.8076	1.369	.7303	.5898
.95	.8134	1.398	.7151	.5817
.96	.8192	1.428	.7001	.5735
.97	.8249	1.459	.6853	.5653
.98	.8305	1.491	.6707	.5570
.99	.8360	1.524	.6563	.5487
1.00	.8415	1.557	.6421	.5403

π radians = 180°, π = 3.14159 26536

1 radian = 57°17′44″.80625 = 57°.29577 95131

1° = 0.01745 32925 19943 radian = 60′ = 3600″

Trigonometric Functions in Radian Measure

Rad	Sin	Tan	Ctn	Cos
1.00	.8415	1.557	.6421	.5403
1.01	.8468	1.592	.6281	.5319
1.02	.8521	1.628	.6142	.5234
1.03	.8573	1.665	.6005	.5148
1.04	.8624	1.704	.5870	.5062
1.05	.8674	1.743	.5736	.4976
1.06	.8724	1.784	.5604	.4889
1.07	.8772	1.827	.5473	.4801
1.08	.8820	1.871	.5344	.4713
1.09	.8866	1.917	.5216	.4625
1.10	.8912	1.965	.5090	.4536
1.11	.8957	2.014	.4964	.4447
1.12	.9001	2.066	.4840	.4357
1.13	.9044	2.120	.4718	.4267
1.14	.9086	2.176	.4596	.4176
1.15	.9128	2.234	.4475	.4085
1.16	.9168	2.296	.4356	.3993
1.17	.9208	2.360	.4237	.3902
1.18	.9246	2.427	.4120	.3809
1.19	.9284	2.498	.4003	.3717
1.20	.9320	2.572	.3888	.3624
1.21	.9356	2.650	.3773	.3530
1.22	.9391	2.733	.3659	.3436
1.23	.9425	2.820	.3546	.3342
1.24	.9458	2.912	.3434	.3248
1.25	.9490	3.010	.3323	.3153
1.26	.9521	3.113	.3212	.3058
1.27	.9551	3.224	.3102	.2963
1.28	.9580	3.341	.2993	.2867
1.29	.9608	3.467	.2884	.2771
1.30	.9636	3.602	.2776	.2675

Rad	Sin	Tan	Ctn	Cos
1.30	.9636	3.602	.2776	.2675
1.31	.9662	3.747	.2669	.2579
1.32	.9687	3.903	.2562	.2482
1.33	.9711	4.072	.2456	.2385
1.34	.9735	4.256	.2350	.2288
1.35	.9757	4.455	.2245	.2190
1.36	.9779	4.673	.2140	.2092
1.37	.9799	4.913	.2035	.1994
1.38	.9819	5.177	.1931	.1896
1.39	.9837	5.471	.1828	.1798
1.40	.9854	5.798	.1725	.1700
1.41	.9871	6.165	.1622	.1601
1.42	.9887	6.581	.1519	.1502
1.43	.9901	7.055	.1417	.1403
1.44	.9915	7.602	.1315	.1304
1.45	.9927	8.238	.1214	.1205
1.46	.9939	8.989	.1113	.1106
1.47	.9949	9.887	.1011	.1006
1.48	.9959	10.983	.0910	.0907
1.49	.9967	12.350	.0810	.0807
1.50	.9975	14.101	.0709	.0707
1.51	.9982	16.428	.0609	.0608
1.52	.9987	19.670	.0508	.0508
1.53	.9992	24.498	.0408	.0408
1.54	.9995	32.461	.0308	.0308
1.55	.9998	48.078	.0208	.0208
1.56	.9999	92.620	.0108	.0108
1.57	1.0000	1255.8	.0008	.0008
1.58	1.0000	−108.65	−.0092	−.0092
1.59	.9998	−52.067	−.0192	−.0192
1.60	.9996	−34.233	−.0292	−.0292

Radians to Degrees, Minutes, and Seconds

Rad		Rad		Rad		Rad		Rad	
1	57°17′44″.8	.1	5°43′46″.5	.01	0°34′22″.6	.001	0° 3′26″.3	.0001	0°0′20″.6
2	114°35′29″.6	.2	11°27′33″.0	.02	1° 8′45″.3	.002	0° 6′52″.5	.0002	0°0′41″.3
3	171°53′14″.4	.3	17°11′19″.4	.03	1°43′07″.9	.003	0°10′18″.8	.0003	0°1′01″.9
4	229°10′59″.2	.4	22°55′05″.9	.04	2°17′30″.6	.004	0°13′45″.1	.0004	0°1′22″.3
5	286°28′44″.0	.5	28°38′52″.4	.05	2°51′53″.2	.005	0°17′11″.3	.0005	0°1′43″.1
6	343°46′28″.8	.6	34°22′38″.9	.06	3°26′15″.9	.006	0°20′37″.6	.0006	0°2′03″.6
7	401° 4′13″.6	.7	40° 6′25″.4	.07	4° 0′38″.5	.007	0°24′03″.9	.0007	0°2′24″.4
8	458°21′58″.4	.8	45°50′11″.8	.08	4°35′01″.2	.008	0°27′30″.1	.0008	0°2′45″.6
9	515°39′43″.3	.9	51°33′58″.3	.09	5° 9′23″.8	.009	0°30′56″.4	.0009	0°3′05″.6

Radians to Degrees

Rad	Degrees	Rad	Degrees	Rad	Degrees
1	57.2958	4	229.1831	7	401.0705
2	114.5916	5	286.4789	8	458.3662
3	171.8873	6	343.7747	9	515.6620

Natural Sines, Cosines, and Tangents

Degs.	Function	0.0°	0.1°	0.2°	0.3°	0.4°	0.5°	0.6°	0.7°	0.8°	0.9°
0	sin	0.0000	0.0017	0.0035	0.0052	0.0070	0.0087	0.0105	0.0122	0.0140	0.0157
	cos	1.0000	1.0000	1.0000	1.0000	1.0000	1.0000	0.9999	0.9999	0.9999	0.9999
	tan	0.0000	0.0017	0.0035	0.0052	0.0070	0.0087	0.0105	0.0122	0.0140	0.0157
1	sin	0.0175	0.0192	0.0209	0.0227	0.0244	0.0262	0.0279	0.0297	0.0314	0.0332
	cos	0.9998	0.9998	0.9998	0.9997	0.9997	0.9997	0.9996	0.9996	0.9995	0.9995
	tan	0.0175	0.0192	0.0209	0.0227	0.0244	0.0262	0.0279	0.0297	0.0314	0.0332
2	sin	0.0349	0.0366	0.0384	0.0401	0.0419	0.0436	0.0454	0.0471	0.0488	0.0506
	cos	0.9994	0.9993	0.9993	0.9992	0.9991	0.9990	0.9990	0.9989	0.9988	0.9987
	tan	0.0349	0.0367	0.0384	0.0402	0.0419	0.0437	0.0454	0.0472	0.0489	0.0507
3	sin	0.0523	0.0541	0.0558	0.0576	0.0593	0.0610	0.0628	0.0645	0.0663	0.0680
	cos	0.9986	0.9985	0.9984	0.9983	0.9982	0.9981	0.9980	0.9979	0.9978	0.9977
	tan	0.0524	0.0542	0.0559	0.0577	0.0594	0.0612	0.0629	0.0647	c.0664	0.0682
4	sin	0.0698	0.0715	0.0732	0.0750	0.0767	0.0785	0.0802	0.0819	0.0837	0.0854
	cos	0.9976	0.9974	0.9973	0.9972	0.9971	0.9969	0.9968	0.9966	0.9965	0.9963
	tan	0.0699	0.0717	0.0734	0.0752	0.0769	0.0787	0.0805	0.0822	0.0840	0.0857
5	sin	0.0872	0.0889	0.0906	0.0924	0.0941	0.0958	0.0976	0.0993	0.1011	0.1028
	cos	0.9962	0.9960	0.9959	0.9957	0.9956	0.9954	0.9952	0.9951	0.9949	0.9947
	tan	0.0875	0.0892	0.0910	0.0928	0.0945	0.0963	0.0981	0.0998	0.1016	0.1033
6	sin	0.1045	0.1063	0.1080	0.1097	0.1115	0.1132	0.1149	0.1167	0.1184	0.1201
	cos	0.9945	0.9943	0.9942	0.9940	0.9938	0.9936	0.9934	0.9932	0.9930	0.9928
	tan	0.1051	0.1069	0.1086	0.1104	0.1122	0.1139	0.1157	0.1175	0.1192	0.1210
7	sin	0.1219	0.1236	0.1253	0.1271	0.1288	0.1305	0.1323	0.1340	0.1357	0.1374
	cos	0.9925	0.9923	0.9921	0.9919	0.9917	0.9914	0.9912	0.9910	0.9907	0.9905
	tan	0.1228	0.1246	0.1263	0.1281	0.1299	0.1317	0.1334	0.1352	0.1370	0.1388
8	sin	0.1392	0.1409	0.1426	0.1444	0.1461	0.1478	0.1495	0.1513	0.1530	0.1547
	cos	0.9903	0.9900	0.9898	0.9895	0.9893	0.9890	0.9888	0.9885	0.9882	0.9880
	tan	0.1405	0.1423	0.1441	0.1459	0.1477	0.1495	0.1512	0.1530	0.1548	0.1566
9	sin	0.1564	0.1582	0.1599	0.1616	0.1633	0.1650	0.1668	0.1685	0.1702	0.1719
	cos	0.9877	0.9874	0.9871	0.9869	0.9866	0.9863	0.9860	0.9857	0.9854	0.9851
	tan	0.1584	0.1602	0.1620	0.1638	0.1655	0.1673	0.1691	0.1709	0.1727	0.1745
10	sin	0.1736	0.1754	0.1771	0.1788	0.1805	0.1822	0.1840	0.1857	0.1874	0.1891
	cos	0.9848	0.9845	0.9842	0.9839	0.9836	0.9833	0.9829	0.9826	0.9823	0.9820
	tan	0.1763	0.1781	0.1799	0.1817	0.1835	0.1853	0.1871	0.1890	0.1908	0.1926
11	sin	0.1908	0.1925	0.1942	0.1959	0.1977	0.1994	0.2011	0.2028	0.2045	0.2062
	cos	0.9816	0.9813	0.9810	0.9806	0.9803	0.9799	0.9796	0.9792	0.9789	0.9785
	tan	0.1944	0.1962	0.1980	0.1998	0.2016	0.2035	0.2053	0.2071	0.2089	0.2107
12	sin	0.2079	0.2096	0.2113	0.2130	0.2147	0.2164	0.2181	0.2198	0.2215	0.2232
	cos	0.9781	0.9778	0.9774	0.9770	0.9767	0.9763	0.9759	0.9755	0.9751	0.9748
	tan	0.2126	0.2144	0.2162	0.2180	0.2199	0.2217	0.2235	0.2254	0.2272	0.2290
13	sin	0.2250	0.2267	0.2284	0.2300	0.2318	0.2334	0.2351	0.2368	0.2385	0.2402
	cos	0.9744	0.9740	0.9736	0.9732	0.9728	0.9724	0.9720	0.9715	0.9711	0.9707
	tan	0.2309	0.2327	0.2345	0.2364	0.2382	0.2401	0.2419	0.2438	0.2456	0.2475
14	sin	0.2419	0.2436	0.2453	0.2470	0.2487	0.2504	0.2521	0.2538	0.2554	0.2571
	cos	0.9703	0.9699	0.9694	0.9690	0.9686	0.9681	0.9677	0.9673	0.9668	0.9664
	tan	0.2493	0.2512	0.2530	0.2549	0.2568	0.2586	0.2605	0.2623	0.2642	0.2661
Degs.	Function	0′	6′	12′	18′	24′	30′	36′	42′	48′	54′

Degs.	Function	0.0°	0.1°	0.2°	0.3°	0.4°	0.5°	0.6°	0.7°	0.8°	0.9°
15	sin	0.2588	0.2605	0.2622	0.2639	0.2656	0.2672	0.2689	0.2706	0.2723	0.2740
	cos	0.9659	0.9655	0.9650	0.9646	0.9641	0.9636	0.9632	0.9627	0.9622	0.9617
	tan	0.2679	0.2698	0.2717	0.2736	0.2754	0.2773	0.2792	0.2811	0.2830	0.2849
16	sin	0.2756	0.2773	0.2790	0.2807	0.2823	0.2840	0.2857	0.2874	0.2890	0.2907
	cos	0.9613	0.9608	0.9603	0.9598	0.9593	0.9588	0.9583	0.9578	0.9573	0.9568
	tan	0.2867	0.2886	0.2905	0.2924	0.2943	0.2962	0.2981	0.3000	0.3019	0.3038
17	sin	0.2924	0.2940	0.2957	0.2974	0.2990	0.3007	0.3024	0.3040	0.3057	0.3074
	cos	0.9563	0.9558	0.9553	0.9548	0.9542	0.9537	0.9532	0.9527	0.9521	0.9516
	tan	0.3057	0.3076	0.3096	0.3115	0.3134	0.3153	0.3172	0.3191	0.3211	0.3230
18	sin	0.3090	0.3107	0.3123	0.3140	0.3156	0.3173	0.3190	0.3206	0.3223	0.3239
	cos	0.9511	0.9505	0.9500	0.9494	0.9489	0.9483	0.9478	0.9472	0.9466	0.9461
	tan	0.3249	0.3269	0.3288	0.3307	0.3327	0.3346	0.3365	0.3385	0.3404	0.3424
19	sin	0.3256	0.3272	0.3289	0.3305	0.3322	0.3338	0.3355	0.3371	0.3387	0.3404
	cos	0.9455	0.9449	0.9444	0.9438	0.9432	0.9426	0.9421	0.9415	0.9409	0.9403
	tan	0.3443	0.3463	0.3482	0.3502	0.3522	0.3541	0.3561	0.3581	0.3600	0.3620
20	sin	0.3420	0.3437	0.3453	0.3469	0.3486	0.3502	0.3518	0.3535	0.3551	0.3567
	cos	0.9397	0.9391	0.9385	0.9379	0.9373	0.9367	0.9361	0.9354	0.9348	0.9342
	tan	0.3640	0.3659	0.3679	0.3699	0.3719	0.3739	0.3759	0.3779	0.3799	0.3819
21	sin	0.3584	0.3600	0.3616	0.3633	0.3649	0.3665	0.3681	0.3697	0.3714	0.3730
	cos	0.9336	0.9330	0.9323	0.9317	0.9311	0.9304	0.9298	0.9291	0.9285	0.9278
	tan	0.3839	0.3859	0.3879	0.3899	0.3919	0.3939	0.3959	0.3979	0.4000	0.4020
22	sin	0.3746	0.3762	0.3778	0.3795	0.3811	0.3827	0.3843	0.3859	0.3875	0.3891
	cos	0.9272	0.9265	0.9259	0.9252	0.9245	0.9239	0.9232	0.9225	0.9219	0.9212
	tan	0.4040	0.4061	0.4081	0.4101	0.4122	0.4142	0.4163	0.4183	0.4204	0.4224
23	sin	0.3907	0.3923	0.3939	0.3955	0.3971	0.3987	0.4003	0.4019	0.4035	0.4051
	cos	0.9205	0.9198	0.9191	0.9184	0.9178	0.9171	0.9164	0.9157	0.9150	0.9143
	tan	0.4245	0.4265	0.4286	0.4307	0.4327	0.4348	0.4369	0.4390	0.4411	0.4431
24	sin	0.4067	0.4083	0.4099	0.4115	0.4131	0.4147	0.4163	0.4179	0.4195	0.4210
	cos	0.9135	0.9128	0.9121	0.9114	0.9107	0.9100	0.9092	0.9085	0.9078	0.9070
	tan	0.4452	0.4473	0.4494	0.4515	0.4536	0.4557	0.4578	0.4599	0.4621	0.4642
25	sin	0.4226	0.4242	0.4258	0.4274	0.4289	0.4305	0.4321	0.4337	0.4352	0.4368
	cos	0.9063	0.9056	0.9048	0.9041	0.9033	0.9026	0.9018	0.9011	0.9003	0.8996
	tan	0.4663	0.4684	0.4706	0.4727	0.4748	0.4770	0.4791	0.4813	0.4834	0.4856
26	sin	0.4384	0.4399	0.4415	0.4431	0.4446	0.4462	0.4478	0.4493	0.4509	0.4524
	cos	0.8988	0.8980	0.8973	0.8965	0.8957	0.8949	0.8942	0.8934	0.8926	0.8918
	tan	0.4877	0.4899	0.4921	0.4942	0.4964	0.4986	0.5008	0.5029	0.5051	0.5073
27	sin	0.4540	0.4555	0.4571	0.4586	0.4602	0.4617	0.4633	0.4648	0.4664	0.4679
	cos	0.8910	0.8902	0.8894	0.8886	0.8878	0.8870	0.8862	0.8854	0.8846	0.8838
	tan	0.5095	0.5117	0.5139	0.5161	0.5184	0.5206	0.5228	0.5250	0.5272	0.5295
28	sin	0.4695	0.4710	0.4726	0.4741	0.4756	0.4772	0.4787	0.4802	0.4818	0.4833
	cos	0.8829	0.8821	0.8813	0.8805	0.8796	0.8788	0.8780	0.8771	0.8763	0.8755
	tan	0.5317	0.5340	0.5362	0.5384	0.5407	0.5430	0.5452	0.5475	0.5498	0.5520
29	sin	0.4848	0.4863	0.4879	0.4894	0.4909	0.4924	0.4939	0.4955	0.4970	0.4985
	cos	0.8746	0.8738	0.8729	0.8721	0.8712	0.8704	0.8695	0.8686	0.8678	0.8669
	tan	0.5543	0.5566	0.5589	0.5612	0.5635	0.5658	0.5681	0.5704	0.5727	0.5750
Degs.	Function	0′	6′	12′	18′	24′	30′	36′	42′	48′	54′

Degs.	Function	0.0°	0.1°	0.2°	0.3°	0.4°	0.5°	0.6°	0.7°	0.8°	0.9°
30	sin	0.5000	0.5015	0.5030	0.5045	0.5060	0.5075	0.5090	0.5105	0.5120	0.5135
	cos	0.8660	0.8652	0.8643	0.8634	0.8625	0.8616	0.8607	0.8599	0.8590	0.8581
	tan	0.5774	0.5797	0.5820	0.5844	0.5867	0.5890	0.5914	0.5938	0.5961	0.5985
31	sin	0.5150	0.5165	0.5180	0.5195	0.5210	0.5225	0.5240	0.5255	0.5270	0.5284
	cos	0.8572	0.8563	0.8554	0.8545	0.8536	0.8526	0.8517	0.8508	0.8499	0.8490
	tan	0.6009	0.6032	0.6056	0.6080	0.6104	0.6128	0.6152	0.6176	0.6200	0.6224
32	sin	0.5299	0.5314	0.5329	0.5344	0.5358	0.5373	0.5388	0.5402	0.5417	0.5432
	cos	0.8480	0.8471	0.8462	0.8453	0.8443	0.8434	0.8425	0.8415	0.8406	0.8396
	tan	0.6249	0.6273	0.6297	0.6322	0.6346	0.6371	0.6395	0.6420	0.6445	0.6469
33	sin	0.5446	0.5461	0.5476	0.5490	0.5505	0.5519	0.5534	0.5548	0.5563	0.5577
	cos	0.8387	0.8377	0.8368	0.8358	0.8348	0.8339	0.8329	0.8320	0.8310	0.8300
	tan	0.6494	0.6519	0.6544	0.6569	0.6594	0.6619	0.6644	0.6669	0.6694	0.6720
34	sin	0.5592	0.5606	0.5621	0.5635	0.5650	0.5664	0.5678	0.5693	0.5707	0.5721
	cos	0.8290	0.8281	0.8271	0.8261	0.8251	0.8241	0.8231	0.8221	0.8211	0.8202
	tan	0.6745	0.6771	0.6796	0.6822	0.6847	0.6873	0.6899	0.6924	0.6950	0.6976
35	sin	0.5736	0.5750	0.5764	0.5779	0.5793	0.5807	0.5821	0.5835	0.5850	0.5864
	cos	0.8192	0.8181	0.8171	0.8161	0.8151	0.8141	0.8131	0.8121	0.8111	0.8100
	tan	0.7002	0.7028	0.7054	0.7080	0.7107	0.7133	0.7159	0.7186	0.7212	0.7239
36	sin	0.5878	0.5892	0.5906	0.5920	0.5934	0.5948	0.5962	0.5976	0.5990	0.6004
	cos	0.8090	0.8080	0.8070	0.8059	0.8049	0.8039	0.8028	0.8018	0.8007	0.7997
	tan	0.7265	0.7292	0.7319	0.7346	0.7373	0.7400	0.7427	0.7454	0.7481	0.7508
37	sin	0.6018	0.6032	0.6046	0.6060	0.6074	0.6088	0.6101	0.6115	0.6129	0.6143
	cos	0.7986	0.7976	0.7965	0.7955	0.7944	0.7934	0.7923	0.7912	0.7902	0.7891
	tan	0.7536	0.7563	0.7590	0.7618	0.7646	0.7673	0.7701	0.7729	0.7757	0.7785
38	sin	0.6157	0.6170	0.6184	0.6198	0.6211	0.6225	0.6239	0.6252	0.6266	0.6280
	cos	0.7880	0.7869	0.7859	0.7848	0.7837	0.7826	0.7815	0.7804	0.7793	0.7782
	tan	0.7813	0.7841	0.7869	0.7898	0.7926	0.7954	0.7983	0.8012	0.8040	0.8069
39	sin	0.6293	0.6307	0.6320	0.6334	0.6347	0.6361	0.6374	0.6388	0.6401	0.6414
	cos	0.7771	0.7760	0.7749	0.7738	0.7727	0.7716	0.7705	0.7694	0.7683	0.7672
	tan	0.8098	0.8127	0.8156	0.8185	0.8214	0.8243	0.8273	0.8302	0.8332	0.8361
40	sin	0.6428	0.6441	0.6455	0.6468	0.6481	0.6494	0.6508	0.6521	0.6534	0.6547
	cos	0.7660	0.7649	0.7638	0.7627	0.7615	0.7604	0.7593	0.7581	0.7570	0.7559
	tan	0.8391	0.8421	0.8451	0.8481	0.8511	0.8541	0.8571	0.8601	0.8632	0.8662
41	sin	0.6561	0.6574	0.6587	0.6600	0.6613	0.6626	0.6639	0.6652	0.6665	0.6678
	cos	0.7547	0.7536	0.7524	0.7513	0.7501	0.7490	0.7478	0.7466	0.7455	0.7443
	tan	0.8693	0.8724	0.8754	0.8785	0.8816	0.8847	0.8878	0.8910	0.8941	0.8972
42	sin	0.6691	0.6704	0.6717	0.6730	0.6743	0.6756	0.6769	0.6782	0.6794	0.6807
	cos	0.7431	0.7420	0.7408	0.7396	0.7385	0.7373	0.7361	0.7349	0.7337	0.7325
	tan	0.9004	0.9036	0.9067	0.9099	0.9131	0.9163	0.9195	0.9228	0.9260	0.9293
43	sin	0.6820	0.6833	0.6845	0.6858	0.6871	0.6884	0.6896	0.6909	0.6921	0.6934
	cos	0.7314	0.7302	0.7290	0.7278	0.7266	0.7254	0.7242	0.7230	0.7218	0.7206
	tan	0.9325	0.9358	0.9391	0.9424	0.9457	0.9490	0.9523	0.9556	0.9590	0.9623
44	sin	0.6947	0.6959	0.6972	0.6984	0.6997	0.7009	0.7022	0.7034	0.7046	0.7059
	cos	0.7193	0.7181	0.7169	0.7157	0.7145	0.7133	0.7120	0.7108	0.7096	0.7083
	tan	0.9657	0.9691	0.9725	0.9759	0.9793	0.9827	0.9861	0.9896	0.9930	0.9965
Degs.	Function	0′	6′	12′	18′	24′	30′	36′	42′	48′	54′

Degs.	Function	0.0°	0.1°	0.2°	0.3°	0.4°	0.5°	0.6°	0.7°	0.8°	0.9°
45	sin	0.7071	0.7083	0.7096	0.7108	0.7120	0.7133	0.7145	0.7157	0.7169	0.7181
	cos	0.7071	0.7059	0.7046	0.7034	0.7022	0.7009	0.6997	0.6984	0.6972	0.6959
	tan	1.0000	1.0035	1.0070	1.0105	1.0141	1.0176	1.0212	1.0247	1.0283	1.0319
46	sin	0.7193	0.7206	0.7218	0.7230	0.7242	0.7254	0.7266	0.7278	0.7290	0.7302
	cos	0.6947	0.6934	0.6921	0.6909	0.6896	0.6884	0.6871	0.6858	0.6845	0.6833
	tan	1.0355	1.0392	1.0428	1.0464	1.0501	1.0538	1.0575	1.0612	1.0649	1.0686
47	sin	0.7314	0.7325	0.7337	0.7349	0.7361	0.7373	0.7385	0.7396	0.7408	0.7420
	cos	0.6820	0.6807	0.6794	0.6782	0.6769	0.6756	0.6743	0.6730	0.6717	0.6704
	tan	1.0724	1.0761	1.0799	1.0837	1.0875	1.0913	1.0951	1.0990	1.1028	1.1067
48	sin	0.7431	0.7443	0.7455	0.7466	0.7478	0.7490	0.7501	0.7513	0.7524	0.7536
	cos	0.6691	0.6678	0.6665	0.6652	0.6639	0.6626	0.6613	0.6600	0.6587	0.6574
	tan	1.1106	1.1145	1.1184	1.1224	1.1263	1.1303	1.1343	1.1383	1.1423	1.1463
49	sin	0.7547	0.7559	0.7570	0.7581	0.7593	0.7604	0.7615	0.7627	0.7638	0.7649
	cos	0.6561	0.6547	0.6534	0.6521	0.6508	0.6494	0.6481	0.6468	0.6455	0.6441
	tan	1.1504	1.1544	1.1585	1.1626	1.1667	1.1708	1.1750	1.1792	1.1833	1.1875
50	sin	0.7660	0.7672	0.7683	0.7694	0.7705	0.7716	0.7727	0.7738	0.7749	0.7760
	cos	0.6428	0.6414	0.6401	0.6388	0.6374	0.6361	0.6347	0.6334	0.6320	0.6307
	tan	1.1918	1.1960	1.2002	1.2045	1.2088	1.2131	1.2174	1.2218	1.2261	1.2305
51	sin	0.7771	0.7782	0.7793	0.7804	0.7815	0.7826	0.7837	0.7848	0.7859	0.7869
	cos	0.6293	0.6280	0.6266	0.6252	0.6239	0.6225	0.6211	0.6198	0.6184	0.6170
	tan	1.2349	1.2393	1.2437	1.2482	1.2527	1.2572	1.2617	1.2662	1.2708	1.2753
52	sin	0.7880	0.7891	0.7902	0.7912	0.7923	0.7934	0.7944	0.7955	0.7965	0.7976
	cos	0.6157	0.6143	0.6129	0.6115	0.6101	0.6088	0.6074	0.6060	0.6046	0.6032
	tan	1.2799	1.2846	1.2892	1.2938	1.2985	1.3032	1.3079	1.3127	1.3175	1.3222
53	sin	0.7986	0.7997	0.8007	0.8018	0.8028	0.8039	0.8049	0.8059	0.8070	0.8080
	cos	0.6018	0.6004	0.5990	0.5976	0.5962	0.5948	0.5934	0.5920	0.5906	0.5892
	tan	1.3270	1.3319	1.3367	1.3416	1.3465	1.3514	1.3564	1.3613	1.3663	1.3713
54	sin	0.8090	0.8100	0.8111	0.8121	0.8131	0.8141	0.8151	0.8161	0.8171	0.8181
	cos	0.5878	0.5864	0.5850	0.5835	0.5821	0.5807	0.5793	0.5779	0.5764	0.5750
	tan	1.3764	1.3814	1.3865	1.3916	1.3968	1.4019	1.4071	1.4124	1.4176	1.4229
55	sin	0.8192	0.8202	0.8211	0.8221	0.8231	0.8241	0.8251	0.8261	0.8271	0.8281
	cos	0.5736	0.5721	0.5707	0.5693	0.5678	0.5664	0.5650	0.5635	0.5621	0.5606
	tan	1.4281	1.4335	1.4388	1.4442	1.4496	1.4550	1.4605	1.4659	1.4715	1.4770
56	sin	0.8290	0.8300	0.8310	0.8320	0.8329	0.8339	0.8348	0.8358	0.8368	0.8377
	cos	0.5592	0.5577	0.5563	0.5548	0.5534	0.5519	0.5505	0.5490	0.5476	0.5461
	tan	1.4826	1.4882	1.4938	1.4994	1.5051	1.5108	1.5166	1.5224	1.5282	1.5340
57	sin	0.8387	0.8396	0.8406	0.8415	0.8425	0.8434	0.8443	0.8453	0.8462	0.8471
	cos	0.5446	0.5432	0.5417	0.5402	0.5388	0.5373	0.5358	0.5344	0.5329	0.5314
	tan	1.5399	1.5458	1.5517	1.5577	1.5637	1.5697	1.5757	1.5818	1.5880	1.5941
58	sin	0.8480	0.8490	0.8499	0.8508	0.8517	0.8526	0.8536	0.8545	0.8554	0.8563
	cos	0.5299	0.5284	0.5270	0.5255	0.5240	0.5225	0.5210	0.5195	0.5180	0.5165
	tan	1.6003	1.6066	1.6128	1.6191	1.6255	1.6319	1.6383	1.6447	1.6512	1.6577
59	sin	0.8572	0.8581	0.8590	0.8599	0.8607	0.8616	0.8625	0.8634	0.8643	0.8652
	cos	0.5150	0.5135	0.5120	0.5105	0.5090	0.5075	0.5060	0.5045	0.5030	0.5015
	tan	1.6643	1.6709	1.6775	1.6842	1.6909	1.6977	1.7045	1.7113	1.7182	1.7251
Degs.	Function	0′	6′	12′	18′	24′	30′	36′	42′	48′	54′

Degs.	Function	0.0°	0.1°	0.2°	0.3°	0.4°	0.5°	0.6°	0.7°	0.8°	0.9°
60	sin	0.8660	0.8669	0.8678	0.8686	0.8695	0.8704	0.8712	0.8721	0.8729	0.8738
	cos	0.5000	0.4985	0.4970	0.4955	0.4939	0.4924	0.4909	0.4894	0.4879	0.4863
	tan	1.7321	1.7391	1.7461	1.7532	1.7603	1.7675	1.7747	1.7820	1.7893	1.7966
61	sin	0.8746	0.8755	0.8763	0.8771	0.8780	0.8788	0.8796	0.8805	0.8813	0.8821
	cos	0.4848	0.4833	0.4818	0.4802	0.4787	0.4772	0.4756	0.4741	0.4726	0.4710
	tan	1.8040	1.8115	1.8190	1.8265	1.8341	1.8418	1.8495	1.8572	1.8650	1.8728
62	sin	0.8829	0.8838	0.8846	0.8854	0.8862	0.8870	0.8878	0.8886	0.8894	0.8902
	cos	0.4695	0.4679	0.4664	0.4648	0.4633	0.4617	0.4602	0.4586	0.4571	0.4555
	tan	1.8807	1.8887	1.8967	1.9047	1.9128	1.9210	1.9292	1.9375	1.9458	1.9542
63	sin	0.8910	0.8918	0.8926	0.8934	0.8942	0.8949	0.8957	0.8965	0.8973	0.8980
	cos	0.4540	0.4524	0.4509	0.4493	0.4478	0.4462	0.4446	0.4431	0.4415	0.4399
	tan	1.9626	1.9711	1.9797	1.9883	1.9970	2.0057	2.0145	2.0233	2.0323	2.0413
64	sin	0.8988	0.8996	0.9003	0.9011	0.9018	0.9026	0.9033	0.9041	0.9048	0.9056
	cos	0.4384	0.4368	0.4352	0.4337	0.4321	0.4305	0.4289	0.4274	0.4258	0.4242
	tan	2.0503	2.0594	2.0686	2.0778	2.0872	2.0965	2.1060	2.1155	2.1251	2.1348
65	sin	0.9063	0.9070	0.9078	0.9085	0.9092	0.9100	0.9107	0.9114	0.9121	0.9128
	cos	0.4226	0.4210	0.4195	0.4179	0.4163	0.4147	0.4131	0.4115	0.4099	0.4083
	tan	2.1445	2.1543	2.1642	2.1742	2.1842	2.1943	2.2045	2.2148	2.2251	2.2355
66	sin	0.9135	0.9143	0.9150	0.9157	0.9164	0.9171	0.9178	0.9184	0.9191	0.9198
	cos	0.4067	0.4051	0.4035	0.4019	0.4003	0.3987	0.3971	0.3955	0.3939	0.3923
	tan	2.2460	2.2566	2.2673	2.2781	2.2889	2.2998	2.3109	2.3220	2.3332	2.3445
67	sin	0.9205	0.9212	0.9219	0.9225	0.9232	0.9239	0.9245	0.9252	0.9259	0.9265
	cos	0.3907	0.3891	0.3875	0.3859	0.3843	0.3827	0.3811	0.3795	0.3778	0.3762
	tan	2.3559	2.3673	2.3789	2.3906	2.4023	2.4142	2.4262	2.4383	2.4504	2.4627
68	sin	0.9272	0.9278	0.9285	0.9291	0.9298	0.9304	0.9311	0.9317	0.9323	0.9330
	cos	0.3746	0.3730	0.3714	0.3697	0.3681	0.3665	0.3649	0.3633	0.3616	0.3600
	tan	2.4751	2.4876	2.5002	2.5129	2.5257	2.5386	2.5517	2.5649	2.5782	2.5916
69	sin	0.9336	0.9342	0.9348	0.9354	0.9361	0.9367	0.9373	0.9379	0.9385	0.9391
	cos	0.3584	0.3567	0.3551	0.3535	0.3518	0.3502	0.3486	0.3469	0.3453	0.3437
	tan	2.6051	2.6187	2.6325	2.6464	2.6605	2.6746	2.6889	2.7034	2.7179	2.7326
70	sin	0.9397	0.9403	0.9409	0.9415	0.9421	0.9426	0.9432	0.9438	0.9444	0.9449
	cos	0.3420	0.3404	0.3387	0.3371	0.3355	0.3338	0.3322	0.3305	0.3289	0.3272
	tan	2.7475	2.7625	2.7776	2.7929	2.8083	2.8239	2.8397	2.8556	2.8716	2.8878
71	sin	0.9455	0.9461	0.9466	0.9472	0.9478	0.9483	0.9489	0.9494	0.9500	0.9505
	cos	0.3256	0.3239	0.3223	0.3206	0.3190	0.3173	0.3156	0.3140	0.3123	0.3107
	tan	2.9042	2.9208	2.9375	2.9544	2.9714	2.9887	3.0061	3.0237	3.0415	3.0595
72	sin	0.9511	0.9516	0.9521	0.9527	0.9532	0.9537	0.9542	0.9548	0.9553	0.9558
	cos	0.3090	0.3074	0.3057	0.3040	0.3024	0.3007	0.2990	0.2974	0.2957	0.2940
	tan	3.0777	3.0961	3.1146	3.1334	3.1524	3.1716	3.1910	3.2106	3.2305	3.2506
73	sin	0.9563	0.9568	0.9573	0.9578	0.9583	0.9588	0.9593	0.9598	0.9603	0.9608
	cos	0.2924	0.2907	0.2890	0.2874	0.2857	0.2840	0.2823	0.2807	0.2790	0.2773
	tan	3.2709	3.2914	3.3122	3.3332	3.3544	3.3759	3.3977	3.4197	3.4420	3.4646
74	sin	0.9613	0.9617	0.9622	0.9627	0.9632	0.9636	0.9641	0.9646	0.9650	0.9655
	cos	0.2756	0.2740	0.2723	0.2706	0.2689	0.2672	0.2656	0.2639	0.2622	0.2605
	tan	3.4874	3.5105	3.5339	3.5576	3.5816	3.6059	3.6305	3.6554	3.6806	3.7062
Degs.	**Function**	0′	6′	12′	18′	24′	30′	36′	42′	48′	54′

Degs.	Function	0.0°	0.1°	0.2°	0.3°	0.4°	0.5°	0.6°	0.7°	0.8°	0.9°
75	sin	0.9659	0.9664	0.9668	0.9673	0.9677	0.9681	0.9686	0.9690	0.9694	0.9699
	cos	0.2588	0.2571	0.2554	0.2538	0.2521	0.2504	0.2487	0.2470	0.2453	0.2436
	tan	3.7321	3.7583	3.7848	3.8118	3.8391	3.8667	3.8947	3.9232	3.9520	3.9812
76	sin	0.9703	0.9707	0.9711	0.9715	0.9720	0.9724	0.9728	0.9732	0.9736	0.9740
	cos	0.2419	0.2402	0.2385	0.2368	0.2351	0.2334	0.2317	0.2300	0.2284	0.2267
	tan	4.0108	4.0408	4.0713	4.1022	4.1335	4.1653	4.1976	4.2303	4.2635	4.2972
77	sin	0.9744	0.9748	0.9751	0.9755	0.9759	0.9763	0.9767	0.9770	0.9774	0.9778
	cos	0.2250	0.2232	0.2215	0.2198	0.2181	0.2164	0.2147	0.2130	0.2113	0.2096
	tan	4.3315	4.3662	4.4015	4.4374	4.4737	4.5107	4.5483	4.5864	4.6252	4.6646
78	sin	0.9781	0.9785	0.9789	0.9792	0.9796	0.9799	0.9803	0.9806	0.9810	0.9813
	cos	0.2079	0.2062	0.2045	0.2028	0.2011	0.1994	0.1977	0.1959	0.1942	0.1925
	tan	4.7046	4.7453	4.7867	4.8288	4.8716	4.9152	4.9594	5.0045	5.0504	5.0970
79	sin	0.9816	0.9820	0.9823	0.9826	0.9829	0.9833	0.9836	0.9839	0.9842	0.9845
	cos	0.1908	0.1891	0.1874	0.1857	0.1840	0.1822	0.1805	0.1788	0.1771	0.1754
	tan	5.1446	5.1929	5.2422	5.2924	5.3435	5.3955	5.4486	5.5026	5.5578	5.6140
80	sin	0.9848	0.9851	0.9854	0.9857	0.9860	0.9863	0.9866	0.9869	0.9871	0.9874
	cos	0.1736	0.1719	0.1702	0.1685	0.1668	0.1650	0.1633	0.1616	0.1599	0.1582
	tan	5.6713	5.7297	5.7894	5.8502	5.9124	5.9758	6.0405	6.1066	6.1742	6.2432
81	sin	0.9877	0.9880	0.9882	0.9885	0.9888	0.9890	0.9893	0.9895	0.9898	0.9900
	cos	0.1564	0.1547	0.1530	0.1513	0.1495	0.1478	0.1461	0.1444	0.1426	0.1409
	tan	6.3138	6.3859	6.4596	6.5350	6.6122	6.6912	6.7720	6.8548	6.9395	7.0264
82	sin	0.9903	0.9905	0.9907	0.9910	0.9912	0.9914	0.9917	0.9919	0.9921	0.9923
	cos	0.1392	0.1374	0.1357	0.1340	0.1323	0.1305	0.1288	0.1271	0.1253	0.1236
	tan	7.1154	7.2066	7.3002	7.3962	7.4947	7.5958	7.6996	7.8062	7.9158	8.0285
83	sin	0.9925	0.9928	0.9930	0.9932	0.9934	0.9936	0.9938	0.9940	0.9942	0.9943
	cos	0.1219	0.1201	0.1184	0.1167	0.1149	0.1132	0.1115	0.1097	0.1080	0.1063
	tan	8.1443	8.2636	8.3863	8.5126	8.6427	8.7769	8.9152	9.0579	9.2052	9.3572
84	sin	0.9945	0.9947	0.9949	0.9951	0.9952	0.9954	0.9956	0.9957	0.9959	0.9960
	cos	0.1045	0.1028	0.1011	0.0993	0.0976	0.0958	0.0941	0.0924	0.0906	0.0889
	tan	9.5144	9.6768	9.8448	10.02	10.20	10.39	10.58	10.78	10.99	11.20
85	sin	0.9962	0.9963	0.9965	0.9966	0.9968	0.9969	0.9971	0.9972	0.9973	0.9974
	cos	0.0872	0.0854	0.0837	0.0819	0.0802	0.0785	0.0767	0.0750	0.0732	0.0715
	tan	11.43	11.66	11.91	12.16	12.43	12.71	13.00	13.30	13.62	13.95
86	sin	0.9976	0.9977	0.9978	0.9979	0.9980	0.9981	0.9982	0.9983	0.9984	0.9985
	cos	0.0698	0.0680	0.0663	0.0645	0.0628	0.0610	0.0593	0.0576	0.0558	0.0541
	tan	14.30	14.67	15.06	15.46	15.89	16.35	16.83	17.34	17.89	18.46
87	sin	0.9986	0.9987	0.9988	0.9989	0.9990	0.9990	0.9991	0.9992	0.9993	0.9993
	cos	0.0523	0.0506	0.0488	0.0471	0.0454	0.0436	0.0419	0.0401	0.0384	0.0366
	tan	19.08	19.74	20.45	21.20	22.02	22.90	23.86	24.90	26.03	27.27
88	sin	0.9994	0.9995	0.9995	0.9996	0.9996	0.9997	0.9997	0.9997	0.9998	0.9998
	cos	0.0349	0.0332	0.0314	0.0297	0.0279	0.0262	0.0244	0.0227	0.0209	0.0192
	tan	28.64	30.14	31.82	33.69	35.80	38.19	40.92	44.07	47.74	52.08
89	sin	0.9998	0.9999	0.9999	0.9999	0.9999	1.000	1.000	1.000	1.000	1.000
	cos	0.0175	0.0157	0.0140	0.0122	0.0105	0.0087	0.0070	0.0052	0.0035	0.0017
	tan	57.29	63.66	71.62	81.85	95.49	114.6	143.2	191.0	286.5	573.0
Degs.	**Function**	0′	6′	12′	18′	24′	30′	36′	42′	48′	54′

Conversion Factors

The customary units of weight and mass are avoirdupois units unless designated otherwise. The symbol (δ) represents the density of a material expressed as a decimal fraction. g equals 980.7 cms. per sec. per sec.

Multiply	by	to obtain
Abamperes	10	amperes.
“	3×10^{10}	statamperes.
abamperes per square cm. .	64.52	amperes per sq. inch.
abampere-turns.	10	ampere-turns.
“ “	12.57	gilberts.
abampere-turns per cm. .	25.40	ampere-turns per inch.
abcoulombs	10	coulombs.
“	3×10^{10}	statcoulombs.
abcoulombs per square cm.	64.52	coulombs per sq. inch.
abfarads	10^9	farads.
“	10^{15}	microfarads.
“	9×10^{20}	statfarads.
abhenries	10^{-9}	henries.
“	10^{-6}	millihenries.
“	$1/9\times10^{-20}$	stathenries.
abmhos per cm. cube . .	$10^5/\delta$	mhos per meter-gram.
“ “ “ “ . . .	1.662×10^2	mhos per mil foot.
“ “ “ “ . . .	10^3	megmhos per cm. cube.
abohms	10^{-15}	megohms.
“	10^{-3}	microhms.
“	10^{-9}	ohms.
“	$1/9\times10^{-20}$	statohms.
abohms per cm. cube . .	10^{-3}	microhms per cm. cube.
“ “ “ “ . . .	6.015×10^{-3}	ohms per mil foot.
“ “ “ “ . . .	$10^{-5}\delta$	ohms per meter-gram.
abvolts	$1/3\times10^{-10}$	statvolts.
“	10^{-8}	volts.
acres	43,560	square feet.
“	6,272,640	square inches.
“	4047	square meters.
“	1.562×10^{-3}	square miles.
“	4840	square yards.
acre-feet	43,560	cubic-feet.
“ “	3.259×10^5	gallons.
amperes	1/10	abamperes.
“	3×10^9	statamperes.
amperes per square cm. .	6.452	amperes per sq. inch.
amperes per square inch .	0.01550	abamperes per sq. cm.
“ “ “ “ . .	0.1550	amperes per sq. cm.
“ “ “ “ . .	4.650×10^8	statamperes per sq. cm.
ampere-turns	1/10	abampere-turns.
“ “	1.257	gilberts.
ampere-turns per cm. . .	2.540	ampere-turns per inch.
ampere-turns per inch . .	0.03937	abampere-turns per cm.
“ “ “ “ . .	0.3937	ampere-turns per cm.
“ “ “ “ . .	0.4950	gilberts per cm.
ares	0.02471	acres.
“	100	square meters.
atmospheres	76	cms. of mercury.
“	29.92	inches of mercury.
“	33.90	feet of water.
“	10,332	kgs. per square meter.
“	14.70	pounds per sq. inch.
“	1.058	tons per sq. foot.

Multiply	by	to obtain
Bars	0.9869	atmospheres.
"	1	dynes per sq. cm.
"	1.020×10^{4}	kgs. per square meter.
"	2,089	pounds per square foot.
"	14.50	pounds per square inch.
board-feet	144 sq. in. $\times$ 1 in.	cubic inches.
British thermal units	778.2	foot-pounds.
" " "	3.930×10^{-4}	horse-power-hours.
" " "	1055	joules.
" " "	0.2520	kilogram-calories.
" " "	107.6	kilogram-meters.
" " "	2.930×10^{-4}	kilowatt hours.
B.t.u. per min.	12.97	foot-pounds per sec.
" " "	0.02358	horse-power.
" " "	0.01758	kilowatts.
" " "	17.58	watts.
B.t.u. per sq. ft. per min.	0.1221	watts per square inch.
bushels	1.244	cubic feet.
"	2150	cubic inches.
"	0.03524	cubic meters.
"	4	pecks.
"	64	pints (dry).
"	32	quarts (dry).
Centares.	1	square meters.
centigrams	0.01	grams.
centiliters	0.01	liters.
centimeters	3.281×10^{-2}	feet.
"	0.3937	inches.
"	0.01	meters.
"	6.214×10^{-6}	miles.
"	10	millimeters.
"	393.7	mils.
"	1.094×10^{-2}	yards.
centimeter-dynes	1.020×10^{-3}	centimeter-grams.
" "	1.020×10^{-8}	meter-kilograms.
" "	7.376×10^{-8}	pound-feet.
centimeter-grams	980.7	centimeter-dynes.
" "	10^{-5}	meter-kilograms.
" "	7.233×10^{-5}	pound-feet.
centimeters of mercury	0.01316	atmospheres.
" " "	0.4461	feet of water.
" " "	136.0	kgs. per square meter.
" " "	27.85	pounds per square foot.
" " "	0.1934	pounds per square inch.
centimeters per second	1.968	feet per minute.
" " "	0.03281	feet per second.
" " "	0.036	kilometers per hour.
" " "	0.6	meters per minute.
" " "	0.02237	miles per hour.
" " "	3.728×10^{-4}	miles per minute.
cms. per sec. per sec.	0.03281	feet per sec. per sec.
" " " " "	0.036	kms. per hour per sec.
" " " " "	0.02237	miles per hour per sec.
circular mils	5.067×10^{-6}	square centimeters.
" "	7.854×10^{-7}	square inches.

Multiply	by	to obtain
circular mils (*cont.*) . . .	0.7854	square mils.
cord-feet	4 ft.×4 ft.×1 ft.	cubic feet.
cords	8 ft.×4 ft.×4 ft.	cubic feet.
coulombs	1/10	abcoulombs.
"	3×10^9	statcoulombs.
coulombs per square inch .	0.01550	abcoulombs per sq. cm.
" " " " .	0.1550	coulombs per sq. cm.
" " " " .	4.650×10^8	statcouls. per sq. cm.
cubic centimeters	3.531×10^{-5}	cubic feet.
" "	6.102×10^{-2}	cubic inches.
" "	10^{-6}	cubic meters.
" "	1.308×10^{-6}	cubic yards.
" "	2.642×10^{-4}	gallons.
" "	10^{-3}	liters.
" "	2.113×10^{-3}	pints (liq.).
" "	1.057×10^{-3}	quarts (liq.).
cubic feet	2.832×10^4	cubic cms.
" "	1728	cubic inches.
" "	0.02832	cubic meters.
" "	0.03704	cubic yards.
" "	7.481	gallons.
" "	28.32	liters.
" "	59.84	pints (liq.).
" "	29.92	quarts (liq.).
cubic feet per minute . . .	472.0	cubic cms. per sec.
" " " " . . .	0.1247	gallons per sec.
" " " " . . .	0.4720	liters per second.
" " " " . . .	62.4	pounds of water per min.
cubic inches	16.39	cubic centimeters.
" "	5.787×10^{-4}	cubic feet.
" "	1.639×10^{-5}	cubic meters.
" "	2.143×10^{-5}	cubic yards.
" "	4.329×10^{-3}	gallons.
" "	1.639×10^{-2}	liters.
" "	1.061×10^5	mil-feet.
" "	0.03463	pints (liq.).
" "	0.01732	quarts (liq.).
cubic meters	10^6	cubic centimeters
" "	35.31	cubic feet.
" "	61,023	cubic inches.
" "	1.308	cubic yards.
" "	264.2	gallons.
" "	10^3	liters.
" "	2113	pints (liq.).
" "	1057	quarts (liq.).
cubic yards	7.646×10^5	cubic centimeters.
" "	27	cubic feet.
" "	46,656	cubic inches.
" "	0.7646	cubic meters.
" "	202.0	gallons.
" "	764.6	liters.
" "	1616	pints (liq.).
" "	807.9	quarts (liq.).
cubic yards per minute . .	0.45	cubic feet per second.
" " " " . .	3.367	gallons per second.
" " " " . .	12.74	liters per second.

Multiply	by	to obtain
Days	24	hours.
"	1440	minutes.
"	86,400	seconds.
decigrams	0.1	grams.
deciliters	0.1	liters.
decimeters	0.1	meters.
degrees (angle)	60	minutes.
" "	0.01745	radians.
" "	3600	seconds.
degrees per second	0.01745	radians per second.
" " "	0.1667	revolutions per minute.
" " "	0.002778	revolutions per second.
dekagrams	10	grams.
dekaliters	10	liters.
dekameters	10	meters.
drams	1.772	grams.
"	0.0625	ounces.
dynes	1.020×10^{-3}	grams.
"	7.233×10^{-5}	poundals.
"	2.248×10^{-6}	pounds.
dynes per square cm.	1	bars.
Ergs	9.480×10^{-11}	British thermal units.
"	1	dyne-centimeters.
"	7.378×10^{-8}	foot-pounds.
"	1.020×10^{-3}	gram-centimeters.
"	10^{-7}	joules.
"	2.389×10^{-11}	kilogram-calories.
"	1.020×10^{-8}	kilogram-meters.
ergs per second	5.688×10^{-9}	B.t. units per minute.
" " "	4.427×10^{-6}	foot-pounds per minute.
" " "	7.378×10^{-8}	foot-pounds per second.
" " "	1.341×10^{-10}	horse-power.
" " "	1.433×10^{-9}	kg.-calories per minute.
" " "	10^{-10}	kilowatts.
Farads	10^{-9}	abfarads.
"	10^{6}	microfarads.
"	9×10^{11}	statfarads.
fathoms	6	feet.
feet	30.48	centimeters.
"	12	inches.
"	0.3048	meters.
"	1.894×10^{-4}	miles.
"	1/3	yards.
feet of water	0.02950	atmospheres.
" " "	0.8826	inches of mercury.
" " "	304.8	kgs. per square meter.
" " "	62.43	pounds per square foot.
" " "	0.4335	pounds per square inch.
feet per minute	0.5080	centimeters per second.
" " "	0.01667	feet per second.
" " "	0.01829	kilometers per hour.
" " "	0.3048	meters per minute.
" " "	0.01136	miles per hour.
feet per second	30.48	centimeters per second.

Multiply	by	to obtain
feet per second (*cont.*) . . .	1.097	kilometers per hour.
" " "	0.5921	knots.
" " "	18.29	meters per minute.
" " "	0.6818	miles per hour.
" " "	0.01136	miles per minute.
feet per 100 feet	1	per cent grade.
feet per second per second	30.48	cms. per sec. per sec.
" " " " "	1.097	kms. per hour per sec.
" " " " "	0.3048	meters per sec. per sec.
" " " " "	0.6818	miles per hour per sec.
foot-pounds	1.285×10^{-3}	British thermal units.
" "	1.356×10^{7}	ergs.
" "	5.050×10^{-7}	horse-power-hours.
" "	1.356	joules.
" "	3.238×10^{-4}	kilogram-calories.
" "	0.1383	kilogram-meters.
" "	3.766×10^{-7}	kilowatt-hours.
foot-pounds per minute . .	1.285×10^{-3}	B.t. units per minute.
" " " " . .	0.01667	foot-pounds per second.
" " " " . .	3.030×10^{-5}	horse-power.
" " " " . .	3.238×10^{-4}	kg.-calories per min.
" " " " . .	2.260×10^{-5}	kilowatts.
foot-pounds per second . .	7.712×10^{-2}	B.t. units per minute.
" " " " . .	1.818×10^{-3}	horse-power.
" " " " . .	1.943×10^{-2}	kg.-calories per min.
" " " " . .	1.356×10^{-3}	kilowatts.
furlongs	40	rods.
Gallons	3785	cubic centimeters.
"	0.1337	cubic feet.
"	231	cubic inches.
"	3.785×10^{-3}	cubic meters.
"	4.951×10^{-3}	cubic yards.
"	3.785	liters.
"	8	pints (liq.).
"	4	quarts (liq.).
gallons per minute	2.228×10^{-3}	cubic feet per second.
" " "	0.06308	liters per second.
gausses	6.452	lines per square inch.
gilberts	0.07958	abampere-turns.
"	0.7958	ampere-turns.
gilberts per centimeter . .	2.021	ampere-turns per inch.
gills	0.1183	liters.
"	0.25	pints (liq.).
grains	1	grains (av.).
"	0.06480	grams.
"	0.04167	pennyweights (troy).
grams	980.7	dynes.
"	15.43	grains.
"	10^{-3}	kilograms.
"	10^{3}	milligrams.
"	0.03527	ounces.
"	0.03215	ounces (troy).
"	0.07093	poundals.
"	2.205×10^{-3}	pounds.
gram-calories (IT)	3.968×10^{-3}	British thermal units.

Multiply	by	to obtain
gram-centimeters	9.297×10^{-8}	British thermal units.
" "	980.7	ergs.
" "	7.235×10^{-5}	foot-pounds.
" "	9.807×10^{-5}	joules.
" "	2.343×10^{-8}	kilogram-calories.
" "	10^{-5}	kilogram-meters.
grams per cm.	5.600×10^{-3}	pounds per inch.
grams per cu. cm.	62.43	pounds per cubic foot.
" " " "	0.03613	pounds per cubic inch.
" " " "	3.405×10^{-7}	pounds per mil-foot.
Hectares	2.471	acres.
"	1.076×10^{5}	square feet.
hectograms	100	grams.
hectoliters	100	liters.
hectometers	100	meters.
hectowatts	100	watts.
hemispheres (solid angle) .	0.5	sphere.
" " " .	4	spherical right angles.
" " " .	6.283	steradians.
henries	10^{9}	abhenries.
"	10^{3}	millihenries.
"	$1/9\times10^{-11}$	stathenries.
horse-power	42.40	B.t. units per min.
" "	33,000	foot-pounds per minute.
" "	550	foot-pounds per second.
" "	1.014	horse-power (metric).
" "	10.68	kg.-calories per minute.
" "	0.7457	kilowatts.
" "	745.7	watts.
horse-power (boiler) . . .	33.520	B.t.u. per hour.
" " " . . .	9.804	kilowatts.
horse-power-hours	2544	British thermal units.
" " "	1.98×10^{6}	foot-pounds.
" " "	2.684×10^{6}	joules.
" " "	641.1	kilogram-calories.
" " "	2.737×10^{5}	kilogram-meters.
" " "	0.7455	kilowatt-hours.
hours	4.167×10^{-2}	days.
"	60	minutes.
"	3600	seconds.
"	5.952×10^{-3}	weeks.
Inches	2.540	centimeters.
"	8.333×10^{-2}	feet.
"	1.578×10^{-5}	miles.
"	10^{3}	mils.
"	2.778×10^{-2}	yards.
inches of mercury	0.03342	atmospheres.
" " "	1.133	feet of water.
" " "	345.3	kgs. per square meter.
" " "	70.73	pounds per square foot.
" " "	0.4912	pounds per square inch.
inches of water	0.002458	atmospheres.
" " "	0.07355	inches of mercury.
" " "	25.40	kgs. per square meter.

Multiply	by	to obtain
inches of water (*cont.*) . .	0.5781	ounces per square inch.
" " "	5.204	pounds per square foot.
" " "	0.03613	pounds per square inch.
Joules (Int.)	9.480×10^{-4}	British thermal units.
" "	10^7	ergs.
" "	0.7378	foot-pounds.
" "	2.389×10^{-4}	kilogram-calories.
" "	0.1020	kilogram-meters.
" "	2.778×10^{-4}	watt-hours.
Kilograms	980,665	dynes.
"	10^3	grams.
"	70.93	poundals.
"	2.205	pounds.
"	1.102×10^{-3}	tons (short).
kilogram-calories	3.968	British thermal units.
" "	3088	foot-pounds.
" "	1.560×10^{-3}	horse-power-hours.
" "	4186	joules.
" "	427.0	kilogram-meters.
" "	1.163×10^{-3}	kilowatt-hours.
kilogram-calories per min.	51.47	foot-pounds per second.
" " " "	0.09358	horse-power.
" " " "	0.06977	kilowatts.
kgs.-cms. squared	2.373×10^{-3}	pounds-feet squared.
" " "	0.3417	pounds-inches squared.
kilogram-meters	9.294×10^{-3}	British thermal units.
" "	9.804×10^7	ergs.
" "	7.233	foot-pounds.
" "	9.804	joules.
" "	2.342×10^{-3}	kilogram-calories.
" "	2.723×10^{-6}	kilowatt-hours.
kilograms per cubic meter	10^{-3}	grams per cubic cm.
" " " "	0.06243	pounds per cubic foot.
" " " "	3.613×10^{-5}	pounds per cubic inch.
" " " "	3.405×10^{-10}	pounds per mil foot.
kgs. per meter	0.6720	pounds per foot.
kgs. per square meter . .	9.678×10^{-5}	atmospheres.
" " " " . .	98.07×10^{-6}	bars.
" " " " . .	3.281×10^{-3}	feet of water.
" " " " . .	2.896×10^{-3}	inches of mercury.
" " " " . .	0.2048	pounds per square foot.
" " " " . .	1.422×10^{-3}	pounds per square inch.
kgs. per square millimeter	10^6	kgs. per square meter.
kilolines	10^3	maxwells.
kiloliters	10^3	liters.
kilometers	10^5	centimeters.
"	3281	feet.
"	3.937×10^4	inches.
"	10^3	meters.
"	0.6214	miles.
"	1094	yards.
kilometers per hour . . .	27.78	centimeters per second.
" " " . . .	54.68	feet per minute.
" " " . . .	0.9113	feet per second.
" " " . . .	0.5396	knots.

Multiply	by	to obtain
kilometers per hour (*cont.*)	16.67	meters per minute.
" " " . . .	0.6214	miles per hour.
kms. per hour per sec. . .	27.78	cms. per sec. per sec.
" " " " " . .	0.9113	ft. per sec. per sec.
" " " " " . .	0.2778	meters per sec. per sec.
" " " " " . .	0.6214	miles per hr. per sec.
kilometers per min. . . .	60	kilometers per hour.
kilowatts	56.88	B.t. units per min.
"	4.427×10^4	foot-pounds per min.
"	737.8	foot-pounds per sec.
"	1.341	horse-power.
"	14.33	kg.-calories per min.
"	10^3	watts.
kilowatt-hours	3413	British thermal units.
" "	2.656×10^6	foot-pounds.
" "	1.341	horse-power-hours.
" "	3.6×10^6	joules.
" "	860	kilogram-calories.
" "	3.672×10^5	kilogram-meters.
knots (length)	6080	feet.
" "	1.853	kilometers.
" "	1.152	miles.
" "	2027	yards.
knots (speed)	51.48	centimeters per second.
" "	1.689	feet per second.
" "	1.853	kilometers per hour.
" "	1.152	miles per hour.
Lines per square cm. . . .	1	gausses.
lines per square inch . . .	0.1550	gausses.
links (engineer's)	12	inches.
links (surveyor's)	7.92	inches.
liters	10^3	cubic centimeters.
"	0.03531	cubic feet.
"	61.02	cubic inches.
"	10^{-3}	cubic meters.
"	1.308×10^{-3}	cubic yards.
"	0.2642	gallons.
"	2.113	pints (liq.).
"	1.057	quarts (liq.).
liters per minute	5.885×10^{-4}	cubic feet per second.
" " "	4.403×10^{-3}	gallons per second.
$\log_{10} N$	2.303	$\log_\epsilon N$ or $ln\ N$.
$\log_\epsilon N$ or $ln\ N$	0.4343	$\log_{10} N$.
lumens per sq. ft.	1	foot-candles.
Maxwells	10^{-3}	kilolines.
megalines	10^6	maxwells.
megmhos per cm. cube .	10^{-3}	abmhos per cm. cube.
" " " " .	2.540	megmhos per inch cube.
" " " " .	$10^2/8$	mhos per meter-gram.
" " " " .	0.1662	mhos per mil foot.
megmhos per inch cube .	0.3937	megmhos per cm. cube.
megohms	10^6	ohms.
meters	100	centimeters.
"	3.281	feet.
"	39.37	inches.

Multiply	by	to obtain
meters (*cont.*)	10^{-3}	kilometers.
"	6.214×10^{-4}	miles.
"	10^{3}	millimeters.
"	1.094	yards.
meter-kilograms	9.807×10^{7}	centimeter-dynes.
" "	10^{5}	centimeter-grams.
" "	7.233	pound-feet.
meters per minute	1.667	centimeters per second.
" " "	3.281	feet per minute.
" " "	0.05468	feet per second.
" " "	0.06	kilometers per hour.
" " "	0.03728	miles per hour.
meters per second	196.8	feet per minute.
" " "	3.281	feet per second.
" " "	3.6	kilometers per hour.
" " "	0.06	kilometers per minute.
" " "	2.237	miles per hour.
" " "	0.03728	miles per minute.
meters per sec. per sec.	3.281	feet per sec. per sec.
" " " " "	3.6	kms. per hour per sec.
" " " " "	2.237	miles per hour per sec.
mhos per meter-gram	$10^{-5}\delta$	abmhos per cm. cube.
" " " "	$10^{-2}\delta$	megmhos per cm. cube.
" " " "	$2.540 \times 10^{-2}\delta$	megmhos per inch cube.
" " " "	$1.662 \times 10^{-3}\delta$	mhos per mil foot.
mhos per mil foot	6.015×10^{-3}	abmhos per cm. cube.
" " " "	6.015	megmhos per cm. cube.
" " " "	15.28	megmhos per in. cube.
" " " "	$601.5/\delta$	mhos per meter-gram.
microfarads	10^{-15}	abfarads.
"	10^{-6}	farads.
"	9×10^{5}	statfarads.
micrograms	10^{-6}	grams.
microliters	10^{-6}	liters.
microhms	10^{3}	abohms.
"	10^{-12}	megohms.
"	10^{-6}	ohms.
"	$1/9 \times 10^{-17}$	statohms.
microhms per cm. cube	10^{3}	abohms per cm. cube.
" " " "	0.3937	microhms per inch cube.
" " " "	$10^{-2}\delta$	ohms per meter-gram.
" " " "	6.015	ohms per mil foot.
microhms per inch cube	2.540	microhms per cm. cube.
microns	10^{-6}	meters.
miles	1.609×10^{5}	centimeters.
"	5280	feet.
"	6.336×10^{4}	inches.
"	1.609	kilometers.
"	1760	yards.
miles per hour	44.70	centimeters per sec.
" " "	88	feet per minute.
" " "	1.467	feet per second.
" " "	1.609	kilometers per hour.
" " "	0.8684	knots.
" " "	26.82	meters per minute.
miles per hour per second	44.70	cms. per sec. per sec.
" " " " "	1.467	feet per sec. per sec.

Multiply	by	to obtain
miles per hr. per sec. (*cont.*)	1.609	kms. per hour per sec.
" " " " "	0.4470	meters per sec. per sec.
miles per minute	2682	centimeters per second.
" " "	88	feet per second.
" " "	1.609	kilometers per min.
" " "	52.10	knots.
" " "	60	miles per hour.
mil-feet	9.425×10^{-6}	cubic inches.
milliers	10^{3}	kilograms.
milligrams	10^{-3}	grams.
millihenries	10^{6}	abhenries.
"	10^{-3}	henries.
"	$1/9 \times 10^{-14}$	stathenries.
milliliters	10^{-3}	liters.
millimeters	0.1	centimeters
"	3.281×10^{-3}	feet.
"	0.03937	inches.
"	6.214×10^{-7}	miles.
"	39.37	mils.
"	1.094×10^{-3}	yards.
mils	2.540×10^{-3}	centimeters.
"	8.333×10^{-5}	feet.
"	10^{-3}	inches.
"	2.540×10^{-8}	kilometers.
"	2.778×10^{-5}	yards.
miner's inches	1.5	cubic feet per min.
minutes	6.944×10^{-4}	days.
"	1.667×10^{-2}	hours.
"	9.921×10^{-5}	weeks.
minutes (angle)	2.909×10^{-4}	radians.
" "	60	seconds (angle).
months	30.42	days.
"	730	hours.
"	43.800	minutes.
"	2.628×10^{6}	seconds.
myriagrams	10	kilograms.
myriameters	10	kilometers.
myriawatts	10	kilowatts.
Ohms	10^{9}	abohms.
"	10^{-6}	megohms.
"	10^{6}	microhms.
"	$1/9 \times 10^{-11}$	statohms.
ohms per meter-gram . . .	$10^{5}/\delta$	abohms per cm. cube.
" " " " . . .	$10^{2}/\delta$	microhms per cm. cube.
" " " " . . .	$39.37/\delta$	microhms per in. cube.
" " " " . . .	$601.5/\delta$	ohms per mil foot.
ohms per mil foot	166.2	abohms per cm. cube.
" " " "	0.1662	microhms per cm. cube.
" " " "	0.06524	microhms per inch cube.
" " " "	$1.662 \times 10^{-3}\delta$	ohms per meter-gram.
ounces	16	drams.
"	437.5	grains.
"	28.35	grams.
"	0.0625	pounds.
ounces (fluid)	1.805	cubic inches.
" "	0.02957	liters.

Multiply	by	to obtain
ounces (troy)	480	grains.
" "	31.10	grams.
" "	20	pennyweights (troy).
" "	0.08333	pounds (troy).
ounces per square inch	0.0625	pounds per square inch.
Pennyweights (troy)	24	grains.
" "	1.555	grams.
" "	0.05	ounces (troy).
perches (masonry)	24.75	cubic feet.
pints (dry)	33.60	cubic inches.
pints (liq.)	473.2	cubic centimeters.
" "	1.671×10^{-2}	cubic feet.
" "	28.87	cubic inches.
" "	4.732×10^{-4}	cubic meters.
" "	6.189×10^{-4}	cubic yards.
" "	0.125	gallons.
" "	0.4732	liters.
poundals	13,826	dynes.
"	14.10	grams.
"	0.03108	pounds.
pounds	444,823	dynes.
"	7000	grains.
"	453.6	grams.
"	16	ounces.
"	32.17	poundals.
pounds (troy)	0.8229	pounds (av.).
pound-feet	1.356×10^{7}	centimeter-dynes.
" "	13,825	centimeter-grams.
" "	0.1383	meter-kilograms.
pounds-feet squared	421.3	kgs.-cms. squared.
" " "	144	pounds-inches squared.
pounds-inches squared	2.926	kgs.-cms. squared.
" " "	6.945×10^{-3}	pounds-feet squared.
pounds of water	0.01602	cubic feet.
" " "	27.68	cubic inches.
" " "	0.1198	gallons.
pounds of water per min.	2.669×10^{-4}	cubic feet per sec.
pounds per cubic foot	0.01602	grams per cubic cm.
" " " "	16.02	kgs. per cubic meter.
" " " "	5.787×10^{-4}	pounds per cubic inch.
" " " "	5.456×10^{-9}	pounds per mil foot.
pounds per cubic inch	27.68	grams per cubic cm.
" " " "	2.768×10^{4}	kgs. per cubic meter.
" " " "	1728	pounds per cubic foot.
" " " "	9.425×10^{-6}	pounds per mil foot.
pounds per foot	1.488	kgs. per meter.
pounds per inch	178.6	grams per cm.
pounds per mil foot	2.306×10^{6}	grams per cubic cm.
pounds per square foot	4.725×10^{-4}	atmospheres.
" " " "	0.01602	feet of water.
" " " "	1.414×10^{-2}	inches of mercury.
" " " "	4.882	kgs. per square meter.
" " " "	6.944×10^{-3}	pounds per square inch.
pounds per square inch	0.06804	atmospheres.
" " " "	2.307	feet of water.
" " " "	2.036	inches of mercury.

Multiply	by	to obtain
pounds per square in. (*cont.*)	703.1	kgs. per square meter.
" " " "	144	pounds per square foot.
Quadrants (angle)	90	degrees.
" "	5400	minutes.
" "	1.571	radians.
quarts (dry)	67.20	cubic inches.
quarts (liq.)	946.4	cubic centimeters.
" "	3.342×10^{-2}	cubic feet.
" "	57.75	cubic inches.
" "	9.464×10^{-4}	cubic meters.
" "	1.238×10^{-3}	cubic yards.
" "	0.25	gallons.
" "	0.9463	liters.
quintals	100	pounds.
quires	25	sheets.
Radians	57.30	degrees.
"	3438	minutes.
"	0.6366	quadrants.
radians per second	57.30	degrees per second.
" " "	9.549	revolutions per minute.
" " "	0.1592	revolutions per second.
radians per sec. per sec.	573.0	revs. per min. per min.
" " " " "	9.549	revs. per min. per sec.
" " " " "	0.1592	revs. per sec. per sec.
reams	500	sheets.
revolutions	360	degrees.
"	4	quadrants.
"	6.283	radians.
revolutions per minute	6	degrees per second.
" " "	0.1047	radians per second.
" " "	0.01667	revolutions per second.
revs. per min. per min.	1.745×10^{-3}	rads. per sec. per sec.
" " " " "	0.01667	revs. per min. per sec.
" " " " "	2.778×10^{-4}	revs. per sec. per sec
revolutions per second	360	degrees per second.
" " "	6.283	radians per second.
" " "	60	revs. per minute.
revs. per sec. per sec.	6.283	radians per sec. per sec.
" " " " "	3600	revs. per min. per min.
" " " " "	60	revs. per min. per sec.
rods	16.5	feet.
Seconds	1.157×10^{-5}	days.
"	2.778×10^{-4}	hours.
"	1.667×10^{-2}	minutes.
"	1.654×10^{-6}	weeks.
seconds (angle)	4.848×10^{-6}	radians.
spheres (solid angle)	12.57	steradians.
spherical right angles	0.25	hemispheres.
" " "	0.125	spheres.
" " "	1.571	steradians.
square centimeters	1.973×10^{5}	circular mils.
" "	1.076×10^{-3}	square feet.
" "	0.1550	square inches.
" "	10^{-4}	square meters.

Multiply	by	to obtain
square centimeters (*cont.*)	3.861×10^{-11}	square miles.
" "	100	square millimeters.
" "	1.196×10^{-4}	square yards.
sq. cms.-cms. sqd.	0.02402	sq. inches-inches sqd.
square feet	2.296×10^{-5}	acres.
" "	1.833×10^{8}	circular mils.
" "	929.0	square centimeters.
" "	144	square inches.
" "	0.09290	square meters.
" "	3.587×10^{-8}	square miles.
" "	1/9	square yards.
sq. feet-feet sqd.	2.074×10^{4}	sq. inches-inches sqd.
square inches	1.273×10^{6}	circular mils.
" "	6.452	square centimeters.
" "	6.944×10^{-3}	square feet.
" "	645.2	square millimeters.
" "	10^{6}	square mils.
" "	7.716×10^{-4}	square yards.
sq. inches-inches sqd. . .	41.62	sq. cms.-cms. sqd.
" " " " . .	4.823×10^{-5}	sq. feet-feet sqd.
square kilometers	247.1	acres.
" "	10.76×10^{6}	square feet.
" "	1.550×10^{9}	square inches.
" "	10^{6}	square meters.
" "	0.3861	square miles.
" "	1.196×10^{6}	square yards.
square meters	2.471×10^{-4}	acres.
" "	10.76	square feet.
" "	1550	square inches.
" "	3.861×10^{-7}	square miles.
" "	1.196	square yards.
square miles	640	acres.
" "	27.88×10^{6}	square feet.
" "	2.590	square kilometers.
" "	3.098×10^{6}	square yards.
square millimeters	1.973×10^{3}	circular mils.
" "	0.01	square centimeters.
" "	1.550×10^{-3}	square inches.
square mils	1.273	circular mils.
" "	6.452×10^{-6}	square centimeters.
" "	10^{-6}	square inches.
square yards	2.066×10^{-4}	acres.
" "	9	square feet.
" "	1296	square inches.
" "	0.8361	square meters.
" "	3.228×10^{-7}	square miles.
statamperes	$1/3 \times 10^{-10}$	abamperes.
"	$1/3 \times 10^{-9}$	amperes.
statcoulombs	$1/3 \times 10^{-10}$	abcoulombs.
"	$1/3 \times 10^{-9}$	coulombs.
statfarads	$1/9 \times 10^{-20}$	abfarads.
"	$1/9 \times 10^{-11}$	farads.
"	$1/9 \times 10^{-5}$	microfarads.
stathenries	9×10^{20}	abhenries.
"	9×10^{11}	henries.
"	9×10^{14}	millihenries.
statohms	9×10^{20}	abohms.

Multiply	by	to obtain
statohms (*cont.*)	9×10^{5}	megohms.
"	9×10^{17}	microhms.
"	9×10^{11}	ohms.
statvolts	3×10^{10}	abvolts.
"	300	volts.
steradians	0.1592	hemispheres.
"	0.07958	spheres.
"	0.6366	spherical right angles.
steres	10^{3}	liters.
Temp. (degs. Cent.)+273	1	abs. temp. (degs. Cent.).
" " " +17.8	1.8	temp. (degs. Fahr.).
temp. (degs. Fahr.)+460	1	abs. temp. (degs. Fahr.).
" " " −32	5/9	temp. (degs. Cent.).
tons (long)	1016	kilograms.
" "	2240	pounds.
tons (metric)	10^{3}	kilograms.
" "	2205	pounds.
tons (short)	907.2	kilograms.
" "	2000	pounds.
tons (short) per sq. ft.	9765	kgs. per square meter.
" " " " "	13.89	pounds per square inch.
tons (short) per sq. in.	1.406×10^{6}	kgs. per square meter.
" " " " "	2000	pounds per square inch.
Volts	10^{8}	abvolts.
"	1/300	statvolts.
volts per inch	3.937×10^{7}	abvolts per cm.
" " "	1.312×10^{-3}	statvolts per cm.
Watts	0.05688	B.t. units per min.
"	10^{7}	ergs per second.
"	44.27	foot-pounds per min.
"	0.7378	foot-pounds per second.
"	1.341×10^{-3}	horse-power.
"	0.01433	kg.-calories per minute.
"	10^{-3}	kilowatts.
watt-hours	3.413	British thermal units.
" "	2656	foot-pounds.
" "	1.341×10^{-3}	horse-power-hours.
" "	0.860	kilogram-calories.
" "	367.2	kilogram-meters.
" "	10^{-3}	kilowatt-hours.
webers	10^{8}	maxwells.
weeks	168	hours.
"	10,080	minutes.
"	604,800	seconds.
Yards	91.44	centimeters.
"	3	feet.
"	36	inches.
"	0.9144	meters.
"	5.682×10^{-4}	miles.
years (common)	365	days.
" "	8760	hours.
years (leap)	366	days.
" "	8784	hours.

Decimal Equivalents of Fractions

Fractions	Decimals	Fractions	Decimals	Fractions	Decimals	Fractions	Decimals
$\frac{1}{64}$	0.015625	$\frac{17}{64}$	0.265625	$\frac{33}{64}$	0.515625	$\frac{49}{64}$	0.765625
$\frac{1}{32}$	0.03125	$\frac{9}{32}$	0.28125	$\frac{17}{32}$	0.53125	$\frac{25}{32}$	0.78125
$\frac{3}{64}$	0.046875	$\frac{19}{64}$	0.296875	$\frac{35}{64}$	0.546875	$\frac{51}{64}$	0.796875
$\frac{1}{16}$	0.0625	$\frac{5}{16}$	0.3125	$\frac{9}{16}$	0.5625	$\frac{13}{16}$	0.8125
$\frac{5}{64}$	0.078125	$\frac{21}{64}$	0.328125	$\frac{37}{64}$	0.578125	$\frac{53}{64}$	0.828125
$\frac{3}{32}$	0.09375	$\frac{11}{32}$	0.34375	$\frac{19}{32}$	0.59375	$\frac{27}{32}$	0.84375
$\frac{7}{64}$	0.109375	$\frac{23}{64}$	0.359375	$\frac{39}{64}$	0.609375	$\frac{55}{64}$	0.859375
$\frac{1}{8}$	0.125	$\frac{3}{8}$	0.375	$\frac{5}{8}$	0.625	$\frac{7}{8}$	0.875
$\frac{9}{64}$	0.140625	$\frac{25}{64}$	0.390625	$\frac{41}{64}$	0.640625	$\frac{57}{64}$	0.890625
$\frac{5}{32}$	0.15625	$\frac{13}{32}$	0.40625	$\frac{21}{32}$	0.65625	$\frac{29}{32}$	0.90625
$\frac{11}{64}$	0.171875	$\frac{27}{64}$	0.421875	$\frac{43}{64}$	0.671875	$\frac{59}{64}$	0.921875
$\frac{3}{16}$	0.1875	$\frac{7}{16}$	0.4375	$\frac{11}{16}$	0.6875	$\frac{15}{16}$	0.9375
$\frac{13}{64}$	0.203125	$\frac{29}{64}$	0.453125	$\frac{45}{64}$	0.703125	$\frac{61}{64}$	0.953125
$\frac{7}{32}$	0.21875	$\frac{15}{32}$	0.46875	$\frac{23}{32}$	0.71875	$\frac{31}{32}$	0.96875
$\frac{15}{64}$	0.234375	$\frac{31}{64}$	0.484375	$\frac{47}{64}$	0.734375	$\frac{63}{64}$	0.984375
$\frac{1}{4}$	0.25	$\frac{1}{2}$	0.5	$\frac{3}{4}$	0.75	1	1

Greek Alphabet

Letters	Name	Letters	Name
Α α	Alpha	Ν ν	Nu
Β β	Beta	Ξ ξ	Xi
Γ γ	Gamma	Ο ο	Omicron
Δ δ	Delta	Π π	Pi
Ε ϵ	Epsilon	Ρ ρ	Rho
Ζ ζ	Zeta	Σ σ	Sigma
Η η	Eta	Τ τ	Tau
Θ θ	Theta	Υ υ	Upsilon
Ι ι	Iota	Φ φ	Phi
Κ κ	Kappa	Χ χ	Chi
Λ λ	Lambda	Ψ ψ	Psi
Μ μ	Mu	Ω ω	Omega

INTERNATIONAL ATOMIC WEIGHTS AND STANDARD COLOR CODE

Standard Color Code of Resistors and Capacitors

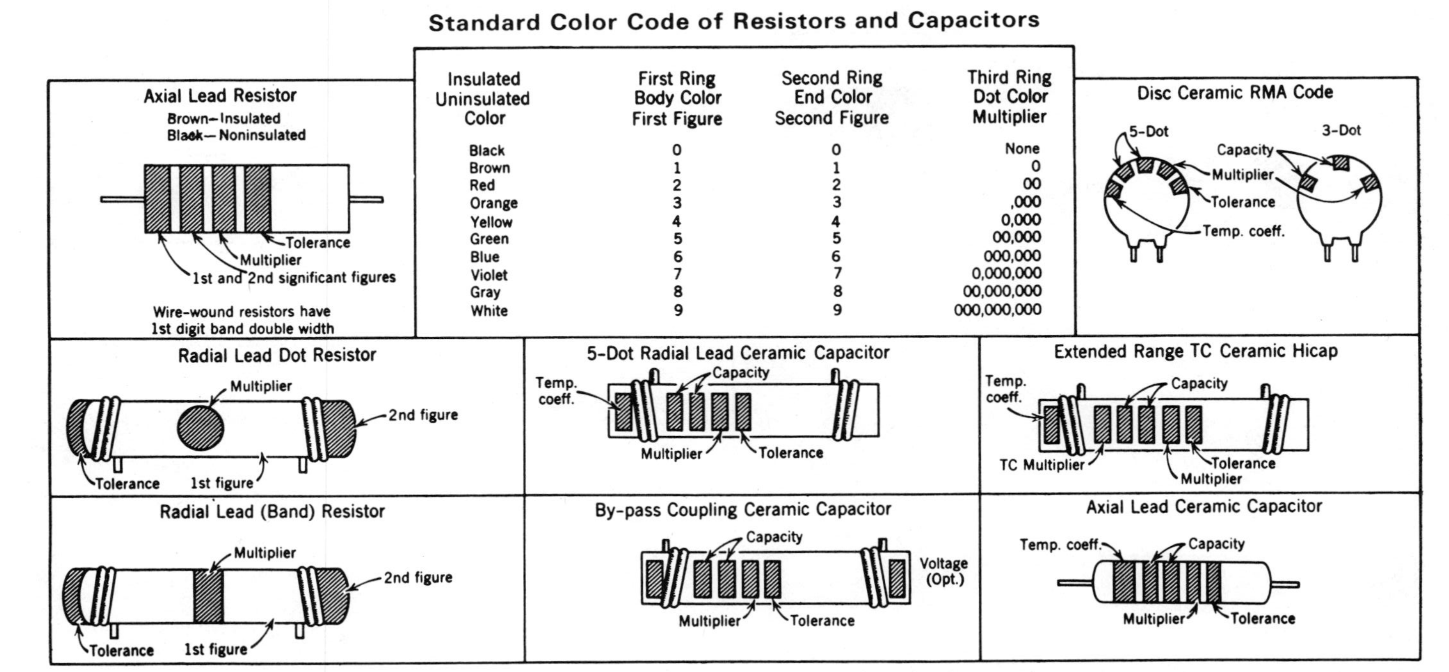

Insulated Uninsulated Color	First Ring Body Color First Figure	Second Ring End Color Second Figure	Third Ring Dot Color Multiplier
Black	0	0	None
Brown	1	1	0
Red	2	2	00
Orange	3	3	,000
Yellow	4	4	0,000
Green	5	5	00,000
Blue	6	6	000,000
Violet	7	7	0,000,000
Gray	8	8	00,000,000
White	9	9	000,000,000

Resistor tolerance is indicated as follows: gold = 5 percent, silver = 10 percent and absence of a fourth band = 20 percent. If the resistor is wire-wound the first band is double width.

Moulded Mica Type Capacitors

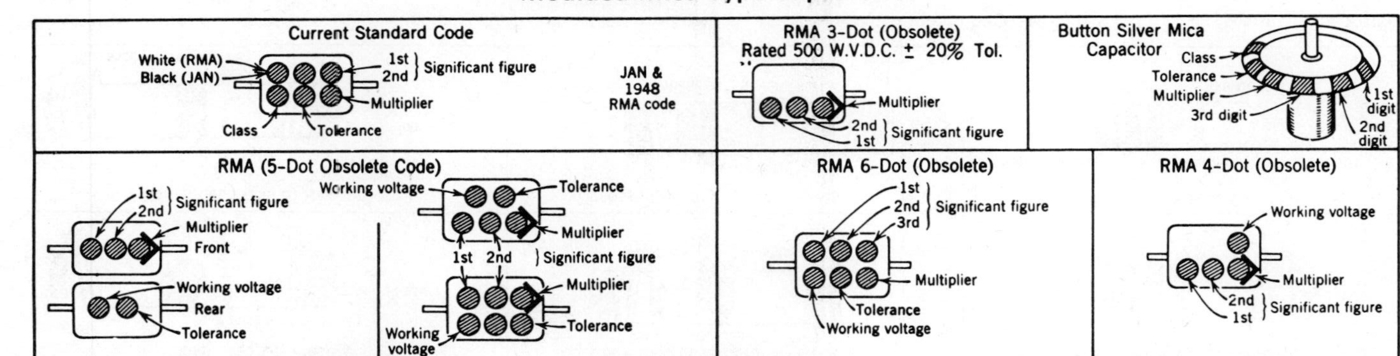

Moulded Paper Type Capacitors

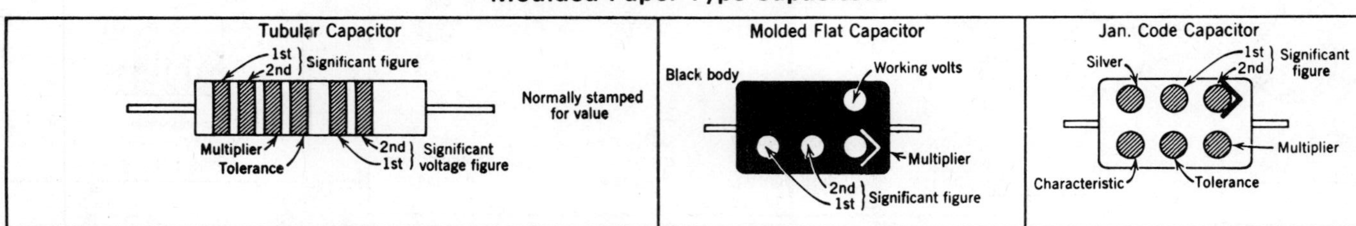

Tolerance ratings of capacitors are read from the color code directly. For example: red = 2 percent, yellow = 4 percent. The voltage rating is found by multiplying the color value by 100.

International Atomic Weights

	Symbol	Atomic No.	Atomic Weight		Symbol	Atomic No.	Atomic Weight
Actinium	Ac	89	(227)	Mercury	Hg	80	200.59
Aluminum	Al	13	26.9815	Molybdenum	Mo	42	95.94
Americium	Am	95	(243)	Neodymium	Nd	60	144.24
Antimony	Sb	51	121.75	Neon	Ne	10	20.183
Argon	Ar	18	39.948	Neptunium	Np	93	(237)
Arsenic	As	33	74.9216	Nickel	Ni	28	58.71
Astatine	At	85	(210)	Niobium	Nb	41	92.906
Barium	Ba	56	137.34	Nitrogen	N	7	14.0067
Berkelium	Bk	97	(249)	Nobelium	No	102	(253)
Beryllium	Be	4	9.0122	Osmium	Os	76	190.2
Bismuth	Bi	83	208.980	Oxygen	O	8	15.9994
Boron	B	5	10.811	Palladium	Pd	46	106.4
Bromine	Br	35	79.909	Phosphorus	P	15	30.9738
Cadmium	Cd	48	112.40	Platinum	Pt	78	195.09
Calcium	Ca	20	40.08	Plutonium	Pu	94	(244)
Californium	Cf	98	(249)	Polonium	Po	84	(210)
Carbon	C	6	12.01115	Potassium	K	19	39.102
Cerium	Ce	58	140.12	Praseodymium	Pr	59	140.907
Cesium	Cs	55	132.905	Promethium	Pm	61	(145)
Chlorine	Cl	17	35.453	Protactinium	Pa	91	(231)
Chromium	Cr	24	51.996	Radium	Ra	88	(226)
Cobalt	Co	27	58.9332	Radon	Rn	86	(222)
Copper	Cu	29	63.54	Rhenium	Re	75	186.2
Curium	Cm	96	(245)	Rhodium	Rh	45	102.905
Dysprosium	Dy	66	162.50	Rubidium	Rb	37	85.47
Einsteinium	Es	99	(254)	Ruthenium	Ru	44	101.07
Erbium	Er	68	167.26	Samarium	Sm	62	150.35
Europium	Eu	63	151.96	Scandium	Sc	21	44.956
Fermium	Fm	100	(252)	Selenium	Se	34	78.96
Fluorine	F	9	18.9984	Silicon	Si	14	28.086
Francium	Fr	87	(223)	Silver	Ag	47	107.870
Gadolinium	Gd	64	157.25	Sodium	Na	11	22.9898
Gallium	Ga	31	69.72	Strontium	Sr	38	87.62
Germanium	Ge	32	72.59	Sulfur	S	16	32.064
Gold	Au	79	196.967	Tantalum	Ta	73	180.948
Hafnium	Hf	72	178.49	Technetium	Tc	43	(99)
Helium	He	2	4.0026	Tellurium	Te	52	127.60
Holmium	Ho	67	164.930	Terbium	Tb	65	158.924
Hydrogen	H	1	1.00797	Thallium	Tl	81	204.37
Indium	In	49	114.82	Thorium	Th	90	232.038
Iodine	I	53	126.9044	Thulium	Tm	69	168.934
Iridium	Ir	77	192.2	Tin	Sn	50	118.69
Iron	Fe	26	55.847	Titanium	Ti	22	47.90
Krypton	Kr	36	83.80	Tungsten	W	74	183.85
Lanthanum	La	57	138.91	Uranium	U	92	238.03
Lead	Pb	82	207.19	Vanadium	V	23	50.942
Lithium	Li	3	6.939	Xenon	Xe	54	131.30
Lutetium	Lu	71	174.97	Ytterbium	Yb	70	173.04
Magnesium	Mg	12	24.312	Yttrium	Y	39	88.905
Manganese	Mn	25	54.9380	Zinc	Zn	30	65.37
Mendelevium	Md	101	(256)	Zirconium	Zr	40	91.22

Value in parenthesis denotes isotope of longest half-life.

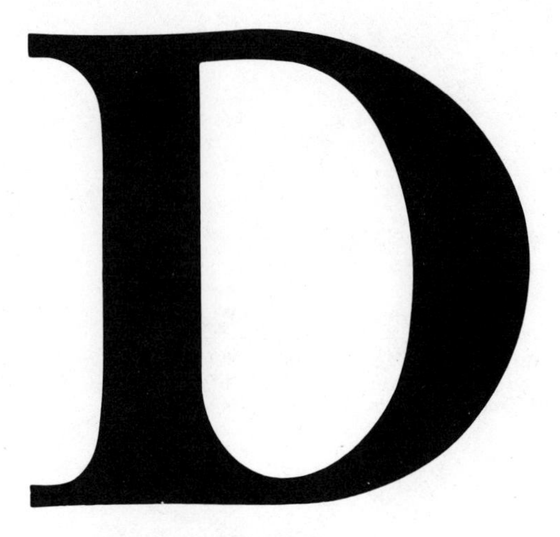

ANSWERS TO REVIEW QUESTIONS

Chapter 1

1. T
2. T
3. electrostatic, magnetic, elastic, chemical
4. F
5. magnetic
6. T
7. (b)
8. T
9. negative, positive
10. ionic, covalent, metallic
11. F
12. T
13. 3

Chapter 2

1. excess, deficit
2. by arbitrary adoption
3. (a)
4. F
5. electric current
6. ions
7. 5 A

8. F
9. (b)
10. voltage, EMF
11. F
12. electrolyte, positive electrode, negative electrode
13. electromotive force
14. joule, coulomb
15. coulomb, second
16. T
17. a closed path which conducts electric current
18. (d)
19. alternating current
20. across

Chapter 3

1. force, distance
2. capacity to do work
3. T
4. (b)
5. F
6. 9.49×10^{-4} BTU/J
7. F
8. meters, kilograms, seconds
9. joule
10. power
11. joule, second
12. increases
13. 550, 746
14. T
15. T
16. 0.5
17. F
18. losses in the motor, heat due to electric resistance, mechanical friction, and other losses
19. smaller
20. damage due to excess heat

Chapter 4

1. electrode potentials
2. chemical reactions
3. 1.5
4. primary
5. F
6. T
7. secondary
8. no gases are formed
9. T
10. specific gravity
11. electrolyte
12. mercury, alkaline, nickel-cadmium
13. A battery is a combination of cells
14. series, parallel
15. T

16. Ah
17. resistance
18. EMF, resistance, current
19. T
20. watt-sec/lb

Chapter 5

1. directly
2. inversely
3. T
4. (a)
5. AWG
6. 20
7. power
8. metal films
9. 1st, 2nd, multiplier, tolerance
10. T
11. orange, white, red
12. 220 kΩ, ±5%
13. increases
14. decreases
15. T
16. negative
17. T
18. conductance
19. mhos

Chapter 6

1. sum
2. T
3. series
4. current
5. positive
6. F
7. T
8. 50
9. smaller
10. voltage
11. T
12. 2 kΩ
13. two
14. conductances
15. parallel
16. T
17. T

Chapter 7

1. parallel
2. series
3. T
4. KCL, KVL
5. F
6. current

7. T
8. furthest

Chapter 8

1. zero
2. infinite
3. T
4. series
5. current, resistance
6. source resistance
7. T
8. without
9. T
10. shorting
11. series, parallel
12. T
13. more than one
14. T
15. currents, voltages
16. F
17. Norton's
18. voltages
19. F
20. T
21. T
22. 50
23. smaller

Chapter 9

1. T
2. reversed
3. galvanometer
4. zero
5. T
5. T
6. F
7. voltage, current
8. series
9. parallel
10. T
11. make before break
12. T
13. parallel
14. series
15. T
16. full scale current
17. T
18. center
19. is not
20. shorting
21. T
22. T
23. smaller

24. high
25. source voltage

Chapter 10

1. mesh, node
2. T
3. element
4. node
5. cannot
6. F
7. closed paths
8. window panes
9. mesh
10. mesh currents
11. T
12. the summation of
13. like, unlike
14. 2 and 3
15. positive, negative
16. clockwise
17. Sect. 10.3
18. T
19. node
20. node voltages
21. T
22. T
23. like, unlike
24. A, B, G_{BA}
25. enter
26. Sect. 10.6
27. $(e-v+1)$, $(v-1)$

Chapter 11

1. attraction
2. like
3. T
4. in motion
5. T
6. F
7. same, opposite
8. magnetized by induction
9. greater
10. T
11. are
12. T
13. right hand rule
14. electromagnet
15. current
16. north
17. $\Phi = \mathcal{F}/\mathcal{R}$
18. F
19. diamagnetic, paramagnetic, larger
20. F

21. hysteresis loop
22. T
23. T
24. torque
25. commutator
26. will not
27. voltage
28. Lenz's
29. T
30. relay

Chapter 12

1. T
2. current
3. directly
4. high
5. Fig. 12.2
6. Fig. 12.2
7. tightly
8. 1.0
9. 0.0
10. F
11. Fig. 12.8
12. are not
13. 1.0
14. 1.0
15. T
16. parallel
17. aiding, bucking
18. coil
19. T
20. falls
21. current
22. 63.2
23. 36.8
24. 5
25. cannot, can

Chapter 13

1. resistance, inductance, capacitance
2. T
3. F
4. always
5. charge
6. inversely
7. perpendicular
8. T
9. 1.0
10. increased
11. F
12. dielectric absorption
13. dielectric strength
14. electrolytic

15. voltage ratings
16. pF
17. resistance
18. series
19. parallel
20. cannot
21. changing
22. R, C
23. 5τ
24. electric field
25. 36.8

Chapter 14

1. periodic
2. F
3. both
4. $\frac{1}{2}$
5. degrees, radians
6. 57.3
7. T
8. sin, cos
9. 2π
11. magnitude, direction
12. alternators, oscillators
13. T
14. lead, lag
15. radians
16. T
17. negative
18. real, imaginary
19. T
20. rectangular, polar
21. F
23. F
24. added
25. $7 + j4$
26. T
27. subtracted from
28. polar

Chapter 15

1. T
2. zero
3. lag
4. ωt
5. resistance
6. reactance
7. negative
8. leads
9. inductive reactance
10. positive
11. T
12. capacitive reactance

13. negative
14. lags
15. F
16. 0, 90
17. 0, −90
18. +90, −90
19. T
20. resonance
21. vectors

Chapter 16

1. T
2. vector
3. current
4. resistances
5. voltage
6. T
7. mhos
8. positive
9. inductive
10. resistance, reactance
11. admittance
12. T
13. leads
14. lags
15. phasors
16. series
17. parallel
18. frequency
20. F
21. impedance
22. T
23. zero
24. do not
25. F
26. T
27. series capacitance comparison
28. inductance
29. sources

Chapter 17

1. T
2. $\frac{1}{2}$
3. watts
4. 0.707
5. 165
6. twice
7. T
8. does not
9. F
10. zero
11. capacitive reactive
12. vars

13. apparent
14. VA
15. negative, positive
16. power factor
17. T
18. leading
19. lagging
20. 1.0
21. zero
22. 1.0
23. parallel
24. series gives high current and voltage
25. complex conjugate
26. single
27. zero
28. 120
29. equal

Chapter 18

1. resonant
2. *R*
3. below, above
4. minimum
5. T
6. zero
7. *Q*
8. F
9. *Q*
10. half-power
11. bandwidth
12. T
13. antiresonant
14. large
15. T
16. infinite
17. T
18. *Q*
19. resistance
20. smaller
21. T
22. T
23. pass
24. F
25. double tuned

Chapter 19

1. iron core
2. T
3. 1.0
4. no
5. leakage
6. resistance
7. core

8. 1.0
9. ideal
10. larger
11. smaller
12. F
13. F
14. equal to
15. divided
16. step-down
17. T
18. F
19. capacitive
20. inductive
21. Q's
22. F
23. k_c
24. greater
25. greater
26. T

Chapter 20

1. Coils on same rotor with 120° angles between starts.
2. T
3. phase sequence
4. star
5. difference
6. T
7. Unbalance causes large circulating currents.
8. T
9. mesh
10. step-up, step-down
11. Y, Δ, single phase
12. T
13. are not
14. T
15. T
16. neutral
17. current, potential
18. F
19. average
20. T
21. upscale, downscale
22. T
23. algebraic

Chapter 21

1. F
2. is not
3. forward
4. open
5. T
6. F
7. period

8. C, R
9. peak
10. zener
11. breakdown
12. T
13. T
14. small
15. small
16. no
17. for zener breakdown
18. T
19. π
20. sinusoidal
21. T
22. high, low
23. dependent
24. less than
25. F

ANSWERS TO SELECTED PROBLEMS

Chapter 1

3. +1
5. (*a*) 8.574×10^8 (*b*) 4.28×10^{-9} (*c*) 1.20×10^{-2} (*d*) 7.35

Chapter 2

3. +2.99 mC
5. 10 mA
7. 0.0001 in./sec, 0.006 in./min, or 0.36 in./h
9. 0.738 sec
11. (*a*) 8 Ω, (*b*) 80 Ω, (*c*) 400 Ω, (*d*) 6.4 kΩ
13. (*a*) 1.00 V, (*b*) 20.0 V, (*c*) 500 V, (*d*) 2000 V
15. 160 Ω
17. (*a*) 21 Ω, (*b*) 2 kΩ, (*c*) 680 kΩ, (*d*) 1.1 MΩ, (*e*) 3.3 kΩ
19. (*a*) 100 kΩ, (*b*) 7,300 μV, (*c*) 3.4 A, (*d*) 4500 Ω

Chapter 3

1. 2.5×10^5 ft-lb
3. 5 keV
5. 549 sec
7. (*a*) 104°F, (*b*) 4.44°C
9. 0.292 A
11. 0.206 W

13. 0.707 mA
15. no; exceeds power rating by factor of two
17. 8.4 ¢
19. 4.2 ¢

Chapter 4

1. 2.01 V, Cu is +
3. Hydrogen ions do not form gas at the copper electrode.
5. minimizing chemical cells
7. recharge
9. 2.77 mA
11. 12 V, 1.2 A
13. 25 hr
15. 94 cells
17. 5.88 V
19. 24.8 Ω

Chapter 5

1. 8.53×10^{-3} Ω
3. 10^{14} Ω
5. 2.3×10^{5} Ω-cm
7. 25.5 Ω
9. 5 ft
11. 0.2 Ω
13. AWG # 10
15. (*a*) 470, (*b*) 560, (*c*) 510
17. −0.05
19. 4 V, 250 Ω

Chapter 6

1. 11.0 mA
3. 2.66 kΩ
5. 1.21 mA, clockwise
7. 2 mA, 27.2 V
13. 1.54 V
15. $P_{R_1} = 26.6$ mW, $P_{R_2} = 182$ mW, $P_{R_3} = 67.8$ mW
17. 120 Ω, 1.41 W
19. 1.5 kΩ
21. 7.5 kΩ and 6 V
23. $I_{R_1} = 400\ \mu A$, $I_{R_2} = 273\ \mu A$, $I_{R_3} = 128\ \mu A$, $I_T = 801\ \mu A$
27. $P_{R_1} = 2.4$ mW, $P_{R_2} = 1.64$ mW, $P_{R_3} = 0.766$ mW, $P_S = 4.8$ mW
29. 300 mA, 20 mA, 13.6 mA, $I_T = 333.6$ mA
31. $V_{R_1} = 6.68$ V, $V_{R_3} = 13.47$ V, $V_{R_5} = 1.85$ V
33. $V_{R_1} = 30.6$ V, $V_{R_3} = 23.6$ V, $V_{R_5} = 13.8$ V, $I_T = 24.5$ mA
35. $I_1 = 4.29$ mA, $I_2 = 2.94$ mA
37. $I_{6k} = I_{1.5k} = 1.33$ mA, $I_{1k} = I_{2.2k} = 2.66$ mA, $V_{S_2}^{+} = 7.35$ V
39. $I_{R_1} = 4/3$ mA, $I_{R_2} = 2/3$ mA, $I_{R_3} = 2$ mA, $I_{R_4} = 2$ mA, $I_{R_5} = 1/2$ mA, $I_{R_6} = 3/2$ mA
41. $I_{R_1} = 1.065$ mA, $I_{R_2} = 1.16$ mA, $I_{R_3} = I_{R_4} = 2.55$ mA, $I_{R_5} = 6.28$ mA
43. $I_{R_1} = 4$ mA, $I_{R_2} = 2$ mA, $I_{R_3} = 2$ mA, $I_{R_4} = I_{R_5} = 1$ mA

Chapter 7

1. 1.54 V
3. zero
5. 14.05 V
7. 1.27 mA
9. 17.7 V, 548 Ω
11. 60.6 V
13. 77.6 V
15. 30 mA
17. 8.25 mA
19. $I_T = 9.03$ mA, $V_{R_1} = 17.5$ V, $V_{R_3} = 17.5$ V
21. $V_{R_4} = 10.2$ V, $V_{R_3} = 3.38$ V, $V_{R_2} = 13.6$ V, $V_{R_1} = 16.3$ V
23. 4.0 mA
25. $V_{R_6} = 4/3$ V, $V_{R_5} = 8/3$ V, $V_{R_4} = 4$ V, $V_{R_3} = 14$ V, $V_{R_2} = 18$ V, $V_{R_1} = 6$ V
27. $I_T = 3/5$ mA, $I_{R_2} = I_{R_3} = 3/10$ mA, $V_{R_1} = 6$ V, $V_{R_2} = 3$ V

Chapter 8

1. 2 mA
3. 83.7 kΩ
5. $V_{TH} = 6$ V, $R_{TH} = 3$ kΩ
7. $V_{TH} = 12$ V, $R_{TH} = 0.0461$ Ω
9. 7 V
11. $I_N = 2$ mA, $R_N = 3$ kΩ
13. $I_N = 5$ mA, $R_N = 100$ kΩ
15. $I_N = 101$ mA, $R_N = R_{TH} = 800$ Ω, $V_{TH} = 90$ V (one solution)
17. 15 V
19. 4.175 mA
25. 120 Ω, 13 V, 1.41 W

Chapter 9

1. 5 mV
3. 50 Ω
5. 5.27 Ω
7. 5 Ω, 50 mV
11. 499.8 kΩ
13. 20 kΩ, 500 kΩ
15. 2 kΩ/V, 1 kΩ/V, 2 kΩ/V
17. 800 kΩ
19. 1.999 MΩ, 3 MΩ, 5 MΩ
21. 0.555 mA, 0.0476 mA
25. No; meter movement too small
27. 20 V, 19.9 V
29. 19.97 V
31. 562 Ω

Chapter 10

1. 6,4,3
3. $V_S = (R_1 + R_2 + R_3)I$, $I = 4/3$ mA
5. $-V_S = (R_1 + R_2 + R_3)I_1 - R_3 I_2$, $V_S = -R_3 I_1 + (R_3 + R_4 + R_5)I_2$
7. $I_1 = -8.45$ mA, $I_2 = 18.6$ mA
9. $I_1 = -0.912$ mA, $I_2 = 0.400$ mA
11. 15.5 mA, 64.8 mA, 13.4 mA
13. $I_1 = -6$ mA, $I_2 = -2$ mA, $I_3 = -2$ mA

15. zero
17. $-I_{S_1} = (1/R_1 + 1/R_2)V_A - (1/R_2)V_B$, $-I_{S_2} = -(1/R_2)V_A + (1/R_2 + 1/R_3)V_B$
19. $V_A = -4.65$ V, $V_B = -4.86$ V, $V_C = -10$ V, $V_D = +5$ V
21. $V_A = -12$ V, $V_B = 0$ V
23. -4.74 V
25. $I_1 = -2.5$ mA, $I_2 = 4.29$ mA, $I_3 = 3.43$ mA

Chapter 11

5. let each line on the drawing equal 10^{-3} Wb of flux
7. 20 μWb
9. same
15. 150 At
17. 2.0 V
19. 25 μWb
21. 6 At
23. 1.79×10^8 At/Wb
25. 1.11×10^9 At/Wb
27. 200 At/m; 0.6 Wb/m^2; 240 μWb; 3×10^{-3} MKS; 4.66×10^5 At/Wb
29. 50 At/in. 33 kilolines/in.^{2}, 33 kilolines, 6.6×10^2 English, 3.03×10^{-2} At/line
31. 10 mA
33. 0.4 Wb/sec
35. (−) on right

Chapter 12

1. 5 mV
3. 89.5 μH
5. 50 μsec
7. 12 μH
9. 3.18 mH
11. 2.83 mH
13. 0.15
15. 50.91 mH
17. 2.0 mH
19. 110 μH
21. $i = 20 \times 10^{-3}(1 - e^{-2\times10^3 t})$
23. 693 μsec
25. 12.6 mA; 17.2 mA; 2.5 msec.
27. $i = 0.25e^{-7.68\times10^2 t}$
29. 76.8 μsec, 34 mA, 384 μsec
31. B: $i = 60 \times 10^{-3}(1 - e^{-66.7t})$ C: $i = 48.7 \times 10^{-3}e^{-213t}$
33. 1.78×10^{-3} J; 76 mW

Chapter 13

3. 2000 V/m
5. 4×10^{-16} newtons
11. 10.6 pF
13. 1125 pF
15. 1.78×10^9 in.^{2}, or 1.236×10^7 ft^2
17. 2950 V
19. 5.45 μC, 2.73 V
21. 2 μF
23. 300 pF

25. $\tau = 2 \times 10^{-3}$; 10 msec; 4 msec.
27. 47 msec; 235 msec.
29. 3164 μF
31. 562×10^{-6} J

Chapter 14

1. yes, no
3. 67°; 63.5°; 90°
5. 9.63; 12.8
7. $\pi/6$; $\pi/4$; $\pi/3$; $13\pi/18$
11. 1.57×10^{6} rad/sec
13. $25 \sin 200\pi t$
15. $6.6 \sin 10^{6}t$ V
17. 159 kHz
19. $10 \sin 1050t$; 167 Hz
21. $30 \sin (10^{6}\pi t + \pi/6)$
23. j; -1; $-j$; $+1$
25. $2.24\angle 63.4^\circ$, $3.6\angle -123.7^\circ$, $5.66\angle 135^\circ$, $3.61\angle -33.7^\circ$
27. $6.06 + j3.5$, $15.3 - j12.9$, $-7.07 + j7.07$, $0 + j5$
29. $4 + j$; $-2 + j3$; $30 - j11$
31. $15.7\angle 93.4^\circ$; $39.6\angle -45^\circ$
34. $0.99\angle 105^\circ$; $3.18\angle -45^\circ$
35. $-2 - j$; $0.62\angle 7.1^\circ$

Chapter 15

3. $5\angle 30^\circ$ Ω
5. $50\angle -45^\circ$ mA
7. $90\angle 25^\circ$ V
9. $(78 - j45)$ kΩ
11. $108\angle 33.7^\circ$
13. 628 Ω
15. 3.18 mH
17. 79.7 kHz
19. $I_P = 74.3\angle -21.8^\circ$ mA; $V_{PR} = 37.2$ V; $V_{PL} = 14.8$ V
21. $V_{PS} = 31.7\angle 28.2^\circ$ V; $X_L = 1.61$ kΩ; $Z = 3.41\angle 28.2^\circ$ kΩ; $I_P = 9.3\angle 0^\circ$ mA
23. 1.25 kΩ; 8000 pF; 15.9 kHz
25. $X_L = 10$ kΩ; $X_C = 8$ kΩ; $Z = 2.83\angle 45^\circ$ kΩ
27. $X_L = 8$ kΩ; $X_C = 10$ kΩ; $Z = 2.83\angle -45^\circ$ kΩ

Chapter 16

1. $(20 - j28)$ kΩ
3. $38.2\angle -47.2^\circ$
5. $4.71\angle 45^\circ$
7. $(0.1 + j0.08)\mho$
9. $0.095 + j0.074\mho$
11. $0.035 + j0.02\mho$
13. $25\angle -30^\circ$ Ω
15. $(0.0125 + j0.0275)\mho$
17. $I_{PR} = 2.5\angle 0^\circ$ mA; $I_{PL} = 2.5\angle -90^\circ$ mA; $I_{PC} = 1.25\angle 90^\circ$ mA; $I_{PT} = 2.76\angle -26.6^\circ$ mA
19. $(1.43 + j1.62)$ A
21. $V_{P1} = 4.4\angle 45^\circ$ V; $V_{P3} = 4.4\angle -45^\circ$ V.

23. $(1.76 - j1.76)$ A
25. $V_{PA} = 7.04\underline{/0^\circ}$, $I_{PZ_2} = 2.49\underline{/-45^\circ}$
27. $I_{P1} = (7 - j6)/4$, $I_{P2} = (-6 + j3)/4$
29. $I_{P1} = -(1 + j12)4$, $I_{P2} = -3(4 + j3)/4$
35. R and C in series
37. 400 pF
39. 600 Ω, 12 μF

Chapter 17

1. $P_P = 0.5$ W; $P = 0.25$ W
3. 19.8 V
5. $v = 8.9 \sin 377t$
7. 1.4 W
9. 292 mW
11. $p = 0.312 \sin 2 \times 10^6 t$, $P_Q = 0.312$ var
13. 96.7 mA
15. 20.1 V
17. 71 VA; 50.2 W; 50.2 var; $\phi = 45^\circ$
19. $P_{QC} = -96.9$ var; $P_{QL} = 77.6$ var; $P = 58.2$ W; $P_Q = -19.3$ var; $P_A = 61.4$ VA
23. 0.89
25. $3833\underline{/-36.9^\circ}$ Ω
29. $600 + j600$
31. 0.0 A
33. zero

Chapter 18

1. 35.6 kHz
3. 5.06 mH
5. 56.3 pF
7. 22.4
9. 12.6 Ω
11. 500 kHz; 25.1; 1.25 V
13. 1.59 kHz
15. 153
17. 290 kHz
19. 8.95
21. 796 kHz; 400; 1.99 kHz
23. 34.6 kHz; 15.3; 2.26 kHz
25. 218 pF
27. 218 pF
29. $L_S = 126$ μH; $L_P = 887$ μH

Chapter 19

1. Fig. 19.1
3. $0.0 \leq k \leq 1.0$
5. 14 μH
7. 2.7 mH
9. 5.98, 0.054, 0.043
11. 928 mA

13. $V_L = 170$ V; $I_L = 85$ mA; $I_P = 1.7$ A; $P = 14.4$ W
15. $Z_P = (600 - j700)\Omega$
17. 1/5
19. 64 Ω
21. $Z_{REF} = 0.0384 - j0.0076$, $Z_{in} \cong 0.0384 + j20$
23. $Z_{in} = 1.4\underline{/43.5^\circ}\ \Omega$, $I_P = 7.14\underline{/43.5^\circ}$ A, $V_S = 3.57\underline{/-133.5^\circ}$ V, $I_S = 0.532\underline{/-196.9^\circ}$ A
25. $19.9\underline{/-168.7^\circ}$, $-4.98\underline{/-177.1^\circ}$
27. 2.53 μF
29. 0.0707
31. 7.14
33. yes, $R \cong 50\ (1/\omega C_P)$

Chapter 20

1. Eq. 20.1–Eq. 20.3 with $V_P = 294$ and $\omega = 377$
3. 120 V
5. Figs. 20.3 and 20.1*c*
7. lighting: *a-n*, *b-n*, *c-n*. motors: a-b-c
9. $P = 9$ kW; $P_{phase} = 3$ kW; $I_{line} = 25$ A in phase with line voltages
11. 29.8 A lagging V_{phase} by 25.8°
13. $I_{phase} = 16.7$ A in phase with V_{line}, $I_{line} = 28.9$ A lagging I_{phase} by 30°
15. $I_{An} = 33.3$ A in phase with V_{An}; $P_A = 4$ kW; $I_{Bn} = 14$ A lagging V_{Bn} by 35.6°; $P_B = 1.36$ kW; $I_{Cn} = 17$ A lagging V_{Cn} by 45°; $P_C = 1440$ W
17. with V_{AB} as the reference: $I_{AB} = 16.7\underline{/0^\circ}$ A; $I_{BC} = 27.9\underline{/-155.5^\circ}$ A; $I_{CA} = 33.9\underline{/75^\circ}$ A; $I_{aA} = 33.8\underline{/-76.3^\circ}$ A; $I_{bB} = 43.7\underline{/-164.5^\circ}$ A; $I_{cC} = 56\underline{/52.4^\circ}$ A
19. 1.81 kW
21. 3.73 kW

Chapter 21

1. 0.128 μA, 9.72 A
3. 12.5 nA
5. Fig. 21.5
7. 12 mV
9. 0.415 $V_{P\text{-}P}$, 24.8 V
11. 33 V
13. Fig. 21.13
19. 0.091
21. 27.8 Ω, 360 mA
23. 50 μA
25. 78
27. 68

INDEX